LIQUID, GASEOUS AND SOLID BIOFUELS - CONVERSION TECHNIQUES

Edited by **Zhen Fang**

Liquid, Gaseous and Solid Biofuels - Conversion Techniques

http://dx.doi.org/10.5772/50479
Edited by Zhen Fang

Contributors

Anne Ruffing, Robert Diltz, Pratap Pullammanappallil, Emad Shalaby, Edilson León Moreno Cárdenas, Deisy Yuliana Cano Quintero, Elkin Alonso Cortés Marín, László Kótai, Armando Tibigin Quitain, Bezergianni, Petra Patakova, Leona Paulova, Mojmir Rychtera, Karel Melzoch, Jean-Michel Lavoie, Charalampos Arapatsakos, Michael Köpke, FungMin Liew, Séan Dennis Simpson, Anli Geng, Fabiana Aparecida Lobo, Fernanda Pollo, Ana Cristina Villafranca, Mercedes De Moraes, Hongjuan Liu, Amar Kumar Mohanty, Singaravelu Vivekanandhan, Nima Zarrinbakhsh, Manjusri Misra, Valeriy Chernyak, Iñaki Gandarias, Pedro L. Arias

Published by InTech

Janeza Trdine 9, 51000 Rijeka, Croatia

Publishing Process Manager Iva Simcic

Technical Editor InTech DTP team

Cover InTech Design team

Additional hard copies can be obtained from orders@intechopen.com

Liquid, Gaseous and Solid Biofuels - Conversion Techniques, Edited by Zhen Fang
p. cm.
ISBN 978-953-51-1050-7

For EU product safety concerns:
IN TECH d.o.o., Prolaz Marije Krucifikse Kozulić 3, 51000 Rijeka, Croatia,
info@intechopen.com or visit our website at intechopen.com.

Meet the editor

Prof. Dr. Zhen FANG is leader and founder of biomass group, Chinese Academy of Sciences, Xishuangbanna Tropical Botanical Garden. He is also an adjunct full Professor of Life Sciences, University of Science and Technology of China. He is the inventor of "fast hydrolysis" process (US patent#: 8268126). His speciality is in thermal/biochemical conversion of biomass, nanocatalyst synthesis and their applications, and pretreatment methods of biomass for biorefineries. He obtained his PhD degrees from China Agricultural University (Ph.D. Biological & Agricultural Engineering, 1991, Beijing) and McGill University (Ph.D. Materials Engineering, 2003, Montreal).

Contents

Preface

Biomass is a renewable, unevenly geographically distributed resource that can be considered sustainable and carbon-neutral if properly managed. It can be converted to high-qualified gaseous, liquid and solid biofuels with many techniques. This book focuses on the latest conversion techniques for the production of liquid and gaseous biofuels that should be of interest to the chemical scientists and technologists.

This book includes 17 chapters contributed by experts around world on conversion techniques. The chapters are categorized into 2 parts: Liquids and Gases and Other Products.

Part 1 (Chapters 1-11) focuses on liquid biofuels. Chapter 1 reviews pathways for the conversion of hemicellulose to biofuels and chemicals. Chapter 2 discusses the production of cellulosic ethanol. Chapter 3 gives the experimental results of ethanol and methanol used in Otto engines. Chapter 4 presents analytic methods to determine trace Cu in ethanol. Chapter 5 reviews gas fermentation process for the production of liquid fuels (e.g., ethanol, butanol and 2,3-butanediol) and other products (e.g., acetic acid and butyric acid). Chapters 6 and 7 overview the production and applications of biobutanol. Chapter 8 describes the metabolic pathways involved in microbial hydrocarbon fuel synthesis and discusses strategies for improving biofuel production using genetic manipulation. Thermal conversion and upgrading techniques (such as catalytic hydroprocessing and microwave irradiation) are introduced in Chapters 9-11.

Part 2 (Chapters 12-17) describes production methods for gases and other products. Chapters 12 and 13 introduce hydrogen production by anaerobic fermentation, and DC and impulse plasma-liquid systems, respectively. Chapter 14 overviews some techniques (e.g., anaerobic digestion, fermentation, lipid extraction and gasification) for the production of biofuels from algae. Chapter 15 briefly introduces the production of biogas, biodiesel and ethanol. Chapter 16 comments on various thermal and biological conversions of oil palm empty fruit bunch to biofuels. Finally, Chapter 17 proposes a biorefinery concept for the co-products of biofuels and value-added biomaterials for sustainable bioeconomy.

This book offers reviews state-of-the-art conversion techniques for biofuels. It should be of interest for students, researchers, scientists and technologists in the engineering and sciences fields.

I would like to thank all the contributing authors for their time and efforts in the careful construction of the chapters and for making this project realizable. It is certain to inspire many young scientists and engineers who will benefit from careful study of these works and that their ideas will lead us to develop even more advances methods for producing liquids and gases from biomass resources.

I am grateful to Ms. Iva Simcic (Publishing Process Manager) for her encouragement and guidelines during my preparation of the book.

Finally, I would like to express my deepest gratitude towards my family for their kind cooperation and encouragement, which help me in completion of this project.

Zhen Fang
Leader of Biomass Group
Chinese Academy of Sciences
Xishuangbanna Tropical Botanical Garden, China

Section 1

Liquids

Chapter 1

Biofuels and Co-Products Out of Hemicelluloses

Ariadna Fuente-Hernández, Pierre-Olivier Corcos,
Romain Beauchet and Jean-Michel Lavoie

Additional information is available at the end of the chapter

http://dx.doi.org/10.5772/52645

1. Introduction

Second generation biofuels are based on the utilisation of non-edible feedstock for the production either of ethanol to be inserted in the gasoline pool or of biodiesel to be inserted in the diesel pool. Ethanol is usually produced out of fermentation of C6 sugars (although other approaches does exist, see [1]) and the latter came, in first generation ethanol, from starch. In second-generation ethanol, the source of carbohydrate considered is usually cellulose, which, in turns, is obtained from lignocellulosic biomass. Recent work by Lavoie*et al.* [2] have depicted an overview of many types of lignocellulosic biomass and in most cases, cellulose, although a major component, is not the only one and is accompanied by lignin, hemicelluloses, extractives and, in case of agricultural biomass, proteins. High grade biomass (as wood chips, sugar cane or even corn) are usually very expensive (more than 100 USD/tonne) because, in most part, of the important demand related to those feedstock in industries and this is why cellulosic ethanol is more than often related to residual biomass. The latter includes but is not limited to residual forest and agricultural biomass as well as energy crops. In all cases, although the feedstock is rather inexpensive (60-80 USD/tonne), it is composed of many different tissues (leaves, bark, wood, stems, etc.) making its transformation rather complex [3]. Industrialisation of second-generation biofuel requires specific pre-treatment that should be as versatile as efficient in order to cope with the economy of scale that has to be implemented in order to make such conversion economical.

The whole economics of cellulosic ethanol relies first on ethanol, which has a commodity beneficiates from a quasi-infinite market as long as prices are competitive. Assuming average cellulose content of 45-55 % (wt) in the lignocellulosic biomass, the ethanol potential of lignocellulosic biomass would range between 313-390 L per tonne of biomass converted.

With an actual market price of 0.48 USD per liter the value of this ethanol would range between 150-187 USD per tonne of biomass processed. Since the latter is more expensive to process (first isolation of cellulose then hydrolysis of cellulose) and considering the fact that the feedstock is itself expensive, there is a necessity to get an added value out of the remaining 55-45 % (wt) content. This residual carbon source is composed mostly of hemicelluloses and of lignin. The latter is a very energetic aromatic-based macromolecule, that has a high calorific value explaining why many processes converting such biomass (as some pulp and paper processes) relies on the combustion of lignin to provide part of the energy for the industry. It could also serve as a feedstock for the production of added-value compounds and although the subject is very pertinent to the field, it is out of the scope of this review, which focuses mostly on C5 sugars derived from hemicelluloses.

Conversion of the carbohydrates is of course an important part of the process although; isolation of hemicellulose for the lignocellulosic matrix is also crucial for such an approach and in consequence should also be briefly assessed. For years now, the pulp and paper industry have worked with lignocellulosic substrates and they have over the year developed many techniques allowing isolation of hemicelluloses. Chemical processes as soda pulping and kraft pulping allows isolation of both lignin and hemicellulose whilst protecting the cellulosic fibres in order to produce the largest amount of pulp possible per ton of biomass. Nevertheless, in both chemical processes previously mentioned, the hemicellulose are rather difficult to reach since they are mixed with a variety of organic and inorganic compounds including lignin as well as the chemicals that were used for the pulping process. During the last decades, the pulp and paper industry have started to look toward other processes that could allow a preliminary removal of hemicelluloses in order to avoid a complicated and expensive isolation after a chemical pulping process.

Amongst the techniques used for prehydrolysis, treatments with hot water catalyzed or not have been investigated in details in literature. As an example, Schild*et al.* [4] performed a preliminary extraction with water (via auto-hydrolysis) or with alkaline water prior to soda pulping in order to recuperate the hemicellulose prior to pulping. Similar testing was also performed on northern spruce with pressurised hot water in the presence of sodium bicarbonate [5]. Hot water extractions were also performed at temperature around 170 °C at different pH (the latter were adjusted with a phthalate buffer) and these experiments showed that control of pH was crucial in order to extract more of the hemicelluloses (up to 8 % wt on original biomass) [6]. Hot water extractions at similar temperature range have also been performed on maple [7] as well as on sugarcane bagasse [8]. Overall the hot water pretreatment may be a very promising approach for isolation of hemicelluloses although reported rates did not go far over 10 % because of the necessity to preserve the cellulosic fibres in order to avoid losses for papermaking. Acid catalyst has also been used as pretreatment to remove hemicellulose prior to pulping as reported by Liu*et al.* [9]. Utilisation of sulphuric acid, although very efficient to remove hemicellulose may also have an impact on cellulose thus reducing the pulp production rates.

Another process that could lead to isolation of hemicellulose is the organosolv process, which is to a certain extent comparable to classical chemical pulping in that sense that the technique allows simultaneous removal both for lignin and hemicelluloses. However, instead of using only an aqueous mixture of ions, the process relies on the utilisation of a combination of ions (usually alkaline) in a 50/50 mixture of aqueous organic solvent. In most cases, the solvent is methanol for obvious economic reasons although other solvents as butanol and certain organic acids have also been investigated to the same purposes. Recent work by Wang*et al.* [10]have shown that in an organosolv process using different solvent as well as different catalyst with poplar, sodium hydroxide was shown to be the best catalyst for hemicellulose removal from the pulp. Recent work by Brosse *et al.* [11] also showed that for *Miscanthus Gigantheus*, an ethanol organosolv process combined with an acid catalyst (sulphuric) lead to removal of most of the hemicelluloses and lignin from the original biomass.

Finally, another approach that could lead to isolation of hemicellulose from a lignocellulosic matrix is steam processes. This technique relies on impregnation of the feedstock with water (either catalyzed or not) then treatment under pressure at temperature ranging from 180-230 °C for a certain period of time after which pressure is relieved suddenly thus creating an "explosion" of the feedstock. Such process could lead, depending on the operating condition, to the isolation of either hemicellulose or lignin in two steps or in a single step. Our team has demonstrated the feasibility of both processes for different substrates [12-14].

Independently of the substrate or the technique used for the isolation of the hemicelluloses, conversion of lignocellulosic biomass, either for the production of paper or for the production of biofuels requires a complete utilization of the carbon compound found in biomass. Once the hemicelluloses are isolated from the original feedstock, they can undergo different types of transformation leading to different added value compounds that could lead to increase the margin of profit for the industries in the field.

Hemicelluloses account for 15-35 % of lignocellulosic biomass dry weight [2] and they are usually composed of different carbohydrates as well as small organic acids as acetic and formic acid. Glucose and xylose are often the most abundant sugars in hemicelluloses hydrolysis although mannose, arabinose and galactose might also be present in lower concentrations. The carbohydrate compositions of some lignocellullosic biomass are shown in Table 1. Whilst the C6 sugars could easily be fermented to ethanol following detoxification of the mixture, C5 sugars remains hard to convert to ethanol, mostly because classical yeasts don't metabolise them and the genetically modified organism that ferment C5 sugars are usually slower than classical organisms used in the production of etanol from C6 sugars. Nevertheless, even if ethanol production may remain a challenge, other alternatives could be considered, both on the chemical and on the microbiological point of view, to allow conversion of C5 sugar into added value products.

Carbohydrates tend to react in acidic, basic, oxidative or reductive mediums and therefore, numerous do arise for the conversion of C5 sugars. Although many options are available, this review will focus solely on 4 different pathways: acid, base, oxidative, and reductive. Each of these pathways could be inserted in an integrated biorefinery process where each of the fractions could be isolated and upgraded to high value compounds (see Figure 1).

Components (wt%)	Energy crops		Agricol residues		Forest residues	Coniferous
	Switchgrass [15]	Miscanthus [16]	Wheat Straw [15]	Corn Stover [17]	Aspen [18]	Loblolly Pine [19]
Glucan	38.5	55.5	39.2	36.2	52.4	36
Xylan	26.3	12.4	24.6	20.1	14.9	7.5
Galactan	1.16	-	-	1.45	2.2	2.5
Mannan	0.13	-	-	-	2.3	8.2
Arabinan	3.41	-	1.9	3.0	0.9	1.6

Table 1. Carbohydrate composition of some lignocellulosic biomass.

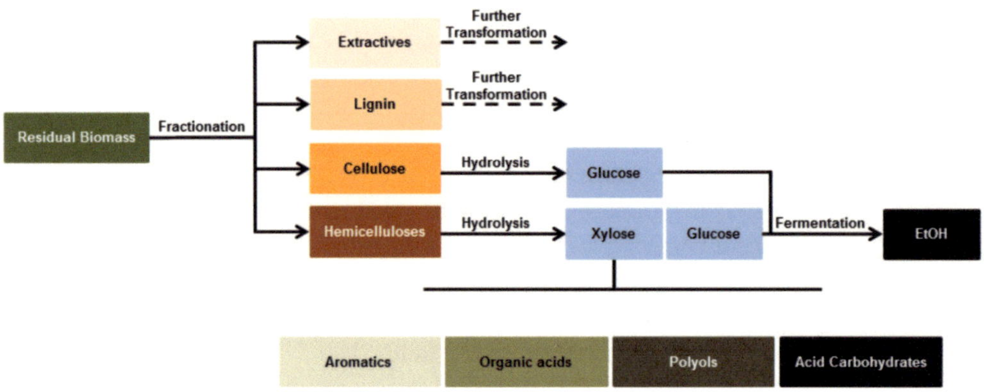

Figure 1. Potential utilization of hemicelluloses in an optimized conversion process for residual lignocellulosic biomass where C6 sugars are converted to ethanol, lignin and extractives to other added value products.

In this review, emphasis will be made on the recent work made for each of these conversion pathways both on the chemical and on the biochemical pathways. The review will focus on these 4 approaches also for their generally simple nature that would make them adaptable to an industrial context. These results will be compared to classical fermentation processes to produce ethanol with different types of organisms that can metabolise C5 sugars.

2. Conversion of xylose under an acid catalyst

2.1. The chemical pathway

Either in cyclic or aliphatic form, xylose then tends to dehydrate thus leading to the production of furfural whilst losing three molecules of water. Although this approach could explain the formation of furfural, it is not the sole options and many detailed reports have shown, by correlating the intermediaries with the actual structure, could be formed by many approaches depending on the reactant as reported by Marcotullio *et al.* [20] using halogen ions and proceeding only via the aliphatic form or as reported by Nimlos *et al.* [21] either via an aliphatic or a cyclic pathway (D-xylopyranose). Many different types of acid catalyst, either Brønsted or Lewis have been tested for the production of furfural. Although most of the acids reported in literature have been efficient so far for the production of the targeted molecule, one of the major side-reaction of furfural is polymerisation which influences the conversion rates and the selectivity of most of the processes reported in literature. An example of the abundance of research on this specific conversion is shown in Table 2 for different dehydration reactions under acid catalyst..

Catalyst	Conversion	Reference
H-Mordenite	98%	[22]
Sulphonic acid/Silica surface	99%	[23]
1-methylimidazole	91%	[24]
KI, KCl (dilute acid)	88%	[20]
NaCl, H_2SO_4	83%	[25]
1-alkyl-3-methylimidazolium	84%	[26]
NaCl, HCl	78%	[27]
Aluminium chloride Hexahydrate	76%	[28]
Amberlyst 70	75%	[29]
Zeolite H-Beta	74%	[30]
MCM-22, ITQ-2	70%	[31]
$FeCl_3$	71%	[32]
Nafion	60%	[33]
Keggin type acids	62%	[34]
Vanadyl pyrophosphate	53%	[35]

Table 2. Molar conversion to furfural in relationship with the catalyst used for the dehydration of xylose to furfural under acid catalyst.

For these reactions, the temperature is generally between 140-240 °C under proportional pressure allowing the mixture to remain liquid. Many researches also use a co-solvent, often toluene in order to isolate furfural from the aqueous mixture. The reason why toluene is so popular to this purpose is mostly related to the fact that toluene has affinity for fufural thus inhibiting its polymerization.

Heterogeneous catalyst has been proven to be very efficient for the process [22,23] although polymerisation tend to reduce the surface activity thus leading to a short-term deactivation of the catalyst. On the other hand, homogeneous catalyst was also shown to be efficient but at this point the whole technique relies on how the organic solvent is dispersed in the aqueous mixture. Reducing the size of the organic solvent particles in water (or vice-versa) to the maximum should allow the best transfer between the aqueous phase to the organic phase, assuming of course that furfural has suitable affinity for the solvent and that the partition coefficient favours the solvent.

Production of furfural itself is of course of significant interest because, amongst many factors, this chemical is commonly used in the industry as a solvent (mostly in oil chemistry). The average world production for furfural is 250 000 t/y and the actual market price evolves around 1000 USD/t [36] with recent market value reported to be closer to 1600 USD/tonne [37]. Furfural can also be a gateway to other products that could be used either as biofuels or as biomolecules. Example of such would be furfuryl alcohol via partial reduction of furfural (see Figure 2 below).

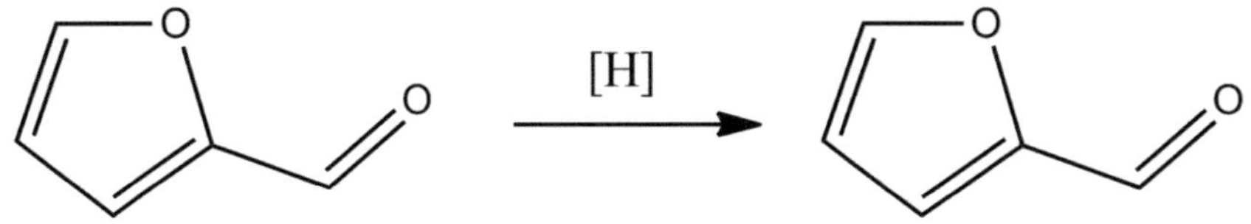

Figure 2. Reduction of furfural to furfuryl alcohol.

Furfuryl alcohol is also of interest since it is used as resins, adhesives and wetting agent, it has been mentioned that most of the 250 Kt/y of the furfural production is oriented toward production of furfuryl alcohol. The market value of this compound has been reported to be around 1800-2000 USD/tonne [38] and many reports in open literature mentions high selectivity for the conversion of furfural with iridium and ruthenium catalyst [39], rhodium [40], iron [41] and with zirconium oxide [42].

Another possible target for the transformation of furfural is for the production of 2-methyltetrahydrofuran (Me-THF) (see Figure 3). The latter is actually accredited as an additive for fuel and therefore, the possible market is virtually very important. It is also used in the petroleum industry to replace tetrahydrofuran (THF) that usually comes from non-renewables.

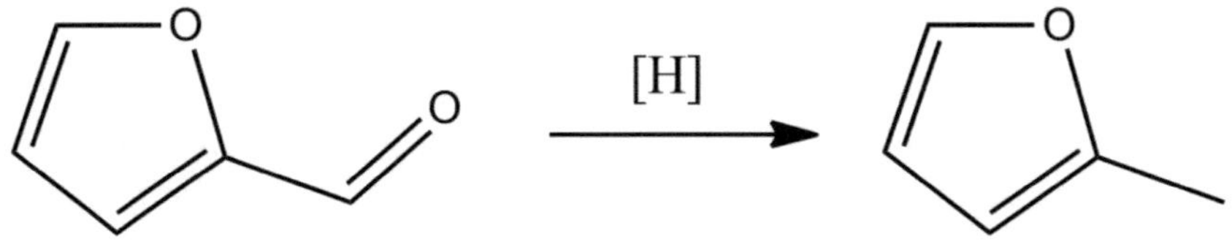

Figure 3. Reduction of furfural to 2-methyltetrahydrofuran.

Reduction of furfural to Me-THF seems to represent an important challenge since there is fewer reports mentioned in literature on the subject, as compared, as an example, to the reduction of furfural to furfuryl alcohol. Wabnitz *et al.* [43, 44] patented a one and two step process allowing conversion of furfural to Me-THF under a palladium-based catalyst and a mixture of palladium and copper oxide and chromium oxide as for the two step process.

Lange [45] patented a process using palladium and titanium oxide whilst Zheng et al. [46] worked with a copper alloy. Value for Me-THF could be estimated from the price of THF which is around 3000 USD/tonne [47] and the gap between the value of furfural and Me-THF could justify the process although hydrogen value can be estimated to be around 4.5 USD/Kg (estimated with the actual price of natural assuming reforming of the latter).

Another potentially interesting approach for a transformation of furfural would be decarboxylation to furan. The general process is depicted in Figure 4 below.

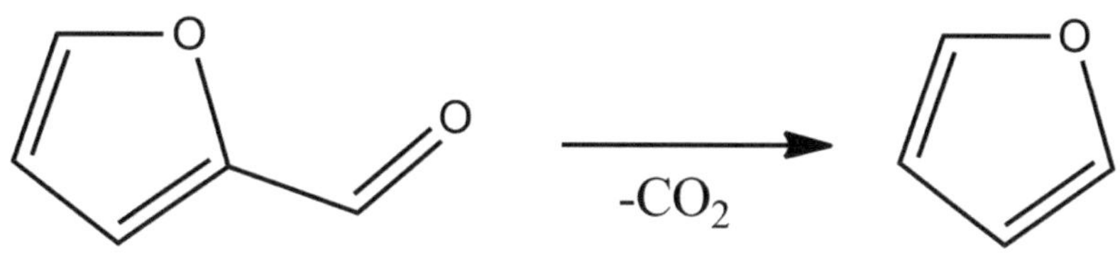

Figure 4. Decarboxylation of furfural to furan.

Many researches have focused on decarboxylation including work by Zhang *et al.* [48] who mentioned decarboxylation with potassium-doped palladium, and Stevens *et al.* [49] who reported conversion with copper chromite in supercritical CO_2.

Results reported in literature show that xylose, under an acid catalyst, tend invariably to dehydrate to furfural thus limiting the possibilities for side-products in such specific conditions. The acids could be Brønsted or Lewis type, all lead to the production of furfural furthermore when temperature are raised above 150 °C.

2.2. The biological pathway

Although furfural is a very common route for the conversion of xylose under an acid catalyst, furfural itself is rarely related to microorganisms in that sense that it is often considered as an inhibitor instead of a metabolite. Nevertheless, to the best of our knowledge, no report mentioned a biological conversion of xylose to furfural.

3. Conversion of xylose under a base catalyst

3.1. The chemical pathway

The interaction between xylose and bases, either Brønsted or Lewis, is rather less reported in the literature when compared to the acid conversion of xylose to furfural indicated in the previous section. Many very different reactions have been reported as in the case of Popoff and Theander [50] that have quantified the cyclic compounds produced after a base-catalyzed reaction of pure D-xylose at 96 °C for 4 hours. The produced compounds are rather peculiars in comparison to other work made on the subject (see Figure 5) since most of the reported compounds are aromatics. The presence of aromatics may be a result that the reaction time was long and the isomerisation that was required in order to induce such reaction was efficient. Johansson and Samuelson [51] tested the effect of alkali treatments (NaOH) on birch xylan and contrarily to the previous research; they found that the treatment led to the production of a variety of organic acids. Testing on untreated xylene showed that most of the organic acids were already obtained from xylans and the most distinctive impact was observed after a 2 day test at 40 °C where the concentrations of L-galactonic and altronic acids increased significantly which could be related to a less severe treatment of xylans that also include C6 sugars.

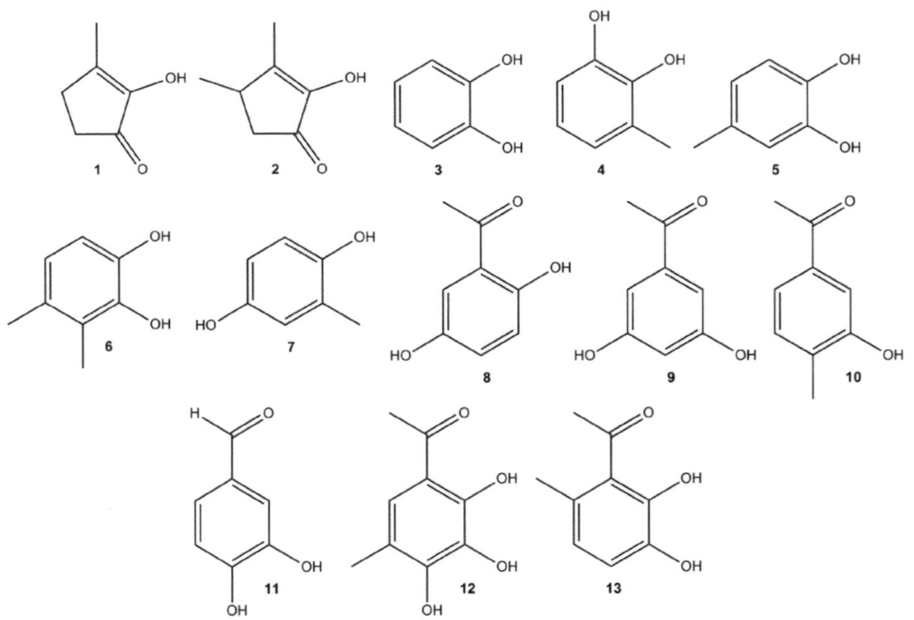

Figure 5. Cyclic and aromatics obtained from the based-catalysed treatment of D-xylose under a sodium hydroxide catalyst where (1) 2-hydroxy-3-methylcyclopent-2-enone; (2) 2-hydroxy-3,4-dimethylcyclopent-2-enone; (3) pyrocatechol; (4) 3-methylbenzene-1,2-diol; (5) 4-methylbenzene-1,2-diol; (6) 3,4-dimethylbenzene-1,2-diol; (7) 2-methylbenzene-1,4-diol; (8) 1-(2,5-dihydroxyphenyl)ethanone; (9) 1-(3,5-dihydroxyphenyl)ethanone; (10) 1-(3,4-dihydroxyphenyl)ethanone; (11) 3,4-dihydroxybenzaldehyde; (12) 1-(2,3,4-trihydroxy -5-methylphenyl)ethanone; (13) 1-(2,3-dihydroxy-6-methylphenyl)ethanone.

El Khadem *et al.* [52] studied the effect of xylose conversion in an alkali medium at low temperatures (room) and for long periods (1-4 weeks) and one of the interesting features of his work was that the process did lead to the epimerization of sugars, but furthermore, it leads to the production of C6 sugars most probably from a reverse aldol reaction. Among the sugars that were formed during the reaction, conversion of xylose was shown to be more efficient to lyxose (18 %) and arabinose (15 %) with a decrease observed for most of the compounds between 1 and 4 weeks (see Figure 6). A vast majority (more than 50 %) of xylose remains on its original form and the reaction leads to the production of 1 % glucose and 2.5 % of sorbose, both are C6 sugars.

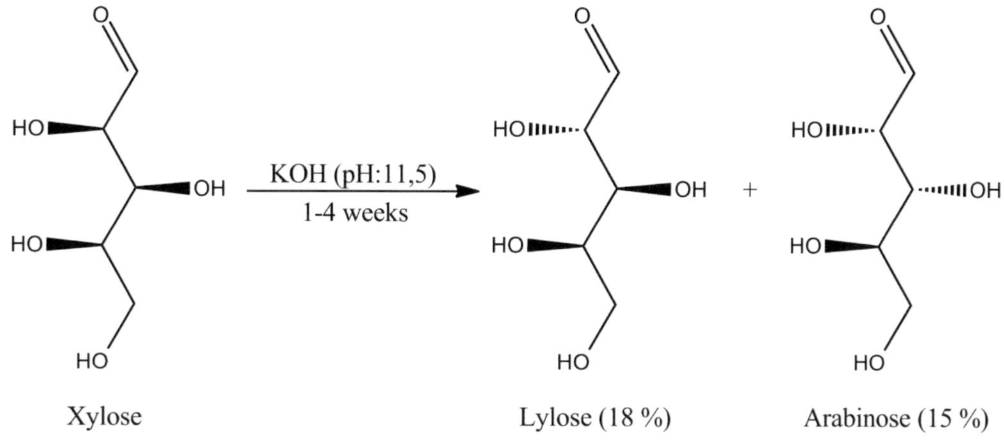

Figure 6. Major epimerisation products from 1-4 week reaction of D-xylose in a pH 11.5 KOH solution at room temperature.

Xylose, as the other carbohydrates, is converted to smaller organic acids when reacted with a strong alkali medium. As an example, Jackson *et al.* [53] have demonstrated that the conversion of xylose to lactic acid could reach 64 % (molar) accompanied by glyceric acid. Although they did not used xylose but rather ribose and arabinose, they were able to reach conversions between 35-43 % into lactic acid using potassium hydroxide as catalyst under microwave irradiation [54]. Rahubadda *et al.* [55] have provided a mechanism for the conversion of xylose to lactic acid under a base catalyst. The simplified pathway is depicted in Figure 7 below.

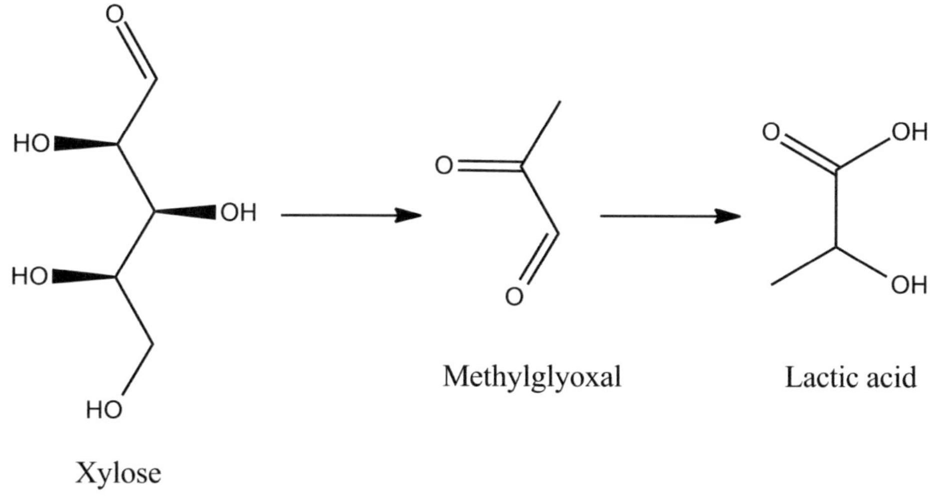

Figure 7. Conversion of D-xylose to lactic acid via the methylglyoxal pathway.

They mentioned in this report that methylglyoxal is most probably derived from glyceraldehyde as depicted in Figure 8 below. The possible reaction leading to methylglyoxal may involve an E2 reaction on C2 leading to removal of the hydroxyl group on C3 then a keto-enol rearrangement to methylglyoxal.

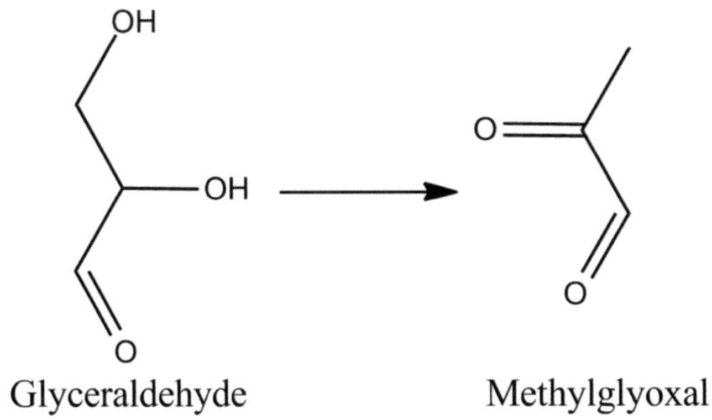

Figure 8. Conversion of glyceraldehyde to methylglyoxal.

Onda *et al.* [56] achieved a conversion rate of more than 20 % when using xylose as a feedstock with a carbon-supported platinum catalyst in alkali solution. In a recent report by Ma *et al.* [57], it was shown that using model compounds, different carbohydrates tend to convert into lactic acid at different levels. Fructose was shown to be more effectively converted to lactic acid than glucose and finally than xylose. The work also showed a correlation between the amount of catalyst (varying from 1-3 % wt.) of NaOH, KOH and $Ca(OH)_2$ respectively. Part of the work by Aspinall *et al.* [58] was aimed at the non-oxidative treatment of xylans from different substrates using sodium hydroxide as solvent. The reaction was performed at room temperature for 25 days and amongst the products that emerged from this reaction, a majority was acidic and lactic acid as well as formic acid were the two major products. Other work by Yang *et al.* [59] showed that higher temperature treatments of xylose (200 °C) in a $Ca(OH)_2$ solution produced about 57 % (mol.) of lactic acid with 2,4-dihydroxybutanoic acid in second with 10 % (mol.). The same conversion patterns were observed by Raharja *et al.* [60] with production rates for lactic acid above 50 %.

3.2. The biological pathway

Amongst the different options for the conversion of xylose reported in the previous chapter, production of lactic acid via the microbial route is a vastly studied field [61-63] since currently, all of the production of lactic acid at an industrial scale in the world is biologically based. Traditionally, the concept evolves around fermenting carbohydrate-based syrup by homolactic organisms, mostly lactic acid bacteria (LAB). The most common carbohydrate-based substrates used to this purpose may be molasses, corn syrup, whey, sugarcane or even beet bagasse. Highly efficient LAB includes *Lactobacillus delbrueckii, L.*

amylophilus, L. bulgaricus and *L. leichmanii*. Mutant *Aspergillus niger* has also been reported to be effective at an industrial scale [64]. LAB have the particularity to possess an homofermentative metabolism producing only lactic acid as extracellular waste product, instead of the heterofermentative pathway yielding by-products such as aldehydes, organic acids and ketones. The catabolic pathway yielding lactic acid is essentially the same across all organisms; the pyruvate intermediate is converted to lactic acid by a lactate dehydrogenase (LDH). Thus for hexose sugars, the theoretical yield is 2 moles of lactate per mole of sugar (or 1g sugar for 1g lactate). This enzymatic catalysis has the advantage over its chemical counterpart to be stereospecific: both L-lactate-dehydrogenase (L-LDH) and D-lactate-dehydrogenase (D-LDH) exist, generating either L-lactate or D-lactate respectively [65]. Both are NAD-dependant (nicotinamide adenine dinucleotide) and may be found alone or together in wild lactate-producing microbial strains. Since optical purity of lactate is a major requirement for the lactate industry, research focuses on stereospecificity as much as yields and productivity [61,66-70].

An efficient lactate producer has to display specific attributes, mainly the adaptability to low-cost substrates, high selectivity of desired enantiomer (L, D or both), high optimal temperature for decreased contamination risks, low pH tolerance and high performances (yield and productivity). LAB display appreciable performances, but lack a low pH tolerance, which implies uses of a pH control apparatus during the fermentation process. LAB optimal pH is near neutral, but the pKa of lactic acid being 3.8, an alkali agent, usually $Ca(OH)_2$, must be used thus generating calcium lactate. After typical batch fermentation, the medium is acidified with H_2SO_4 therefore regenerating and purifying the lactic acid [64]. Another drawback of LAB is their requirement for a complex growth medium, since they are auxotroph for certain amino acids and vitamins [71]. In order to overcome this problem, many fungi were also investigated for lactate production. Strains of *Rhizopus, Mucor* and *Monilla sp.* have shown potential whilst other fungi even displayed amylolytic activity, which could lead to a direct starch-to-lactate conversion [72-74].

Most researches still focuses on hexose conversion, and research group have optimized strains and process strategies in order to obtain high lactate titers, yields and productivities. Ding and Tan [75] developed a glucose fed-batch strategy using *L. casei* and generating up to 210 g/L of lactic acid with a 97 % yield. Chang *et al.* [76] proposed a continuous high cell density reactor strategy yielding a titer of 212.9 g/L and productivity of 10.6 g/L/h with *Lb. rhamnosus*. Dumbrepatil *et al.* [77] created a *Lb. delbrueckii* mutant by ultraviolet (UV) mutagenesis producing 166 g/L with productivity of 4.15 g/L/h in batch fermentation. Genetically engineered non-LAB biocatalysts yet have to match the performances of highly efficient wild LAB. In fact, *C. glutamicum, S. cerevisiae* and *E. coli* recombinant have been developed, but with limited success [61].

The search for lignocellulose-to-lactate biocatalysts have led to the discovery of many strains of pentose-utilizing LAB. *Lb. pentosus ATCC8041* [78, 79], *Lb. bifermentans DSM20003* [80], *Lb. brevis* [81], *Lb. Plantarum* [82], *Leuconostoc lactis* [83, 84], and *E. mundtii QU 25* [85, 86]. Lactic acid produced from xylose per say has been investigated by few [84,85, 87, 88], but with mitigated results, mainly due to the fact that the pentose-utilizing

LAB do not perform as well in pentoses as in hexoses-rich metabolism. This phenomenon is most likely due to the fact that pentoses are metabolized in the PK pathway (phosphoketolase), thus for a given strain, even if hexoses are fermented through an homofermentative route, pentose will yield heterofermentative products (i.e. acetic and lactic acid) [78, 89]. Nevertheless, Tanaka *et al.*[84] have shown that in addition to the PK, *L. lactis* could metabolize xylulose-5-phosphate (X5P), an intermediate pentose catabolite, through the pentose phosphate pathway (PPP). The theoretical yield through the PPP is 5 moles of lactate for 2 moles of pentoses, but through the PK it decreases to 1:1 [61], thus, the conversion advantage of the PPP is obvious. Okano *et al.* [87,89] demonstrated this approach by creating a pentoses-utilizing *Lb. plantarium* recombinant in which the native L-lactate dehydrogenase (L-LDH) gene was disrupted, leaving only the homologous D-lactate dehydrogenase (D-LDH) active. However, this strain produced both acetic and D-lactic acid; hence the PK gene (*xpk1*) was substituted by a heterologous transketolase (*tkt*) from *L. lactis*, thereby shifting heterolactic fermentation to a homolactic one.

Modification of yeast strains in order to achieve xylose-to-lactate conversion has also been investigated, as an example Ilmen *et al.* [90] expressed the L-LDH gene from *L. helveticus* in *P. stipitis* and was able to reach a titer of 58 g/L of lactate with a yield of 58 %. These results were obtained despite the fact that no effort had been made to silence the native PDC/ADH (pyruvate decarboxylase/alcohol dehydrogenase) ethylic pathway, consequently 4.5 g/L of ethanol was simultaneously produced as the endogenous PDC rivalled against the recombinant L-LDH for pyruvate. Tamakawa *et al.* [88] went further by transforming *C. utilis*, disrupting the native *pdc1* gene, and expressing heterologous LDH, XR (xylose reductase), XDH (xylitol dehydrogenase) and XK (xylulokinase) enzymes. Furthermore, to prevent the redox imbalance, they increased the XR's NADH (reduced nicotinamide adenine dinucleotide) affinity by site-directed mutagenesis. In batch culture this recombinant was able to yield titers up to 93.9 g/L of lactate at a yield of 91 %. Table 3 shows the most recent and most efficient strains developed for lactic acid production, both from hexoses and pentoses.

Strain	**Gen Eng Str**	**Medium**	**Process**	**LA (g/L)**	**Tf (h)**	**Yield (g/g)**	**Prd (g/L/h)**	**Ref**
E. mundtii QU 25	-	Cellobiose	Batch	119	106	0.83	1.12	[86]
	-	Xylose	Batch	86.7		0.84	0.9	[85]
	-	Glucose/ cellobiose	Batch	35.1	15	0.91	2.99	[86]
Lactobacillus sp. RKY2	-	Wood hydrolysates*	Continuous w/cell recycling	27	-	0.9	6.7	[91]

Strain	Gen Eng Str	Medium	Process	LA (g/L)	Tf (h)	Yield (g/g)	Prd (g/L/h)	Ref
Lb. bifermentas DSM 20003	-	Wheat bran hydrolysates	Batch	62.8	60	0.83	1.17	[80]
Lb. casei NCIMB 3254	-	Cassava bagasse	Batch SSF**	83.8	60	0.96	1.4	[92]
Lb. delbrueckii Uc-3	UV mutagenesis	Cellobiose	Batch	90	40	0.9	2.25	[93]
		Molasse	Batch	166	40	0.95	4.15	[77]
Lb. lactis RM 2-24	UV mutagenesis	Cellobiose	Batch	80	48	0.8	1.66	[94]
		Cellulose	Batch SSF	73	48	0.73	1.52	
Lb. plantarum ΔldhL1-xpk1::tkt	Disruption of endogenous LDH gene. Replacment of endogenous PK (*xpk1*) gene with heterologous *tkt* to redirect the PK pathway to the PPP.	Arabinose	Batch	38.6	28	0.82	1.37	[89]
Lb. plantarum ΔldhL1-xpk1::tkt-Δxpk2	Idem as above. Disruption of 2nd PK gene (*xpk2*) to terminate acetate production.	Xylose	Batch	41.2	60	0.89	0.67	[87]
Lb. rhamnosus ATCC 7469	-	Paper Sludge	Batch SSF	73	168	0.97	0.45	[95]
Lb. rhamnosus ATCC 9595 (CECT288)	-	Apple pomace	Batch	32.5	6	0.88	5.41	[96]
L. lactis IO-1	-	Xylose	Batch	33.3	-	0.68	-	[84]
S. cerevisiae recombinant	Replacement of native *pdc1* and *pdc5* by heterologous bovine L-LDH gene.	Glucose	Batch	82.3	192	0.83	0.43	[97]

Strain	Gen Eng Str	Medium	Process	LA (g/L)	Tf (h)	Yield (g/g)	Prd (g/L/h)	Ref
S. cerevisiae recombinant	Disruption of *pcd1* and *adh1* genes. Expression of bovine L-LDH.	Glucose	Batch	71.8	65	0.74	1.1	[98]
K. lactis	Disruption of PDC and PDH genes. Expression of bovine L-LDH gene.	Glucose	Semi-Batch	60	500	0.85	0.12	[99]
C. utilis	Disruption of endogenous PDC gene.Expression of heterologous LDH, XR, XDH and XK. XR gene site-specific mutation for preferential NADH cofactor utilization	Xylose	Batch	66.7	78	0.79	1.62	[88]
			Batch	93.9	78	0.91	2.18	
P. stipitis	Expression of LDH from *L. helveticus.*	Xylose	Batch	58	147	0.58	0.39	[90]

* *No xylose consumption occurred*

***SSF = simultaneous saccharification and fermentation*

Table 3. Lactic acid concentration (LA), time of fermentation (Tf), yield and production rate for the most common microorganisms used for the biological conversion of xylose to lactic acid

Lactic acid seems to be, on the biological as well as on the chemical point of view the best possible compound that could be derived from a based-catalysed reaction of xylose. Racemic mixtures of lactic acid (most probably derived from chemical synthesis) can be evaluated to 1150 USD/tonne [100] whilst the pure isomer was reported to have a price market around 1750 USD/tonne [101]. As in many cases, the price will vary proportionally with purity of the compound. Utilisation of lactic acid on the market is mostly related to polymers, food, pharmaceutical and detergents. The annual world demand for the compound should reach a little more than 367 Ktonnes/year by 2017 [102].

4. Conversion of xylose under reducing conditions

4.1. The chemical pathway

Xylose, as all the other carbohydrates that can be isolated from lignocellulosic biomass, has a carbonyl function that is susceptible to transformations, including reduction. One of the most common compounds that can be derived from xylose is xylitol, a pentahydroxy chiral compound as depicted in Figure 9.

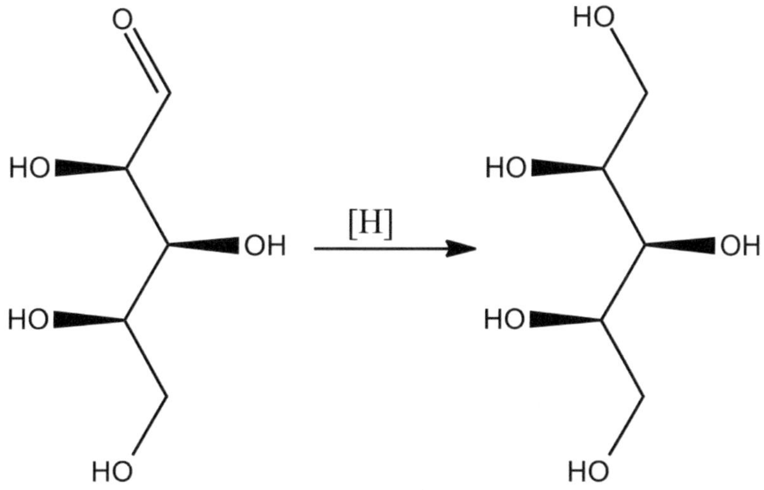

Figure 9. Simplified conversion of D-xylose to D-xylitol.

Amongst the most reported catalysts in the literature are nickel and Raney nickel. According to Wisniak *et al.* [103] they are good catalysts for the production of xylitol from xylose with total conversion at 125 °C and 515 psi. In the same year, the authors published the use of ruthenium, rhodium and palladium for the reduction of xylose [104] concluding that the efficiency of those metals was declining in the order Ru>Rh>Pd at temperatures around 100-125 °C under pressure. Mikkola *et al.* [105, 106] also used nickel as a catalyst by ultrasonic process that generated close to 50 % conversion of xylose to xylitol. From this process was reported that an important problem was the deactivation of the catalyst. Utilisation of nickel also led to the publication of two patents, one in 2003 [107] and another in 2007 [108]. In the case of the first, the concept relied on the isomerization of D-xylose to L-xylose prior to catalytic reduction under a nickel catalyst.

Ruthenium as well as ruthenium-based compounds has also been reported as catalysts for the reduction of xylose to xylitol. Ruthenium has been operated at temperatures between 90 °C and 110 °C under pressure using ruthenium supported either on silica [109] or on carbon [110]. Conversion rates for the latter have been reported to reach 35 % to xylitol for the latter with coproduction of glycerol and ethylene glycol. Ruthenium chloride ($RuCl_3$) has also been reported as a catalyst for the reduction of xylose to xylitol [111, 112].

Treatment of carbohydrates at a higher severity leads to the hydrogenolysis, implying not only the carbonyl compounds being reduce to alcohol but a breakage of the carbon-carbon bonds in the original carbohydrate. Recent work [113] shows that temperature above 250 °C and pressure between 600-1000 psi, can lead to conversion of xylose to ethylene glycol, propylene glycol and glycerol, as depicted in Figure 10 below.

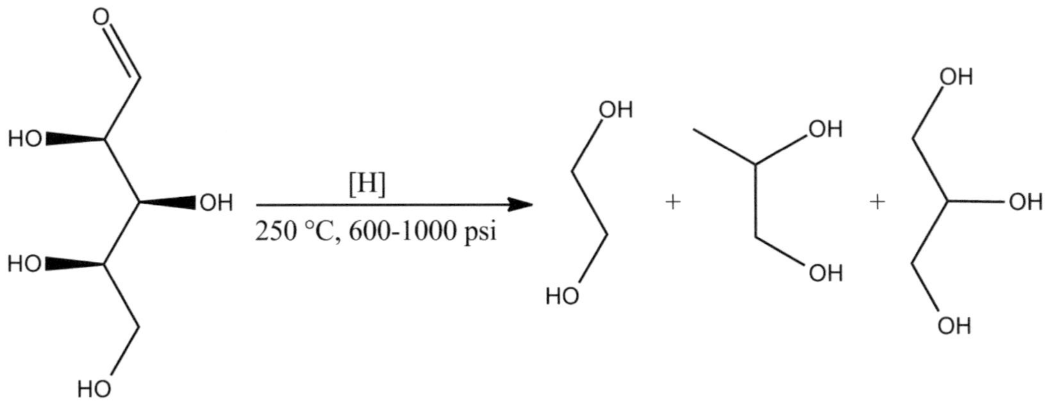

Figure 10. Simplified conversion of D-xylose to ethylene glycol, propylene glycol and glycerol as reported by Crabtree *et al.* [113].

Production of ethylene glycol and glycerol has also been reported by Guha *et al.* [110] as a side product of their xylitol production. Hydrogenolysis of xylitol is a logical suite for reduction of xylose and specific work has been reported using different catalytic systems and experimental setups. As an example, it was recently reported [114] that xylitol could be converted into a mixture of polyols and different other products as formic acid and lactic acid as well as xylitol, which, according to the previously mentioned work in this chapter, is given when xylose is submitted to a noble metal catalyst under hydrogen. In this specific case, the catalyst was platinum supported on carbon under a base-catalyzed matrix. Chopade *et al.* [115] also presented a patent reporting the conversion of carbohydrates (including xylose) into polyols using a ruthenium catalyst as did Dubeck and Knapp in 1984 [116].

In 2010 it was reported the use of nickel as a catalyst for hydrogenolysis of xylose [117] whilst Kasehagen [118] reported hydrogenolysis of carbohydrates under a nickel-iron-copper catalyst using a matrix of alkali salts with glycerol as the main product. The effects of nickel was studied by Wright [119] but this time using tungsten as a co-catalyst. Finally, there is a report about hydrogenolysis of carbohydrates under a rhenium catalyst [120].

4.2. The biological pathway

Only a few bacteria have been shown to naturally produce xylose as a metabolite. It has been showed [121] that a bacteria belonging to the genus *Gluconobacter* was able to produce xylitol from arabitol by way of a membrane-bound D-arabitol deshydrogenase (AraDH), followed by a soluble XDH. Rangaswamy *et al.* [122] isolated strains of *Serratia*, *Cellulomonas* and *Corynebacterium* species that were able both to grow and produce xylitol with xylose as sole carbon source, although the reported yields were very low. In early work [123, 124], it was found that both *Corynebacterium* and *Enterobacter liquefaciens* strains were able to grow and produce xylitol from xylose although gluconate had to be present as cosubstrate. Nevertheless, studies using wild bacterial strains for xylitol production are scares [122, 125-127]. In most metabolic pathways, bacteria go through direct xylose to xylulose conversion via isomerisation, bypassing the xylitol intermediate. Subsequently, xylulose is phosphorylated in X5P and can be metabolized by most prokaryotes and eukaryotes via the PPP, or the PK pathway in the case of heterolactic bacteria (Figure 11) [128].

Although the fact that yeast and fungi are generally more efficient xylitol producers than bacteria is widely recognized [129], certain highly productive species such as *Candida* are actually known for their pathogenic nature [130]. Moreover, construction of recombinant yeasts by introduction of xylose reduction pathway in GRAS species such as *S. cerevisiae* have been accomplish, although these recombinant still have to match the productivities found using non-GMO organisms (genetically modified *organisms*) [131-134]. Bacterial species on the other hand present high yields, fast metabolism and many GRAS (generally recognized as safe) species with recombinant strains often display higher efficiencies than their non-altered counter-part [135].

It was found that the catabolic rate of xylose is usually enhanced by the presence of a co-substrate such as glucose [136, 137]. However, most organisms preferentially use glucose to any other sugars due to allosteric competition in sugar transport and/or repression of other carbon catabolites [138, 139]. Thus, a suitable biocatalyst would have to simultaneously metabolize both substrates. This functionality was achieved in *E. coli* [140]by replacing the putative cAMP-dependent receptor protein (CRP) with a cAMP-independent mutant, which also expressed a plasmid-based xylose transporter. Similarly, some authors [125] used this approach as well as inserting the heterologous XR gene and silencing the endogenous xylose isomerase (XI). Alternatively, heterologous XR and XDH may be introduced and the putative XK (*xylB* gene) silenced.

Other well suited candidates for such a bioconversion would be LAB, offering the advantage of an energy metabolism completely independent of their limited biosynthetic activity, thus their glycolysis pathways may be engineered without disturbing other key structural pathways [129]. By introduction of yeast XR gene, as well as a heterologous xylose transporter in *L. lactis,* they showed that bacterial productivity and yield might reach those of the best yeasts. Even if all xylose is not consumed when in high initial concentration, the non-pathogenic and anaerobic nature of *L. lactis* is a notable advantage.

Early work done on *Corynebacterium glutamicum* showed another alternative for the production of xylitol but the necessity of inserting gluconate as co-substrate for NADPH (nicotinamide adenine dinucleotide phosphate) regeneration rendered the application non economical [122,124]. Sasaki *et al.* [141] developed a *C. glutamicum* recombinant achieving simultaneous co-utilization of glucose/xylose. This was done by introducing the pentose transporter area in *C. glutamicum* chromosomal DNA (deoxyribonucleic acid). *C. glutamicum* is a noticeable candidate for its non-pathogenic and gram-positive nature, as well as its ex-

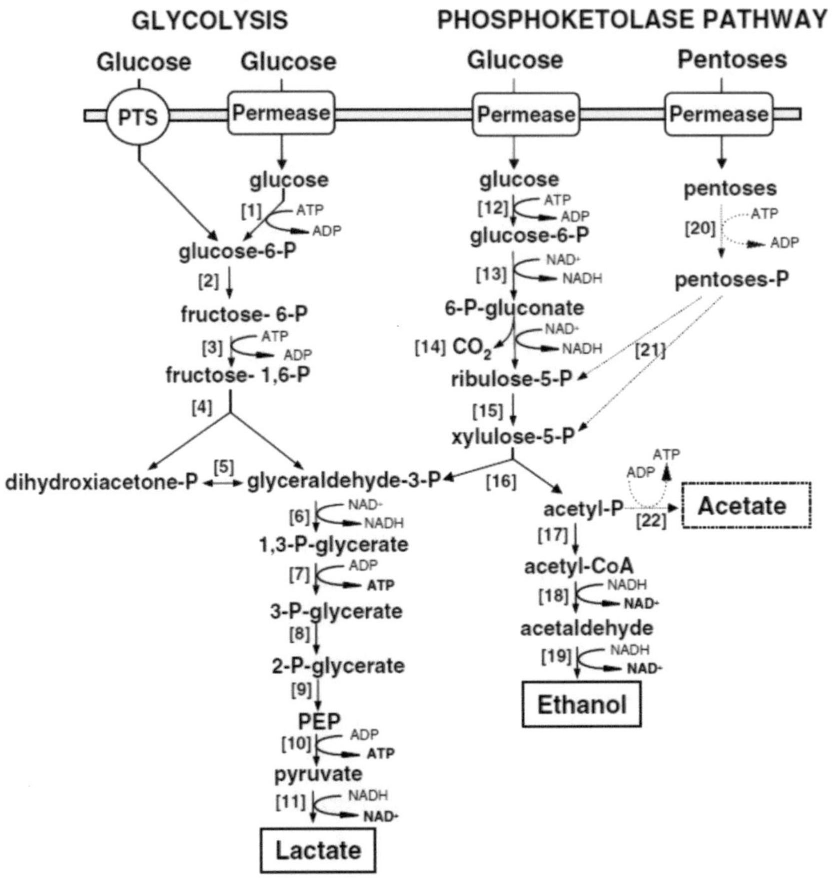

Figure 11. Glycolysis and phosphoketolase (pentose phosphate) pathways in lactic acid bacteria (1) glucokinase, (2) phosphoglucose isomerase, (3) phosphofructokinase, (4) fructose 1,6-bisP aldolase, (5) triose-phosphate isomerase, (6) glyceraldehyde-3P dehydrogenase, (7) phosphoglycerate kinase, (8) phosphoglycerate mutase, (9) enolase, (10) pyruvate kinase, (11) lactate dehydrogenase, (12) hexokinase, (13) glucose-6P dehydrogenase, (14) 6-phosphogluconate dehydrogenase, (15) ribulose-5P 3-epimerase, (16) xylulose-5P phosphoketolase, (17) phosphotransacetylase, (18) acetaldehyde dehydrogenase, (19) alcohol dehydrogenase; (20) pentose kinase, (21) pentose phosphate epimerase or isomerase, (22) acetate kinase. *CoA* coenzyme A.

tensive use for amino and nucleic acid industrial synthesis [142, 143]. It was established [135] that xylitol productivity may be improved by disabling the xylitol import system (ptsF gene) and suggested that more work done on xylitol export system and redox balance may yield further improvements. Nevertheless, their CtXR7 *C. glutamicum* recombinant attained a productivity of 7.9 g/L/h and final xylitol concentration of 166 g/L after 21 h (see Table 4). This was achieved by (to date this is considered the best xylitol bacterial producer):

- introduction homologous pentose transporter (*araE*);
- disruption of the native lactate deshydrogenase (*ldhA*);
- expression of single-site mutant XR from C. *tenuis*;

- disruption of XK native gene (*xylB*);
- disruption of phosphoenolpyruvate-dependent fructose phosphotransferase (*ptsF* gene; PTS^{fru}).

Strain	Genetic Engineering Strategy	Yield g/g	Xylose g/L	Xylitol (g/L)	Tf (h)	Prd (g/l/h)	Process Strategy	Reference
Candida athensensis SB18	-	83%	250	207.8	175	1.15	Batch limited O_2	[144]
		87%	300	256.5	250	0.97	Fed Batch limited O_2	
		79%	200	151.71	156	0.97	Batch limited O_2	
C. tropicalis ASM III	-	93%	200	130	120	1.08	Batch limited O_2	[145]
Candida sp. 559-9	-	99%	200	173	121	1.44	Batch limited O_2	[146]
C. tropicalis KCTC 10457	-	87%	200	172	48	3.66	Batch limited O_2	[147]
C. tropicalis KFCC 10960	-	93%	270	251	55	4.56	Fed Batch	[148]
C. tropicalis KCTC 10457	-	90%	260	234	48	4.88	Fed Batch	[147]
C. guilliermondii	-	73%	250	-	-	-	Fed Batch limited O_2	[149]
C. tropicalis	-	82%	750	189	58	4.94	Fed Batch/ Cell recylcing/ Glucose cosubstrate/ limited O_2	[150]
C. tropicalis	-	69%	100	-	-	5.7	Cell recycling/ limited O_2	[151]
C. tropicalis	-	85%	214	182	15	12	cell recycling/ limited O_2	[147]
S. cerevisiae	Expression heterologous XR gene from *P. stipitis*.	95%	190	-	-	0.4	Fed batch/ Glucose cosubstrate	[152]
Corynebacterium glutamicum CtXR7	Expression of *araE* pentose transporter gene. Disruption of *ldhA*. Single site mutation of heterologous XR gene. Disruption of *xylB*& PTS^{fru} genes.	-	120	166	21	7.9	Fed batch/ Glucose cosubstrate/ 40g/L dry cell	[135]
D. hansenii NRRL Y-7426	-	38%	45	19.7	72	0.274	Batch/ Detoxified grape marc hydrolysates	[153]
S. cerevisiae	Overexpression ALD6 & ACS1 genes.Expression of *P. stipitis* XR gene.	~100%	20	91.3	60	1.76	Fed batch/ Glucose cosubstrate	[154]
Lactobacillus brevis NZ9800	Expression of *P. stipitis* XR gene.Expression of *Lb. brevisxylT* symporteur.	~100%	160	75	41	2.72	Fed batch/ Glucose cosubstrate	[129]

Strain	Genetic Engineering Strategy	Yield g/g	Xylose g/L	Xylitol (g/L)	Tf (h)	Prd (g/l/h)	Process Strategy	Reference
C. tropicalis	-	83%	80	96.5	120	1.01	Fed batch/ Corn Cob hydrolysates/limited O_2	[155]
C. tropicalisSS2	Xylitol-assimilation deficient strain by chemical mutagenesis.	93%	100	220	70	3.3	Fed batch/ aerobic	[156]
C. trolpicalis JH030	-	71%	45	31.1	80	0.44	Batch/ Rice straw hydrolysates	[157]

Table 4. Overview of the different strains allowing conversion of xylose to xylitol including yields, fermentation time (Tf), production (Prd) and the process strategy.

As previously discussed for ethanol, the redox imbalance that often occurs from XR/XDH preferential use of NADPH/NAD+ cofactors is a key factor for xylitol accumulation in the cell. In most yeast studied, it has been shown that XR has a marked preference for NADPH, while XDH has a quasi-unique specificity for NAD+ [126]. The main exception being *P. stipitis* who shows a nearly by-specificity for NAD(P)(H) for its XR and *P. tannophilus* whose XDH shows a higher activity with NADP+ than NAD+ [158] proposed a theoretical maximum xylitol yield in yeasts of 0.905 mol of xylitol per mol of xylose when NADH was efficiently used as cofactor by the XR or under aerobic condition where the NADH can be oxidized back to NAD+ in the respiratory chain. Otherwise, under anaerobic conditions, the theoretical yield drops to 0.875. These yields follow the equations (1) and (2) below respectively:

$$126 \text{ xylose} + 3\, O_2 + 6 \text{ ADP} + 6\, P_i + 48\, H_2O \rightarrow 114 \text{ xylitol} + 6 \text{ ATP} + 60\, CO_2 \quad (1)$$

$$48 \text{ xylose} + 15\, H_2O \rightarrow 42 \text{ Xylitol} + 2 \text{ ethanol} + 24\, CO_2 \quad (2)$$

Owing the better yield both in xylitol and ATP (adenosine triphosphate) under oxygen-limited xylitol production, aeration is a crucial parameter. As a general trend, xylitol production increases when oxygen is allowed in the medium under a certain threshold concentration [159]. This preference is yeast specific since for *P. stipitis* it is reported that the absence of dissolved oxygen is needed for optimal xylitol production; while *P. tannophilus* reaches maximum yields under anoxic conditions [160, 161].

Many strains of *S. cerevisiae* have been transformed for xylose utilization in the early 90's. As for xylose-to-xylitol, Hallborn *et al.* [152] reported a highly efficient conversion of xylose to xylitol (95 % of theoretical). It has been suggested that the incapacity of *S. cerevisiae* to rapidly replenish its NADPH pool from its PPP during xylose metabolism is what causes the metabolic bottleneck [162, 163]. This is mainly due to the fact that xylose is a non-preferred carbon substrate for *S. cerevisiae* and do not provide sufficient energy for growth and metabolism [164].

C. tropicalis is a candidate of choice for xylitol production among the few native strains reported as the best xylitol producers to date (see Table 4) and this research for native strains and genetically engineered recombinant is still under way today [155-157]. As in *S. cerevisiae*, the PPP is the major NADPH biosynthesis pathway and efforts have been made to increase its flux. Ahmad *et al.* [165] recently successfully increased the metabolic flux toward PPP for NADPH regeneration, thereby enhancing xylitol production of the original strain by 21 %. This was done by disrupting XDH putative gene, and over-expressing homologous glucose-6-phosphate dehydrogenase (G6PDH) and 6-phosphogluconate dehydrogenase (6-PGDH). Table 4 summarize the best xylitol producing strains found in the literature up to date.

Reduction of xylose either at low or at high severity thus producing either xylitol or polyols (including glycerol) is a process driven by the price of hydrogen. On the other hand, the market for small polyols as ethylene or propylene glycol may generate more opportunity than the xylitol market. Xylitol market value is between 3650 and 4200 USD/tonne [166] whilst ethylene glycol is reported at a market price of 980-1500 USD/tonne [167] and propylene glycol at 1500-1700 USD/tonne [168]. The market for each of the previously mentionned compound is around 100 Ktonnes/y for xylitol [169], 19 Mtonnes/y for ethylene glycol [170] and 1.4 Mtonnes/y for propylene glycol. Although the market for smaller polyols may seem to be larger, as an example conversion of xylose to ethylene glycol and propylene glycol would require 3 times as much hydrogen if compared to xylitol. Since the price for hydrogen can be estimated roughly at 4.5-5 USD/Kg, the very concept of polyols production relies on the efficiency of the hydrogenolysis process therefore explaining why many of the reported litterature in this chapter are patents.

5. Conversion of xylose under oxidizing conditions

5.1. The chemical pathway

Oxidation of xylose has been numerously reported in the literature although focus interest, both on the biological as well as chemical point of view has been focused toward a simple oxidation of xylose to xylonic acid (see Figure 12).

Oxidation of xylose has been reported for a variety of different metallic catalyst including gold for high conversion rates [171]. Using a process performed a little higher than room temperature in a basic pH for 1 hour, they were able to reach a 78 % conversion of xylose to xylonic acid. Using comparable catalyst, Pruesse *et al.* [172] were able to reach 99 % selectivity with a conversion rate of 21 mmol/min/g (Au) in a continuous reactor. Nevertheless, contrarily to Bonrath, Pruesse and co-worker used a mixture of gold and palladium to perform this oxidation and temperature slightly higher (60 °C as compared to 40 °C).

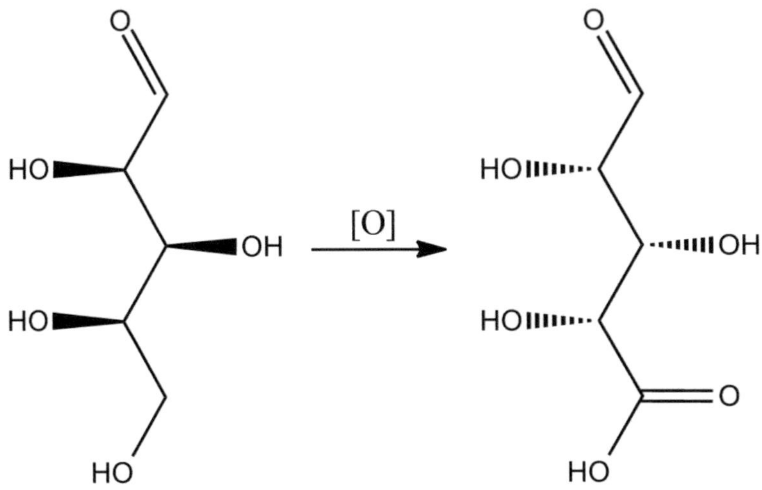

Figure 12. Simplified conversion of xylose to xylonic acid

Copper has also been indirectly investigated for the conversion of xylose to xylonic acid in that sense that Van der Weijden *et al.* [173] used C5 sugars (including xylose) for the reduction of copper sulfate in wastewater with very promising results. Although emphasis was not put on the carbohydrate itself, results showed that the reduction of copper from (II) to elemental was possible yet economical at larger scale. Xylonic acid was also observed as by-product of xylose oxidation using chlorine, as a side reaction of lignin oxidation. In this work [174], the concentration of xylonic acid increased by a factor of 40 after the chlorination process. Interesting enough, the xylitol concentration also increased, which might lead to the conclusion that oxidation, was probably not the sole factor here and that side reactions as the Cannizarro reaction between two xylose molecules could have been occurring. Jokic *et al.* [175] showed that it was possible up to an efficiency of 80 % to convert xylose simultaneously to xylonic acid and xylitol using electrotechnologies. Such process could be to a certain extent compared to the Cannizarro reaction where the original aldehyde is acting as redox reagent.

Further oxidation of xylose leads to a trihydroxydiacid, more specifically xylaric acid as depicted in Figure 13 below.

Conversion of C5 sugars and to a smaller extent of xylose into aldaric acids has been described in literature in a few reports. Kiely *et al.* [176] reported that a conversion up to 83 % xylose into 2,3,4-trihydroxyglutaric acid was achievable in a reaction mixture composed of nitric acid and $NaNO_2$. The side product of this reaction was reported to be disodium tetrahydroxysuccinate. Conversion of xylose to xylaric adic was also reported [177] using oxygen under a platinum catalyst all of this in an alkali promoted medium. Comparable conversion process [178] was obtained without any alkali, though still performed the reaction in water at 90 °C under 75 psi of oxygen. The conversion for this process was 29 %. Fleche *et al.* [179] reported a maximum conversion of 58% once again using platinum supported on alumina.

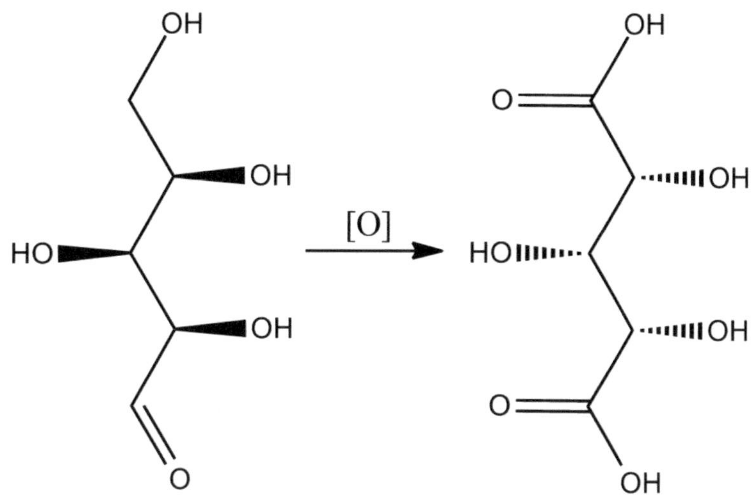

Figure 13. Simplified scheme for the conversion of xylose xylaric acid

Severer oxidizing conditions leads to a breakage of the carbon-carbon bonds in the carbohydrate molecule leading to the production, mostly, of small organic acids as formic and acetic acid on glucose [180]. A simplified scheme of such a reaction is presented in Figure 14 below:

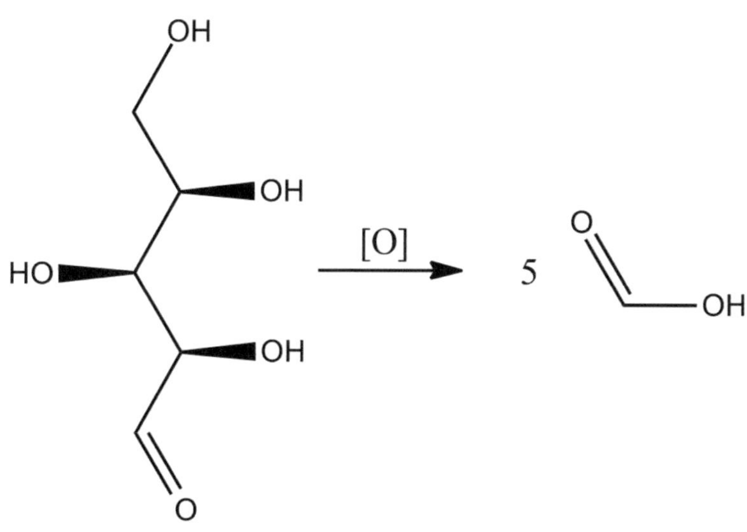

Figure 14. Simplified scheme for the conversion of xylose to formic acid under more severe oxidizing conditions.

An example of sever oxidation of xylose in a mixture of hydrogen peroxide and ammonium hydroxide have been recently reported [181] with a conversion of 96 % at room temperature for 1 h. Similar conversion of xylose was reported [182] for a process using oxygen and a molybdenum and vanadium catalyst. The reaction was done for 26 h at 353 K and 30 bar for a conversion of up to 54 % into formic acid with carbon dioxide as by-product.

5.2. The biological pathway

Xylonic acid synthesis from xylose has been reported for *Acetobacter* sp. [183], *Enterobacter cloacea* [184], *Erwinia* sp. [185, 186], *Fusarium lini* [187], *Micrococcus* sp. [188], *Penicillium corylophilum, Pichiaquer cuum* [185], *Pseudomonas* sp. [189, 190], *Pullularia pullulans* [191], *Gluconobacter* and *Caulobacter* [192, 193].

In metabolic pathways, xylose is converted to xylonate via 2 key enzymes. First, a xylose dehydrogenase (XD) oxidizes xylose to D-xylono-1,4-lactone (xylonolactone) using either NAD + or NADP+ as cofactor. This reaction is followed by the hydrolysis of xylonolactone to xylonate either spontaneously or by an enzyme with lactonase activity [194, 195]. It is hypothesized that *Pseudomonas* and *Gluconobacter sp.* both carry a membrane-bound pyrroloquinoline quinine (PQQ)-dependent XD and a cytoplasmic one [195, 196]. Stephens *et al.* [193] recently proposed a full xylose catabolic pathway for *C. crescentus*. Note that a similar pathway was proposed for arabinose yielding L-arabonate [197]. As shown in Figure 15, the proposed metabolic pathway for *C. crescentus* shows that xylonate is an intermediate in catabolic reactions that is quite different from the XI or XR/XDH previously discussed which were more intensively studied.

Researches on highly efficient microbial xylonic acid production are scarce compared to biofuels or xylitol. Even if the identification of xylonate producing species began as early as 1938 [187], the first attempt to isolate a possible industrial biocatalyst was done by Buchert *et al.* [185], who identified *P. fragi ATCC4973* as a potentially high efficiency xylonate producer (92 % of initial sugar converted to xylonic acid with initial xylose concentration of 100 g/L). In further work, *P. fragi* and *G. oxydans* showed yields of over 95 % but the low tolerance of those native strains to inhibitors tends to be problematic for industrial uses [192]. As discussed above, the metabolic pathways implied by xylonate have been investigated in the recent years [193,196]. The first recombinant microorganism engineered for the industrial production of xylonate was done by Toivari *et al.* [198]. By introducing the heterologous *Trichoderma reesei xyd1* gene (coding for the NADP+ dependant XD) in *S. cerevisiae*, they were able to obtain up to 3.8 g/L xylonate with 0.036 g/L/h productivity and 40 % yield. Nygard *et al.* [195] engineered *K. lactis* by introducing *T. reesei xyd1* and deleting the putative *xyl1* gene coding for the XR. Up to 19 g/L xylonate where produced when grown on a xylose (40 g/L) and galactose (10.5 g/L) medium. The native ability of fast xylose uptake was an advantage, but high intracellular xylonate concentration was observed, which may indicate difficulties with product export. Liu *et al.* [199] used similar approach engineering *E. coli* by disrupting the native xylose metabolic pathways of XI and XK (as shown in Figure 16). The native pathway of xylonate was also blocked by disrupting xylonic acid dehydratase genes. The XD from *C. crescentus* was introduced and 39.2 g/L of xylonate from 40 g/L of xylose in minimal medium was obtained at high productivity 1.09 g/L/h. From these results it is clear that research is at its genesis and significant efforts will be required for the creation of a highly productive and effective xylonate production biocatalyst.

At this point it is rather hard to verify the potential or the economic value of oxidation products from xylose. Complete oxidation to formic acid could be the most suitable approach at this point since the market for xylonic and xylaric acid is not as well defined as for the simple methanoic acid with its actual market value between 750-950 USD/tonne [200] and an annual world demand suspected to reach 573 Ktonnes in 2012 [201]. Conversion of xylaric acid into glutaric acid (pentanedioic acid) would lead to a very interesting market as a plasticizer but dehydration or reduction of the three central hydroxyl groups may be a challenge that could be winning at lab scale although a multiple synthesis pathway would be very difficult to reach economic at an industrial level.

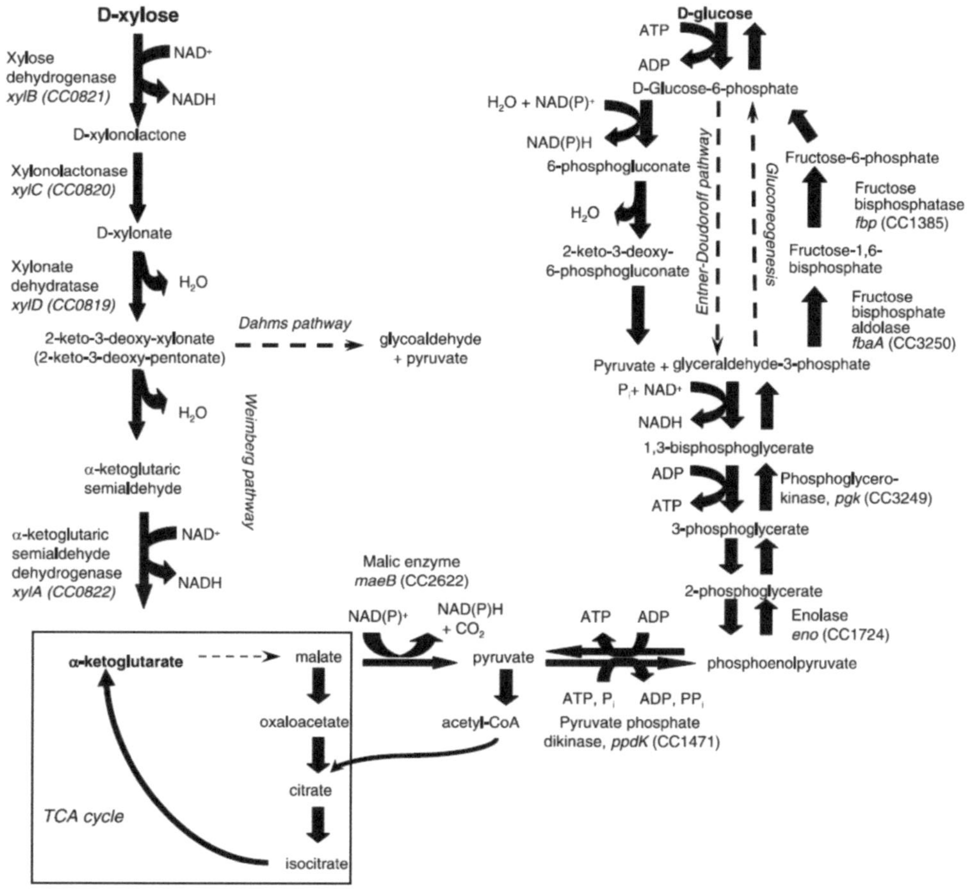

Figure 15. Proposed pathway ford-xylose metabolism in *C. crescentus* [193].

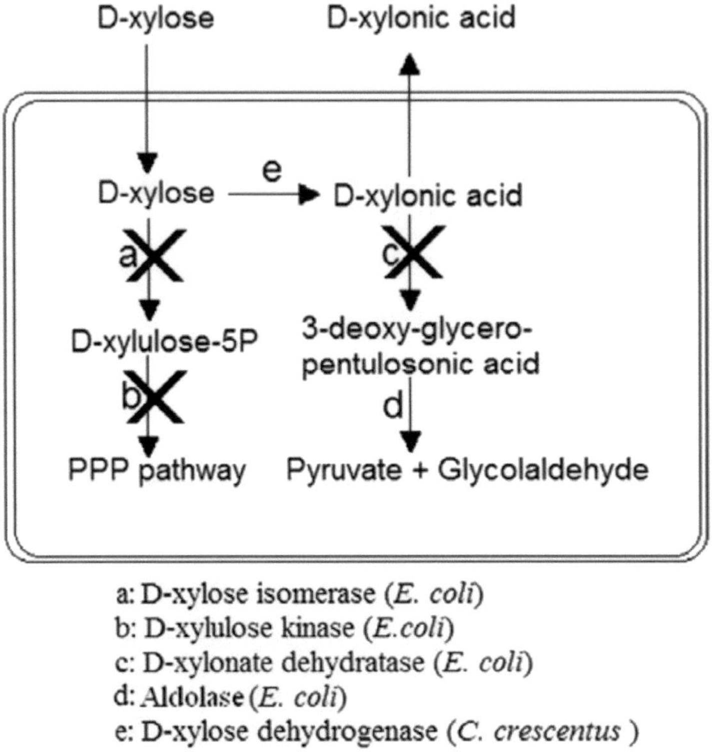

Figure 16. D-xylose and D-xylonic acid metabolic pathways in *E. coli*. The symbol X denotes that the gene is disrupted.

6. Conclusion

Second-generation ethanol or "cellulosic ethanol" relies on the utilisation of lignocellulosic biomass as a source of carbohydrates via the "bio" conversion route (keeping in mind that other pathway, as thermocatalytic pathways, may also lead to cellulosic ethanol). Production of ethanol thus requires isolation of cellulose from lignocellulosic matrix, then hydrolysis of cellulose to glucose prior to fermentation. Both of the previously mentioned steps represent challenges for industry, but the whole economic of the process is perhaps the most challenging part of cellulosic ethanol production. Cellulose is usually available in lignocellulosic biomass in the 45-60 % range which, assuming a perfect conversion implies production of 300-400 L/tonne of lignocellulosic biomass processed. At an actual price of 0.48 USD/L, each ton of biomass has a potential value of about 150-200 USD/tonne of biomass processed.

The conversion of lignocellulosic biomass is rather more complex and to a certain extent more expensive than starch-based feedstock as corn and therefore, one can assume that the conversion price is going to be higher than classical or first generation ethanol production. Keeping that fact in mind, the conversion of cellulose to glucose itself is a major technological challenge since it either requires enzymes, ionic liquids or strong acids that are rather expensive to buy or expensive to recycle and since it is of outmost importance for the production of the ethanol, technology is to a certain extent limited by this reality.

The remaining carbon content of lignocellulosic biomass is also an important factor to be considered. Since the maximum production of ethanol from the total feedstock could vary around 300-400 L per tonne, there is at this point a necessity to generate co-products from the biomass in order to make this whole process economic at the end thus coping for technological problem as conversion of cellulose to glucose. Lignin is one of the most abundant macromolecule on earth bested only by cellulose. The aromatic nature of lignin is a challenge for ethanol production but not for added value compounds as aromatic monomers that could displace actual monomers used in the polymer industry that are usually obtained from non-renewable materials.

Hemicelluloses are also an important part of the lignocellulosic biomass. Hemicelluloses, contrarily to cellulose that is characterized by an amorphous and a crystalline part, are highly ramified and easy to hydrolyse. Usually, a simple diluted alkali solution, acidic solution or even hot water can allow conversion of hemicellulose to simple sugars. The major problem with hemicellulose is the heterogeneous composition including but not limited to small acids and a variety of C6 and C5 sugars. Whilst the C6 sugars could be easily fermented to ethanol, pending reduction of the organic acids and other inhibitors, the C5 sugars require speciality yeasts for fermentation.

Other than the classical fermentative pathway, C5 sugars can as well be converted, biologically as well as chemically into a wide variety of added value products and "green" compounds. In this paper, we have identified 4 pathways for the conversion of C5 sugars but more specifically xylose, a common carbohydrate in biomass hemicelluloses.

Reaction of xylose under an acid catalyst is probably one of the most investigated fields in this domain. The target for this conversion being furfural, a well-known chemical as well as precursor for other compound as furan, Me-THF, THF and furfuryl alcohol, a reactant used in the polymer industry. The best approach for the conversion of xylose furfural, to the best of our knowledge, is chemical as no microorganism allowing conversion of C5 sugars to furfural has been identified so far. The conversion of xylose to furfural was reported to reach more than 95 % for both heterogeneous and homogeneous catalyst. On the other hand, the selectivity toward furfural is not always as efficient since the latter undergoes polymerisation in acidic medium, which often also leads to deactivation of the catalyst.

A basic catalyst leads to a conversion of C5 sugars to lactic acid although this pathway as not been deeply investigated in the literature. Lactic acid is a compound well in demand on the market but the limitations for the chemical transformation is the lack of stereospecificity of the products. Conversion of xylose under a base catalyst leads to the production of a racemic mixture of D- and L-lactic acid and thus reducing the market value of the product, particularly if the polymer industry is targeted. On the other hand, the biological conversion of xylose to lactic acid is a well-known and extensively reported process for which the production was reported to reach 6.7 g/L/d for genetically modified organisms as, in this specific case, *Lactobacillus sp. RKY2*. According to the reports, the production of lactic acid would be more efficient by the biological approach since it can lead to a stereospecific and a higher market value.

Reduction of xylose can lead to many different products including xylitol for lower severity up to diols as ethylene glycol and propylene glycol at higher severity. It is ambiguous to determine at this point if either the chemical or the biological pathway is more efficient for the production of xylitol since reports on both pathways have shown promising results. The main problem with the xylitol market is that although it is increasing, it is fairly small and therefore it is harder to fit in a new production of xylitol. On the other hand, a more severe reduction of xylose, leading to diols, could be a very interesting opportunity for the production of ethylene glycol and propylene glycol, two very important products in the chemical industry. The downside of this approach would be the production of glycerol as a side-product.

Finally, oxidation of xylose is, at this point, the approach with the lower potential for a rapid commercialisation since the market for xylonic acid and xylaric acid is hard to size at present. The conversion process, both chemical and biological seems to have significant potential in terms of scalability but the end usage is not well defined at this point. The best option would be to produce glucaric acid from xylaric acid, which could be used as a plasticizer. On the other hand, such a process, overall rather complicated, would add a significant cost for a product that would land in the commodity range.

Acknowledgement

We would like to acknowledge Enerkem, Greenfield Ethanol, CRB Innovations and the Ministry of Natural Resources of Quebec for financial support of the Industrial Chair in Cellulosic Ethanol.

Author details

Ariadna Fuente-Hernández, Pierre-Olivier Corcos, Romain Beauchet and
Jean-Michel Lavoie*

*Address all correspondence to: jean-michel.lavoie2@usherbrooke.ca

Industrial Research Chair on Cellulosic Ethanol (CRIEC), Département de Génie Chimique et de Génie Biotechnologique, Université de Sherbrooke, Sherbrooke, Québec, Canada

References

[1] Lavoie JM, Marie-Rose S, Lynch D. Non-homogeneous residual feedstocks to biofuels and chemicals via the methanol route.Biomass Conversion and Biorefinery 2012; 1-6, ISSN: 2190-6823.

[2] Lavoie JM, Beauchet R, Berberi V, Chornet M. Biorefining Lignocellulosic Biomass via the Feedstock Impregnation Rapid and Sequential Steam Treatment. Biofuel's engineering process technology.Intech publisher 2011; 685-714, ISBN: 978-953-307-480-1.

[3] Lavoie JM, Chornet M, Chornet E. Biomass refineries: relationships between feedstock and conversion approach. Energy and sustainability II, WIT press 2009; 35-48, DOI:10.2495/ESUS090041.

[4] Schild G, Sixta H. Sulfur-free dissolving pulps and their application for viscose and lyocell.Cellulose 2011;18(4) 1113-1128, DOI:10.1007/s10570-011-9532-0.

[5] Song T, Pranovich A, Bjarne H. Characterisation of Norway spruce hemicelluloses extracted by pressurised hot-water extraction (ASE) in the presence of sodium bicarbonate. Holzforschung 2011;65(1) 35-42, ISSN: 0018-3830.

[6] Song T, Pranovich A, Holmbom B. Effects of pH control with phthalate buffers on hot-water extraction of hemicelluloses from spruce wood. Bioresource technology 2011 ; 102 (22) 10518-10523;ISSN: 0960-8524.

[7] Duarte GV, Ramarao BV, Amidon TE. Polymer induced flocculation and separation of particulates from extracts of lignocellulosic materials. Bioresource technology 2010;101(22) 8526-8534, PMID: 20605092.

[8] Peng F, Ren JL, Xu F, Bian J, Peng P, Sun RC. Comparative study of hemicelluloses obtained by graded ethanol precipitation from sugarcane bagasse. Journal of agricultural and food chemistry 2009;57(14) 6305-6317, DOI:10.1021/ jf900986b.

[9] Liu W, Hou Q, Mao C, Yuan Z, Li K. Hemicelluloses Prior to Aspen Chemithermomechanical Pulping: Pre-Extraction, Separation, and Characterization. Journal of Agricultural and Food Chemistry 2012;60(19) 4880-4885, DOI: 10.1021/jf300787b.

[10] Wang K, Yang H, Guo S, Tang Y, Jiang J, Xu F, Sun RC. Organosolv fractionation process with various catalysts for improving bioconversion of triploid poplar. Process Biochemistry 2012;47(10) 1503-1509, DOI: 10.1016/j.procbio.2012.06.002.

[11] Brosse N, El Hage R, Sannigrahi P, Ragauskas A. Dilute sulphuric acid and ethanol organosolv pretreatment of Miscanthus x Giganteus. Cellulose Chemistry and Technology 2010;44(1-3) 71-78,ISSN:05769787.

[12] Lavoie JM, Capek-Menard E, Gauvin H, Chornet E. Production of pulp from Salix viminalis energy crops using the FIRSST process. Bioresource Technology 2010;101(13) 4940-4946, DOI:10.1016/j.biortech.2009.09.021.

[13] Lavoie JM, Beauchet R, Berberi V, Chornet M. Biorefining quasi homogeneous biomass via the Feedstock Impregnation Rapid and Sequential Steam Treatment. Biofuel's engineering process technology 2011; Intech publishing 685-715, ISBN: 978-953-307-480-1.

[14] Lavoie JM, Beauchet R. Biorefinery of Cannabis sativa using one-and two-step steam treatments for the production of high quality fibres. Industrial Crops and Products 2012;37(1) 275-283, DOI:10.1016/j.indcrop.2011.11.016.

[15] Pasangulapati V, Ramachandriya KD, Kumar A, Wilkins MR, Jones CL, Huhnke RL. Effects of cellulose, hemicellulose and lignin on thermochemical conversion characteristics of the selected biomass. Bioresource Technology. 2012;114 663–669, ISSN: 0960-8524.

[16] Sathitsuksanoh N, Zhu Z, Zhang YHP. Cellulose solvent- and organic solvent-based lignocellulose fractionation enabled efficient sugar release from a variety of lignocellulosic feedstocks. Bioresource Technology 2012;117 228–233, ISSN: 0960-8524

[17] Xu J, Zhang X, Cheng JJ. Pretreatment of corn stover for sugar production with switchgrass-derived black liquor. Bioresource Technology 2012;111 255–260, ISSN: 0960-8524

[18] Jun A, Tschirner UW, Tauer Z. Hemicellulose extraction from aspen chips prior to kraft pulping utilizing kraft white liquor. Biomass and bioenergy 2012;37 229-236, ISSN:09619534.

[19] Rana D, Rana V, Ahring BK. Producing high sugar concentrations from loblolly pine using wet explosion pretreatment. Bioresource Technology 2012;121 61–67, ISSN: 0960-8524

[20] Marcotullio G, de Jong W. Furfural formation from D-xylose: the use of different halides in dilute aqueous acidic solutions allows for exceptionally high yields. Carbohydrate Research 2011;346(11) 1291-3, ISSN:1873-426X.

[21] Nimlos MR, Qian X, Davis M, Himmel ME, Johnson DK. Energetics of xylose decomposition as determined used quantum mechanics. J. Phys. Chem. A.2006;110(42) 11824-11838, ISSN:1089-5639.

[22] Lessard J, Morin JF, Wehrung JF, Magnin D, Chornet E. High Yield Conversion of Residual Pentoses into Furfural via Zeolite Catalysis and Catalytic Hydrogenation of Furfural to 2-Methylfuran. Top Catal. 2010;53(15-18) 1231-1234, ISSN:10225528.

[23] Bugrayev A, Al-Haq N, Okopie RA, Qazi A, Suggate M, Sullivan AC, Wilson JRH. Covalently linked ethylmercaptophenyl sulfonic acid and ethylmercaptobenzyl sulfonic acid silica materials - Synthesis and catalytic activity. Journal of Molecular Catalysis A: Chemical 2008;280(1-2) 96-101, ISSN: 1381-1169.

[24] Tao F, Song H, Chou L. Efficient process for the conversion of xylose to furfural with acidic ionic liquid. Canadian Journal of Chemistry 2011;89(1) 83-87, DOI: 10.1139/V10-153.

[25] Rong C, Ding X, Zhu Y, Li Y, Wang L, Qu Y, Ma X, Wang Z. Production of furfural from xylose at atmospheric pressure by dilute sulfuric acid and inorganic salts.Carbohydrate Research 2012;350(1) 77-80, ISSN:1873-426X.

[26] Lima S, Neves P, Antunes MM, Pillinger M, Ignatyev N, Valente AA. Conversion of mono/di/polysaccharides into furan compounds using 1-alkyl-3-methylimidazolium ionic liquids. Appl. Catal. A: Gen 2009;363(1-2)93-99, DOI: 10.1016/j.apcata.2009.04.049.

[27] Gürbüz EI, Wettstein SG, Dumesic JA. Conversion of Hemicellulose to Furfural and Levulinic Acid using Biphasic Reactors with Alkylphenol Solvents. ChemSusChem 2012;5(2) 383-387, DOI:10.1002/cssc.201100608.

[28] Yang Y, Hu CW, Abu-Omar MM. Synthesis of Furfural from Xylose, Xylan, and Biomass Using AlCl3 6H2O in Biphasic Media via Xylose Isomerization to Xylulose. ChemSusChem 2012;5(2) 405-410, DOI:10.1002/cssc.201100688.

[29] Agirrezabal-Telleria I, Larreategui A, Requies J, Güemez MB, Arias PL. Furfural production from xylose using sulfonic ion-exchange resins (Amberlyst) and simultaneous stripping with nitrogen. Bioresource Technology 2011;102(16) 7478-7485, DOI: 10.1016/j.biortech.2011.05.015.

[30] Lima S, Antunes MM, Fernandes A, Pillinger M, Ribeiro MF, Valente AA. Catalytic cyclodehydration of xylose to furfural in the presence of zeolite H-Beta and a micro/mesoporous Beta/TUD-1 composite material. Applied Catalysis, A: General 2010;388(1-2) 141-148, ISSN: 0926-860X.

[31] Antunes MM, Lima S, Fernandes A, Pillinger M, Ribeiro MF,Valente AA. Aqueous-phase dehydration of xylose to furfural in the presence of MCM-22 and ITQ-2 solid acid catalysts. Applied Catalysis, A: General 2012; 417-418 (29) 243-252, DOI:10.1016/j.apcata.2011.12.046.

[32] vom Stein T, Grande PM, Leitner W, Dominguez de Maria P. Iron-Catalyzed Furfural Production in Biobased Biphasic Systems: From Pure Sugars to Direct Use of Crude Xylose Effluents as Feedstock. ChemSusChem 2011;4(11) 1592-1594, DOI:10.1002/cssc.201100259.

[33] Lam E, Majid E, Leung AC, Chong JH, Mahmoud KA, Luong JH. Synthesis of Furfural from Xylose by Heterogeneous and Reusable Nafion Catalysts. ChemSusChem2011;4(4) 535-541, DOI:10.1002/cssc.201100023.

[34] Dias AS, Pillinger M, Valente AA. Liquid phase dehydration of D-xylose in the presence of Keggin-type heteropolyacids. Appl. Catal. A. 2005;285(1-2) 126-131, DOI: 10.1016/j.apcata.2005. 02.016.

[35] Sádaba I, Lima S, Valente AA, López Granados M. Catalytic dehydration of xylose to furfural: vanadyl pyrophosphate as source of active soluble species. Carbohydrate Research 2011;346(17) 2785-2791, DOI:10.1016/j.carres. 2011.10.001.

[36] Win DT. Furfural-Gold from Garbage. AU J.T. 2005;8(4) 185-190, http://www.journal.au.edu/au_ techno/2005/apr05/vol8no4_abstract04.pdf.

[37] Jinan ZZ International Trade Co., Ltd.

[38] Zibo Pulisi Trading Co., Ltd.

[39] Zhou Z, Ma Q, Zhang A, Wu M. Synthesis of water-soluble monotosylated ethylenediamines and their application in ruthenium and iridium-catalyzed transfer hydrogenation of aldehydes. Applied Organometallic Chemistry 2011;25(12)856-861, DOI: 10.1002/aoc.1824.

[40] Ajjou AN, Pinet JL. The biphasic transfer hydrogenation of aldehydes and ketones with isopropanol catalyzed by water-soluble rhodium complexes. Journal of Molecular Catalysis A: Chemical 2004;214(2)203-206, ISSN:1381-1169.

[41] Li H, Luo H, Zhuang L, Dai W, Qiao M. Liquid phase hydrogenation of furfural to furfuryl alcohol over the Fe-promoted Ni-B amorphous alloy catalysts. Journal of Molecular Catalysis A: Chemical 2003;203(1-2)267-275, DOI:10.1016/S1381-1169(03)00368-6.

[42] Inada K, Shibagaki M, Nakanishi Y, Matsushita H. The catalytic reduction of aldehydes and ketones with 2-propanol over silica-supported zirconium catalyst. ChemInform 1994;25(15) no pages, DOI: 10.1002/chin.199415090.

[43] Wabnitz T, Breuninger D, Heimann J, Backes R, Pinkos R. Method for one-step synthesis of 2-methyltetrahydrofuran from furfural and catalyst 2009; PCT Int. Appl., WO/ 2009/003882 A1 20090108.

[44] Wabnitz T, Breuninger D, Heimann J, Backes R, Pinkos R. Production of 2-methyltetrahydrofuran from furfural over two catalysts in a structured bed 2009; PCT Int. Appl., WO/2009/003881 A1 20090108.

[45] Lange JP. Process for the hydrogenolysis of furfuryl derivatives PCT Int. Appl.,2009; PCT/EP2009/057899, WO/2009/156439 A1 20091230, (June 2009).

[46] Zheng HY, Zhu YL, Teng BT, Bai ZQ, Zhang CH, Xiang HW, Li YW. Towards understanding the reaction pathway in vapor phase hydrogenation of furfural to 2-methylfuran. Journal of Molecular Catalysis A: Chemical 2006;246(1-2) 18-23, DOI:10.1016/j.molcata.2005.10.003.

[47] TPFTZ Qingshan Int'l Business Technology Co., Ltd.

[48] Zhang W, Zhu Y, Niu S, Li Y. A study of furfural decarbonylation on K-doped Pd/Al2O3 catalysts. Journal of Molecular Catalysis A: Chemical 2011;335(1-2) 71-81, ISSN:1873-314X.

[49] Stevens JG, Bourne RA, Twigg MV, Poliakoff M. Real-Time Product Switching Using a Twin Catalyst System for the Hydrogenation of Furfural in Supercritical CO2. Angewandte Chemie International Edition 2010; 49(47), ISSN:1433-7851.

[50] Popoff T, Theander O. Reactions of D-xylose and D-glucose in alkaline aqueous solutions.Carbohydrate Research1976;48(1) 13-21, DOI:10.1016/S0008-6215(00)83510-7.

[51] Johansson MH, Samuelson O. Reducing end groups in brich xylan and their alkaline degradation. Wood Science and Technology 1977; 11(4) 251-163, ISSN:0043-7719.

[52] El Khadem HS, Ennifar S, Isbell HS. Contribution of the reaction pathways involved in the isomerization of monosaccharides by alkali. Carbohydrate Research 1987;169 13-21, ISSN: 0008-6215.

[53] Jackson JE, Miller DJ, Marincean S. Process for the preparation of lactic acid and glyceric acid from aqueous solutions of hexoses and pentoses in the presence of an excess of a strongly basic anionic exchange resin 2007; U.S. Pat. Appl. Publ., 20070066844.

[54] Epane G, Laguerre JC, Wadouachi A, Marek D. Microwave-assisted conversion of D-glucose into lactic acid under solvent-free conditions. Green Chemistry 2010;12(3) 502-506, DOI:10.1039/B922286C.

[55] Rahubadda A, Montoya A, Haynes BS. C5 sugar decomposition products under hot compressed water conditions [online]. In:CHEMECA 2011: Engineering a Better World: Sydney Hilton Hotel, NSW, Australia, 2011;18-21 Barton, A.C.T.: Engineers Australia, 2011: 698-706. Conference paper. Availability: http://search.informit.com.au/documentSummary;dn=168748429919269;res=IELENGISBN: 9780858259676.

[56] Onda A, Ochi T, Kajiyoshi K, Yanagisawa K. A new chemical process for catalytic conversion of D-glucose into lactic acid and gluconic acid. Applied Catalysis, A: General 2008;343(1-2) 49-54, DOI:10.1016/j.apcata.2008.03.017.

[57] Ma C, Jin F,Cao J,Wu B. Hydrothermal Conversion of Carbohydrates into Lactic Acid with Alkaline Catalysts: 4th International Conference onBioinformatics and Biomedical Engineering (iCBBE), June 18-20, 2010, Chengdu, China, 1-4, DOI 10.1109/ICBBE.2010.5516468.

[58] Aspinall GO,Greenwood CT,Sturgeon RJ. The degradation of xylans by alkali. J. Chem. Soc. 1961;3667-3674, DOI 10.1039/JR9610003667.

[59] Yang BY. Montgomery R. Alkaline degradation of fructofuranosides.Carbohydrate Research 1996;280(1) 47-57, DOI: 10.1016/0008-6215(95)00233-2.

[60] Raharja S, Rigal L, Barre L, VidalPF. Design of a continuous process for the alkaline treatment of xylose into lactic acid. The Canadian Journal of Chemical Engineering 1997;75(5)913-920, DOI: 10.1002/cjce.5450750511.

[61] Okano K, Tanaka T, Ogino C, Fukuda H, Kondo A. Biotechnological production of enantiomeric pure lactic acid from renewable resources: Recent achievements, perspectives, and limits. Applied Microbiology and Biotechnology 2010;85(3)413-423, DOI:10.1007/s00253-009-2280-5.

[62] Yadav AK, Chaudhari AB, Kothari RM. Bioconversion of renewable resources into lactic acid: An industrial view. Critical reviews in biotechnology 2011;31(1) 1-19, PMID: 20476870.

[63] Abdel-Rahman MA, Tashiro Y, Sonomoto K. Lactic acid production from lignocellulose-derived sugars using lactic acid bacteria: Overview and limits. Journal of Biotechnology 2011;156(4) 286-301, DOI:10.1016/j.jbiotec. 2011.06.017.

[64] Datta R, Henry M. Lactic acid: Recent advances in products, processes and technologies-A review. Journal of Chemical Technology and Biotechnology 2006;81(7) 1119-1129, DOI: 10.1002/jctb.1486.

[65] Garvie EI. Bacterial lactate dehydrogenases. Microbiological reviews 1980;44(1) 106-139. PMCID: PMC373236

[66] John RP, Nampoothiri KM, Pandey A. Fermentative production of lactic acid from biomass: An overview on process developments and future perspectives. Applied Microbiology and Biotechnology 2007;74(3)524-534, DOI:10.1007/s00253-006-0779-6.

[67] Wee Y, Kim J, Ryu H. Biotechnological production of lactic acid and its recent applications. Food Technology and Biotechnology 2006;44(2) 163-172, ISSN: 1330-9862.

[68] Zhang ZY, Jin B, Kelly JM. Production of lactic acid from renewable materials by Rhizopus fungi. Biochemical engineering journal 2007;35(3) 251-263, ISSN:1369-703X.

[69] Zhao B, Wang L, Li F, Hua D, Ma C, Ma Y, Xu P. Kinetics of D-lactic acid production by Sporolactobacillus sp. strain CASD using repeated batch fermentation. Bioresource technology 2010;101(16) 6499-6505, DOI: 10.1016/j.biortech.2010.03.069.

[70] Zhao B, Wang L, Ma C, Yang C, Xu P, Ma Y. Repeated open fermentative production of optically pure L-lactic acid using a thermophilic Bacillus sp. strain. Bioresource technology 2010;101(16) 6494-6498, DOI: 10.1016/j.biortech.2010.03.051.

[71] Salminen S, Ramos P, Fonden R. Substrates and lactic acid bacteria. Food Science and Technology (New York, NY, United States) 1993;58295-306. ISBN:0-8247-8907-5.

[72] Hang YD. Direct fermentation of corn to L(+)-lactic acid by Rhizopus oryzae. Biotechnology Letters 1989;11(4) 299-300, DOI:10.1007/BF01031581.

[73] Krištofíková L, Rosenberg M, Vlnová A, Šajbidor J, Čertík M. Selection of Rhizopus strains for L(+)-lactic acid and gamma-linolenic acid production. Folia Microbiologica (Prague, Czech Republic) 1991;36(5) 451-455, DOI:10.1007/BF02884065.

[74] Miura S, Arimura T, Itoda N, Dwiarti L, Feng JB, Bin CH, Okabe M. Production of L-lactic acid from corncob. Journal of Bioscience and Bioengineering 2004;97(3) 153-157, ISSN: 1389-1723.

[75] Ding S, Tan T. L-lactic acid production by Lactobacillus casei fermentation using different fed-batch feeding strategies. Process Biochemistry (Amsterdam, Netherlands) 2006;41(6) 1451-1454, ISSN: 1359-5113.

[76] Chang HN, Kim N, Kang J, Jeong CM, Choi J, Fei Q, Kim BJ, Kwon S, Lee SY, Kim J. Multi-stage high cell continuous fermentation for high productivity and titer. Bioprocess and Biosystems Engineering 2011;34(4) 419-431, ISSN 16157591.

[77] Dumbrepatil A, Adsul M, Chaudhari S, Khire J, Gokhale D. Utilization of molasses sugar for lactic acid production by Lactobacillus delbrueckii subsp. delbrueckii mutant Uc-3 in batch fermentation. Applied and Environmental Microbiology 2008;74(1) 333-335, DOI:10.1128/AEM.01595-07.

[78] Bustos G, de la Torre N, Moldes AB, Cruz JM, Dominguez JM. Revalorization of hemicellulosic trimming vine shoots hydrolyzates trough continuous production of lactic acid and biosurfactants by L. pentosus. Journal of Food Engineering 2006;78(2) 405-412, ISBN:0260-8774.

[79] Zhu Y, Lee YY, Elander RT. Conversion of aqueous ammonia-treated corn stover to lactic acid by simultaneous saccharification and cofermentation. Applied Biochemistry and Biotechnology 2007;137-140(1-2) 721-738, ISSN:0273-2289.

[80] Givry S, Prevot V, Duchiron F. Lactic acid production from hemicellulosic hydrolyzate by cells of Lactobacillus bifermentans immobilized in Ca-alginate using response surface methodology. World Journal of Microbiology & Biotechnology 2008;24(6) 745-752, DOI:0.1007/s11274-007-9534-0.

[81] Chaillou S, Bor Y, Batt CA, Postma PW, Pouwels PH. Molecular cloning and functional expression in Lactobacillus plantarum 80 of xylT, encoding the D-xylose-H+ symporter of Lactobacillus brevis. Applied and Environmental Microbiology 1998;64(12) 4720-4728, PMCID:PMC90914.

[82] Helanto M, Kiviharju K, Leisola M, Nyyssölä A. Metabolic engineering of Lactobacillus plantarum for production of L-ribulose. Applied and Environmental Microbiology 2007;73(21) 7083-7091, DOI:10.1128/AEM.01180-07.

[83] Ohara H, Owaki M, Sonomoto K. Xylooligosaccharide fermentation with Leuconostoc lactis. Journal of Bioscience and Bioengineering 2006;101(5) 415-420, ISSN: 1389-1723.

[84] Tanaka K, Komiyama A, Sonomoto K, Ishizaki A, Hall SJ, Stanbury PF. Two different pathways for D-xylose metabolism and the effect of xylose concentration on the yield coefficient of L-lactate in mixed-acid fermentation by the lactic acid bacterium Lactococcus lactis IO-1. Applied Microbiology and Biotechnology 2002;60(1-2) 160-167, ISSN:0175-7598.

[85] Abdel-Rahman M, Tashiro Y. Zendo T, Sonomoto K. Effective (+)-Lactic Acid Production by Co-fermentation of Mixed Sugars. Journal of Biotechnology 2010;150(Supplement) 347-348, DOI:10.1016/j.jbiotec.2010.09.384.

[86] Abdel-Rahman MA, Tashiro Y, Zendo T, Hanada K, Shibata K, Sonomoto K. Efficient homofermentative L-(+)-lactic acid production from xylose by a novel lactic acid bac-

terium, Enterococcus mundtii QU 25. Applied and Environmental Microbiology 2011;77(5) 1892-1895, DOI: 10.1128/AEM.02076-10.

[87] Okano K, Yoshida S, Yamada R, Tanaka T, Ogino C, Fukuda H, Kondo A. Improved production of homo-D-lactic acid via xylose fermentation by introduction of xylose assimilation genes and redirection of the phosphoketolase pathway to the pentose phosphate pathway in L-lactate dehydrogenase gene-deficient Lactobacillus plantarum. Applied and Environmental Microbiology 2009;75(24) 7858-7861, DOI:10.1128/AEM.01692-09.

[88] Tamakawa H, Ikushima S, Yoshida S. Efficient production of L-lactic acid from xylose by a recombinant Candida utilis strain. Journal of Bioscience and Bioengineering 2012;113(1) 73-75, ISSN:1389-1723.

[89] Okano K, Yoshida S, Tanaka T, Ogino C, Fukuda H, Kondo A. Homo-D-lactic acid fermentation from arabinose by redirection of the phosphoketolase pathway to the pentose phosphate pathway in L-lactate dehydrogenase gene-deficient Lactobacillus plantarum. Applied and Environmental Microbiology 2009;75(15)5175-5178, ISSN: 0099-2240.

[90] Ilmen M, Koivuranta K, Ruohonen L, Suominen P, Penttila M. Efficient production of L-lactic acid from xylose by Pichia stipitis. Applied and Environmental Microbiology 2007;73(1)117-123, DOI:10.1128/AEM.01311-06.

[91] Wee Y, Ryu H. Lactic acid production by Lactobacillus sp. RKY2 in a cell-recycle continuous fermentation using lignocellulosic hydrolyzates as inexpensive raw materials. Bioresource technology2009;100(18)4262-4270, ISSN:0960-8524.

[92] John RP, Nampoothiri KM, Pandey A. Simultaneous saccharification and fermentation of cassava bagasse for L-(+)-lactic acid production using Lactobacilli. Applied Biochemistry and Biotechnology 2006;134(3)263-272, ISSN:0273-2289.

[93] Adsul M, Khire J, Bastawde K, Gokhale D. Production of lactic acid from cellobiose and cellotriose by Lactobacillus delbrueckii mutant Uc-3. Applied and Environmental Microbiology 2007;73(15) 5055-5057, DOI:10.1128/AEM.00774-07.

[94] Singhvi M, Joshi D, Adsul M, Varma A, Gokhale D. D-(-)-Lactic acid production from cellobiose and cellulose by Lactobacillus lactis mutant RM2-24. Green Chemistry 2010;12(6)1106-1109, DOI:10.1039/b925975a.

[95] Marques S, Santos JA. L Gírio, FM, Roseiro JC. Lactic acid production from recycled paper sludge by simultaneous saccharification and fermentation. Biochemical engineering journal 2008;41(3)210-216, DOI:10.1016/j.bej.2008.04. 018.

[96] Gullon B, Yáñez R, Alonso JL, Parajó JC. L-Lactic acid production from apple pomace by sequential hydrolysis and fermentation. Bioresource technology 2008;99(2) 308-319, ISSN:0960-8524.

[97] Ishida N, Saitoh S, Onishi T, Tokuhiro K, Nagamori E, Kitamoto K, Takahashi H. The effect of pyruvate decarboxylase gene knockout in Saccharomyces cerevisiae on L-

lactic acid production. Bioscience, biotechnology, and biochemistry 2006;70(5)1148-1153, PMID: 16717415.

[98] Tokuhiro K, Ishida N, Nagamori E, Saitoh S, Onishi T, Kondo A, Takahashi H. Double mutation of the PDC1 and ADH1 genes improves lactate production in the yeast Saccharomyces cerevisiae expressing the bovine lactate dehydrogenase gene. Applied Microbiology and Biotechnology 2009;82(5)883-890, ISSN:0175-7598.

[99] Bianchi MM, Brambilla L, Protani F, Liu C, Lievense J, Porro D. Efficient homolactic fermentation by Kluyveromyces lactis strains defective in pyruvate utilization and transformed with the heterologous LDH gene. Applied and Environmental Microbiology 2001;67(12)5621-5625, ISSN: 0099-2240.

[100] Tianjin Tiger International Trade Co., Ltd.

[101] Qingdao Abel Technology Co., Ltd.

[102] PRWeb Online visibility on Vocus, Global Lactic Acid Market to Reach 367.3 Thousand Metric Tons by 2017, According to New Report by Global Industry Analysts, Inc. http://www.prweb.com/releases/lactic_acid/polylactic_acid/prweb9369473.htm

[103] Wisniak J, Hershkowitz M, Stein S. Hydrogenation of Xylose to Xylitol, Ind. Eng Chem., Prod. Res. Develop. 1974;13(1)75, DOI:10.1021/i360049a015.

[104] Wisniak J, Hershkowitz M, Stein S. Hydrogenation of Xylose over Platinum Group Catalysts. Ind. Eng. Chem. Prod. Res. Dev. 1974;13(4)232-236,DOI:10.1021/i360052a004.

[105] Mikkola JP, Salmi T. Three-phase catalytic hydrogenation of xylose to xylitol - prolonging the catalyst activity by means of on-line ultrasonic treatment. Catalysis Today 2001;64(3-4)271-277, DOI: 10.1016/S0920-5861(00)00530-7.

[106] Mikkola JP, Kubicka D, Kuusisto J, Granholm N, Salmi T, Holmbom B. Non-traditional three-phase reactor setup for simultaneous acoustic irradiation and hydrogenation. Journal of Chemical Technology and Biotechnology 2003;78(2-3)203-207, DOI: 10.1002/jctb.739.

[107] Heikkila H, Ojamo H, Tylli M, Ravanko V, Nurmi J, Haimi P, Alen R, Koivikko H. Preparation of L-xylose and its use for the production of xylitol. U.S. Pat. Appl. Publ., 20030097029, (May 2003).

[108] Kuusisto J, Heikkilae H, Tylli M, Golde M, Riihimaeki T. Catalytic hydrogenation of sugar to sugar alcohol, using increasing reaction temperature and/or addition of monocarboxylic acid to maintain catalyst activity.Brit. UK Pat. Appl. GB 2437517(October2007).

[109] Vanoppen D, Maas-Brunner M, Kammel U, Arndt JD. Preparation and use of silicon dioxide-supported ruthenium catalysts for saccharide hydrogenation, Ger. Offen. DE10128205(December 2002).

[110] Guha SK, Kobayashi H, Hara K, Kikuchi H, Aritsuka T, Fukuoka A. Hydrogenolysis of sugar beet fiber by supported metal catalyst. Catalysis Communications 2011;12(11) 980-983, DOI:10.1016/j.catcom.2011.02.017.

[111] Kwak BS, Lee BI, Kim TY, Kim JW, Lee SI. Process for preparation of sugar alcohols by ruthenium-catalyzed hydrogenation of sugars. PCT Int. Appl., WO 2005021475 A1 20050310 (2005).

[112] Kwak BS, Lee BI, Kim TY, Kim JW, Lee SI. Method for preparing sugar alcohols using ruthenium- and zirconia- catalyzed hydrogenation of monosaccharides. PCT Int. Appl., WO 2006093364 A1 20060908 (2006).

[113] Crabtree SP, Tyers DV. Hydrogenolysis of sugar feedstock. PCT Int. Appl., PCT/GB2004.004391, WO/2005/051874, (June 2005).

[114] Sun J, Liu H. Selective hydrogenolysis of biomass-derived xylitol to ethylene glycol and propylene glycol on supported Ru catalysts. Green Chemistry 2011;13(1)135-142, DOI:10.1039/c0gc00571a.

[115] Chopade SP, Miller DJ, Jackson JE, Werpy TA, Frye Jr JG, Zacher AH. Catalysts and process for hydrogenolysis of sugar alcohols to polyols. US Patent 6291725, (January 2001).

[116] Dubeck M, Knapp GG. Two stage hydrogenolysis of carbohydrate to glycols using sulfide modified ruthenium catalyst in second stage. US Patent 4,476,331, (September 1984).

[117] Holcomb DE. Catalyst and method for production of polyols by hydrogenolysis of carbohydrates. US Patent 7,692,001, (April 2010).

[118] Kasehagen L. Hydrogenolysis of reducible sugars to obtaini a high percentage of glycerol. US Patent 3,396,199, (August 1968).

[119] Wright LW. Hydrogenation and hydrogenolysis of carbohydrates with tungsten oxide promoted supported nickel catalyst. US Patent 3,965,199, (June 1976).

[120] Werpy TA, Frye Jr JG, Zacher AH, Miller DJ. Hydrogenolysis of 6-carbon sugars and other organic compounds. US Patent 6,841,085, (January 2005).

[121] Suzuki S, Sugiyama M, Mihara Y, Hashiguchi K, Yokozeki K. Novel enzymatic method for the production of xylitol from D-arabitol by Gluconobacter oxydans. Bioscience, biotechnology, and biochemistry 2002;66(12)2614-2620, DOI: 10.1271/bbb.66.2614.

[122] Rangaswamy S, Agblevor FA. Screening of facultative anaerobic bacteria utilizing D-xylose for xylitol production. Applied Microbiology and Biotechnology 2002;60(1-2)88-93, ISSN:0175-7598.

[123] Yoshitake J, Ohiwa H, Shimamura M, Imai T. Production of polyalcohol by a Corynebacterium species. I. Production of pentitol from aldopentose. Agricultural and Biological Chemistry 1971;35(6) 905-911.

[124] Yoshitake J, Ishizaki H, Shimamura M, Imai T. Xylitol production by an Enterobacter species. Agricultural and Biological Chemistry 1973;37(10) 2261-2266.

[125] Cirino PC, Chin JW, Ingram LO. Engineering Escherichia coli for xylitol production from glucose-xylose mixtures. Biotechnology and bioengineering 2006;95(6)1167-1176, ISSN:0006-3592.

[126] Parajó JC, Domínguez H, Domínguez JM. Biotechnological production of xylitol. Part 1: Interest of xylitol and fundamentals of its biosynthesis. Bioresource technology 1998;65(3)191-201, DOI:10.1016/S0960-8524(98)00038-8.

[127] Izumori K, Tuzaki K. Production of xylitol from D-xylulose by Mycobacterium smegmatis. Journal of Fermentation Technology 1988;66(1) 33-36, DOI: 10.1016/0385-6380(88)90126-4.

[128] Monedero V, Pérez-Martínez G, Yebra MJ. Perspectives of engineering lactic acid bacteria for biotechnological polyol production. Applied Microbiology and Biotechnology 2010;86(4)1003-1015, ISSN:0175-7598.

[129] Nyyssoelae A, Pihlajaniemi A, Palva A, von Weymarn N, Leisola M. Production of xylitol from D-xylose by recombinant Lactococcus lactis. Journal of Biotechnology 2005;118(1)55-66, ISSN:0168-1656.

[130] Fridkin SK, Jarvis WR. Epidemiology of nosocomial fungal infections. Clinical microbiology reviews 1996;9(4)499-511, ISSN: 0893-8512.

[131] Meinander NQ, Hahn-Haegerdal B. Fed-batch xylitol production with two recombinant Saccharomyces cerevisiae strains expressing XYL1 at different levels, using glucose as a cosubstrate: a comparison of production parameters and strain stability. Biotechnology and bioengineering 1997;54(4)391-399, ISSN:0006-3592.

[132] Parajó JC, Domínguez H, Domínguez JM. Biotechnological production of xylitol. Part 2: Operation in culture media made with commercial sugars. Bioresource technology 1998;65(3)203-212, DOI:10.1016/S0960-8524(98)00036-4.

[133] Lee WJ, Ryu YW, Seo JH. Characterization of two-substrate fermentation processes for xylitol production using recombinant Saccharomyces cerevisiae containing xylose reductase gene. Process Biochemistry (Oxford) 2000;35(10)1199-1203, DOI:10.1016/S0032-9592(00)00165-5.

[134] Govinden R, Pillay B, Van Zyl WH, Pillay D. Xylitol production by recombinant Saccharomyces cerevisiae expressing the Pichia stipitis and Candida shehatae XYL1 genes. Applied Microbiology and Biotechnology 2001;55(1)76-80, ISSN:0175-7598.

[135] Sasaki M, Jojima T, Inui M, Yukawa H. Xylitol production by recombinant Corynebacterium glutamicum under oxygen deprivation. Applied Microbiology and Biotechnology 2010;86(4)1057-1066, ISSN:0175-7598.

[136] Akinterinwa O, Khankal R, Cirino PC. Metabolic engineering for bioproduction of sugar alcohols. Current opinion in biotechnology 2008;19(5)461-467, ISSN:0958-1669.

[137] Nigam P, Singh D. Processes for fermentative production of xylitol - a sugar substitute. Process Biochemistry (Oxford) 1995;30(2)117-124, DOI: 10.1016/0032-9592(95)80001-8.

[138] Dien BS, Cotta MA, Jeffries TW. Bacteria engineered for fuel ethanol production: current status. Applied Microbiology and Biotechnology 2003;63(3)258-266, ISSN: 0175-7598.

[139] Hahn-Hagerdal B, Karhumaa K, Fonseca C, Spencer-Martins I, Gorwa-Grauslund M. Towards industrial pentose-fermenting yeast strains. Applied Microbiology and Biotechnology 2007;74(5)937-953, ISSN:0175-7598.

[140] Khankal R, Chin JW, Cirino PC. Role of xylose transporters in xylitol production from engineered Escherichia coli. Journal of Biotechnology 2008;134(3-4)246-252, ISSN:0168-1656.

[141] Sasaki M, Jojima T, Kawaguchi H, Inui M, Yukawa H. Engineering of pentose transport in Corynebacterium glutamicum to improve simultaneous utilization of mixed sugars. Applied Microbiology and Biotechnology 2009;85(1)105-115, ISSN:0175-7598.

[142] Kinoshita S. Glutamic acid bacteria. In: Demain AL, Solomon NA (eds) Biology of industrial microorganisms. Benjamin/Cummins, London 1985;6 115-142.

[143] Terasawa M, Yukawa H. Industrial production of biochemicals by native immobilization. Bioprocess technology 1993;16, Ind. Appl. Immobilized Biocatal.37-52.

[144] Zhang J, Geng A, Yao C, Lu Y, Li Q. Xylitol production from D-xylose and horticultural waste hemicellulosic hydrolysate by a new isolate of Candida athensensis SB18. Bioresource technology 2012;105134-141, ISSN: 0960-8524.

[145] Lopez F, Delgado OD, Martinez MA, Spencer JFT, Figueroa LIC. Characterization of a new xylitol-producer Candida tropicalis strain. Antonie van Leeuwenhoek 2004;85(4)281-286, ISSN:0003-6072.

[146] Ikeuchi T, Azuma M, Kato J, Ooshima H. Screening of microorganisms for xylitol production and fermentation behavior in high concentrations of xylose. Biomass and Bioenergy 1999;16(5)333-339, DOI:10.1016/S0961-9534(99)00005-7.

[147] Kwon S, Park S, Oh D. Increase of xylitol productivity by cell-recycle fermentation of Candida tropicalis using submerged membrane bioreactor. Journal of Bioscience and Bioengineering 2006;101(1)13-18, ISSN:1389-1723.

[148] Oh DK, Kim SY. Increase of xylitol yield by feeding xylose and glucose in Candida tropicalis. Applied Microbiology and Biotechnology 1998;50(4)419-425, ISSN: 0175-7598.

[149] Ojamo H. Yeast xylose metabolism and xylitol production. VTT Publications 1994;176, ISBN:9513844145.

[150] Choi J, Moon K, Ryu Y, Seo J. Production of xylitol in cell recycle fermentations of Candida tropicalis. Biotechnology Letters 2000;22(20)1625-1628, DOI:10.1023/A:1005693427389.

[151] Granström TB. Biotechnological production of xylitol with Candida yeasts. Aalto University, Library Publication Archive, 160, ISBN:951-22-5997-4.

[152] Hallborn J, Walfridsson M, Airaksinen U, Ojamo H, Hahn-Hagerdal B, Penttila M, Kerasnen S. Xylitol production by recombinant Saccharomyces cerevisiae. Biotechnology (Nature Publishing Company) 1991;9(11)1090-1095, PMID 1367625

[153] Salgado JM, Rodríguez N, Cortés S, Domínguez JM. Effect of nutrient supplementation of crude or detoxified concentrated distilled grape marc hemicellulosic hydrolysates on the xylitol production by Debaryomyces hansenii. Preparative biochemistry & biotechnology 2012;42(1) 1-14, ISSN: 1082-6068.

[154] Oh E, Bae Y, Kim K, Park Y, Seo J. Effects of overexpression of acetaldehyde dehydrogenase 6 and acetyl-CoA synthetase 1 on xylitol production in recombinant Saccharomyces cerevisiae. Biocatalysis and Agricultural Biotechnology 2012;1(1)15-19, DOI:10.1016/j.bcab.2011.08.011.

[155] Li M, Meng X, Diao E, Du F. Xylitol production by Candida tropicalis from corn cob hemicellulose hydrolysate in a two-stage fed-batch fermentation process. Journal of Chemical Technology and Biotechnology 2012;87(3)387-392, DOI:10.1002/jctb.2732.

[156] Jeon YJ, Shin HS, Rogers PL. Xylitol production from a mutant strain of Candida tropicalis. Letters in applied microbiology 2011;53(1)106-113, ISSN:0266-8254.

[157] Huang C, Jiang Y, Guo G, Hwang W. Development of a yeast strain for xylitol production without hydrolysate detoxification as part of the integration of co-product generation within the lignocellulosic ethanol process. Bioresource technology 2012;102(3)3322-3329, ISSN:0960-8524.

[158] Barbosa MFS, de Medeiros MB, de Mancilha IM, Schneider H, Lee H. Screening of yeasts for production of xylitol from D-xylose and some factors which affect xylitol yield in Candida guilliermondii. Journal of industrial microbiology 1988;3(4)241-251, DOI:10.1007/BF01569582.

[159] Roseiro JC, Peito MA, Gírio FM, Amaral-Collaco MT. The effects of the oxygen transfer coefficient and substrate concentration on the xylose fermentation by Debaryomyces hansenii. Archives of Microbiology 1991;156(6) 484-490, DOI:10.1007/BF00245396.

[160] Ligthelm ME, Prior BA, du Preez JC. The oxygen requirements of yeasts for the fermentation of D-xylose and D-glucose to ethanol. Applied Microbiology and Biotechnology 1988;28(1)63-68, ISSN:0175-7598.

[161] Debus D, Methner H, Schulze D, Dellweg H. Fermentation of xylose with the yeast Pachysolen tannophilus. European Journal of Applied Microbiology and Biotechnology 1983;17(5) 287-291, DOI:10.1007/BF00508022.

[162] Kötter P, Ciriacy M. Xylose fermentation by Saccharomyces cerevisiae. Applied Microbiology and Biotechnology 1993;38(6)776-783, DOI:10.1007/BF00167144.

[163] Hallborn J, Gorwa MF, Meinander N, Penttilä M, Keränen S, Hahn-Haegerdal B. The influence of cosubstrate and aeration on xylitol formation by recombinant Saccharomyces cerevisiae expressing the XYL1 gene. Applied Microbiology and Biotechnology 1994;42(2-3)326-333, ISSN: 0175-7598.

[164] Sonderegger M, Jeppsson M, Hahn-Hagerdal B, Sauer U. Molecular Basis for Anaerobic Growth of Saccharomyces cerevisiae on Xylose, Investigated by Global Gene Expression and Metabolic Flux Analysis. Applied and Environmental Microbiology 2004;70(4)2307-2317, DOI:10.1128/AEM.70.4.2307-2317.

[165] Ahmad I, Shim WY, Jeon WY, Yoon BH, Kim J. Enhancement of xylitol production in Candida tropicalis by co-expression of two genes involved in pentose phosphate pathway. Bioprocess and Biosystems Engineering 2012;35(1-2)199-204, ISSN: 1615-7591.

[166] Shanghai Yanda Biotechnology Co., Ltd.

[167] Wuhan Xuyadi Chemicals Co., Ltd.

[168] Tianjin Yuanlong Chemical Industry Co., Ltd.

[169] http://english.jl.gov.cn/Investment/Opportunities/Majorproject/201104/t20110401_971857.html

[170] http://www.azom.com/news.aspx?newsID=24757

[171] Bonrath W, Fischesser J. Novel reaction with a gold catalyst. WO2008148549 (December 2008).

[172] Pruesse U, Heidkamp K, Decker N, Herrmann M, Vorlop KD. Processing of raw materials from renewable resources by selective oxidation with gold catalysts. Chemie Ingenieur Technik 2010;82(8) 1231-1237.

[173] Van der Weijden RD, Mahabir J, Abbadi A. Reuter, MA. Copper recovery from copper (II) sulfate solutions by reduction with carbohydrates. Hydrometallurgy 2002;64(2)131-146, DOI:10.1016/S0304-386X(02)00031-2.

[174] Ericsson T, Samuelson O. Treatment of Birch Xylan with Chlorine in Aqueous Solution. Acta Chemica Seandinavica B,Summary citation from AGRICOLA, the online catalog of the National Agricultural Library (NAL)1975;29(3) 309-314, DOI: 10.3891/acta.chem.scand.29b-0309.

[175] Jokic A, Ristic N, Jaksic MM, Spasojevic M, Krstajic N. Simultaneous electrolytic production of xylitol and xylonic acid from xylose. Journal of applied Electrochemistry 1991;21(4) 321-326, ISSN: 0021-891X.

[176] Kiely DE, Hash KR. Process for the oxidization of monosaccharides to glycosyl carboxylic acids using oxygen and nitric acid as co-oxidants. U.S. Pat. Appl. Publ, US 20080033205(February 2008).

[177] Venema FR, Peters JA, van Bekkum H. Platinum-catalyzed oxidation of aldopentoses to aldaric acids. Journal of Molecular Catalysis 1992;77(1)75-85; DOI: 10.1016/0304-5102(92)80186-K.

[178] Boussie TR, Dias EL, Fresco ZM, Murphy VJ. Production of glutaric acid and derivatives from carbohydrate-containing materials. PCT Int. Appl., AU-A-2010259935, WO 2010/144871, (December 2010).

[179] Fleche G. Process for the manufacture of xylaric acid and uses thereof US Patent 5731467, (March 1998).

[180] Isbell HS, Frush HL, Martin ET. Reactions of carbohydrates with hydroperoxides: Part I. Oxidation of aldoses with sodium peroxide. Carbohydrate Research 1973;26(2)287-295, DOI:10.1016/S0008-6215(00)84515-2.

[181] Pullanikat P, Jung SJ, Yoo KS, Jung KW. Oxidative degradation of reducing carbohydrates to ammonium formate with H2O2 and NH4OH. Tetrahedron Letters 2010;51(47)6192-6194, ISSN:0040-4039.

[182] Wölfel R,Taccardi N,Bösmann A,Wasserscheid P. Selective catalytic conversion of biobased carbohydrates to formic acid using molecular oxygen. Green Chemistry 2011;13(10)2759-2763, DOI:10.1039/C1GC15434F.

[183] Bernhauer K, Riedl-Tumova E. Oxidations by acetic acid bacteria. X. Oxidation of enantiomorphic forms of xylose and arabinose. Biochemische Zeitschrift 1950;321 26-30.

[184] Ishizaki H, Ihara T, Yoshitake J, Shimamura M, Imai T. D-Xylonic acid production by Enterobacter cloacae. Nippon Nogei Kagaku Kaishi 1973;47(12) 755-761, DOI: 10.1271/nogeikagaku1924.47.755.

[185] Buchert J, Viikari L, Linko M, Markkanen P. Production of xylonic acid by Pseudomonas fragi. Biotechnology Letters 1986;8(8) 541-546, DOI:10.1007/BF01028079.

[186] Uchida K, Suzuki Y. Physiology of plant pathogens. 3. Production of D-galactonic acid, 2-keto-D-galactonic acid, L-arabonic acid, and D-xylonic acid by Erwinia milletiae. Nippon Nogei Kagaku Kaishi 1975;49(5) 257-262.

[187] Hayasida A. Demonstration of pentose oxidase and catalase in fusaria. Biochemische Zeitschrift 1938;298169-178.

[188] Ohsugi M, Tochikura T, Ogata K. Production of D-xylonic acid by Micrococcus species. Agricultural and Biological Chemistry 1970;34(3)357-363.

[189] Lockwood LB, Nelson GEN. The oxidation of pentoses by Pseudomonas. Journal of Bacteriology1946;52(5)581-586.

[190] Weimberg R. Pentose oxidation by Pseudomonas fragi. Journal of Biological Chemistry 1961;236(3)629-635, PMID 13783864.

[191] Kiessling H, Lindberg B, McKay J. Some products of the metabolism of D-xylose by Pullularia pullulans. Acta chemica Scandinavica 1962;16 1858-1862, DOI: 10.3891/acta.chem.scand.16-1858.

[192] Buchert J, Puls J, Poutanen K. Comparison of Pseudomonas fragi and Gluconobacter oxydans for production of xylonic acid from hemicellulose hydrolyzates. Applied Microbiology and Biotechnology 1988;28(4-5)367-372, DOI:10.1007/BF00268197.

[193] Stephens C, Christen B, Fuchs T, Sundaram V, Watanabe K, Jenal U. Genetic analysis of a novel pathway for D-xylose metabolism in Caulobacter crescentus. Journal of Bacteriology 2007;189(5)2181-2185, ISSN: 0021-9193.

[194] Yamanaka K, Gino M, Kaneda R. A specific NAD-D-xylose dehydrogenase from Arthrobacter sp. Agricultural and Biological Chemistry 1977;41(8)1493-1499, ISSN: 0002-1369.

[195] Nygard Y, Toivari MH, Penttila M, Ruohonen L, Wiebe MG. Bioconversion of d-xylose to d-xylonate with Kluyveromyces lactis. Metabolic engineering 2011;13(4)383-391, ISSN:1096-7176.

[196] Meijnen J, de Winde JH, Ruijssenaars HJ. Establishment of oxidative D-xylose metabolism in Pseudomonas putida S12. Applied and Environmental Microbiology 2009;75(9)2784-2791, ISSN: 0099-2240.

[197] Novick NJ, Tyler ME. (1982). L-Arabinose metabolism in Azospirillum brasiliense. Journal of Bacteriology 1982;149(1)364-367, PMCID PMC216631.

[198] Toivari MH, Ruohonen L, Richard P, Penttilae M, Wiebe MG. Saccharomyces cerevisiae engineered to produce D-xylonate. Applied Microbiology and Biotechnology 2010;88(3)751-760, ISSN:0175-7598.

[199] Liu H, Valdehuesa KNG, Nisola GM, Ramos KRM, Chung W. High yield production of D-xylonic acid from D-xylose using engineered Escherichia coli. Bioresource technology 2012;115244-248, ISSN:0960-8524.

[200] Henan Premtec Enterprise Corporation

[201] http://www.prweb.com/releases/formic_acid/leather_textiles/prweb1569064.htm

Chapter 2

Production of 2nd Generation of Liquid Biofuels

Leona Paulová, Petra Patáková,
Mojmír Rychtera and Karel Melzoch

Additional information is available at the end of the chapter

http://dx.doi.org/10.5772/53492

1. Introduction

Fluctuations in the price of oil and projections on depletion of accessible oil deposits have led to national and international efforts to enhance the proportion of energy derived from renewable sources (bioenergy) with special emphasis on the transport sector (e.g. according to Directive 2009/28 EC, by 2020, 20% of energy in EU-27 should be met from renewable sources and 10% should be used in transportation). To fulfil the legal requirements, wider exploitation of biofuels made from renewable feedstocks, as a substitute for traditional liquid fuels, will be inevitable; e.g. the demand for bioethanol in the EU is expected to reach 28.5 billion litres by 2020 [1], while in America 36 billion gallons of ethanol must be produced by 2022 [2]. Bioethanol, which has a higher octane level then petrol but only contains 66% of the energy yield of petrol, can be used as blend or burned in its pure form in modified spark-ignition engines [2]. This will improve fuel combustion, and will contribute to a reduction in atmospheric carbon monoxide, unburned hydrocarbons, carcinogenic emissions and reduce emissions of oxides of nitrogen and sulphur, the main cause of acid rain [2]. Butanol-gasoline blends might outcompete ethanol-gasoline ones because they have better phase stability in the presence of water, better low-temperature properties, higher oxidation stability during long term storage, more favourable distillation characteristics and lower volatility with respect to possible air pollution. Recently performed ECE 83.03 emission tests [3] have shown negligible or no adverse effects on air pollution by burning butanol-gasoline blends (containing up to 30% v/v of butanol) in spark ignition engines of Skoda passenger cars.

Although most of the world's bioethanol is currently produced from starch or sugar raw materials, attention is increasingly turning to 2nd generation biofuels made from lignocellulose, e.g. agriculture and forest wastes, fast growing trees, herbaceous plants, industrial wastes or wastes from wood and paper processing. The concept of ethanol production from lignocellulose sugars is not new. Probably the first technical attempt to degrade polysaccharides in wood was carried out by the French scientist Henri Braconnot in 1819 using 90% sulfuric acid [4]. His findings were exploited much later, in 1898, with the opening of the first cellulosic ethanol plant in Germany, followed by another one in 1910 in the US [5, 6]. During World War II, several industrial plants were built to produce fuel ethanol from cellulose (e.g. in Germany, Russia, China, Korea, Switzerland, US), but since the end of the war, most of these have been closed due to their non-competitiveness with synthetically produced ethanol [7]. In spite of all the advantages of lignocellulosic as a raw material (e.g. low and stable price, renewability, versatility, local availability, high sugar content, noncompetitiveness with food chain, waste revaluation) and extensive efforts of many research groups to reduce bottlenecks in technology of lignocellulosic ethanol production (e.g. energy intensive pretreatment, costly enzymatic treatment, need for utilization of pentose/hexose mixtures, low sugar concentration, low ethanol concentration), large scale commercial production of 2nd generation bioethanol has not been reopened yet [8], although many pilot and demonstration plants operate worldwide [9]. Identically, only first generation biobutanol is produced in China (approx. annual amount 100 000 t) and Brazil (approx. annual amount 8 000 t) [10]. At the 2012 London Olympic Games, British Petrol introduced its three most advanced biofuels i.e. cellulosic ethanol, renewable diesel and biobutanol. At a demonstration plant at Hull UK, biobutanol, produced by Butamax (joint venture of BP and DuPont) was blended at 24 % v/v with standard gasoline and used in BMW-5 series hybrids without engine modifications [11]. As the final price of both ethanol and 1-butanol produced by fermentation is influenced mostly by the price of feedstock, the future success of industrial ABE fermentation is tightly linked with the cost of pre-treatment of lignocellulosic material into a fermentable substrate.

2. Characterization of 2nd generation feedstock

Plant biomass can be used as a sustainable source of organic carbon to create bioenergy, either directly in the form of heat and electricity, or as liquid biofuels produced by thermochemical or biochemical methods or their combination [12]. In contrast to fossil energy sources, which are the result of long-term transformation of organic matter, plant biomass is created via photosynthesis using carbon dioxide as a source of carbon and sunlight as a source of energy and therefore is rapidly produced. The world annual production of biomass is estimated to be 146 billion metric tons [13], which could contribute 9-13% of the global energy supply yielding 45±10 EJ per year [14, 15].

Lignocellulose, which is stored in plant cell walls makes up a significant part of biomass representing 60-80% of woody tissue of stems, 15-30% of leaves or 30-60% of herbal stems [16]. Since it is not digestible for human beings, its use as a feedstock for bioprocesses does not compete with food production as in the case of sugar or starch raw materials.

All lignocellulose consist of three main polymeric components – cellulose, non-cellulosic carbohydrates (predominantly represented by hemicellulose) and lignin; its proportion and

http://dx.doi.org/10.5772/53492

structure differs for different types of biomass (Table1) and it is also influenced by variety, climatic conditions, cultivation methods and location. Minor components of the cell wall are represented by proteoglycans, pectin, starch, minerals, terpenes, resins tannins and waxes.

Biomass	Cellulose	Hemicellulose	Lignin	Reference
Hardwood	45-47	25-40	20-55	[17, 18]
Softwood	40-45	25-29	30-60	[17, 18]
Wheat straw	30-49	20-50	8-20	[19-22]
Rye straw	30.9	21.5	25.3	[21]
Corn fibre	15	35	8	[23]
Corn cobs	35-45	35-42	5-15	[22, 23]
Corn stover	39-42	19-25	15-18	[22, 23]
Corn straw	42.6	21.3	8.2	[20]
Rice straw	32-47	15-27	5-24	[20, 22, 23]
Rice hulls	24-36	12-19	11-19	[22]
Sugarcane bagasse	40	24-30	12-25	[20, 22, 23]
Switchgrass	30-50	10-40	5-20	[17, 23, 24]
Bermuda grass	25-48	13-35	6-19	[22, 23]
Cotton seed hairs, flax	80-95	5-20	0	[18, 22]
Municipal solid waste – separated fibre	49	16	10	[25]
Primary municipal sludge	29.3	not identified	not identified	[26]
Thickened waste activated sludge	13.8	not identified	not identified	[26]
Sawdust	45.0	15.1	25.3	[22]
Waste paper from chemical pulps	50-70	12-20	6-10	[17]
Newspaper	40-55	25-40	18-20	[1, 17]
Used office paper	55.7	13.9	5.8	[1]
Magazine	34.3	27.1	14.2	[1]
Cardboard	49.6	15.9	14.9	[1]
Paper sludge	33-61	14.2	8.4-15.4	[27, 28]
Chemical pulps	60-80	20-30	2-10	[18]

Table 1. Overview and composition of lignocellulosic biomass and other lignocellulosic sources

Cellulose is a homopolymer of 500-1 000 000 D-glucose units (e.g. 10 000 units in wood, 15 000 in native cotton) linked by β-1,4-glycosidic bonds [19, 26, 29]; the cellulose chains (200-300) are grouped together to form cellulose fibres. The strong inter-chain hydrogen bonds between hydroxyl groups of glucose residues in radial orientation and the aliphatic hydrogen atoms in axial positions creates a semi-crystalline structure resistant to enzymatic hydrolysis; weaker hydrophobic interactions between cellulose sheets promote the formation of a water layer near the cellulose surface, which protects cellulose from acid hydrolysis [30]. Cellulose originating from different plants has the same chemical structure, but it differs in crystalline structure and inter-connections between other biomass components. Microfibrils made of cellulose are surrounded by covalently or non-covalently bound hemicellulose, which is a highly branched heteropolymer made from 70-300 monomers units of pentoses (xylose, arabinose), hexoses (galactose, glucose, mannose) and acetylated sugars (e.g. glucuronic, galacturonic acids). Unlike cellulose, hemicellulose is not chemically homogenous and its composition depends on the type of material - hardwood contains predominantly xylans while softwood consists mainly of glucomannans [17, 23, 29,31]. Lignin, an amorphous heteropolymer of three phenolic monomers of phenyl propionic alcohols, namely p-coumaryl, coniferyl and sinapylalcohol, creates a hydrophobic filler, which is synthesized as a matrix displacing water in the late phase of plant fibre synthesis, and forms a layer encasing the cellulose fibres. Its covalent crosslinking with hemicellulose and cellulose forms a strong matrix, which protects polysaccharides from microbial degradation, makes it resistant to oxidative stress, and prevents its extraction by neutral aqueous solvents [31]. Forest biomass has the highest content of lignin (30-60% and 30-55 % for softwoods and hardwoods, respectively), while grasses and agricultural residues contain less lignin (10-30% and 3-15% respectively) [17].

There are several groups of lignocellulosic plant biomasses that can be exploited as a feedstock for bioprocessing. Woody biomass is represented mainly by hardwoods (angiosperm trees, e.g. poplar, willow, oak, cottonwood, aspen) and softwoods (conifers and gymnosperm trees e.g. pine, cedar, spruce, cypress, fir, redwood) together with forest wastes such as sawdust, wood chips or pruning residues. Nowadays the trend in this area is to use fast growing trees (poplar, willow) with genetically changed wood structures e.g. lower lignin content [32]. The advantage of forest biomass is its flexible harvesting time, thus avoiding long storage periods, and its high density, contributing to cost-effective transportation. Agricultural residues are represented mainly by corn stover or stalks, rice and wheat straw or sugarcane bagasse. The world's annual production of rice straw, wheat straw and corn straw that can be exploited for bioethanol production is 694.1, 354.3 and 203.6 million tons, respectively [20]. In the USA, 370 million and 350-450 million tons of forest biomass and agricultural wastes respectively are produced per year [17]. Although agrowastes are partly reutilized, e.g. as animal fodder, bedding, domestic fuel, used for cogeneration of electricity or reused in agriculture, a large fraction is still disposed as waste and is left in the fields; this can be utilized as a raw material for biofuels production. Sugarcane is nowadays one of the most important feedstocks for production of 1st generation bioethanol and also one of the plants with the highest photosynthetic efficiency, yielding around 55 tons of dry matter per hectare annually (approx. 176 kg/ha/day). Sugar cane bagasse, the fibrous lignocellulosic material remaining as waste is mostly used as a solid fuel in sugar mills or distilleries but due to its high cellulose content (Table 1) it can be reutilized as a feedstock for production of 2nd generation bioethanol. In the sugarcane season of 2010/11, the total sugar cane crop reached almost 1.627 billion tons (on 23 million hectares), which corresponds to 600 million tons of wet sugar cane bagasse [33]. Minor, but also important residues are the leaves, called sugarcane trash, amounting to 6-8 tons per hectare of sugarcane crop [34]. Another group of lignocellulosic biomass, herbaceous energy crops and grasses, which are represented predominantly by switch grass, alfalfa, sorrel or miscanthus [24], are interesting due to their low demands on soil quality, low-cost investments, fast growth, low moisture content, high yield per hectare (e.g. 20 t/ha for miscanthus) and high carbohydrate content (Table 1). Besides lignocellulosic plant materials, other low-cost large volume feedstocks such as municipal solid waste, municipal wastewater, food-processing waste or waste from the paper industry can be utilized for bioethanol production. Mixed municipal recovery solid waste (MSW) consists of approximately 55% mineral waste, 6% of metallic waste, 5% animal and vegetable waste (food residues, garden waste), 3% of paper and cardboard waste and 31% of others [35]. In the EU alone, the annual production of municipal wastes amounts 2.6 million tons, 65% of which is derived from renewable resources [35, 36]. The main challenge in its bioprocessing is its heterogeneous composition. To be used for ethanol production, degradable fractions of MSW should be separated after sterilization; cellulosic material (paper, wood or yard waste) represents approximately 60% of the dry weight of typical MSW as shown in Table 1 [25, 37]. Beside the solid wastes, lignocellulose extracted from municipal wastewater treatment processes can also be used as low-cost feedstock for biofuel production [26]. In Canada, 6.22 Mt of sugar could be annually produced using municipal sludge/biosolids and livestock manures [26]. Municipal wastewaters, which include faecal materials, scraps of toilet paper and food residues, should be pre-treated to separate solid and liquid fractions, the former of which is processed further to gain simple sugars. Primary sludge contains more cellulose compared to activated sludge (Table 1) because it is consumed in the activated sludge process and is further degraded by anaerobic digestion processes in the sewage disposal plant [26]. When talking about industrial wastes as 2nd generation raw materials for biofuels, wastes from cellulose/paper production cannot be neglected. Paper sludge is waste solid residue from wood pulping and papermaking processes and is represented by poor-quality paper fibres, which are too short to be used in paper machines. It is attractive as a raw material for bioprocessing mainly due to its low cost (it is currently disposed of in landfills or burned), its high carbohydrate content (Table 1) and its structure, which doesn´t require any pretreatment [8, 27, 28]. Another waste is represented by sulphite waste liquor (SWL), a solution of monomeric sugars formed during the sulfite pulping process by dissolution of lignin and most hemicelluloses. About 1 ton of solid waste is dissolved in SWL (11-14% solids) per ton of pulp and its annual production is around 90 billion litres [38]. SWL is usually burned after its concentration and evaporation, but since it's main components are sugars and lignosulfonates, its use as a raw material for bioethanol production has potential. Chemical composition of SWL (a spectrum of fermentable sugars, inhibitors, nutrients and minerals) differs significantly with the type of wood and technological procedures, e.g. concentration of the main sugars in SWL (% of dry matter) ranges for xylose from

3 to 5 % in soft wood (spruce, western hemlock) up to 21 % in eucalyptus, the highest concentration of galactose and glucose around 2.5 % is in soft wood SWL, content of mannose can reach values of almost 15 % in soft wood SWL [39-42]. SWL cannot be fermented without careful pretreatment - stripping off free sulfur dioxide and simultaneous concentration, steaming, removing inhibitors, adding nutrients, and adjusting the pH [43].

Although lignocellulose biomass is cheap and predominantly comprises waste material, the logistics, handling, storage and transportation dramatically increases its cost and therefore its use directly on site is preferred over to its processing in a central plant [8]. Further price increases occur due to the character of material - most lignocelluloses mentioned above are not fermentable by common ethanol producers and must be decomposed and hydrolysed into simple sugars before fermentation is carried out.

3. Biomass disruption in pretreatment process

A prerequisite for ethanol production from lignocellulose is to break recalcitrant structure of material by removal of lignin, and to expose cellulose, making it more accessible to cellulolytic enzymes by modifying its structure; this happens in the pretreatment process. Basically, lignocellulose processing into fermentable sugars occurs in two steps: a) pretreatment yielding a liquid fraction that is mostly derived from hemicellulose and lignin and a solid fraction rich in cellulose, b) further enzymatic or chemical hydrolysis of the solid (wet) cellulose fraction to yield fermentable sugars.

Delignification (extraction of lignin by chemicals) is an essential prerequisite for enzymatic digestion of biomass; it disrupts the lignin polymeric structure, leading to biomass swelling and increase in its surface area and enables contact of cellulolytic enzymes with cellulose fibres. Although some pretreatment methods do not lead to a significant decrease in lignin content, all of them alter its chemical structure making biomass more digestible even though it may contain the same amount of lignin as non-pretreated biomass [29]. Hemicellulose is often dissolved during pretreatment because it is thermosensitive and easily acid-hydrolysed due to its amorphous branched structure; the liquid fraction obtained after pretreatment thus contains mainly pentose sugars (D-xylose, D-arabinose) originating from hemicelluloses, and strains fermenting pentose sugars must be used for its processing into ethanol as discussed later. The solid wet fraction obtained after pretreatment contains predominantly cellulose and needs further processing to yield fermentable sugars.

The conversion of lignocellulose into fermentable sugars is more difficult to achieve than conversion of starch; starchy material is converted from a crystalline to an amorphous structure at temperatures of 60-70°C, while lignocellulose is more resistant - a temperature of 320°C and a pressure of 25 MPa is needed to achieve its amorphous structure in water [17]. Therefore complete decomposition of cellulose is rarely attainable. Although lignocellulose pretreatment is an energy-intensive process, which contributes significantly to the price of the final product (18-20% of the total cost of lignocellulosic bioethanol is attributed to pretreatment) [8], it is a necessary expense because enzymatic hydrolysis of non-pretreated material provides less than 20% of the theoretical maximum yield of fermentable sugars for the majority of lignocellulose feedstocks [44]. The resistance of biomass to enzymatic attack is characterized by a number of physical variables such as lignin content, crystallinity index (ratio of crystalline to amorphous composition of cellulose), degree of polymerization, chain length, specific surface area, pore volume or particle size [31], which are material specific; e.g. pretreatment of woody biomass differs considerably from agriculture biomass, while paper sludge doesn´t need any processing.

Efficient pretreatment of biomass is characterized by an optimum combination of variables which leads to effective disruption of the complex lignocellulosic structure, removes most of the lignin, reduces cellulose crystallinity and increases the surface area of cellulose that is accessible to enzymatic attack. At the same time, it should minimize the loss of sugars, limit the formation of toxic compounds, enable the recovery of valuable components (e.g. lignin or furfural), use high solids loading, be effective for many lignocellulosic materials, reduce energy expenses, minimize operating costs and maximize the sugar yield in the subsequent enzymatic processing [45-47]. Pretreatment efficiency is usually assessed as: a) total amount of recoverable carbohydrates analysed as concentration of sugars released in the liquid and solid fraction after pretreatment, b) conversion of cellulose, expressed as the amount of sugars released by enzymatic hydrolysis of the solid phase, c) fermentability of released sugars, expressed as the amount of ethanol produced in the subsequent fermentation or d) its toxicity (concentration of inhibitory compounds released by sugar and lignin decomposition) analysed by HPLC or measured as the ability of test strains to grow.

Although it might seem that the problem of lignocellulose pretreatment has been solved by the chemical pulping process, which has been used commercially for a long time to produce various paper products, the opposite is true; despite most lignin is removed in these processes, they have been optimized to maintain the strength and integrity of cellulose fibres that are used for papermaking or as chemical feedstock and thus they are not easily hydrolysed by enzymes. The traditional sulfite pulping process was first reported in 1857 where treatment of wood with a mixture of sulfur dioxide in hot water considerably softened the wood; in 1900 the sulfurous acid process was patented [6]. Nowadays chemical pulp production based on the sulphite method [38] use sulfurous acid and its salts (Ca^{2+}, Mg^{2+}, Na^+ and NH_4^+) in combination with SO_2 as a cooking liquor at temperatures of 120 - 150 °C. Sulfurous acid is an impregnation agent, improving the penetration of hydrolytic chemicals inside the wood structure [48], and importantly, promotes sulfonation of lignin leading to formation of lignosulfonic acid and its salts, that are soluble [49, 50]. Combinations of salts and cooking conditions produce different qualities of cellulose and different compositions of the sulfite waste liquors. Possibility to optimize old sulphite pulping process to obtain higher degree of saccharification of hard and softwoods had led to various modifications of process condition [48, 51-54]. So called SPORL technique is based on application of solution of bisulphate salts and sulfur dioxide (sulfurous acid) on biomass; sulfuric acid can also be added depending on lignin content (the higher amount of sulfuric acid is necessary for biomass with higher content of lignin, e.g. softwood, eucalyptus).

Many other processes have been investigated over the last decades in order to intensify lignocellulose pretreatment process by exploiting various physical, chemical and biological methods or their combination as reviewed elsewhere [29, 47] and summarized in Table 2.

Pretreatment	Condition	Advantages	Disadvantages	Reference
	Physical pretreatment			
Mechanical (chipping, shredding, milling, grinding)	Normal temperature and pressure	Decreased cellulose crystallinity, increased surface area, decreased degree of polymerization	High energy demand, no lignin removal	[29, 47]
	Biological pretreatment			
Biological pretreatment – soft, brown or white rot fungi	Normal temperature and pressure	Low cost, low energy consumption, degradation of lignin and hemicellullose	Low efficiency, loss of carbohydrates (consumed by fungi), long residence times (10-14 days), need for carefully controlled growth condition, big space	[29, 45, 47]
	Chemical pretreatment			
Dilute acid pretreatment (H_2SO_4, HCl, H_3PO_4, HNO_3),	Concentration<4%, temperature 140-215 °C, pressure 0.5 MPa, reaction time seconds to minutes	High reaction rates, lignin disruption, increased accessibility of cellulose, improved digestibility, moderate temperatures	Little lignin removed, hemicellulose dissolved, sugar decomposition (inhibitors), need for acid recycling and pH adjustment	[29, 31, 45, 55]
Concentrated acid hydrolysis (H_2SO_4, H_3PO_4)	Concentration 70-77%, temperature 40-100 °C	Crystalline structure of cellulose completely destroyed, amorphous cellulose achieved, low temperature	Hemicellulose dissolved, equipment corrosion, sugar decomposition (inhibitors), need for acid regeneration, pH adjustment, environmental concerns	[45]
Alkali pretreatment (NaOH, KOH, $Ca(OH)_2$)	Temperature 25-130 °C	Decreased crystallinity of cellulose, decreased polymerization, lignin removal, few inhibitors	Hemicellulose dissolved, pH adjustment	[29, 55]
Ammonia pretreatment	Temperature 25-60 °C, reaction time several days	High delignification, cellulose swelling, high volatility of ammonia, low cost, ammonia	Cellulose crystallinity not reduced, environmental concerns	[45, 47]

Pretreatment	Condition	Advantages	Disadvantages	Refe-rence
		recycle, continuous process, short residence times		
Ozonolysis	Room temperature, normal pressure, reaction time - hours	Lignin degradation, no inhibitors, ambient temperature	Hemicellulose dissolved	[21]
Combined acid and alkali pretreatment (formic acid-aqueous ammonia, dilute sulphuric acid-sodium hydroxide)		Cellulose digestion, fractionation of lignocellulose, most of non-cellulosic components removed, high loading		[45]
Combined acid and organic solvent (concentrated H_3PO_4 + aceton),	Moderate temperatures	Cellulose crystalline structure disrupted, high yield of amorphous cellulose, lignin removed, reduced enzyme loading	Hemicellulose dissolved	[45]
Ionic liquid (IL)pretreatment	Temperature <100 °C, cellulose recovered by addition of water, ethanol or acetone	Lignin extraction, low temperature, high biomass loading, high lignin solubility, cellulose dissolution, solvents recovered and reused, environmentally friendly	Cellulose recovered by addition of acetone, deionized water or alcohol, IL denaturates enzymes, IL must be washed before reused	[29, 44, 45]
	Physicochemical pretreatment			
Steam explosion	Temperature 160-240 °C, pressure 0.7-4.8 MPa, reaction time 1-10 min followed by biomass explosion	Extensive redistribution of lignin, high cellulose digestibility, cellulose swelling, limited use of chemicals	Little lignin removed, incomplete destruction of biomass matrix, sugar decomposition (inhibitors), hemicellulose dissolved, high energy consumption	[29, 31, 45, 55]
Acid-catalyzed steam explosion	Steam explosion catalysed by addition of H_2SO_4 or SO_2	Decreased time and temperature compared to steam explosion	Inhibitors formation, hemicellulose dissolved, high temperature	[45]

Pretreatment	Condition	Advantages	Disadvantages	Reference
Liquid hot water pretreatment	Temperature 180-230 °C, elevated pressure, pH 4-7, reaction time up to 15 min	Increased accessibility of cellulose, no inhibitors, no chemicals added, no need for pH adjustment and washing	Hemicellulose dissolved, lower loading	[45, 56]
Ammonia fiber explosion (AFEX)	Anhydrous liquid ammonia, temperature 60-120° C, pressure above 3 MPa, reaction time 30-60 min, followed decompression	Decreased crystallinity of cellulose, expanded fibre structure, increased accessible surface area, lignin depolymerisation and removal, low inhibitor concentrations, low temperature	Not suitable for softwood, hemicellulose dissolved, cost of ammonia, environmental concerns	[29, 31, 45, 55]
Ammonia recycle percolation	Aqueous ammonia (5-15%), temperature 150-180 °C, reaction time 10-90 min, flow 1-5 ml/min	Lignin removed, decreased crystallinity, low inhibitor concentrations, moderate temperatures	Hemicellulose dissolved, environmental concerns	[29]
Organosolv pretreatment	Organic (ethanol, methanol, ethylene glycol, glycerol, DMSO) or organic-aqueous mixtures, with catalyst at temperature >180 °C (HCl, H_2SO_4), temperature 100-250°C	Biomass fractionalization, pure cellulose, selectivity, effective for high-lignin biomass, organic solvents easily recovered (distillation) and reused, less energy	Hemicellulose dissolved, high cost of chemicals, inhibitors formation, need for containment vessels, explosion hazard, environmental concerns	[29]
Carbon dioxide explosion treatment	Supercritical CO_2, pressure 7-28 MPa, temperature 200 °C, time – several minutes	Increased surface area, low cost chemical, no inhibitors, high solid loading	Effectivity increased with moisture content, costly equipment	[29]
Wet oxidative pretreatment	Addition of oxidizing agent (oxygen, water, hydrogen peroxide)	Low concentration of inhibitors	High pressure and temperature, costly equipment and chemicals (oxygen)	[29]

Table 2. Overview and main characteristics of methods leading to biomass pretreatment

Acid treatments lead mainly to hydrolysis of hemicelluloses (pentose and hexose fractions) while alkaline treatments bring about lignin removal. Concentrated acids such as sulphuric or hydrochloric have been used as powerful agents to treat lignocelluloses, but due to their toxicity, corrosivity and necessity of recovery after hydrolysis, attention has shifted to milder conditions e.g. 0.5 % (v/v) sulfuric acid [57]. To improve cellulose hydrolysis in dilute acid processes, higher temperatures are favoured [58] since at a moderate temperature, direct saccharification resulted in low yields. As demonstrated by Candido et al. [59] for bagasse, dilute acid hydrolysis is greatly influenced by reaction time; at 100°C in 10% v/v sulfuric acid, the loss of mass and hemicellulose content decreased with time while soluble lignin concentration increased. Several modifications of the dilute acid hydrolysis method have been reported, e.g. acid hydrolysis with 1 % H_2SO_4 to remove hemicellulose and lignin followed by an alkaline step to increase the yield of cellulose. Methods based on the use of organosolv, wet oxidation, steam explosion or steam enriched with various impregnating agents (SO_2, CO_2, NH_3) are also often used for lignocellulose pretreatment as summarized in Table 2. The principle of the organosolv is mild hydrolysis of lignocellulose catalysed by sulfuric acid or sodium hydroxide in the reactor followed by extraction into ethanol at temperatures around 175 °C. Taking sugar cane bagasse as an example, the solid to liquid ratio can vary from 1 to 5 kg/l or lower, and solubilized lignin and hemicellulose appear in the liquid phase [34]. Wet oxidation is widely used in research and development technologies. Martín et al. [60] compared wet oxidation of bagasse, which was mixed with water (ca. 6 % w/v dry bagasse) in a special autoclave under slightly alkaline conditions, with steam explosion. In the wet oxidation procedure, slightly lower solubilisation of lignin, higher solubilisation of hemicellulose and higher cellulose content in the solid phase (approx. 60 % w/w) was achieved in comparison with steam explosion (45 % w/w). The effect of steam enrichment with CO_2 or SO_2 proved promising results as for enzymatic hydrolysis of cellulose and the low content of inhibitors, especially 2-furalaldehyde and 5-hydroxymethyl-2-furalaldehyde.

In summary, biomass pretreatment is a key bottleneck in the bioprocessing of lignocellulose biomass and even though all methods have distinct advantages, as summarized in Table 2, the main problems are high energy consumption and low substrate loading, leading to low sugar recovery. However, increasing the biomass concentration leads to high solid slurries which are very viscous, with a pasta-like behaviour, creating a challenge for mixing, pumping and handling; this increases energy demands reflected in a higher price for the ethanol as well as concentrates toxic compounds, thus counteracting any potential benefits [61].

Although the pretreatment process disrupts the complex structure of the material and causes partial hydrolysis of cellulose, the content of fermentable sugars is still very low; further enzymatic degradation of the cellulose polymeric chain must be carried out to increase the concentration of glucose, which is utilized (optimally together with hemicellulose-derived monomers) in fermentation as shown in Figure 1.

Most commercial enzyme preparations (the largest producers are Genencor, Novozymes or Spezyme) are produced by cultivation of *Trichoderma resei* as mixtures of enzymes with endo-1,4-ß-D-glucanase (EC 3.2.1.4, hydrolysis of (1→4) glucosidic linkages inside the chain), exo-1,4- ß-glucosidase (EC 3.2.1.74, hydrolysis of (1→4) linkage in (1→4)-β-D-glucans to remove successive glucose units), ß-glucosidase (EC 3.2.1.21, hydrolysis of terminal non-reducing β-D-glucosyl residues with release of β-D-glucose) and β-1,4-glucan cellobiohydrolase (EC 3.2.1.91, hydrolysis of (1→4)-β-D-glucosidic linkages in cellulose and cellotetraose releasing cellobiose from non-reducing ends of the chains) activities working in synergy.

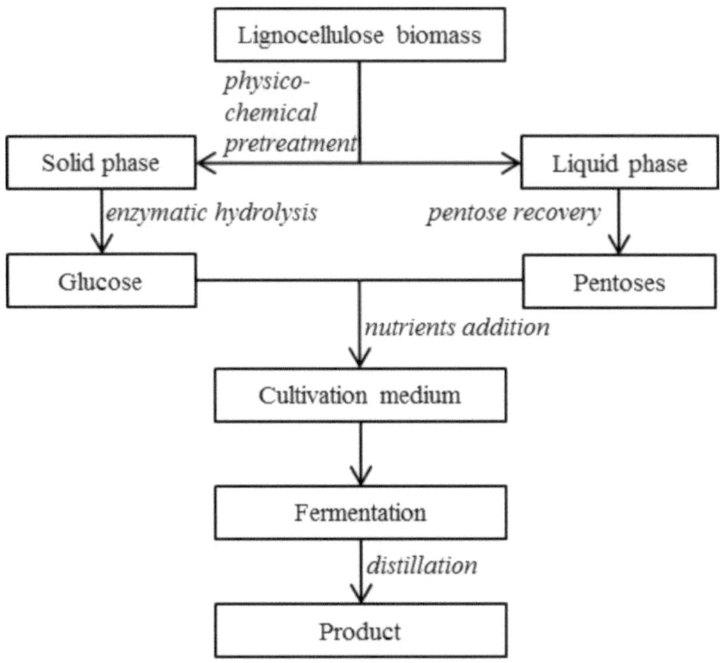

Figure 1. Simplified diagram of production of liquid biofuels from lignocellulose biomass

In recent years, the efficiency of commercial enzyme mixtures has rapidly increased and permits high conversions of cellulose to glucose; e.g. 85% and 91% yields of glucose were reported for ionic liquid pretreated poplar and switchgrass [62] and 85% and 83% yields were achieved for acid pretreated poplar and rice straws respectively [17, 63, 64]. Although the differential between the price of amylolytic and cellulolytic enzymes is currently reduced, the major difference is in dosing; about 40 -100 times more enzyme (based on protein weight) is required to breakdown cellulose compared to starch [29]. According to economic analyses, the conversion of biomass into fermentable sugars, which includes enzyme production and enzymatic hydrolysis together with indispensable pretreatment of biomass, comprises 33 % of the total cost [8, 17] and the estimated cost of cellulases is 50 cents per gallon (3.785 l) of ethanol, which is often comparable to the purchase cost of the feedstock [65]. For this reason attention has turned to further improvement of the composition and activity of enzyme cocktails, e.g. by constructing tailor-made multienzyme systems. It was shown that addition of xylanase and pectinase to alkali-pretreated biomass can reduce the negative effect of hemicellulose and pectin, which can restrict access of cellulases to the cellulose surface, while β-xylosidase can decompose xylobiose and polymerized xylooligomers to avoid inhibition of cellulolytic enzymes [22, 45]. Unfortunately, improved enzyme cocktails are not generally applicable, e.g. an enzyme complex enriched with β-mannanase and amyloglucosidase improved digestibility of dried distillers grains, but this was not required for corn stover [22]. Furthermore, the rate and efficiency of enzymatic hydrolysis can be affected by enzyme adsorption to non-cellulolytic substrates, e.g. lignin through phenolic groups and hydrophobic interactions, which limits the accessibility of cellulose to cellulases [45, 47]. To reduce this effect, "designer cellulosomes" have been recently constructed [45]. The cellulosome is a large complex of cellulolytic enzymes, originally produced by anaerobic bacteria [66], and has been engineered to comprise a recombinant chimeric scaffolding protein and many bound protein hybrids that have low lignin binding affinity. A different approach is represented by the addition of non-catalytic additives, e.g. surfactants (e.g. Tween, polyethylene glycol), polymers or proteins (bovine serum albumin, gelatine), which compete with cellulolytic enzymes for adsorption sites of lignin and thus prevent non-productive enzyme binding and can also facilitate enzyme recycling. Addition of expansins (plant proteins), expansin-like proteins or swollenin (fungal protein) promotes enhanced enzymatic hydrolysis by disrupting hydrogen bonding between cellulose and other cell-wall polysaccharides [45]. Recycling of enzymes, e.g. by ultrafiltration, re-adsorbtion onto fresh substrate, enzyme immobilization onto various materials e.g. chitosan-alginate composite, chitosan-clay composite, Eupergit C, mesoporous silicates, silicagel or kaolin are other approaches to reduce pretreatment costs [45].

The activity of cellulolytic enzymes can be reduced not only by ineffective binding, but also by feedback inhibition by glucose and cellobiose released by hydrolysis of cellulose as reviewed by Andric et al. [67] and by inhibitory effects of toxic products that may be released during pretreatment (type and concentration depends on biomass and process conditions) and can affect not only the rate and yield of saccharification but also substrate fermentability.

4. Toxic compounds released in pretreatment process

Toxic products can generally be divided into three main groups – aliphatic acids, furan derivatives and phenolic compounds [68-70] released by degradation of carbohydrates, and compounds arising from lignin. In acidic solutions, cellulose and hemicellulose are broken down into hexose and pentose sugars, which are further decomposed at high temperatures into furan derivatives represented mainly by 2-furaldehyde (furfural, FF) and 5-hydroxymethyl-2-furaldehyde (hydroxymethylfurfural, HMF). Free aliphatic acids, represented mainly by acetic, formic or levulinic acids, are created by substituents cleaved from lignin and hemicelluloses within the pretreatment, or are produced by cells during fermentation, while phenolic derivatives (4-hydroxybenzoic acid, 3,4-dihydroxybenzoic acid or vanilin) arise mainly from lignin decomposition in alkaline solution [71]. About 40 lignocellulose degradation products have been identified in various hydrolysates [71], the type and amount depending on type of biomass and pretreatment conditions [68]; e.g. furfural, hydroxymethylfurfural and levulinic acid occur in higher concentrations at low pH combined with high temperature and pressure [68, 71], while vanilin, vanilic, benzoic and 4-hydroxy-coumaric acids are formed under alkaline conditions at elevated temperatures and acetic acid is produced in significant concentrations independent of the process and type of biomass [71]. Although many studies on the effect of inhibitors on cellulolytic enzymes have been published, a general conclusion is not easy to draw because it is influenced not only by the type and origin of the enzyme preparation, but also by its dosing and the concentration of inhibitors. However, in general, compounds exhibiting higher hydrophobicity tend to be more inhibitory to cellulolytic enzymes, the greatest inhibitory effect being caused by acetic and formic acids [72-74], while the activity of enzymes is not practically influenced by levulinic acid [73]. On the other hand, the presence of inhibitory compounds also affects ethanol productivity in the subsequent fermentation by influencing metabolic functions of ethanol producing strains. Inhibitory effects are described by type and concentration of toxic compounds (their effect is intensified when present in combination) and the strain used for ethanol production, but generally, fermentation is mainly influenced by the presence of furan derivatives together with phenolic compounds and weak acids (at low pH). As reviewed elsewhere [70, 75], low molecular weight compounds are able to penetrate the cell, while inhibitors with high molecular weights affect expression and activity of sugar and ion transporters. Growth and rate of ethanol production by *Saccharomyces cerevisiae*, the main ethanol producing strain, is significantly inhibited by furfural, while ethanol yield is almost not influenced [75] due to its ability to detoxify the broth by reduction of furfural to furfuryl alcohol, which is less toxic.

Surprisingly, in butanol production process, *C.beijerinckii* BA101, *C. acetobutylicum* P260, *C. acetobutylicum* ATCC 824, *Clostridium saccharobutylicum* 262 and *Clostridium butylicum* 592 were not sensitive towards sugar degradation products like furfural or hydroxymethylfurfural (up to concentrations of 2-3 g/l) but its growth and solvent production were inhibited by ϱ-coumaric and ferulic acids present at a concentration of 0.3 g/l [76-78]. Solvent productivity and final solvent concentration in *C. beijerinckii* P260 were stimulated by addition of furfural or hydroxy methylfurfural (or both compounds) to the fermentation medium, at concentrations of up to 1 g/l [79]. *C. acetobutylicum* ATCC 824 metabolized furfural and hydroxymethyl furfural into furfuryl alcohol and 2,5-bis-hydroxymethylfuran, respectively and these compounds positively influenced solvent production up to a concentration of 2 g/l. It was hypothesised that this biotransformation step, independent of initial furfural and HMF concentrations, might increase solventogenesis via an increased rate of regeneration of NAD^+ [80]. Another possible inhibitor of phenolic origin, syringaldehyde, caused inhibition of solvent production by *C. beijerinckii* NCIMB 8052 over the whole range tested (0.2-1 g/l). This inhibition was probably caused by decreased expression and activity of coenzyme A transferase, which participated in utilization of butyric and acetic acids, because these acids accumulated in the medium [81].

The inhibitory effects of toxic compounds released by sugars and lignin degradation can be reduced in several ways, e.g. optimization of pretreatment conditions to minimize the formation of inhibitors, use of specific detoxification methods, e.g. precipitation by calcium hydroxide (overliming) alone or in combination with sulphite addition, adsorption on charcoal, evaporation of the volatile fraction, extraction with ethyl acetate or diethyl ether, ion extraction, treatment with peroxidase (E.C. 1.11.7) and laccase (EC 1.10.3.2), or use of microbial strains with increased resistance to inhibitors (achieved by adaptation or prepared by genetic modification) [75, 82, 83]. Lignin degradation products, ϱ-coumaric, ferulic and vanillic acids, together with vanillin, were effectively removed from a model solution of phenolic compounds by treatment with 0.01μM peroxidase (E.C. 1.11.7), resulting in improved growth and butanol production by *C. beijerinckii* NCIMB 8052 [84]. Sulphuric acid-hydrolysed corn fiber was treated with XAD-4 resin, resulting in an improvement of butanol yield achieved with *C. beijerinckii* BA101 [85]. Another popular approach for detoxification of acid hydrolysates for butanol production is "overliming" i.e. addition of $Ca(OH)_2$ in excess to hydrolysate [78, 85]. Although this detoxification method has been known for a long time, its mode of action, especially in the case of butanol production, is not completely clear. Addition of $Ca(OH)_2$ to an acid hydrolysate decreases furfural and HMF concentrations [86, 87] but does not affect acid concentrations; thus it is only possible to assume a beneficial neutralization effect. Furthermore it may be useful to treat hydrolysates with activated carbon [88].

5. Fermentation of lignocellulosic substrates

5.1. Ethanol fermentation

Fermentation of lignocellulose hydrolysates is more complicated compared to fermentation of 1st generation feedstock (sugar cane juice, molasses, grains) for several reasons: a) pentose sugars (predominantly xylose) are present along with hexoses (mainly glucose, mannose, galactose) in the hydrolysate, b) toxic compounds released during pretreatment can influence metabolic activity of the fermentation strain, c) low concentrations of fermentable sugars hamper the attainment of a high ethanol concentration. Because lignocellulose hydrolysates are poor in some nutrients (phosphorus, trace elements, and vitamins) they are usually supplemented, e.g. by addition of corn steep or yeast extract before being used as a substrate for fermentation. For an efficient process it is necessary to identify a strain that utilizes both pentose and hexose sugars, produces ethanol with a high yield and productivity and is tolerant to both inhibitors and ethanol. One of the main challenges is to simultaneously co-ferment pentose and hexose sugars, but neither yeast *S. cerevisiae* nor the bacterium *Z. mobilis*, which are usually used for ethanol production, contain genes for expression of xylose reductase and xylitol dehydrogenase [89]. In order to enhance process effectiveness, co-fermentation or sequential fermentation of hexoses and pentoses has been examined by combining good ethanol producers with strains naturally utilizing pentoses e.g. *Pichia stipitis, Candida shehatae, Pachysolen tannophillus, Klebsiella oxytoca*. However, xylose utilization is the rate limiting step due to catabolite repression by hexoses and the low availability of oxygen, and inhibition of pentose-utilizing strains by ethanol [90, 91]. Moreover, the yield of ethanol by co-fermentation is usually lower than with separate processes, e.g. yields of 0.5 g ethanol per g glucose (98% of theoretical) and 0.15 g/g xylose (29% theoretical) were achieved by separate cultivation of *Z. mobilis* and *P. tannophillus* respectively, but in optimized co-fermentation, the yield was just 0.33 g ethanol/g sugar. The same yield was obtained in a 5-reactor process combining *P. stipitis* and *S. cerevisiae* [92], but it was enhanced to 0.49 g/g sugars (96% theoretical) by cultivation of an adapted co-culture of *S. cerevisiae, P. tannophilis* and recombinant *E. coli* in dilute-acid softwood hydrolysate [93]. In a subsequent process employing *P. stipilis* and *S. cerevisiae,* which was inactivated before *Pichia* inoculation to avoid oxygen competition, 75% of theoretical ethanol yield was achieved [94]. A different approach is represented by the use of a recombinant strain prepared either by cloning genes encoding xylose utilization into good ethanol producers or to construct synthetic pathways for ethanol production in pentose-utilizing hosts. Wild type yeasts can be genetically modified to utilize xylose by introducing fungal genes encoding xylose reductase and xylitol dehydrogenase or bacterial/fungal genes for xylose isomerase [95]. Yeast *S. cerevisiae* was transformed with the *xylA* gene from *Thermus thermophiles* and *Piromyces sp.* to produce xylose isomerase, but unfortunately, this enzyme was inhibited by xylitol, favouring instead, its formation. Recently a recombinant strain of *S. cerevisiae* expressing a heterologous *xylA* gene produced 0.42 g/g of ethanol from xylose [96]. A strategy using *xyl1* and *xyl2* genes from *P. stipitis* introduced into *S. cerevisiae* produced transformants that exclusively consumed xylose, but produced significant amounts of xylitol [97]. On the other hand, with recombinant *Z. mobilis,* which carried *E. coli* genes encoding for xylose isomerase, xylulokinase, transketolase and transaldolase, 86% ethanol yield from xylose was achieved. Another strain of *Z. mobilis,* expressing genes *araABD* from *E. coli,* encoding L-arabinose isomerase, L-ribulokinase, L-ribulose-5-P-4 epimerase together with genes for transketolase and transaldolase, was able to grow on arabinose with 98% ethanol yield. *E. coli,* which naturally utilizes a wide range of substrates including pentoses, was transformed by genes encoding pyruvate decarboxylase and alcohol dehydrodenase, resulting in enhanced ethanol production [96]. Adaptation of recombinant strains to inhibitors can further increase the yield of ethanol, e.g. the ethanol yield achieved with a genetically engineered strain of *S. cerevisiae* grown on bagasse hydrolysate was increased from 0.18 g/g to 0.38 g/g after adaptation [98]. Recombinant strains that not only consume pentoses but also hydrolyse hemicelluloses by co-expressing endoxylanase, β-xylosidase and β-glucosidase activities has recently been constructed [95] and yields of 0.41 g/g of ethanol were obtained from total sugars in a rice straw hydrolysate.

In addition to the wide range of sugars, their low concentration in hydrolysates is problematic. Since ethanol recovery by distillation is only economically viable on the industrial scale for yields greater than 4% (w/w), which for most hydrolysates requires a dry mass concentration greater than 20% [45], the use of high substrate loading is needed. Effect of substrate concentration (unbleached hardwood pulp and organosolve pretreated poplar) on glucose concentration resulting from the enzyme hydrolysis was studied in [52] and [99]. In laboratory scale after 48 h of enzymatic hydrolysis 158 g/l glucose in the hydrolyzate was reached, ethanol concentration after fermentation ranged between 50.4 and 63.1 g/l. The general problem for this kind of conversions is that high load of the pulp or pretreated lignocelulosic material gives rise to high viscosity and thus also to mixing and transport problems. These extremely high yields of glucose can be attributed to a very efficient peg mixer. Problems connected with use of such high viscosity slurries can be overcome by various strategies, e.g. maximizing dry matter by removing most hemicellulose and lignin, utilizing alternative bioreactors with novel mixing modes (e.g. peg mixer, shaking, gravitational tumbling, hand stirring) or gradual dosing of substrate into the bioreactor (fed-batch), which enables the use of more substrate and thus increases the yield of ethanol above values achievable in batch mode. Moreover, the actual concentration of toxic substrates is reduced and yield and/or productivity is enhanced by controlled dosing of substrate and prolonged cultivation time, thus shortening unprofitable periods between batches [45, 89]. Feed rates should reflect the type of hydrolysate and strain. Continuous cultures usually using immobilized cells (to prevent their wash out from the bioreactor at high dilution rates) is another strategy to increase process productivity [89].

Integration strategies, which replace classical separate hydrolysis and fermentation processes (SHF) by combining several process steps in one vessel represents another approach for lignocellulosic ethanol production. Simultaneous saccharification and fermentation (SSF), which combines enzymatic hydrolysis and fermentation in one step, permits an increased rate of cellulose hydrolysis by elimination of product inhibition (the released glucose is consumed by the microbial strain), an increased rate of sugar consumption, reduced contamination due to the presence of ethanol and a reduced number of reactors. However, SSF is constrained by different temperature optima for each process (the cellulase optimum is usually 40-50 °C, whereas the fermentation temperature usually cannot exceed 35 °C for most ethanol producers) and carbon source limitation in the early stages of the process. Several modifications of SSF to ease the problems and increase productivity have been published. These include the use of thermotolerant ethanol producers [100, 101], application of a presaccharification step [102] or the use of recombinant strains consuming both hexose and pentose sugars (a simultaneous saccharification and co-fermentation process (SSCF)) [103] in batch or fed-batch mode [104]. Consolidated bioprocessing (CBP), which combines cellulase production, cellulose hydrolysis and fermentation into a single step have been investigated as a way of reducing the cost of cellulolytic enzymes, increasing volumetric productivity and reducing capital investment [105]. Some biofuel companies (e.g. Mascoma and Qteros) have been founded based on this concept [105]. CBP microorganisms should combine high cellulase production and secretory capability, the ability to utilize a broad range of sugars, tolerance to high concentrations of salts, solvents and inhibitors, high ethanol productivity and yield, have a known genomic DNA sequence and developed recombinant technologies and ideally be usable as feed protein after fermentation [105]. There is a lack of native organisms that combine the ability to produce cellulolytic enzymes and be homoethanolic with high titres and yields. Although some thermophilic anaerobic bacteria e.g. *Clostridium thermocellum,* are high cellulase producers and utilize both pentose and hexose sugars, they have a low tolerance to ethanol ~30 g/l [106] and an insufficient yield ~0.2 g/g [107]. Therefore recombinant strains have been prepared by engineering cellulolytic microorganisms (e.g. *C. thermocellum, C. phytofermentans, C. cellulolyticum, T. reesei or F.oxysporum*) to produce ethanol. Knockout mutants of *Thermoanaerobacterium saccharolyticum* that lack lactic and acetic acid production exhibited an ethanol yield from xylose of 0.46 g/g [108], while recombinant *Geobacillus thermoglucosidasius* produced 0.42-0.47 g/g of ethanol from hexoses [65]. Another attempt, to create a recombinant cellulose-utilizing microorganism using non-cellulolytic strains with high ethanol production have not been very successful; although some recombinant ethanologenic strains secreting some active cellulases have been prepared [106, 109, 110], their requirement for a nutrient rich medium and often sensitivity to end-product inhibition hamper their use [105].

5.2. ABE (acetone-butanol-ethanol) fermentation

Different, so-called solventogenic species of the genus *Clostridium*, like *Clostridium acetobutylicum*, *Clostridium beijerinckii* or *Clostridium saccharoperbutylacetonicum*, can be used for 1-butanol production by ABE fermentation. The fermentation usually proceeds in two steps; at first butyric and acetic acids, along with hydrogen and carbon dioxide, are formed and then metabolic switching leads to the formation of solvents (mainly 1-butanol and acetone) and the cessation/slowdown of acid and gas production (for recent reviews see [111-114]). Industrial fermentative ABE (butanol) production, which has quite a long and impressive history connected with both World Wars, is nowadays carried out only in China and Brazil (estimated annual production of 100 000 t and 8 000 t from corn starch and sugar cane juice, respectively) [10]. However, many corporations such as BP, DuPont, Gevo, Green Biologics, Cobalt Technologies and others have declared their interest in this field. A unique example of the use of lignocellulosic hydrolysate on an industrial scale is the former Dukshukino plant (operated in the Soviet Union up to 1980s) producing acetone and butanol by fermentation. The plant was based on current "very modern" biorefinery concepts which assumed the conversion of complex feedstocks (hydrolysates of agricultural waste + molasses or corn) into many valuable products i.e. in addition to solvents (acetone, butanol and ethanol), it was possible to produce liquid CO_2, dry ice, H_2, fodder yeast, vitamin B_{12} and biogas [115].

The most interesting approach to fermentation of any lignocellulosic substrate is probably consolidated bioprocessing (CBP) i.e. a method in which a single microorganism is used for both substrate decomposition and fermentation to produce the required metabolites. Although some clostridial species such as *Clostridium thermocellum* can utilise cellulosic substrates and produce ethanol [116, 117], the ABE fermentation pattern unfortunately cannot be produced using clostridia. However, *C.acetobutylicum* ATCC 824 possesses genes for various cellullases and a complete cellulosome [118-120]. But even if production of some cellulases by *C. acetobutylicum* ATCC 824 was induced by xylose or lichenan [118], cellulose utilization was not achieved, possibly because of insufficient or deficient synthesis of an unknown specific chaperone that could be responsible for correct secretion of cellulases [119]. Nevertheless as solventogenic *Clostridium* species are soil bacteria that differ significantly in fermentative abilities and genome sizes, it is not excluded that in the future, some solventogenic species with cellulolytic activity will be isolated from an appropriate environment. Recently, a new strain of *Clostridium saccharobutylicum* with hemicellulolytic activity and ABE fermentation pattern was found amongst 50 soil-borne, anaerobic, sporulating isolates [121].

Substrate	Pretreatment	Microbial strain	ABE concentration (g/l) /yield (%) /productivity (g/l/h)	Reference
Wheat straw	Diluted sulphuric acid+ enzyme	*C. beijerinckii* P260	13/25/0.14	[139]
Wheat bran	Diluted sulphuric acid	*C. beijerinckii* ATCC 55025	12/32/0.16	[140]
Corn fiber	Diluted sulphuric acid+ XAD-4 resin treatment + enzyme	*C. beijerinckii* BA101	8/32/0.11	[127]
Corn cobs	Steam explosion + enzyme	*C. acetobutylicum*	21/31/0.45	[131]
Rice straw	Alkali + $(NH_4)_2SO_4$ precipitation + activated carbon treatment + enzyme	*C. saccharoperbutylacetonicum* ATCC 27022	13/28/0.15	[88]
Sugar cane bagasse	Alkali + $(NH_4)_2SO_4$ precipitation + activated carbon treatment + enzyme	*C. saccharoperbutylacetonicum* ATCC 27022	14/30/0.17	[88]
Cassava bagasse	Heat + enzyme	*C. acetobutylicum* JB200	34/39/0.63	[130]
Domestic organic waste	Steam explosion, lyophilization + enzyme +4 fold concentration of released sugars	*C. acetobutylicum* DSM 792	9/26/0.08	[132]
Dried distiller`s grain and solubles	Diluted acid + overliming +enzyme hot water + overliming+ enzyme AFEX +overliming+ enzyme	*C. saccharobutylicum* 260 *C. butylicum 592* *C. butylicum 592*	12/35/0.20 13/32/0.20 12/32/0.20	[78]
Sweet sorghum stem	Diluted acetic acid	*C. acetobutylicum* ABE 0801	19/32/0.10	[141]

All fermentations were run in SHF mode i.e. sugar release and fermentation were separate processes.

AFEX stands for ammonium fiber expansion process.

Table 3. Selection of batch ABE fermentations in laboratory scale using lignocellulosic hydrolysates as a substrate

Until now, lignocellulosic substrates must be prehydrolysed for the ABE process. In the case of fermentation of lignocellulosic hydrolysate, usually containing low concentrations of fermentable sugars, one of the main bottlenecks in the ABE process, the low final titre of butanol (caused by severe butanol toxicity towards bacterial cells), is of minor importance. In fact hydrolysates are very good substrates for clostridia that express extensive fermentative abilities [122, 123]and can utilise not only cellulose-derived glucose but also hemicellulose monomers (xylose, arabinose, galactose, mannose). Co-fermentation of various sugar mixtures was described for *Clostridium beijerinckii* SA-1 (ATCC 35702) [124], *Clostridium acetobutylicum* DSM 792 [125], *C. acetobutylicum* ATCC 824 [126] and *C.beijerinckii* P260 [127] however, at the same time, catabolic repression of xylose utilization in the presence of glucose was demonstrated in *C. acetobutylicum* ATCC 824 [128, 129].

An overview of fermentation parameters achieved in batch ABE fermentations of different hydrolysates is presented in Table 3. The most promising results were obtained by Lu et al. [130] using cassava bagasse and a mutant strain, *C.acetobutylicum* JB200; the results of Marchal et al. [131] were unique at the scale used (48 m^3) as shown in Table 3. A frequent problem of lignocellulosic hydrolysates is a low final concentration of fermentable sugars caused by low density of the original substrate. This can be overcome by evaporation of the hydrolysate [132](see Table 3) or by addition of glucose and/or other carbohydrates present in the hydrolysate (this is only possible in laboratory scale experiments) [85,133-135]. In the case of glucose supplemented corn stover and switchgrass hydrolysates, final ABE concentrations of 26 and 15 g/l were achieved [135]. With *C.beijerinckii* P260, use of diluted and $Ca(OH)_2$ treated barley straw hydrolysate supplemented with glucose resulted in a solvent concentration of 27 g/l, a yield of 43% and productivity of 0.39 g/l/h [133]. In addition to materials presented in Table 3, other substrates like diluted sulfite spent liquor supplemented with glucose [134], palm empty fruit bunches [136, 137] or hardwood [138] were used in the ABE process but in these cases, additional optimizations were necessary.

In addition to a batch fermentation arrangement, semi-continuous fermentation of enzymatically hydrolyzed SO_2 pretreated pine wood using *C.acetobutylicum* P262 resulted in 18 g/l of solvents, a yield of 36% and solvent productivity of 0.73 g/l/h [142]. Further, fed-batch fermentation of wheat straw hydrolysate supplemented with varying concentrations of hydrolysate sugars (glucose, xylose, arabinose and mannose) using *C.beijerinckii* P260 yielded a solvent productivity of 0.36 g/l/h if gas stripping was used [127]. In the cases shown in Table 3, enzyme hydrolysis preceeded fermentation, however simultaneous saccharification and fermentation (SSF) was also tested. In SSF of acid pre-hydrolyzed wheat straw using *C.beijerinckii* P262 and solvent removal by gas stripping, 21 g/l of ABE was produced with a productivity of 0.31 g/l/h [127] Nevertheless, the solvent yield from hardwood using SSF was rather low, at 15% [138].

6. Conclusion

Intensive research over the last decades on lignocellulose-derived ethanol have focused mainly on intensification of biomass pretreatment, production of cellulolytic enzymes, and strain and process improvements, and have eliminated some of the main technological bottlenecks. Although a number of projects on 2nd generation bioethanol ended with the opening of pilot and demonstration plants around the world (production capacity in millions of gallons for the year 2012 given in brackets) e.g. the POET demonstration plant in Iowa (0.02 from corn stover and cobs), Abengoa in Kansas (0.01 from corn stover), Blue Sugarsin Wyoming (1.3 from stover and cobs), Chempolis in Finland (3.7 from paper waste), Fiberight in Iowa (6.0 MSW), Iogen in Canada (0.48 from stover), Praj MATRIX in India (0.01 from cellulose), UPM-Kymemene/Mesto in Finland (0.68 from mixed cellulose) and in spite of several proclamations, none of them is operating at the industrial scale [9]. To make this possible, further reductions in processing costs will be necessary to achieve a product that is competitive with 1st generation bioethanol. Further process integration is required, including decreased energy demand during pretreatment, increased sugar concentration, higher enzyme activity and strain recycling. By-products, e.g. lignin separated after pretreatment procedure can be used to generate energy for ethanol plant operations (lignin has higher caloric value (25.4 MJ/kg) then the biomass itself [8]) or used as a dispersant and binder in concrete admixtures, as an alternative to phenolic and epoxy resins, or as the principal component in thermoplastic blends, polyurethane foams or surfactants [143]. A combination of 1st and 2nd generation feedstocks (e.g. corn cobs together with stover) can eliminate bottlenecks and lead to product competitiveness. Higher bioethanol production costs can also be compensated for by political and economic instruments such as tax incentives (e.g. tax exemption on biofuels and higher excise taxes for fossil fuels) and legislation (mandatory blends) to enable ready access of 2nd generation biofuels to the market [30]. Butanol, as a second generation biofuel, might be produced via fermentation and used as an excellent fuel extender in addition to ethanol if the technological bottleneck of a low final concentration, yield and productivity could be overcome, and the assumption that suitable cheap waste pretreatments were possible.

Acknowledgement

The. review was performed thanks to financial support of the projects Kontakt ME10146 of Ministry of Education, Youth and Sport of Czech Republic and BIORAF No.TE01020080 of the Technological Agency of the Czech Republic.

Author details

Leona Paulová, Petra Patáková, Mojmír Rychtera and Karel Melzoch

*Address all correspondence to: Leona.paulova@vscht.cz

Department of Biotechnology, Institute of Chemical Technology Prague, Prague, Czech Republic

References

[1] Wang L., Sharifzadeh M., Templer R., Murphy R. J. Technology performance and economic feasibility of bioethanol production from various waste papers. Energy & Environmental Science 2012;5, 5717-5730.

[2] Nigam P. S., Singh A. Production of liquid biofuels from renewable resources. Progress in Energy and Combustion Science 2011;37(1), 52-68.

[3] Patáková P., Maxa D., Rychtera M., Linhová M., Fribert P., Mužíková Z., Lipovský J., Paulova L., Pospíšil M., Šebor G., Melzoch K. Perspectives of biobutanol production and use. In: Biofuel's Engineering Process Technology. Bernandes M., (ed.). Bernandes M. ed. Rijeka, Croatia: InTech; 2011. p. 243-266.

[4] Braconnot H. Hydrolysis of cellulose into sugar. Gilbert´s Annalen der Physik 1819;63, 348.

[5] U.S. Department of Energy - Energy Efficiency and Renewable Energy Biomass Program. Dilute acid hydrolysis. http://www1.eere.energy.gov/biomass/printable_versions/dilute_acid.html#background (accessed 3 August 2012).

[6] Sherrard E. C., Kressman F. W. Review of processes in the United States prior to World War II. Industrial & Engineering Chemistry Research 1945;37(1), 5-8.

[7] Lin Y., S. T. Ethanol fermentation from biomass resources: Current state and perspectives. Applied Microbiology and Biotechnology 2006;69, 627-642.

[8] Banerjee S., Mudliar S., Sen R., Giri B. Commercializing lignocellulosic bioethanol: Technology bottlenecks and possible remedies. Biofuels Bioproducts & Biorefining 2010;4, 77-93.

[9] Biofuel Digest. Free database download:207 advanced fiofuels&chems projets.http://www.biofuelsdigest.com/bdigest/2011/11/16/free-database-download-207-advanced-biofuels-chems-projects (accessed 13 August 2012).

[10] Green E. M. Fermentative production of butanol - the industrial perspective. Current Opinion in Biotechnology 2011;22(3), 1-7.

[11] The 2012 London Olympics, biofuels-style: BP to showcase its hhree most advanced biofuels. http://www.biofuelsdigest.com/bdigest/2012/07/23/the-2012-london-olympics-biofuels-style-bp-to-showcase-its-three-most-advanced-biofuels (accessed 8 August 2012).

[12] Gray K. A., Zhao L., Emptage M. Bioethanol. Current Opinion in Chemical Biology 2006;10(2), 141-146.

[13] Balat M., Ayar G. Biomass energy in the world, use of biomass and potential trends. Energy Sources 2005(29), 931-940.

[14] Kim S., Dale B. E. Global potential bioethanol production from wasted crops and crop residues. Biomass and Bioenergy 2004(26), 361-375.

[15] Ladanai S., Vinterback J. Global potential of sustainable biomass for energy. Uppsala: Swedish University of Agricultural Sciences; 2009. p. 29.

[16] Moller R., Toonen M., Van Beilen J., Salentijn E., Clayton D. Crop platforms for cell wall biorefining: lignocellulose feedstocks, Outputs from the EPOBIO project: CPLpress Science publishers; 2007.

[17] Limayem A., Ricke S. C. Lignocellulosic biomass for bioethanol production: Current perspectives, potential issues and future prospects. Progress in Energy and Combustion Science 2012(38), 449-467.

[18] Balat M. Production of bioethanol from lignocellulosic materials via the biochemical pathway: a review. Energy Conversion and Management 2011(52), 858-875.

[19] Volynets B., Dahman Y. Assesment of pretreatments and enzymatic hydrolysis of wheat straw as a sugar source for bioprocess industry. International journal of energy and environmnet 2011;2(3), 427-446.

[20] Sarkar N., Ghosh S. K., Bannerjee S., Aikat K. Bioethanol production from agricultural wastes: an overview. Renewable Energy 2012(37), 19-27.

[21] García-Cubero T., Gonzáles-Benito G., Indacoechea I., Coca M., Bolado S. Effect of ozonolysis pretreatment on enzymatic digestibility of wheat and rye straw. Bioresour Technol 2009(100), 1608-1613.

[22] Van Dyk J. S., Pletschke B. I. A review of lignocellulose bioconversion using enzymatic hydrolysis and synergist cooperation between enzymes - factors afftecting enzymes, conversion and synergy. Bitechnology Advances 2012.

[23] Saha B. C. Hemicellulose bioconversion. Journal Indian Microbiology and Biotechnology 2003;30, 279-291.

[24] McKendry P. Energy production from biomass (part 1): Overview of biomass. Bioresour Technol 2002;83, 37-46.

[25] Li S., Zhang X., Andresen J. M. Production of fermentable sugars from enzymatic hydrolysis of pretreated municipal solid waste after autoclave process. Fuel 2011.doi. 10.1016/j.fuel.2011.07.012

[26] Champagne P., Li C. Enzymatic hydrolysis of cellulosic municipal wastewater treatment process residuals as feedstocks for the recovery os simple sugars. Bioresour Technol 2009;100 (4), 5700-5706.

[27] Peng L., Chen Y. Conversion of paper sludge to ethanol by separate hydrolysis and fermentation (SHF) using Saccharomyces cerevisiae. Biomass and Bioenergy 2011(35), 1600-1606,Yamashita Y., Kurosumi A., Sasaki C., Nakamura Y. Ethanol production from paper sludge by immobilized Zymomonas mobilis. Biochemical Engineering Journal 2008(42), 314-319.

[28] Yamashita Y., Kurosumi A., Sasaki C., Nakamura Y. Ethanol production from paper sludge by immobilized Zymomonas mobilis. Biochemical Engineering Journal. 2008(42):314-319.

[29] Agbor V. B., Cicek N., Sparling R., Berlin A., Levin D. B. Biomass pretreatment: Fundamentals toward application. Biotechnology Advances 2011(29), 675-685.

[30] Bessou C., Ferchaud F., Gabrielle B., Mary B. Biofuels, greenhouse gases and climate change. A review. Agronomy for Sustainable Development 2011;31(1), 1-79.

[31] Vidal B. C., Jr., Dien B. S., Ting K. C., Singh V. Influence of feedstock particle size on lignocellulose conversion--a review. Appl Biochem Biotechnol 2011;164(8), 1405-1421.

[32] El Bassam N. Handbook of Bioenergy Crops: A complete reference to species, development and applications: Routledge; 2010.

[33] OECD Agriculture outlook. http://www.keepeek.com/Digital-Asset-Management/oecd/agriculture-and-food/oecd-fao-agricultural-outlook-2011_agr_outlook-2011-en (accessed 24 July 2012).

[34] Singh P., Suman A., Tiwari P., Arya N., Gaur A., Shivastava A. K. Biological pretreatment of sugarcane trash for its conversion to fermentable sugars. World Journal of Microbiology and Biotechnology 2008;24, 667-673.

[35] Fodor Z., Klemeš J. J. Waste as alternative fuel - minimising emissions and effluents by advanced design. Process safety and environmental protection 2012;90, 263-284.

[36] Jensen J. W., Felby C., Jorgensen H., Ronch G. O., Norholm N. D. Enzymatic processing of municipal solid waste. Waste Management 2010;30, 2497-2503.

[37] Ballesteros M., Sáez F., Ballesteros I., Manzanares P., Negro M. J., Martínez J. M., Castaneda R., Dominguez J. M. O. Ethanol production from the organic fraction obtained after thermal pretreatment of municipal solid waste. Appl Biochem Biotechnol 2009;161, 423-431.

[38] Fernandes D. L. A., Pereira S. R., Serafim L. S., Evtugin D. V., Xavier A. M. R. B. Bioethanol. Rijeka: InTech; 2012.

[39] Sjorostrom E. Wood Chemistry. Fundamentals and Applications. New York, USA: Academic Press; 1993.

[40] Rodsrud G. Biorefining – expanding the sulfite pulping biorefinery concept. World Biofuels Markets; March 22 – 24, 2011; Rotterdam, NL2011.

[41] Mulvany P. K., Agar H. D., Peniston Q. P., Mccarthy J. L. A chromatographic study of sulfite waste liquor sugars. 118th Meeting of the American Chemical Society; Sept. 3 – 8, 1950; Chicago, Illinois1950.

[42] Lai L. X. Bioproducts from sulfite pulping: Bioconversion of sugar streams from pulp, sludge and spent sulfite liquor: University of Washington; 2010.

[43] Harris E. E. Food-yeast production from wood-processing byproducts. Yeast in Feeding Symposium; Nov. 8 – 10, 1948; Milwaukee, Wisconsin, USA1948.

[44] Vancov T., Alston A.-S., Brown T., Mcintosh S. Use of ionic liquids in converting lignocellulosic material to biofuels. Renewable Energy 2012;45, 1-6.

[45] Huang R., Su R., Qi W., He Z. Bioconversion of lignocellulose into bioethanol: process intensification and mechanism research. Bioenergy Resources 2011;4, 225-245.

[46] Modenbach A. A., Nokes S. E. The use of high-solids loadings in biomass pretreatment - a review. Biotechnology and Bioengineering 2012;109(6), 1430-1442.

[47] Chandra R. P., Bura R., Mabee W. E., Berlin A., Pan X., Saddler J. N. Substrate pretreatment: the key to effective enzymatic hydrolysis of lignocellulosic? Advanced Biochemical Engineering/Biotechnology 2007;108, 67-93.

[48] Wang G. S., Pan X. J., Zhu J. Y., Gleisner R., Rockwood D. Sulfite pretreatment to overcome recalcitrance of lignocellulose sport for robust enzymatic saccharification of hardwoods. Biotechnology Progress 2009;25(4), 1086-1093.

[49] Nikitin N. I. The chemistry of cellulose and wood: Israel Program for Scientific Translations; 1966.

[50] Sjorostrom E. Wood chemistry. Fundamental and applications. New York, USA: Academic Press; 1993.

[51] Zhu J. Y., Pan X. J., Wang G. S., Gleisner R. Sulfite pretreatment (SPORL) for robust enzymatic saccharification of spruce and red pine. Bioresource Technology 2009;100, 2411-2418

[52] Zhang X., Qin W., Paice M. G., Saddler J. N. High consistency enzymatic hydrolysis of hardwood substrates. Bioresource Technology 2009;100, 5890-5897.

[53] Zhang X., Tu M., Paice M., Sacciadis G., Jiang Z., Jemaa N., Thibault A. Bioconversion of knot rejects from a sulphite pulp mill to ethanol. BioResources. 2010;5(1):23-42

[54] Helle S. S., Petretta R. A., Duff S. J. B. Fortifying spent sulfite pulping liquor with hydrolyzed recect knots. Enzyme and Microbial Technology. 2007(41):44-50.

[55] Modenbach A. A., Nokes S. E. The use of high-solids loadings in biomass pretreatment - a review. Biotechnology and Bioengineering 2012;102(6), 1430-1442.

[56] Mosier N., Hendrickson R., Ho N., Sedlak M., Ladish M. Optimization of pH controlled liquid hot water pretreatment of corn stover. Bioresour Technol 2005(96), 1986-1993.

[57] Hahn-Hagerdal B., Galbe M., Gorwa-Grauslund M. F., Lidén G., Zacchi G. Bio-ethanol - the fuel of tomorrow from the residues of today. Trends in Biotechnology 2006;24(12), 549-556.

[58] Sun Y., Cheng J. Hydrolysis of lignocellulosic materials for ethanol production: a review. Bioresource Technology 2002;8, 1-11.

[59] Candido R. G., Godoy G. G., Concalves A. R. Study of sugarcane bagasse pretreatment with sulfuric acid as a step of cellulose obtaining. World Academy of Science, Engineering and Technology. 2012; 61, 101-105.

[60] Martín C., Marcet M., Thomsem A. B. Comparison between wet oxidation and steam explosion as pretreatment method for enzymatic hydrolysis of sugarcane bagasse. BioResources 2008;3(3), 670-683.

[61] Modenbach A. A., Nokes S. E. The use of high-solids loadings in biomass pretreatment-a review. Biotechnololgy and Bioengineering 2012;109(6), 1430-1442.

[62] Samayam I. P., Schall C. A. Saccharification of ionic liquid pretreated biomass with commercial enzyme mixtures. Bioresource Technology 2010;101, 561-3566.

[63] Hsu T.-C., Guo G.-L., Chen W.-H., Hwang W.-S. Effect of dilute acid pretreatment of rice straw on structural properties and enzymatic hydrolysis. Bioresource Technology 2009;101, 4907-4913.

[64] Kumar R., Wyman C. E. Effects of cellulase and xylanase enzymes on the deconstruction of solids from pretreatment of poplar by leading technologies. Biotechnology Progress. 2009;25(2):302-314.

[65] Olson D. G., Mcbride J. E., Shaw A. J., Lynd L. R. Recent progress in consolidated bioprocessing. Current Opinion in Biotechnology 2012;23, 396-405.

[66] Doi R. H., Kosugi A. Cellulosomes: plant-cell-wall degrading enzyme complexes. Nature reviews Microbiology 2004;2, 541-551.

[67] Andric P., Meyer A. S., Jensen P. A., Dam-Johansen K. Reactor design for minimizing product inhibition during enzymatic lignocellulose hydrolysis: I. Significance and mechanism of cellobiose and glucose inhibition on cellulolytic enzymes. Biotechnology Advances 2010;28, 308-324.

[68] Almeida J. R. M., Modig T., Petersson A., Hahn-Hagerdal B., Lidén G., Gorwa-Grauslund M. F. Increased tolerance and conversion of inhibitors in lignocellulosic hydrolysates by Saccharomyces cerevisiae. Journal of Chemical Technology and Biotechnology 2007;82, 340-349.

[69] Palmqvist E., Hahn-Hagerdal B. Fermentation of lignocellulosic hydrolysates. I:Inhibition and detoxification. Bioresource Technology 2000;74, 17-24.

[70] Palmquist E., Hahn-Hagerdal B. Fermentation of lignocellulosic hydrolysates. II: Inhibitors and mechanisms of inhibitions. Bioresource Technolology 2000;74, 25-33.

[71] Du B., Sharma L. N., Becker C., Chen S.-F., Mowery R. A., Van Walsum G. P., Chambliss C. K. Effects of variying feedstock-pretreatment chemistry combinantions on the formation and accumulation of potentially inhibitory degradation products in biomass hydrolysates. Biotechnology and Bioengineering 2010;107(3), 430-440.

[72] Panagiotou G., Olsoon L. Effect of compounds released during pretreatment of wheat straw on microbial growth and enzymatic hydrolysis rates. Biotechnology and Bioengineering 2007;96(2), 250-258.

[73] Cantarella M., Cantarella L., Gallifuoco A., Spera A., Alfani F. Effect of inhibitors released during steam-explosion treatment of poplar wood on subsequent enzymatic hydrolysis and SSF. Biotechnology Progress 2004;20, 200-206.

[74] Ximenes E., Kim Y., Mosier N., Dien B. S., Ladish M. Deactivation of cellulases by phenols. Enzyme and Microbial Technology. 2011(48):54-60.

[75] Klinke H. B., Thomsen A. B., Ahring B. K. Inhibition of ethanol-producing yeast and bacteria by degradation products produced during pre-treatment of biomass. Applied Microbiology and Biotechnology 2004;66, 10-26.

[76] Ezeji T., Qureshi N., Blaschek H. P. Butanol production from agricultural residues: impact of degradation products on *Clostridium beijerinckii* growth and butanol fermentation. Biotechnology and Bioengineering 2007;97(6), 1460-1469.

[77] Ezeji T., Blaschek H. P. Butanol production from lignocellulosic biomass: Blackwell Publishing; 2010.p.19-37.

[78] Ezeji T., Blaschek H. P. Fermentation of dried distillers' grains and solubles (DDGS) hydrolysates to sovents and value-added products by solventogenic *Clostridia*. Bioresour Technol 2008;99(12), 5232-5242.

[79] Qureshi N., Bowman M. J., Saha B. C., Hector R., Berhow M. A., Cotta M. A. Effect of cellulosic sugar degradation products (furfural and hydroxymethyl furfural) on acetone-butanol-ethanol (ABE) fermentation using *Clostridium beijerinckii* P260. Food and Bioproducts Processing 2012;90(3), 533-540.

[80] Zhang Y., Han B., Ezeji T. C. Biotransformation of furfural and 5-hydroxymethyl furfural (HMF) by *Clostridium acetobutylicum* ATCC 824 during butanol fermentation. New Biotechnology 2012;29(3), 345-351.

[81] Richmond C., Ujor V., Ezeji T. C. Impact of syringaldehyde on the growth of *Clostridium beijerinckii* NCIMB 8052 and butanol production. Biotech 2012;2(2), 159-167.

[82] Palmqvist E., Hahn-Hagerdal B. Fermentation of lignocellulosic hydrolysates. II: inhibitors and mechanism of inhibition. Bioresource Technology 2000(74), 25-33.

[83] Palmqvist E., Hahn-Hagerdal B. Fermetation of lignocellulosic hydrolysates. I. Inhibition and detoxification. Bioresource Technology 2000;74:17-24.

[84] Cho D. H., Lee Y. J., Um Y., Sang B.-I., Kim Y. H. Detoxification of model phenolic compounds in lignocellulosic hydrolysates with peroxidase for butanol production from *Clostridium beijerinckii*. Applied and Microbial Biotechnology 2009;83(6), 1035-1043.

[85] Qureshi N., Ezeji T. C., Ebener J., Dien B. S., Cotta M. A., Blaschek H. P. Butanol production by *Clostridium beijerinckii*. Part I: Use of acid and enzyme hydrolyzed corn fiber. Bioresource Technology 2008;99(13), 5915-5922.

[86] Larsson S., Reimann A., Nilvebrandt N.-O., Jönsson L. J. Comparison of different methods for the detoxification of lignocellulose hydrolyzates of spruce. Applied Biochemistry and Biotechnology 1999;77(1-3).

[87] Martinez A., Rodriguez M. E., York S. W., Preston J. F., Ingram L. O. Effects of Ca(OH)2 treatments ("overliming") on the composition and toxicity of bagasse hemicellulose hydrolysates. Biotechnology and Bioengineering. 2000;69(5):526-536.

[88] Soni B. K., Das K., Ghose T. K. Bioconversion of agro-wastes into acetone butanol. Biotechnology Letters 1982;4(1), 19-22.

[89] Taherzadeh M. J., Karimi K. Acid-based hydrolysis processes for ethanol from lignocellulosic material: a review. BioResources 2007;2(3), 472-499.

[90] Fu N., Peiris P. Co-fermentation of a mixture of glucose and xylose to ethanol by *Zymomonas mobilis* and *Pachysolen tannophilus*. World Journal of Microbiology and Biotechnology 2008(24), 1091-1097.

[91] Lebeau T., Jouenne T., Junter G. A. Simultaneous fermentation of glucose and xylose by pure and mixed cultures of Saccharomyces cerevisiae and Candida shehatae immobilized in a two-chambered bioreactor. Enzyme and Microbial Technology. 1997(21):265-272.

[92] Groothen D. R. J., Jansen M. L., Van Der Lans R. G. J. M., Luyben K. C. a. M. Reactors in series for the complete converstion of glucose/xylose mixtures by *Pichia stipitis* and *Saccharomyces cerevisiae*. Enzyme and Microbial Technology 1991;13(10), 828-833.

[93] Qian M., Tian S., Li X., Zhang J., Pan Y., Yang X. Ethanol production from dilute-acid softwood hydrolysate by co-culture. Applied Microbiology and Biotechnology 2006;134(3), 27-283.

[94] Li Y., Park J.-Y., Shiroma R., Tokuyasu K. Bioethanol production from rice straw by a sequential use of *Saccharomyces cerevisiae* and *Pichia stipitis* with heat inactivation of Saccharomyces cerevisiae cells prior to xylose fermentation. Journal of Bioscience and Bioengineering 2011;111(6), 682-686.

[95] Sakamoto T., Hasunuma T., Hori Y., Yamada R., Kondo A. Direct ethanol production from hemicellulosic materials of rice straw by use of an engineered yast strain codisplaying three types of hemicellulolytic enzymes on the surface of xylose-utilizing *Saccharomyces cerevisiae* cells. Journal of Biotechnology 2012(158), 203-210.

[96] Dellomonaco C., Fava F., Gonzales R. The path to next generation biofuels: successes and challenges in the era of synthetic biology. Microbial Cell Factories 2010;9(3).

[97] Scalcinati G., Otero J. M., Van Vleet J. R. H., Jeffries T. W., Olsson L., Nielsen J. Evolutionary engineering of *Saccharomyces cerevisiae* for efficient aerobic xylose consumption. FEMS Yeast Research 2012;12(5), 585-597.

[98] Martín C., Marcet M., Almazán O., Jonsson L. J. Adaptation of a recombinant xylose-utilizing Saccharomyces cerevisiae strain to sugarcane bagasse hydrolysate with high concentration of fermentable inhibitors. Bioresource Technology 2007;98, 1767-1773.

[99] Quin W. High consistency enzymatic hydrolysis of lignocelluloses. Vancouver: The University of British Columbia; 2010.

[100] Suryawati L., Wilkins M. R., Bellmer D. D., Huhnke R. L., Maness N. O., Banat I. M. Simultaneous saccharification and fermentation of kanlow switchgrass pretreated by hydrothermolysis using *Kluyveromyces marxianus* IMB4. Biotechnology and Bioengineering 2008;101(5), 894-902.

[101] García-Aparicio M. P., Oliva J. M., Manzanares P., Ballesteros M., Ballesteros I., González A., Negro M. J. Second-generation ethanol production from steam exploded barley straw by *Kluyveromyces marxianus* CECT 10875. Fuel. 2011;90,1624-1630.

[102] de Souza C. J. A., Costa D. A., Rodrigues M. Q. R. B., Dos Santos A. F., Lopes M. R., Abrantes A. B. P., Dos Santos Costa P., Silveira W. B., Passos F. M. L., Fietto L. G. The influence of presaccharification, fermentation temperature and yeast strain on ethanol production from sugarcane bagasse. Bioresource Technology 2012;109, 63-69.

[103] Chen M.-L., Wang F.-S. Optimization of a fed-batch simultaneous saccharification and cofermentation process from lignocellulose to ethanol. Industrial & Engineering Chemistry Research 2010;49, 5775-5789.

[104] Olofsson K., Bertilsson M., Lidén G. A short review on SSF - an interesting process option for ethanol production from lignocellulosic feedstocks. Bitechnology for Biofuels 2008;1(7), 1-14.

[105] Zhang X.-Z., Heng Y., Zhang P. One-step production of biocommodities from lignocellulosic biomass by recombinant cellulolytic *Bacillus subtilis*: Opportunities and chalenges. Engineering in Life Sciences 2010(5), 398-406.

[106] Lee R. Lynd W. H. V. Z., John E. Mcbride, Mark Laser. Consolidated bioprocessing of cellulosic biomass: an update. Current Opinion in Biotechnology 2005(16), 577-583.

[107] Chang T., Yao S. Thermophilic, lignocellulolytic bacteria for ethanol production: current state and perspectives. Applied Microbiology and Biotechnology 2011(92), 13-27.

[108] Shaw A. J., Podkaminer K. K., Desai S. G., Bardsely J. S., Rogers S. R., Thorne P. G., Hogsett D. A., Lynd L. R. Metabolic engineering of a thermophilic bacterium to produce ethanol at high yield. Proceedings of the National Academy of Sciences 2008;105(37), 1769-13774.

[109] Haan R. D., Rose S. H., Lynd L. R., Van Zyl W. H. Hydrolysis and fermentation of amorphous cellulose by recombinant *Saccharomyces cerevisiae* Metabolic Engineering 2007(9), 87-94.

[110] Fujita Y., Ito J., Ueda M., Fukuda H., Kondo A. Synergistic saccharification, and direct fermentation to ethanol, of amorphous cellulose by use of an engineered yeast strain codisplaying three types of cellulolytic enzyme Applied and Environmental Microbiology. 2004; 20, 688-691.

[111] Patáková P., Linhová M., Rychtera M., Paulová L., Melzoch K. Novel and neglected issues of acetone–butanol–ethanol (ABE) fermentation by *Clostridia*: *Clostridium* metabolic diversity, tools for process mapping and continuous fermentation systems. Biotechnology Advances. 2013; 31(1), 58-67.

[112] Kumar M., Gayen K. Developments in biobutanol production: new insights. Applied Energy. 2011;88(6), 1999–2012

[113] Tashiro Y., Sonomoto K. Advances in butanol production by *Clostridia*. In: Méndez-Vilas A . (ed.) Current Research, Technology and Education Topics in Applied Microbiology and Microbial Biotechnology. Bajadoz, Spain Formatex Research Center; 2010. p. 1383-1394.

[114] Lee S. Y., Park J. H., Jang S. H., Nielsen L. K., Kim J., Jung K. S. Fermentative Butanol Production by *Clostridia*. Biotechnology and Bioengineering. 2008;101(2):209-228.

[115] Zverlov V. V., Berezina O., Velikodvorskaya G. A., Schwarz W. H. Bacterial acetone and btanol poduction by idustrial frmentation in the Soviet Union: Use of hydrolyzed agricultural waste for biorefinery. Applied Microbiology and Biotechnology 2006;71(5), 587-597.

[116] Raman B., Mckeown C. K., Rodriguez M., Brown S. D., Mielenz J. R. Transcriptomic analysis of *Clostridium thermocellum* ATCC 27405 cellulose fermentation. BMC Microbiology. 2011; 11, 134.

[117] BMC Microbiology 2011 2011;11(134),Carere C. R., Sparling R., N. C., Levin D. B. Third generation biofuels via direct cellulose fermentation. Journal of Molecular Sciences 2008;9(7), 1342-1360.

[118] López-Contreras A. M., Gabor K., Martens A. A., Renckens B. a. M., Claasen P. a. M., Van Der Oost J., De Vos W. M. Substrate-induced production and secretion of cellulases by *Clostridium acetobutylicum*. Applied and Environmental Microbiology 2004;70(9), 5238-5243.

[119] Mingardon F., Chanal A., Tardif C., Fierobe H.-P. The issue of secretion in heterologous expression of *Clostridium cellulolyticum* cellulase-encoding genes in *Clostridium acetobutylicum* ATCC 824. Applied and Environmental Microbiology 2011;77(9), 2831-2838.

[120] Cui Z., Li Y., Xiao Y., Feng Y., Cui Q. Resonance assignments of cohesin and dockerin domains from *Clostridium acetobutylicum* ATCC 824. Biomolecular NMR Assign 2012.

[121] Berezina O. V., Brandt A., Yarotsky S., Schwartz W. H., Zverlov V. V. Isolation of a new butanol-producing *Clostridium* strain: High level of hemicellulosic activity and structure of solventogenesis genes of a new *Clostridium saccharobutylicum* isolate. Systematic and Applied Microbiology 2009;2(7), 449-459.

[122] Tracy B. P., Jones S. W., Fast A. G., Indurthi D. C., Papoutsakis E. T. *Clostridia*: the importance of their exceptional substrate and metabolic diversity for biofuel and biorefinery applications. Current Opinion in Biotechnology. 2012;23(3), 364-381.

[123] Shi Y., Li Y.-X., Li Y. Y. Large number of phosphotransferase genes in the *Clostridium beijerinckii* NCIMB 8052 genome and the study on their evolution. BMC Bioinformatics 2010(11 (Suppl 11)), S9,Tracy B. P., Jones S. W., Fast A. G., Indurthi D. C., Papoutsakis E. T. *Clostridia*: the importance of their exceptional substrate and metabolic diversity for biofuel and biorefinery applications. Current Opinion in Biotechnology 2012;23(3), 364-381.

[124] Heluane H., Evans M. R., Dagher S. F., Bruno-Bárcena J. M. Nutritional requirements of solventogenic *Clostridia* growing under butanol stress and co-utilization of D-glucose/D-xylose.A meta-analysis and a functional validation. Applied and Environmental Microbiology 2011;77(13), 4473-4485.

[125] Survase S. A., Heiningen A., Granström T. Continuous bio-catalytic conversion of sugar mixture to aceton-butanol-ethanol by immobilized *Clostridium acetobutylicum* DSM 792. Applied Microbiology and Biotechnology 2012;93(6), 2309-2316.

[126] Wang L., Chen H. Increased fermentability of enzymatically hydrolyzed steam-exploded cornstover for butanol production by removal of fermentation Inhibitors. Process Biochemistry 2011;46(2), 604-607.

[127] Qureshi N., Saha B. C., Cotta M. A. Butanol production from wheat straw by simultaneous saccharification and fermentation using *Clostridium beijerinckii*: Part II-Fed-batch fermentation. Biomass and Bioenergy 2008;32(2), 176–183.

[128] Ren C., Gu Y., Hu S., Wu Y., Wang P., Yang Y., Yang S., Jiang W. Identification and inactivation of pleiotropic regulator CcpA to eliminate glucose repression of xylose utilization in *Clostridium acetobutylicum*. Metabolic Engineering 2010;12(5), 446-454.

[129] Grimmler C., Held C., Liebl W., Ehrenreich A. Transcriptional analysis of catabolite repression in *Clostridium acetobutylicum* growing on mixtures of D-glucose and D-xylose. Journal of Bacteriology. 2010;150(3):315-323.

[130] Lu C., Zhao J., Yang S.-T., Wei D. Fed-batch fermentation for n-butanol production from cassava bagasse hydrolysate in a fibrous bed bioreator with continuous gas stripping. Bioresource Technology 2012;104, 80-387.

[131] Marchal R., Ropars M., Pourquié J., Fayolle F., Vandecasteele J. P. Large-scale enzymatic hydrolysis of agricultural lignocellulosis biomass. Part 2: Conversion into acetone-butanol. Bioresource Technology 1992;42(3), 205-217.

[132] Claassen P. a. M., Budde M. a. W., López-Contreras A. M. Acetone, butanol and dthanol production from domestic organic waste by solventogenic *Clostridia*. Journal of Molecular and Microbiological Biotechnology 2000;2(1), 39-44.

[133] Qureshi N., Saha B. C., Dien B., Hector R. E., Cotta M. A. Production of butanol (a biofuel) from agricultural Residues: Part I- Use of barley straw hydrolysate. Biomass and Bioenergy 2010;34(4), 559-565.

[134] Survase S. A., Sklavounos E., Jurgens G., Heiningen A., Granström T. Continuous acetone-butanol-ethanol fermentation using SO2-ethanol-water spent liquor from spruce. Bioresource Technology 2011;102(23), 10996-11002.

[135] Qureshi N., Saha B. C., Hector R. E., Dien B., Hughes S., Liu S., Iten L., Bowman M. J., Sarath G., Cotta M. A. Production of butanol (a biofuel) from agricultural residues: Part II- use of corn stover and switchgrass hydrolysates. Biomass and Bioenergy 2010;34(4), 566-571.

[136] Noomtim P., Cheirsilp B. Production of Butanol from Palm Empty Fruit Bunches Hydrolyzate by *Clostridium acetobutylicum*. Energy Procedia 2011(9), 140-146.

[137] Ibrahim M. F., Abd-Aziz S., Razak M. N., Phang L. Y., Hassan M. A. Oil palm empty fruit bunch as alternative substrate for acetone-butanol-ethanol production by *Clostridium butyricum* EB6. Appl Biochem Biotechnol. 2012;166(7):1615-1625.

[138] Shah M. M., Song S. K., Lee Y. Y., Torget R. Effect of pretreatment on simultaneous saccharification and fermentation of hardwood into acetone/butanol. Appl Biochem Biotechnol 1991;28-29(1), 99-109.

[139] Qureshi N., Saha B. C., Hector R. E., Hughes S. R., Cotta M. A. Butanol production from wheat straw by simultaneous saccharification and fermentation using *Clostridium beijerinckii*: Part I- Batch Fermentation Biomass and Bioenergy 2008;32(2), 176-183.

[140] Liu Z., Ying Y., Li F., Ma C., Xu P. Butanol production by *Clostridium beijerinckii* ATCC 55025 from wheat bran. Journal of Industrial Microbiology and Biotechnology 2010;37(5), 495-501.

[141] Yu J., Zhang T., Zhong J., Zhang X., Tan T. Biorefinery of sweet sorghum stem. Biotechnology Advances 2012;30(4), 811-816.

[142] Parekh S. R., Wayman M. Ethanol and butanol production by fermentation of enzymatically saccharified SO2-prehydrolysed lignocellulosics. Enzyme and Microbial Technology 1988;10(11), 660-668.

[143] Lora J. H., Gasser W. G. Recent industrial applications of lignin: a sustainable alternative to nonrenewable materials. Journal of Polymers and the Environment 2002;10(1/2), 39-48.

Chapter 3

Biofuels Ethanol and Methanol in OTTO Engines

Charalampos Arapatsakos

Additional information is available at the end of the chapter

http://dx.doi.org/10.5772/52772

1. Introduction

Today, humanity faces many environmental problems, one of which is atmospheric pollution that leads to greenhouse effect, ozone formation and to many health problems to human beings. Also, many countries around the world face the problem of energy shortage. At the same time we must not forget the need for clean air, clean fuel and biodegradable, renewable materials. Hazardous pollutants that lead to atmospheric pollution have many sources and automobile's exhaust emission is one of these. Petroleum-based products that have been used as fuels produce dangerous gas emissions. In order to decrease environmental impacts, scientists and many governments turned their attention to renewable fuels as alternatives to conventional fossil fuels and as oxygenates [1]. The beginning of the 21st century finds humans more familiar with the concept of sustainable development. We must prevent the degradation of our environment focusing in more friendly technologies. This need lead scientists to the use of other energy sources that can be used with the same efficiency but won't have damaging effect to the environment. The increased vehicle number that usually uses petroleum-based fuels results to dangerous emissions production such as carbon monoxide (CO), carbon dioxide (CO_2), hydrocarbons (HC), nitrogen oxides (NOx) and others. These emissions besides the fact that lead to environmental degradation they also constitute a threat for human health. People's concern about the risks associated with hazardous pollutants results to an increased demand for renewable fuels as alternatives to fossil fuels [1,2]. Ethanol and methanol are alcohols that can be used as fuels instead of gasoline in automobile engines. For better understanding of the use of these two alcohols we must examine them separately. Fuel ethanol is an alternative fuel that is produced from biologically renewable resources that it can also be used as an octane enhancer and as oxygenate. Ethanol (ethyl alcohol, grain alcohol, ETOH) is a clear, colorless liquid alcohol with characteristic odor and as alcohol is a group of chemical compounds whose molecules contain a hydroxyl group, -OH, bonded to a carbon atom. Is produced with the process of fermentation of grains such as wheat, barley, corn, wood, or sugar cane. In the United States ethanol is made by the fermentation of corn [1-3]. By the reaction of fermentation simple sugars change into ethanol and carbon dioxide with the presence of zymase, an enzyme from yeast. Ethanol can also be made from cellulose that is obtained from agricultural residue and waste paper [1]. It is a high-octane fuel with high oxygen content (35% oxygen by weight) and when blended properly in gasoline produces a cleaner and more complete combustion. Ethanol is used as an automotive fuel either by itself or in blends with gasoline, such as mixtures of 10% ethanol and 90%gasoline, or 85% ethanol and 15% gasoline [3-6]. Many countries around the world use ethanol as fuel. For example, in Brazil ethanol is produced using as raw material sugarcane and many vehicles use ethanol as fuel. Also in Canada and in Sweden ethanol is highly promoted as fuel because of the many environmental benefits that ethanol has. When gasoline is used as fuel hydrocarbons (HC) escape to the atmosphere. Many hydrocarbons are toxic and some, such as benzene, cane cause cancer to humans. If ethanol is used as fuel hydrocarbons are not being produced because ethanol is an alcohol that does not produce HC when is burned. The reaction of hydrocarbons and nitrogen oxides that are produced from the gasoline burning, in the presence of sunlight leads to the formation of photochemical smog. The use of ethanol as fuel can contribute to the decrease of photochemical smog since it does not produces hydrocarbons [5-8]. Vehicles that burn petroleum fuels produce carbon monoxide (CO) because these fuels do not contain oxygen in their molecular structure. Carbon monoxide is a toxic gas that is formed by incomplete combustion. When ethanol, which contains oxygen, is mixed with gasoline the combustion of the engine is more complete and the result is CO reduction [9-11].

Using renewable fuels, such as ethanol, there is also a reduction of carbon dioxide (CO_2) in the atmosphere. Carbon dioxide is non-toxic but contributes to the greenhouse effect. Because of the fact that plants absorb carbon dioxide and give off oxygen, that balances the amount of CO_2 that is formed during combustion absorbed by plants used to produce ethanol. That is why the use of ethanol will partially offset the greenhouse effect that is formed by carbon dioxide emissions of burning gasoline [11-13]. Ethanol, as an octane enhancer, can substitute benzene and other benzene-like compounds, which are powerful liver carcinogens, and reduce their emissions to the atmosphere. Besides the environmental benefits, production and use of ethanol, which is a renewable fuel, increases economic activity, creates job openings, stabilizes prices and can increase farm income. That is why ethanol as an automotive fuel has many advantages.

Methanol (CH_3OH) is an alcohol that is produced from natural gas, biomass, coal and also municipal solid wastes and sewage. It is quite corrosive and poisonous and has lower volatility compared to gasoline, which means that is not instantly flammable. Usually methanol is used as a gasoline-blending compound, but it can be used directly as an automobile fuel with some modifications of the automobile engine.

Although there are many feedstocks that are being used for the production of methanol, natural gas is more economic. Methanol is produced from natural gas with a technology of steam reforming. By this method natural gas is transformed to a synthesis gas that is fed to a reactor vessel to produce methanol and water at the presence of a catalyst. The reactions(equation 1,2) that represent methanol production are the following [4]:

$$2CH_4 + 3H_2O \rightarrow CO + CO_2 + 7H_2 \text{ - Synthesis gas} \quad (1)$$

$$CO + CO_2 + 7H_2 \rightarrow 2CH_3OH + 2H_2 + H_2O \quad (2)$$

The main advantage of methanol as fuel is that is being produced from resources that can be found globally, while a large percentage of petroleum is located in Middle East. Furthermore, the materials needed for methanol production such as natural gas or biomass are renewable. This means that methanol can also be cheaper and more economically attractive than gasoline. When fossil fuels are used in automobiles produce exhaust emissions of hydrocarbons, carbon dioxide and other gases that contribute to the greenhouse effect. Methanol can give lower HC and CO emissions and besides that the vehicles that use methanol emit minimum particulate matter compared to gasoline, which usually has damaging effect to humans. In addition, methanol has high-octane content that promotes better the process of combustion. Another advantage of methanol is that if it does ignite can cause less severe fires to the vehicle because is less flammable than gasoline [4]. Some disadvantages that methanol has are the lower energy content compared to gasoline, the fact that is not volatile enough for easy cold starting and can damage plastic and rubber fuel system components. The vehicle that uses methanol for fuel must have a large storage tank because pure methanol burns faster than gasoline, and corrosion resistant, materials must be used for the storage equipment [14-16]. Renewable fuels such as ethanol and methanol will probably replace petroleum-based fuels in the near future because petroleum reserves are not sufficient enough to last many years. Also, the severe environmental problems around the world will eventually lead to the use of more environmentally friendly technologies. The question that is examined in this chapter is how the mixtures of gasoline-ethanol and gasoline-methanol behave in a four-stroke engine from the aspect of emissions and fuel consumption.

2. Experimental part

The experimental measurements were carried out on a four-stroke, air-cooled engine. This is a one-cylinder engine with 123cm^3 displacement that is connected with a phase single alternative generator (230V/50Hz) with maximum electrical load approximately 1KW(picture 1). The engine according to the manufacturer uses as fuel gasoline. The engine functioned without load and under full load conditions (1KW) using different fuel mixtures: gasoline, gasoline-10%ethanol, gasoline-20%ethanol, gasoline-30%ethanol, gasoline-40%ethanol, gasoline-50%ethanol, gasoline-60%ethanol, gasoline-70%ethanol gasoline-80%ethanol gasoline-90%ethanol and 100% ethanol, gasoline-10% methanol, gasoline-20%methanol, gasoline-30%methanol, gasoline-40%methanol, gasoline-50%methanol, gasoline-60%methanol, gasoline-70%methanol. During the tests, exhaust gases measurements, were also monitored for every fuel mixture and for every load conditions. Also, during the function of the engine the consumption was recorded for every fuel. There was lack of engine regulation concerning the stable air/fuel ratio. For this purpose, data Acquisition cart was used with the terminal wiring board with on-board Cold Junction The data acquisition card was installed at a PC. This particular measuring system and software completed a scanning cycle per channel every 0.1 second approximately. This measuring speed was considered adequate for the purpose of the experiment and the sampling capabilities of the chemical sensors. For the exhaust gas (CO and HC) measurements a analyzer was used.

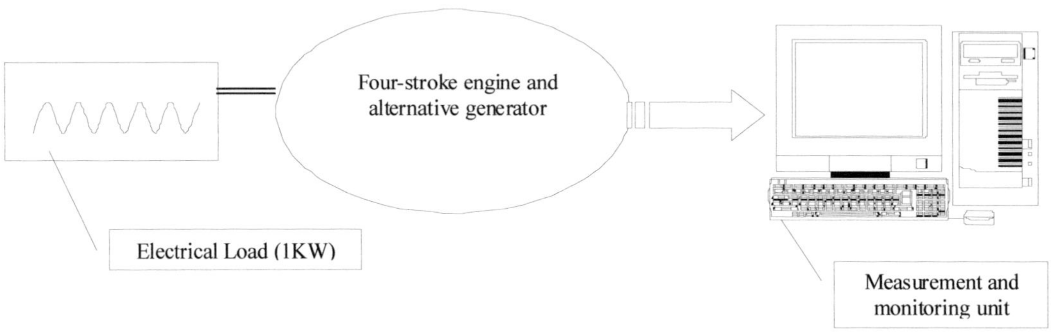

Figure 1. The illustration of the experimental unit

The figures of CO and HC emissions, for every fuel and for every load conditions, are represented below [4-7]:

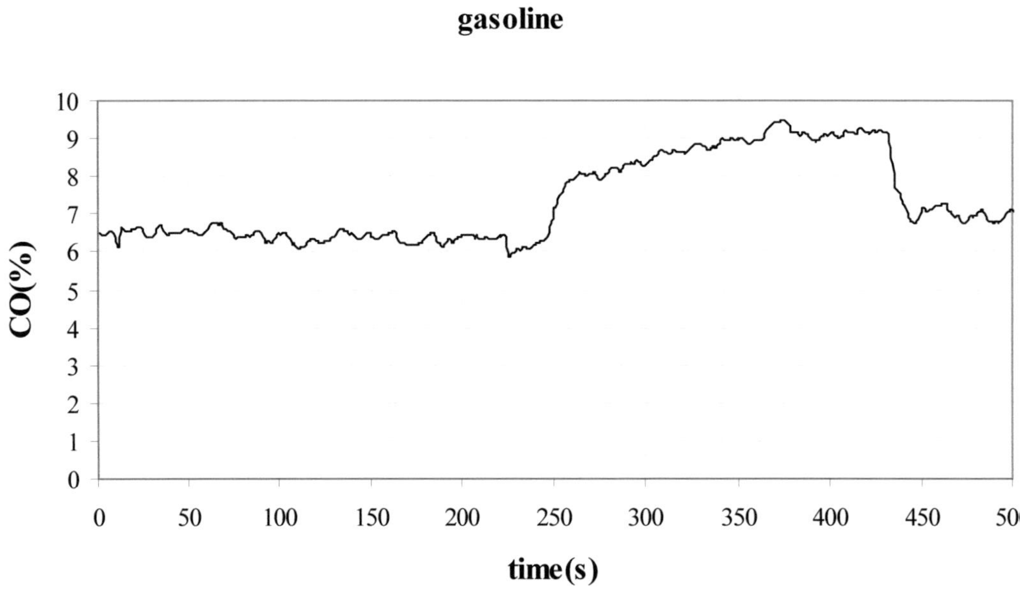

Figure 2. The CO variation when gasoline is used as fuel

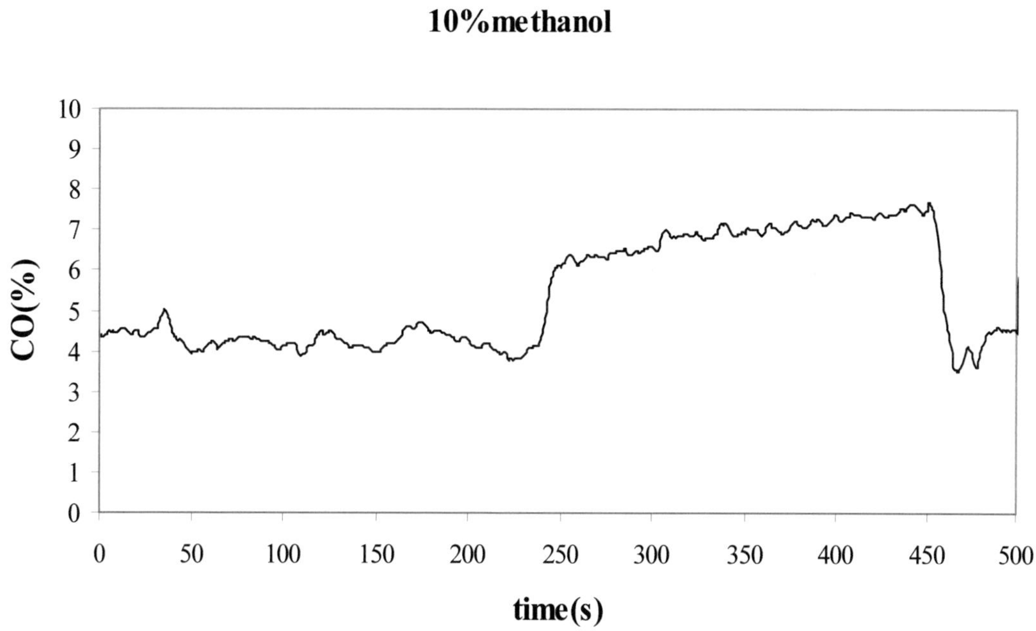

Figure 3. The CO variation when mixture of gasoline-10%methanol is used as fuel

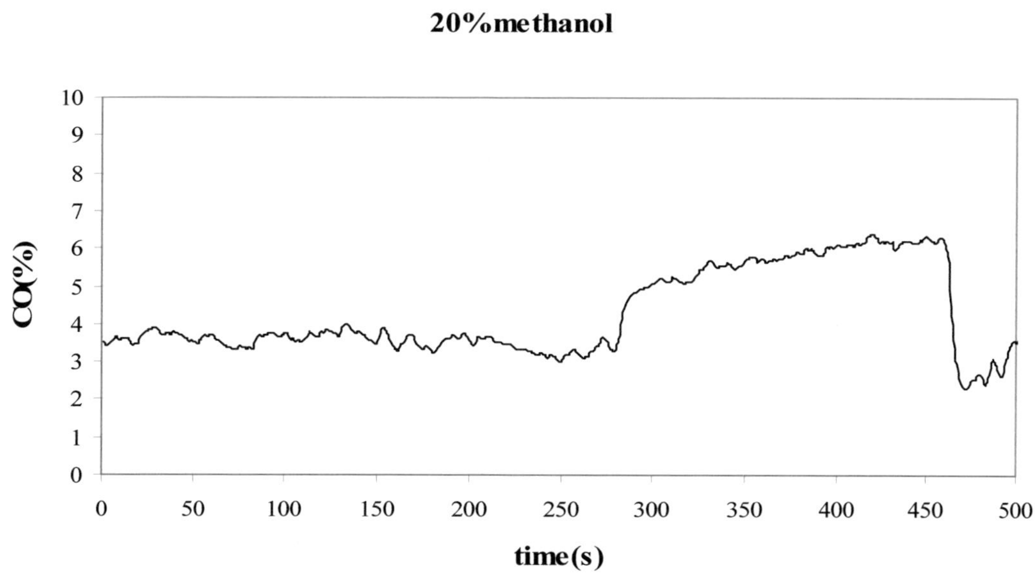

Figure 4. The CO variation when mixture of gasoline-20%methanol is used as fuel

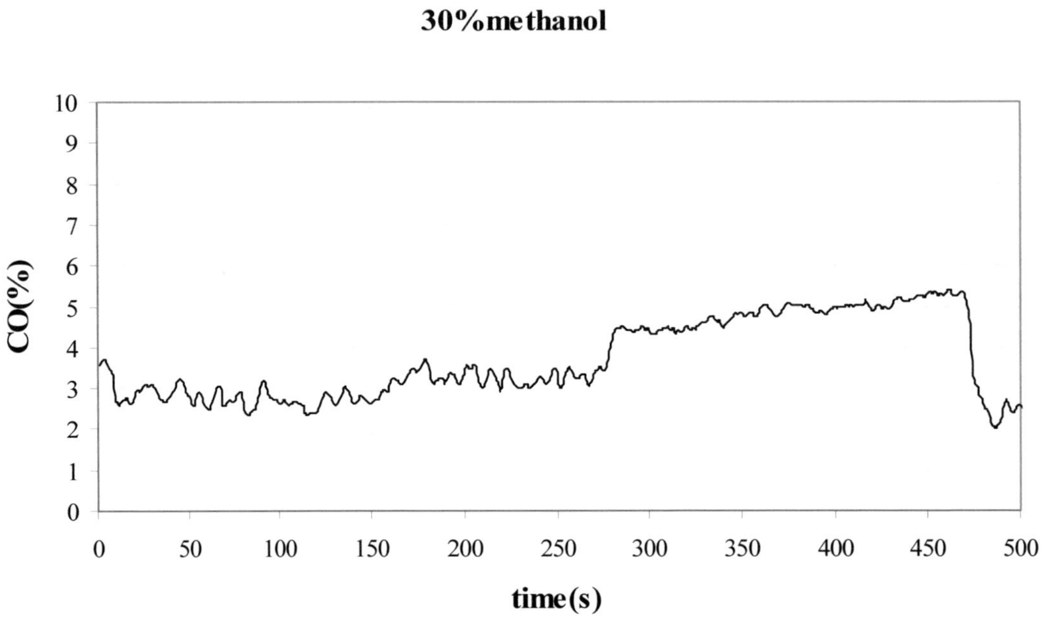

Figure 5. The CO variation when mixture of gasoline-30%methanol is used as fuel

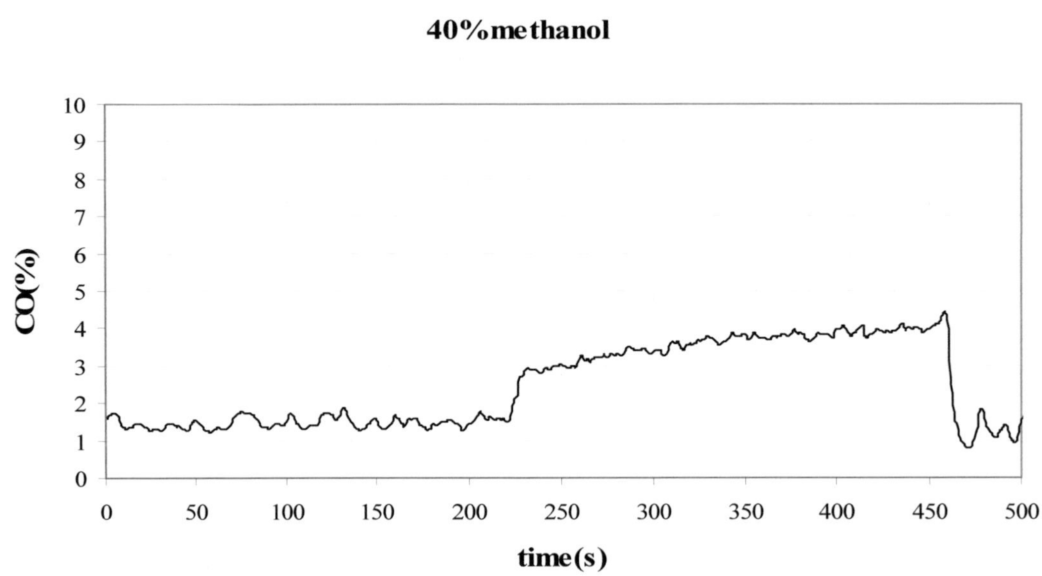

Figure 6. The CO variation when mixture of gasoline-40%methanol is used as fuel

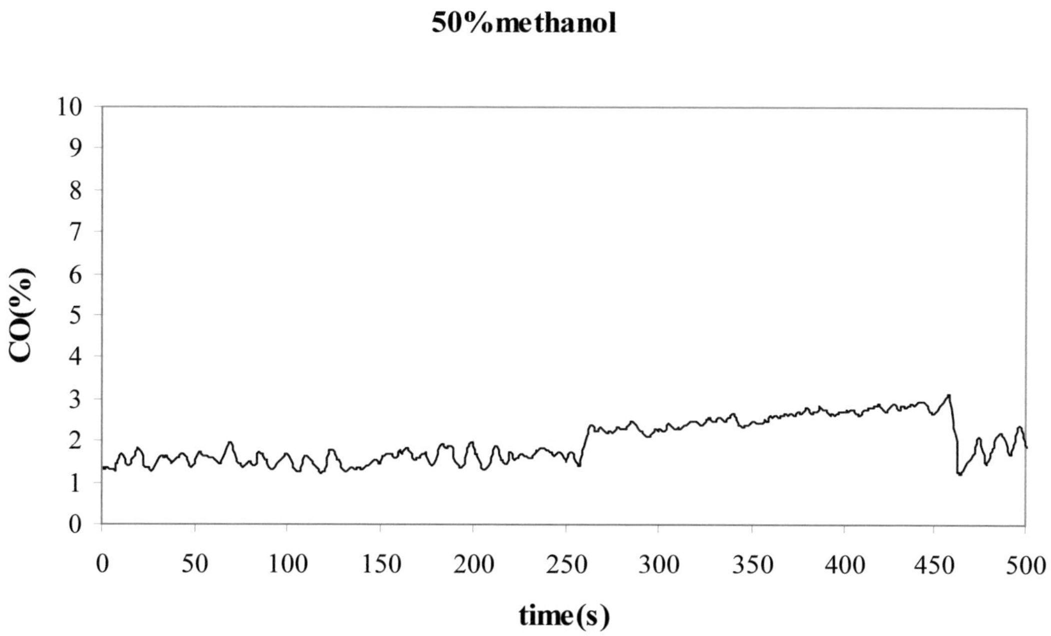

Figure 7. The CO variation when mixture of gasoline-50%methanol is used as fuel.

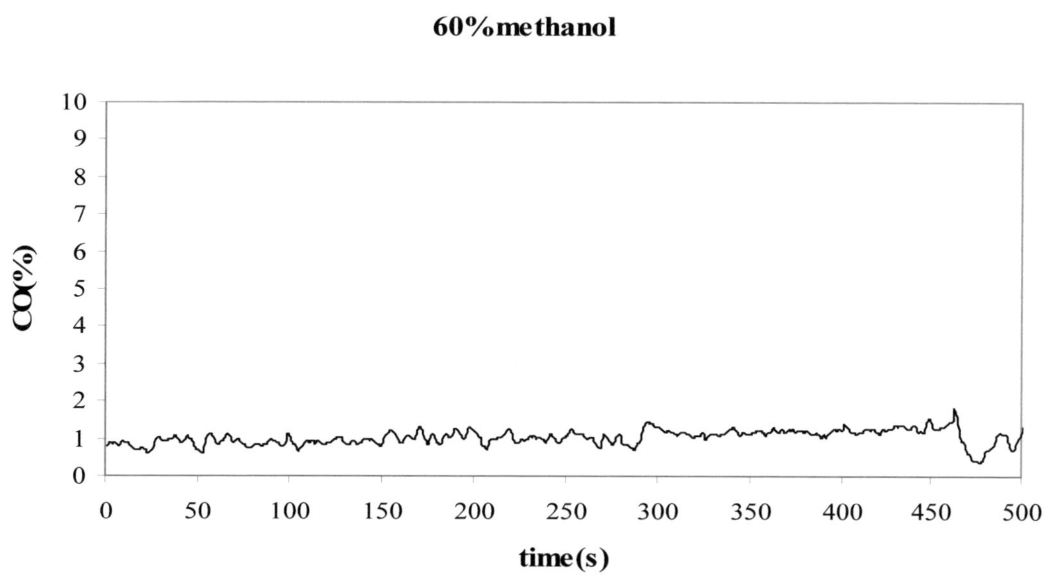

Figure 8. The CO variation when mixture of gasoline-60%methanol is used as fuel

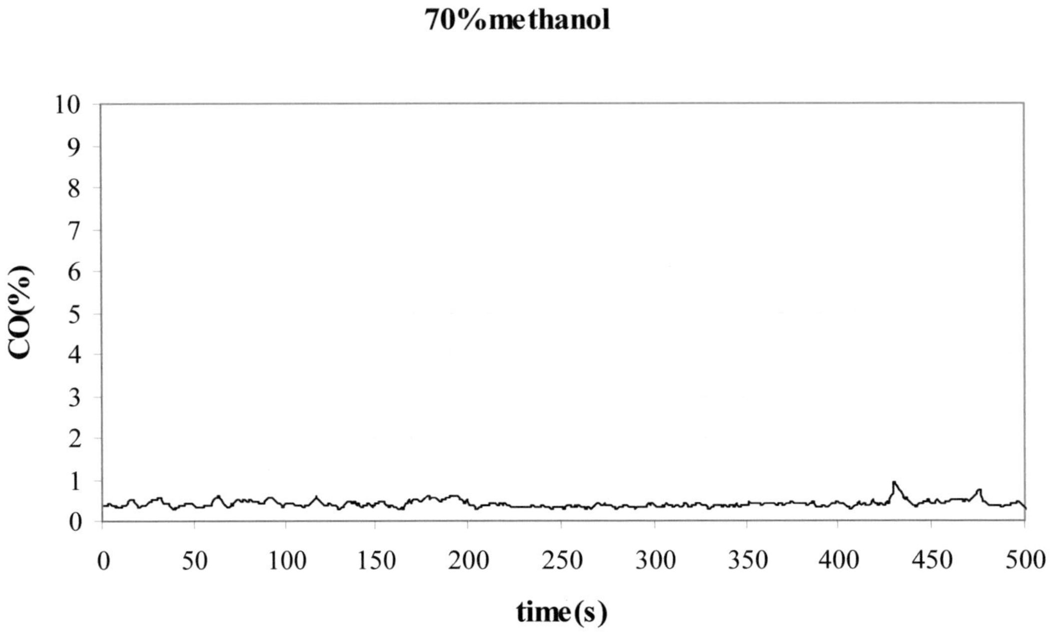

Figure 9. The CO variation when mixture of gasoline-70%methanol is used as fuel.

Figure 2 represents CO emissions when the fuel that is used is gasoline. The engine functions without load at first and then (after 250s) functions under full load conditions (1KW). The average value of CO emissions during the function of the engine without load is 6,41%, while at full load conditions the average value of CO emissions is 8,7%. Following, a mixture of gasoline with 10% methanol is used (fig. 3) and the same test is conducted with this mixture. From figure 3 it is being observed that the average value of CO emissions without load conditions of the engine is 4,87%, while at full load conditions the percentage of CO emissions is 6,9%. The same tests are conducted while increasing the percentage of the methanol in the fuel, using the mixtures: gasoline-20%methanol(fig. 4), gasoline-30%methanol(fig. 5), gasoline-40%methanol(fig. 6), gasoline-50%methanol(fig. 7), gasoline-60%methanol(fig. 8), and gasoline-70%methanol(fig. 9).

The HC emissions when the fuel that is used is gasoline are represented at figure 10. As it was mentioned above, the engine functioned without load at first and then (after 250s approximately) functioned under full load conditions (1KW). During the function of the engine without load the average value of HC emissions is 1091ppm, while at full load conditions the average value of HC emissions is 730ppm. The mixture of gasoline with 10% methanol is illustrated at figure 11. At this figure is being observed that the average value of HC emissions without load conditions of the engine is 496ppm, while at full load conditions the HC emissions is 613ppm. When the percentage of the methanol in the fuel increases: gasoline-20%methanol(fig. 11), gasoline-30%methanol(fig. 13), gasoline-40%methanol(fig. 14), gasoline-50%methanol(fig. 15), gasoline-60%methanol(fig. 16), and gasoline-70%methanol(fig. 17).

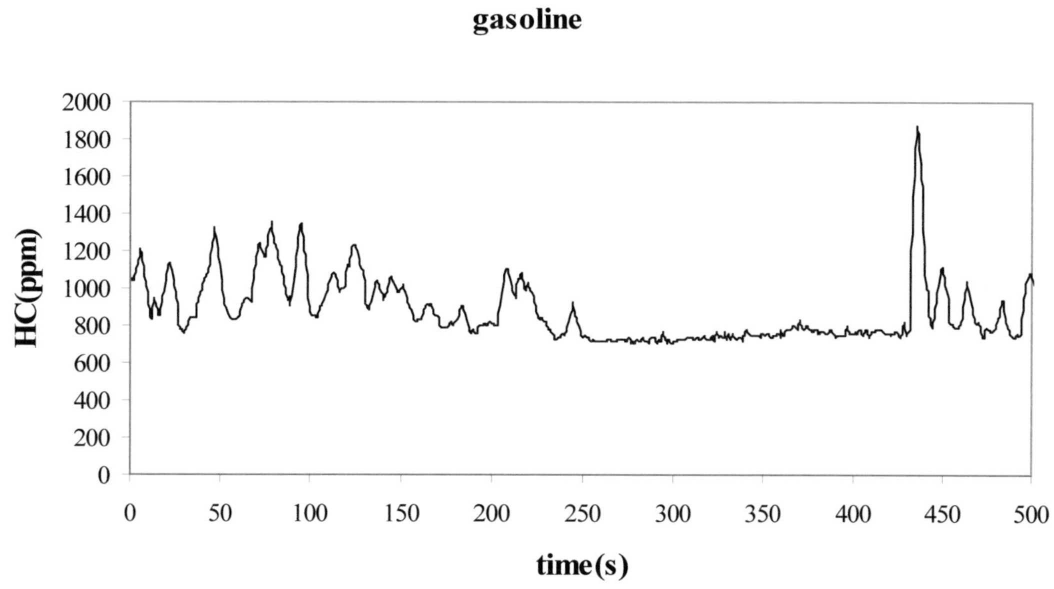

Figure 10. The HC variation when gasoline is used as fuel.

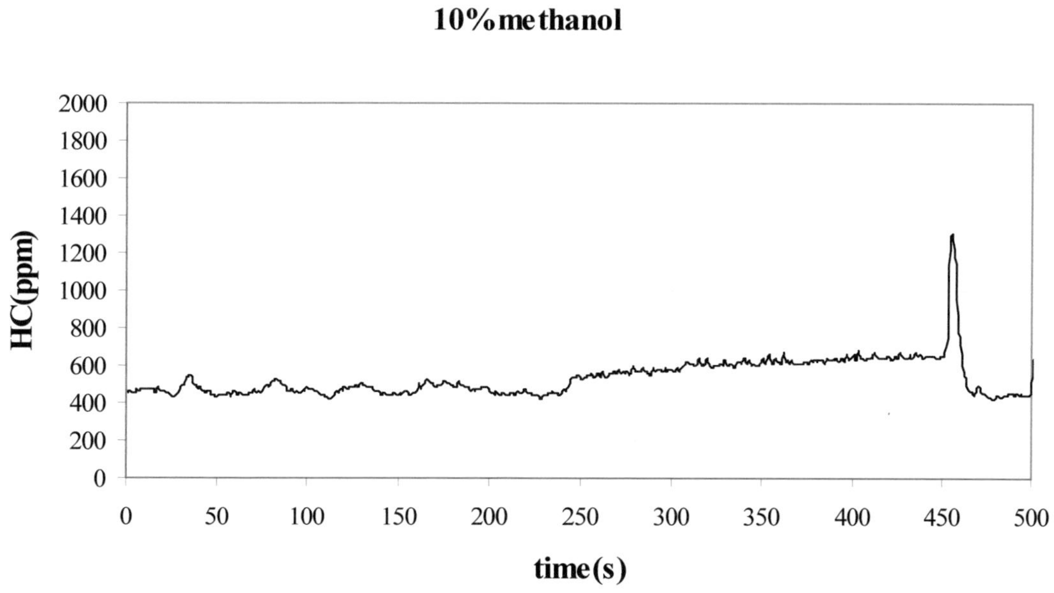

Figure 11. The HC variation when mixture of gasoline-10%methanol is used as fuel

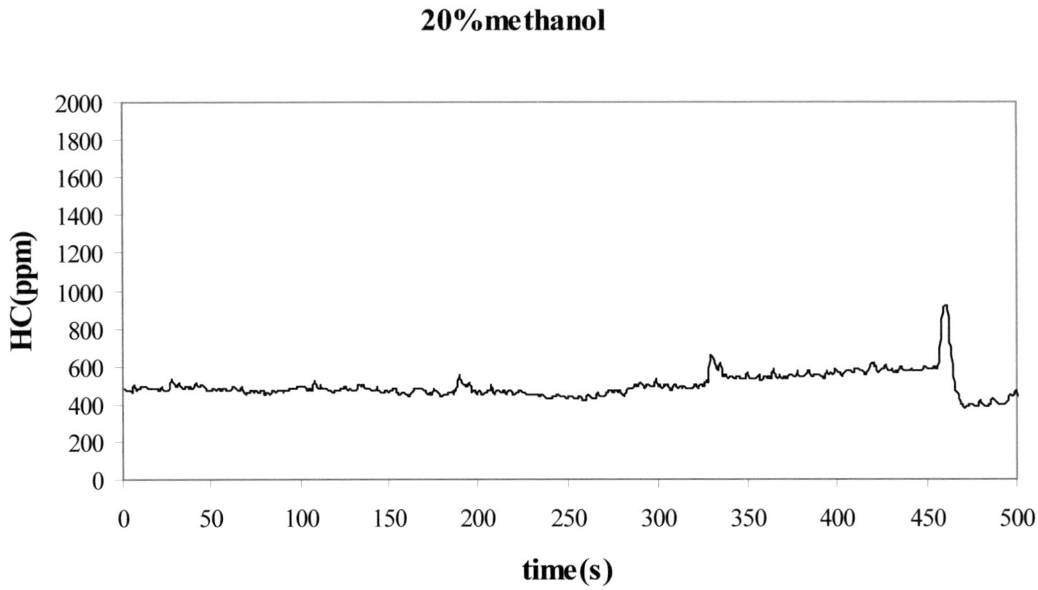

Figure 12. The HC variation when mixture of gasoline-20%methanol is used as fuel.

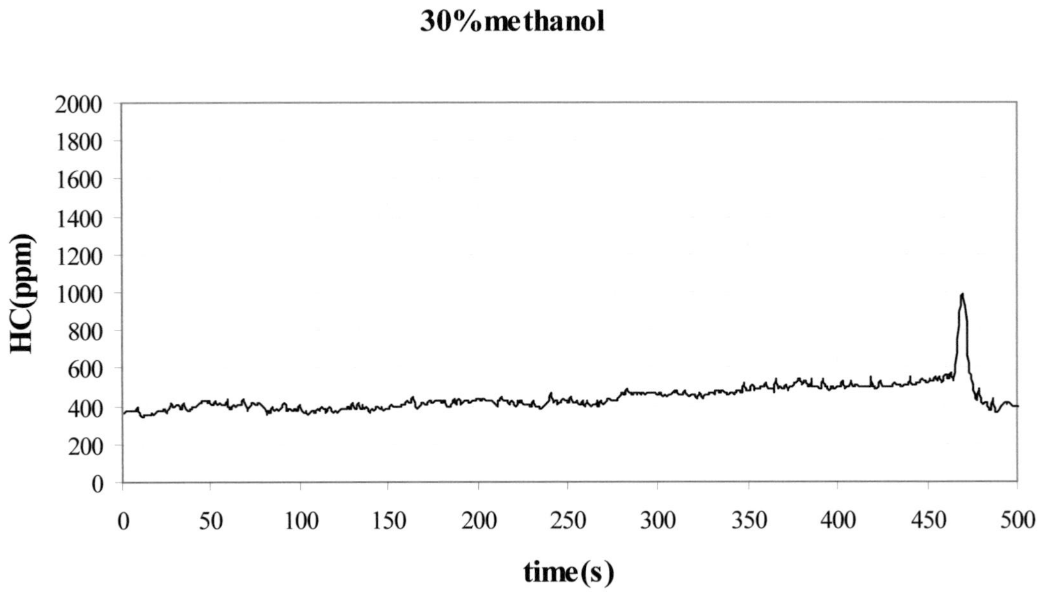

Figure 13. The HC variation when mixture of gasoline-30%methanol is used as fuel

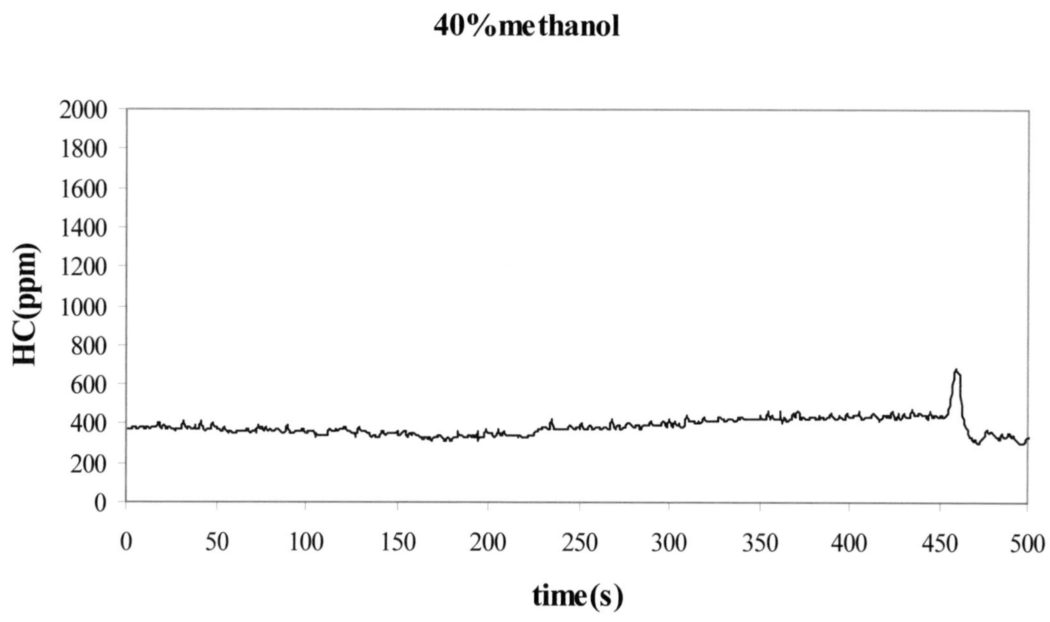

Figure 14. The HC variation when mixture of gasoline-40%methanol is used as fuel.

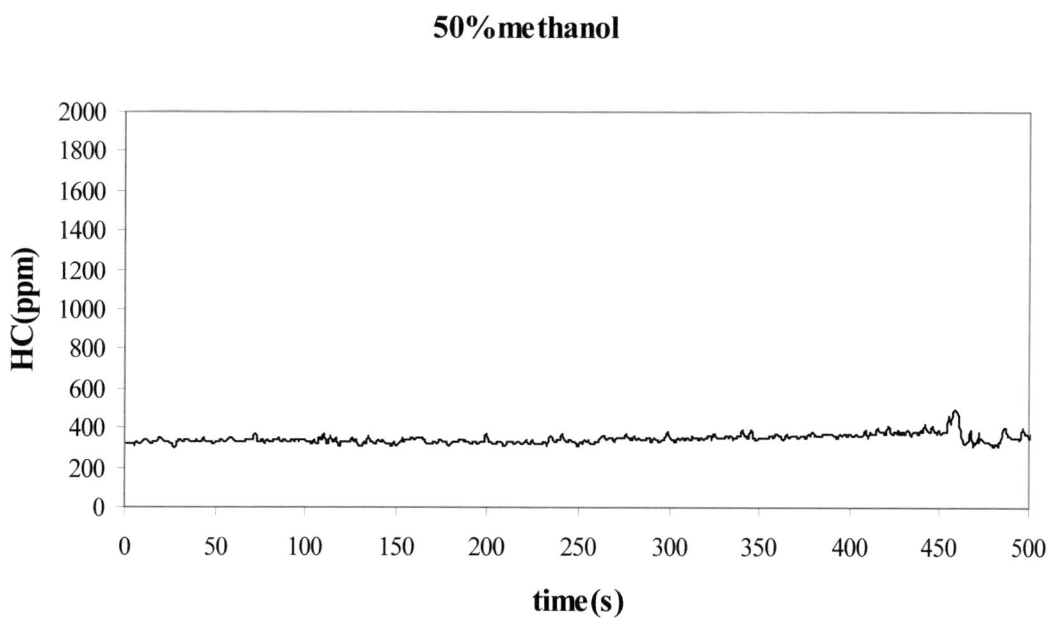

Figure 15. The HC variation when mixture of gasoline-50%methanol is used as fuel.

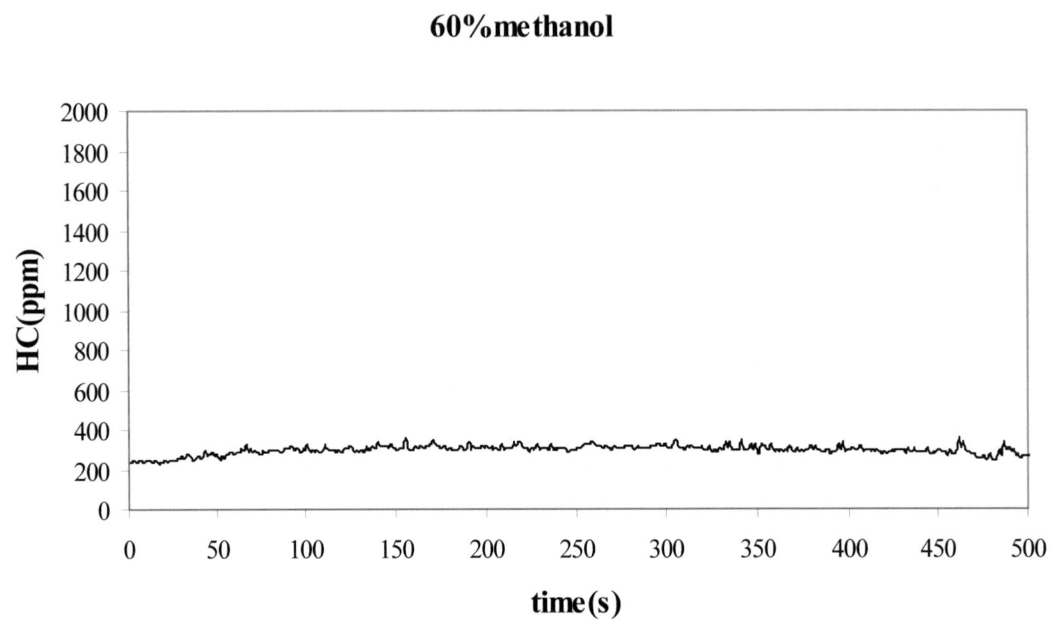

Figure 16. The HC variation when mixture of gasoline-60%methanol is used as fuel.

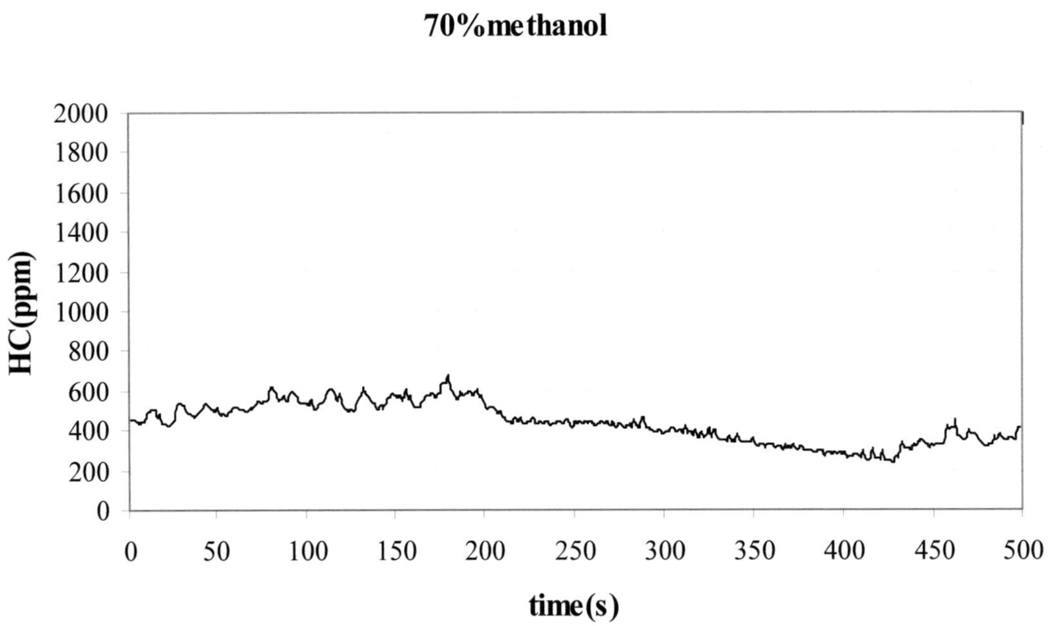

Figure 17. The CO variation when mixture of gasoline-70%methanol is used as fuel.

In the case of HC emissions there is also a decrease of emissions when the percentage of methanol in the fuel increases at idle and under full load conditions. There is an exception at the mixture gasoline-70%methanol where the average value of HC without load is 534ppm and under full load is 367ppm. These values are higher than the values that correspond to the mixture of gasoline-60%methanol (295ppm, 298ppm). This is explained by mentioning the fact that during the use of the mixture gasoline-70%methanol there was a malfunction of the engine that was cause by the bad mixture of the air with the fuel(gasoline-70%methanol), since the engine was not regulated(ratio air/fuel) for every mixture maintaining the adjustments for gasoline. Also it must reported that the addition of methanol in the fuel led to HC decrease for the same mixture but for different load conditions. When gasoline was used HC emissions were higher at no load conditions than at full load conditions(1KW), while during the use of gasoline-methanol mixtures this was reversed. This is due to the better combustion under full load conditions because methanol has higher octane number than gasoline [4-7].

It is important to mention that when mixture gasoline-80%methanol was tested the engine could not function properly.

The CO and HC emissions are represented in the figures below, for the mixtures: gasoline, gasoline-ethanol, for every fuel and for every load conditions. For these mixtures the average values of the emissions (CO, HC) are presented at the figures below. From the average values, the variation of those emissions can be better understood.

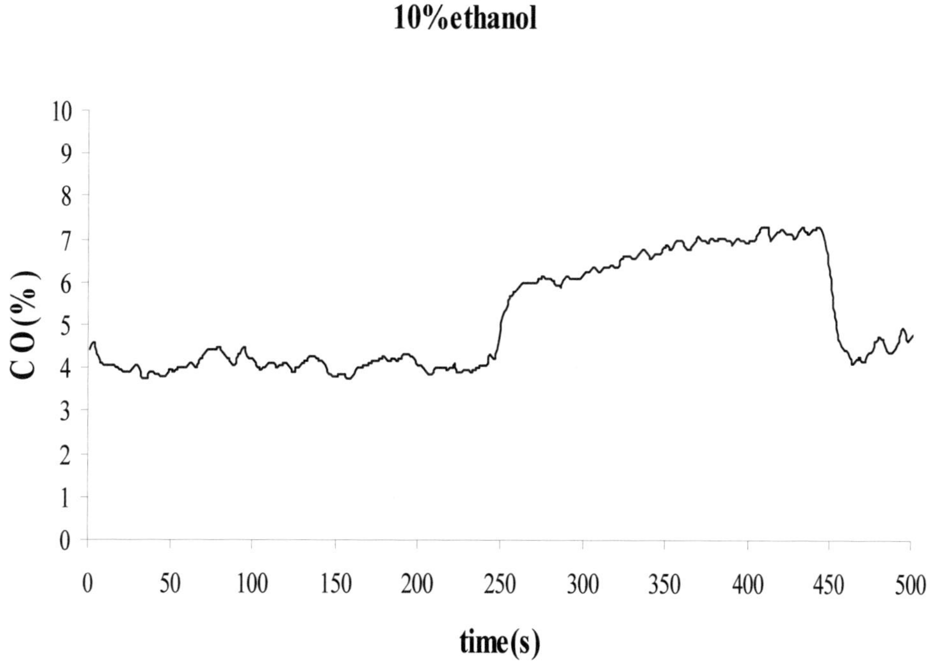

Figure 18. The CO variation for the gasoline-10% ethanol mixture is used as fuel

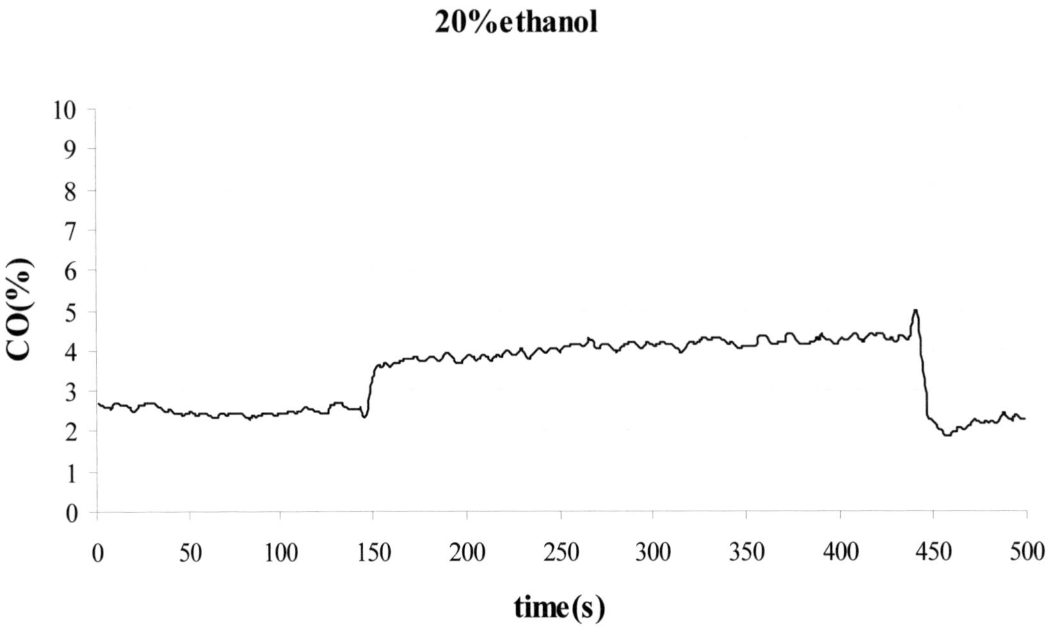

Figure 19. The CO variation for the gasoline-20% ethanol mixture is used as fuel

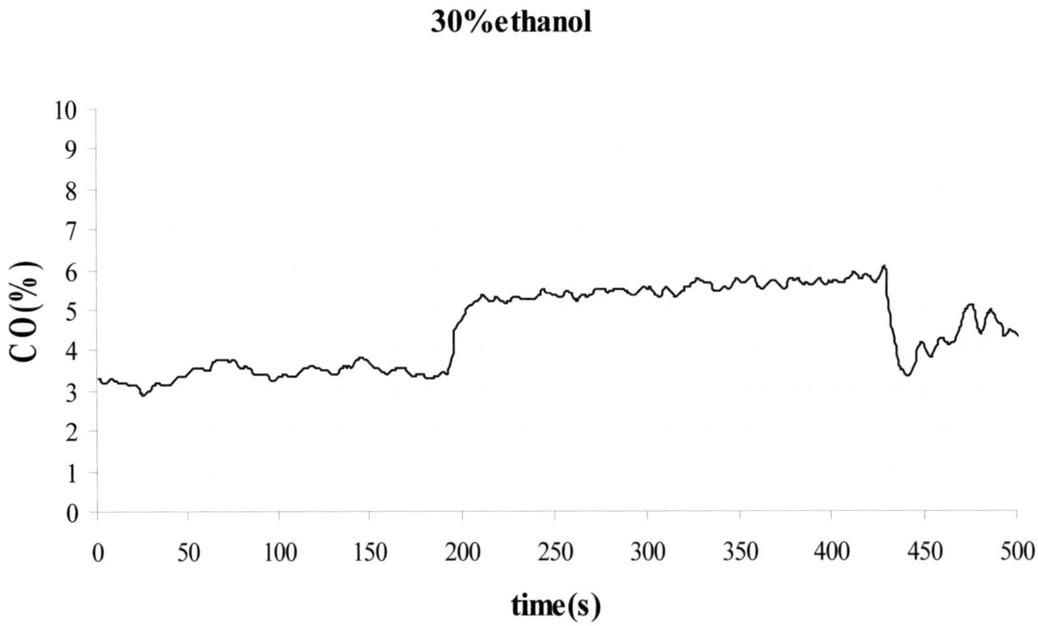

Figure 20. The CO variation for the gasoline-30% ethanol mixture is used as fuel

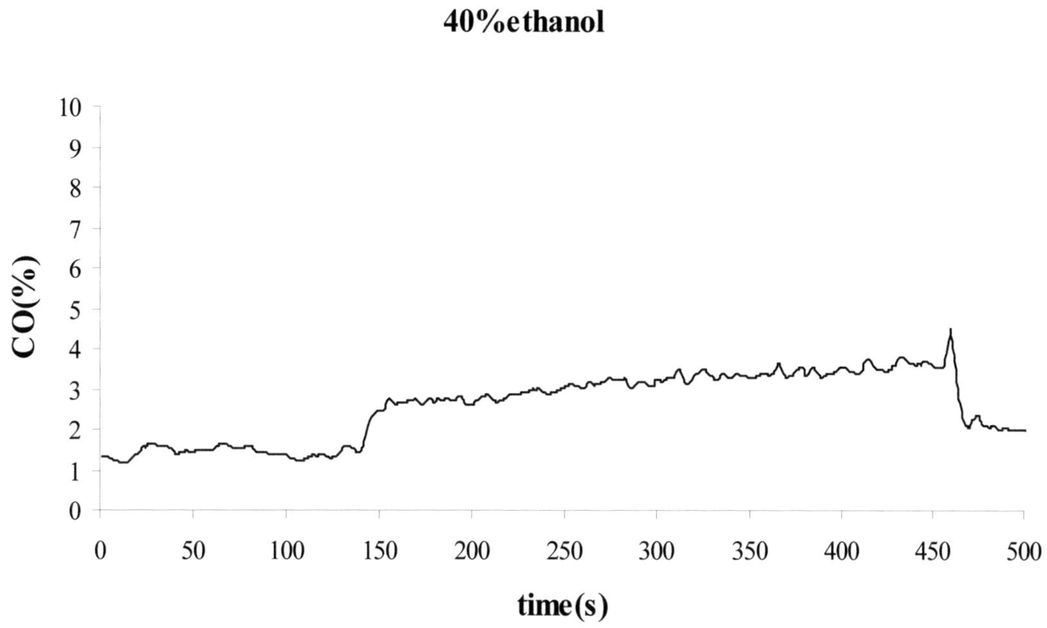

Figure 21. The CO variation for the gasoline-40% ethanol mixture is used as fuel

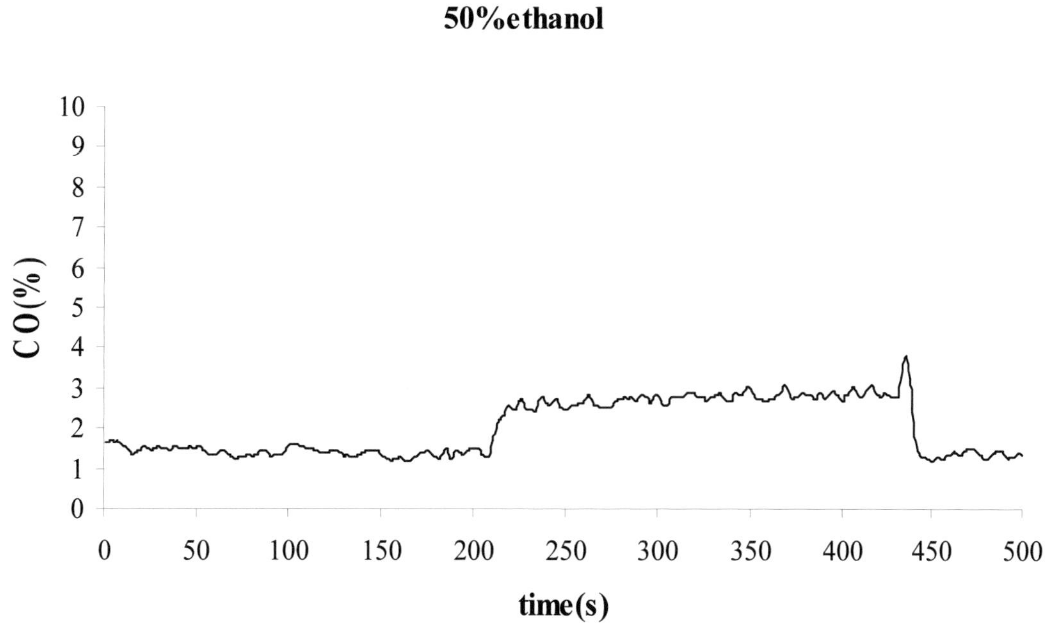

Figure 22. The CO variation when gasoline-50% ethanol mixture used as fuel

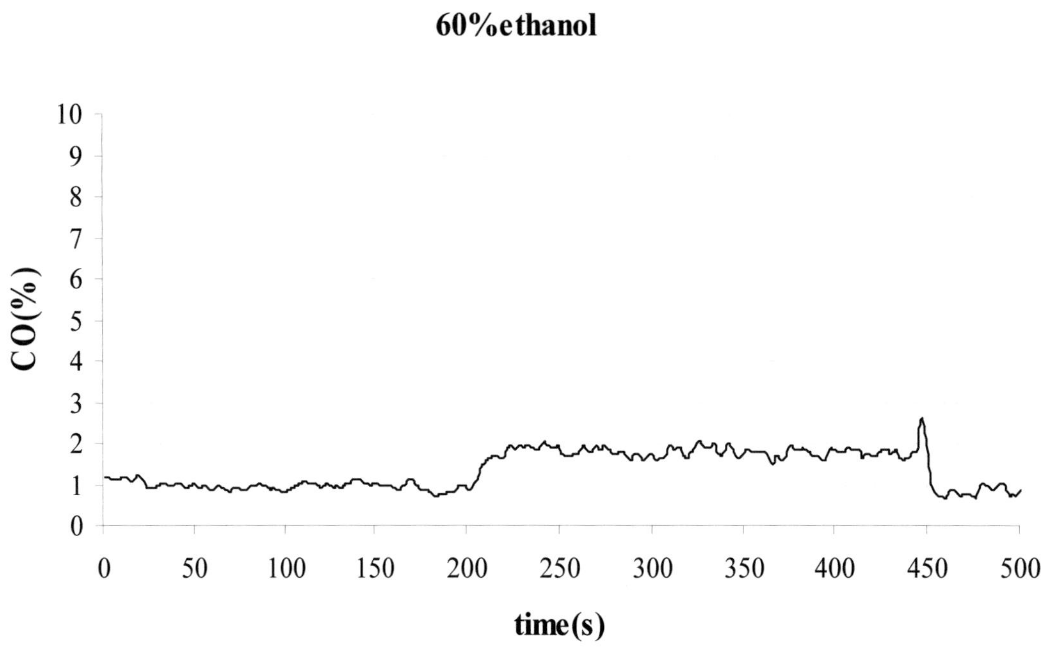

Figure 23. The CO variation when gasoline-60% ethanol mixture used as fuel

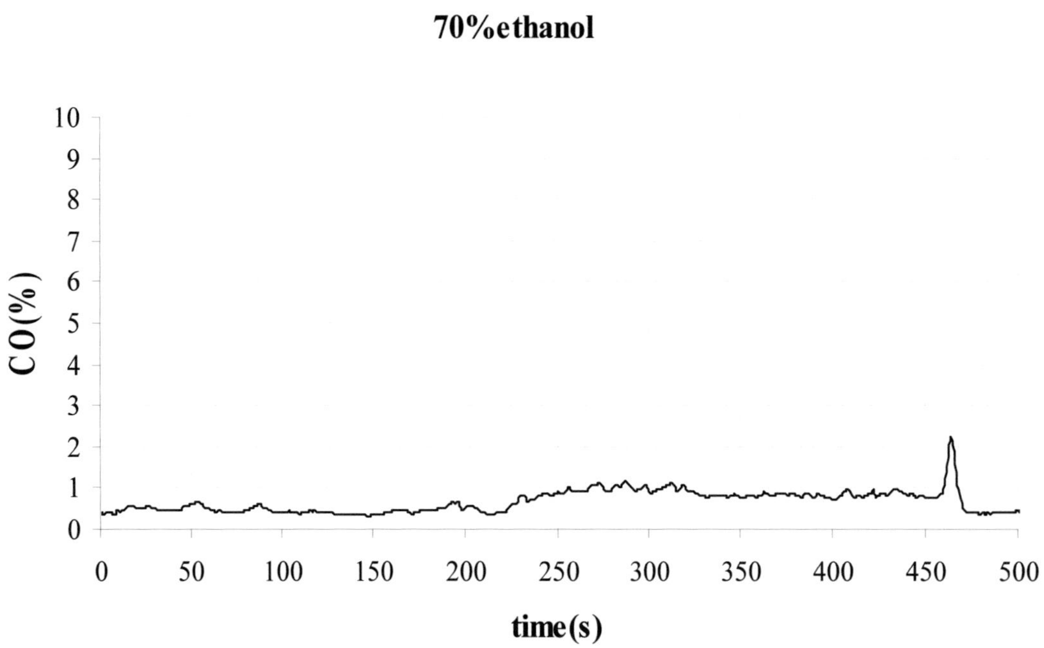

Figure 24. The CO variation when gasoline-70% ethanol mixture used as fuel

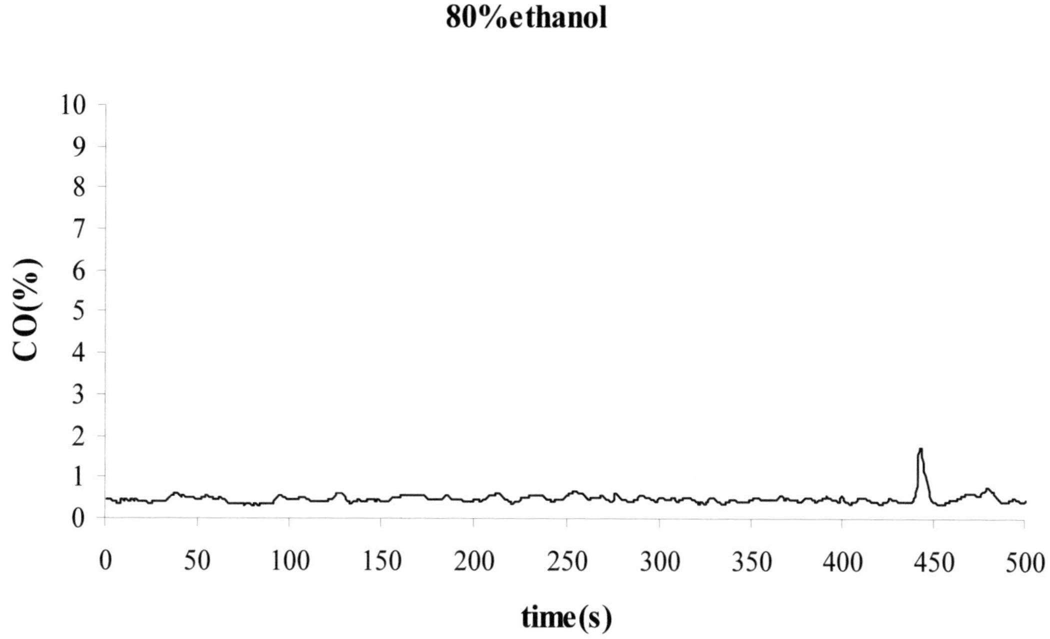

Figure 25. The CO variation when gasoline-80% ethanol mixture used as fuel

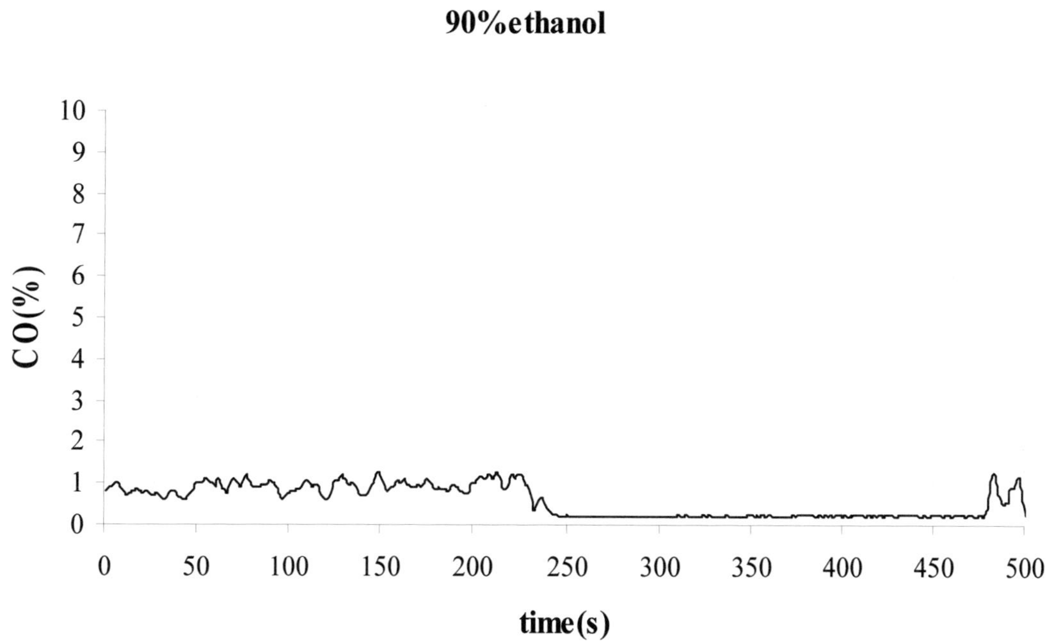

Figure 26. The CO variation when gasoline-90% ethanol mixture used as fuel

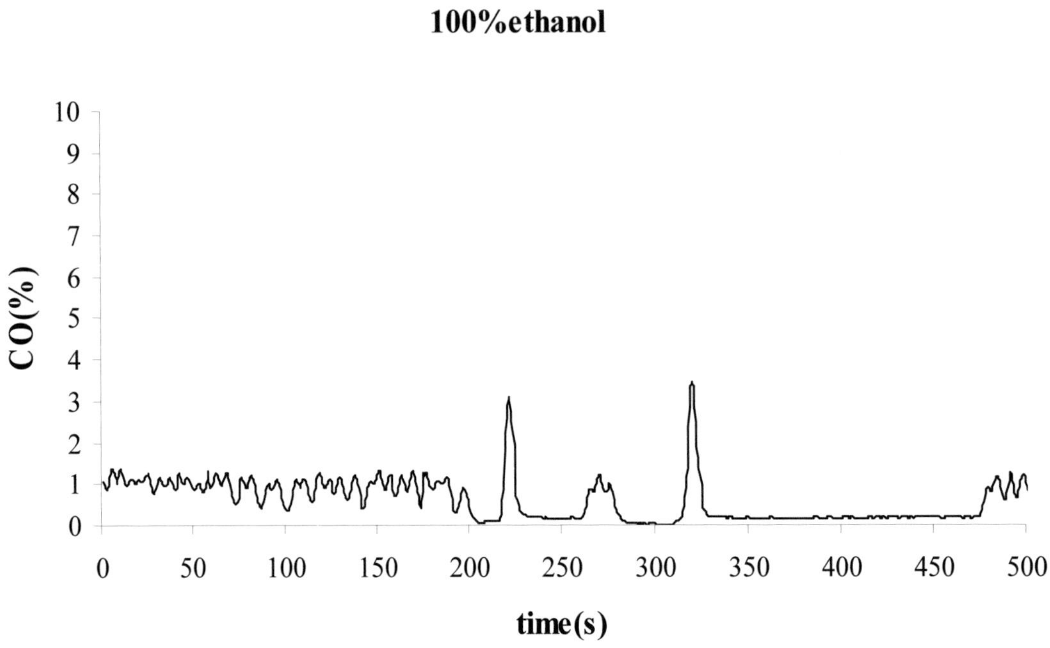

Figure 27. The CO variation when 100% ethanol used as fuel

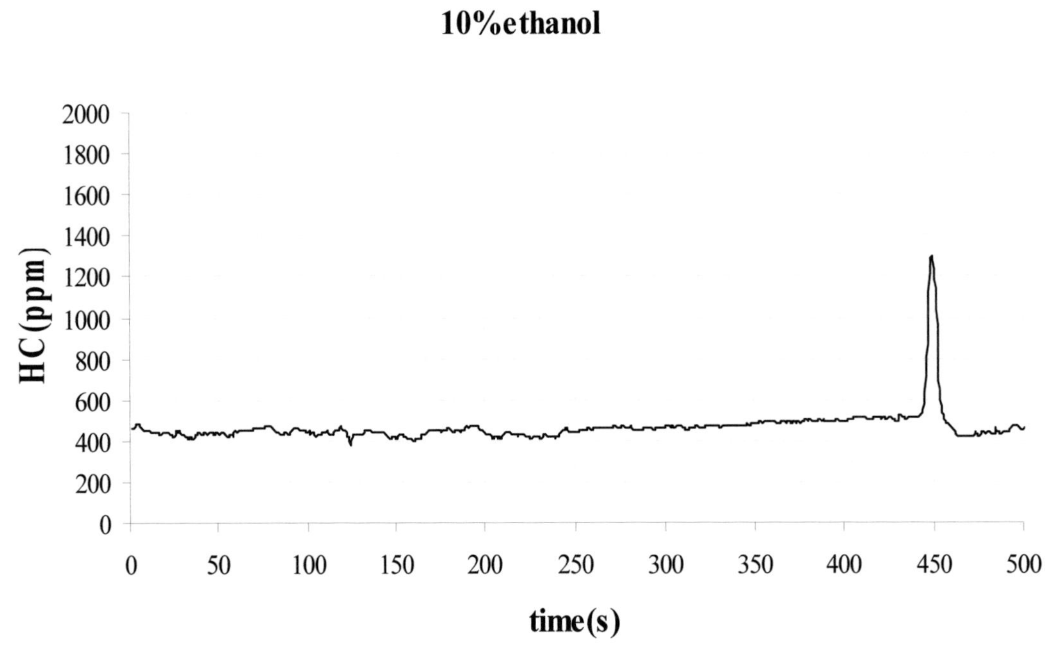

Figure 28. The HC variation when gasoline-10% ethanol mixture used as fuel

http://dx.doi.org/10.5772/52772

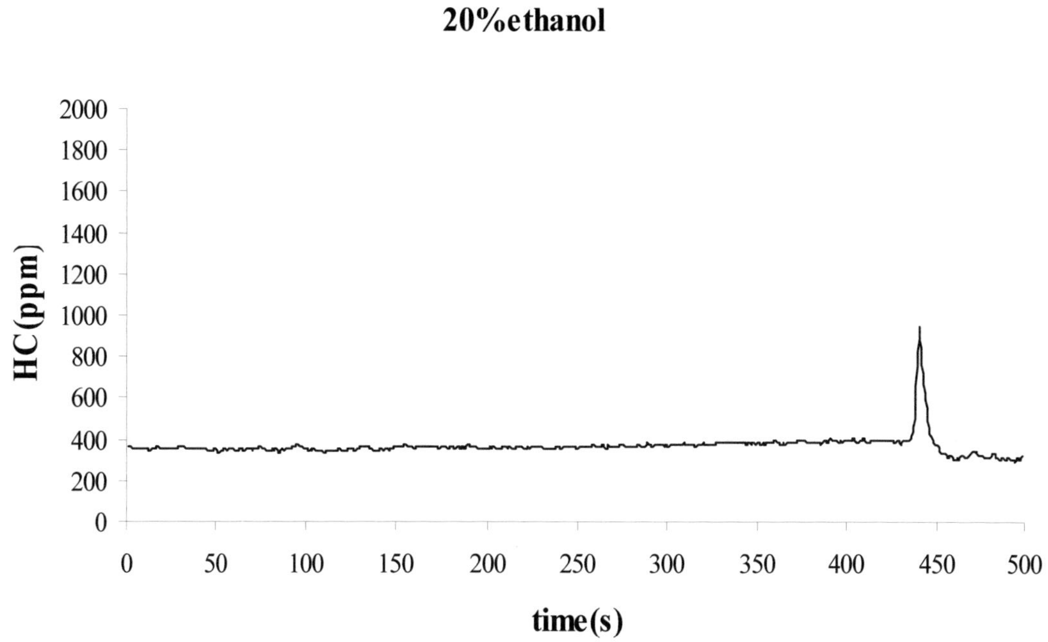

Figure 29. The HC variation when gasoline-20% ethanol mixture used as fuel

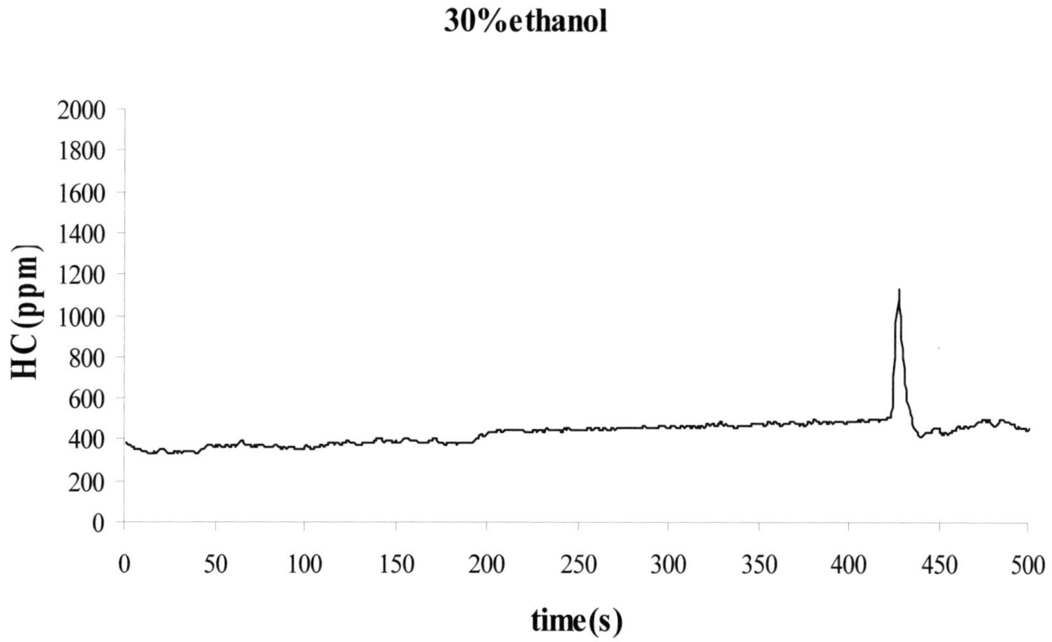

Figure 30. The HC variation when gasoline-30% ethanol mixture used as fuel

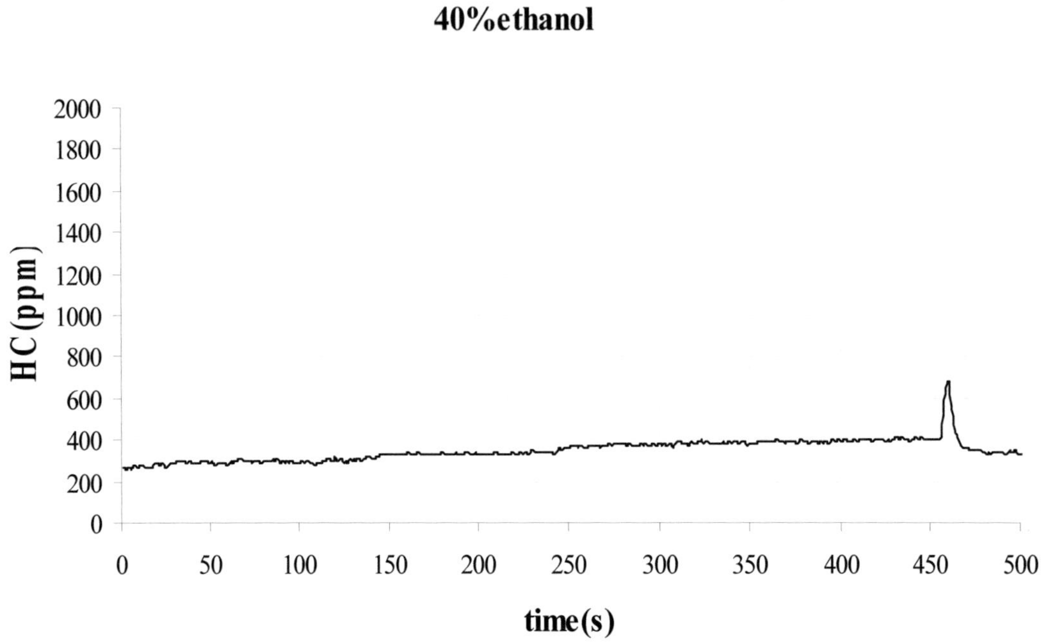

Figure 31. The HC variation when gasoline-40% ethanol mixture used as fuel

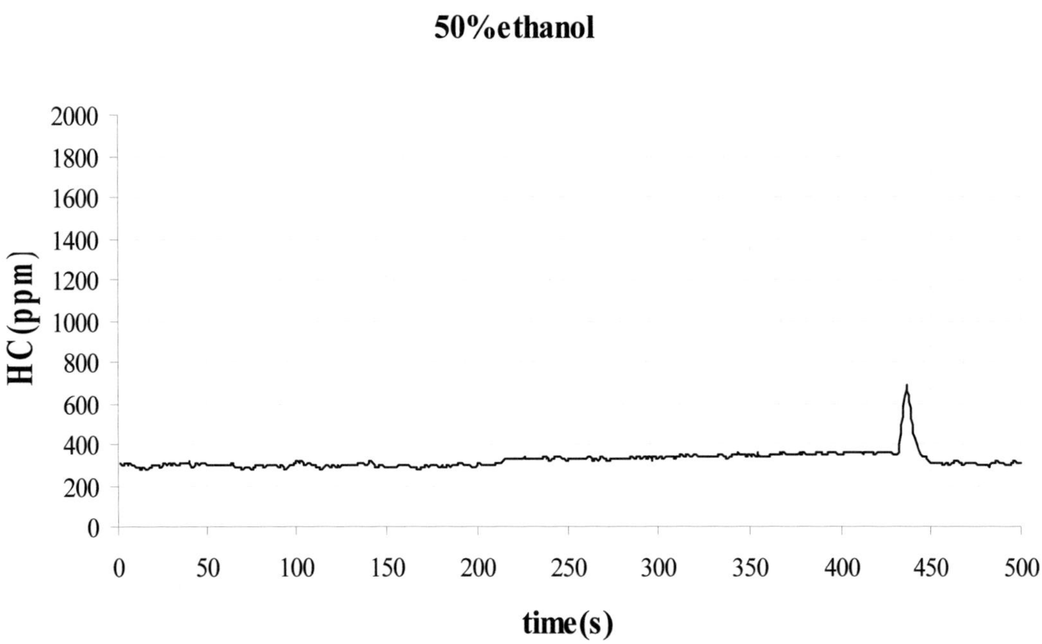

Figure 32. The HC variation when gasoline-50% ethanol mixture used as fuel

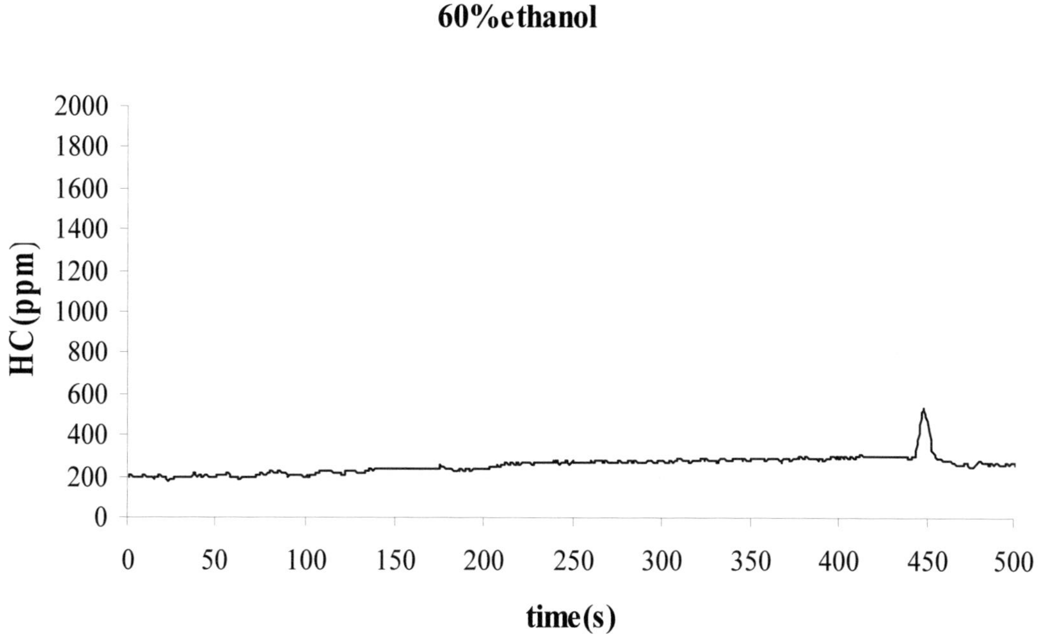

Figure 33. The HC variation when gasoline-60% ethanol mixture used as fuel

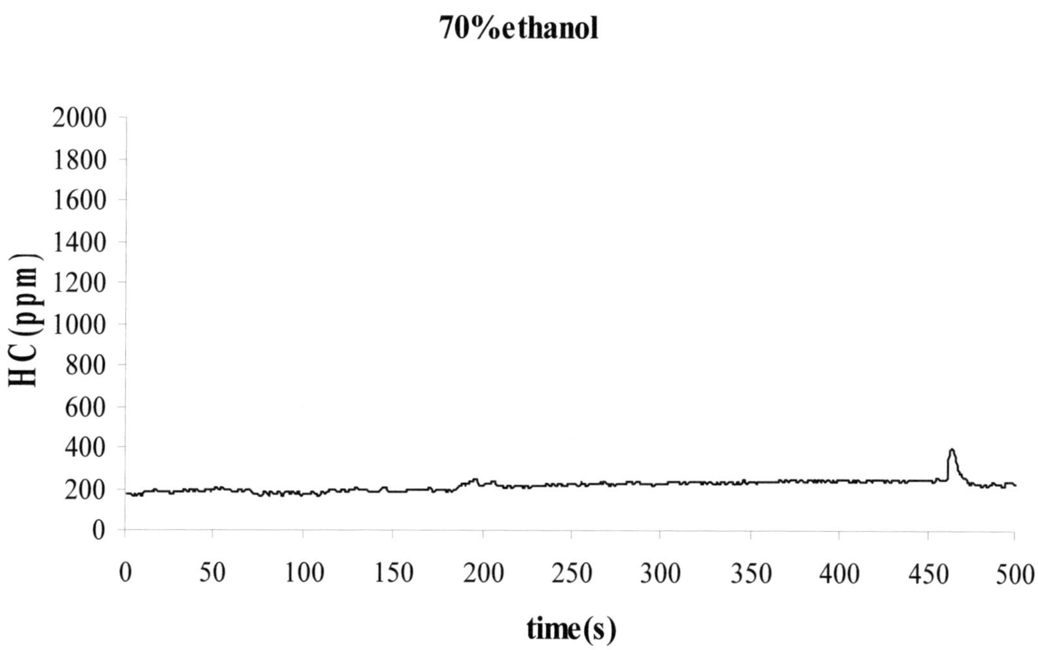

Figure 34. The HC variation when gasoline-70% ethanol mixture used as fuel

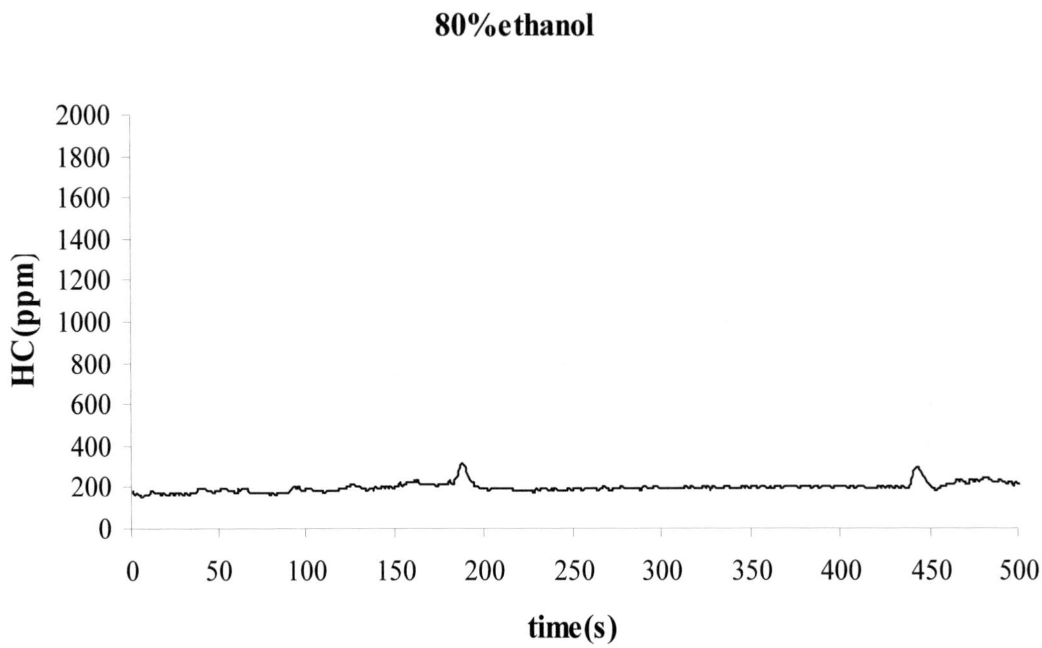

Figure 35. The HC variation when gasoline-80% ethanol mixture used as fuel

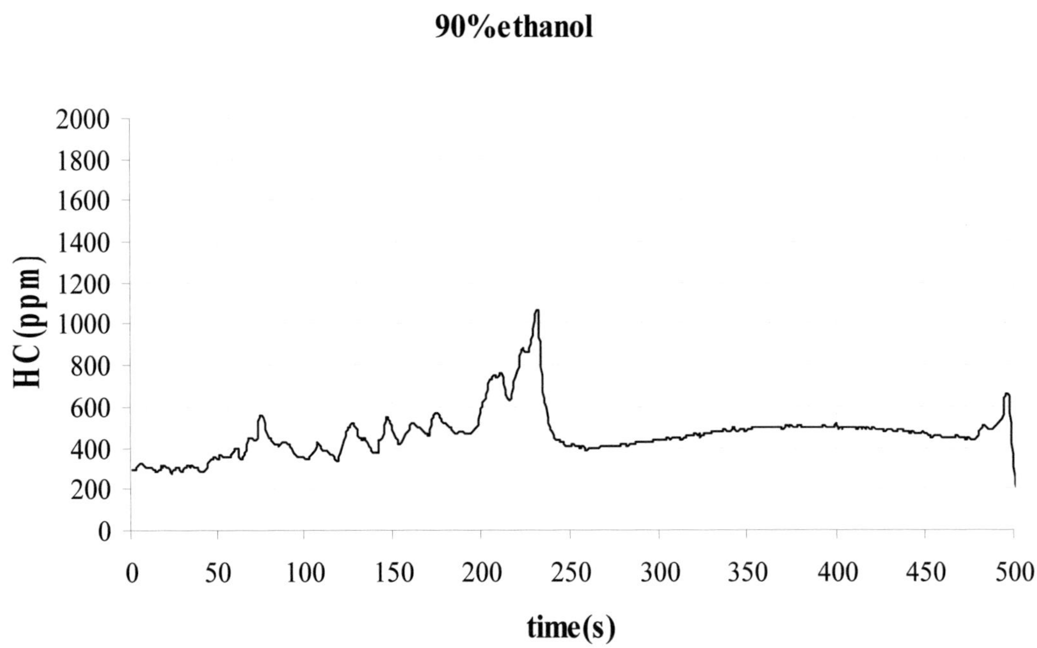

Figure 36. The HC variation when gasoline-90% ethanol mixture used as fuel

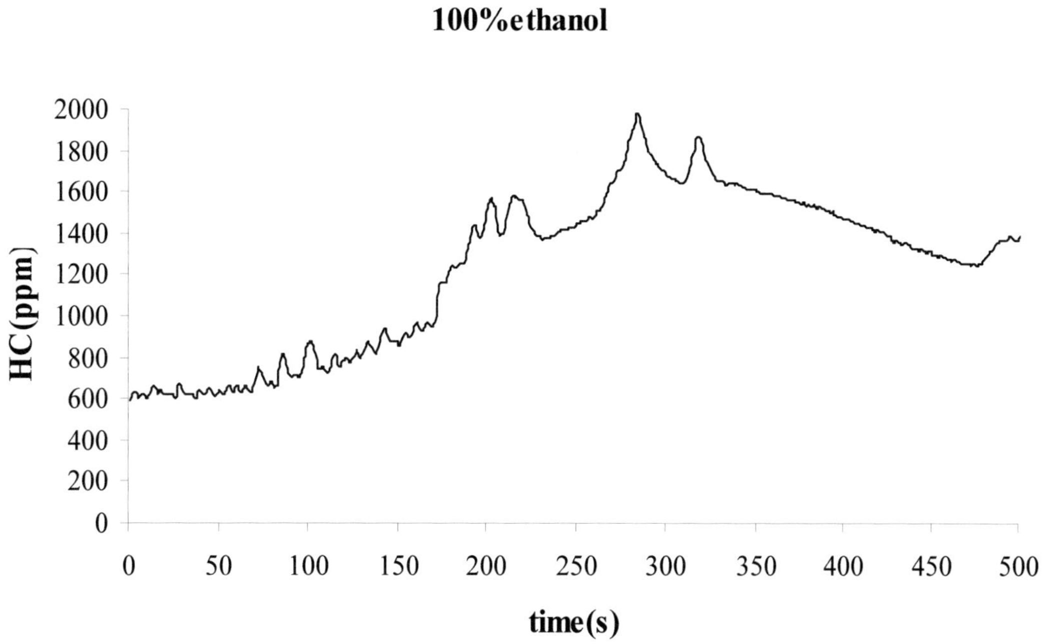

Figure 37. The HC variation when 100% ethanol used as fuel

Figures 18 - 28 present the CO variation when as fuel is used gasoline –ethanol and gasoline –methanol mixtures when the engine functioned without load and under full load conditions(1KW). From these figures is observed lower CO emissions when gasoline-ethanol mixtures are used compared to the mixtures gasoline-methanol, until the mixture of 70% ethanol and methanol. Over the 70% percentage of methanol the engine could not function and that is why there is no further presentation of comparative curves of CO emissions. It must also be mentioned that for the mixtures of gasoline –70% methanol, gasoline –90%ethanol and 100%ethanol the engine malfunctioned. The average values of CO emissions for the above mixtures and for both load conditions are presented in the figure 38 below [4-7]:

In the figures 28 - 37 is observed higher decrease of HC in the case were methanol is used, with exception of the use of gasoline –70%methanol mixture where the HC are higher compared to the mixture gasoline-70%ethanol. This is due to the malfunction that occurred during the use of gasoline-70%methanol mixture. There was also malfunction of the engine when the mixtures of gasoline-90%ethanol and 100%ethanol were used, which had as result the HC increase during the use of those mixtures. These observations are presented more clearly in the figure 38 below [4-7]:

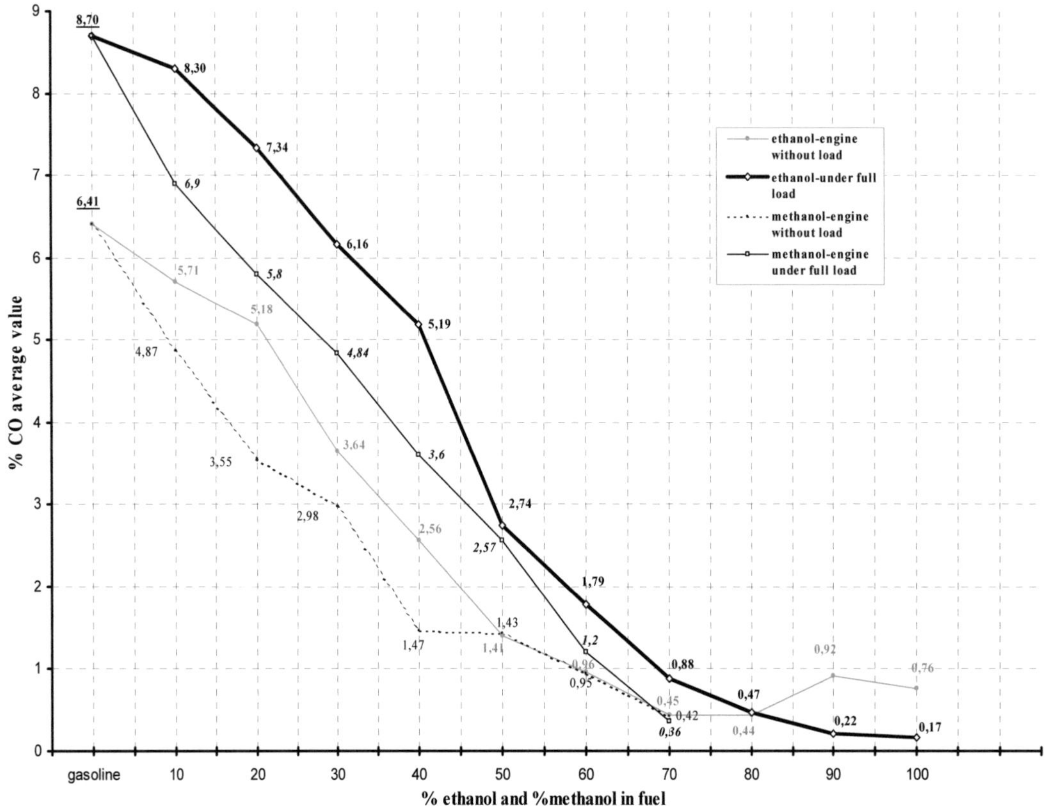

Figure 38. The CO emission average value for every gasoline-ethanol and gasoline-methanol mixture

Figure 39 shows the average values of HC for every mixture, when the engine functions without load and under full load conditions. It is being observed grater decrease of HC during the use of methanol in the fuel contrary to the use of ethanol.

Also is shown HC emissions decrease compared to gasoline, while the percentage of methanol and ethanol in the fuel increases without load and under full electrical load conditions (1KW). At higher percentage of ethanol in the fuel 90%ethanol and 100%ethanol it is observed HC emissions increase, which is due to incomplete combustion. Indeed, during the tests of the mixtures: gasoline-70%methanol, gasoline – 90%ethanol and 100%ethanol, there was an engine malfunction mostly at without electrical load, as it was mentioned above. This malfunction is showed from the rounds per minute recording in the figures below:

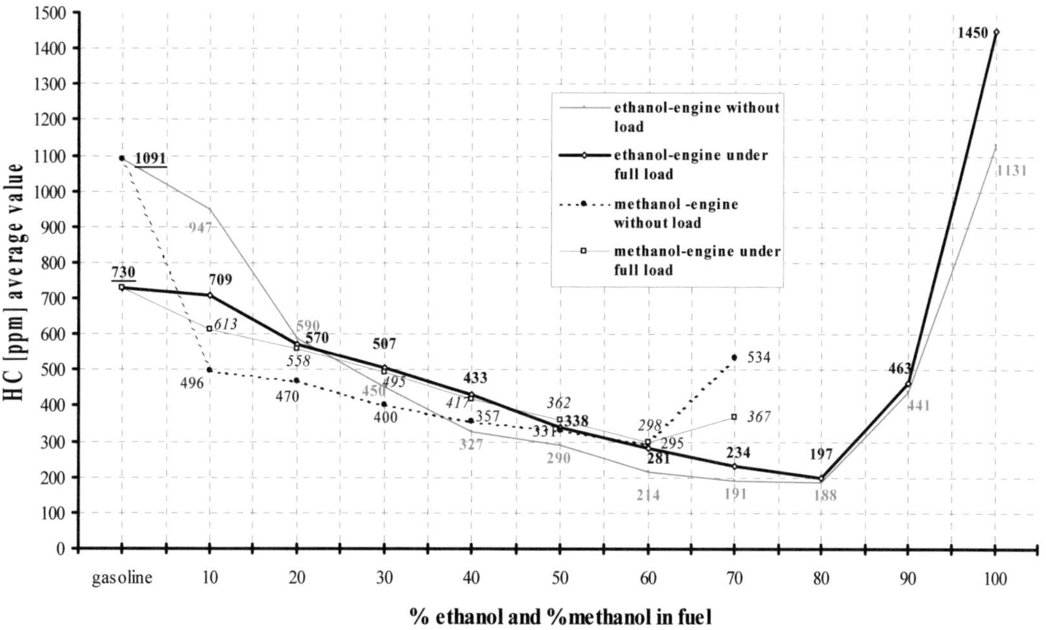

Figure 39. The HC emission average value for every gasoline-ethanol and gasoline-methanol mixtures

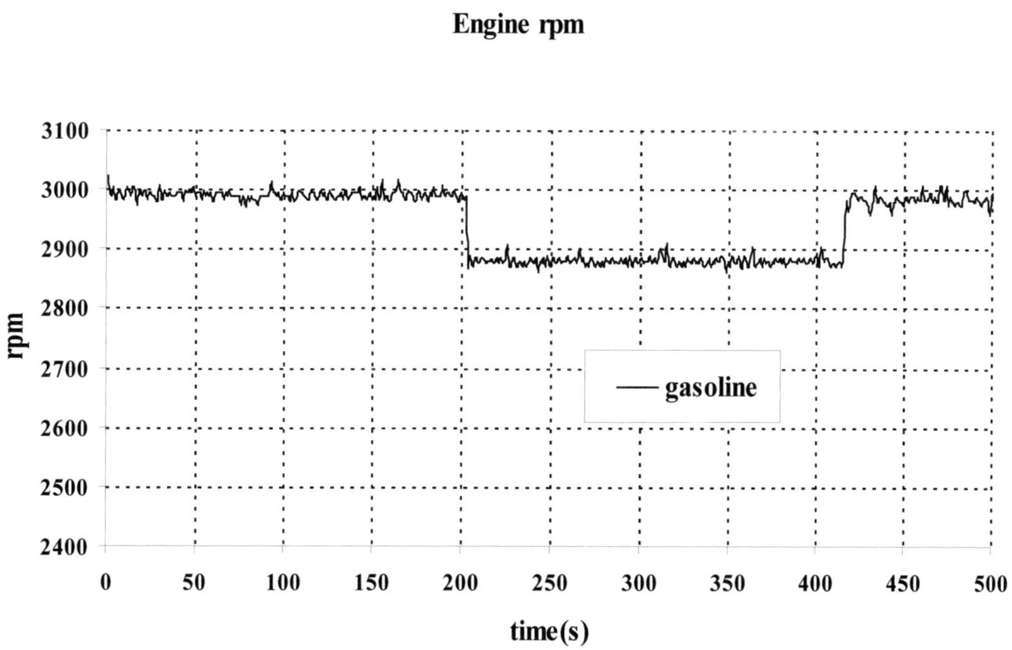

Figure 40. The rpm variation when used fuel gasoline

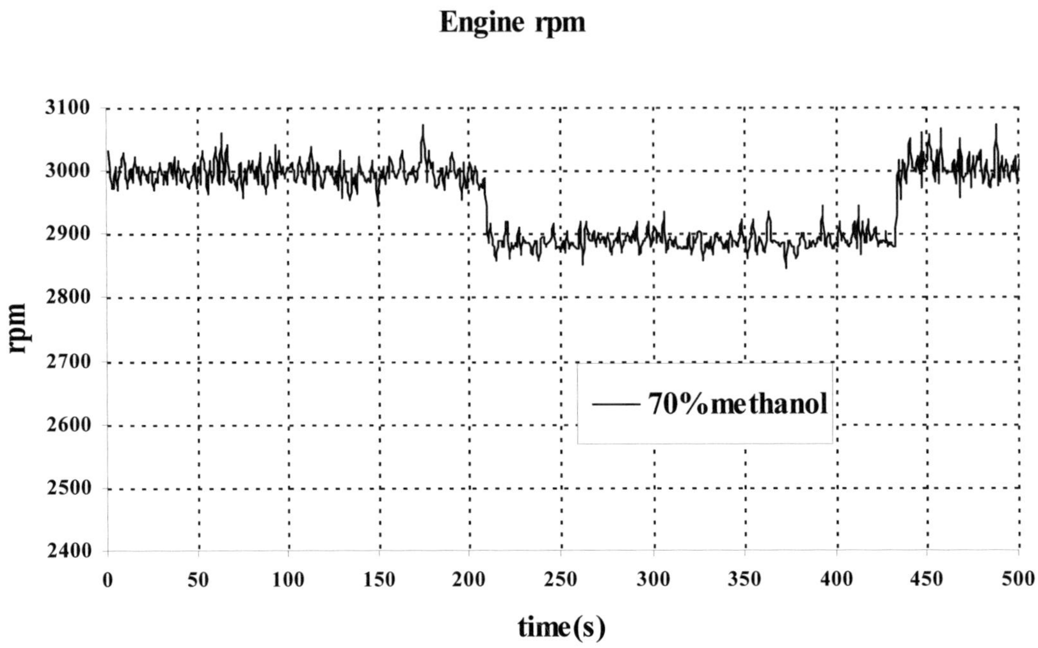

Figure 41. The rpm variation when used fuel gasoline –70%methanol

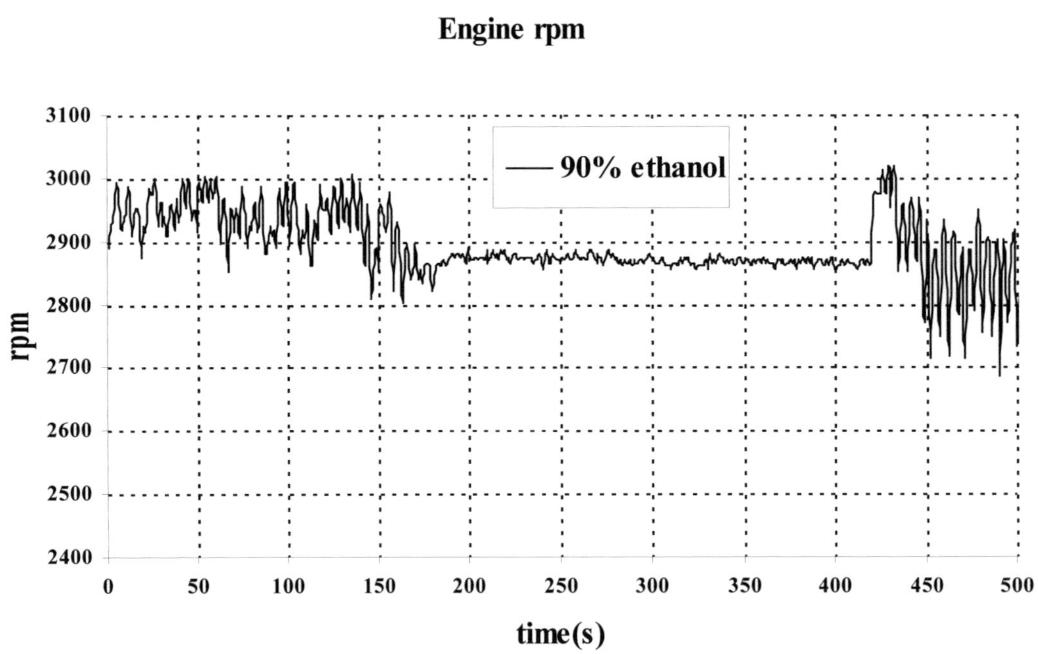

Figure 42. The rpm variation when used fuel gasoline –90%ethanol

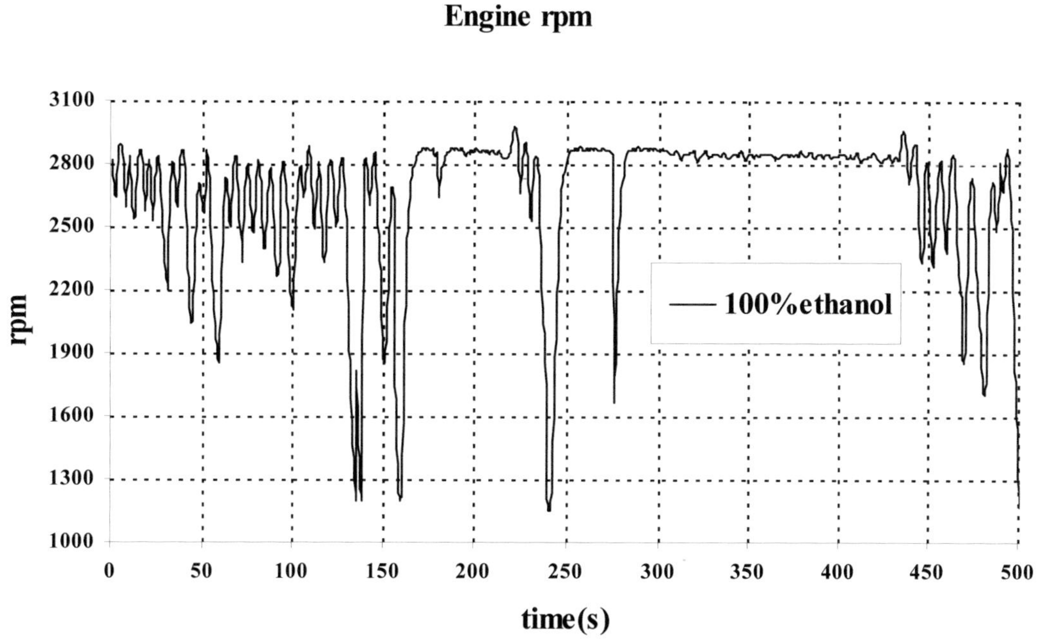

Figure 43. The rpm variation when used fuel 100%ethanol

During the tests the rounds per minute of the engine were recorded as it was mentioned above. The normal variation of the engine rpm appears in figure 40. The same variation that is illustrated in this figure corresponds to the mixtures gasoline until the mixtures gasoline-90% ethanol and gasoline-60%methanol, without any change. As it is presented in figure 39, the average value of the engine rpm without load (0-200s and 420-500s) is approximately 2990rpm while at full load conditions (200-420s) the average value of the engine rpm is 2880rpm. It must be noted that the engine has a round stabilizer. In figures 41, 42 and 43 the mixtures gasoline-70%methanol, gasoline-90%ethanol and 100%ethanol are illustrated and irregular variation of the engine rpm is presented, which is caused from the engine malfunction. Higher irregular variation is observed at without load condition, and lower at full load conditions in the case of use ethanol. This malfunction is due to the smaller calorific value of methanol and ethanol than the gasoline, and to the fact that there is no adjustment of the air/fuel ratio during the use of gasoline-methanol and gasoline-ethanol mixtures. The initial adjustment that corresponds to gasoline as fuel is maintained [6,7].

Furthermore, during the tests the consumption of the fuel was recorded for every mixture separately and for every load conditions. The results of the consumption recording are illustrated in the figure below:

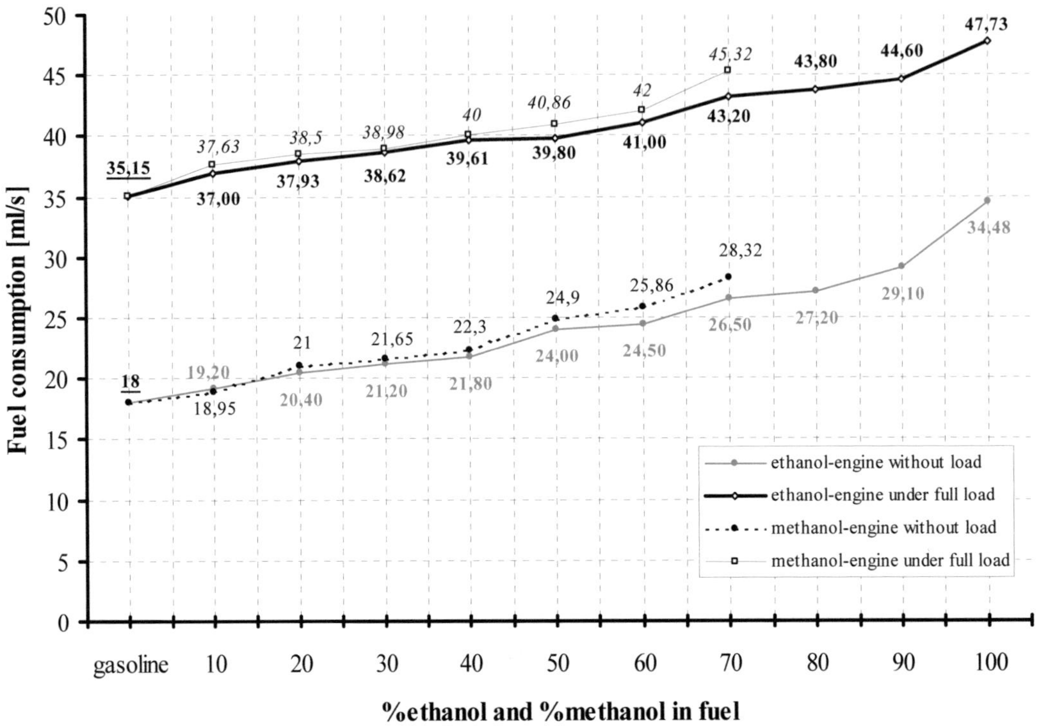

Figure 44. The fuel consumption

Figure 44 shows an increase of fuel consumption when the percentage of methanol and ethanol in the fuel increases than gasoline. Also, between the use of the mixtures of methanol and ethanol is observed small increase during the use of methanol because of the smaller calorific value that methanol has compared to ethanol. The smaller calorific value of methanol and ethanol compared to gasoline and also the lack of regulation (ratio air/fuel) of the engine, results to the consumption increase contrary to the use gasoline. This increase of consumption happens automatically for the rounds regulator that the engine has, for the maintaining of the rounds constant.

3. Conclusion

From the observations above is appeared that methanol and ethanol as mixture with gasoline results in an emissions (CO and HC) decrease when the engine functions without load and under full load conditions. There is an exception in the use of the mixtures: gasoline-70%methanol, gasoline –90% ethanol and 100%ethanol where there is observed an HC emissions increase because of the incomplete combustion and consequently due to engine malfunction. Also, it must be mentioned that the adjustment of the engine (air/fuel ratio) was that which referred to the use of gasoline as fuel. From the aspect of consumption, there was a consumption compared to gasoline increase when the percentage of the methanol and ethanol in the fuel was increased in both load conditions. Between the use of methanol and ethanol mixtures is observed higher increase of consumption when the mixtures of methanol are used due to the fact that methanol has lower calorific value compared to ethanol. From the aspect of emissions, when the mixtures of gasoline with methanol and ethanol are compared, there is grater reduction of emissions in the case where methanol is used. It can be said that this is caused because of the smaller carbon chain of the methanol molecule, which results to the better combustion of methanol. It is also observed that the engine functions with the mixtures of methanol until the use of 70%methanol mixture with gasoline, while with ethanol mixtures until 100% ethanol as fuel (with the initial adjustment of the air/fuel ratio that is made for gasoline). This is due to the fact that ethanol has higher octane number compared to methanol. Finally, it is important the fact that methanol and ethanol are a renewable fuels, which present emissions decrease compared to gasoline, when they are used, in a time period where petroleum reservations are depleted and the environmental pollution is one of the most important problems that humanity faces [4-7].

Author details

Charalampos Arapatsakos*

Department of Production and Management Engineering, Democritus University of Thrace, Xanthi, Greece

References

[1] Arapatsakos I. C, Sparis D. P., "Testing the Two Stroke Engine Using Mixtures of Gasoline - Ethanol" International Journal of Heat & Technology, Vol. 16, pp. 57-63, 1998.

[2] Arapatsakos I. C, "Air and Water Influence of Two Stroke Outboard Engine Using Gasoline -Ethanol Mixtures" Transactions of SAE 2000-01-2973, Book SP-1565, 2000.

[3] Arapatsakos I. Charalampos, "Testing The Tractor Engine Using Diesel –Ethanol Mixtures Under Full Load Conditions" International Journal of Heat & Technology, Vol. 19, n.1, 2001.

[4] Arapatsakos I. Charalampos, Karkanis N. Anastasios, Sparis D. Panagiotis, "Gas Emissions And Engine Behavior When Gasoline-Alcohols Mixtures Are Used" Journal of Environmental Technology, Vol. 24, 2003.

[5] Arapatsakos I. Charalampos, Karkanis N. Anastasios, Sparis D. Panagiotis, "Tests On A Small Four Engine Using As Fuel Gasoline-Bioethanol Mixtures" Transactions of WIT, ISSN 13569-5886, 2003.

[6] Charalampos Arapatsakos, Anastasios Karkanis, Panagiotis Sparis, "Gasoline – Ethanol, Methanol Mixtures And A Small Four-Stroke Engine" International journal of heat and technology Vol. 22, n.2 2004.

[7] Charalampos Arapatsakos, Anastasios Karkanis, Panagiotis Sparis, "Environmental Pollution From The Use Of Alternative Fuels In A Four-Stroke Engine" International journal of Environment and Pollution Vol 21,n.4 2004.

[8] Charalampos Arapatsakos, Anastasios Karkanis, Panagiotis Sparis, "Environmental Contribution Of Gasoline –Ethanol Mixtures" WSEAS Transactions on Environment and Development Vol. 2, Is. 7, 2006.

[9] Charalampos Arapatsakos, Anastasios Karkanis, Panagiotis Sparis "Methanol Blends As Motor Fuels" journal of wseas transactions on environment and development 2008.

[10] C. Arapatsakos, K. Papastavrou, "Experimental Measurements Of Bioethanol Use In Four Stroke Engine" International journal of heat and technology, Issue 1, 2009.

[11] Arapatsakos I. C., Sparis D. P., "Testing A Low Output Two Stroke Outboard Engine Using Gasoline-Ethanol Mixtures" International Conference on The Naval Technology for the 21st Century, pp.305-310. Hellenic Naval Academy Pireus-Greece, 1998.

[12] C. Arapatsakos, "Air Pollution From Four Stroke Engine That Use Gasoline – Ethanol Blends" 2nd International Conference on Renewable Energy Sources(RES'08), Corfu, 2008.

[13] C. Arapatsakos, "The Gas Emissions From The Use Of Bioethanol International Conference on Energy and Environment", Cambridge 2009.

[14] C. Arapatsakos, "Influence Of Gasoline – Methanol Mixtures In A Two Stroke Engine International Conference on Energy and Environment", Cambridge 2009.

[15] C. Arapatsakos, "Engine Behavior Using Mixtures of Gasoline-Ethanol"" 5th International Conference on Energy and Environment, Cambridge 2010.

[16] C. Arapatsakos, M. Moschou "The Influence of Air and Water When It Is Used as Fuel the Mixtures of Gasoline -Alcohols" 5th International Conference on Energy and Environment, Cambridge 2010.

Chapter 4

Analytical Methodology for Determination of Trace Cu in Hydrated Alcohol Fuel

Fabiana Aparecida Lobo, Fernanda Pollo,
Ana Cristina Villafranca and Mercedes de Moraes

Additional information is available at the end of the chapter

http://dx.doi.org/10.5772/52514

1. Introduction

Many private and governmental initiatives have been established worldwide to identify viable alternatives to petroleum derivatives [1,2].The goals are to reduce dependence on imported energy from non-renewable sources, while mitigating environmental problems caused by petroleum products, and to develop national technologies in the alternative energy field.

Ethyl alcohol (ethanol) is considered to be a highly viable alternative fuel. Its production from biomass means that it can provide a source of energy that is both clean and renewable. The inclusion of ethanol as a component of gasoline can help to reduce problems of pollution in many regions, since it eliminates the needto use tetraethyl lead (historically notorious as a highly toxic trace component of the atmosphere in major cities) as an anti-knock additive.

The quantitative monitoring of metal elements in fuels (including gasoline, alcohol, and diesel) is important from an economic perspective in the fuel industry as well as in the areas of transport and environment. The presence of metalspecies (ions or organometallic compounds) in automotive fuels can cause engine corrosion, reduce performance, and contribute to environmental contamination [2-5].

The low concentrations of metals in fuels typically require the use of sensitive spectrometric analytical techniques for the purposes of quality control. Atomic absorption spectrometry (AAS) can be applied for the quantitative determination of many elements (metals and semi-metals) in a wide variety of media including fuels, foodstuffs, and biological, environmental, and geological materials, amongst others. The principle of the technique is based on measurement of the absorption of optical radiation, emitted from a source, by ground-state atoms in the gas phase. Atomization can be achieved using a flame, electrothermal heating, or specific chemical reaction (such as the generation of Hg cold vapor). Electrothermalatomizers include graphite tubes, tungsten filaments, and quartz tubes (for atomization of hydrides), as well as metal or ceramic tubes. Flame atomic absorption spectrometry (FAAS) is mostly used for elemental analysis at higher concentration levels, of the order of mg L^{-1}[3-5]. Table 1 lists some of the published studies concerning the application of AAS for determination of metals in fuels.

Technique	Matrix	Sample preparation	Reference
Graphite furnace atomic absorption spectrometry (GFAAS)	Fuel oil	Microwave digestion	6
Flame atomic absorption spectrometry	Fuel oil	Wet digestion	7
Atomic fluorescence spectrometry with cold vapor generation	Gasoline and other petroleum derivatives	Wet digestion	8
Graphite furnace atomic absorption spectrometry	Fuel oil	Microwave digestion	9
Graphite furnace atomic absorption spectrometry	Crude fuel oil	Dilution in xylene/I_2, wet and microwave digestion	10
Flame atomic absorption spectrometry	Engine oil	Wet digestion	11
Graphite furnace atomic absorption spectrometry	Fuel oils and naphtha	Oil-water emulsion	12
Inductively coupled plasma mass spectrometry (ICP-MS)	Gasoline	Emulsion with Triton X-100 surfactant	13
Flame atomic absorption spectrometry	Ethyl alcohol	Dilution in HNO_3	5
Graphite furnace atomic absorption spectrometry	Ethyl alcohol	Dilution in HNO_3	3,4
Graphite furnace atomic absorption spectrometry	Ethyl alcohol	Dilution in HNO_3	2
Graphite furnace atomic absorption spectrometry	Ethyl alcohol and acids	Dilution in HNO_3	14
Graphite furnace atomic absorption spectrometry with a high-resolution continuum source (HR-CS-GFAAS)	Crude oil	Oil-water emulsion	15
Atomic absorption spectrometry with thermal nebulization in a tube heated in a flame (TS-FF-AAS)	Ethyl alcohol fuel and gasoline	Dilution in HNO_3/oil-water emulsion	16
Graphite furnace atomic absorption spectrometry with a high-resolution continuum source	Crude oil	Oil-water emulsion	17

Technique	Matrix	Sample preparation	Reference
Atomic absorption spectrometry	Gasoline, diesel, ethyl alcohol and biodiesel	- (Review)	18
Graphite furnace atomic absorption spectrometry	Ethyl alcohol and acids	Direct determination	19
Graphite furnace atomic absorption spectrometry	Biodiesel	Microwave digestion and emulsion	1
Flame atomic absorption spectrometry	Ethyl alcohol	Solid phase extraction	20
Graphite furnace atomic absorption spectrometry	Biodiesel	Emulsion	21
Atomic absorption spectrometry with vapor generation (VP-FAAS)	Ethyl alcohol	Treatment with acid under UV irradiation	22

Table 1. Analytical methods for the determination of inorganic contaminants in fuels.

The thermospray (TS) technique was originally developed by Vestal et al.in 1978 [23]as an interface between liquid chromatography and mass spectrometry. In atomic absorption spectrometry, the tube was heated electrically in order to maintain a constant temperature, which restricted use of the method to only a few elements. However, Gáspárand Berndt (2000) proposed the TS-FF-AAS procedure, in which a metal tube is positioned above the flame of the atomic absorption spectrometer, as a reactor. The sample solution is transported through a metal capillary, connected to the tube, and heated simultaneously by the flame. On reaching the hot tip of the capillary, the liquid partially vaporizes, forming an aerosol. In turn, the aerosol is vaporized within the tube, producing an atomic cloud that absorbs the radiation emitted by the lamp.

The TS-FF-AAS method was used as an interface between high performance liquid chromatography (HPLC) and FAAS, employing a flow injection system [25-60].

The objective of this work is to describe the analysis of Cu present in hydrated ethyl alcohol fuel (HEAF) using the technique of atomic absorption spectrometry with thermal nebulization in a tube heated in a flame (TS-FF-AAS). The atomizers used were a metal tube (Ni-Cr alloy) and a ceramic tube (Al_2O_3).

2. Experimental procedures

2.1. Instruments and accessories

The instrumentation consisted of an atomic absorption spectrometer fitted with a flame atomizer (Perkin-Elmer, model AAnalyst 100), a hollow cathode Cu lamp (λ = 324.8 nm, slit width = 0.7 nm,i = 15 mA), with an air/acetylene (4:2 ratio) flame gas mixture, and background correction using a deuterium lamp. Other equipment comprised an analytical balance (Sartorius BL 2105) and a peristaltic pump (Ismatec, model ICP 8).

The TS-FF-AAS assembly employed a Rheodyne RE9725 injection valve, PEEK tubing, and a ceramic thermocouple insulator capillary (OMEGATITE450, OMEGA, USA). The capillary wascomposed of Al_2O_3(>99.8%), resistant to temperatures up to 1900 °C, with $Ø_{ext}$= 1.6 mm and two orifices with $Ø_{int}$ = 0.4 mm (this capillary provided better results than a stainless steel HPLC capillary, with less noise in the absorbance signal). The atomization tubes were a metal tube composed of Ni-Cr super-alloy (Inconel, length 100 mm, $Ø_{int}$ = 10.0 mm, $Ø_{ext}$ = 12.0 mm, 6 orifices with Ø = 2.5 mm, perpendicular to an orifice with Ø = 2.0 mm), and a ceramic tube (99.9% Al_2O_3, length 100 mm, $Ø_{int}$ = 10.0 mm, $Ø_{ext}$ = 12.0 mm, 6 orifices with Ø = 2.5 mm, perpendicular to an orifice with Ø = 2.0 mm).

Data acquisition employed the software MQDOS (Microquímica), and the absorbance values were proportional to the height of the transient signals.

The temperature in the interior of the atomization tube was measured in two ways. The first method employed a thermocouple with an earthed connection, positioned adjacent to the metal tube, oriented towards the orifice where the ceramic capillary used to introduce the sample into the atomizer was located. The temperature measured for the metal tube was 983 ± 1°C. Secondly, the thermocouple with connection exposed was positioned adjacent to the ceramic capillary within the tube, where a temperature of between 1030 °C and 1060°C was measured, at which the tube glowed ruby-red above the flame [16,40,45,54].

When 50 μL of HNO_3 (~0.1 mol L^{-1}) was injected at a rate of approximately 1.5 mL min^{-1},there was a temperature reduction of around 50°C, due to cooling of the tube by the solution, followed by a rapid return to the maximum temperature range.

2.2. Reagents, solutions and samples

Working standard solutions were prepared from a stock 1000 mg L^{-1} copper standard solution (spectroscopic grade), by dilution in 0.14 mol L^{-1} HNO_3 (Synth).

The HEAF samples were prepared by mixing the fuel with an equal volume of 0.14 mol L^{-1} HNO_3,with final volumes of 50 mL [3-5]. Subsequent quantification employed the standard additions procedure.

2.3. Assembly of the TS-FF-AAS system

A schematic diagram of the TS-FF-AAS system is shown in Figure 1.

It is recommended that the Inconel tube should only be positioned above the burner head after lighting the flame, to avoid the possibility of an explosion within the tube due to gas accumulation. The TS-FF-AAS system was therefore first assembled, after which the spectrometer flame was ignited immediately after opening the gas valves to avoid any explosion risk. This procedure facilitated the positioning of the tube above the burner head, which was performed while the flame was extinguished. All analyses employed a fixed volume of sample, injected into the flow of air as the carrier, since previous work has shown that injection using carrier solutions results in greater sample dilution and dispersion [40,45,54,58].

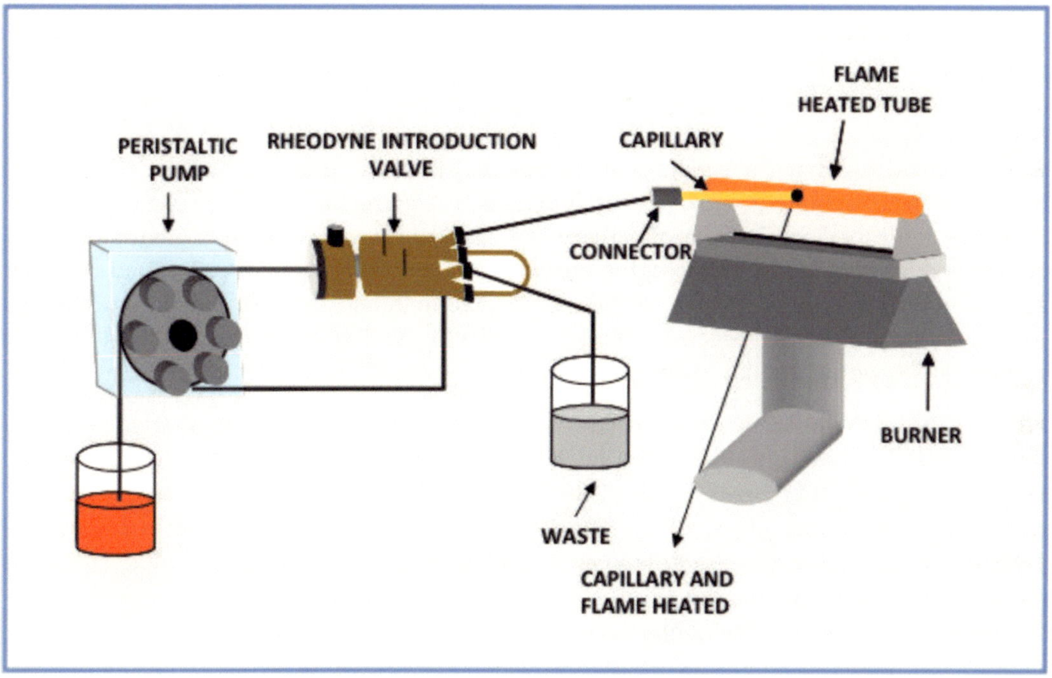

Figure 1. Schematic arrangement of the TS-FF-AAS system (adapted from Davies & Berndt 2003[59].

The sample was introduced into the system using a manual Rheodyne valve (Figure 1), after which it was transported to the ceramic capillary in the flow of air. Since the capillary was heated simultaneously with the metal or ceramic reactor tube, the liquid was partially vaporized, forming a thermospray, and atomization occurred on arrival in the tube, generating a transient signal that was captured and stored by the software. The determination employed the height of the transient signal peak.

2.4. Optimization of carrier flow rate and sample volume

The influences of the carrier flow rate (in the range 9.0-18.0 mL min^{-1}) and the sample volume (50, 100, and 200µL) were evaluated using a standard of 200µg Cu L^{-1}.

2.5. Construction of analytical curve

After optimization of the system, analytical curves were constructed in the concentration range 0.1-0.4 mg Cu L^{-1} in 0.14mol L^{-1} HNO_3. Additions of analyte were made to the sample mixed with an equal volume of 0.14 mol L^{-1} HNO_3. The detection limit (DL) was calculated from 12 blank readings for each type of tube (metal or ceramic).

3. Results and discussion

3.1. Optimization of carrier flow rate and sample volume

Good peak reproducibility was achieved when samples were injected into the air flow as carrier. When samples were injected into 0.14 mol L^{-1} HNO_3, used as the carrier, there was a rise in the baseline (as expected, due to increase of the blank), followed by a fall due to cooling of the metal or ceramic tubes. This cooling was significant, since no transient signals were obtained following injection of standards, indicating that the temperature within the tubes was insufficient to atomize the analyte, which remained dispersed in the carrier solution. This confirmed the findings of earlier work that the use of air (or other gas) as the carrier avoids dilution and dispersion of the sample. Here, all analyses were performed using air as the carrier, not only because it was less expensive than use of a solution, and minimized waste generation, but also because it enabled the TS-FF-AAS system to be used to determine copper, which would not have been possible using a solution as the carrier.

Figure 2 shows the influence of the carrier (air) flow rate, in the range 9.0-18.0 mL min^{-1}, on the absorbance values obtained using 50 μL of a standard of 200 μg Cu L^{-1}in 0.14 mol L^{-1} HNO_3, using both tubes.In the case of the metal tube, lower absorbance values were obtained at low flow rates, because the sample arrived slowly at the atomizer,increasing the measurement duration and resulting in an unpredictable and erratic vaporization. Hence, as the flow rate was increased, the absorbance also increased due to a more homogeneous vaporization of the sample [23,27,58].

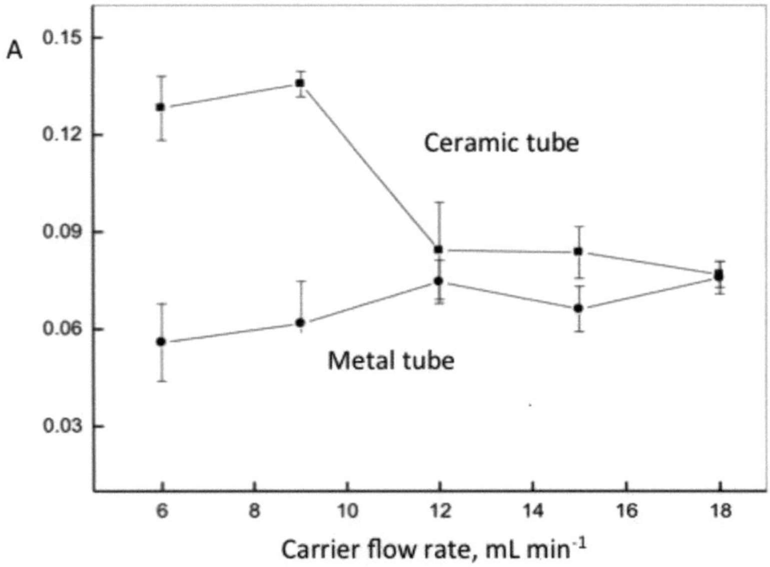

Figure 2. Influence of carrier (air) flow rate on the absorbance obtained for 50 μLof a solution of200 μg Cu L^{-1} in 0.14mol $L^{-1}HNO_3$, using the metal and ceramic tubes.

This increase proceeded up to a carrier flow rate of 12.0 mL min^{-1}, above which there was no significant variation in absorbance. The highest absorbance value was obtained at a flow rate of 18.0 mL min^{-1}, which was therefore selected as the best flow rate to use with the metal tube.

When the ceramic tube was used, maximum absorbance was achieved at a carrier flow rate of 9.0 mL min^{-1}. At higher flow rates, the residence time of the liquid in the heated section of the ceramic capillary was considerably diminished, reducing the time available for evaporation of the liquid, so that the sample was not delivered in the form of vapor/aerosol, but rather as a flow of liquid. The temperature within the tube decreased, and the color of the tube changed from ruby-red to opaque grey. It was also possible to see droplets emerging from the atomizer tube. Hence, the absorbance values did not increase, while greater variability in the signal resulted in elevated standard deviation values. A flow rate of 3.0 mL min^{-1}was selected, at which the absorbance signal was maximized, and the standard deviation was minimized.

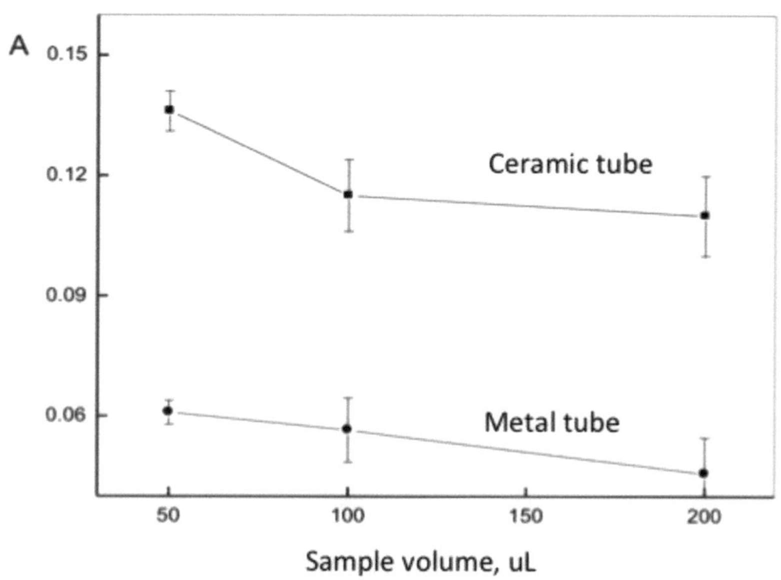

Figure 3. Influence of sample volume on the absorbances obtained for a solution of 200 µg CuL^{-1}in0.14mol L^{-1}HNO_3, using carrier flow rates of 9.0 and 18.0 mL min^{-1}for the ceramic and metal tubes, respectively.

The sample volume was varied between 50 and 200 µL, using carrier flow rates of 18.0 and 9.0 mL min^{-1} for the metal and ceramic tubes, respectively. The results (Figure 3) revealed that for both tubes a sample volume of 50 µL generated the highest absorbance value, with a low standard deviation, reflecting good repeatability in the experimental measurements. When 100 µL of sample was used, there was a slight cooling of the ceramic capillary, and consequently of the atomization tubes, while there was no increase in the absorbance values. At a sample volume of 200 µL, the ceramic capillary and the tube were substantially cooled, and there was no homogeneous thermospray formation, with erratic generation of droplets that acted to disperse the light radiation (probably to a large degree, since the deuterium lamp was unable to fully correct the resulting background signal). The unpredictable atomization resulted in very high standard deviation values. Using air as the carrier, a sample volume of 50 μL was selected for the subsequent measurements, due to greater atomization homogeneity, satisfactory absorbance for a 30 mg Cu L^{-1} standard, and a low SD value.

3.2. Construction of analytical curves

Figure 4 illustrates the results obtained for the analytical curve in the concentration range 0.1-0.4 μg Cu L^{-1} in 0.14 mol $L^{-1}HNO_3$, using the optimized conditions of the TS-FF-AAS system. The transient signals were repeatable, and (for both tubes) the curve was linear in the concentration range studied. A two-fold greater sensitivity was achieved using the ceramic tube.

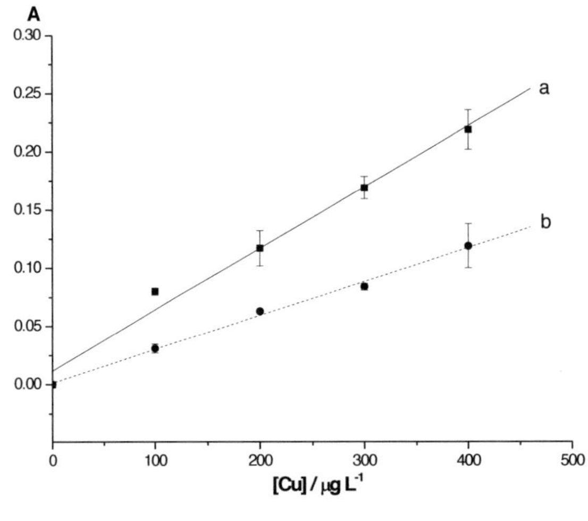

Figure 4. Regression lines fitted to the analytical curves of Cu obtained using the ceramic tube (a) and the metal tube (b) Equations of the lines: $A = 1.16x10^{-2} + 5.27x10^{-4}(Cu)$ (ceramic tube);$A = 1.20x10^{-3} + 2.91x10^{-4}(Cu)$ (metal tube).

Figure 5 illustrates the results obtained for the analytical curves constructed using concentrations of Cu in the range 100-400 μg L^{-1}, with additions of analytein 0.14 mol L^{-1} HNO_3 to equal volumes of sample, under the optimized TS-FF-AAS system conditions. The presence of 75.8 μg Cu L^{-1} in the sample was calculated from curve (a), obtained using the ceramic tube. This value was slightly above the detection limit (Table 1), although below the concentration of the first point of the analytical curve. In the case of the metal tube (curve (b)), a Cu concentration of 80.0 μg L^{-1} was below the detection limit for this tube, but was nevertheless in agreement with the result obtained for the ceramic tube.

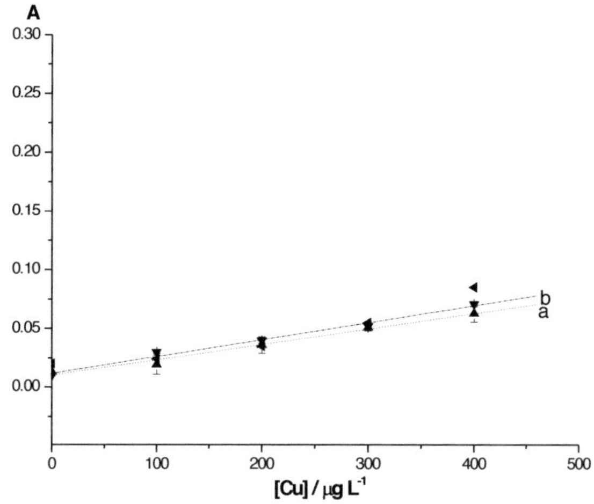

Figure 5. Regression lines fitted to the analytical curves of Cu in 1:1 mixtures of fuel samples and standards prepared in 0.14 mol L^{-1} HNO_3, obtained using the ceramic tube (a) and the metal tube (b) Equations of the lines: $A = 1.00x10^{-2} + 1.32x10^{-4}$ (Cu) (ceramic tube); $A = 1.16x10^{-2} + 1.45x10^{-4}$ (Cu) (metal tube).

	Analytical characteristics	**Ceramic tube**	**Metal Tube**
HNO_3	Detection limit, DL (µg L^{-1})	55.6	56.0
	Characteristic concentration, C_o (µg L^{-1})	8.35	15.1
	Analytical curve interval (µg L^{-1})	100 – 400	
	Correlation coefficient (r)	0.9930	0.9978
	Analytical frequency (h^{-1})	26	100
HEAF	Detection limit, DL (µg L^{-1})	64.5	128
	Characteristic concentration, C_o (µg L^{-1})	33.3	30.3
	Analytical curve interval (µg L^{-1})	100 – 400	
	Correlation coefficient (r)	0.9918	0.9927
	Analytical frequency (h^{-1})	53	82

Table 2. Analytical characteristics for determination of Cu using the TS-FF-AAS system with ceramic and metal tubes.

The analytical parameters obtained for the determination of Cu under the optimized conditions of the TS-FF-AAS system are provided in Table 2. The analytical curves were linear for a concentration range of 100-400 µg Cu L^{-1} in 0.14 mol L^{-1} HNO_3. The system could be considered to be sensitive, with characteristic concentrations of 8 and 15 µg Cu L^{-1} for the ceramic and metal tubes, respectively, and analysis frequencies (using HNO_3 medium) of 26 and 100 determinations per hour, respectively. Better analytical performance of the system was achieved using the ceramic tube, compared to the metal tube. The data showed that the TS-FF-AAS technique was more sensitive than FAAS, with nine-fold (ceramic tube) and five-fold (metal tube) increases in sensitivity, relative to FAAS with pneumatic nebulization, for which the characteristic concentration was 77 $\mu g\ L^{-1}$. The increase in power of detectionobtained using the ceramic tube was around twice that for the metal tube. The sensitivity for determination of copper using the ceramic tube was therefore two-fold that obtained using the metal tube.

4. Conclusions

The TS-FF-AAS system can be used to determine copper at low concentrations, using either metal (Inconel) or ceramic (Al_2O_3) tubes as atomizers. Following optimization considering the most important experimental variables affecting atomization, these systems provided significantly improved detection limits for Cu determination, with nine-fold (ceramic tube) and five-fold (metal tube) increases in sensitivity, compared to traditional FAAS with pneumatic nebulization. The TS-FF-AAS technique is simple, fast, effective, and inexpensive. It requires low volumes of sample (as little as 50 μL) and reagents, and reduces waste generation. The method offers a useful new alternative for the determination of copper in alcohol.

Acknowledgments

The authors thank UFOP and CNPq for financial assistance.

Author details

Fabiana Aparecida Lobo[1], Fernanda Pollo[2], Ana Cristina Villafranca[2] and Mercedes de Moraes[2]

1 UFOP - Universidade Federal de Ouro Preto, Brazil

2 UNESP – Universidade Estadual Paulista, Brazil

References

[1] Lobo, FA, Gouveia, D, Rosa, AH, Oliveira, AP, Pereira-Filho, ER, Fraceto, LF, Dias-Filho, N L.Comparison of the univariate and multivariate methods in the optimization of experimental conditions for determining Pb, Cu, Ni and Cd in biodiesel by GFAAS. Fuel, 2009;88:1907-14.

[2] Saint'pierre T, Aucélio R Q, Curtius A J. Trace Elemental determination in alcohol automotive fuel by electrothermal atomic absorption spectrometry.Microchemical Journal 2003;75: 59-97.

[3] Oliveira, AP, Gomes Neto, JA, Moraes, M, Lima, EC. Direct determination of Al, As, Cu, Fe, Mn and Ni in fuel ethanol by simultaneous GFAAS using integrated platforms pretreated with W-Rh permanent modifier together with Pd+Mg modifier.AtomicSpectroscopy 2002;23:190-5.

[4] Oliveira, AP, Gomes Neto, JA., Moraes, M, Lima, EC. Simultaneous determination of al, As, Cu, Fe, Mn and Ni in Fuel Ethanol by GFAAS.AtomicSpectroscopy 2002;23: 39-43.

[5] Oliveira, AP, Okumura, LL, Gomes Neto, JA, Moraes M. Avaliação do método das adições de analitopara a determinação de sódioemálcoolcombustívelporespectrometria de emissãoatômicaemchama.Eclética Química 2002;27:285-91.

[6] Bettinelli, M, Spezia, S, Baroni, U, Bizzarri, G. The use of reference materials in the fossil fuels quality control. Mikrochim. Acta 1996;123:217-30.

[7] Iantcheva, M. Atomicabsorptionspectrophotometrydeterminationofaluminiumandvanadiuminnavyoilsandrelatedrawmaterials. Anal. Lab 1996;5:31-5

[8] Liang, L, Horvat, M, Danilchik, P. A novel analytical method for determination of picogram levels of total mercury in gasoline and other petroleum based products. Sci. Total Envir. 1996;187: 57-64

[9] Misselwitz, R, Sengutta, U. Determination of elements in oily matrices. Laborpraxis 1997;21:41-8

[10] Kowalewska Z, Bulska E, Hulanicki A. Organic palladium and palladium-magnesium chemical modifiers in direct determination of lead in fractions from distillation of crude oil by electrothermal atomic absorption analysis. Spectrochimica. Acta, Part B 1999;54:835-43

[11] Zięba-Palus J, Kościelniak P. An analysis of the similarity of motor oils on the basis of their elemental composition. Forensic Science Internationalt 2000;112:81-90.

[12] Meeravali NN, Kumar, SJ. The utility of a W-Ir permanent chemical modifier for the determination of Ni and V in emulsified fuel oils and naphta by transverse heated electrothermal atomic absoption spectrometer. Journal. Analytica. Atomic. Spectrometry 2001;16:527-32.

[13] Sant'pierre TD, Dias, LF, Pozebon D, Aucélio RQ, Curtius AJ, WelzB.Determination of Cu, Mn, Ni and Sn in gasoline by electrothermal vaporization inductively coupled plasma mass spectrometry, and emulsion sample introduction. Spectrochim. Acta Part B 2002;57:1991-2001.

[14] Giacomelli MBO, da Silva JBB, Sain't Pierre TD, Curtius A. Use of iridium plus rhodium as permanent modifier to determine As, Cd and Pb in acids and ethanol by electrothermal atomic absorption spectrometry. Microchemical Journal 2004;77: 151-6

[15] Vale MGR, Damin ICF, Klassen A, Silva MM, Welz B, Silva AF, Lepri FG. Method development for the determination of nickel in petroleum using line-source and hight-resolution continuum-source graphite furnace atomic absorption spectrometry. Microchemical J. 2004;77:131-40

[16] Villafranca AC. Avaliação da espectrometria de absorçãoatômicacomnebulizaçãotérmicaemtuboaquecidoemchama (TS-FF-AAS) paradeterminação de Cd, Cu, Pb e Znemálcoolcombustível e óleodiesel. 2004. 144f. Tese (DoutoradoemQuímica) – Instituto de Química, UniversidadeEstadualPaulista, Araraquara.

[17] Lepri FG, Welz B, Borges DLG, Silva AF, Vale MGR, Heitmann U. Speciation analysis of volatile and non-volatile vanadium compounds in Brazilian crude oils using high-resolution continuum source graphite furnace atomic absorption spectrometry. AnalyticaChimicaActa2006;558:195-200.

[18] Korn MGA, do Santos, DSS, Welz, B, Vale MGR, Teixeira AP, Lima DC, Ferreira SLC. Atomic spectrometric methods for the determination of metals and metalloids in automotive fuels - A review. Talanta 2007;73:1-11

[19] Amorim Filho VR, Caldas NM, Raposo Junior JL, Flumignan DL, de Olivreira JE, Gomes Neto JA. Simultaneous Determination of Ba, Cr, Mo (Group 1), and Cu, Fe, Ni, and Pb (Group 2) in Commercial Fuel Ethanol by Graphite Furnace AAS.Atomic Spectroscopy, 2008;29:230-35.

[20] Alves VN, Mosquetta R, Melo Coelho NM, Bianchin JN, DiPietroRoux KC, Martendal E, Carasek E. Determination of cadmium in alcohol fuel using Moringaoleifera seeds as a biosorbent in an on-line system coupled to FAAS. Talanta 2010;80:1133-8.

[21] [21]Lobo FA, Gouveia D, Oliveira AP, Romão LP, Fraceto LF, Dias Filho NL, Rosa A H. Development of a method to determine Ni and Cd in biodiesel by graphite furnace atomic absorption spectrometry.Fuel 2011;90:142–6.

[22] Silva CS, Oreste EQ, Nunes AM, Vieira MA, RibeiroAS.Determination of mercury in ethanol biofuel by photochemical vapor generation. J. Anal. Atomic Spectrom. 2012;27:689-94.

[23] Gáspar A, Berndt H. Thermospray flame furnace atomic absorption spectrometry (TS-FF-AAS) — a simple method for trace element determination with microsamples in the μg/l concentration range.SpectrochimicaActaPart B: AtomicSpectroscopy2000.;55:587-97

[24] Gáspar A, Sézales E, Berndt H. Analysis of submicroliter samples using thermospray flame furnace atomic absorption spectroscopy.Anal. andBioanal. Chem. 2002;372: : 136-40

[25] Zeng C, Hu, Y, Luo J. Ionic liquid-based hollow fiber supported liquid membrane extraction combined with thermospray flame furnace AAS for the determination of cadmium. MicrochimicaActa 2012;177:53-8

[26] Miranda K, Dionísio AGG, Pessoa Neto OD, Pereira-Filho ER. Determination of Cd levels in smoke condensate of Brazilian and Paraguayan cigarettes by Thermospray Flame Furnace Atomic Absorption Spectrometry (TS-FF-AAS). Microchem. J. 2012;100:27-30

[27] Tarley CRT, Arruda MAZ. A sensitive method for cadmium determination using an on-line polyurethane foam precooncentratio system and thermospray flame furnace atomic absorption Spectroscopy.Analytical Sciences 2004;20:961-6.

[28] [28]Zeng C, Yang F, Zhou N. Hollow fiber supported liquid membrane extraction coupled with thermospray flame furnace atomic absorption spectrometry for the speciation of Sb(III) and Sb(V) in environmental and biological samples. Microchem. J. 2011;98:307-11

[29] Rezende HC, Nascentes CC, Coelho NMM. Cloud point extraction for determination of cadmium in soft drinks by thermospray flame furnace atomic absorption spectrometry. Microchem. J. 2011;97:118-21

[30] Sepulveda EG, Fuentes AA, Hinojosa JN. On-line preconcentration system coupled to thermospray flame furnace-AAS for cadmium determination at sub mu g L-1 levels in seawater. Química Nova 2011;34:700-703.

[31] Bezerra MA, Lemos VA, Garcia JS, da Silva, DG, Araujo AS, Arruda MAZ. Thermospray generation directly into a flame furnace-An alternative to improve the detection power in atomic absorption spectrometry.Talanta, . 2010;82:437-443

[32] Miranda K, Pereira-Filho ER. Potentialities of thermospray fame furnace atomic absorption spectrometry (TS-FF-AAS) in the fast sequential determination of Cd, Cu, Pb and Zn. Analytical Methods 2009;1: 215-219

[33] Gomes MS, Pereira-Filho ER. Ti and Ni tubes combined in thermospray flame furnace atomic absorption spectrometry (TS-FF-AAS) for the determination of copper in biological samples. Microchemical Journal 2009;93:93-8

[34] Da-Col JA, Domene SMA, Pereira-Filho ER. Fast Determination of Cd, Fe, Pb, and Zn in Food using AAS. Food Anal Methods 2009;2:110-115

[35] Wu P, He S, Luo B, Hou X. Flame Furnace Atomic Absorption Spectrometry: A Review. Applied Spectroscopy Reviews 2009;44:411-37

[36] Andrade FP, Nascentes CC, Costa LM, Pereira-Filho ER. Cadmium and Lead Cloud Point Preconcentration and Determination in Tobacco Samples by Thermospray Flame Furnace Atomic Absorption Spectrometry.J. Brazilian Chem. Soc. 2009;20:1460-6

[37] Brancalion ML, Sabadini E, Arruda MAZ. Thermospray nebulization for flame furnace atomic absorption spectrometry - Correlations between spray formation and

cadmium analytical sensitivity. SpectrochimicaActaPart B-AtomicSpectroscopy, . 2009;64:89-94

[38] Lemos VA, Bezerra MA, Amorim FAC. On-line preconcentration using a resin functionalized with 3,4-dihydroxybenzoic acid for the determination of trace elements in biological samples by thermospray flame furnace atomic absorption spectrometry. Journal of HarzadousMaterials 2008;157:613-9

[39] Schiavo D, Nobrega JA. Interferences in Thermospray Flame Furnace AAS: Co and Mn Behavior. Spectroscopy Letters 2008;41:354-60

[40] Lobo FA, Villafranc, AC, de Oliveira AP, de Moraes M. Evaluation and Application of TS-FF-AAS System Using Ceramic Tube as Atomization Cell for Tin Determination in Canned Foods. Atomic Spectroscopy 2008;29:186-92

[41] Matos GD, Arruda MAZ. Improvements in cobalt determination by thermospray flame furnace atomic absorption spectrometry using an on-line derivatization strategy. Talanta 2008;76:475-8

[42] Coelho LM, Bezerra MA, Arruda MAZ, Bruns RE, Ferreira SLC. Determination of Cd, Cu, and Pb after cloud point extraction using multielemental sequential determination by thermospray flame furnace atomic absorption spectrometry (TS-FF-AAS).Separation Science and Technology 2008;43:815-27

[43] Wu P, Gao Y, Cheng G, Yang W, Lv Y, Hou X. Selective determination of trace amounts of silver in complicated matrices by displacement-cloud point extraction coupled with thermospray flame furnace atomic absorption spectrometry. Journal of Analytical Atomic Spectrometry 2008;23:752-7

[44] Schiavo D, Neira JY, Nobrega JA. Direct determination of Cd, Cu and Pb in wines and grape juices by thermospray flame furnace atomic absorption spectrometry. Talanta 2008;76:1113-8

[45] Lobo FA, Oliveira AP, Villafranca AC, Moraes M. Comparison of metallic and ceramic tubes as atomization cells for tin determination by TS-FF-AAS. Atomic Spectroscopy 2007;28:17-23

[46] Rosini F, Nascentes CC, Neira JY, Nobrega JA. Evaluation of selenium behavior in thermospray flame furnace atomic absorption spectrometry.Talanta2007;73:845-9

[47] Brancalion ML, Sabadini E, Arruda MAZ. Description of the thermospray formed at low flow rate in thermospray flame furnace atomic absorption Spectrometry based on high-speed images. Analytical Chemistry 2007;79:6527-33

[48] Amorim FAC, Bezerra MA. Online preconcentration system for determining ultratrace amounts of Cd in vegetal samples using thermospray flame furnace atomic absorption spectrometry.MicrochimicaActa 2007;159:183-9

[49] Petrucelli GA, Poppi RJ, Mincato RL, Pereira-Filho ER. TS-FF-AAS and multivariate calibration: A proposition for sewage sludge slurry sample analyses. Talanta 2007;71:620-6

[50] Wu P, Zhang Y, Lv Y, Hou X. Cloud point extraction-thermospray spectrometry for determination of flame quartz furnace atomic absorption ultratrace cadmium in water and urine. SpectrochimicaActa Part B Atomic Spectroscopy 2006;61:1310-4

[51] Petrucelli GA, Stocco, PK, Bueno MIMS, Pereira-Filho ER. Tube atomizers in thermospray flame furnace atomic absorption spectrometry: characterization using X-ray fluorescence, scanning electron microscopy and chemometrics. Journal of Analytical Atomic Spectrometry 2006;21:1298-1304

[52] Lobo FA, Villafranca AA, Oliveira AP, Moraes M. Sistema TS-FF-AAS comchamaacetileno-arcomoalternativaemrelação à chamaacetileno-óxidonitrosoem FAAS nadeterminação de estanho. Eclética Química 2005;30:61-68.

[53] González E, Ramón A, Medina V, Neira J, Urcesino G. Espectrofotometria de Absorción Atómica con Tubo en La Llama: Aplicación en la Determinación Total de Cadmio, Plomo y Zinc en Aguas Frescas de Mar y Sedimentos Marino. Química Nova 2004; 27:873-7

[54] Donati GL, Nascentes CC, Nogueira ARA, Arruda MAZ, Nobrega JS. Acid extraction and cloud point preconcentration as sample preparation strategies for cobalt determination in biological materials by thermospray flame furnace atomic absorption spectrometry. Microchemical Journal 2006;82:189-95

[55] Pereira MG, Pereira-Filho ER, Berndt H, Arruda, MAZ. Determination of cadmium and lead at low levels by using preconcentration at fullerene coupled to thermospray flame furnace atomic absorption spectroscopy.SpectrochimicaActaPart B Atomic-Spectroscopy 2004;59:515-21

[56] Nascentes C, Arruda MAZ, Nogueira ARA, Nobrega JA. Direct determination of Cu and Zn in fruit juices and bovine milk by thermospray flame furnace atomic absorption spectroscopy. Talanta 2004;64:912-7.

[57] Davies J, Berndt H. Improvements in thermospray flame furnace atomic absorption spectroscopy. AnalyticaChimicaActa 2003;479:215-23

[58] Pereira-Filho ER, Berndt H, Arruda MAZ. Simultaneous sample digestion and determination of Cd, Cu and Pb in biological samples using thermospray flame furnace atomic absorption spectroscopy (TS-FF-AAS) with slurry sample introduction. Journal of Analytical Atomic Spectrometry 2002;17:1308-15

Chapter 5

Gas Fermentation for Commercial Biofuels Production

Fung Min Liew, Michael Köpke and
Séan Dennis Simpson

Additional information is available at the end of the chapter

http://dx.doi.org/10.5772/52164

1. Introduction

With diminishing global reserves of crude oil and increasing demand, especially from developing countries, the pressure on oil supply will grow. Although the 2007-2010 financial crisis brought down the price of crude oil (per barrel) from a record peak of US $145 in July 2008, factors such as recovering global economies and political instability in the Middle East have restored the price of crude oil to the US$100 mark. At current rate of consumption, the global reserves of petroleum are predicted to be exhausted within 50 years [1, 2]. This, coupled with the deleterious environmental impacts that result from accumulating atmospheric CO_2 from the burning of fossil fuels, the development of affordable, and environmentally sustainable fuels is urgently required. Many countries have responded to this challenge by legislating mandates and introducing policies to stimulate research and development (R&D) and commercialization of technologies that allow the production of low cost, low fossil carbon emitting fuels. For instance, the European Union (EU) has mandated member countries to a target of deriving 10% of all transportation fuel from renewable sources by 2020 [3]. Between 2005 and 2010, renewable energies such as solar, wind, and biofuels have been increasing at an average annual rate of 15-50% [4]. Renewable energy accounted for an estimated 16% of global final energy consumption in 2009 [4].

Biofuels have been defined as solid (bio-char), liquid (bioethanol, biobutanol, and biodiesel) and gaseous (biogas, biosyngas, and biohydrogen) fuels that are mainly derived from biomass [5]. Liquid biofuels provided a small but growing contribution towards worldwide fuel usage, accounting for 2.7% of global road transport fuels in 2010 [4]. The world's largest producer of biofuels is the United States (US), followed by Brazil and the EU [4]. In 2009, US and Brazil accounted for approximately 85% of global bioethanol production while Europe generated about 85% of the world's biodiesel [6]. The global market for liquid biofuels (bioethanol and biodiesel) increased dramatically in recent years, reaching US$83 billion in 2011 and is projected to US$139 billion by 2021 [7].

The use and production of biofuels has a long history, starting with the inventors Nikolaus August Otto and Rudolph Diesel, who already envisioned the use of biofuels such as ethanol and natural oils when developing the first Otto cycle combustion and diesel engines [6]. While fermentative production of ethanol has been used for thousands of years, mainly for brewing beer starting in Mesopotamia 5000 B.C., fermentative production of another potential biofuel butanol, has only been discovered over the last century, but had significant impact. During the World War 1, Chaim Weizmann successfully applied a process called ABE (acetone-butanol-ethanol) fermentation using *Clostridium acetobutylicum* to generate industrial scale acetone (for cordites, the propellant of cartridges and shells) from starchy materials [6, 8]. His contribution was later recognised in the Balfour declaration in 1917 and he became the first President of the newly founded State of Israel [6, 8]. Intriguingly, the enormous potential of butanol produced at that time was not realized and the substance was simply stored in huge containers [6]. ABE fermentation became the second biggest ever biotechnological process (after the ethanol fermentation process) ever performed, but the low demand of acetone following the conclusion of the war led to closure of all the plants [8]. Although ABE fermentation briefly made a comeback during the Second World War, increasing substrate costs and increasing stable supply of low cost crude oil from the Middle East rendered the technology economically unviable. Recently, a resurgence of the technology is underway as some old plants are reopened and new plants are being built or planned in China, the US, the United Kingdom (UK), Brazil, France and Austria [6, 8].

Traditionally sugar substrates derived from food crops such as sugar cane, corn (maize) and sugar beet have been the preferred feedstocks for the production of biofuels. However, world raw sugar prices have witnessed significant volatility over the last decade or so, ranging from US$216/ton in year 2000 to a 30 year high of US$795/ton in February 2011 due to global sugar deficits and crop shortfall [9]. This has created uncertainty and raised sustainability issues about its use as a feedstock for large scale biofuel production. This review aims to shed light on the use of syngas and industrial waste gas as feedstocks, and the emerging field of gas fermentation to generate not only biofuels, but also other high-value added products. The advantages of gas fermentation over conventional sugar-based fermentation and thermochemical conversions, and their flexibility in utilizing a spectrum of feedstocks to generate syngas will be discussed. The biochemistry, genetic and energetic background of the microorganisms that perform this bioconversion process will be critically examined, together with recent advances in systems biology and synthetic biology that offer growing opportunities to improve biocatalysts in terms of both the potential products that can be produced and their process performance. The key processes such as gasification, bioreactor designs, media formulation, and product recovery will be analysed. Finally, the state of commercialization of gas fermentation will be highlighted and an outlook will be provided.

http://dx.doi.org/10.5772/52164

2. Advantages of gas fermentation

The production of first generation biofuels relies on food crops such as sugar beet, sugar cane, corn, wheat and cassava as substrates for bioethanol; and vegetable oils and animal fats for biodiesel. Although years of intense R&D have made methods of bioethanol production (typically using the yeast *Saccharomyces cerevisiae*) technologically mature, there remain some serious questions regarding its sustainability. The use of food crops as a source of carbohydrate feedstocks by these processes requires high-quality agricultural land. The inevitable conflict between the increasing diversion of crops or land for fuel rather than food production has been highlighted as one of the prime causes of rising global food prices. Furthermore, corn ethanol producers in the US, have historically enjoyed a 45-cent-a-gallon federal tax credit for years (which ended in early 2012), costing the government US$30.5 billion between 2005 to 2011, raising questions about its economic competitiveness with gasoline [10, 11].

These arguments have stimulated the search for so-called second generation biofuels, which utilize non-food lignocellulose biomass such as wood, dedicated energy crops, agricultural residues and municipal solid wastes as feedstocks. Biomass consists of cellulose, hemicellulose and lignin, and the latter of which is extremely resistant to degradation. One approach to unlocking the potential in this abundant feedstock is to separate the lignin from the carbohydrate fraction of the biomass via extensive pre-treatment of the lignocellulose involving, for example, steam-explosion and/or acid hydrolysis. These pre-treatments are designed to allow the carbohydrate portion of the biomass to be broken down into simple sugars, for example by enzymatic hydrolysis using exogenously added cellulases to release fermentable sugars [12]. Such approaches have been found to be expensive and rate limiting [6, 12, 13]. Alternatively, processes using cellulolytic microorganisms (such as *C. cellulolyticum*, *C. thermocellum*, and *C. phytofermentans*) to carry out both the hydrolysis of lignocelluloses and sugar fermentation in a single step, termed 'Consolidated Bioprocessing Process (CBP)' [12] have been proposed, however the development of these is still at an early stage, and again low conversion rates seem to be a major limitation that needs to be overcome.

Microorganisms such as acetogens, carboxytrophs and methanogens are able to utilize the CO_2 + H_2, and/or CO available in such syngas as their sole source of carbon and energy for growth as well as the production of biofuels and other valuable products. However, only acetogens are described to synthesize metabolic end products that have potentials as liquid transportation fuels. While biological processes are generally considered slower than chemical reactions, the use of these microbes to carry out syngas fermentation offers several key advantages over alternative thermo-chemical approaches such as the Fischer-Tropsch' process (FTP). First, microbial processes operate at ambient temperatures and low pressures which offer significant energy and cost savings. Second, the ambient conditions and irreversible nature of biological reactions also avoid thermodynamic equilibrium relationships and allow near complete conversion efficiencies [14, 15]. Third, biological conversions are commonly more specific due to high enzymatic specificities, resulting in higher product yield with the formation of fewer by-products. Fourth, unlike traditional chemical catalysts which require a set feed gas composition to yield desired product ratios or suite, microbial processes have freedom to operate for the production ofthe same suite of products across a wider range of $CO:H_2$ ratios in the feed gas [16]. Fifth, biocatalysts exhibit a much higher tolerance to poisoning by tars, sulphur and chlorine than inorganic catalysts [6, 16]. However, some challenges have been identified for syngas fermentation to be commercialized, including gas mass transfer limitations, long retention times due to slow cell growth, and lower alcohol production rates and broth concentrations. Recent progress and development to remedy these issues will be highlighted in this review.

3. Feedstock and gasification

Due to the flexibility of the microbes to ferment syngas with diverse composition, virtually any carbonaceous materials can be used as feedstock for gasification. Non-food biomass that can be employed as feedstock for gasification includes agricultural wastes, dedicated energy crops, forest residues, and municipal organic wastes, or even glycerol and feathers [16-20]. Biomass is available on a renewable basis, either through natural processes or anthropogenic activities (e.g. organic wastes). It has been estimated that out of a global energy potential from modern biomass of 250 EJ per year in 2005, only 9 EJ (3.6%) was used for energy generation [18]. The use of existing waste streams such as municipal organic waste also differentiate itself from other feedstocks such as dedicated energy crops because these wastes are available today at economically attractive prices, and they are often already aggregated and require less indirect land use. Alternatively, gasification of non-biomass sources such as coal, cokes, oil shale, tar sands, sewage sludge and heavy residues from oil refining, as well as reformed natural gas are commonly applied as feedstocks for the FTP and can also be used for syngas fermentation [15, 21]. Furthermore, some industries such as steel manufacturing, oil refining and chemical production generate large volume of CO and/or CO_2 rich gas streams as wastes. Tapping into these sources using microbial fermentation process essentially convert existing toxic waste gas streams into valuable commodities such as biofuels. The overall process of gas fermentation is outlined in Figure 1.

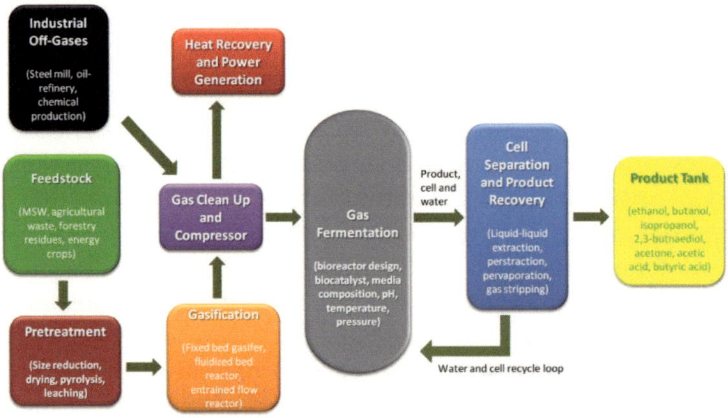

Figure 1. Overview of gas fermentation process

Prior to gasification, biomass generally needs to go through a pre-treatment process encompassing drying, size reduction (e.g. chipping, grinding and chopping), pyrolysis, fractionation and leaching depending on the gasifier configuration [22, 23]. This upstream pre-treatment process can incur significant capital expense and add to the overall biomass feedstock cost, ranging from US$16-70 per dry ton [22]. Gasification is a thermo-chemical process that converts carbonaceous materials to gaseous intermediates at elevated temperature (600-1000°C), in the presence of an oxidizing agent such as air, steam or oxygen [16, 22]. The resulting syngas contains mainly CO, CO_2, H_2 and N_2, with varying amounts of CH_4, water vapour and trace amount of impurities such as H_2S, COS, NH_3, HCl, HCN, NO_x, phenol, light hydrocarbons and tar [17, 22, 24]. The composition and amount of impurities of syngas depends on the feedstock properties (e.g. moisture, dust and particle size), gasifier type and operational conditions (e.g. temperature, pressure, and oxidant) [17, 22]. Table 1 summarizes typical composition of syngas and other potential gas streams derived from various sources.

4. Fixed bed gasifier

Depending on the direction of the flows of carbonaceous fuel and oxidant (air or steam), fixed bed gasifier can be further categorized into updraft or downdraft reactor. In the updraft (counter-current) version of the fixed bed gasifier, biomass enters from the top while gasifying agent from the bottom. The biomass moves down the reactor through zones of drying (100°C), pyrolysis (300°C), gasification (900°C) and finally oxidation zone (1400°C) [18]. Although this mode of gasifier is often associated with high tar content in the exit gas, recent advances in tar cracking demonstrated that very low tar level is achievable [31]. The direct heat exchange of the oxidizing agent with the entering fuel feed results in low gas exit temperature and hence high thermal efficiency [18, 23]. The downdraft (co-current) gasifier has very similar design as the updraft reactor, except the carbonaceous fuel and oxidizing agent flow in the same direction. In comparison to the updraft gasifier, the downdraft reactor has lower tar content in the exit gas but exhibit lower thermal efficiency [23]. Due to the size limitation in the constriction (where most of the gasification occurs) of the reactor, this mode of gasifier is considered unsuitable for large scale operation [18].

5. Fluidized bed reactor

In fluidized bed reactor, the carbonaceous fuel is mixed together with inert bed material (e.g. silica sand) by forcing fluidization medium (e.g. air and/or steam) through the reactor. The inert bed facilitates better heat exchange between the fuel materials, resulting in nearly isothermal operation conditions and high feedstock conversion efficiencies [18, 22]. The maximum operating temperature of the gasifier is typically around 800 - 900°C, which is limited by the melting point of the bed material [18]. Furthermore, the geometry of the reactor and excellent mixing properties also means that fluidized bed reactors are suitable for up-scaling [18, 22]. Due to these properties, fluidized bed reactor is currently the most commonly used gasifier for biomass feedstock [32]. However, this mode of gasifier is not suitable for feedstocks with high levels of ash and alkali metals because the melting of these components causes stickiness and formation of bigger lumps, which ultimately negatively affect the hydrodynamics of the reactor [18].

	Composition vol%, dry basis						Ref.
	CO	CO_2	H_2	N_2	CH_4	Other	
Non-biomass source							
Coal gasification	59.4	10	29.4	0.6	0	0.6	[25]
Coke oven gas	5.6	1.4	55.4	4.3	28.4	4.9	[25]
Partial oxidation of heavy fuel oil	47	4.3	46	1.4	0.3	1	[26]
Hardwood chips + 20 wt%liquid crude glycerol	19.73	11.67	19.38	NR*	3.82	NR*	[19]
Steam reforming of natural gas	15.5	8.1	75.7	0.2	0.5	0	[25]
Steam reforming of Naphtha	6.7	15.8	65.9	2.6	6.3	2.7	[25]
Water gas	30	3.4	31.7	13.1	12.2	9.6	[25]
Steel Mill	44	22	2	32	0	0	[27]
Biomass and organic waste source							
Demolition wood + sewage sludge	10.53	15.02	8.02	60.46	3.19	2.78	[28]
Cacao shell	8	16.02	9.02	61.45	2.34	3.17	[28]
Dairy biomass	8.7	15.7	18.6	56	0.6	0.4	[29]
Switchgrass	14.7	16.5	4.4	56.8	4.2	3.4	[13]
Kentucky bluegrass straw	12.9	17.4	2.6	64.2	2.1	0.8	[30]
Willow	9.4	17.1	7.2	60.42	3.3	2.58	[28]

Note: NR, not reported

The factors that determine which type of gasifier to employ are scale of operation, feedstock size and composition, tar yield and sensitivity towards ash [18]. Currently, three main types of gasifier are commercially employed: fixed bed, fluidized bed and entrained flow reactors [18].

Table 1. Typical composition of syngas and other potential gas streams from various sources

6. Entrained flow reactor

Entrained flow reactor is the preferred route for large scale gasification of coal, petcoke and refinery residues because of high carbon conversion efficiencies and low tar production [22]. This mode of gasifier does not require inert bed material but relies on feeding the feedstocks co-currently with oxidizing agent at high velocity to achieve a pneumatic transport regime

[18]. At operating temperature of 1200-1500°C, this method is able to convert tars and methane, resulting in better syngas quality [18]. Importantly this technology requires the feedstocks to be pulverised into fine particles of ~50 μm before feeding, which is not a major issue for coal but very difficult and costly for biomass sources [18, 22].

7. Microbes and biochemistry of gas fermentation

Acetogens are defined as obligate anaerobes that utilize the reductive acetyl-CoA pathway for the reduction of CO_2 to the acetyl moiety of acetyl-coenzyme A (CoA), for the conservation of energy, and for the assimilation of CO_2 into cell carbon [33]. In addition to the reductive acetyl-CoA pathway, four other biological pathways are known for complete autotrophic CO_2 fixation: the Calvin cycle, the reductive tricarboxylic acid (TCA) cycle, the 3-hydroxypropionate/malyl-CoA cycle and the 3-hydroxypropionate/4-hydroxybutyrate cycle [34]. Since the earlier atmosphere of earth was anoxic and the acetyl-CoA pathway is biochemically the simplest among the autotrophic pathways (the only linear pathway, whereas the other four pathways are cyclic), it has been postulated to be the first autotrophic process on earth [35, 36]. The reductive acetyl-CoA pathway is also known as the 'Wood-Ljungdahl' pathway, in recognition of the two pioneers, Lars G. Ljungdahl and Harland G. Wood, who elucidated the chemical and enzymology of the pathway using *Moorella thermoacetica* (formerly: *Clostridium thermoaceticum*) [35] or CODH/ACS pathway after the key enzyme of the pathway Carbon Monoxide dehydrogenase/Acetyl-CoA synthase. This ancient pathway is diversely distributed among at least 23 different bacterial genera: *Acetitomaculum, Acetoanaerobium, Acetobacterium, Acetohalobium, Acetonema, Alkalibaculum, "Bryantella", "Butyribacterium",Caloramator, Clostridium, Eubacterium, Holophaga, Moorella, Natroniella, Natronincola, Oxobacter, Ruminococcus, Sporomusa, Syntrophococcus, Tindallia, Thermoacetogenium, Thermoanaerobacter*, and *Treponema* [33]. A selection of mesophilic and thermophilic acetogens are presented in Table 2. Acetogens are able to utilize gases CO_2 + H_2, and/or CO to produce acetic acid and ethanol according to the following stoichiometries:

$$2\,CO_2 + 4\,H_2 \Rightarrow CH_3COOH + 2\,H_2O \qquad \Delta H = -75.3\ kJ/mol \tag{1}$$

$$2\,CO_2 + 6\,H_2 \Rightarrow C_2H_5OH + 3\,H_2O \qquad \Delta H = -97.3\ kJ/mol \tag{2}$$

$$4\,CO + 2\,H_2O \Rightarrow CH_3COOH + 2\,CO_2 \qquad \Delta H = -154.9\ kJ/mol \tag{3}$$

$$6CO + 3\,H_2O \Rightarrow C_2H_5OH + 4\,CO_2 \qquad \Delta H = -217.9\ kJ/mol \tag{4}$$

The Acetyl-CoA pathway is essentially a terminal electron-accepting process that assimilates CO_2 into biomass [35]. It constitutes an *Eastern* (or Carbonyl) branch and a *Western* (or Meth-

Species	Substrate	Product(s)	T_{opt} (°C)	pH_{opt}	Genome Status	Ref.
Mesophilic Microorganisms						
Acetobacterium woodii	H_2/CO_2, CO	Acetate	30	6.8	Available	[41, 42]
Acetonema longum	H_2/CO_2	Acetate, butyrate	30-33	7.8		[43]
Alkalibaculum bacchi	H_2/CO_2, CO	Acetate, ethanol	37	8.0-8.5		[44, 45]
Blautia producta	H_2/CO_2, CO	Acetate	37	7		[46]
Butyribacterium methylotrophicum	H_2/CO_2, CO	Acetate, ethanol, butyrate, butanol	37	6		[47-49]
Clostridium aceticum	H_2/CO_2, CO	Acetate	30	8.3	Under construction	[50-52]
Clostridium autoethanogenum	H_2/CO_2, CO	Acetate, ethanol, 2,3-butanediol, lactate	37	5.8-6.0		[27, 53]
Clostridium carboxidivorans or "P7"	H_2/CO_2, CO	Acetate, ethanol, butyrate, butanol, lactate	38	6.2	Draft	[54, 55]
Clostridium drakei	H_2/CO_2, CO	Acetate, ethanol, butyrate	25-30	5.8- 6.9		[55-57]
Clostridium formicoaceticum	CO	Acetate, formate	37	NR		[50, 58, 59]
Clostridium glycolicum	H_2/CO_2	Acetate	37-40	7.0-7.5		[60, 61]
Clostridium ljungdahlii	H_2/CO_2, CO	Acetate, ethanol, 2,3-butanediol, lactate	37	6	Available	[27, 62, 63]
Clostridium magnum	H_2/CO_2	Acetate	30-32	7.0		[64, 65]
Clostridium mayombei	H_2/CO_2	Acetate	33	7.3		[66]
Clostridium methoxybenzovorans	H_2/CO_2	Acetate, formate	37	7.4		[67]
"Clostridium ragsdalei" or "P11"	H_2/CO_2, CO	Acetate, ethanol, 2,3-butanediol, lactate	37	6.3		[68]
Clostridium scatologenes	H_2/CO_2, CO	Acetate, ethanol, butyrate	37-40	5.4-7.5		[55, 56]
Eubacterium limosum	H_2/CO_2, CO	Acetate	38-39	7.0-7.2	Available	[41, 69]
Oxobacter pfennigii	H_2/CO_2, CO	Acetate, butyrate	36-38	7.3		[70]
Thermophilic Microorganisms						
Moorella thermoacetica	H_2/CO_2, CO	Acetate	55	6.5-6.8	Available	[71-73]
Moorella thermoautotrophica	H_2/CO_2, CO	Acetate	58	6.1		[74]
Thermoanaerobacter kiuvi	H_2/CO_2	Acetate	66	6.4		[72]
Notes: NR, not reported						

Table 2. Acetogens

yl) branch (Figure 2.). The *Western* branch employs a series of enzymes to carry out a six-electron reduction of CO_2 to the methyl group of acetyl-CoA, starting from the conversion of CO_2 to formate by formate dehydrogenase. Formyl-H_4folate synthase then condenses formate with H_4folate to form 10-formyl-H_4folate, which is then converted to 5,10-methenyl-H_4folate by a cyclohydrolase. This is followed by a dehydrogenase that reducesmethenyl- to 5,10-methylene-H_4hydrofolate, before (6S)-5-CH_3-H_4folate is formed by methylene-H_4folate reductase [37]. A B12-depedent methyltransferase (MeTr) then transfer the methyl group of (6S)-5-CH_3-H_4folate to corrinoid iron-sulphur protein (CoFeSP) of the bi-functional carbon monoxide dehydrogenase/acetyl-CoA synthase (CODH/ACS) complex [37]. The bi-functional CODH/ACS enzyme complex is formed by two autonomous proteins, an $\alpha_2\beta_2$ tetramer (CODH/ACS) and a $\gamma\delta$ heterodimer (CoFeSP), and the genes are often arranged in an operon, together with MeTr [37, 38]. In the *Eastern* branch, the CODH component catalyzes the reduction of CO_2 to CO. The central molecule, acetyl-CoA, is finally generated when CO, methyl group (bound to CoFeSP) and CoASH are condensed by ACS. Given the pivotal role of CODH/ACS, it is unsurprising that this complex was found to be the most highly expressed transcripts under autotrophic conditions in *C. autoethanogenum* [27], and can represent up to 2% of the soluble cell protein of an acetogen [39]. CODH/ACS is not unique to acetogenic bacteria, as it is also present in sulphate-reducing bacteria, desulfitobacteria, and Archaea (methanogens and Archaeoglobus) [38, 40].

The reducing equivalents required for fixation of CO_2 carbon into acetyl-CoA come from the oxidation of molecular hydrogen under chemolithoautotrophic growth, or NADH and reduced ferredoxin under heterotrophic growth [75]. An extensive review by Calusinska *et al.* (2010) highlighted the diversity of ubiquitous hydrogenases that Clostridia possess although only one acetogen *C. carboxidivorans* was included in this study [76], which catalyze the reversible oxidation of hydrogen:

$$H_2 \Leftrightarrow 2\,H^+ + 2\,e^- \tag{5}$$

The direction of the hydrogenase reaction is directed by the redox potential of the components able to interact with the enzyme. Hydrogen evolution occurs when electron donor is available, whereas the presence of electron acceptor results in hydrogen oxidation [77]. Hydrogenases can be classified into three phylogenetically distinct classes of metalloenzymes: [NiFe]-, [FeFe]-, and [Fe]-hydrogenases [76]. In *Methanosarcina barkeri*, the Ech hydrogenase, a [NiFe]-hydrogenase, was demonstrated to oxidize H_2 to reduce ferredoxin [78]. During acetoclastic methanogenesis, Ech hydrogenase oxidize ferredoxin to generate H_2 [78]. Although genome analysis revealed the presence of Ech-like hydrogenase in *C. thermocellum, C. phytofermentans*, *C. papyrosolvens*, and *C. cellulolyticum*, their physiological roles remained unknown [76]. Clostridia harbour multiple distinct [FeFe]-hydrogenases, perhaps reflecting their ability to respond swiftly to changing environmental conditions [76]. The monomeric, soluble [FeFe]-hydrogenase of *C. pasteurianum* is one of the best studied. It transfer electrons from reduced ferredoxins or flavodoxins to protons, forming H_2 [79]. A trimeric [FeFe]-hydrogenase found in *C. difficile, C. beijerinckii*, and *C. carboxidivorans* were hypothesized to couple formate oxidation to reduce protons into H_2 [76]. In *Thermotoga maritima*, an electron bifurcating, trimeric [FeFe]-hydrogenase was identified, that was shown to simultaneously oxidize reduced ferredoxin and NADH to evolve hydrogen under low H_2 partial pressure [80]. Under high H_2 partial pressure, the authors hypothesized that the NADH is oxidized to produce ethanol. *In silico* analysis revealed homologs of this bifurcating hydrogenase in a few Clostridia including *C. beijerinckii* and *C. thermocellum* [80]. In addition to classical hydrogenases, CODH/ACS and pyruvate:ferredoxin oxidoreductase (PFOR) from *M. thermoacetica* were shown to have hydrogen evolving capability, possibly as a mean of disposing excess reducing equivalents when electron carriers are limited and/or CO concentration is sufficient to inhibit conventional hydrogenases [81].

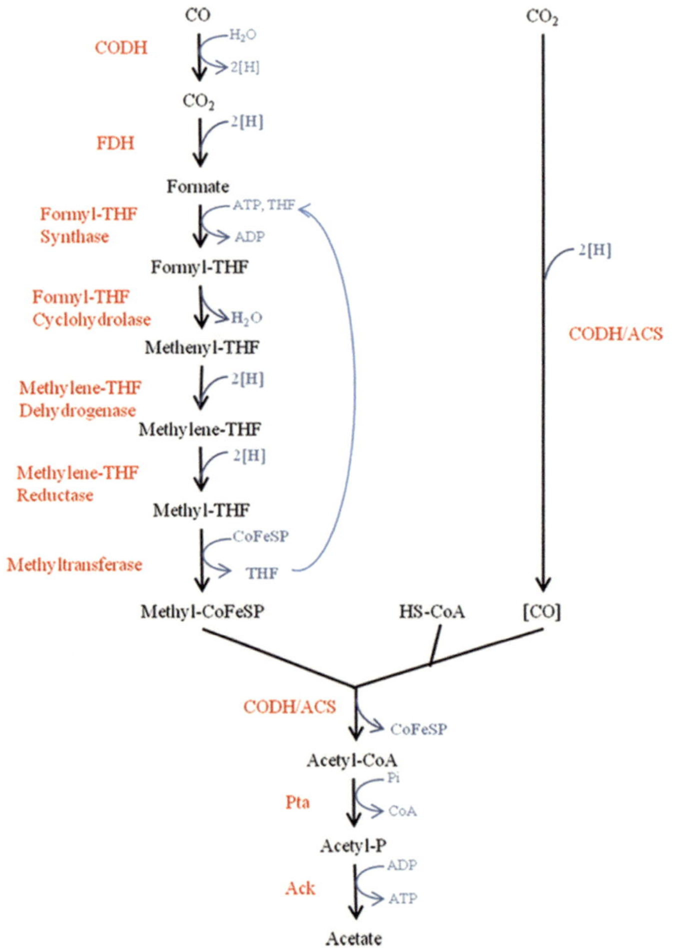

Figure 2. Wood-Ljungdahl Pathway. Ack, acetate kinase; ACS, acetyl-CoA synthase; CODH, carbon monoxide dehydrogenase; CoFeSP, corrinoid iron sulfur protein; FDH, formate dehydrogenase; Pta, phosphotransacetylase; THF, tetrahydrofolate.

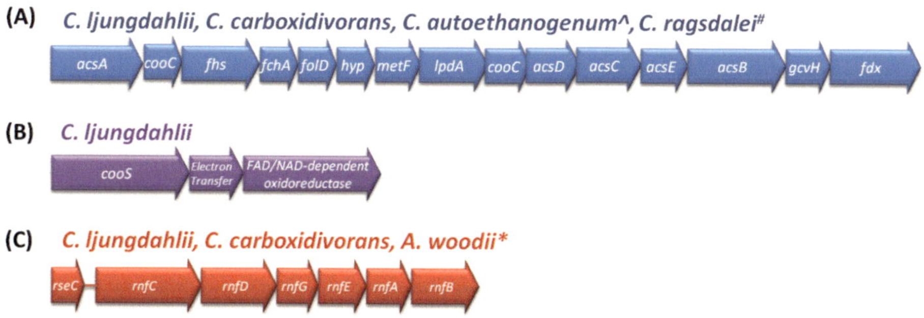

Figure 3. The organization of genes involved in acetogenesis and energy conservation from sequenced key acetogens. (A) Wood-Ljungdahl cluster; (B) carbon monoxide dehydrogenase (CODH) cluster; (C) Rnf complex cluster. *acsA*, CODH subunit; *acsB*, ACS subunit; *acsC*, corrinoid iron-sulfur protein large subunit; *acsD*, corrinoid iron-sulfur protein small subunit; *acsE*, methyltransferase subunit; *cooC*, gene for CODH accessory protein; *cooS*, CODH; *fchA*, formimidotetrahydrofolate cyclodeaminase; *fdx*, ferredoxin; *fhs*, formyl-tetrahydrofolate synthase; *folD*, bifunctional methylenetetrahydrofolate dehydrogenase/formyl-tetrahydrofolate cyclohydrolase; *gcvH*, gene for glycine cleavage system H protein; *hyp*, hypothetical protein; *lpdA*, dihydrolipoamide dehydrogenase; *metF*, methylene-tetrahydrofolate reductase; *rnfA, rnfB, rnfC, rnfD, rnfE, rnfG*, electron transport complex protein subunits; *rseC*, sigma E positive regulator. ^, truncated *acsA*. #, truncated fdx. *, lack *rseC*.

Most acetogens are also able to utilize another gas carbon monoxide (CO). In contrast to CO_2, CO can serve as both a source of carbon btut also as source of electrons such that hydrogen is not necessarily required. With a CO_2/CO reduction potential of -524 to -558mV, CO is approximately 1000-fold more capable of generating extremely low potential electrons than NADH, capable of reducing cellular electron carriers such as ferredoxin and flavodoxin [38, 82]. The reducing equivalents generated from CO oxidation can be coupled to reduction of CO_2 into acetate, butyrate and/or methane, evolution of molecular hydrogen from protons, reduction of nitrate/nitrite, reduction of sulfur species and reduction of aldehydes into alcohols [35, 83]. However, relatively few microorganisms are able to utilize CO as sole carbon and energy source, probably due to growth inhibition from sensitivity of their metalloproteins and hydrogenases towards CO [38, 83]. During exponential growth of *Pseudomonas carboxydovorans* (an aerobic carboxydotroph), it was demonstrated via immunological localization studies that 87% of the key enzyme CODH is associated with the inner cytoplasmic membrane, but this association was lost at the end of the exponential growth phase and a reduction in CO-dependent respiration rate was observed [84, 85]. It should be mentioned that aerobic and anaerobic CODH enzymes are structurally very different. CODH has been reported to be a very rapid and efficient CO oxidizer at rates between 4,000 and 40,000 s^{-1}, and reduces CO_2 at $11s^{-1}$ [86, 87]. Other electron donors commonly used by acetogens include formate, CH_3Cl, lactate, pyruvate, alcohols, betaine, carbohydrate, acetoin, oxalate and citrate [88]. CODH is able to split water in a biological water-gas shift reaction into hydrogen and electron according to the stoichiometry:

$$CO + H_2O \Rightarrow CO_2 + 2H^+ + 2e^- \qquad (6)$$

The operation of this water gas shift reaction is the biochemical basis for the tremendous flexibility that acetogens have in terms of input gas composition. Via this reaction these organisms can flexibly use CO or H_2 as a source of electrons.Recently, some acetogens such as *C. ljungdahlii*, *C. aceticum*, *M. thermoacetica*, *Sporomusa ovata*, and *S. sphaeroides* have additionally been shown to utilize electrons derived from electrodes to reduce CO_2 into organic compounds such as acetate, formate, fumarate, caffeine, and 2-oxo-butyrate [89]. Termed microbial electrosynthesis, this nascent concept offers another route for acetogens to harvest the electrons generated from sustainable sources (e.g. solar and wind) to reduce CO_2 into useful multi-carbon products such as biofuels [90].

Under chemolithoautotrophic conditions, acetogenesis must not only fix carbon but also conserve energy. Approximately 0.1 mol of ATP is required for generation of 1g of dry biomass in anaerobes [82]. Acetyl-CoA is an energy rich molecule that through the combined actions of Pta (phosphotransacetylase) and Ack (acetate kinase), one ATP can be generated via substrate level phosphorylation (SLP). However, the activation of formate to 10-formyl-H4folate in the methyl-branch of Acetyl-CoA pathway consumes one ATP so no net gain in ATP is achieved via this mechanism [35, 75]. Furthermore, the reduction of CO_2 to the carbonyl group also requires energy, estimated at one third of ATP equivalent [35]. Recent advances indicated that other modes of energy conservation such as electron transport phosphorylation (ETP) or chemiosmotic processes that are coupled to the translocation of protons or sodium ions are implicated in acetogens. Acetogens such as *M. thermoacetica* harbour membrane-associated electron transport system containing cytochrome, menaquinones, and oxidoreductases that translocate H^+ out of the cell [33]. For acetogens that lack such membranous electron transport system, such as *Acetobacterium woodii* and *C. ljungdahlii*, a membrane-bound corrinoid protein is hypothesized to facilitate extrusion of Na^+ or protons during the transfer of methyl group from methyl-H_4F to CODH/ACS [75]. However, all enzymes involved are predicted to be soluble rather than membrane bound. Recent evidence suggested coupling to an Rnf complex in *A. woodii*, and *C. ljungdahlii* (Figure 3) which acts as ferredoxin:NAD^+-oxidoreductase [62, 91-93]. The Rnf complex is also found in other Clostridia (but not in ABE model organism *C. acetobutylicum*) and bacteria, and was originally discovered in *Rhodobacter capsulatus* where it is involved in nitrogen fixation [93]. Using reduced ferredoxin (Fd^{2-}) generated from CO oxidation, carbohydrate utilization and/or hydrogenase reactions, this membrane-bound electron transfer complex is predicted to reduce NAD^+ with concomitant translocation of Na^+/ H^+. The ion gradient generated from the above processes is harvested by H^+- or Na^+- ATP synthase to generate ATP [33, 93]. The recent genome sequencing of *A. woodii* revealed that Rnf complex is likely to be the only ion-pumping enzyme active during autotrophic growth and the organism's entire catabolic metabolism is optimized to maximize the Fd^{2-}/NAD^+ ratio [42]. Recently, a third mechanism of energy conservation which involves bifurcation of electrons by hydrogenases was proposed for anaerobes [94] and demonstrated for enzymes hydrogenase (see above; [80]), butyryl-CoA dehydrogenase [94, 95], or an iron-sulfur flavoprotein Nfn [96]. A similar mechanism has also been proposed for the methylene-THF reductase of the reductive acetyl-CoA pathway, which would enable this highly exergonic reduction step ($\Delta G^{0'} = -22$ kJ/mol) to be coupled with the Rnf complex for additional energy conservation [62]. However, no experimental proof to support this hypothesis has been published to date.

In an attempt to generate an autotrophic *E. coli*, the genes encoding MeTr, the two subunits of CODH/ACS, and the two subunits of CoFeSP from *M. thermoacetica* were cloned and heterologously expressed in *E. coli* [97]. Although the MeTr was found to be active, the other subunits misassembled hence no active enzymes were found [97]. Autotrophic capability is clearly a very complex process that involves many genes other than the CODH/ACS complex and tetrahydrofolate pathway, including compatible cofactors, electron carriers, specific chaperones and energy conservation mechanisms. For instance, more than 200 genes are predicted to be involved in methanogenesis and energy conservation from CO_2 and H_2 in methanogens [98]. A recent patent application described the introduction of three Wood-Ljungdahl pathway genes encoding MeTr, CoFeSP subunit α and β from *C. difficile* into *C. acetobutylicum* [99]. The recombinant strain was shown to incorporate more CO_2 into extracellular products than wild-type [99].

8. Products of gas fermentation

Acetyl-CoA generated via the Wood-Ljungdahl pathway serves as key intermediate for synthesis of cell mass as well as products. All acetogens are described to produce acetate, in order to gain energy via SLP to compensate for the energy invested in activating formate in the *Western* branch of the reductive acetyl-CoA pathway. Acetate and ATP are formed via acetyl-phosphate through the successive actions of Pta and Ack. *pta* and *ack* are arranged in the same operon and they were reported to be constitutively expressed [100]. With CO_2 and H_2 as substrate, only acetate has been observed as major product [44], with minor amounts of ethanol produced in rare cases with *C. ljungdahlii* [101], *C. autoethanogenum* [53], or "*Moorella sp.*" [102, 103]. Using the more reduced substrate CO, production of a range of other products have been reported, such as ethanol, butanol, butyrate, 2,3-butanediol [104], and lactate (Figure 4.) [105]. From a biofuel perspective, ethanol and butanol are of particular interest. Ethanol and butanol have even been described as the main fermentation products over acetate in some acetogens under specific conditions. Ethanol producers include *C. ljungdahlii* [62, 63], *C. autoethanogenum* [53], "*C. ragsdalei*" ("*Clostridium* strain P11") [106, 107], "*Moorella sp.*" [102, 103], *Alkalibaculum bacchii* [44], *C. carboxidivorans* ("*Clostridium* strain P7") [54, 55], and *B. methylotrophicum* [49, 108]. The latter two have also been described to produce butanol.

Due to historical roles in ABE fermentation, organisms like *C. acetobutylicum*, *C. beijerinckii*, *C. saccharobutylicum*, and *C. saccharoperbutylacetonicum* have been much more extensively characterized than acetogenic Clostridia [95]. Since *C. acetobutylicum* was the first *Clostridium* to be fully sequenced [109] and it remains the most commonly used species for industrial production of solvents to date [110], it provides a model for study of solventogenesis. Although sugar- and starch-utilizing ABE Clostridia and acetogens exhibit clear distinctions in substrate utilization and thus metabolism, they share some similarities in the biochemical pathway and genetic organization of product synthesis and can be used as model for comparison. Structure of key genes and operons (except for the absence of acetone biosynthetic genes) have been found to be very similar in sequenced acetogen *C. carboxidivorans* [54], and in respect of acetate and ethanol genes to some extent also in *C. ljungdahlii* [62]. For instance, the operon structure of *pta-ack*, *ptb-buk* and the *bcs* cluster of acetogen *C. carboxidivorans* are highly similar to starch-utilizing *C. acetobutylicum* and *C. beijerinckii* [54, 109] (Figure 5). Due to these reasons, solventogenic genes from starch-utilizing Clostridia are ideal targets for heterologous expression in acetogens for improvement of product yield and expansion of product range.

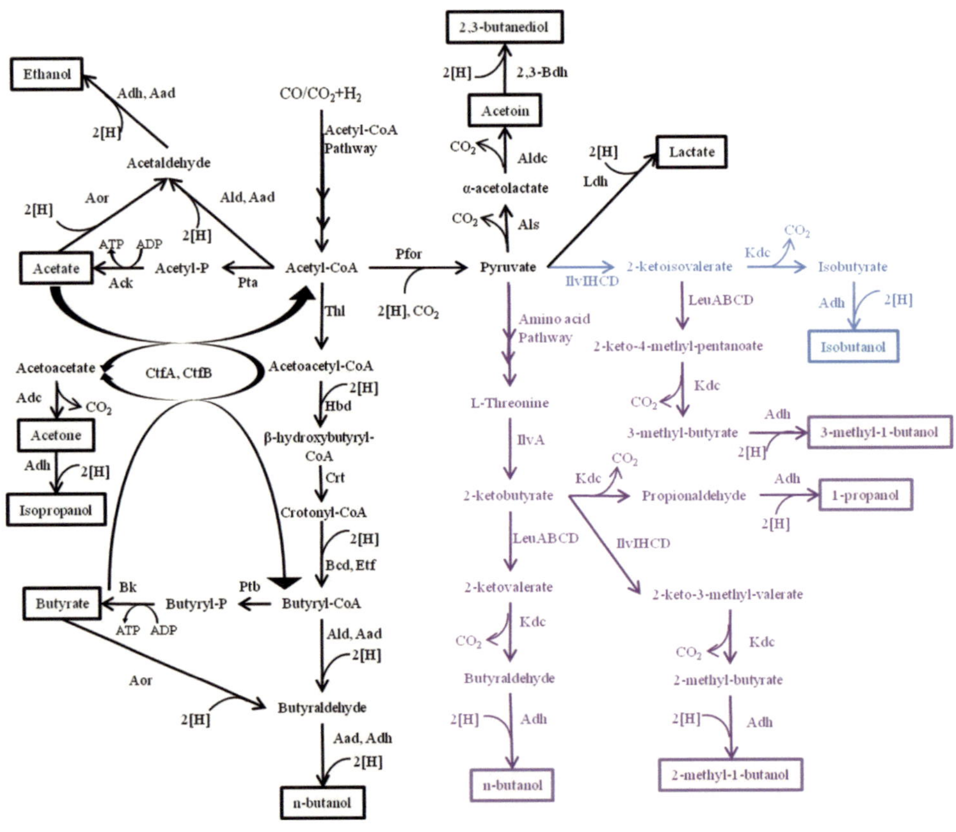

Figure 4. Scheme of metabolite production from gas fermentation using native and genetically modified Clostridia. Black denotes well-characterized pathways in Clostridia. Blue shows demonstrated heterologous pathways that have been engineered into Clostridia. Purple designates hypothetical pathways that can be engineered into Clostridia. Products are highlighted in boxes. Aad, aldehyde/alcohol dehydrogenase; Ack, acetate kinase; Adc, acetoacetate decarboxylase; Adh, alcohol dehydrogenase; Ald, aldehyde dehydrogenase; Aldc, acetolactate decarboxylase; Aor, aldehyde oxidoreductase; Bcd, butyryl-CoA dehydrogenase; Bk, butyrate kinase; Crt, crotonase; CtfA & CtfB, CoA transferase A & B; Etf, electron-transferring flavoprotein; Hbd, hydroxybutyryl-CoA dehydrogenase; IlvA, threonine deaminase; IlvIHCD, valine and isoleucine biosynthesis; Kdc, 2-ketoacid decarboxylase; Ldh, lactate dehydrogenase; LeuABCD, leucine and norvaline biosynthesis; Pfor, Pyruvate ferredoxin oxidoreductase; Pta, phosphotransacetylase; Ptb, phosphotransbutyrylase; Thl, thiolase; 2,3-Bdh, 2,3-butanediol dehydrogenase.

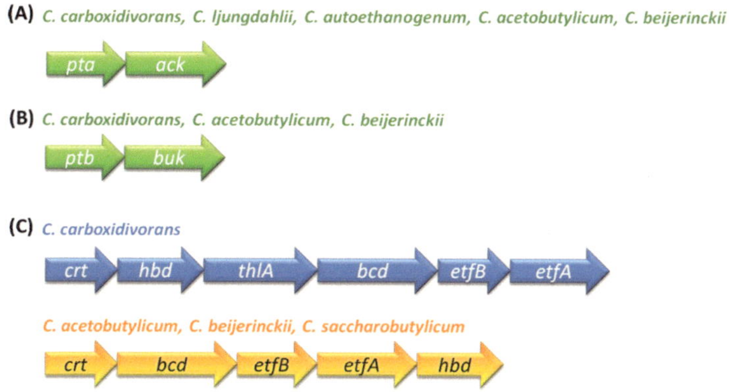

Figure 5. Similarity of acidogenesis and butanol formation gene clusters of acetogens and sugar-utilizing Clostridia. (A) Acetate-forming operon; (B) butyrate-forming operon; (C) butanol-forming operon. *ack*, acetate kinase; *buk*, butyrate kinase; *bcd*, butyryl-CoA dehydrogenase; *crt*, crotonase; *etfA*, electron-transferring flavoprotein subunit A; *etfB*, electron-transferring flavoprotein subunit B; *hbd*, 3-hydroxybutyryl-CoA dehydrogenase; *ptb*, phosphotransbutyrylase; *thlA*, thiolase.

Similar to sugar- and starch-utilizing ABE Clostridia, acetogens such as *C. carboxidivorans* [111, 112], *C. ljungdahlii* [113], and *C. autoethanogenum* [27] also typically undergo biphasic fermentation under autotrophic conditions. The first phase involves the production of carboxylic acids (acidogenic), H_2 and CO_2 during exponential growth. This is followed by the solventogenic phase in which part of the produced acids are reassimilated or reduced into solvents, which usually occurs during stationary growth phase [114]. This shift from acidogenesis to solventogenesis is of industrial importance and several transcriptional analysis on *C. acetobutylicum* [100, 115], and *C. beijerinckii* [116] have been performed to shed light on this process. In both organisms, the onset of solventogenesis coincides with an increase in expression of master sporulation/solventogenesis regulator gene *spo0A*, solventogenic genes such as *ald*, *ctfA-ctfB*, and *adc*, as well as down-regulation of chemotaxis/motility genes [100, 115, 116]. Physiologically, the signals that induce solventogenesis were hypothesized to involve temperature, low pH, high concentrations of undissociated acetic and butyric acids, limiting concentrations of sulphate or phosphate, ATP/ADP ratio and/or NAD(P)H levels [117].

For Clostridia such as acetogen *C. carboxidivorans* [54], which harbour the genes thiolase (*thlA*), 3-hydroxybutyryl-CoA dehydrogenase (*hbd*), crotonase (*crt*) and butyryl-CoA dehydrogenase (*bcd*), the two carbon acetyl-CoA can be converted to four carbon butyryl-CoA [95]. ThlA compete with the activities of Pta, Ald (aldehyde dehydrogenase), and PFOR to condense two acetyl-CoA into one acetoacetyl-CoA, and plays a key role in regulating the C2:C4 acid ratio [110, 118]. Since the formation of acetate yields twice as much ATP per mole of acetyl-CoA relative to butyrate formation, thiolase activity indirectly affects ATP yield [118]. Under physiological conditions, Crt catalyzes dehydration of β-hydroxybutyryl-CoA to crotonyl-CoA [119]. Bcd was shown to require a pair of electron transfer flavoproteins (EtfA and EtfB) to convert crotonyl-CoA to butyryl-CoA [120]. Furthermore, the Bcd was demonstrated to form a stable complex with EtfA and EtfB, and they were shown to couple the reduction of crotonyl-CoA to butyryl-CoA with concomitant generation of reduced ferredoxins, which can be used for energy conservation via Rnf complex [94, 119]. Subsequent actions of phosphotransbutyrylase (*ptb*) and butyrate kinase (*buk*) then generate ATP and butyrate from butyryl-CoA [118].

Under low extracellular pH of 4-4.5, the secreted undissociated acetic acid (pK_a 4.79) and/or butyric acid (pK_a 4.82) diffuse back into cell cytoplasm and then dissociate into the respective salts and protons because of the more alkaline intracellular conditions. Without further interventions, the result of this is abolishment of the proton gradient and inevitable cell death [95]. The conversion of acetate and butyrate into solvents increase the pH, thus provide some time for the organism to sporulate and secure long term survival. However, the solvents produced are toxic because they increase membrane fluidity and disrupt critical membrane-associated functions such as ATP synthesis, glucose uptake and other transport processes [114, 121]. In *C. acetobutylicum*, it has been demonstrated that the addition of 7-13 g/l of butanol, or up to 40 g/l of acetone and ethanol resulted in 50% growth inhibition [122]. The bacterium is likely to experience a different cytotoxic effect from endogenously produced solvents because the organism has time to adapt to increasing amount of solvents.

The reassimilation of acetate and butyrate into the respective acyl-CoA and acetoacetate is catalyzed by acetoacetyl-CoA:acetate/butyrate CoA transferase (CtfA and CtfB) [110, 117, 118]. Acetoacetate is deconstructed by acetoacetate decarboxylase (Adc) into acetone and CO_2. This enzyme is missing in acetogenic *C. carboxidivorans* compared to the ABE strains [54, 123]. Some ABE strains such as *C. beijerinckii* NRRL B593 also possess a primary/secondary alcohol dehydrogenase that converts acetone to isopropanol [124]. In acetogenic "*C. ragsdalei*", reduction of acetone to isopropanol was also observed although the mechanism of this reduction is as yet unknown [124, 125]. Again, *C. carboxidivorans* lacks this activity [125]. The recycled acetyl-CoA and butyryl-CoA can be converted to ethanol and butanol through the actions of coenzyme A-acylating aldehyde dehydrogenase (Ald) and alcohol dehydrogenase (Adh) [110, 118]. Ald converts acyl-CoA into aldehydes, and the enzyme has been purified from *C. beijerinckii* NRRL B593 and was shown to be NADH-specific, exhibit higher affinity with butyraldehyde than acetaldehyde, but possess no Adh activity [126]. In *C. ljungdahlii*, two variants of aldehyde:ferredoxin oxidoreductases (AOR) are present in the genome, and they are hypothesized to couple reduced ferredoxin from CO oxidation via the CODH (see above) to perform the reversible reduction of acetate into acetaldehyde, which can be further reduced into ethanol [62].

The final step of solventogenesis utilizes Adh to reduce acetaldehyde and butyraldehyde into ethanol and butanol, respectively. For ethanol synthesis, transposon mutagenesis and enzymatic assay in *C. acetobutylicum* showed the involvement of a specific Ald that does not interact with butyryl-CoA, and a NAD(P)H-dependent Adh [127, 128]. The production of butanol by *C. acetobutylicum* is mainly due to the action of butanol dehydrogenase A and B (BdhA and BdhB), and bifunctional butyraldehyde/butanol dehydrogenase 1 and 2 (AdhE1 and AdhE2) [95]. In *C. carboxidivorans* [54] and *C. ljungdahlii* [62] both *adhE1* and *adhE2* are arranged in tandem and separated by a 200bp gap which contains a putative terminator [62,

111]. This is likely the result of gene duplication [62]. qRT-PCR analysis from *C. carboxidivorans* fed with syngas showed that the two *adhE* showed differential expression, and the more abundant *adhE2* was significantly upregulated over 1000 fold in a time span that coincided with the greatest rate of butanol production [111].

Pyruvate is a central molecule for anabolism and it is predominantly generated from glycolysis during heterotrophic growth. But under autotrophic growth, this four carbon molecule can be synthesized by PFOR and potentially also the pyruvate-formate lyase (PFL). Two variants of PFOR were reported in *C. autoethanogenum*, and transcriptional analysis showed that they were differentially expressed when grown using industrial waste gases (containing CO, CO_2 and H_2) [104]. Unlike PFL from most other microorganisms that only catalyze the lysis of pyruvate into formate and acetyl-CoA, clostridial PFL (*C. kluyveri, C. butylicum,* and *C. butyricum*) were reported to readily catalyze the reverse reaction (i.e. pyruvate formation) [129]. Apart from roles in anabolism, pyruvate is also a precursor to other products such as lactic acid and 2,3-butanediol. Small amounts of lactic acid are converted from pyruvate in acetogens, a reaction which is catalyzed by lactate dehydrogenase (Ldh) [104, 118]. Recently, Köpke *et al.* (2011) reported the production of 2mM 2,3-butanediol from acetogenic bacteria (*C. autoethanogenum, C. ljungdahlii,* and *C. ragsdalei*) using industrial waste gases (containing CO, CO_2 and H_2) as feedstock [104]. Pyruvate is first converted into α-acetolactate by the enzyme acetolactate synthase, followed by acetolactate decarboxylase which split acetolactate into acetoin and CO_2, before a final reduction of acetoin into 2,3-butanediol by 2,3-butanediol dehydrogenase [104] (Figure 4).

9. Strain improvement and metabolic engineering

The genomes of several solventogenic Clostridia, including gas fermenting species, have been sequenced since 2001 [54, 62, 109, 119, 123, 130], and an array of transcriptomic [100, 116, 121, 131, 132], proteomic [132] and systems analysis [133, 134] are being made increasingly available. However, the generation of stable recombinant Clostridia has been severely hindered by the difficulties encountered introducing foreign DNA into cells and a lack of established genetic tools for this genera of bacteria. In comparison to starch-utilizing Clostridia, very little information is available for metabolic engineering of acetogens. Although this section describes recent advances in the development of genetic tools for mostly sugar-utilizing Clostridia, these techniques are highly relevant and applicable to the closely related acetogenic Clostridia for biofuels or chemical production via gas fermentation.

The ideal microbial catalyst for industrial scale gas fermentation might exhibit the following traits: high product yield and selectivity, low product inhibition, no strain degeneration, asporogenous, prolonged cell viability, strong aero-tolerance, high biomass density and efficient utilization of gas substrates. These can be achieved by directed evolution, random mutagenesis and/or targeted genetic engineering. Traditionally, chemical mutagenesis [135-137] and adaption strategies [138, 139] have been deployed to select for these traits. However, these strategies are limited and often come with the expense of unwanted events.

First attempts of targeted genetic modification of Clostridia were made in the early 1990s by the laboratory of Prof. Terry Papoutsakis [140-142]. While these pioneering efforts relied on use of plasmids for (over)expression of genes in *C. acetobutylicum,* more sophisticated tools were later developed for a range of solventogenic and pathogenic Clostridia.

Antisense RNA (asRNA) has been employed to down-regulate genes. Here, single stranded RNA binds to a complementary target mRNA and prevents translation by hindering ribosome-binding site interactions [143]. For instance, this method has been used to knockdown *ctfB* resulting in production of 30 g/l solvents with significantly suppressed acetone yield in *C. acetobutylicum* ATCC 824 [144, 145].

Several homologous recombination methods have been developed for integration or knock-out of genes in a range of sugar-utilizing Clostridia. In early stage, knockout mutants were almost exclusively generated from single crossover events that could revert back to wild-type [146-152], with stable double crossovers only observed in rare cases [153, 154]. For *C. acetobutylicum* [155] and cellulolytic *C. thermocellum* [156] counter selectable markers have been developed to allow more efficient screening for the rare second recombination event.

ClosTron utilizes the specificity of mobile group II intron Ll.*ltrB* from *Lactoccocus lactis* to propagate into a specified site in the genome via a RNA-mediated, retro-homing mechanism which can be used to disrupt genes [157]. This technique has initially been developed by InGex and Sigma-Aldrich under the name 'TargeTron™' and successfully adapted to a range of solventogenic and pathogenic Clostridia including *C. acetobutylicum, C. difficile, C. sporogenes, C. perfringens,* and *C. botulinum* [158-160] by the laboratory of Prof. Nigel Minton.

The same laboratory recently also developed another method for integration of DNA into the genome. Termed Allele-Coupled Exchange (ACE), this approach does not employ a counter selective marker to select for the rare second recombination event. Rather, it utilizes the activation or inactivation of gene(s) that result in a selectable phenotype, and asymmetrical homology arms to direct the order of recombination events [161]. Remarkably, the whole genome of phage lambda (48.5kb minus a 6kb region) was successfully inserted into the genome of *C. acetobutylicum* ATCC 824 in three successive steps using this genetic tool. This technique was also demonstrated in *C. difficile* and *C. sporogenes* [161].

For reverse engineering, mainly transposon mutagenesis has been utilized. Earlier efforts of transposon mutagenesis were demonstrated in *C. acetobutylicum* P262 (now: *C. saccharobutylicum* [162]), *C. acetobutylicum* DSM792, *C. acetobutylicum* DSM1732, and *C. beijerinckii* NCIM 8052, but issues with multiple transposon insertions per mutant, and non-random distribution of insertion were reported [163, 164]. Recent developments have seen the successful generation of mono-copy random insertion of transposon *Tn1545* into cellulolytic *C. cellulolyticum* [165] and mariner transposon *Himar1* into pathogenic *C. difficile* [166].

While there is still a lack of some other essential metabolic engineering tools such as efficient inducible promoters, the array of available tools that enabled significant improvements to the ABE process and cellulolytic Clostridia fermentations as summarized in Table 3.

Organism	Genetic modification	Phenotypes/Effects	Ref
Acetogens			
C. ljungdahlii	Plasmid overexpression of butanol biosynthetic genes from *C. acetobutylicum* (*thlA, crt, hbd, bcd, adhE* and *bdhA*)	Produced 2 mM butanol from syngas	[62]
C. autoethanogenum	Plasmid overexpression of butanol biosynthetic genes from *C. acetobutylicum* (*thlA, crt, hbd, bcd, etfA, & etfB*)	Produced 26 mM butanol using steel mill gas	[167]
C. autoethanogenum	Plasmid expression of native *groES* and *groEL*	Increased alcohol tolerance	[168]
C. aceticum	Plasmid overexpression of acetone operon from *C. acetobutylicum* (*adc, ctfAB, thlA*)	Produced up to 140 μM acetone using gas	[169, 170]
Acidogenesis and Solventogenesis			
C. acetobutylicum	Inactivation of *buk* and overexpression of *aad*	Produced same amount of butanol as control but relatively more ethanol, corresponding to a total alcohol tolerance of 21.2 g/l	[171]
C. acetobutylicum	Inactivation of *hbd* using ClosTron	Produced 716 mM ethanol by diverting C4 products	[172]
C. acetobutylicum	Inactivation of *ack* using ClosTron	Reduction in acetate kinase activity by more than 97% resulted in 80% less acetate produced but similar final solvent amount	[173]
C. tyrobutylicum	Inactivation of *ack* and plasmid overexpression of *adhE2* from *C. acetobutylicum*	Produced 216 mM butanol	[174]
C. thermocellum	Inactivation of *ldh* and *pta* via homologous recombination	Showed 4 fold increase in ethanol yield (122 mM instead of 28 mM)	[156]
C. cellulolyticum	Inactivation of*ldh* and *mdh* (malate dehydrogenase) using ClosTron	Generated 8.5 times higher ethanol yield (56.4 mM) than wild type (6.5 mM)	[175]
C. acetobutylicum	Plasmid overexpression of a synthetic acetoneoperon (*adc, ctfA, ctfB*) and primary/secondary *adh* from *C. beijerinckii* NRRL B593	Produced 85 mM isopropanol	[176]

C. acetobutylicum	Genome insertion of *adh* gene from *C.beijerinckii* NRRL B593 using Allele-Coupled Exchange	Converted acetone into 28 mM isopropanol without affecting the yield of other fermentation products	[161]
Biosynthesis of New Products			
C. cellulolyticum	Plasmid overexpression of *kivD, yqhD, alsS, ilvC* and *ilvD*	Produced 8.9 mM isobutanol by diverting 2-ketoacid intermediates	[177]
C. acetobutylicum	Plasmid expression of native *ribGBAH* operon and mutated PRPP amidotransferase	Produced 70 mg/l riboflavin and 190 mM butanol	[178]
Solvent- and Aero-tolerance			
C. acetobutylicum	Plasmid overexpression of glutathione *gshA* and *gshB* from *E. coli*	Improved aero- and solvent-tolerance	[179]
C. acetobutylicum	Plasmid overexpression of chaperone *groESL*	Showed 85% decrease in butanol inhibition and 33% increase in solvent yield	[180]
Substrate Utilization			
C. acetobutylicum	Plasmid expression of *acsC, acsD* and *acsE* from *C. difficile*	Increased incorporation of CO_2 into extracellular products	[99]
C. saccharoperbutylacetonicum strain N1-4	Knockdown hydrogenase *hupCBA* expression using siRNA delivered from plasmid	Significantly reduced hydrogen uptake activity to 13% (relative to control strain)	[181]

Table 3. Genetically modified solventogenic Clostridia

In contrast, to date only a limited number of acetogenic Clostridia have been successfully modified. Pioneering work in this area has been undertaken in the laboratory of Prof. Peter Dürre. *C. ljungdahlii*, a species that does not naturally produce butanol, was modified with butanol biosynthetic genes (*thlA, hbd, crt, bcd, adhE* and *bdhA*) from *C. acetobutylicum* ATCC 824 resulting in production of up to 2 mM of butanol using synthesis gas as sole energy and carbon source [62]. By delivering a plasmid with acetone biosynthesis genes *ctfA, ctfB, adc,* and *thlA* in *C. aceticum,* production of up to 140 µM acetone was demonstrated from various gas mixes (80% H_2/20% CO_2 and 67% H_2/33% CO_2) [169, 170]. Recent patent filings by Lanza-Tech describe the production of butanol as main fermentation product and increased alcohol tolerance in genetically engineered acetogens. Up to 26 mM butanol were produced with genetically modified *C. ljungdahlii* and *C. autoethanogenum* using steel mill gas (composition 44% CO, 32% N_2, 22% CO_2, and 2% H_2) as the only source of carbon and energy when the butanol biosynthetic genes *thlA, hbd, crt, bcd, etfA,* and *etfB* were heterologously expressed [167]. Overexpression of native *groESL* operon in *C. autoethanogenum* resulted in a strain that displayed higher alcohol tolerance relative to wild-type when challenged with ethanol [168].

Besides the classical Clostridial butanol pathway (which constitutes genes *thlA, crt, hbd, bcd, etfA and etfB*; see earlier section), a non-fermentative approach has been described and demonstrated in *E. coli* for branched chain higher alcohol production [182]. This alternative approach requires a combination of highly active amino acid biosynthetic pathway and artificial diversion of 2-keto acid intermediates into alcohols by introduction of two additional genes: broad substrate range 2-keto-acid decarboxylase (*kdc*) which converts 2-keto acids into aldehydes, followed by *Adh* to form alcohols [182]. Engineered strains of *E. coli* have been shown to produce alcohols such as isobutanol, n-butanol, 2-methyl-1-butanol, 3-methyl-1-butanol and 2-phenylethanol via this strategy [182]. For instance, the overexpression of *kivD* (KDC from *Lactococcus lactis*), *adh2, ilvA,* and *leuABCD* operon, coupled with deletion of *ilvD* gene and supplementation of L-threonine, increased n-butanol yield to 9 mM while producing 10 mM of 1-propanol [182].An even more remarkable yield of 300 mM isobutanol was achieved through introduction of k*ivD, adh2, alsS* (from *B. subtilis*), and *ilvCD* into *E. coli* [182]. Like butanol, isobutanol exhibits superior properties as a transportation fuel when compared to ethanol [177]. By applying similar strategy into *C. cellulolyticum,* 8.9 mM isobutanol was produced from cellulose when k*ivD, yqhD, alsS, ilvC,* and *ilvD* were overexpressed [177]. This result suggests that such non-fermentative pathway is suitable target for metabolic engineering of acetogens for the biosynthesis of branched chain higher alcohols. Via synthetic biology and metabolic engineering, production of additional potential liquid transportation fuels like farnesese or fatty acid based fuels has successfully been demonstrated in *E. coli* or yeast from sugar [183, 184]. Given the unsolved energetics in acetogens, it is unclear if production of such energy dense liquid fuels could be viable via gas fermentation.

10. Fermentation and product recovery

10.1. Bioreactor design

An optimum gas fermentation system requires efficient mass transfer of gaseous substrates to the culture medium (liquid phase) and microbial catalysts (solid phase). Gas-to-liquid mass transfer has been identified as the rate-limiting step and bottleneck for gas fermentation because of the low aqueous solubility of CO and H_2, respectively at only 77% and 68% of that of oxygen (on molar basis) at 35°C [185]. Hence, a bioreactor design that delivers sufficient gas-to-liquid mass transfer in an energy-efficient manner at commercial scale for gas fermentation represents a significant engineering challenge. A brief overview of reactor configurations reported in gas fermentation operations is given below.

10.2. Continuous Stirred Tank Reactor (CSTR)

In continuous stirred tank reactor (CSTR), gas substrates are continuously fed into the reactor and mechanically sheared by baffled impellers into smaller bubbles, which has greater interfacial surface area for mass transfer [16]. In addition, finer bubbles have a slower rising velocity and a longer retention time in the aqueous medium, resulting in higher gas-to-liquid mass transfer [24]. Fermentation reactions using *C. ljungdahlii* have been successfully maintained in a 2 litre CSTR under autotrophic conditions for more than a month, while achieving peak ethanol level of 6.5 g/l and CO conversion rate of 93% [186]. The production of 49 g/l of ethanol from gas substrates using *C. ljungdahlii* was demonstrated using CSTR [113]. In another example, a 100 litre stirred tank reactor was demonstrated to produce up to 24.57 g/l ethanol, 9.25 g/l isopropanol and 0.47 g/l n-butanol during a 59-day semi-batch gas fermentation using "*C. ragsdalei*" strain P11 as biocatalysts [112]. An improved version of CSTR incorporates microbubble sparger to generate finer bubbles to achieve higher mass transfer coefficient [187]. Although CSTR offers complete mixing and uniform distribution of gas substrates to the microbes, the high power per unit volume required to drive the stirrer are thought to make this approach economically unviable for commercial scale gas fermentation systems [187].

10.3. Bubble column reactor

In contrast to CSTR, gas mixing in bubble column reactor is achievable by gas sparging, without mechanical agitation. This reactor configuration has fewer moving parts, and consequently has a lower associated capital and operational costs while exhibiting good heat and mass transfer efficiencies, making it a good candidate for large scale gas fermentation [17]. However, excessive level of gas inflow for enhanced mixing have been cited as an issue that leads to heterogeneous flow and back-mixing of the gas substrates [16, 17]. *C. carboxidivorans* strain P7 was cultured in a 4 litre bubble column reactor for 20 days using a combination of producer gas and synthetic syngas, generating a peak ethanol concentration of 6 g/l [13].

10.4. Immobilized cell column reactor

One of the key challenges of gas fermentation is cell density. Immobilization of microbes through crosslinking or adsorption to insoluble biosupport materials and the subsequent packing within the column offers a range of benefits [14]. These include high cell densities, plug flow operation, high mass transfer rate via direct contact between microbe and gas, reduction of retention time, and operation without mechanical agitation [14, 16]. However, channelling issues may arise when the microbe overgrows and completely fill the interstitial space. Due to limitations in column dimensions and packing, this reactor configuration lacks flexibility to operate in various gas fermentation conditions [14, 16].

10.5. Trickle-bed reactor

Trickle-bed reactor is a gas- or liquid- continuous reactor consisting of packed bed, which liquid culture trickles down through packing media containing suspended or immobilized cells [16, 24, 187]. The gas substrate is delivered either co-currently or counter-currently to the liquid flow, and no mechanical agitation is required [187].In this reactor format, low gas and liquid flow rates are typically applied, generating relatively low pressure drops [187]. Trickle-bed reactor was found to exhibit excellent gas conversion rates and higher productivities than CSTR and bubble column reactor [15].

11. Gas fermentation parameters

11.1. Gas composition

The gas composition and its impurities can have an impact on the productivity of the gas fermentation process. Greater molar ratio of H_2:CO allows greater efficiency in the conversion of the carbon from CO into products such as ethanol, because reducing equivalents are generated from oxidation of H_2 (rather than CO). However, CO is also a known inhibitor of hydrogenase which can affect utilization of H_2 during fermentation. In *B. methylotrophicum*, H_2 utilization was inhibited until CO was exhausted [108]. When CO is consumed, acetogens are able to grow using CO_2 and H_2. Common impurities from biomass gasification or other waste gases are tar, ash, char, ethane, ethylene, acetylene, H_2S, NH_3 and NO [17, 22, 24, 188].These have been shown to cause cell dormancy, inhibition of hydrogen uptake, low cell growth and shift between acidogenesis and solventogenesis in acetogens [13, 188]. For instance, NH_3 from the feed gas readily convert into NH_4^+ in the culture media and these ions were recently shown to inhibit hydrogenase and cell growth of acetogen "*C. ragsdalei*" [189]. A number of strategies to mitigate the impact of such impurities have been proposed, for example installing 0.025 mm filters, or the use of gas scrubbers or cyclones, and improvement in gasification efficiency and scavenging for contaminants in the gas stream using agents such as potassium permanganate, sodium hydroxide or sodium hypochlorite [24, 190-192]. H_2S does not have a negative effect on acetogens such as *C. ljungdahlii* up to 5.2% (v/v) [193].

11.2. Substrate pressure

The partial pressure of syngas components have a major influence on microbial growth and product profiles because the enzymes involved are sensitive to substrate exposure [194]. Due to the low solubility of CO and H_2 in water, the growth of dense bacterial cell cultures can face mass transfer limitations, so increasing the partial pressure of gaseous substrates can help alleviate this problem. For instance, studies in which the CO partial pressure (P_{CO}) increased from 0.35 to 2.0 atm showed that this resulted in a 440% increase in maximum cell density, a significant increase in ethanol productivity and a decrease in acetate production in *C. carboxidivorans* strain P7 [195]. In another study involving *C. ljungdahlii*, the increase of P_{CO} from 0.8 to 1.8 atm had a positive effect on ethanol production, and the microbe did not exhibit any substrate inhibition at high P_{CO} [196].In less CO-tolerant microorganisms, the effect of increasing P_{CO} partial pressure range from non-appreciable in the case of *Rhodospirillum rubrum* [197], to negative impact on doubling time of *Peptostreptococcus productus* (now: *Blautia product*) [194] and *Eubacterium limosum* [198]. Similar to CO, the increase in partial pressure of H_2 (pH_2) to 1700 mbar enhanced acetate productivity of *A. woodii* to 7.4g acetate/ l/day [199].

11.3. Medium formulation

Although acetogens are able of utilizing CO and CO_2/H_2 as carbon and energy source, other constituents such as vitamins, trace metal elements, minerals and reducing agents are also required for maintenance of high metabolic activity [16, 113]. Studies indicated that formation of ethanol in solventogenic Clostridiais non-growth associated and limitation of growth by reducing availability of carbon-, nitrogen- and phosphate- nutrients shift the balance from acidogenesis to solventogenesis [113, 200, 201]. Optimization of medium formulation for *C. ljungdahlii* through reduction of B-vitamin concentrations and elimination of yeast extract significantly enhanced the final ethanol yield to 48 g/l in a CSTR with cell recycling (23 g/l without cell recycling) [113]. Another study by Klasson *et al.* showed thatthe replacement of yeast extract with cellobiose not only increased maximum cell concentration, but also enhanced ethanol yield by 4-fold [14]. Media formulation for *C. autoethanogenum* was investigated using Plackett-Burman and central composite designs, but only low ethanol yield was recorded overall [202]. In an attempt to reduce the cost of fermentation medium and improve process economics, 0.5 g/l of cotton seed extract without other nutrient supplementation was shown to be a superior medium for *C. carboxidivorans* strain P7 in producing ethanol from syngas fermentation [203]. A recent study showed that increasing concentrations of trace metal ions such as Ni^{2+}, Zn^{2+}, SeO_4^-, WO_4^-, Fe^{2+} and elimination of Cu^{2+} from medium improved enzymatic activities (FDH, CODH, and hydrogenase), growth and ethanol production in "*C. ragsdalei*" under autotrophic conditions [107].

A low redox potential is necessary for strict anaerobes to grow, hence reducing agents such as sodium thioglycolate, ascorbic acid, methyl viologen, benzyl viologen, titanium (III)–citrate, potassium ferricyanide, cysteine-HCl and sodium sulfide are commonly added to fermentation medium [14, 16, 204]. Furthermore, the addition of reducing agent directs the electron and carbon flow towards solventogenesis by enhancing the availability of reducing equivalents to form NADH for alcohol production [16, 205]. Excessive addition of reducing agents can cause slower microbial growth due to reduced ATP formation from acetogenesis so it is important to determine the optimum concentration of reducing agents [14, 16]. The sulfur containing gases (e.g. H_2S) present in syngas are toxic to chemical catalysts but can be beneficial for microbial catalysts by reducing medium redox potential, stimulate redox sensitive enzymes such as CODH, and promote alcohol formation [206, 207].

11.4. Medium pH

Like other organisms, acetogens have a limited range of pH for optimal growth so the pH of the fermentation medium needs to be closely controlled. The extracellular pH directly influences the intracellular pH, membrane potential, proton motive force, and consequently substrate utilization and product profile [208, 209]. In most studies, lowering pH medium divert carbon and electron flow from cell and acid formation towards alcohol production [113, 209-211]. By applying this knowledge, Gaddy and Clausen performed a two-stage CSTR syngas fermentation systems using *C. ljungdahlii* where they set the first reactor at pH 5 to promote cell growth, and pH 4 - 4.5 in the second reactor to induce ethanol production [212]. One recent study with *C. ljungdahlii* showed conflicting results in which cell density and ethanol production were both higher at pH 6.8 when compared to pH 5.5 [213].

11.5. Temperature

The optimum temperature for mesophilic acetogens are between 30-40°C, while thermophilic acetogens grow best between 55 and 58°C. The fermentation temperature not only affects substrate utilization, growth rate and membrane lipid composition of the acetogens, but also gas substrate availability because gas solubility increases with decreasing temperature [24, 211]. "*C. ragsdalei*"was reported to produce more ethanol at 32°C than at the optimum growth temperature of 37°C [211].

12. Cell separation and product recovery

To retain high cell densities in reactor, microbes can be grown as biofilm attached to carrier material. Planktonic cells can be retained in the fermentation broth by installing solid/liquid separators such as membranous ultra-filtration units, spiral wound filtration systems, hollow fibres, cell-recycling membranes and centrifuges [214-216]. The concentrations of solvents from gas fermentation rarely exceed 6% [w/v] so a cost- and energy- efficient product recovery process is required. Furthermore, acetogens also exhibit low resistance towards solvents like ethanol [217, 218] and butanol [219, 220] so an *in situ*/online product recovery system can enhance solvent productivity by decreasing solvent concentrations (and hence toxicity) in the fermentation broth. Distillation has been the traditional method of product recovery but the associated high energy costs have led to the development of alternative methods such as liquid-liquid extraction, pervaporation, perstraction, and gas stripping [24, 221].

12.1. Liquid-liquid extraction

In liquid-liquid extraction, a water-insoluble organic extractant is mixed with the fermentation broth [222]. Because solvents are more soluble in the organic phase than in the aqueous phase, they get selectively concentrated in the extractant.Although this technique does not remove water or nutrients from the fermentation broth, some gaseous substrates might be removed because CO and H_2 have much higher solubility in organic solvents than water [222, 223]. Oleyl alcohol has been the extractant of choice due to its relatively non-toxicity [224].

12.2. Perstraction

Liquid-liquid extraction is associated with several problems including toxicity to the microbes, formation of emulsion, and the accumulation of microbes at the extractant and fermentation broth interphase [222]. In an attempt to remediate these problems, perstraction was developed and this technique employs membrane to separate the extractant from the fermentation broth. This physical barrier prevent direct contact between the microbe and the toxicity of extractant, but it can also limit the rate of solvent extraction and is susceptible to fouling [219, 221]

12.3. Pervaporation

In a product recovery technique termed pervaporation, a membrane that directly comes in contact with fermentation broth is used to selectively remove volatile compounds such as ethanol and butanol [219, 222]. The volatile compounds diffuse through the membrane as vapour and are then collected by condensation. To facilitate volatilization of permeates into vapour, a partial pressure difference across the membrane is usually maintained by applying a vacuum or inert gas (e.g. N_2) across the permeate side of the membrane [219]. Polydimethylsiloxane (PDMS) is the current material of choice for the membrane, but other materials such as poly(1-trimethylsilyl-1-propyne) (PTMSP), hydrophobic zeolite membranes, and composite membranes have also been investigated [225].

12.4. Gas stripping

Gas stripping is an attractive product recovery method for gas fermentation because the exit gas stream from the bioreactor can be used for *in situ*/online product recovery [219]. Following product recovery via condensation, the effluent and gas can be recycled back into the bioreactor. In sugar-based fermentation using *C. beijerinckii* mutant strain BA101, *in situ* gas stripping was shown to improve ABE productivity by 200%, complete substrate utilization and also complete acid conversion into solvents, when compared to non-integrated process [226].

13. Commercialization

The growing commercial interests in using gas fermentation as a platform for biofuels production is evident in the recent spike in patent fillings within the field [105]. A 2009 report compared mass and energy conversion efficiencies from a process engineering standpoint between enzymatic hydrolysis fermentation of lignocellulose, syngas fermentation and FTP [227]. The authors concluded that while syngas fermentation offers a range of advantages such as low pretreatment requirement and low energy requirement for bioconversion, the technology is severely limited by low ethanol productivity [227]. Another report documented the techno-economic analysis of gas fermentation and concluded that the selling price of ethanol using this technology would still be significantly higher than gasoline in 2009 [228]. In contrast, Griffin and Schultz recently compared the production of ethanol from CO-rich gas using thermo-chemical route and biological gas fermentation route [22]. The authors concluded that gas fermentation offers superior fuel yield per volume of biomass feed, carbon conversion to fuel, energy efficiency and lower carbon emissions relative to the thermo-chemical approach to bioethanol production.

Ethanol and butanol are the most attractive fuel products from current gas fermentation but other by-products such as 2,3-butanediol, acetic acid and butyric acid are also valuable commodities that have the potential to provide significant additional revenue streams, setting off costs for biofuel production. 2,3-butanediol is a high value commodity which can be used to synthesize chemical products such as 1,3-butanediane, methyl ethyl ketone, and gamma butyrolactone, with a combined potential market value of $43 billion [104]. Acetic acid is an important precursor for synthesis of polymers while butyric acid can be used as a flavouring agent in the food industry [229, 230]. With the development of advanced genetic tools for expansion of product range, the industry might witness an increasing emphasis on the production of high-value commodities in addition to biofuels.

Several companies are actively engaged in the development of the gas fermentation technology and some are approaching commercialization. Bioengineering Resources Inc (BRI) founded by Prof. James Gaddy of University of Arkensas, Fayetteville, an early pioneer in the investigation of gas fermentation at scale, was the first company to explore the potential of gas fermentation for industrial bioethanol production. BRI was acquired by chemical company INEOS and rebranded as INEOS Bio (www.ineosbio.com). A pilot-scale facility in Arkansas has been operated since 2003 using several isolates of *C. ljungdahlii* [231] and is building a US$130 million commercial facility in Florida with its joint venture partner New Planet Energy Florida [232]. The commercial facility is expected to start operation in the second quarter of 2012 and is aiming to generate 8 million gallon of cellulosic ethanol per annum and 6 MW of power to the local communities [232]. INEOS Bio also announced design of a second plant, the Seal Sands Biorefinery in Teeside, UK [233].

Founded in 2006, Coskata Inc. (www.coskata.com) is a US-based company that has reported achieving ethanol yields of 100 gallons per dry ton of wood biomass in a semi-commercial facility in Pennsylvania [234]. The company licensed several microbial strains from the University of Oklahoma [235], which has filed patents and journal publications for acetogens such as "*C. ragsdalei*" [211, 236, 237] and *C. carboxidivorans* [55, 112]. A patent documenting a new ethanologenic species, "*C. coskatii*" was also recently filed by Coskata [238]. Backed by a conditional US$250 million loan guarantee from the US Department of Agriculture (USDA), Coskata has announced that it is planning to build a commercial plant with the capacity to produce 55 million gallon fuel grade ethanol per annum in Alabama [234, 239]. While the initial strategy saw biomass as feedstock, the company recently announced its first commercial plant will be switched to 100% natural gas as feedstock [240]. A planned IPO with the aim to tap into private investors to finance the plant was put on hold [241]. In 2012, Coskata and INEOS Bio were involved in a trade secret dispute which culminated in a settlement that see INEOS Bio receiving US$2.5 million cash payment, shares and right to receive 2.5% of future ethanol royalties from Coskata [242].

LanzaTech is a NZ/US based company that has developed a gas fermentation technology to utilize industrial off-gases from steel making and other sources, as well as syngas produced from biomass as feedstocks. The company has reported the development of a proprietary Clostridial biocatalyst that is able to convert the CO-rich waste gas with minimal gas conditioning into bioethanol and the platform chemical 2,3-butanediol. The use of industrial off-gases as feedstock not only helps to reduce the carbon footprint of the steel-making operations but also allows the production of valuable commodities without the costs associated with feedstock gasification. The company has estimated that up to 30 billion gallon of bioethanol per year can be produced from the CO-rich off gases produced through steel manufacturers globally [243]. Founded in 2005, LanzaTech has successfully demonstrated bioethanol production at a pilot plant at BlueScope Steel in Glenbrook, NZ, since 2008 and the company has recently started operating its 100,000 gallon bioethanol per year demonstration facility in Shanghai, China, using waste gas collected from an adjacent steel mill plant owned by its partner Baosteel Group [243, 244]. LanzaTech is planning to build a commercial facility with the capacity to produce 50 million gallon of bioethanol per annum in China by 2013 [243]. The recent acquisition of a biorefinery facility developed by the US-based gasification technology company Range Fuels in Georgia, and a milestone signing of its first commercial customer, Concord Enviro Systems (India), highlighted LanzaTech's intention to utilize MSW and lignocellulosic waste as feedstocks for biofuel and chemical production [243, 244].

14. Conclusion

One of the fundamental factors that govern the environmental and economical sustainability of biofuel production is feedstock. Through gasification, a spectrum of renewable non-food feedstock such as agricultural wastes, dedicated energy crops, forest residues, and MSW can be converted into syngas. This article presents a detailed examination of gas fermentation technology in capturing the carbon and energy from syngas and produce biofuels and chemicals. In comparison to indirect fermentation of lignocellulose via enzymatic hydrolysis, and thermo-chemical FTP, gas fermentation offers several advantages such as good product yield and selectivity, operation in ambient conditions, high tolerance to gas impurities, and elimination of expensive pre-treatment steps and costly enzymes. Furthermore, some industries such as steel mill, natural gas steam reforming, oil refining and chemical production generate large volumes of CO-rich off-gas. Gas fermentation can access these existing feedstocks and generate valuable products from these while reducing carbon emissions. Pivotal to gas fermentation is acetogens such as *C. ljungdahlii*, *C. carboxidivorans*, "*C. ragsdalei*" and *C. autoethanogenum*, which are able to metabolize CO, and CO_2/H_2 into a range of products such as ethanol, butanol, isopropanol, acetone, 2,3-butanediol, acetic acid and butyric acid. Sustained effort in studying the physiology and biochemistry using advanced molecular techniques such as genomics, transcriptomics, proteomics, metabolomics and systems biology are essential to further the understanding of these microbes. Furthermore, recent advances in Clostridial genetic tools offer endless opportunities to engineer strains that have improved product yield, substrate utilization, no strain degeneration, and synthesis of new products.

The main challenges associated with commercialization of gas fermentation have been identified as gas-to-liquid mass transfer limitation, product yield, substrate utilization efficiency, low biomass density and product recovery. Further development of bioreactor is necessary to improve the availability of gas substrates and maintain high cell density for higher productivity. Improvement in integrated product recovery technology is also essential to lower the costs of product recovery and alleviate product inhibition. Gas fermentation appears to be mature enough for commercialization since several companies have already demonstrated their technologies at pilot scale and are moving towards commercialization in the near future.

Author details

Fung Min Liew, Michael Köpke and Séan Dennis Simpson

LanzaTech NZ Ltd., Parnell, Auckland, New Zealand

References

[1] Vasudevan PT, Gagnon MD, Briggs MS. Environmentally sustainable biofuels – The case for biodiesel, biobutanol and cellulosic ethanol. In: Singh OV, Harvey SP, editors. Sustainable Biotechnology: Sources of Renewable Energy: Springer Science +Business Media B.V.; 2010. p. 43-62.

[2] Appenzeller T. The end of cheap oil. National Geographic Magazine. 2004.

[3] European_Union. Directive 2009/28/EC of the European Parliament and of the Council of 23 April 2009 on the promotion of the use of energy from renewable sources and amending and subsequently repealing Directives 2001/77/EC and 2003/30/EC. 2009.

[4] REN21. Renewables 2011 Global Status Report. Paris: REN21, 2011.

[5] Demirbas A. Political, economic and environmental impacts of biofuels: A review. Applied Energy. 2009 Nov;86:S108-S17. PubMed PMID: WOS:000271170300013.

[6] Köpke M, Noack S, Dürre P. The past, present, and future of biofuels – Biobutanol as Promising Alternative. In: dos Santos Bernades MA, editor.: InTech; 2011. p. 451-86.

[7] Pernick R, Wilder C, Winnie T. Clean Energy Trends 2012. Clean Edge Inc., 2012.

[8] Dürre P. Biobutanol: an attractive biofuel. Biotechnology journal. 2007;2(12):1525-34.

[9] OECD-FAO. OECD-FAO Agricultural Outlook 2011-2020. 2011.

[10] Murse T. Understanding the ethanol subsidy 2011. Available from: http://usgovinfo.about.com/od/moneymatters/a/The-Federal-Ethanol-Subsidy.htm.

[11] Pear R. After three decades, tax credit for ethanol expires. The New York Times. 2nd January 2012.

[12] Carere CR, Sparling R, Cicek N, Levin DB. Third generation biofuels via direct cellulose fermentation. International Journal of Molecular Sciences. 2008 Jul;9(7):1342-60. PubMed PMID: WOS:000257986800014.

[13] Datar RP, Shenkman RM, Cateni BG, Huhnke RL, Lewis RS. Fermentation of biomass-generated producer gas to ethanol. Biotechnology and bioengineering. 2004;86(5):587-94.

[14] Klasson KT, Ackerson MD, Clausen EC, Gaddy JL. Bioreactor design for synthesis gas fermentations. Fuel. 1991 May;70(5):605-14. PubMed PMID: WOS:A1991FL49400009.

[15] Klasson KT, Ackerson MD, Clausen EC, Gaddy JL. Bioconversion of synthesis gas into liquid or gasesous fuels. enzyme and microbial technology. 1992 Aug;14(8):602-8. PubMed PMID: WOS:A1992JE44900001.

[16] Mohammadi M, Najafpour GD, Younesi F, Lahijani P, Uzir MH, Mohamed AR. Bioconversion of synthesis gas to second generation biofuels: A review. Renewable & Sustainable Energy Reviews. 2011 Dec;15(9):4255-73. PubMed PMID: WOS: 000298764400005.

[17] Abubackar HN, Veiga MC, Kennes C, Coruña L. Biological conversion of carbon monoxide: rich syngas or waste gases to bioethanol. Biofuels, Bioproducts & Biorefining. 2011:93-114.

[18] Siedlecki M, de Jong W, Verkooijen AHM. Fluidized bed gasification as a mature and reliable technology for the production of bio-syngas and applied in the production of liquid transportation fuels-A review. Energies. 2011 Mar;4(3):389-434. PubMed PMID: WOS:000288788000002.

[19] Wei L, Pordesimo LO, Haryanto A, Wooten J. Co-gasification of hardwood chips and crude glycerol in a pilot scale downdraft gasifier. Bioresource Technology. 2011 May; 102(10):6266-72. PubMed PMID: WOS:000291125800097.

[20] Dudynski M, Kwiatkowski K, Bajer K. From feathers to syngas - Technologies and devices. Waste Management. 2012 Apr;32(4):685-91. PubMed PMID: WOS: 000302822500007.

[21] Sipma J, Henstra AM, Parshina SM, Lens PN, Lettinga G, Stams AJM. Microbial CO conversions with applications in synthesis gas purification and bio-desulfurization. Critical reviews in biotechnology. 2006;26(1):41-65.

[22] Griffin DW, Schultz MA. Fuel and chemical products from biomass syngas: A comparison of gas fermentation to thermochemical conversion routes. Environmental Progress & Sustainable Energy. 2012 Jul;31(2):219-24. PubMed PMID: WOS: 000302794000008.

[23] McKendry P. Energy production from biomass (part 3): gasification technologies. Bioresource Technology. 2002 May;83(1):55-63. PubMed PMID: WOS:000175355300006.

[24] Munasinghe PC, Khanal SK. Biomass-derived syngas fermentation into biofuels: Opportunities and challenges. Bioresour Technol. 2010;101(13):5013-22.

[25] Bartish CM, Drissel GM. Carbon monoxide. In: Othmer K, editor. Encyclopedia of Chemical Technology. New York: John Wiley and Sons; 1978.

[26] van Houten RT, Lettinga G. Biological sulphate reduction with synthesis gas: Microbiology and technology. Immobilized Cells: Basics and Applications. 1996;11:793-9. PubMed PMID: WOS:A1996BF78D00106.

[27] Köpke M, Mihalcea C, Liew FM, Tizard JH, Ali MS, Conolly JJ, et al. 2,3-butanediol production by acetogenic bacteria, an alternative route to chemical synthesis, using industrial waste gas. Applied and Environmental Microbiology. 2011 Aug;77(15): 5467-75. PubMed PMID: WOS:000293224500045.

[28] van der Drift A, van Doorn J, Vermeulen JW. Ten residual biomass fuels for circulating fluidized-bed gasification. Biomass & Bioenergy. 2001;20(1):45-56. PubMed PMID: WOS:000167001400006.

[29] Gordillo G, Annamalai K. Adiabatic fixed bed gasification of dairy biomass with air and steam. Fuel. 2010 Feb;89(2):384-91. PubMed PMID: WOS:000271295400017.

[30] Boateng AA, Banowetz GM, Steiner JJ, Barton TF, Taylor DG, Hicks KB, et al. Gasification of Kentucky bluegrass (Poa pratensis l.) straw in a farm-scale reactor. Biomass & Bioenergy. 2007 Feb-Mar;31(2-3):153-61. PubMed PMID: WOS:000244170800006.

[31] Bridgwater T. Biomass for energy. Journal of the Science of Food and Agriculture. 2006 Sep;86(12):1755-68. PubMed PMID: WOS:000240412500004.

[32] Swanson RM, Platon A, Satrio JA, Brown RC. Techno-economic analysis of biomass-to-liquids production based on gasification. Fuel. 2010 Nov;89:S2-S10. PubMed PMID: WOS:000282368600002.

[33] Drake HL, Gössner AS, Daniel SL. Old acetogens, new light. Annals of the New York Academy of Sciences. 2008;1125:100-28.

[34] Thauer RK. Microbiology - A fifth pathway of carbon fixation. Science. 2007 Dec; 318(5857):1732-3. PubMed PMID: WOS:000251616800025.

[35] Drake HL, Küsel K, Matthies C. Acetogenic prokaryotes. In: Dworkin M, Falkow S, Rosenberg E, Schleifer K-H, Stackebrandt E, editors.: Springer New York; 2006. p. 354-420.

[36] Lindahl PA, Chang B. The evolution of acetyl-CoA synthase. Origins of Life and Evolution of the Biosphere. 2001;31(4-5):403-34. PubMed PMID: WOS:000170975700006.

[37] Ragsdale SW. Enzymology of the Wood-Ljungdahl pathway of acetogenesis. Incredible Anaerobes: from Physiology to Genomics to Fuels. 2008;1125:129-36. PubMed PMID: WOS:000255395600007.

[38] Ragsdale SW. Life with carbon monoxide. Critical Reviews in Biochemistry and Molecular Biology. 2004 May-Jun;39(3):165-95. PubMed PMID: WOS:000223849300003.

[39] Ragsdale SW, Clark JE, Ljungdahl LG, Lundie LL, Drake HL. Properties of purified carbon monoxide dehydrogenase from Clostridium thermoaceticum, a nickel, iron-sulfur protein. Journal of Biological Chemistry. 1983;258(4):2364-9. PubMed PMID: WOS:A1983QD12900054. English.

[40] Morsdorf G, Frunzke K, Gadkari D, Meyer O. Microbial growth oncarbon monoxide. Biodegradation. 1992;3:61-82.

[41] Genthner BRS, Bryant MP. Additional characteristics of one-carbon-compound utilization by Eubacterium limosum and Acetobacterium woodi. Applied and Environmental Microbiology. 1987 Mar;53(3):471-6. PubMed PMID: WOS:A1987G206800001.

[42] Poehlein A, Schmidt S, Kaster AK, Goenrich M, Vollmers J, Thurmer A, et al. An ancient pathway combining carbon dioxide fixation with the generation and utilization of a sodium ion gradient for ATP synthesis. Plos One. 2012 Mar;7(3). PubMed PMID: WOS:000304523400022.

[43] Kane MD, Breznak JA. Acetonema longum gen nov SP-nov, an H2/CO2 acetogenic bacterium from the termite, Pterotermes occidentis. Archives of Microbiology. 1991 Aug;156(2):91-8. PubMed PMID: WOS:A1991GB27400002.

[44] Allen TD, Caldwell ME, Lawson Pa, Huhnke RL, Tanner RS. Alkalibaculum bacchi gen. nov., sp. nov., a CO-oxidizing, ethanol-producing acetogen isolated from livestock-impacted soil. International journal of systematic and evolutionary microbiology. 2010;60(Pt 10):2483-9.

[45] Liu K, Atiyeh HK, Tanner RS, Wilkins MR, Huhnke RL. Fermentative production of ethanol from syngas using novel moderately alkaliphilic strains of Alkalibaculum bacchi. Bioresource Technology. 2012 Jan;104:336-41. PubMed PMID: WOS: 000301155800045.

[46] Lorowitz WH, Bryant MP. Peptostreptococcus productus strain that grows rapidly with CO as the energy source. Applied and Environmental Microbiology. 1984;47(5): 961-4. PubMed PMID: WOS:A1984SQ91300014.

[47] Lynd L, Kerby R, Zeikus JG. Carbon monoxide metabolism of the methylotrophic acidogen Butyribacterium methylotrophicum. Journal of bacteriology. 1982;149(1): 255-63.

[48] Zeikus JG, Lynd LH, Thompson TE, Krzycki Ja, Weimer PJ, Hegge PW. Isolation and characterization of a new, methylotrophic, acidogenic anaerobe, the marburg strain. Current Microbiology. 1980;3(6):381-6.

[49] Grethlein AJ, Worden RM, Jain MK, Datta R. Evidence for production of n-butanol from carbon monoxide by Butyribacterium methylotrophicum. Journal of Fermentation and Bioengineering. 1991;72(1):58-60.

[50] Lux MF, Drake HL. Reexamination of the metabolic potentials of the acetogens Clostridium aceticum and Clostridium formicoaceticum - chemolithoautotrophic and aromatic-dependent growth. Fems Microbiology Letters. 1992 Aug;95(1):49-56. PubMed PMID: WOS:A1992JJ52500008.

[51] Adamse AD. New isolation of Clostridium aceticum (Wieringa). Antonie Van Leeuwenhoek Journal of Microbiology. 1980;46(6):523-31. PubMed PMID: WOS:A1980LA02100002.

[52] Braun M, Mayer F, Gottschalk G. Clostridium aceticum (Wieringa), A microorganism producing acetic acid from molecular hydrogen and carbon dioxide. Archives of Microbiology. 1981;128(3):288-93. PubMed PMID: WOS:A1981LB03200006.

[53] Abrini J, Naveau H, Nyns EJ. Clostridium autoethanogenum, sp. nov., an anaerobic bacterium that produces ethanol from carbon monoxide. Archives of microbiology. 1994;161(4):345-51.

[54] Bruant G, Lévesque M-J, Peter C, Guiot SR, Masson L. Genomic analysis of carbon monoxide utilization and butanol production by Clostridium carboxidivorans strain P7. PloS one. 2010;5(9):e13033.

[55] Liou JSC, Balkwill DL, Drake GR, Tanner RS. Clostridium carboxidivorans sp nov., a solvent-producing clostridium isolated from an agricultural settling lagoon, and reclassification of the acetogen Clostridium scatologenes strain SL1 as Clostridium drakei sp nov. International Journal of Systematic and Evolutionary Microbiology. 2005 Sep;55:2085-91. PubMed PMID: WOS:000232239600055.

[56] Küsel K, Dorsch T, Acker G, Stackebrandt E, Drake HL. Clostridium scatologenes strain SL1 isolated as an acetogenic bacterium from acidic sediments. International journal of systematic and evolutionary microbiology. 2000;50 Pt 2:537-46.

[57] Gössner AS, Picardal F, Tanner RS, Drake HL. Carbon metabolism of the moderately acid-tolerant acetogen Clostridium drakei isolated from peat. FEMS microbiology letters. 2008;287(2):236-42.

[58] Andreese.Jr, Gottscha.G, Schlegel HG. Clostridium formicoaceticumnov spec - isolation, description and distinction from C aceticum and C thermoaceticum. Archiv Fur Mikrobiologie. 1970;72(2):154. PubMed PMID: WOS:A1970G704100007.

[59] Diekert GB, Thauer RK. Carbon monoxide oxidation by Clostridium thermoaceticum and Clostridium formicoaceticum. Journal of bacteriology. 1978;136(2):597-606.

[60] Kusel K, Karnholz A, Trinkwalter T, Devereux R, Acker G, Drake HL. Physiological ecology of Clostridium glycolicum RD-1, an aerotolerant acetogen isolated from sea grass roots. Applied and Environmental Microbiology. 2001 Oct;67(10):4734-41. PubMed PMID: WOS:000171237700045.

[61] Ohwaki K, Hungate RE. Hydrogen utilization by Clostridia in sewage sludge. Applied and Environmental Microbiology. 1977;33(6):1270-4.

[62] Köpke M, Held C, Hujer S, Liesegang H, Wiezer A, Wollherr A, et al. Clostridium ljungdahlii represents a microbial production platform based on syngas. Proceedings of the National Academy of Sciences of the United States of America. 2010;107(29): 13087-92.

[63] Tanner RS, Miller LM, Yang D. Clostridium ljungdahlii sp. nov., an acetogenic species in clostridial rRNA homology group I. International journal of systematic bacteriology. 1993;43(2):232-.

[64] Bomar M, Hippe H, Schink B. Lithotrophic growth and hydrogen metabolism by Clostridium magnum. Fems Microbiology Letters. 1991;83(3):347-50.

[65] Schink B. Clostridium magnum sp nov, a non-autotrophic homoacetogenic bacterium. Archives of Microbiology. 1984;137(3):250-5.

[66] Kane MD, Brauman A, Breznak JA. Clostridium mayombei sp nov, an H2/CO2 acetogenic bacterium from the gut of the african soil-feeding termite, Cubitermes speciosus. Archives of Microbiology. 1991;156(2):99-104.

[67] Mechichi T, Labat M, Patel BKC, Woo THS, Thomas P, Garcia JL. Clostridium methoxybenzovorans sp nov., a new aromatic o-demethylating homoacetogen from an olive mill wastewater treatment digester. International Journal of Systematic Bacteriology. 1999 Jul;49:1201-9. PubMed PMID: WOS:000081448000032.

[68] Huhnke R, Lewis RS, Tanner RS, inventors. Isolation and characterization of novel Clostridial species. 2008.

[69] Genthner BRS, Davis CL, Bryant MP. Features of rumen and sewage sludge strains of Eubacterium limosum, a methanol-utilizing and H2-CO2-utilizing species. Applied and Environmental Microbiology. 1981;42(1):12-9. PubMed PMID: WOS:A1981LX19300003.

[70] Krumholz LR, Bryant MP. Clostridium pfennigii sp nov uses methoxyl groups of monobenzenoids and produces butyrate. International Journal of Systematic Bacteriology. 1985;35(4):454-6. PubMed PMID: WOS:A1985ASV9200009.

[71] Pierce E, Xie G, Barabote RD, Saunders E, Han CS, Detter JC, et al. The complete genome sequence of Moorella thermoacetica (f. Clostridium thermoaceticum). Environmental microbiology. 2008;10(10):2550-73.

[72] Daniel SL, Hsu T, Dean SI, Drake HL. Characterization of the H2-dependent and CO-dependent chemolithotrophic potentials of the acetogens Clostridium thermoaceticum and Acetogenium kivui. Journal of Bacteriology. 1990 Aug;172(8):4464-71. PubMed PMID: WOS:A1990DR47100043.

[73] Kerby R, Zeikus JG. Growth of Clostridium thermoaceticum on H2/CO2 or CO as energy source. Current Microbiology. 1983;8(1):27-30. PubMed PMID: WOS:A1983QB38000006.

[74] Savage MD, Wu ZR, Daniel SL, Lundie LL, Drake HL. Carbon monoxide-dependent chemolithotrophic growth of Clostridium thermoautotrophicum. Applied and Environmental Microbiology. 1987 Aug;53(8):1902-6. PubMed PMID: WOS:A1987J392000033.

[75] Müller V. Energy conservation in acetogenic bacteria. Applied and environmental microbiology. 2003;69(11):6345-53.

[76] Calusinska M, Happe T, Joris B, Wilmotte A. The surprising diversity of clostridial hydrogenases: a comparative genomic perspective. Microbiology-Sgm. 2010 Jun; 156:1575-88. PubMed PMID: WOS:000279798900001.

[77] Vignais PM, Colbeau A. Molecular biology of microbial hydrogenases. Current Issues in Molecular Biology. 2004;6:159-88. PubMed PMID: WOS:000228691500013.

[78] Vignais PM. Hydrogenases and H^+-reduction in primary energy conservation. In: Schafer G, Penefsky HS, editors. Bioenergetics: Energy Conservation and Conversion. 45: Springer Berlin / Heidelberg; 2007. p. 223-52.

[79] Demuez M, Cournac L, Guerrini O, Soucaille P, Girbal L. Complete activity profile of Clostridium acetobutylicum FeFe -hydrogenase and kinetic parameters for endogenous redox partners. Fems Microbiology Letters. 2007 Oct;275(1):113-21. PubMed PMID: WOS:000249428700015.

[80] Schut GJ, Adams MWW. The iron-hydrogenase of Thermotoga maritima utilizes ferredoxin and NADH synergistically: a new perspective on anaerobic hydrogen production. Journal of bacteriology. 2009;191(13):4451-7.

[81] Menon S. Unleashing hydrogenase activity in carbon monoxide dehydrogenase/ acetyl-CoA synthase and pyruvate: ferredoxin oxidoreductase. Biochemistry. 1996;2960(96):15814-21.

[82] Thauer RK, Jungermann K, Decker K. Energy conservation in chemotrophic anerobic bacteria. Bacteriological Reviews. 1977;41(1):100-80. PubMed PMID: WOS:A1977DC99200003.

[83] Oelgeschläger E, Rother M. Carbon monoxide-dependent energy metabolism in anaerobic bacteria and archaea. Archives of microbiology. 2008;190(3):257-69.

[84] Rohde M, Mayer F, Meyer O. Immunocytochemical localization of carbon-monoxide oxidase in Pseudomonas carboxydovorans - The enzyme is attached to the inner aspect of the cytoplasmic membrane. Journal of Biological Chemistry. 1984;259(23): 4788-92. PubMed PMID: WOS:A1984TV66900080.

[85] Rohde M, Mayer F, Jacobitz S, Meyer O. Attachement of CO dehydrogenase to the cytoplasmic membrane is limiting the respiratory rate of Pseudomonas carboxydovorans. Fems Microbiology Letters. 1985;28(2):141-4. PubMed PMID: WOS:A1985AMV6400003.

[86] Kumar M, Lu WP, Ragsdale SW. Binding of carbon disufide to the site of acetyl-CoA synthesis by the nickel-iron-sulfur protein, carbon monoxide dehydrogenase, from Clostridium thermoaceticum. Biochemistry. 1994 Aug;33(32):9769-77. PubMed PMID: WOS:A1994PC54500048.

[87] Lindahl PA, Munck E, Ragsdale SW. CO dehydrogenase from Clostridium thermoaceticum - EPR and electrochemical studies in CO2 and argon atmosphere. Journal of Biological Chemistry. 1990 Mar;265(7):3873-9. PubMed PMID: WOS:A1990CQ48000047.

[88] Ragsdale SW, Pierce E. Acetogenesis and the Wood-Ljungdahl pathway of CO(2) fixation. Biochimica et biophysica acta. 2008;1784(12):1873-98.

[89] Nevin KP, Hensley SA, Franks AE, Summers ZM, Ou JH, Woodard TL, et al. Electrosynthesis of organic compounds from carbon dioxide is catalyzed by a diversity of acetogenic microorganisms. Applied and Environmental Microbiology. 2011 May; 77(9):2882-6. PubMed PMID: WOS:000289773100009.

[90] Nevin KP, Woodard TL, Franks AE, Summers ZM, Lovley DR. Microbial electrosynthesis : Feeding microbes electricity to convert carbon dioxide and water to multicarbon extracellular organic compounds. mBio. 2010;1(2).

[91] Biegel E, Schmidt S, Müller V. Genetic, immunological and biochemical evidence for a Rnf complex in the acetogen Acetobacterium woodii. Environmental microbiology. 2009;11(6):1438-43.

[92] Li F, Hinderberger J, Seedorf H, Zhang J, Buckel W, Thauer RK. Coupled ferredoxin and crotonyl coenzyme A (CoA) reduction with NADH catalyzed by the butyryl-CoA dehydrogenase/Etf complex from Clostridium kluyveri. Journal of bacteriology. 2008;190(3):843-50.

[93] Müller V, Imkamp F, Biegel E, Schmidt S, Dilling S. Discovery of a ferredoxin:NAD+-oxidoreductase (Rnf) in Acetobacterium woodii: a novel potential coupling site in acetogens. Annals of the New York Academy of Sciences. 2008;1125:137-46.

[94] Herrmann G, Jayamani E, Mai G, Buckel W. Energy conservation via electron-transferring flavoprotein in anaerobic bacteria. Journal of Bacteriology. 2008 Feb;190(3): 784-91. PubMed PMID: WOS:000252616800002.

[95] Dürre P. Handbook on Clostridia. Dürre P, editor: CRC Press Taylor & Francis Group; 2005.

[96] Wang SN, Huang HY, Moll J, Thauer RK. NADP(+) reduction with reduced ferredoxin and NADP(+) reduction with NADH are coupled via an electron-bifurcating enzyme complex in Clostridium kluyveri. Journal of Bacteriology. 2010 Oct;192(19): 5115-23. PubMed PMID: WOS:000281866900032.

[97] Roberts DL, Jameshagstrom JE, Garvin DK, Gorst CM, Runquist JA, Baur JR, et al. Cloning and expression of the gene-cluster encoding key proteins involved in acetyl-CoA synthesis in Clostridium thermoaceticum - CO dehydrogenase, the corrinoid Fe-S protein, and methyltransferase. Proceedings of the National Academy of Sciences of the United States of America. 1989 Jan;86(1):32-6. PubMed PMID: WOS:A1989R820200008.

[98] Kaster AK, Goenrich M, Seedorf H, Liesegang H, Wollherr A, Gottschalk G, et al. More than 200 genes required for methane formation from H-2 and CO2 and energy conservation are present in Methanothermobacter marburgensis and Methanothermobacter thermautotrophicus. Archaea-an International Microbiological Journal. 2011. PubMed PMID: WOS:000292403100001.

[99] Papoutsakis ET, AL-Hinai MA, Jones SW, Indurthi DC, Mitchell DK, Fast A, inventors. Recombinant Clostridia that fix CO2 and CO and uses thereof. U.S.2012.

[100] Alsaker KV, Papoutsakis ET. Transcriptional program of early sporulation and stationary-phase events in Clostridium acetobutylicum. Journal of bacteriology. 2005;187(20):7103-18.

[101] Köpke M. Genetische Veränderung von Clostridium ljungdahlii zur produktion von 1-butanol aus synthesegas.: Ulm University; 2009.

[102] Sakai S, Nakashimada Y, Yoshimoto H, Watanabe S, Okada H, Nishio N. Ethanol production from H-2 and CO2 by a newly isolated thermophilic bacterium, Moorella sp HUC22-1. Biotechnology Letters. 2004 Oct;26(20):1607-12. PubMed PMID: WOS: 000225691400013.

[103] Sakai S, Nakashimada Y, Inokuma K, Kita M, Okada H, Nishio N. Acetate and ethanol production from H-2 and CO2 by Moorella sp using a repeated batch culture. Journal of Bioscience and Bioengineering. 2005 Mar;99(3):252-8. PubMed PMID: WOS:000229052300009.

[104] Köpke M, Mihalcea C, Liew FM, Tizard JH, Ali MS, Conolly JJ, et al. 2,3-butanediol production by acetogenic bacteria, an alternative route to chemical synthesis, using industrial waste gas. Applied and Environmental Microbiology. 2011 Aug;77(15): 5467-75. PubMed PMID: WOS:000293224500045.

[105] Köpke M, Mihalcea C, Bromley JC, Simpson SD. Fermentative production of ethanol from carbon monoxide. Current Opinion in Biotechnology. 2011 Jun;22(3):320-5. PubMed PMID: WOS:000292228300003.

[106] Kundiyana DK, Wilkins MR, Maddipati P, Huhnke RL. Effect of temperature, pH and buffer presence on ethanol production from synthesis gas by "Clostridium ragsdalei". Bioresource technology. 2011;102(10):5794-9.

[107] Saxena J, Tanner RS. Effect of trace metals on ethanol production from synthesis gas by the ethanologenic acetogen, Clostridium ragsdalei. Journal of Industrial Microbiology & Biotechnology. 2011 Apr;38(4):513-21. PubMed PMID: WOS: 000288673700004.

[108] Heiskanen H, Virkajärvi I, Viikari L. The effect of syngas composition on the growth and product formation of Butyribacterium methylotrophicum. Enzyme and Microbial Technology. 2007;41(3):362-7.

[109] Noelling J. Genome sequence and comparative analysis of the solvent-producing bacterium Clostridium acetobutylicum. Journal of. 2001;183(16):4823-38.

[110] Huang H, Liu H, Gan Y-R. Genetic modification of critical enzymes and involved genes in butanol biosynthesis from biomass. Biotechnology advances. 2010;28(5): 651-7.

[111] Ukpong MN, Atiyeh HK, De Lorme MJM, Liu K, Zhu X, Tanner RS, et al. Physiological response of Clostridium carboxidivorans during conversion of synthesis gas to

solvents in a gas-fed bioreactor. Biotechnology and bioengineering. 2012;Accepted article.

[112] Kundiyana DK, Huhnke RL, Wilkins MR. Syngas fermentation in a 100-L pilot scale fermentor: Design and process considerations. Journal of Bioscience and Bioengineering. 2010 May;109(5):492-8. PubMed PMID: WOS:000279089600011.

[113] Phillips JR, Klasson KT, Claussen EC, Gaddy JL. Biological production of ethanol from coal synthesis gas. Applied Biochemistry And Biotechnology. 1993;39(1):0-7.

[114] Lee SY, Park JH, Jang SH, Nielsen LK, Kim J, Jung KS. Fermentative butanol production by clostridia. Biotechnology and Bioengineering. 2008 Oct;101(2):209-28. PubMed PMID: WOS:000259512900001.

[115] Sauer U, Durre P. Differential induction of genes related to solvent formation during the shift from acidogenesis to solventogenesis in continuous culture of Clostridium acetobutylicum. Fems Microbiology Letters. 1995 Jan;125(1):115-20. PubMed PMID: WOS:A1995QA11200018.

[116] Shi Z, Blaschek HP. Transcriptional analysis of Clostridium beijerinckii NCIMB 8052 and the hyper-butanol-producing mutant BA101 during the shift from acidogenesis to solventogenesis. Applied and Environmental Microbiology. 2008 Dec;74(24): 7709-14. PubMed PMID: WOS:000261513700031.

[117] Köpke M, Dürre P. Biochemical production of biobutanol. In: Luque R, Campelo J, Clark JH, editors. Camebridge, UK: Woodhead Publishing Ltd; 2011. p. 221-57.

[118] Gheshlaghi R, Scharer JM, Moo-young M, Chou CP. Metabolic pathways of clostridia for producing butanol. Biotechnology advances. 2009;27(6):764-81.

[119] Seedorf H, Fricke WF, Veith B, Bruggemann H, Liesegang H, Strittimatter A, et al. The genome of Clostridium kluyveri, a strict anaerobe with unique metabolic features. Proceedings of the National Academy of Sciences of the United States of America. 2008 Feb;105(6):2128-33. PubMed PMID: WOS:000253261900064.

[120] Inui M, Suda M, Kimura S, Yasuda K, Suzuki H, Toda H, et al. Expression of Clostridium acetobutylicum butanol synthetic genes in Escherichia coli. Applied microbiology and biotechnology. 2008;77(6):1305-16.

[121] Tomas CA, Beamish J, Papoutsakis ET. Transcriptional analysis of butanol stress and tolerance in Clostridium acetobutylicum. Journal of Bacteriology. 2004 Apr;186(7): 2006-18. PubMed PMID: WOS:000220363200011.

[122] Jones DT, Woods DR. Acetone-butanol fermentation revisited. Microbiological reviews. 1986;50(4):484-524.

[123] Paul D, Austin FW, Arick T, Bridges SM, Burgess SC, Dandass YS, et al. Genome sequence of the solvent-producing bacterium Clostridium carboxidivorans strain P7T. Journal of bacteriology. 2010;192(20):5554-5.

[146] Green EM, Boynton ZL, Harris LM, Rudolph FB, Papoutsakis ET, Bennett GN. Genetic manipulation of acid formation pathways by gene inactivation in Clostridium acetobutylicum ATCC 824. Microbiology-Uk. 1996 Aug;142:2079-86. PubMed PMID: WOS:A1996VB82900020.

[147] Green EM, Bennett GN. Inactivation of an aldehyde/alcohol dehydrogenase gene from Clostridium acetobutylicum ATCC 824. Applied biochemistry and biotechnology. 1996;57-58:213-21.

[148] Wilkinson SR, Young M. Targeted integration of genes into the Clostridium acetobutylicum genome. Microbiology-Uk. 1994 Jan;140:89-95. PubMed PMID: WOS:A1994NE68400013.

[149] Shimizu T, Bathein W, Tamaki M, Hayashi H. The virr gene, a member of a class of 2-component response regulators, regulates the production of perfringolysin-o, collagenase, and hemagglutinin in Clostridium perfringens. Journal of Bacteriology. 1994 Mar;176(6):1616-23. PubMed PMID: WOS:A1994NB99400008.

[150] Liyanage H, Kashket S, Young M, Kashket ER. Clostridium beijerinckii and Clostridium difficile detoxify methylglyoxal by a novel mechanism involving glycerol dehydrogenase. Applied and Environmental Microbiology. 2001 May;67(5):2004-10. PubMed PMID: WOS:000168488400003.

[151] O'Connor JR, Lyras D, Farrow KA, Adams V, Powell DR, Hinds J, et al. Construction and analysis of chromosomal Clostridium difficile mutants. Molecular Microbiology. 2006 Sep;61(5):1335-51. PubMed PMID: WOS:000239701400020.

[152] Wilkinson SR, Young DI, Morris JG, Young M. Molecular genetics and the initiation of solventogenesis in Clostridium beijerinckii (formerly Clostridium acetobutylicum) NCIMB-8052. Fems Microbiology Reviews. 1995 Oct;17(3):275-85. PubMed PMID: WOS:A1995TA64700007.

[153] Harris LM, Welker NE, Papoutsakis ET. Northern, morphological, and fermentation analysis of spo0A inactivation and overexpression in Clostridium acetobutylicum ATCC 824. Journal of Bacteriology. 2002 Jul;184(13):3586-97. PubMed PMID: WOS: 000176237400020.

[154] Sarker MR, Carman RJ, McClane BA. Inactivation of the gene (cpe) encoding Clostridium perfringens enterotoxin eliminates the ability of two cpe-positive C-perfringens type A human gastrointestinal disease isolates to affect rabbit ileal loops. Molecular Microbiology. 1999 Sep;33(5):946-58. PubMed PMID: WOS:000082779700005.

[155] Soucaille P, Rainer F, Christian C, inventors. Process for chromosomal integration and DNA sequence replacement in clostridia. EU. 2008.

[156] Argyros DA, Tripathi SA, Barrett TF, Rogers SR, Feinberg LF, Olson DG, et al. High ethanol titers from cellulose by using metabolically engineered thermophilic, anaerobic microbes. Applied and Environmental Microbiology. 2011 Dec;77(23):8288-94. PubMed PMID: WOS:000297164100012.

[157] Karberg M, Guo HT, Zhong J, Coon R, Perutka J, Lambowitz AM. Group II introns as controllable gene targeting vectors for genetic manipulation of bacteria. Nature Biotechnology. 2001 Dec;19(12):1162-7. PubMed PMID: WOS:000172524400030.

[158] Heap JT, Pennington OJ, Cartman ST, Carter GP, Minton NP. The ClosTron: A universal gene knock-out system for the genus Clostridium. Journal of Microbiological Methods. 2007 Sep;70(3):452-64. PubMed PMID: WOS:000249573800008.

[159] Heap JT, Kuehne Sa, Ehsaan M, Cartman ST, Cooksley CM, Scott JC, et al. The ClosTron: Mutagenesis in Clostridium refined and streamlined. Journal of microbiological methods. 2010;80(1):49-55.

[160] Camiade E, Peltier J, Bourgeois I, Couture-Tosi E, Courtin P, Antunes A, et al. Characterization of Acp, a Peptidoglycan hydrolase of Clostridium perfringens with N-acetylglucosaminidase activity That is implicated in cell separation and stress-induced autolysis. Journal of Bacteriology. 2010 May;192(9):2373-84. PubMed PMID: WOS:000276685800009.

[161] Heap JT, Ehsaan M, Cooksley CM, Ng YK, Cartman ST, Winzer K, et al. Integration of DNA into bacterial chromosomes from plasmids without a counter-selection marker. Nucleic Acids Res. 2012 Apr 1;40(8):e59. PubMed PMID: 22259038. Pubmed Central PMCID: 3333862. Epub 2012/01/20. eng.

[162] Keis S, Shaheen R, Jones DT. Emended descriptions of Clostridium acetobutylicum and Clostridium beijerinckii, and descriptions of Clostridium saccharoperbutylacetonicum sp nov and Clostridium saccharobutylicum sp nov. International Journal of Systematic and Evolutionary Microbiology. 2001 Nov;51:2095-103. PubMed PMID: WOS:000172374400020.

[163] Bertram J, Durre P. Conjugal transfer and expression of Streptococcal transposons in Clostridium acetobutylicum. Archives of Microbiology. 1989 May;151(6):551-7. PubMed PMID: WOS:A1989AA14000015.

[164] Woolley RC, Pennock A, Ashton RJ, Davies A, Young M. Transfer of Tn1545 and Tn916 to Clostridium acetobutylicum. Plasmid. 1989 Sep;22(2):169-74. PubMed PMID: WOS:A1989CK56900010.

[165] Blouzard JC, Valette O, Tardif C, de Philip P. Random mutagenesis of Clostridium cellulolyticum by using a Tn1545 derivative. Applied and Environmental Microbiology. 2010 Jul;76(13):4546-9. PubMed PMID: WOS:000279082800049.

[166] Cartman ST, Minton NP. A mariner-based transposon system for in vivo random mutagenesis of Clostridium difficile. Appl Environ Microbiol. 2010 Feb;76(4):1103-9. PubMed PMID: 20023081. Pubmed Central PMCID: 2820977. Epub 2009/12/22. eng.

[167] Koepke M, Liew FM, inventors. Recombinant microorganism and methods of production thereof. U.S.2011.

[168] Simpson SD, Koepke M, Liew FM, inventors. Recombinant microorganisms and methods of use thereof. U.S.2011.

[169] Lederle S. Heterofermentative acetonproduktion: University of Ulm, Germany; 2011.

[170] Becker U, Grund G, Orschel M, Doderer K, Löhden G, Brand G, et al., inventors. Zellen und Verfahren zur Herstellung von Aceton. Germany. 2009.

[171] Harris LM, Desai RP, Welker NE, Papoutsakis ET. Characterization of recombinant strains of the Clostridium acetobutylicum butyrate kinase inactivation mutant: need for new phenomenological models for solventogenesis and butanol inhibition? Biotechnology and bioengineering. 2000;67(1):1-11.

[172] Lehmann D, Lutke-Eversloh T. Switching Clostridium acetobutylicum to an ethanol producer by disruption of the butyrate/butanol fermentative pathway. Metabolic Engineering. 2011 Sep;13(5):464-73. PubMed PMID: WOS:000294291200002.

[173] Kuit W, Minton NP, Lopez-Contreras AM, Eggink G. Disruption of the acetate kinase (ack) gene of Clostridium acetobutylicum results in delayed acetate production. Applied Microbiology and Biotechnology. 2012 May;94(3):729-41. PubMed PMID: WOS: 000302283700015.

[174] Yu MR, Zhang YL, Tang IC, Yang ST. Metabolic engineering of Clostridium tyrobutyricum for n-butanol production. Metabolic Engineering. 2011 Jul;13(4):373-82. PubMed PMID: WOS:000291471500003.

[175] Li YC, Tschaplinski TJ, Engle NL, Hamilton CY, Rodriguez M, Liao JC, et al. Combined inactivation of the Clostridium cellulolyticum lactate and malate dehydrogenase genes substantially increases ethanol yield from cellulose and switchgrass fermentations. Biotechnology for Biofuels. 2012 Jan;5. PubMed PMID: WOS: 000299821600001.

[176] Lee J, Jang YS, Choi SJ, Im JA, Song H, Cho JH, et al. Metabolic engineering of Clostridium acetobutylicum ATCC 824 for isopropanol-butanol-ethanol fermentation. Applied and Environmental Microbiology. 2012 Mar;78(5):1416-23. PubMed PMID: WOS:000300537400012.

[177] Higashide W, Li YC, Yang YF, Liao JC. Metabolic Engineering of Clostridium cellulolyticum for Production of isobutanol from cellulose. Applied and Environmental Microbiology. 2011 Apr;77(8):2727-33. PubMed PMID: WOS:000289459300020.

[178] Cai XP, Bennett GN. Improving the Clostridium acetobutylicum butanol fermentation by engineering the strain for co-production of riboflavin. Journal of Industrial Microbiology & Biotechnology. 2011 Aug;38(8):1013-25. PubMed PMID: WOS: 000293002000015.

[179] Zhu LJ, Dong HJ, Zhang YP, Li Y. Engineering the robustness of Clostridium acetobutylicum by introducing glutathione biosynthetic capability. Metabolic Engineering. 2011 Jul;13(4):426-34. PubMed PMID: WOS:000291471500008.

[180] Tomas CA, Welker NE, Papoutsakis ET. Overexpression of groESL in Clostridium acetobutylicum results in increased solvent production and tolerance, prolonged me-

tabolism, and changes in the cell's transcriptional program. Applied and Environmental Microbiology. 2003 Aug;69(8):4951-65. PubMed PMID: WOS: 000184672500078.

[181] Nakayama S-i, Kosaka T, Hirakawa H, Matsuura K, Yoshino S, Furukawa K. Metabolic engineering for solvent productivity by downregulation of the hydrogenase gene cluster hupCBA in Clostridium saccharoperbutylacetonicum strain N1-4. Applied Microbiology and Biotechnology. 2008 Mar;78(3):483-93. PubMed PMID: WOS: 000253132500012.

[182] Atsumi S, Hanai T, Liao J. Non-fermentative pathways for synthesis of branched-chain higher alcohols as biofuels. Nature. 2008;451(7174):86-9.

[183] Clomburg JM, Gonzalez R. Biofuel production in Escherichia coli: the role of metabolic engineering and synthetic biology. Applied Microbiology and Biotechnology. 2010 Mar;86(2):419-34. PubMed PMID: WOS:000274707600003.

[184] Peralta-Yahya PP, Keasling JD. Advanced biofuel production in microbes. Biotechnology Journal. 2010 Feb;5(2):147-62. PubMed PMID: WOS:000275275500005.

[185] Kapic A, Jones ST, Heindel TJ. Carbon monoxide mass transfer in a syngas mixture. Industrial & Engineering Chemistry Research. 2006 Dec;45(26):9150-5. PubMed PMID: WOS:000242786200049.

[186] Mohammadi M, Younesi H, Najafpour G, Mohamed AR. Sustainable ethanol fermentation from synthesis gas by Clostridium ljungdahlii in a continuous stirred tank bioreactor. Journal of Chemical Technology and Biotechnology. 2012 Jun;87(6): 837-43. PubMed PMID: WOS:000304190300018.

[187] Bredwell MD, Srivastava P, Worden RM. Reactor design issues for synthesis-gas fermentations. Biotechnology Progress. 1999 Sep-Oct;15(5):834-44. PubMed PMID: WOS:000082936200007.

[188] Ahmed A, Cateni BG, Huhnke RL, Lewis RS. Effects of biomass-generated producer gas constituents on cell growth, product distribution and hydrogenase activity of Clostridium carboxidivorans P7T. Biomass and Bioenergy. 2006;30(7):665-72.

[189] Xu D, Lewis RS. Syngas fermentation to biofuels: Effects of ammonia impurity in raw syngas on hydrogenase activity. Biomass and Bioenergy. 2012.

[190] Ahmed A, Lewis RS. Fermentation of biomass-generated synthesis gas: Effects of nitric oxide. Biotechnology and Bioengineering. 2007 Aug;97(5):1080-6. PubMed PMID: WOS:000247960300008.

[191] Brogren C, Karlsson HT, Bjerle I. Absorption of NO in an alkaline solution of KMnO4. Chemical Engineering & Technology. 1997 Aug;20(6):396-402. PubMed PMID: WOS:A1997XX28400006.

[192] Chu H, Chien TW, Li SY. Simultaneous absorption of SO2 and NO from flue gas with KMnO4/NaOH solutions. Science of the Total Environment. 2001 Jul;275(1-3): 127-35. PubMed PMID: WOS:000169969700011.

[193] Klasson KT, Ackerson MD, Clausen EC, Gaddy JL. Biological conversion of coal and coal-derived synthesis gas. Fuel. 1993 Dec;72(12):1673-8. PubMed PMID: WOS:A1993MJ23700016.

[194] Vega JL, Holmberg VL, Clausen EC, Gaddy JL. Fermentation parameters of Peptostreptococcus productus on gaseous substrates (CO, H2/CO2). Archives of Microbiology. 1989;151(1):65-70. PubMed PMID: WOS:A1989R169500012.

[195] Hurst KM, Lewis RS. Carbon monoxide partial pressure effects on the metabolic process of syngas fermentation. Biochemical Engineering Journal. 2010;48(2):159-65.

[196] Najafpour G, Younesi H. Ethanol and acetate synthesis from waste gas using batch culture of Clostridium ljungdahlii. Enzyme and Microbial Technology. 2006 Jan; 38(1-2):223-8. PubMed PMID: WOS:000234199300033.

[197] Gaddy JL, Chen GJ, inventors. Bioconversion of waste biomass to useful products. 1998.

[198] Genthner BRS, Bryant MP. Growth of Eubacterium limosum with carbon monoxide as the energy source. Applied and Environmental Microbiology. 1982;43(1):70-4. PubMed PMID: WOS:A1982MX20500009.

[199] Demler M, Weuster-Botz D. Reaction Engineering analysis of hydrogenotrophic production of ccetic ccid by Acetobacterium woodii. Biotechnology and Bioengineering. 2011 Feb;108(2):470-4. PubMed PMID: WOS:000285393000024.

[200] Cotter JL, Chinn MS, Grunden AM. Ethanol and acetate production by Clostridium ljungdahlii and Clostridium autoethanogenum using resting cells. Bioprocess Biosyst Eng. 2009;32(3):369-80.

[201] Kanchanatawee S, Maddox IS. Effect of phosphate and ammonium ion concentrations on solvent production in defined medium by Clostridium acetobutylicum. Journal of Industrial Microbiology. 1990 Jul;5(5):277-82. PubMed PMID: WOS:A1990DU68500002.

[202] Guo Y, Xu JL, Zhang Y, Xu HJ, Yuan ZH, Li D. Medium optimization for ethanol production with Clostridium autoethanogenum with carbon monoxide as sole carbon source. Bioresource Technology. 2010 Nov;101(22):8784-9. PubMed PMID: WOS: 000281262900044.

[203] Kundiyana DK, Huhnke RL, Maddipati P, Atiyeh HK, Wilkins MR. Feasibility of incorporating cotton seed extract in Clostridium strain P11 fermentation medium during synthesis gas fermentation. Bioresource Technology. 2010 Dec;101(24):9673-80. PubMed PMID: WOS:000282201200037.

[204] Sim JH, Kamaruddin AH. Optimization of acetic acid production from synthesis gas by chemolithotrophic bacterium - Clostridium aceticum using statistical approach.

Bioresource Technology. 2008 May;99(8):2724-35. PubMed PMID: WOS: 000254072200003.

[205] Vega JL, Prieto S, Elmore BB, Clausen EC, Gaddy JL. The biological production of ethanol from synthesis gas. Applied Biochemistry and Biotechnology. 1989 Jan-Aug; 20-1:781-97. PubMed PMID: WOS:A1989U642400058.

[206] Hu P, Jacobsen LT, Horton JG, Lewis RS. Sulfide assessment in bioreactors with gas replacement. Biochemical Engineering Journal. 2010 May;49(3):429-34. PubMed PMID: WOS:000277782900018.

[207] Do YS, Smeenk J, Broer KM, Kisting CJ, Brown R, Heindel TJ, et al. Growth of Rhodospirillum rubrum on synthesis gas: Conversion of CO to H-2 and poly-beta-hydroxyalkanoate. Biotechnology and Bioengineering. 2007 Jun;97(2):279-86. PubMed PMID: WOS:000246434700008.

[208] Devi MP, Mohan SV, Mohanakrishna G, Sarma PN. Regulatory influence of CO2 supplementation on fermentative hydrogen production process. International Journal of Hydrogen Energy. 2010 Oct;35(19):10701-9. PubMed PMID: WOS: 000283977100086.

[209] Grethlein AJ, Worden RM, Jain MK, Datta R. Continuous production of mixed alcohol and acids from carbon monoxide. Applied Biochemistry and Biotechnology. 1990 Spr-Sum;24-5:875-84. PubMed PMID: WOS:A1990DE57300079.

[210] Girbal L, Croux C, Vasconcelos I, Soucaille P. Regulation of metabolic shifts in Clostridium acetobutylicum ATCC 824. Fems Microbiology Reviews. 1995 Oct;17(3): 287-97. PubMed PMID: WOS:A1995TA64700008.

[211] Kundiyana DK, Wilkins MR, Maddipati P, Huhnke RL. Effect of temperature, pH and buffer presence on ethanol production from synthesis gas by "Clostridium ragsdalei". Bioresource Technology. 2011 May;102(10):5794-9. PubMed PMID: WOS: 000291125800033.

[212] Gaddy JL, Clausen EC, inventors. Clostridium ljungdahlii, an anaerobic ethanol and acetate producing microorganism. 1992.

[213] Cotter JL, Chinn MS, Grunden AM. Influence of process parameters on growth of Clostridium ljungdahlii and Clostridium autoethanogenum on synthesis gas. Enzyme and Microbial Technology. 2009 May;44(5):281-8. PubMed PMID: WOS: 000264908000006.

[214] Frohlich H, Villian L, Melzner D, Strube J. Membrane technology in bioprocess science. Chemie Ingenieur Technik. 2012 Jun;84(6):905-17. PubMed PMID: WOS: 000304438600024.

[215] Rushton A, Ward AS, Holdich RG. Solid-liquid filtration and separation technology. Germany: VCH Publishers, Inc.; 1996.

[216] Kargupta K, Datta S, Sanyal SK. Analysis of the performance of a continuous membrane bioreactor with cell recycling during ethanol fermentation. Biochemical Engineering Journal. 1998 Jan;1(1):31-7. PubMed PMID: WOS:000077894400004.

[217] Brown SD, Guss AM, Karpinets TV, Parks JM, Smolin N, Yang S, et al. Mutant alcohol dehydrogenase leads to improved ethanol tolerance in Clostridium thermocellum. Proceedings of the National Academy of Sciences of the United States of America. 2011;108(33):13752-7.

[218] Shao XJ, Raman B, Zhu MJ, Mielenz JR, Brown SD, Guss AM, et al. Mutant selection and phenotypic and genetic characterization of ethanol-tolerant strains of Clostridium thermocellum. Applied Microbiology and Biotechnology. 2011 Nov;92(3):641-52. PubMed PMID: WOS:000295673800020.

[219] Ezeji T, Milne C, Price ND, Blaschek HP. Achievements and perspectives to overcome the poor solvent resistance in acetone and butanol-producing microorganisms. Applied Microbiology and Biotechnology. 2010 Feb;85(6):1697-712. PubMed PMID: WOS:000273978400007.

[220] Borden JR, Papoutsakis ET. Dynamics of genomic-library enrichment and identification of solvent tolerance genes for Clostridium acetobutylicum. Applied and environmental microbiology. 2007;73(9):3061-8.

[221] Ezeji TC, Li Y. Advanced product recovery technologies. Hoboken: Wiley; 2009.

[222] Ezeji TC, Qureshi N, Blaschek HP. Bioproduction of butanol from biomass: from genes to bioreactors. Current opinion in biotechnology. 2007;18(3):220-7.

[223] Jauregui-Haza UJ, Pardillo-Fontdevila EJ, Wilhelm AM, Delmas H. Solubility of hydrogen and carbon monoxide in water and some organic solvents. Latin American Applied Research. 2004 Apr;34(2):71-4. PubMed PMID: WOS:000220679200001.

[224] Karcher P, Ezeji TC, Qureshi N, Blaschek HP. Microbial production of butanol: product recovery by extraction. New Delhi: IK International Publishing House Pvt Ltd; 2005.

[225] Ezeji TC, Li Y. Advanced product recovery technologies. Hoboken: Wiley; 2009.

[226] Ezeji TC, Qureshi N, Blaschek HP. Production of acetone, butanol and ethanol by Clostridium beijerinckii BA101 and in situ recovery by gas stripping. World Journal of Microbiology & Biotechnology. 2003 Aug;19(6):595-603. PubMed PMID: WOS: 000184717400007.

[227] Wei L, Pordesimo LO, Igathinathane C, Batchelor WD. Process engineering evaluation of ethanol production from wood through bioprocessing and chemical catalysis. Biomass & Bioenergy. 2009 Feb;33(2):255-66. PubMed PMID: WOS:000263995200011.

[228] Piccolo C, Bezzo F. A techno-economic comparison between two technologies for bioethanol production from lignocellulose. Biomass & Bioenergy. 2009 Mar;33(3): 478-91. PubMed PMID: WOS:000264433600016.

[229] Yoneda N, Kusano S, Yasui M, Pujado P, Wilcher S. Recent advances in processes and catalysts for the production of acetic acid. Applied Catalysis a-General. 2001 Nov;221(1-2):253-65. PubMed PMID: WOS:000173000700019.

[230] Zigova J, Sturdik E, Vandak D, Schlosser S. Butyric acid production by Clostridium butyricum with integrated extraction and pertraction. Process Biochemistry. 1999 Oct;34(8):835-43. PubMed PMID: WOS:000083166300009.

[231] Gaddy JL, inventor. Biological production of ethanol from waste gases with Clostridium ljungdahlii. US. 2000.

[232] Guzman DD. INEOS Bio looks to start US cellulosic ethanol in 2012. ICIS Chemical Business. 2011:22-3.

[233] Bio I. INEOS Bio - Pioneering waste-to-biofuel technology attracts cross-party support in UK 2011. Available from: http://www.ineosbio.com/76-Press_releases-16.htm.

[234] Coskata. Coskata, Inc.'s semi-commercial facility demonstrates two years of successful operation 2011. Available from: http://www.coskata.com/company/media.asp?story=504B571C-0916-474E-BFFA-ACB326EFDB68.

[235] Coskata. About us 2012. Available from: http://www.coskata.com/company/.

[236] Huhnke RL, Lewis RS, Tanner RS, inventors. Isolation and characterization of novel Clostridial species. US. 2010.

[237] Kundiyana DK, Huhnke RL, Wilkins MR. Effect of nutrient limitation and two-stage continuous fermentor design on productivities during "Clostridium ragsdalei" syngas fermentation. Bioresource Technology. 2011 May;102(10):6058-64. PubMed PMID: WOS:000291125800069.

[238] Zahn JA, Saxena J, inventors. Novel ethanologenic Clostridium species, Clostridium coskatii. US. 2012.

[239] Kuo I. Coskata grabs biggest slice of USDA's $405M in biofuels loan guarantees: Reuters; 2011. Available from: http://www.reuters.com/article/2011/01/21/idUS126007139820110121.

[240] Lane J. Coskata switches focus from biomass to natural gas; to raise $100M in natgas-oriented private placement: Biofuels Digest; 2012. Available from: http://www.biofuelsdigest.com/bdigest/2012/07/20/coskata-switches-from-biomass-to-natural-gas-to-raise-100m-in-natgas-oriented-private-placement/.

[241] Wernau J. Biofuel firm Coskata drops initial public offering plans amid volatile IPO market: Chicago Tribute Business; 2012. Available from: http://www.chicagotribune.com/business/ct-biz-0721-ipos-20120721,0,1092013.story.

[242] Lane J. The intrepid investor: How are biofuels IPOs performing and why? : Biofuels Digest; 2012. Available from: http://www.biofuelsdigest.com/bdigest/2012/02/09/the-intrepid-investor-part-i/.

[243] Guzman DD, Chang J. Bio-based 2,3 BDO set for 2014 sales. ICIS Chemical Business. 2012:24.

[244] LanzaTech. Media Releases. Available from: http://lanzatech.com/media/media-releases.

Chapter 6

The Promising Fuel-Biobutanol

Hongjuan Liu, Genyu Wang and Jianan Zhang

Additional information is available at the end of the chapter

http://dx.doi.org/10. 5772/52535

1. Introduction

In recent years, two problems roused peoples' concern. One is energy crisis caused by the depleting of petroleum fuel. The other is environmental issues such as greenhouse effect, global warming, etc. Therefore, renewable sources utilization technology and bioenergy production technology developed fast for solving such two problems. Bioethanol as one of the biofuel has been applied in automobiles with gasoline in different blending proportions (Zhou and Thomson, 2009; Yan and Lin, 2009). Biobutanol is one of the new types of biofuel. It continuously attracted the attention of researchers and industrialists because of its several distinct advantages.

1.1. Property of butanol

Butanol is a four carbon straight chained alcohol, colorless and flammable. Butanol can be mixed with ethanol, ether and other organic solvent. Butanol can be used as a solvent, in cosmetics, hydraulic fluids, detergent formulations, drugs, antibiotics, hormones and vitamins, as a chemical intermediate in the production of butyl acrylate and methacrylate, and additionally as an extract agent in the manufacture of pharmaceuticals. Butanol has a 4-carbon structure and the carbon atoms can form either a straight-chain or a branched structure, resulting in different properties. There exist different isomers, based on the location of the–OH and carbon chain structure. The different structures, properties and main applications are shown as Table 1.

Although the properties of butanol isomers are different in octane number, boiling point, viscosity, etc., the main applications are similar in some aspects, such as being used as solvents, industrial cleaners, or gasoline additives. All these butanol isomers can be produced from fossil fuels by different methods, only n-butanol, a straight-chain molecule structure can be produced from biomass.

	n-Butanol	2-Butanol	iso-Butanol	tert-Butanol
Molecular structure	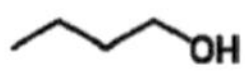	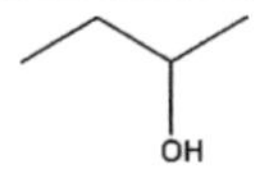	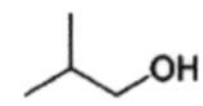	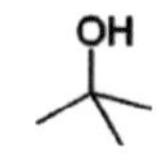
Density (g/cm3)	0. 81	0. 806	0. 802	0. 789
Boiling point(°C)	118	99. 5	108	82. 4
Melting point(°C)	-90	-115	-108	25-26
Refractive index(n20D)	1. 399	1. 3978	1. 3959	1. 3878
Flash point(°C)	35	22-27	28	11
Motor octane number	78	32	94	89
Main applications	Solvents-for paints, resins, dyes, etc. Plasticizers- improve a plastic material processes Chemical intermediate -for butyl esters or butyl ethers, etc. Cosmetics- including eye makeup, lipsticks, etc. Gasoline additive	Solvent Chemical intermediate- for butanone, etc. Industrial cleaners -paint removers Perfumes or in artificial flavors	Solvent and additive for paint Gasoline additive Industrial cleaners -paint removers Ink ingredient	Solvent Denaturant for ethanol Industrial cleaners- paint removers Gasoline additive for octane booster and oxygenate Intermediate for MTBE, ETBE, TBHP, etc.

Table 1. Structures, properties and main applications of n-butanol, 2-Butanol, iso-Butanol and tert-Butanol

1.2. Advantages of butanol as fuel

Except the use of solvent, chemical intermediate and extract agent, butanol also can be used as fuel, which attracted people's attention in recent years. Because of the good properties of high heat value, high viscosity, low volatility, high hydrophobicity, less corrosive, butanol has the potential to be a good fuel in the future. The properities of butanol and other fuels or homologues are compared as Table 2. (Freeman et al., 1988; Dean, 1992)

Fuel	Octane number	Cetane number	Evaporation heat (MJ/kg)	Combustion energy(MJ/dm³)	Flammability limits (%vol)	Saturation pressure (kPa) at 38°C
Gasoline	80-99	0-10	0. 36	32	0. 6-0. 8	31. 01
Methanol	111	3	1. 2	16	6-36. 5	31. 69
Ethanol	108	8	0. 92	19. 6	4. 3-19	13. 8
Butanol	96	25	0. 43	29. 2	1. 4-11. 2	2. 27

Table 2. Properities of butanol and other fuels

Butanol appeared the good properties compared with it's homologues such as 2-butanol, iso-butanol and tert-butanol and other fuels such as Gasoline and ethanol. Actually, when ethanol is mixed with gasoline (less than 10%), there exists some disadvantages. Firstly, the heating value of ethanol is one sixth of gasoline. The fuel consumption will increase 5% if the engine is not retrofitted. Secondly, acetic acid will be produced during the burning process of ethanol, which is corrosive to the materials of vehicle. The preservative must be added when the ethanol proportion upper than 15%. Thirdly, ethanol is hydroscopic and the liquid phase separation may be occurring with high water proportion. Furthermore, ethanol as fuel cannot be preserved easily and it is more difficult in the process of allocation, storage, transition than that of gasoline.

Compared with ethanol, butanol overcomes above disadvantages and it shows potential advantages. For example, Butanol has higher energy content and higher burning efficiency, which can be used for longer distance. The air to fuel ratio and the energy content of butanol are closer to gasoline. So, butanol can be easily mixed with gasoline in any proportion. Butanol is less volatile and explosive, has higher flash point, and lower vapor pressure, which makes it safer to handle and can be shipped through existing fuel pipelines. In addition, Butanol can be used directly or blended with gasoline or diesel without any vehicle retrofit (Durre, 2007; Pfromm et al., 2010).

Actually, the first-time synthesis of biobutanol at laboratory level was reported by Pasteur in 1861 (Durre, 1998) and the industrial synthesis of biobutanol was started during 1912–1914 by fermentation (Jones and Woods, 1986). However, before 2005, butanol was mainly used as solvent and precursor of other chemicals due to the product inhibition and low butanol productivity. To bring awareness to butanol's potential as a renewable fuel, David Ramey drove his family car from Ohio to California on 100% butanol (http://www.consumerenergyreport.com /2011/02/09/reintroducing-butanol/). And then, two giant companies DuPont and BP have declared to finance development of a modernize production plant supported by research and development. (http://biomassmagazine.com/articles/2994 /eu-approves-bp-dupont-biobutanol-venture) The economy of biobutanol production also was revaluated. The research of a continuous fermentation pilot plant operating in Austria in the 1990s introduced new technologies and proved economic feasibility with agricultural waste potatoes. (Nimcevic and Gapes, 2000).

2. Production methods of butanol

Butanol can be obtained using chemical technologies, such as Oxo-synthesis and aldol condensation. It is also possible to produce butanol in the process of fermentation by bacteria and butanol as one of the products called biobutanol. The most popular bacteria species used for fermentation is *Clostridium acetobutylicum*. Because the main products of this process containing acetone, butanol and ethanol, the fermentation is called ABE fermentation (Qureshi and Maddox, 1995).

2.1. Chemical process

Butanol can be produced by chemical synthesis. One process is Oxo-synthesis, which involves the reaction of propylene with carbon monoxide and hydrogen in the presence of cobalt or rhodium as the catalyst. The mixture of n-butyraldehyde and isobutyraldehyde are obtained and then the mixture can be hydrogenated to the corresponding n-butanol and isobutyl alcohols (Park, 1996).The reactions are as following:

$$CH_3CH\text{–}CH_2 + CO + H_2 \rightarrow CH_3CH_2CH_2CHO + (CH_3)_2CHCHO \tag{1}$$

$$\begin{array}{ll} CH_3CH_2CH_2CHO + H_2 \rightarrow CH_3CH_2CH_2CH_2OH & (a) \\ (CH_3)_2CHCHO + H_2 \rightarrow (CH_3)_2CHCH_2OH & (b) \end{array} \tag{2}$$

When using cobalt as the catalyst, the reaction processes at 10~20MPa and 130~160°°C, the products ratio of n-butyraldehyde and isobutyraldehyde is 3. Rhodium as the catalyst used in industry from 1976 and the reaction processes at 0.7-3MPa and 80-120°°C.The products ratio of n-butyraldehyde and isobutyraldehyde can reach 8-16. Hydrogenaration processes by using the catalyst of nickel or copper in gaseous phase or nickel in liquid phase. Some by-products can be transferred into butanol at high temperature and high pressure that will enhance the product purity.

Another route is aldol condensation, which involves the reaction of condensation and dehydration from two molecules of acetic aldehyde. And then, the product crotonaldehyde was transformed into n-butanol by hydrogenation at 180°°C and 0.2MPa. The reaction is as following: $CH_3CH{=}CHCHO + 2H_2 \longrightarrow CH_3CH_2CH_2CH_2OH$

Comparing the two processes, Oxo-synthesis route has the advantages of materials easily obtained, comparable moderate reaction conditions, enhanced ratio of n-butanol to isobutyl alcohol. So, Oxo-synthesis process is the main industrial route for n-butanol production. There are also some other fossil oil derived raw materials such as ethylene, propylene and triethylaluminium or carbon monoxide and hydrogen are used in butanol production (Zverlov, et al., 2006).

2.2. Biological process

Except the chemical ways, butanol can also be obtained from biological ways with the renewable resources by the microorganism through fermentation. The *Clostridia* genus is very common for butanol synthesis under anaerobic conditions, and the fermentation products are often the mixture of butanol, acetone and ethanol. A few kinds of *Clostridium* can utilize cellulose and hemicellulose with the ability of cellulolytic activities (Mitchell et al., 1997; Berezina et al. 2009).

Compared with the chemical ways for butanol production, biological ways has the distinct advantages. For example, it can utilize the renewable resources such as wheat straw, corn core, switch grass, etc. Furthermore, biological process has high product selectivity, high security, less by-products. Furthermore, the fermentation condition of butanol production is milder than that of chemical ways and the products are easier to separate. The process of biobutanol production with Lignocellulosic feedstocks is as following (Fig. 1):

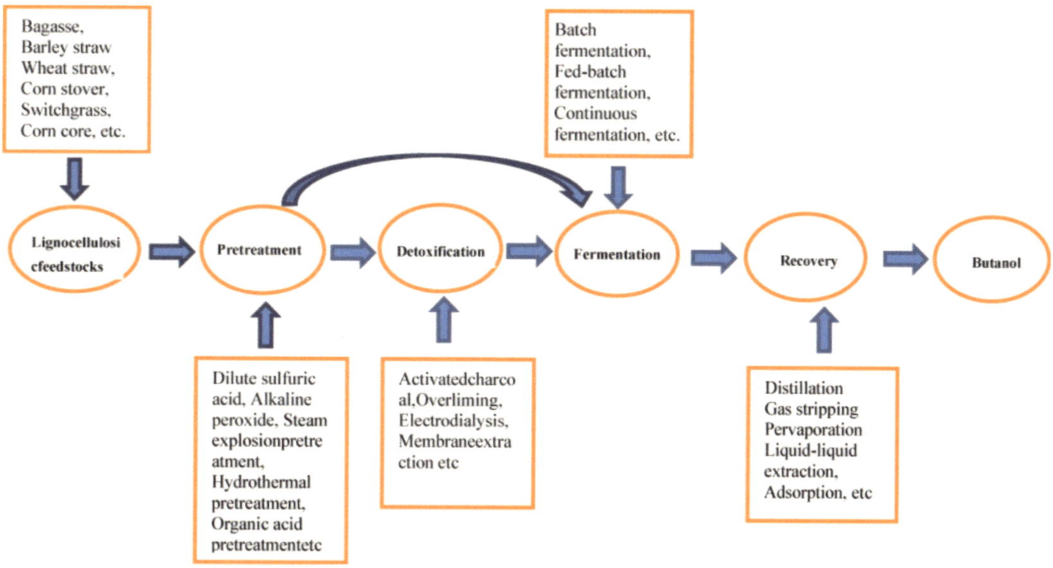

Figure 1. Butanol production process from lignocellulosic feedstocks

For the first step, biomass containing lignocellulosics should be pretreated before they were used as the substrate for the fermentation, except for a few high cellulase activity strains (Ezeji and Blaschek, 2008). The pretreatment methods are different according to the different types of biomass used. There often use dilute sulfuric acid pretreatment, alkaline peroxide pretreatment, steam explosion pretreatment, hydrothermal pretreatment, organic acid pretreatment etc. Some inhibitors such as acetic acid, furfural, 5- hydroxymethyl furfural, phenols etc. that need to be further detoxified. The ordinary detoxification methods are using activated charcoal (Wang et al., 2011), overliming (Sun and Liu, 2012; Park et al., 2010), electrodialysis (Qureshi et al., 2008c), membrane extraction (Grzenia et al., 2012) to remove the inhibitors. This step is determined by different feed stock and different pretreatment methods. After the fermentation, the desired product is recovered and purified in the downstream process. Biological ways has been set up for many years while it was inhibited for industrial application for economic reasons. So, as an alternative fuel, biomass feedstock for biobutanol production must be widely available at low cost (Kent, 2009). Therefore, by using agricultural wastes for butanol production such as straw, leaves, grass, spoiled grain and fruits etc are much more profitable from an economic point of view. Recently, other sources such as algae culture (Potts et al., 2012; Ellis et al., 2012) also is studied as one substrate for butanol production.

3. Biobutanol production by fermentation

3.1. Microbes

Clostridium is a group of obligate, Gram positive, endospore-forming anaerobes. There are lots of strains used for ABE fermentation in different culture collections, such as ATCC (American Type Culture Collection), DSM (German Collection of Microorganisms, or Deutsche Sammlung Von Mikroorganismen), NCIMB (National Collections of Industrial & Marine Bactria Ltd), and NRRL (Midwest Area National Center for Agriculture Utilization Research, US Department of Agriculture). The different strains share similar phonotype such as main metabolic pathway and end products. Molecular biology technology offers efficient method for classification. The butanol-producing clostridium can be assigned to four groups according to their genetic background, named *C. acetobutylicum, C. beijerinckii, C. saccharoperbutyl acetonicum*, and *C. saccharobutylicum*, respectively. *C. acetobutylicum* is phylogenetically distinct from the other three groups.

The common substrate for the solvent production by these strains is soluble starch. The original starch-fermenting strains belong to *C. acetobutylicum*. A recently isolated butanol-producing strain *C. saccharobutylicum* showed high hemicellulotic activity (Berezina et al., 2009). All of the four group strains can ferment glucose-containing medium to produce solvent. In 4% glucose TYA medium, *C. beijerinckii* gave the lowest solvent yield (28%), while the solvent yield was upper than 30% compared to the other three groups (Shaheen et al., 2000). In standard supplement maize medium (SMM), *C. acetobutylicum* is the best strain for maize fermentation, and the total solvent concentration can reach 19g/L. The solvent yield was 16, 14, and 11 for that of *C. beijerinckii, C. saccharoperbutyl acetonicum*, and *C. saccharobutylicum* respectively. However, *C. acetobutylicum* can't ferment molasses well and it produces bright yellow riboflavin in milk, which is different from other groups and easy identified. The best molasses-fermenting strains belong to *C. saccharobutylicum* and *C. beijerinckii* (Shaheen et al., 2000). *C. saccharoperbutyl acetonicum* can utilize sugar, molasses and maize. Comparing to *C. acetobutylicum, C. beijerinckii* was more tolerant to acetic acid and formic acid (Cho et al., 2012), which suggests the advantage when using lignocellulosic hydrolysate treated with acetic and formic acid as substrate.

There are also some *C. beijerinckii* strains produce isopropanol instead of acetone (George et al., 1983). Some microorganisms can produce biobutanol from carbon monoxide (CO) and molecular hydrogen (H_2), including acetogens, *Butyribacterium methylotrophicum, C. autoethanogenum, C. ljungdahlii* and *C. carboxidiworans*. The *C. carboxidivorans strain P7(T)* genome possessed a complete Wood-Ljungdahl pathway gene cluster which is responsible for CO, hydrogen fixation and conversion to acetyl-CoA(Fig.2) (Bruant et al., 2010).

http://dx.doi.org/10. 5772/52535

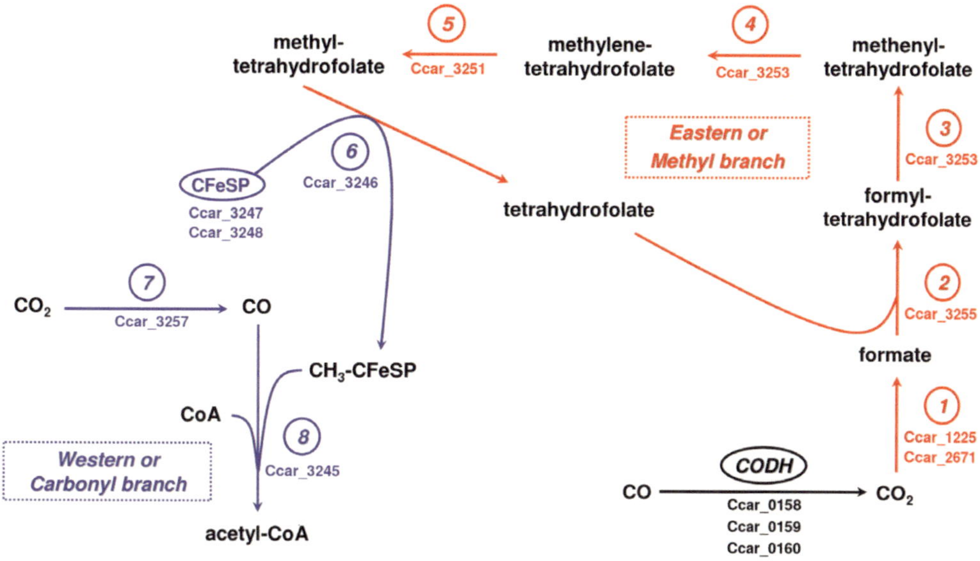

Figure 2. Wood-Ljungdahl pathway in Carboxdivorans Strain P7T. (Bruant et al. 2010, http://creativecommons. org/licenses/by/3. 0/)Wood-Ljungdahl pathway key enzymes and protein identified in C. carboxidivorans strain P7T. 1, formate dehydrogenase; 2, formate-tetrahydrofolate ligase; 3 and 4, bifunctionalmethenyl-tetrahydrofolatecyclohydrolase/methylene-tetrahydrofolate dehydrogenase (NADP+); 5, 5, 10-methylene-tetrahydrofolate reductase; 6, 5-methyl-tetrahydrofolate:- corrinoid iron-sulfur protein methyltransferase; 7, carbon monoxide dehydrogenase; 8, acetyl-CoA synthase; CFeSP, corrinoid iron-sulfur protein; CODH, additional carbon monoxide dehydrogenase complex. Reactions from the western branch are indicated in blue, those from the eastern branch are indicated in red. The corresponding genes in strain P7T genome are indicated below the enzyme.

3.2. Metabolic pathway

The ABE producing strains can hydrolyze starch to glucose or other hexose by amylases. Glucose was firstly converted to pyruvate through the Embden-Meyerhoff pathway (EMP, or glycolysis). Pyruvate was then cleaved to acetyl-CoA by pyruvate ferredoxin oxidoreductase. Acetyl-CoA is the common precursor of all the fermentation intermediate and end products. The enzyme activity and the coding genes have been widely assayed and described in butanol-producing strains (Dürre et al., 1995; Gheshlaghi et al., 2009).

The ABE fermentation process can be divided into two successive and distinct phase as acidogenesis phase and solvetogenesis phase. The acidogenesis phase is accompanied with cell exponential growth and pH drop, accumulation of acetate and butyrate. Solventogenesis phase begins with endospore forming and the cells entering stationary state. The products of acidogenesis phase include acetate and butyrate. Acetate forms from Acetyl-CoA, which is catalyzed by two enzymes, phosphotransacetylase (PTA, or phosphate acetyltransferase, endoced by *pta* gene) and acetate kinase (AK, encoded by *ak* gene). The butyrate synthesis is a little complicated with more steps. At first, two molecular of acetyl-CoA is catalyzed by thiolase (thl, or acetyl-CoA acetyltransferase, encoded by *thl* gene) and transforms into one molecular C4 unit acetoacetyl-CoA, which is another important node and precursor of butyrate, acetone, and butanol synthesis. The acetoacetyl-CoA is subjected to three enzymes in turn and another C4 unit butyryl-CoA is the intermediate product. The three enzymes are hydroxybutyryl-CoA dehydrogenase (encoded by *hbd* gene) (Youngleson et al., 1995), crotonase (CRT, or hydroxybutyryl-CoA dehydrolase, encoded by *crt* gene), and butyryl-CoA dehydrogenase (BCD, encoded by *bcd* gene). Accordingly, three encoded genes coexist in the BCS operon with additional two genes coding for the α and β subunit of electron transfer protein (Bennett and Rudolph, 1995). Butyryl-CoA was then catalyzed by phosphotransbutylase (PTB, or phosphate butyltransferase, encoded by *ptb* gene) and butyrate kinase (BK, encoded by *bk* gene) to form butyrate during acidogenesis phase.

As the organic acid accumulation, pH drop to the lowest point during the fermentation. This leads to the switch of acidogenesis phase to solventogenesis phase. Acetate and butyrate are reassimilated and participate in the solvent formation. Under the catalyzing of CoA transferase (CoAT, two unit encoded by *ctfα* and *ctfβ*), acetate and butyrate was transformed into acetyl-CoA and butyryl-CoA respectively again. The alcohols formation share the same key enzymes, NAD(P)H dependent aldehyde/alcohol dehydrogenases (encoded by *adh*1 and *adh*2 gene) (Chen, 1995). In addition, Butanol owns its unique butanol dehydrogenase (encoded by *bdh* gene) (Welch et al., 1989). The formation of acetone from acetoacetyl-CoA is a two-step reaction. Acetoacetyl-CoA is catalyzed to acetoacetate by CoA transferase. Acetone is produced after a molecular CO_2 released from acetoacetate by decarboxylase (AADC, encoded by *aadc* gene) (Janati-Idrissi et al., 1988; Cary et al., 1993). Both acid reassimilation and acetone formation utilize CoA transferase, however, the butyrate uptake was not concomitant with the production of acetone (Desai et al., 1999). The metabolic pathway accompanied by electron transfer and reduction force forming. The main ABE fermentation pathway was illustrated in Fig.3.

Solventogenic genes *aad, ctfA, ctfB* and *adc* constitute the *sol* operon (Durre et al., 1995). In some conditions, butanol producing strains lose the ability to produce solvents after repeated subculturing, called as degenerated (DGN) strain. In *C. acetobutylicum* ATCC 824, the plasmid pSOL1 carrying the *sol* operon was found missing during degenerating process (Cornillot et al., 1997). For *C. saccharoperbutyl acetonicum* strain N1-4, the *sol* genes maintained in degenerated DGN3-4 strain, while the *sol* operon was hardly induced during solventogenesis. Extract from the culture supernatants of wild-type N1-4 is enough to induce the transcription of the *sol* operon in DGN3-4 (Kosaka et al., 2007). It suggested that the degeneration maybe caused by the incompetence of the induction mechanism of the *sol* operon. The transcription of *sol* operon may be under the control of the quorum-sensing mechanism in *C. saccharoperbutyl acetonicum*.

Though the metabolic pathway is clear, the underlying regulation mechanism is poorly understood, such as the phase switch of fermentation, the relationship between solventogenesis and sporulation. Answering these questions is critical to improve the efficiency of butanol producing fundamentally. Proteomics and transcriptomics can provide more unknown details, which will be helpful for solving these problems (Sivagnanam et al., 2011; Sivagnanam et al., 2012).

http://dx.doi.org/10. 5772/52535

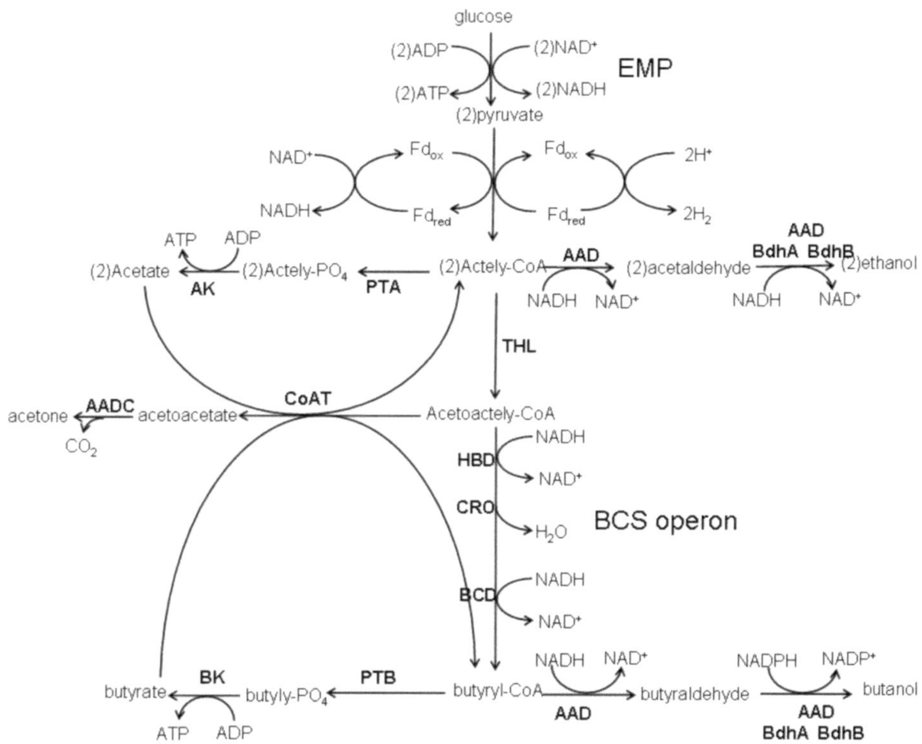

Figure 3. Metabolic pathway of Acetone-butanol-ethanol fermentation. EMP: Embden-Meyerhoff pathway (glycolysis); AK, acetate kinase; PTA, phosphotransacetylase; CoAT, CoA transferase; AADC, acetoacetate decarboxylase; THL, thiolase; BK, butyrate kinase; PTB, phosphotransbutylase; HBD, hydroxybutyryl-CoA dehydrogenase; CRO, crotonase; BCD, butyry-CoA dehydrogenase; AAD, aldylde/ alcohol dehydrogenase; BdhA, butyryl-CoA dehydrogenase A; BdhB, butyryl-CoA dehydrogenase B.

3.3. Metabolic engineering

The increasing genetic knowledge provides feasible technique for the strain modification. Many efforts have been made to construct the strain with high butanol tolerance, superior butanol yield, productivity and less byproduct. The process can be classified into pathway-based construction and regulation-based construction.

Except butanol, acetone and ethanol are main products in ABE fermentation. The byproduct, especially acetone is low valuable and undesirable. Blocking the expression key enzyme gene for acetone is thought perfect to decrease the split flux and enhance butanol yield. However, the results were not ideal as expected. Knocking out the C. acetobutylicum EA 2018 *adc* gene, the acetone is still produced in low level (Jiang et al., 2009). In *C. beijerinckii* 8052, the strain with *adc* gene disruption produced similar acetone with the original wild type strain (Han et al., 2011). To block acetate and acetone pathway by knocking out gene *adc* and *ctfA* reduced solvent production (Lehmann et al., 2012). These results demonstrated that the butanol metabolic mechanism is more complicated than expected.

Acetate and butyrate are produced during acidogenesis, and then they are transformed into acetyl-CoA and butyl-CoA to participate the solvent formation during solventogenesis phase. It seems an ineffective loop. In fact, the "inefficiency" loop is necessary for acid accumulation and switching to solventogenesis, at the same time, energy and reduction force were reserved. Disruption of acetate and butyrate pathway didn't enhance butanol production. Knocking out acetate biosynthetic pathway gene by Clos Tron had no significant influence on the metabolite distribution (Lehmann et al., 2012). Disruption of *ptb* gene blocked the butyrate synthesis and led to acetate and lactate accumulation. Some mutant strain without *bk* gene even can't survival (Sillers et al., 2008). It indicated that the pathways seeming useless were necessary for butanol synthesis. What's more, it is not possible to improve performance by decrease acid formation.

The genes participate in butanol synthesis including of *thl*, BCS operon, and *add, bdh*. Overexpression these genes are thought useful to increase the butanol yield. Overexpression of *aad* gene alone could enhance butanol production (Nair and Papoutsakis, 1994; Tummala et al., 2003). Transformed strain M5 (*sol* operon deficient because of lose of plasmid pSOL) with a plasmid carrying *aad* gene restored butanol-producing capability (Nair and Papoutsakis, 1994). Overexpression of *aad* gene and down-regulated *ctf* gene increased the butanol and ethanol production. To boost the butyryl-CoA pool, the strain with both *thl* and *aad* overexpression was constructed. However, butyrate and acetone concentration were increased, not butanol. The *thl* overexpression with *ctf* knock down didn't change the product significantly (Sillers et al., 2009). So, the metabolic is more complicated than it seems. Theoretical analyses also suggested alteration single solvent-associated gene is not sufficient to increase butanol yield (Haus et al., 2011).

Low butanol tolerance of the strains is another problem of butanol production. Although butanol synthesis is spontaneous in clostridium, the wild type strains can't endure high butanol concentration upper than 2%. Butanol stress influence gene expression of amino acid, nucleotide, glycerolipid biosynthesis and the cytoplasmic membrane composition (Janssen et al., 2012). Cells have heat shock response system will protect it from heat or other stress (Bahl, Müller et al. 1995). Overexpression of grosESL improved the strain tolerance and butanol titer (Tomas et al., 2003).

The utilization of xylose and other carbon sources was inhibited by glucose is a phenomenon called as Carbon catabolite repression (CCR). CCR limited the efficiency of butanol fermentation with lignocellulosic material as substrate. The utilization rate of pentose was improved efficiently by knocking out pleiotropic regulator gene *ccpA*, *glcG* (responsibility for phosphoenoopyruvate-dependent phophotransferase system, PTS) and overexpressing the genes of xylose utilization (Ren et al., 2010; Xiao et al., 2012). By heterogonous expression transaldolase gene talA in ATCC 824, the xylose utilization was improved significantly (Gu et al., 2009). Knocking out xylose repressor gene *XylR* also increased the fermentation efficiency (Xiao et al., 2012).

There also some strategies aim at the upstream regulation. Global transcription machinery engineering (gTME) is thought to be a promising method to improve the butanol-producing performance (Alper et al., 2006; Papoutsakis, 2008). By regulating the transcription factor, the gTME strategy is thought to be able to change the metabolic strength and direction. gTME has been shown an efficient solution to improve substrate utilization, product tolerance, and production in yeast (Alper et al. 2006) and *E. coli* (Chen et al., 2011). In butanol-producing *Clostridium*, the metabolic pathway have been described clearly, however, the mechanism of metabolism regulation is still not fully understood. This situation keeps the gTME strategy away from butanol-producing strains. Much effort should be devoted on the proteomics and transcriptomics etc. that will increase more details behind the appearance of ABE fermentation. A true gTME strategy will bring fresh and effective innovation to the butanol fermentation.

The concept of metabolic engineering is to develop strains as "cell factory" which is efficient for desired products production from renewable sources (Na et al., 2010). Some microbes attracted interests because they are more tolerant to butanol than *Clostridium*, although these bacteria haven't natural solvent-producing ability. Some kinds of Lactic acid bacteria can grow in 3-4% butanol (Liu et al., 2012) after long term adaption, that makes them promising host for butanol producing. The synthetic biology strategy has been implemented by constructing the whole butanol-producing pathway in *Escherichia coli, Bacillus subitilis, Saccharomyces cerevisiae* and *Pseudomonas putida* (Shen and Liao, 2008; Nielsen et al., 2009). This strategy deserves further attempts in spite of the poor final butanol concentration.

3.4. Fermentation application

ABE fermentation can be conducted as batch, fed-batch, and continuous under anaerobic conditions. Batch fermentation is the simplest mode. The substrate is typical 40-80g/L and the efficiency decreased as substrate concentration upper than 80g/L (Shaheen et al, 2000). With optimized physiological and nutritional parameters, 20g/L n-butanol was obtained by *C. beijerinckii* ATCC 10132 in 72h (Isar and Rangaswamy, 2012). Fed-batch fermentation was adopted to avoid substrate inhibition. However, because of product inhibition, the substrate feeding seems ineffective. The solvent must be removed from the broth to decrease the product toxicity. The solvent can be removed by several ways such as liquid-liquid extraction, perstraction, gas-stripping, and pervaporation etc. (Qureshi and Maddox, 1995; Qureshi and Blaschek, 2001b). The whole systemic technique of high productivity was constructed by continuous feeding combined with product removal (Qureshi et al., 1992), such as using membrane reactor (Qureshi et al., 1999a). With these techniques, the fermentation can be continuing for a long time and resulting in higher productivity. To improve the utilization efficiency of cells, the immobilization system is used (Huang et al., 2004; Qureshi et al., 2000; Lienhardt et al., 2002). Comparing with the free cell system, the immobilization system is easier to separate cells from product, can reach high cell concentration and productivity, and can decrease nutrient depletion and product inhibition.

Co-culture is another important way for butanol fermentation (Abd-Alla and El-Enany, 2012). *C. beijerinckii* NCIMB 8052 was entangled with ATCC 824 and thought as *C. acetobuty-*

licum before the 16S rDNA based method was exploited (Johnson and Chen, 1995). These data implied that they could be cocultured before isolation. A microflora of four strain isolated from hydrogen-forming sludge of sewage performed a little high solvent yield (Cheng et al., 2012). Different strains possess various advantages, either with larger carbon substrate, higher butanol yield, or with high substrate and product tolerance. The co-culture should possess potential benefits and be harnessed fully after all the details are disclosed for each individual strain.

4. Separation of butanol product

Because butanol has a higher boiling point than water, therefore, distillation is not suitable for butanol recovery. Other processes such as adsorption, pervaporation, membrane pertraction, reverse osmosis and gas stripping have been developed to improve recovery performance and reduce costs (Oudshoorn et al., 2009; Ezeji et al., 2004b).

4.1. Adsorption process

Adsorption is the technology operating easily for the butanol separation. Butanol can be adsorpted by the adsorbents in the fermenter and then the butanol was obtained by desorption. A variety of materials can be used as adsorbents for butanol recovery and silicalite is the common one used (Qureshi et al., 2005b; Ezeji et al., 2007). Silicalite is a form of silica with a zeolite-like structure and hydrophobic properties, it can selectively adsorb small organic molecule like C1–C5 alcohols from dilute aqueous solutions (Zheng et al., 2009). However, adsorption separation process is not suitable on an industrial or semi-technical scale because the capacity of adsorbent is very low.

4.2. Butanol recovery by membrane reactor

Immobilization of microorganisms in the membrane or using membrane reactors is another option of butanol removal. The productivity can be enhanced obviously by this way. Huang et al. reported the continuous ABE fermentation by immobilized *C. acetobutylicum* cells with the fibrous as carrier and a productivity of 4.6 g/L/h was obtained (Huang et al., 2004). Qureshi et al. studied the butanol fermentation by immobilized *C. beijerinckii* cells with different carriers such as clay brick, the reactor productivity was enhanced to 15.8 g/(lh) (Qureshi and Blaschek, 2005a). Although the butanol productivity increased by using immobilized cell fermentation, leakage of cells from the matrices is a frequent problem for the industrial application. There still some other problems such as poor mechanical strength and increase mass transfer resistance etc.

4.3. Butanol recovery by gas stripping

Gas stripping seems to be a promising technique that can be applied to butanol recovery combined with ABE fermentation. When the gas (ordinary N_2 or CO_2) are bubbled through the fermentation broth, it captures the solvents. The solvents then condensed in the condenser and are collected in a receiver. Ezeji applied gas stripping on the fed-batch fermentation, 500 g glucose was consumed and 233 g/l solvent was produced with the productivity of 1.16 g/(Lh) and the yield of 0.47 g/g.When combined with continuous fermentation with gas stripping, 460g/l solvent was obtained with 1163g glucose consuming (Ezeji et al., 2004a; Ezeji et al., 2004b).

4.4. Butanol recovery by pervaporation

Pervaporation is a membrane-based process that allows selective removal of volatile compounds from fermentation broth. The membrane is placed in contact with the fermentation broth and the volatile liquids or solvents diffuse through the membrane as a vapor which is recovered by condensation. A vacuum applied to the side of permeate. Polydimethylsiloxane membranes and silicon rubber sheets are generally used for the pervaporation process. Selection of a suitable polymer forming the active part of the membrane is a key factor in this case. In the batch fermentation, Evans and Wang increased the solvent concentration and productivity from 24.2g/l and 0.34g/(lh) to 32.8g/l and 0.5g/(lh) with pervaporation (Evans and Wang, 1988). Groot et al. applied pervaporation on the fed-batch fermentation and the solvent productivity and concentration reached 0.98g/lh and 165.1g/l (Groot et al., 1984). The Reverse osmosis is another recovery technique that based on membranes. Before the reverse osmosis is carried out, the suspended vegetative organisms must be removed using the hollow-fiber ultra-filter. After the pretreatment, reverse osmosis starts to dewater the fermentation liquor by rejecting solvents but allowing water to pass through the membrane. And then, the products are concentrated (Zheng et al., 2009).

4.5. Liquid–liquid extraction

Liquid–liquid extraction can be used to remove solvents from the fermentation broth. In this process, the water-insoluble organic extractant is mixed with the fermentation broth. Butanol is more soluble in the organic (extractant) phase than in the aqueous (fermentation broth) phase. So, butanol can be selectively concentrated in the organic phase. As the extractant and fermentation broth are immiscible, the extractant can easily be separated from the fermentation broth after butanol extraction. (Qureshi and Blaschek, 1999a). However, there still some problems with liquid–liquid extraction such as toxicity of extractant, extraction solvent losing, the formation of an emulsion, etc. Oleyl alcohol as a good extractant with relatively low-toxic has been used widely by the researchers (Karcher et al., 2005; Ezeji, 2006).

4.6. Application of ionic liquids

The butanol extraction process using conventional solvents may be useful, but the solvents used are often volatile, toxic and dangerous. In recent years, a growing interest in ionic liquids(IL) which also can be used in butanol recovery. Ionic liquids are organic salts present in the liquid state at room conditions, have very low vapor pressure and low solubility in water. Hence, Ionic liquids is valuable solvent in the extraction process from aqueous solutions (Fadeev and Meagher, 2001; Garcia-Chavez et al., 2012). Ionic liquids as the non-volatile, environment friendly solvents have been used in various chemical processes. With the development of the technology, ionic liquids extraction would be more promising for butanol recovery.

5. Biobutanol production from renewable resources

Biobutanol is no doubt a superior candidate renewable energy facing the exhausted fossil-energy. The clostridium can incorporate simple and complex soluble sugar, such as corn, molasses, cassava, and sugar beet. The ABE fermentation is also a solution to deal with agriculture residue, spoilage material, and domestic organic waste (Table 3). Additionally, using renewable resources is also ideal for environment problem solving.

Raw materials	Bacterial strain	Fermentation process	ABE concentration (g/l)	ABE Yield (g/g)	ABE productivity (g/lh)	References
Barley straw	*C. beijerinckii*	Batch fermentation	26.64	0.43	0.39 g/lh	Qureshi et al., 2010a
Wheat straw	*C.beijerinckii*	simultaneous saccharification and fermentation combined with gas stripping	21.42	0.41	0.31	Qureshi et al., 2008a
Corn fiber	*C. beijerinckii*	Batch fermentation	9.3	0.39	0.10	Qureshi et al., 2008b
Corn stover	*C. beijerinckii*	Batch fermentation	26.27	0.44	0.31	Qureshi et al., 2010b
Rice straw	*C. saccharoperbutylacetonicum*	Batch fermentation	13	0.28	0.15	Soni et al.(1982)
Bagasse	*C. saccharoperbutylacetonicum*	Batch fermentation	18.1	0.33	0.3	Soni et al.(1982)
Switch grass (*Panicum virgatum*)	*C. beijerinckii*	Batch fermentation	14.61	0.39	0.17	Qureshi et al., 2010a
Domestic organic waste	*C. acetobutylocum*	Batch fermentation	9.3	0.38	0.08	Claassen et al., 2000
Sago	*C. saccharobutylicum*	Batch fermentation Continuous Fermentation ($D=0.11h^{-1}$)	16.38 7.74±0.55	0.33 0.29	0.59 0.85	Liew et al., 2005
Defibrated-sweet potato-slurry (DSPS)	*C. acetobutylocum*	Batch fermentation Continuous Fermentation, immobilized cell ($D=0.129\ h^{-1}$)	5.87 7.73	0.29 0.195	0.12 1	Badr et al., 2001
Cassava	Co-culture of *B. Subtilis* and *C. butylicum*	Batch fermentation	9.71	~0.21	0.135	Tran et al., 2010

Table 3. Butanol production with different raw materials

Food-based substrate arouses many problems. The cost of butanol from glucose was four fold higher than that from sugarcane and cellulose materials (Kumar et al., 2012). For the cellulose-based substrate, the crystal structure of cellulose is hard to use for normal ABE fermentation clostridium. The pretreatment of cellulose is costly, complex, and often leads to new environment problems. For example, using corn as substrate, the cost is 0.44-0.55 US $/kg butanol by the hyper-butanol producing strain *C. beijerinckii* BA101 (Qureshi and Blaschek, 2000) by continuous fermentation combined with butanol separation. The cost reached 0.73-1.07 US$/kg when grass-rooted plant was used as substrate (Qureshi and Blaschek, 2001a). A promising solution is co-culture of butanol-producing and cellulolytic strains. However, many obstacles must be cleared before the system is constructed. It's difficult for different strains to play a role in turn in the substrate medium. Firstly, strain with high hydrolysis activity must be obtained. Secondly, the procedure must also be optimized.

Some strains can use CO_2, H_2, and CO as substrate (Tracy et al., 2012). The celluloses substrate can be transformed into CO ($_2$) and H_2 firstly. The simple substrates then are used by *C. carboxidivorans* to produce butanol. The more simple and feasible process is still need to be further explored for different substrates.

6. The promising application and prospect of biobutanol

Due to the excessive exploitation, the fossil fuels are facing scarce and they cannot be generated. On the other hand, most of the carbon emissions result from fossil fuel combustion. Reducing the use of fossil fuels will considerably reduce the amount of carbon dioxide and other pollutants produced. Renewable energy has the potential to provide energy services with low emissions of both air pollutants and greenhouse gases. Currently, renewable energy sources supply over 14% of the total world energy demand. Biofuels as the important renewable energy are generally considered as sustainability, reduction of greenhouse gas emissions, regional development, social structure and agriculture, and security of supply (Reijnders, 2006). Biodiesel and bioethanol are presently produced as a fuel on an industrial scale, including ETBE partially made with bioethanol, these fuels make up most of the biofuel market (Antoni et al., 2007).

Biobutanol also has a promising future for the excellent fuel properties. It has been demonstrated that n-butanol can be used either 100% in unmodified 4-cycle ignition engines or blended up with diesel to at least 30% in a diesel compression engine or blended up with kerosene to 20% in a jet turbine engine in 2006 (Schwarz et al., 2006). The production of biobutanol from lignocellulosic biomass is promising and has been paid attention by many companies. Dupont and BP announced a partnership to develop the next generation of biofuels, with biobutanol as first product (Cascone, 2007). In 2011, Cobalt Technologies Company and American Process Inc. (API) have been partnering to build an industrial-scale cellulosic biorefinery to produce biobutanol. Additionally, the companies agreed to jointly market a GreenPower+ biobutanol solution to biomass power facilities and other customers worldwide. The facility is expected to start ethanol production in early 2012 and switch to biobutanol in mid-2012. The annual production of biobutanol is estimated to 470, 000 gallons. (http://www.greencarcongress.com/2011/04/cobalt-20110419.html, http://www.renewableenergyfocususa.com/view/17558/cobalt-and-api-cooperate-on-biobutanol/) Gevo, Inc. signed a Joint Development Agreement with Beta Renewables, a joint venture between Chemtex and TPG, to develop an integrated process for the production of bio-based isobutanol from cellulosic, non-food biomass, such as switch grass, miscanthus, agriculture residues and other biomass will be readily available. (http://www.greencarcongress.com/biobutanol/). Syntec company also is currently developing catalysts to produce bio-butanol from a range of waste biomass, including Municiple Solid Waste, agricultural and forestry wastes. (http://www.syntecbiofuel.com/butanol.php). Utilization the waste materials improve the economy of butanol production that makes biobutanol great potential to be the next new type of biofuel in spite of the existing drawbacks.

7. Conclusions

Biobutanol production has only recent years booming again after long time of silence. Quite a lot of progress has been made with the technology development of metabolic engineering in enhancing solvent production, increasing the solvent tolerance of bacteria, improving the selectivity for butanol. Fortunately, *Clostridia* have been tested being able to consume lignocellulosic biomass for ABE fermentation. The complex regulation mechanism of butanol synthesis is still need to be further study. For the strain improvement, for example, constructing better butanol tolerance strains, more suitable hosts and genetic methods are required to be set up. Furthermore, more efficient techniques for removing the inhibitors in the lignocellulosic hydrolysate need to be developed. In addition, from the economic point of view, the integrated system of hydrolysis, fermentation, and recovery process also are important to be further developed to reduce the operation cost of butanol synthesis.

Author details

Hongjuan Liu*, Genyu Wang and Jianan Zhang

*Address all correspondence to: liuhongjuan@tsinghua.edu.cn; zhangja@tsinghua. edu. cn

Institute of Nuclear and New Energy Technology, Tsinghua University, Beijing, P. R., China

References

[1] Abd-Alla MH, El-Enany AWE. Production of acetone-butanol-ethanol from spoilage date palm (Phoenix dactylifera L.) fruits by mixed culture of Clostridium acetobutylicum and Bacillus subtilis. Biomass Bioenergy. 2012, 42: 172-178.

[2] Alper H, Moxley J, Nevoigt E, Fink GR, Stephanopoulos G. Engineering yeast transcription machinery for improved ethanol tolerance and production. Science. 2006, 314(5805): 1565-1568.

[3] Antoni D, Zverlov VV, Schwarz WH. Biofuels from microbes. ApplMicrobiolBiotechnol. 2007, 77:23–35

[4] Badr HR, Toledo R, Hamdy MK. Continuous acetone ethanol butanol fermentation by immobilized cells of Clostridium acetobutylicum. Biomass Bioenergy. 2001, 20:119–132

[5] Bennett GN, Rudolph FB. The central metabolic pathway from acetyl-CoA to butyryl-CoA in Clostridium acetobutylicum. FEMS Microb Rev. 1995, 17(3): 241-249.

[6] Berezina OV, Brandt A, Yarotsky S, Schwarz WH, ZverlovVV. Isolation of a new butanol-producing Clostridium strain: High level of hemicellulosic activity and structure of solventogenesis genes of a new Clostridium saccharobutylicum isolate. SystApplMicrobiol. 2009, 32(7): 449-459.

[7] Bruant G, Levesque MJ, Peter C, Guiot SR, Masson L. Genomic Analysis of Carbon Monoxide Utilization and Butanol Production by Clostridium carboxidivorans Strain P7(T). Plos One. 2010, 5(9).

[8] Cary JW, Petersen DJ, Papoutsakis ET, Bennett GN. Sequence and arrangement of genes encoding enzymes of the acetone-production pathway of Clostridium acetobutylicum ATCC 824. Gene. 1993, 123(1): 93-97.

[9] Cascone, R. Biofuels: What is beyond ethanol and biodiesel? Hydrocarbon. 2007, 86(9)95-109.

[10] Chen JS. Alcohol dehydrogenase: multiplicity and relatedness in the solvent-producing clostridia. FEMS Microb Rev. 1995, 17(3): 263-273.

[11] Chen T, Wang J, Yang R, Li J, Lin M, Lin Z. Laboratory-evolved mutants of an exogenous global regulator, IrrE from Deinococcus radiodurans, enhance stress tolerances of Escherichia coli. PLoS One. 2011, 6(1): e16228.

[12] Cho, DH, Shin SJ, Kim YH. Effects of acetic and formic acid on ABE production by Clostridium acetobutylicum and Clostridium beijerinckii. BiotechnolBioproc E. 2012, 17(2): 270-275.

[13] ClaassenPAM, BuddeMAW, López-Contreras AM. Acetone, butanol and ethanol production from domestic organic waste by solventogenic clostridia. J MolMicrob Biotech. 2000, 2(1): 39-44.

[14] CornillotE, NairRV, Papoutsakis ET, Soucaille P. The genes for butanol and acetone formation in Clostridium acetobutylicum ATCC 824 reside on a large plasmid whose loss leads to degeneration of the strain. J Bacteriol. 1997, 179(17): 5442-5447.

[15] Dean JA. Lange's handbook of chemistry. 14th edition. New York: McGraw-Hill;1992

[16] Desai RP, Harris LM, Welker NE, Papoutsakis ET. Metabolic Flux Analysis Elucidates the Importance of the Acid-Formation Pathways in Regulating Solvent Production by Clostridium acetobutylicum. Metablic Eng. 1999, 1(3): 206-213.

[17] Dürre P, Fischer RJ, Kuhn A, Lorenz K, Schreiber W, Stürzenhofecker B, Ullmann S, Winzer K, Sauer U. Solventogenic enzymes of Clostridium acetobutylicum: catalytic

properties, genetic organization, and transcriptional regulation. " FEMS Microb Rev. 1995, 17(3): 251-262.

[18] Dürre P. Biobutanol: an attractive biofuel. Biotechnol J. 2007, 2:1525–1534.

[19] Dürre P. New insights and novel developments in clostridial acetone/butanol/isopropanefermentation. ApplMicrobBiotechnol, 1998, 49:639–648.

[20] Ellis JT, Hengge NN, Sims RC, Miller CD. Acetone, butanol, and ethanol production from wastewater algae. Bioresource Technol. 2012, 111:491-495.

[21] Evans PJ, Wang HY. Enhancement of butanol formation by Clostridium acetobutylicum in the presence of decanol-oleyl alcohol mixed extractants. Appl Environ Microbiol. 1988, 54:1662–1667.

[22] Ezeji T, Blaschek HP. Fermentation of dried distillers' grains and solubles (DDGS) hydrolysates to solvents and value-added products by solventogenic clostridia. Bioresource Technol. 2008, 99(12): 5232-5242.

[23] Ezeji TC, Qureshi N, Blaschek HP. Acetone butanol ethanol (ABE) production from concentrated substrate: reduction in substrate inhibition by fed-batch technique and product inhibition by gas stripping. ApplMicrobiolBiotechnol. 2004a, 63:653–8.

[24] Ezeji TC, Qureshi N, Blaschek HP. Bioproduction of butanol from biomass: from genes to bioreactors. CurrOpinBiotechnol, 2007, 18:220-227.

[25] Ezeji TC, Qureshi N, Blaschek HP. Butanol fermentation research: Upstream and downstream manipulations. Chem Rec. 2004b, 4:305–314.

[26] Ezeji TC, Qureshi N, Karcher P, Blaschek HP. Butanol production from corn. In Alcoholic Fuels: Fuels for Today and Tomorrow. Edited by Minteer SD. New York, NY: Taylor & Francis, 2006:99-122.

[27] Fadeev AG, Meagher MM. Opportunities for ionic liquids in recovery of biofuels. ChemCommun. 2001, 295-296.

[28] Freeman J, Williams J, Minner S, Baxter C, DeJovine J, Gibbs L, Lauck J, Muller H,. Saunders H. Alcohols and ethers: a technical assessment of their application as fuels and fuel components, API publication 4261. 2nd ed. New York: American Institute of Physics; 1988.

[29] Garcia-Chavez LY, Garsia CM, Schuur B, de Haan AB. Biobutanol Recovery Using Nonfluorinated Task-Specific Ionic Liquids. Ind Eng Chem Res. 2012, 51(24): 8293-8301.

[30] George HA, Johnson JL, Moore WE, Holdeman LV, Chen JS. Acetone, Isopropanol, and Butanol Production by Clostridium beijerinckii (syn. Clostridium butylicum) and Clostridium aurantibutyricum. Appl Environ Microbiol. 1983, 45(3): 1160-1163.

[31] GheshlaghiR, Scharer JM, Moo-Young M, Chou CP. Metabolic pathways of clostridia for producing butanol. Biotechnol Adv. 2009, 27(6): 764-781.

[32] Groot WJ, Oever CE van den, Kossen NWF. Pervaporation for simultaneous product recovery in the butanol/isobutanol batch fermentation. BiotechnolLett. 1984, 6:709–714.

[33] Grzenia DL, Schell DJ, Wickramasinghe SR. Membrane extraction for detoxification of biomass hydrolysates. Bioresource Technol. 2012, 111:248-254.

[34] Gu Y, Li J, Zhang L, Chen JNiu LX, Yang YL, Yang S, Jiang WH. Improvement of xylose utilization in Clostridium acetobutylicum via expression of the talA gene encoding transaldolase from Escherichia coli. J Biotechnol. 2009, 143(4): 284-287.

[35] Haus S, Jabbari S, Millat T, Janssen H, Fischer RJ, Bahl H, King JR, Wolkenhauer O. A systems biology approach to investigate the effect of pH-induced gene regulation on solvent production by Clostridium acetobutylicum in continuous culture. BMC Syst Biol. 2011, 5: 10.

[36] Huang WC, Ramey DE, Yang ST. Continuous production of butanol by Clostridium acetobutylicum immobilized in a fibrous bed reactor. ApplBiochemBiotechnol. 2004, 113:887-898.

[37] Isar J, Rangaswamy V. Improved n-butanol production by solvent tolerant Clostridium beijerinckii. " Biomass Bioenerg. 2012, 37: 9-15.

[38] Janati-IdrissiR, JunellesAM, Petitdemange H, Gay R. Regulation of coenzyme a transferase and acetoacetate decarboxylase activities in clostridium acetobutylicum. " Annales de l'Institut Pasteur / Microbiologie. 1988, 139(6): 683-688.

[39] Janssen H, Grimmler C, Ehrenreich A, Bahl H, Fischer RJ. A transcriptional study of acidogenic chemostat cells of Clostridium acetobutylicum—Solvent stress caused by a transient n-butanol pulse. J Biotec. http://dx. doi. org/10. 1016/j. jbiotec. 2012. 03. 027.

[40] Jiang Y, Xu CM, Dong F, Yang YL, Jiang WH, Yang S. Disruption of the acetoacetate decarboxylase gene in solvent-producing Clostridium acetobutylicum increases the butanol ratio. Metab Eng. 2009, 11(4–5): 284-291.

[41] Johnson JL, Chen JS. Taxonomic relationships among strains of clostridium-acetobutylicum and other phenotypically similar organisms. FEMS Microbiol Rev. 1995, 17(3): 233-240.

[42] Jones DT, Woods DR. Acetone-Butanol fermentation revisited. Microbiol Rev 1986, 50(4):484–524.

[43] Karcher P, Ezeji TC, Qureshi N, Blaschek HP. Microbial production of butanol: product recovery by extraction. In Microbial Diversity: Current Perspectives and Potential Applications. Edited by Satyanarayana T, Johri BN. New Delhi: IK International Publishing House Pvt. Ltd; 2005, 865-880.

[44] Kent SK. Biofuels in the U. S. —challenges and opportunities. Renew Energy 2009, 34:14–22.

[45] Kosaka T, Hirakawa H, Matsuura K, Yoshino S, Furukawa K. Characterization of the sol operon in butanol-hyperproducing Clostridium saccharoperbutylacetonicum strain N1-4 and its degeneration mechanism. Biosci Biotech Bioch. 2007, 71(1): 58-68.

[46] Kumar M, Goyal Y, Sarkar A, Gayen K. Comparative economic assessment of ABE fermentation based on cellulosic and non-cellulosic feedstocks. Appl Energy. 2012, 93: 193-204.

[47] Lehmann D, Hönicke D, Ehrenreich A, Schmidt M, Weuster-Botz D, Bahl H. Modifying the product pattern of Clostridium acetobutylicum: physiological effects of disrupting the acetate and acetone formation pathways. Appl Microbiol Biotechnol. 2012, 94(3): 743-754.

[48] Lienhardt J, Schripsema J, Qureshi N, BlaschekHP. Butanol production by Clostridium beijerinckii BA101 in an immobilized cell biofilm reactor - Increase in sugar utilization. Appl Biochem Biotechnol. 2002, 98: 591-598.

[49] Liew ST, Arbakariya A, Rosfarizan M, Raha AR. Production of solvent (acetonebutanol- ethanol) in continuous fermentation by Clostridium saccharobutylicum DSM 13864 using gelatinised sago starch as a carbon source. Malays J Microbiol. 2005, 2(2): 42–45

[50] Liu S, Wilkinson BJ, Bischoff KM, Hughes SR, Rich JO, Cotta MA. Adaptation of lactic acid bacteria to butanol. "BiocatalAgriBiotechnol. 2012, 1(1): 57-61.

[51] Mitchell WJ. Physiology of Carbohydrate to Solvent Conversion by Clostridia. Adv-Microb Physiol. R. K. Poole, Academic Press. 1997, 39: 31-130.

[52] Na D, Kim TY, Lee SY. Construction and optimization of synthetic pathways in metabolic engineering. CurrOpinMicrobiol. 2010, 13(3): 363-370.

[53] Nair RV, Papoutsakis ET. Expression of plasmid-encoded aad in Clostridium acetobutylicum M5 restores vigorous butanol production. J Bacteriol. 1994, 176(18): 5843-5846.

[54] NielsenDR, LeonardE, Yoon SH, Tseng HC, Yuan CJ, Prather KJ. Engineering alternative butanol production platforms in heterologous bacteria. Metab Eng. 2009, 11(4–5): 262-273.

[55] Nimcevic D, Gapes JR. The acetone–butanol fermentation in pilot plant and pre-industrial scale. JMolMicrobiolBiotechnol. 2000, 2:15–20.

[56] Oudshoorn A, Van der Wielen LAM, Straathof AJJ. Assessment of options for selective 1-butanol recovery from aqueous solution. IndEngChem Res, 2009, 48:7325-7336.

[57] Papoutsakis ET. Engineering solventogenicClostridia. CurrOpin Biotech. 2008, 19(5): 420-429.

[58] Park CH. Pervaporativebutanol fermentation using a new bacterial strain. Biotechnol Bioprocess Eng 1996, 1:1–8.

[59] Park J, Shiroma R, Al-Haq MI, Zhang Y, Ike M, Arai-Sanoh Y, Ida A, Kondo M, Tokuyasu K. A novel lime pretreatment for subsequent bioethanol production from rice straw – Calcium capturing by carbonation (CaCCO) process. Bioresour. Technol. 2010, 101(17): 6805-6011.

[60] Pfromm PH, Boadu VA, Nelson R, Vadlani P, Madl R. Bio-butanol vs. bio-ethanol: a technical and economic assessment for corn and switch grass fermented by yeast or Clostridium acetobutylicum. Biomass Bioenerg. 2010, 34(4):515-524.

[61] Potts T, Du JJ, Paul M, May P, Beitle R, Hestekin J. The production of butanol from Jamaica bay macro algae. Environ Prog Sustain Energy. 2012, 31(1):29-36.

[62] Qureshi N, Blaschek HP. ABE production from corn: a recent economic evaluation. " J IndMicrobiolBiot. 2001a, 27(5): 292-297.

[63] Qureshi N, Blaschek HP. Economics of butanol fermentation using hyper-butanol producing Clostridium beijerinckii BA101. Food Bioprod Process 2000, 78(C3): 139-144.

[64] QureshiN, MeagherMM, HutkinsRW. Recovery of butanol from model solutions and fermentation broth using a silicalite silicone membrane. J MembranE Sci. 1999a, 158(1-2): 115-125.

[65] Qureshi N, Blaschek HP. Production of acetone butanol ethanol (ABE) by a hyper-producing mutant strain of Clostridium beijerinckii BA101 and recovery by pervaporation. BiotechnolProg 1999b, 15:594–602.

[66] Qureshi N, Saha BC, Dien B, Hector RE, Cotta MA. Production of butanol (a biofuel) from agricultural residues: part I—use of barley straw hydrolysate. Biomass Bioenerg. 2010a, 34:559–565

[67] Qureshi N, Blaschek HP. Recovery of butanol from fermentation broth by gas stripping. " Renewable Energy. 2001b, 22(4): 557-564.

[68] Qureshi N, Blaschek HP: Butanol production from agricultural biomass. In Food Biotechnology. Edited by Shetty K, Pometto A, Paliyath G. Boca Raton, FL: Taylor & Francis Group plc; 2005a, 525-551.

[69] Qureshi N, Saha BC, Hector RE, Hughes SR, Cotta MA. Butanol production from wheat straw by simultaneous saccharification and fermentation using Clostridium beijerinckii: Part I – Batch fermentation. Biomass Bioenerg. 2008a, 32, 168-175.

[70] Qureshi N, Ezeji TC, Ebener J, Dien BS, Cotta MA, Blaschek HP. Butanol production by Clostridium beijerinckii. Part I: use of acid and enzyme hydrolyzed corn fiber. Bioresourse Technol. 2008b, 99:5915–5922.

[71] Qureshi N, Hughes S, Maddox IS, Cotta MA. Energy-efficient recovery of butanol from model solutions and fermentation broth by adsorption. Bioprocess Biosyst Eng. 2005b, 27(4):215-222.

[72] QureshiN, Maddox IS, Friedl A. Application of continuous substrate feeding to the abe fermentation - relief of product inhibition using extraction, perstraction, stripping, and pervaporation. Biotechnol Progr. 1992, 8(5): 382-390.

[73] Qureshi N, Maddox IS. Continuous production of acetone-butanol-ethanol using immobilized cells of Clostridium acetobutylicum and integration with product removal by liquid-liquid extraction. J Ferment Bioeng. 1995, 80(2):185-189.

[74] Qureshi N, Saha BC, Hector RE, Cotta MA. Removal of fermentation inhibitors from alkaline peroxide pretreated and enzymatically hydrolyzed wheat straw: Production of butanol from hydrolysate using Clostridium beijerinckii in batch reactors. Biomass Bioenerg. 2008c, 32(12):1353-1358.

[75] Qureshi N, Saha BC, Hector RE, Dien B, Hughes S, Liu S, Iten L, Bowman MJ, Sarath G, Cotta MA. Production of butanol (a biofuel) from agricultural residues: part II - use of corn stover and switchgrasshydrolysates. Biomass Bioenerg. 2010b, 35:559–669

[76] Reijnders L. Conditions for the sustainability of biomass based fuel use. Energy Policy. 2006, 34:863–876.

[77] Ren C, Gu Y, Hu SY, Wu Y, Wang P, Yang YL, Yang C, Yang S, Jiang WH. Identification and inactivation of pleiotropic regulator CcpA to eliminate glucose repression of xylose utilization in Clostridium acetobutylicum. Metab Eng. 2010, 12(5): 446-454.

[78] Schwarz WH, Gapes JR, Zverlov VV, Antoni D, Erhard W, Slattery M. Personal communication and demonstration at the TU Muenchen (Campus Garching and Weihenstephan) in June 2006

[79] Shaheen R, Shirley M, Jones DT. Comparative fermentation studies of industrial strains belonging to four species of solvent-producing Clostridia. J Mol Microbiol Biotechnol. 2000, 2(1): 115-124.

[80] Shen CR, Liao JC. Metabolic engineering of Escherichia coli for 1-butanol and 1-propanol production via the keto-acid pathways. Metab Eng. 2008, 10(6): 312-320.

[81] Sillers R, Al-Hinai MA, Papoutsakis ET. Aldehyde-alcohol dehydrogenase and/or thiolase overexpression coupled with CoA transferase downregulation lead to higher alcohol titers and selectivity in Clostridium acetobutylicum fermentations. Biotechnol Bioeng. 2009, 102(1): 38-49.

[82] Sillers R, Chow A, Tracy B, Papoutsakis ET. Metabolic engineering of the non-sporulating, non-solventogenic Clostridium acetobutylicum strain M5 to produce butanol without acetone demonstrate the robustness of the acid-formation pathways and the importance of the electron balance. Metab Eng. 2008, 10(6): 321-332.

[83] Sivagnanam K, Raghavan VGS, Shah M, Hettich RL, Verberkmoes NC, Lefsrud MG. Comparative shotgun proteomic analysis of Clostridium acetobutylicum from butanol fermentation using glucose and xylose. " Proteome Sci 2011, 9:66

[84] Sivagnanam K, Raghavan VGS, Shah M, Hettich RL, Verberkmoes NC, Lefsrud MG. Shotgun proteomic monitoring of Clostridium acetobutylicum during stationary

phase of butanol fermentation using xylose and comparison with the exponential phase. " J Ind Microb Biotechnol. 2012, 39(6): 949-955.

[85] Soni BK, Das K, Ghose TK. Bioconversion of agro-wastes into acetone butanol. Biotechnology Letters. 1982, 4(1):19-22.

[86] Sun ZJ, Liu SJ. Production of n-butanol from concentrated sugar maple hemicellulosic hydrolysate by Clostridia acetobutylicum ATCC824. BIOMASS & BIOENERGY. 2012, 39(SI):39-47.

[87] Tomas CA, Welker NE, Papoutsakis ET. Overexpression of groESL in Clostridium acetobutylicum results in increased solvent production and tolerance, prolonged metabolism, and changes in the cell's transcriptional program. " Appl Environ Microbiol. 2003, 69(8): 4951-4965.

[88] Tracy BP, Jones SW, Fast AG, Indurthi DC, Papoutsakis ET. Clostridia: the importance of their exceptional substrate and metabolite diversity for biofuel and biorefinery applications. Curr Opin Biotechnol. 2012, 23(3): 364-381.

[89] Tran HTM, Cheirsilp B, Hodgson B, Umsakul K. Potectial use of Bacillus subtilis in a co-culture with Clostridium butylicum for acetone-butanol-ethanol production from cassava starch. Biochem Eng. 2010, 48:260–267

[90] TummalaSB, Welker NE, Papoutsakis ET. Design of antisense RNA constructs for downregulation of the acetone formation pathway of Clostridium acetobutylicum. J Bacteriol. 2003, 185(6): 1923-1934.

[91] Wang L, Chen HZ. Increased fermentability of enzymatically hydrolyzed steam-exploded corn stover for butanol production by removal of fermentation inhibitors. Process Biochem. 2011, 46(2):604-607.

[92] Welch RW, Rudolph FB, Papoutsakis E. Purification and characterization of the NADH-dependent butanol dehydrogenase from Clostridium acetobutylicum (ATCC 824). Arch Biochem Biophys. 1989, 273(2): 309-318.

[93] XiaoH, LiZ, Jiang Y, Yang Y, Jiang W, Gu Y, Yang S. Metabolic engineering of d-xylose pathway in Clostridium beijerinckii to optimize solvent production from xylose mother liquid. Metab Eng. 2012, DOI: 10. 1016/j. ymben. 2012. 05. 003

[94] Yan J, Lin T. Biofuels in Asia. Appl Energy. 2009, 86:1–10.

[95] Youngleson JS, Lin FP, Reid SJ, Woods DR. Structure and transcription of genes within the β-hbd-adh1 region of Clostridium acetobutylicum P262. FEMS Microb Lett. 1995, 125(2–3): 185-191.

[96] Zheng YN, Li LZ, Xian M, Ma YJ, Yang JM, Xu X, He DZ. Problems with the microbial production of butanol. J IndMicrobiolBiotechnol. 2009, 36:1127-1138.

[97] Zhou A, Thomson E. The development of biofuels in Asia. Appl Energy 2009, 86:11–20.

[98] Zverlov VV, Berezina O, Velikodvorskaya GA, Schwarz WH. Bacterial acetone and butanol production by industrial fermentation in the Soviet Union: use of hydrolyzed agricultural waste for biorefinery. ApplMicrobiolBiotechnol. 2006, 71:587–97.

Chapter 7

Biobutanol from Renewable Agricultural and Lignocellulose Resources and Its Perspectives as Alternative of Liquid Fuels

László Kótai, János Szépvölgyi, Mária Szilágyi, Li Zhibin, Chen Baiquan, Vinita Sharma and Pradeep K. Sharma

Additional information is available at the end of the chapter

http://dx.doi.org/10.5772/52379

1. Introduction

Biobutanol (n-C_4H_9OH, available as fermentation product of various carbohydrate derivatives obtained from different resources of agricultural production such as crops and wastes) is one of the most promising biofuels in the near future. It can be produced by the so-called ABE (acetone-butanol-ethanol) type anaerobic fermentation discovered by Pasteur [1, 2] and industrialized by Weizmann [3]. Main problems associated with industrial production of biobutanol include high energy demand for processing of dilute ferment liquors and high volume of wastewater. A bioreactor with a volume of 100 m^3 produces at 90% filling ratio 1053 kg of butanol, 526 kg of acetone and 175 kg of ethanol together with 2900 kg of carbon dioxide, 117 kg of hydrogen and 84150 kg of wastewater. Efforts to increase productivity and decrease production costs resulted in many new methods. This chapter summarizes some selected results on methods of biobutanol production.

2. History of industrial biobutanol production

During investigations aimed at discovering cheaper sources of acetone and butanol for chemical industry, Weizmann [3] isolated an organism which could ferment a fairly concentrated corn mash with good yields of acetone and butanol. In 1915 the British Admirality took over the research and carried out large-scale tests in an improvised apparatus but without providing proper conditions for laboratory testing. Thus the experiment failed due to lack of strict sterility throughout the system. Later on, British Acetones Ltd. undertook the initiation to duplicate laboratory bacteriological conditions on a commercial scale using corn meal. Butanol and acetone were produced from April 1916 to November 1919 in a total of 3458 runs of 24,000 gals of mash each. There was no run unfit for distillation [4]. By modifying the raw materials and technological conditions an explosion-like development of acetone-butanol fermentation technologies took place. Beesch [5] collected the available knowledge about industrial acetone-butanol fermentation process details, including usability of raw materials, problems of contaminations, infections, treatment of the by-products and recovering the end-products. During World War II the ABE fermentation became the most voluminous industrial biochemical process. However, the cheap petrochemical-based butanol production withdrawn it almost completely in the USA and Europe later on. In China, however, the ABE fermentation industry started only in the early 1950s in Shanghai and expanded rapidly thereafter. At its peak, there were about 30 plants all over the country and the total annual production of solvents reached 170,000 tons [6]. The success of the ABE industry in China had special features like development of continuous fermentation technologies such as in Russia, where the AB plants were the only full-scale industrial plants which used hydrolyzates of lignocellosic wastes for butanol fermentation and the process was finally run in a continual mode [7]. In China, the main strategic considerations were as follows: maintaining maximal growth and acid production phase, adoption of multiple stages in the solvent phase to allow gradual adaptation to increasing solvent, and incorporation of stillage to offer enough nutrients to delay cell degeneration. A biorefinery concept for the use of all byproducts has been elaborated and was partially put into practice. Due to the tremendous national demand for solvents, China has begun a new round of ABE fermentation research. It is expected that a new era in the ABE industry is on the horizon [6].

3. Basic principles of biobutanol production

The ABE fermentation is a complicated multistage process with a series of consecutive and parallel reactions influenced by a series of technological factors. The presence or absence of natural constituents or contaminations in the used raw materials has important influence on the productivity and product distribution. The ABE fermentation is controlled by intracellular redox processes which is influenced by a variety of technological conditions.

3.1. General mechanism of ABE fermentation

The ABE fermentation is a two-stage process: first, an acid-producing and then a solvent producing process takes place, but the solvent producing metabolic pathway could be observed only above 20 g/L starting sugar concentration [8]. Key factors in starting of solventogenesis are the undissociated intracellular butyric acid concentration and the summarized amount of the undissociated butyric and acetic acids within the cells. These are in relationships with the pH and the concentration of butyric and acetic acids in the ferment mash of course, and a boundary condition is that glucose concentration should be above 15 g/L at the moment of the final consumption of butyric acid, because a high glucose flux is required to generate as much amount of ATP as is enough to supply the energy demand of the butyric acid-butanol transformation [9]. Hartmanis et al. studied the pathway for uptake of acids during the solvent formation phase of ABE fermentation by C. acetobutylicum using ^{13}C NMR [10]. Actively metabolizing cells showed that butyrate can be taken up from the medium and quantitatively converted to butanol without accumulation of intermediates. The activities of acetate phosphotransacetylase, acetate kinase, and phosphate butyryltransferase rapidly decreased to very low levels when the organism began to form solvents. This indicates that the uptake of acids does not occur via a reversal of these acid-forming enzymes. No short-chain acyl-CoA synthetase activity could be detected. Apparently, an acetoacetyl-CoA:acetate (butyrate) CoA-transferase is solely responsible for uptake and activation of acetate and butyrate in C. acetobutylicum. The transferase exhibits broad carboxylic acid specificity. The key enzyme in the uptake is acetoacetate decarboxylase which is induced late in the fermentation and pulls the transfer reaction towards formation of acetoacetate. The major implication is that it is not feasible to obtain a batch-wise BuOH fermentation without acetone formation and retention of a good yield of BuOH [10]. Ferredoxin enzymes also play important role in the ABE processes, thus the presence of iron in the appropriate form and concentration is essential factor in the appropriate solvent production. When Clostridium acetobutylicum was grown in batch culture under Fe limitation (0.2 mg/L) at pH 4.8, glucose was fermented to BuOH as the major fermentation end product, and small quantities of HOAc were produced. The final conversion yield of glucose into BuOH could be increased from 20% to 30% by Fe limitation. The BuOH-acetone ratio was changed from 3.7 (control) to 11.8. Hydrogenase specific activity was decreased by 40% and acetoacetate decarboxylase specific activity by 25% under Fe limitation. Thus, Fe limitation affects C and electron flow in addition to hydrogenase [11].

Terracciano and Kashket investigated the intracellular physiological conditions associated with the induction of butanol-producing enzymes in Clostridium acetobutylicum. During the acidogenic phase of growth, the internal pH decreased in parallel with decrease in the external pH, but the internal pH did not go below 5.5 throughout batch growth. Butanol was found to dissipate the proton motive force of fermenting C. acetobutylicum cells by decreasing the transmembrane pH gradient, whereas the membrane potential was affected only slightly. In growing cells, the switch from acid to solvent production occurred when the internal undissociated butyric acid concentration reached 13 mM and the total intracellular undissociated acid concentration (acetic plus butyric acids) was at least 40 to 45 mM [12]. C. acetobutylicum ATCC 824 cells harvested from a phosphate-limited chemostat culture maintained at pH 4.5 had intracellular concentrations of acetate, butyrate and butanol which were 13-, 7- and 1.3-fold higher, respectively, than the corresponding extracellular concentrations. Cells from a culture grown at pH 6.5 had intracellular concentrations of acetate and butyrate, which were only 2.2-fold higher than the respective external concentrations. The highest intracellular concentrations of these acids were attained at pH 5.5. When cells were suspended in anaerobic citrate-phosphate buffer at pH 4.5, exogenous acetate and butyrate caused a concentration-dependent decrease in the intracellular pH, while butanol had relatively little effect until the external concentration reached 150 mM. Acetone had no effect at concentrations ≤200 mM. These data demonstrate that acetate and butyrate are concentrated within the cell under acidic conditions and thus tend to lower the intracellular pH. The high intracellular butyrate concentration presumably leads to induction of solvent production thereby circumventing a decrease in the intracellular pH great enough to be deleterious to the cell [13].

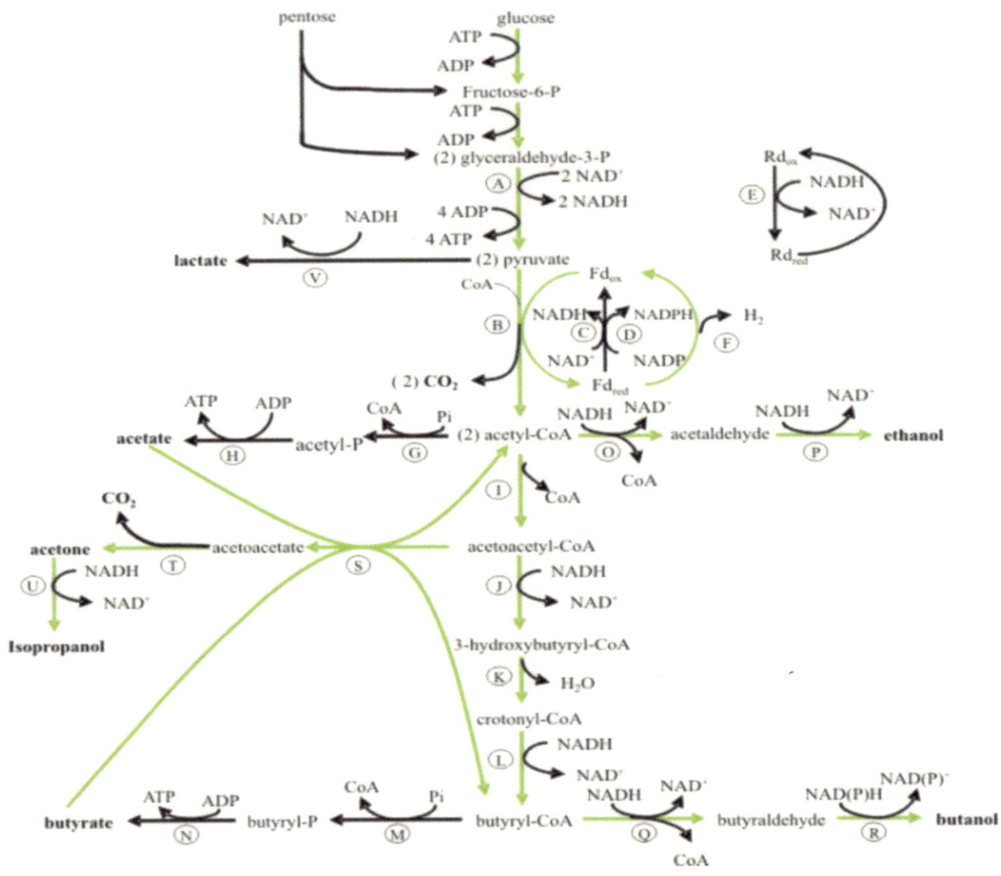

Figure 1. Mechanism of acetogenesis

Harris et al. suggested [14] that butyryl phosphate (BuP) is a regulator of solventogenesis in Clostridium acetobutylicum. Determination of BuP and acetyl phosphate (AcP) levels in various C. acetobutylicum strains (wild(WT), M5, a butyrate kinase (buk) and a phosphotransacetylase (pta) mutant) showed that the buk mutant had higher levels of BuP and AcP than the wild strain; the BuP levels were high during the early exponential phase, and there was a peak corresponding to solvent production [15]. Consistently with this, solvent formation was initiated significantly earlier and was much stronger in the buk mutant than in all other strains. For all strains, initiation of butanol formation corresponded to a BuP peak concentration that was more than 60 to 70 pmol/g (dry wt.), and higher and sustained levels corresponded to higher butanol formation fluxes. The BuP levels never exceeded 40 to 50 pmol/g (dry wt.) in strain M5, which produces no solvents. The BuP profiles were bimodal, and there was a second peak midway through solventogenesis that corresponded to carboxylic acid reutilization. AcP showed a delayed single peak during late solventogenesis corresponding to acetate reutilization. As expected, in the pta mutant AcP levels were very low, yet this strain exhibited strong butanol prodn. These data suggest that BuP is a regulatory mol. that may act as a phosphodonor of transcriptional factors. DNA array-based transcriptional anal. of the buk and M5 mutants demonstrated that high BuP levels corresponded to downregulation of flagellar genes and upregulation of solvent formation and stress genes [15].

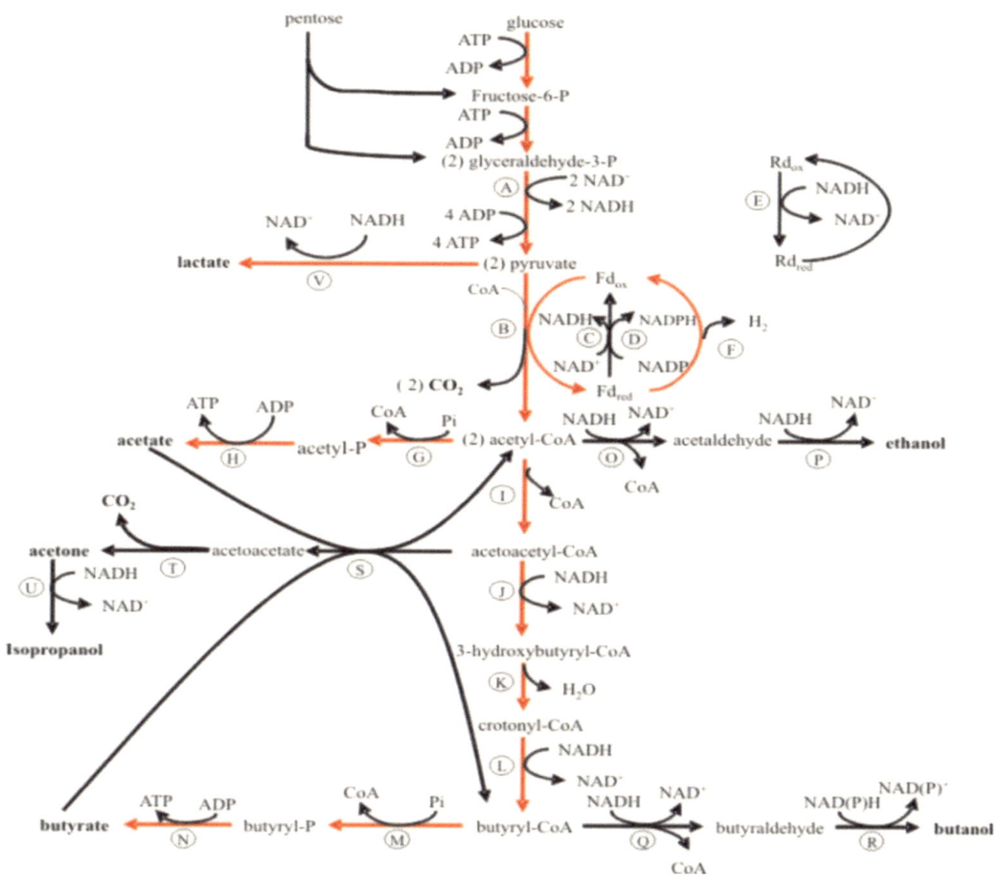

Figure 2. Mechanism of solventogenesis

3.2. Basic reasons for autoinhibition or butanol and intermediate acids toxicity

The toxicity of accumulated butanol and the intermediates is a very important feature of the ABE fermentation. Costa studied [16] the growth rates of Clostridium acetobutylicum in presence of BuOH, EtOH, Me_2CO, acetate and butyrate. Acetate and butyrate were the most toxic compounds, with concentrations of 5 and 8.5 g/L, respectively, stopped the cell growth. An EtOH concentration of 51 g/L or 11 g BuOH/L reduced cell growth by

50%. Acetone did not inhibit cell growth at 29 g/L, thus ethanol and acetone were non-toxic at a normal fermentation. Some mutant strains, however, more tolerant towards butanol, for example Lin and Bladchek [17] obtained a derivative of C. acetobutylicum ATCC 824 which grew at concentrations of BuOH that prevented growth of the wild-type strain at a rate which was 66% of the uninhibited control. This strain produced consistently higher concentrations of BuOH (5-14%) and lower concentrations of acetone (12.5-40%) than the wild-type strain in 4-20% extruded corn broth. Characterization of the wild-type and the mutant strain demonstrated the superiority of the latter in terms of growth rate, time of onset of BuOH production, carbohydrate utilization, pH resistance, and final BuOH concentration in the fermentation broth [17]. Moreira et al. [18] initiated a fundamental study attempting to elucidate the mechanism for BuOH toxicity in the acetone-BuOH fermentation by Clostridium acetobutylicum. Butanol as a hydrophobic compound inserted into the membrane increases the passive proton flux, forms a "hole" for proton on the membrane. This eliminates hydrogen ions form the cell and the intracellular pH increases. The strains which are able to decrease the membrane fluidity are more resistant towards butanol. The cells have deacidifying mechanism to keep the intracellular pH value at 6 when the pH value of the ferment liquor is located between 4 and 5 can reduce acids into alcohols, which increases their butanol producing ability. Lepage et al [19] studied the changes in membrane lipid composition of C. acetobutylicum during ABE fermentation. Large changes were found in phospholipid composition and in fatty acid composition, the latter characterized mainly by a decrease in the unsaturated/saturated fatty acid (U/S) ratio.

Compound	50 % inhibiton	100 % inhibition
Acetic acid	2.7	5.0
Butyric acid	4.1	8.5
Ethanol	51.0	69.0
Butanol	11.0	15.0

Table 1. Inhibitory concentrations (g L^{-1}) of ABE solvents and intermediates on fermentations carried out by C. acetobutylicum

Effects of the addition of alcohols (EtOH, BuOH, hexanol, and octanol) and of acetone were also studied. In all cases, large changes were observed in the U/S ratio, but with differences which were related to the chain length of the alcohols. The effect of solvents appears to account for a large part of changes in lipid composition observed during the fermentation. The pH was also important, a decrease in pH resulting in a decrease in the U/S ratio and in an increase in cyclopropane fatty acids. The effect of increasing temperature was mainly to increase fatty acid chain lengths [19].

4. General conditions of the ABE fermentation

Optimal conditions of ABE fermentation strongly depend on many factors such as the selected raw materials or their composition, and are essentially influenced by the selected strain as well. Furthermore, a series of important factors can decrease or increase the yield and changes the distribution of the ABE solvents even with the same raw material or bacterium strain. Some selected pieces of information are summarized below.

4.1. Selection of raw materials

A wide scale of monosaccharide or disaccharide or other sugar-based oligomeric or polymeric substrates can be used as starting material in ABE fermentations. Compere and Griffith [20] studied the effect of substrate types on the yield and distribution of valuable ABE products with different strains of Clostridia. Different concentrations (2.5-10%) of arabinose, ribose, xylose, xylane, fructose, glucose, cellobiose, lactose, sucrose, dextrin and cellulose were tested and the amount and the distribution of acetone, ethanol, butanol and the H_2/CO_2 ratio strongly depended on the type of strain, the type of sugar and their concentrations as well. Ounine et al. [21] studied the fermentation of glucose, arabinose and xylose with C. Acetobutylicum and found conversion into solvents in 32, 29, and 28%, respectively. Growth yields were similar on the all sugars, but glucose or arabinose was consumed in preference to xylose and with faster growth. It means that not only corn, cereals, molasses, but wastes of dairy products [22,23] or agricultural byproducts as corn stalk, corncob, cellulose wastes and other raw materials can also be utilized [24,25]. By using wastes, however, it is an important key factor that the contaminants can prevent the ABE fermentation. For example, when alkaline peroxide pretreated wheat straw was hydrolyzed using cellulolytic and xylanolytic enzymes, and the hydrolyzate was used to produce butanol using Clostridium beijerinckii P260, the culture produced less than 2.59 g L-1 ABE solvents, but after removal of the formed inhibitor salts with electrodialysis, 22.17 g/L of ABE solvents were formed. This was higher value than the ABE solvent concentration (21.37 g/L) given from glucose. A comparison of use of different substrates (corn fiber, wheat straw) and different pretreatment techniques (dilute sulfuric acid, alkaline peroxide) suggested that generation of inhibitors was substrate and pretreatment specific [26]. Selection of raw material for ABE process cannot be independent from the selection of the bacteria strain. For example, cassava, due to its high starch content and low cost, is a promising candidate substrate for large-scale ABE fermentation processes. However, the solvent yield from the fermentation of cassava reaches only 60% of that achieved by fermenting corn. Addition of ammonium acetate (CH_3COONH_4) to the cassava medium significantly promotes solvent production with a high butanol ratio C. Acetobutylicum mutant (EA 2018). When cassava medium was supplemented with 30 mM ammonium acetate, the acetone, butanol and total solvent prodn. reached 5.0, 13.0 and 19.4 g/l, respectively, after 48 h of fermentation which level of solvent production is comparable to that obtained from corn medium. Both ammonium (NH_4^+) and acetate (CH_3COO-) were required for increased solvent synthesis [27]

4.2. Fermenting microorganisms

Depending on the composition and properties of raw materials, the selection and conditioning of the appropriate bacterium strain are essential. In order to improve the economic efficacy of ABE fermentation, the butanol ratio is to be increased by eliminating the production of other byproducts such as acetone and specific mutants are to be developed which show high butanol tolerance, high productivity or other advantageous properties.

Harada [28, 29] isolated a new strain of Clostridium (Cl. Madisonii) which produced BuOH amounting to 28.7% of the initial total sugar and the fermented broth included 1.38% BuOH. The age of the culture also plays important role in the productivity. By using older inoculated bacteria, the production of acetone increased and the ratio of BuOH to Me_2CO decreased from 2.24 to 1.88 [30]. Harada [31] concluded that the seed culture at the last stage of the acid-decreasing phase gave the best yield as inoculum in the main fermentation. Butanol-resistant mutants have been isolated by Hermann from soil which produced significantly higher solvent concentrations (about 30%) than the wild-type strain [32]. The sporulation-deficient (spo) early-sporulation Clostridium acetobutylicum P262 mutants produced higher solvent yields than did the spoB mutant which was a late-sporulation one. In conventional batch fermentation, the wild-type strain produced 15.44 g L^{-1} of solvents after 50 h at a productivity of 7.41 g L^{-1} d^{-1} of solvents. The spoA2 mutant produced 15.42 g L^{-1} of solvents at a productivity of 72.4 g L^{-1} d^{-1} of solvents with a retention time of 2.4 h in a continuous immobilized cell system employing a fluidized bed reactor [33].

Using two different types of Clostridia to improve the productivity of each (acidogenic and solventogenic) phase is also known. Bergstroem and Foutch [34] improved the BuOH production from sugars by combining two cultures of Clostridium: one that produces butyric acid, and another that converts butyrate to BuOH. Thus, C. butylicum NRRL B592 and C. pasteurianum NRRL B598 were cultured together in thioglycolate medium containing 2.5% added glucose and a $CaCO_3$ chip to maintain pH, at 37 °C under anaerobic conditions. The yield of BuOH was 20 % more as compared to the value when C. butylicum was cultured alone.

Initiation of gene-structure changes by destructive methods such as irradiations or chemicals followed by selection is a well known method in the production of highly effective Cl. Acetobutylicum strains. Yasuda [35] heated ABE producing microorganisms at 100 °C to destroy all vegetative forms except spores which were kept at –10 °C, then treated with electric discharge in vacuum by using 50,000 V and 0.002 A DC for stimulation. High-yield butanol producing Clostridium strain was prepared through irradiation of the wild strain with ^{60}Co γ-rays at an irradiation dosage of 100-1,000 Gy and a dosage rate of 3-5 Gy/min [36].

Chemical mutation with N-methyl-N′-nitrosoguanidine is one of the most frequently used method to produce excellent ABE fermenting strains. Hermann et al [37] prepared a strain of C. acetobutylicum that hyperproduces acetone and BuOH by mutation of C. acetobutylicum IFP903. A new mutant (CA101) of C. pasteurianum prepared in this way could produce 2.1 g BuOH/L in 2 days. By using the parental strain, the production of BuOH was only 0.6 g/L [38]. The C. acetobutylicum strain 77 was isolated from the parent strain ATCC 824 with the abovementioned method in the presence of butanol. The mutant grew more rapidly (μ = 0 69

h^{-1}) than the parent strain (μ = 0 27 h^{-1}) and, at the stationary phase, the cell dry weight of mutant strain was about 50% higher than that of the parent strain. Strain 77 metabolised glucose faster than wild strain and solvent production started earlier with higher specific production rates than the parent strain. From 65 g of glucose, 20 g L^{-1} of solvents (butanol, 14 5 g; acetone, 3 5 g; ethanol, 2 g) were formed by the wild strain in 53 h, whereas the mutant used 75 g of glucose and excreted nearly 24 g L^{-1} of solvents (butanol, 15.6 g; acetone, 4.5 g; ethanol, 3.7 g) in 44 h [39]. A frequently used chemical to inititate mutation in C. Acetobutylicum strains is methanesulfonic acid ethyl ester (EMS). EMS is effective in inducing mutants resistant to ampicillin, erythromycin, and butanol (15 g/l). Optimal mutagenesis occurs at 85–90% kill corresponding to a 15 minute exposure to 1.0% (v/v) EMS at 35 C. At optimal conditions, the frequency of resistant mutant CFU/ total CFU plated increases 100–200 fold [40].

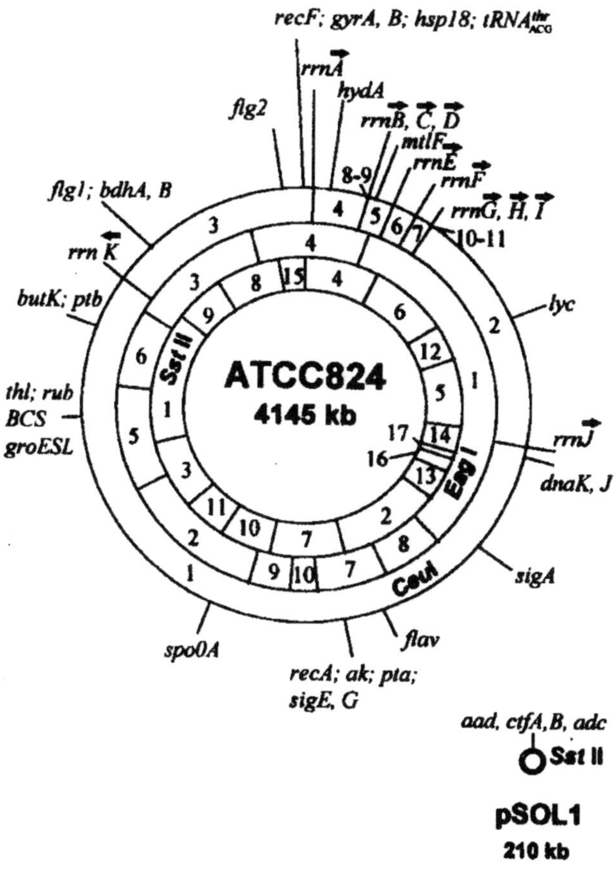

Figure 3. Physical and genetic map of the C. acetobutylicum ATCC 824 genome [242].

Genetical engineering opened unlimited perspectives in the preparation of ABE fermenting microorganisms. Genetically modified C. Acetobutylicum, E. Coli and S. Cereviase and other microorganisms play important role in the future production of ABE solvents under more convenient conditions than in classical ABE fermentation. The acetoacetate decarboxylase gene (adc) in the hyperbutanol-producing industrial strain Clostridium acetobutylicum EA 2018 was disrupted when the butanol ratio was increased from 70 to 80.05%, while acetone production decreased to approx. 0.21 g/L in the adc-disrupted mutant (2018adc). Regulation of the electron flow by addition of methylviologen altered the carbon flux from acetic acid production to butanol production in strain 2018adc, which resulted in an increased butanol ratio of 82% and a corresponding improvement in the overall yield of butanol from 57 to 70.8% [41].

Larossa and Smulski found genes involved in a complex that is a three-component proton motive force-dependent multidrug efflux system to be involved in E. coli cell response to butanol by screening of transposon random insertion mutants. Reduced production of the AcrA and/or AcrB proteins of the complex confers increased butanol tolerance [42]. Green and Bennett subcloned the genes coding for enzymes involved in butanol or butyrate formation into a novel Escherichia coli-Clostridium acetobutylicum shuttle vector constructed from pIMP1 and a chloramphenicol acetyl transferase gene [43]. The resulting replicative plasmids, referred to as pTHAAD (aldehyde/alcohol dehydrogenase) and pTHBUT (butyrate operon), were used to complement C. acetobutylicum mutant strains, in which genes encoding aldehyde/alcohol dehydrogenase (aad) or butyrate kinase (buk) had been inactivated by recombination with Emr constructs. Complementation of strain PJC4BK (buk mutant) with pTHBUT restored butyrate kinase activity and butyrate production during exponential growth. Complementation of strain PJC4AAD (aad mutant) with pTHAAD restored NAD(H)-dependent butanol dehydrogenase activity, NAD(H)-dependent butyraldehyde dehydrogenase activity and butanol production during solventogenic growth [43]. Shen and Liao constructed an Escherichia coli strain that produces 1-butanol and 1-propanol from glucose [44]. First, the strain converts glucose to 2-ketobutyrate, a common keto-acid intermediate for isoleucine biosynthesis. Then, 2-ketobutyrate is converted to 1-butanol via chemicals involved in the synthesis of the unnatural amino acid norvaline. The synthesis of 1-butanol is improved through deregulation of amino-acid biosynthesis and elimination of competing pathways. The final strain demonstrated a production titre of 2 g/L with nearly 1:1 ratio of butanol and propanol [44]. Green et al [45] made recombinant thermophilic bacteria of the family Bacillaceae which have been engineered to produce butanol and/or butyrate. The Bacillaceae is preferably of the genus Geobacillus or Ureibacillus [45]. Young et al described a method of modifying prokaryotic and eukaryotic hosts for the fermentation production of aliphatic alcohols. Elements of the gene for a CAAX proteinase (prenylated protein-processing C-terminal proteinase) are used to increase alcohol tolerance. This can be used in combination with other changes to increase alcohol tolerance [46]. Fermenting with modified eukaryotic cells in a suitable fermentation broth, wherein butanol and ethanol are produced at a ratio between 1:2 to 1:100, is described by Dijk et al. [47]. Since fermentations with yeasts do not require sterile environment, genetically modified yeasts are very prosperous micrørganisms in ABE fermentation. Yeast cells capable of producing butanol and comprising a nucleotide sequence encoding a butyryl-CoA dehydrogenase and at least one nucleotide sequence encoding an electron transfer flavoprotein were described by Mueller et al. [48].

4.3. Effect of medium composition, temperature and nitrogen sources

The appropriate temperature for optimal fermentation ability of C. Acetobutylicum strains strongly depends not only on the type of strain, but on the composition of the medium and raw materials as well, and is strongly influenced by a series of factors such as presence or absence of additives, sugar concentration, pH, and others. McNeil and Christiahsen studied the effect of temperature on the solvent production by C. acetobutylicum in the range 25 to 40°C [49]. It was found that the solvent yield decreased with increasing temperature. Considering total solvent yield and productivity only, the optimum fermentation temperature was found to be 35°C [49]. Comparison of the solvent production by using strains of C. acetobutylicum and C. butylicum from whey showed that higher yields of solvents were observed at 37 °C or 30 °C, respectively [50]. Oda and Yamaguchi [51] concluded that temperature control played important role in the solvent yield and the optimal temperatures were not found to be the same during different stages of the process. Harada [52] concluded that the yield of BuOH was increased from 18.4-18.7% to 19.1-21.2% by lowering the temperature from 30 °C to 28 °C when the growth of the bacteria reached a maximal rate.

Fouad et al. [53] studied fourteen different media in the fermentative production of acetone and butanol. The highest total yields were achieved in medium containing potato starch and soluble starch as C sources. Compositon and pH of the medium have important influence on ABE fermentation. The contaminants in the media have decisive effect on the ABE fermentation. For example, hydrolysates obtained by enzymatic saccharification of wheat straw or cornstover pretreated by steam explosion in classical or acidic conditions, were found non-fermentable into acetone-butanol. A simple treatment involving heating the hydrolysates in presence of calcium or magnesium compounds such as $Ca(OH)_2$ or $MgCO_3$ at neutral pH values restored normal fermentability to these hydrolysates [54]. Sugar concentration of the media also influences the ABE fermentation. Fond et al. [55] studied growing of C. acetobutylicum in fed-batch cultures at different feeding rates of glucose. The sugar conversion to BuOH and Me_2CO increased with increasing the glucose flow whereas, on the contrary, conversion to butyric acid was highest at slow glucose feeding rate. The AcOH concentration was constant at different flows of glucose and the solventogenesis was not inhibited at high flow of sugar [55].

The amount and chemical form of inorganic and organic nitrogen sources basically affect on the ABE process. They influence also strongly depends on presence or absence of other important additives. Among studied inorganic nitrogen compounds, ammonium nitrate and urea could stop the fermentation in the middle, $(NH_4)HSO_4$, NH_4Cl, and $(NH_4)_2HPO_4$ resulted acetone-rich fermentation, while $(NH_4)_2CO_3$ and NH_4OH gave BuOH-rich fermentation [56]. Baghlaf et al [57] studied the effect of different concentrations of corn steep liquor, fodder yeast, soybean meal, corn bran, rice bran, and KH_2PO_4 in the ABE fermenatation, and the organism preferred utilization of natural organic sources. The best concentration of KH_2PO_4, favouring the ABE production was found to be 2 g/L. Oda [58,59] occurred a little effect of adding $(NH_4)_2SO_4$ to EtOH-extracted soybean meal in the yield of solvents, however, cane molasses and dried yeasts were good supplements to the same soybean meal. Addition of asparagine retarded the fermentation. When used as the sole N source, soybean press cake and egg white were good; the others tested were, in the order of decreasing suitability, EtOH-extracted soybean meal, casein, fish protein, zein, gluten, yeast protein, and gelatine. With peanut cake as the N source, Ca salts were not desirable. The stimulants tested were mostly effective: they were, in the order of decreasing effect, liver (best), rice bran-clay, α-alanine, α-methylphenethylamine-H_2SO_4, β-alanine, p-aminobenzoic acid, naphthaleneacetic acid, and cane molasses-clay (the last two were slightly worse than the control without stimulant). Doi et al. [60] could occur that growth-promoting amino acids in the casein acid-hydrolyzate can be divided into three groups: the bacteria required isoleucine, valine, and glutamic acid; asparagine, serine, threonine, alanine, and glycine accelerated fermentation. Leucine, phenylalanine, methionine, tryptophan, proline, lysine, histidine, and arginine were not required for growth and cystine and tyrosine inhibited fermentation.

4.4. Acetate and butyrate additives

Since both acetic and butyric acids are intermediate products of ABE fermentation, and butyrate is almost completely consumed during the solventogenic phase, addition of these intermediates to increase the yield of butanol has already been studied in detail. Beneficial effects were observed with addition of AcOH (completely reduced to Me_2CO), butyric acid (50-80% recovery as BuOH), and sodium acetate NaOAc (60% recovery as Me_2CO), while bad results were obtained with addition of formic acid and calcium acetate [61,62]. Nakhmanocivh and Shcheblikina [63] used a 4% glucose medium, with corn gluten or flour mash, and additon of 0.1N $Ca(OAc)_2$ raised acetone yield by 20-24% and 0.1 N calcium butyrate raised BuOH yield by 45-60% in C. acetobutylicum fermentations. Though $Ca(OAc)_2$ accelerated the fermentation, it was only 40-50% fermented itself. Utilization of $Ca(OOCPr)_2$ goes further (above 70%), mostly by conversion to $Ca(OAc)_2$.

Tang concluded [64] that addition of 1.5 g/L acetic acid increased the cell growth and enhanced acetone production in ABE fermentation. The final concentration of acetone was 21.05%, and the butanol production was not improved. Similarly, addition of 1.0 g/L butyric acid increased the cell growth and enhanced butanol production, the final concentration of butanol was 24.32% while the acetone production was not improved. Additon of acetic acid and butyric acid together (10 mM each) to C. acetobutylicum grown on glucose (2%) in a pH-controlled minimal medium caused rapid induction of acetone and butanol synthesis (within 2 h) [65]. The specific growth rate of the culture and the rate of H_2 production decreased gradually from the onset of the experiment, whereas the rate of CO_2 production remained unchanged. No correlation was found between solvent production and sporulation of the culture [65]. A 32 % conversion rate of the glucose into solvents took place when the same fermentation was carried out on a synthetic medium (BuOH:acetone:EtOH was 0.6:1.9:6). This was changed to 34 and 35 % (BuOH:acetone:EtOH was 5:3:6 or 0.8:2.4:6) by adding HOAc or butyric acid, respectively [66].

Fond et al. [67] studied the effect of HOAc and butyric acid additon in the fermentation of various kinds of carbohydrates using fed-batch fermentations. Different specific rates of carbohydrate utilisation were obtained by variations in feeding rates of sugar. At low catabolic rates of sugar addition of acetic acid or butyric acid, alone or together, increased the rate of metabolic transition by a factor 10 to 20, the amount of solvents by a factor 6 and the percentage of fermented glucose to solvents by a factor 3. The same results were obtained with both glucose and xylose fermentations. Depending on the rates of growth, butanol production began at acid levels of 3–4 g L^{-1} for fast metabolism and at acid levels of 8–10 g L^{-1} for slow metabolism. Associated with slow metabolism, reassimilation of acids required values as high as 6.5 g L^{-1} of acetic acid and 7.5 g L^{-1} of butyric acid. At a high rate of metabolism, acetic and butyric acids were reassimilated at concentrations of 4.5 g L^{-1} [67]. Significant increases in acetone and BuOH production could be observed by Yu and Saddler [68] by growing C. acetobutylicum on xylose in presence of added HOAc or butyric acid. Increased yields could not be accounted for by conversion of the low amounts of acetic or butyric acid added. The effect was greater when the acid was added before, rather than during fermentation, so pH change alone is probably not responsible and enzyme induction may be involved in this process. Addition of acetate or butyrate ensures fermentation at neutral pH conditions as well. Holt et al. used C. acetobutylicum NCIB 8052 (ATCC 824) and monitored a batch culture at 35 °C in a glucose (2%) minimal medium. At pH 5, good solvent production was obtained in the unsupplemented medium, although addition of acetate plus butyrate (10 mM each) caused solvent production to be initiated at a lower biomass concentration. At pH 7, although a purely acidogenic fermentation was maintained in the unsupplemented medium, low concentrations of acetone and n-butanol were produced when the glucose content of the medium was increased (to 4% [wt./vol.]). Substantial solvent concentrations obtained at pH 7 in a 2% glucose medium supplemented with high concentrations of acetate plus butyrate (100 mM each, supplied as their K salts). Thus, C. acetobutylicum NCIB 8052, like C. beijerinckii VPI 13436, are able to produce solvents at neutral pH, although good yields are obtained only when adequately high concentrations of acetate and butyrate are supplied. Supplementation of the glucose minimal medium with propionate (20 mM) at pH 5 led to production of some n-propanol as well as acetone and n-butanol; the final culture medium was virtually acid free. At pH 7, supplementation with propionate (150 mM) again led to formation of n-propanol but also provoked production of some acetone and n-butanol, although in considerably smaller amounts than those obtained when the same basal medium had been fortified with acetate and butyrate at pH 7 [69].

4.5. Effect of carbon monoxide, carbon dioxide and hydrogen

From technological viewpoint, fermentation can be divided into two well separable phases: acid formation phase and, after reaching an autoinhibition limit value of the acids, solvent formation phase. These steps can be performed in separated technological environments as well [70].

Hydrogen formation takes place in the acidogenic phase, so the composition of the gases (CO_2, H_2) changes during the fermentation process. The larger part of the carbon dioxide is formed in the pathway of acetone formation. Presence of hydrogen and carbon dioxide has large influence on each metabolic step. The effect of H_2 and CO_2 as product gases on solvent production was studied in a continuous culture of alginate-immobilized C. acetobutylicum. Fermentations were carried out at various dilution rates. With 10% H_2 and 10% CO_2 in the sparging gas, a dilution rate of 0.07 h^{-1} was found to maximize volumetric productivity (0.58 $g \times L^{-1} \times h^{-1}$), while maximal specific productivity of 0.27 $g^{-1} \times h^{-1}$ occurred at 0.12 h^{-1}. Continuous cultures with vigorous sparging of N_2 produced only acids. It was concluded that in the case of continuous fermentation H_2 is essential for good solvent production, although good solvent production is possible in an H_2-absent environment in case of batch fermentations. When the fermentation was carried out at atmospheric pressure under H_2-enriched conditions, presence of CO_2 in the sparging gas did not slow down glucose metabolism; rather it changed the direction of the phosphoroclastic reaction and, as a result, increased the butanol/acetone ratio [71].

Klei et al. [72] studied the effect of pure CO_2 on the second phase of ABE fermentation. CO_2 pressures up to 100 psig were used in a batch fermentor using glucose as substrate. Maximal solvent production occurred near 25 psig CO_2 at the expense of cell growth. In addition, the BuOH:Me_2CO ratio changed sharply at 40 psig from 5:1 to 20:1 and EtOH production was eliminated at >50 psig. As the pressure increased, both conversion rates of organic acids to solvents and the utilization rate of substrate glucose decreased.

Pressurization of the fermentation vessel with H_2 appeared to decrease, rather than increase, the formation of neutral solvents in batch fermentations [73]. However, increasing H_2 partial pressure increased BuOH and EtOH yields from glucose by an average of 18% and 13%, respectively, and the yields of acetone and of endogenous H_2 decreased by an average of 40% and 30%, respectively, and almost no effect was observed on the growth of the culture. The BuOH-to-acetone ratio and the fraction of BuOH in the total solvents also increased with H_2 partial pressure. There were no major differences in the observed pattern of change with pressurization at either t = 0 or t = 18 h [74].

Redox active additives such as carbon monoxide have important influence on the ABE fermentation processes. Addition of CO inhibited the hydrogenase activity of cell extracts and viable metabolizing cells. Increasing the partial pressure of CO (2 to 10%) in unshaken anaerobic culture tube headspaces significantly inhibited (90% inhibition at 10% CO) both growth and H_2 production. The growth was not sensitive to low partial pressures of CO (~15%) in pH-controlled fermentors (pH 4.5). CO addition dramatically altered the glucose fermentation balance of C. acetobutylicum by diverting carbon and electrons away from H_2, CO_2, acetate and butyrate production and towards production of EtOH and BuOH. The BuOH concentration increased from 65 to 106 mM and the BuOH productivity (the ratio of BuOH produced/total acids and solvents produced) increased by 31% when glucose fermentation was maintained at pH 4.5 in presence of 85% N_2-15% CO vs. N_2 alone [75]. Carbon monoxide sparged into batch fermentations of C. acetobutylicum inhibited production of H_2 and enhanced production of solvents by making available larger amounts of NAD(P)H_2 to the cells. CO also inhibited biomass growth and acid formation as well. Its effect was mostly pronounced under fermentation conditions of excess carbon- and nitrogen-source supply [76]. When continuous, steady-state, glucose-limited cultures of Clostridium acetobutylicum were sparged with CO, complete or almost complete acidogenic fermentations became solventogenic. Alcohol (butanol and ethanol) and lactate production at very high specific production rates were initiated and sustained without acetone, and little or no acetate and butyrate formation. In one fermentation strong butyrate uptake without acetone formation was observed. Growth could be sustained even with 100% inhibition of H_2 formation. Although CO gasing inhibited growth up to 50%, and H_2 formation up to 100%, it enhanced the rate of glucose uptake up to 300%. These results support the hypothesis that solvent formation is triggered by an altered electron flow [77]. The metabolic modulation by CO was particularly effective when organic acids such as acetic and butyric acid were added to the fermentation as electron sinks. The uptake of organic acids was enhanced, and increase in butyric acid uptake by 50-200% over control was observed. H_2 production could be reduced by 50% and the ratio of solvent could be controlled by CO modulation and organic acid addition. Acetone production could be eliminated if desired. BuOH yield could be increased by 10-15%. Total solvent yield could be increased by 1-3% and the electron efficiency to acetone-BuOH-EtOH solvents could be increased from 73% for controls to 80-85% for CO- and organic acid-modulated fermentations. The dynamic nature of electron flow in this fermentation was elucidated and mechanisms for metabolic control were hypothesized [78].

4.6. Other factors

Wyne [79] studied the inhibition of ABE fermentation of maize mash by C. acetobutylicum influenced by 30 representative inorganic and organic acids. Several acids caused complete inhibition when the initial reaction was between pH 3.90 and 3.65, the following being included: HCl, HNO_3, H_2SO_4, H_3PO_4, succinic, maleic, malonic, levulinic, crotonic, glycolic, p-hydroxybutyric, formic, acetic, propionic, butyric and isobutyric. The toxic effects are probably associated with a critical C_{H+} in the cell interior, closely approximating the observed extracellular C_{H+} associated with an inhibitory effect. All three chloroacetic acids are much more toxic than acetic acid, but hydroxy derivatives of the lower fatty acids are not more toxic than the corresponding normal acids. Pyruvic, lactic and glyceric acids are tolerated at higher C_{H+} levels. In the lower fatty acids the inhibiting C_{H+} was appreciably lower with each successive higher homolog. On the basis of molar concentration the order of effectiveness of inhibition was as follows: nonylic > caprylic > heptylic > formic > caproic = isocaproic > valeric = isovaleric > isobutyric = butyric ≥ propionic = acetic. Capillary activity has relatively little effect with formic, acetic, propionic and butric acids, but was very marked with higher homologs [79]. Inhibitory effect of these acids can easily be removed by neutralization [80]. When the ABE fermentation is over, the culture medium may be treated by blowing NH_3 to neutralize most of organic acids and, after distilling out the solvents, the residue can be treated with non-N-containing nutrients, e.g. dried sweet potatoes, and the fermentation may be repeated in the same way in order to save the quantity of nutrient and to increase the yield [81].

The effect of agitation speed and pressure was studied by Doremus et al [82]. Batch fermentations were run at varying agitation rates and were either pressurized to 1 bar or nonpressurized. Agitation and pressure both affect the level of dissolved H_2 in the media which, in turn, influence solvent production. In nonpressurized fermentations volumetric productivity of BuOH increased as the agitation rate decreased. While agitation had no significant effect on BuOH productivity under pressurized conditions, overall BuOH productivity increased over that obtained in nonpressurized runs. Maximal butyric acid productivity, however, occurred earlier and increased as agitation increased. Peak H_2 productivity occurred simultaneously with peak butyric acid productivity. The proportion of reducing equivalents used in forming the above products was determined using a redox balance based on the fermentation stoichiometry. An inverse relationship between the final concentrations of acetone and acetoin was found in all fermentations studied [82]. Using shear activation of C. acetobutylicum by pumping the cells through capillaries, the cell growth, glucose consumption and product formation rates are considerably increased. Shear-activated continuous cell culture can be used as an inoculum with a well-defined fermentation activity for batch cultures. Different runs of such batch cultivation yield well-reproducible results which could not be obtained from inocula of other cultures or even of heat-shocked spores. The cells can attain a growth rate higher than 1.6 h^{-1}. The shear-activated continous culture growth is affected already at a butanol concentration lower than 1.6 g L^{-1}[83], Afschar et al (1986) [80]. The effect of viscosity on the ABE fermentation was studied by Korneeva et al. [84]. Viscosity of the medium was a limiting factor in ABE production by C. acetobutylicum during fermentation with starch and grains such as wheat and rye flour. Various concentrations of agar-agar (0.1, 0.5 and 0.8%) were added to the medium which showed that elevation of viscosity reduces saccharification, increases the concentration of nonfermented sugars, and decreases the yield of solvents. Prior treatment of the substrate with α-amylase reduced the viscosity of the medium and improved fermentation and solvent yields [84].

Although the ABE fermentation is a strictly anaerobic process, [2] Nakhmanovich and Kochkina [85] could increase the BuOH yield by 3.4-9.1% by short periodical aeration of the medium. Redox potential was measured before and after bubbling and decreased sharply by aeration. In batch and continuous cultivations of C. acetobutylicum ATCC 824 on lactose, a strong relationship was observed between redox potential of broth and cellular metabolism [86]. The specific productivity of BuOH and of butyric acid was maximal at a redox potential of -250 mV. The specific production rate of butyric acid decreased rapidly at higher and lower redox potentials. For BuOH, however, it achieved a lower but stable value. This was true for both dynamic and steady states. Continuous fermentations involving lactose exhibited sustained oscillation at low dilution rates. Such oscillation appears to be related to BuOH toxicity to the growth of cells. At higher dilution rates, where BuOH concentrations were relatively low, no such oscillation was observed. Broth redox potential apparently is an excellent indicator of the resulting fermentation product partitioning [86]. Some selected examples are given in Table 2 and 3.

5. General considerations on developments of ABE fermentation

5.1. Immobilization

Immobilization of C. Acetobutylicum strains prevents bacteria from existing in the ferment mash and is a very essential facility in a variety of integrated solvent recovery methods. Haeggstroem and Molin [87] concluded that immobilized vegetative cells of C. Acetobutylicum have a similar product formation pattern when incubated in a simple glucose-salts solution as ordinary growing cells. If vegetative cells of the organism are immobilized in the solvent production phase, solvents are continuously produced on extended incubation. By immobi1izing spores of the organism, the disturbance of the cells metabolic activity during the immobilization procedure was avoided. After the outgrowth of viable cells within the gel, the washed gel preparation retained at a high production capacity in the non-growth stage and the results indicate that continuous production might be fully possible. The butanol productivity was also found to be higher with immobilized cells than in a normal batch process. Haeggstroem [88] used immobilized spores of Clostridium acetobutylicum in a calcium alginate gel. The productivity of the system was 67 g BuOH/L-day and with immobilized cells it was possible to achieve continuous BuOH production for 1000 h. Foerberg et al. [89] developed a technique for maintaining constant activity during continuous production with immobilized, non-growing cells. A single stage continuous system with alginate-immobilized C. Acetobutylicum, was mainly fed with a glucose medium that supported fermentation of acetone-BuOH but did not permit microbial growth. The inactivation that occurred during these conditions was prevented by pulse-wise addition of nutrients to the reactor. By using this technique, the ratio of biomass to BuOH was reduced to 2% compared to 34% in a traditional batch culture. At steady state conditions BuOH was the major end product with yield coefficients of 0.20 (g/g glucose). The productivity of BuOH was 16.8 g L^{-1} d^{-1} during these conditions. In a corresponding system with immobilized growing cells the ratio of biomass to BuOH was 52-76% and the formation of butyric and acetic acid increased thereby reducing the yield coefficients for BuOH to 0.11 (g g^{-1}). With the intermittent nutrient dosing technique, const. activity from immobilized non-growing cells has been achieved for 8 weeks.

Characteristics of the process	**Yield**	**Content Productivity**		**Ref.**
	g g^{-1}	**g L^{-1}**	**g $L^{-1}h^{-1}$**	
Complex medium, yeast extract, glucose, *Cl.Acetobutylicum ATCC 824,* continuous	0.26	12.0	2.50	[244]
Synthetic P-limited medium, two-stage reactorglucose, continuous, *Cl. AcetobutylicumATCC 824*	0.42	18.0	0.54	[70]
Complex medium, continuous, glucose*Cl. Acetobutylicum ATCC 824*	0.32	13.0	0.75	[245]
Synthetic medium, yeast extract, two-stage, cell recycling, *Cl. Acetobutylicum ATCC 824*glucose, continuous	0.30	7.0	4.50	[108]
Synthetic medium, glucose, cell recycling, *Cl. Acetobutylicum ATCC 824,* continuous	0.29	13.0	6.50	[111]
Complex medium, yeast extract, immbolized, intermittent feeding, glucose, continuous,*Cl. Acetobutylicum*	0.20	1.0	0.70	[89]
Complex medium, yeast extract, two-stage, Immobilized *Cl. Acetobutylicum DSM 792,* Glucose, continuous	0.21	3.9	4.02	[94]
Complex medium, glucose, yeast extract, two-stage, *Cl. Acetobutylicum DSM 792,* continuous	0.25	15.4	1.93	[94]

Table 2. Comparison of maximum solvent productivities, yields and concentrations with glucose as sugar source

Several carriers have been tested for production of ABE solvents by immobilized local strain of C. acetobutylicum. Thus, both batch and continuous fermentations were performed by using sodium alginate, polyacrylamide, activated carbon, and silica gel carriers. Calcium alginate was found to be the most suitable with batch culture techniques where the total solvent production was 19.55 g L^{-1} after 4 days. On the other hand, higher solvent yields with continuous fermentation was noticed with silica gel G-60 (0.063-0.2 mm) with 13.06 g L^{-1} solvent production. In all cases, the tested solid supports were of inferior effect for solvent production under the exptl. conditions used as compared with Ca-alginate [90]. High–strength carriers were also tested for C. acetobutylicum ATCC 824 in batch fermentation. Coke, kaolinite and montmorillonite clay appeared to have a beneficial effect on the fermentation, although the effectiveness appeared to be dependent on the medium used. One of the least expensive materials, coke, was suitable for use in continuous culture. Steady state conditions could be maintained for more than 30 days with total solvent productivity and a yield of 12 g L^{-1}, 1.12 g L^{-1} h^{-1} and 0.3 g total solvent/g glucose used, respectively [91]. Entrapment of C.acetobutylicum AS 1.70 with PVA as the base and by means of absorption in the corncob as the carrier is recommended. Experiments have been done to produce acetone and butanol in a statical way in batches and by changing the corn as medium circulatingly [92]. The vegetative cells of C. acetobutylicum AS 1.70 were also immobilized onto CR (ceramic ring) carriers by adsorption. The continuous production of acetone-BuOH from 8% corn mash concentration was carried out for 90 days in a system of 3-stage packed column reactor (total vol. 5.18 L). The maximal concentration of solvent (acetone, BuOH, and EtOH) was 21.9 g L^{-1} and the productivity of the column was 24.73 g L^{-1} d^{-1}. The residual starch concentration was 0.43% and the conversion efficiency of starch was 40.5% [93]. ABE solvent production was also carried out with C. acetobutylicum DSM 792 (ATCC 824) in a two-stage stirred tank cascade using free and immobilized cells. The cells were immobilized by alginate, κ-carrageenan or chitosan. The cell-containing pellets were dried or chemically treated to improve their long-term stability. Dried calcium alginate yielded the best matrix system. It remained stable after a fermentation time of 727 h in stirred tank reactors. The solvent (sum of acetone, butanol and ethanol) productivity of 1.93 g L^{-1} h^{-1} at a solvent concentration of 15.4 g L^{-1} with free cells was increased to 4.02 g L^{-1} h^{-1} at a solvent concentration of 4.0 g L^{-1} h^{-1} with calcium alginate-immobilized cells (25% cell loading, 12 g L^{-1} pellet concentration, 3 g L^{-1} wet cell mass concentration). With pellet diameter of 0.5 mm, the biocatalyst efficiency was <50% [94]. Immobilized cells of C. saccharoperbutylacetonicum N1-4 (ATCC 13564) were tested in an anaerobic batch culture system. Two different methods of immobilization, active immobilization in alginate and passive immobilization by employing stainless steel scrubber, nylon scrubber, polyurethane with uniform pore's size, polyurethane with different pore's size and palm oil empty fruit bunch fiber were studied. Immobilization in alginate was carried out on the effect of cell's age, initial culture pH and temperature on the production of ABE. Immobilized solventogenic cells (18 h) produced the highest total solvents concentration as compared to other phases with productivity of 0.325 g L^{-1} h^{-1}. The highest solvents production by active immobilization of cells was obtained at pH 6.0 with 30 °C with productivity of 0.336 g L^{-1} h^{-1}. Polyurethane with different pore's size is significantly better than other materials tested for solvents productivity and YP/S at 3.2 times and 1.9 times, respectively, compared to free cells after 24 h fermentation. We concluded that passive immobilization technique increases the productivity (215.12 %) and

YP/S (88.37 %) of solvents by C. saccharoperbutyl-acetonicum N1-4 [95]. C.beijerinckii was immobilized in calcium alginate to produce BuOH continuously from glucose. Two different alginate geometries (beads and coated wire-netting) were used for continuous experiments and two mathematical models (sphere and flat plate) were developed. Calculations revealed that no glucose limitation was present in both cases. Furthermore, the biomass build-up in the alginate was probably a surface process [96].

Cells of C.acetobutylicum immobilized on bonechar were used for the production of ABE solvents from whey permeate. When the process was performed in packed bed reactors operated in a vertical or inclined mode, solvent productivities up to 6 kg m^{-3} h^{-1} were obtained. However, the systems suffered from blockage due to excess biomass production and gas hold-up. These problems were less apparent when a partially-packed bed reactor was operated in horizontal mode. A fluidized bed reactor was the most stable of the systems investigated, and a productivity of 4.8 kg m^{-3} h^{-1} was maintained for 2000 h of operation. The results demonstrate that this type of reactor may have a useful future role in the ABE fermentation [97]. Schoutens determined the optimal conditions necessary for the continuous BuOH production from whey permeate with C. beyerinckii LMD 27.6 immobilized in calcium alginate beads. The influence of three parameters on the BuOH production was investigated: fermentation temperature, dilution rate (during start-up and at steady state) and concentration of Ca^{2+} in the fermentation broth. Both a fermentation temperature of 30 °C and a dilution rate of ≤0.1 h-1 during the start-up phase are required to achieve continuous BuOH production from whey permeate. BuOH can be produced continuously from whey permeate in reactor productivities 16-fold higher than those found in batch cultures with free C. beyerinckii cells on whey media [98]. Fermentation of cane sugar molasses by immobilized C. acetobutylicum cells was greatly affected by inoculum size, calcium alginate concentration and molar ammonium nitrogen to molasses ratios. The pH value of the medium and incubation temperature both influenced the ABE production. The maximum total solvent content reached 22.54 g L^{-1} at inoculum size 6% (w/w), molasses concentration 140 g/l, sodium alginate amount 3 %, and molar ammonium nitrogen to molasses ratios 0.48, pH 5.5. Attempts to recyclize the fermentation process by using immobilized spores of C. acetobutylicum afforded total solvent contents of 22.54, 20.64, 19.31 g L^{-1} during the first 3 runs, respectively [99].

5.2. Continuous fermentation

Continuous fermentation is a preferred operational mode to decrease cost of production and increase efficiency. It can easily be performed with using cascade reactors with suppressing butanol concentration below the inhibition limit. The butanol concentration supressing can be performed by dilution or with various methods of recovery with adsorption, extraction, stripping, membrane techniques or with combination of these methods. Increase of the active biomass amount in the mash by cell recycling plays key role in continuous ABE fermentation processes, as well.

Dyr et al. [100] observed formation of neutral solvents in continuous ABE fermentation process by means of C. acetobutylicum without morphological adaptation due to the altered way of cultivation. The results obtained leave no doubt as to the possibility of employing the continuous method for acetone-butanol fermentation [101]. A cascade type continuous ABE fermentation method was developed from soluble starch by building an equipment consistsing a battery of 11 fermenting tanks [102]. The first tank is used as an incubator and an activator for the culture. In the remaining tanks, the actual fermentation is carried out. The feed liquor is continuously supplied. The continuous fermentation process for Me2CO-BuOH production is a 1st-order reaction. A continuous ABE fermentation process was developed and adopted in plants using starch raw materials by Yarovenko [103]. The basis for a continuous process is knowledge of laws of continuous mixing of liquids in batteries of connected vessels which are discussed by Yarovenko [104]. The length of fermentation considerably influences the acidity of the fermented mixture at the end of the process. Owing to differences in mash composition and duration of process, acid level is mostly higher in continuous fermentation than in a discontinuous one. With continuous acetone-butanol process fermentation, speed could be raised 1.58 times compared to the semicontinuous method. In the continuous fermentation, it is useful to operate with 2-5 parallel batteries and to cultivate bacteria in separated vessels. The carbohydrates produced by saccharification under different conditions were studied as they were of great importance on length and course of fermentation. Operation of the battery's head fermentor has a great influence on the whole process, the amount of inoculum, acid production, and fermentation speed. To provide an adequate microorganism concentration and to reduce the risk of infection in the battery's head fermentor, mash from the 2nd vessel is recycled. The acidity increase was evident primarily in the last tank. Optimum concentration of the cells to be inoculated at the start of the fermentation 7×10^9/ml for C. acetobutylicum and physiologically mature cells should comprise about 80% of the total inoculum. The flow rate into the main fermentor should be harmonized with the utilization rate of carbohydrate in the battery. Bacteria in the main vessel must be maintained at their respective stationary phase of growth. The continuous ABE fermentation increased productivity efficiency 20%. The carbohydrate utilization was improved by 2.4%, along with the characteristics of the beer [103,104]. The Japanese K.F. Engineering [105] described an apparatus for production of Me_2CO and BuOH by immobilized ABE-producing microorganisms, where the immobilized microorganisms are first exposed to a batch process until active gas formation is observed, and then, a continuous production process was performed.

The availability and demand of biosynthetic energy (ATP) is an important factor in the regulation of solvent production in steady state continuous cultures of C. acetobutylicum. The effect of biomass recycle at a variety of dilution rates and recycle ratios on product yields and selectivities was determined. Under conditions of non-glucose limitation, when the ATP supply is not growth-limiting, a lower growth rate imposed by biomass recycle leads to a reduced demand for ATP and substantially higher acetone and butanol yields. When the culture is glucose limited, however, biomass recycle results in lower solvent and higher acid yields [106]. Wijjeswarapu et al. studied continuous BuOH fermentation by C. acetobutylicum in a stirred tank reactor. The results of glucose fermentation with cell recycling revealed the formation of small amounts of EtOH, moderate amounts of Me_2CO and BuOH, and large amounts of AcOH and butyric acid. Without cell recycling overall BuOH production was decreased by a factor of 3.5 [107]. Afshar et al. used a cascade system and cell recycling. At a dry cell mass concentration of 8 g/L and a dilution rate of $D=0.64\ h^{-1}$, a solvent productivity of $5.4\ g/L^{-1}\ h^{-1}$ could be attained. To avoid degeneration of the culture which occurs with high concentrations of ABE solvents a 2-stage cascade with cell recycling and turbidostatic cell concentration control was used as optimal solution, the 1st stage of which was kept at relatively low cell and product concentrations. A solvent productivity of 3 and 2.3 g L^{-1} h^{-1}, respectively, was achieved at solvent concentrations of 12 and 15 g L^{-1} [108]. Huang and Ramey [109] determined the influence of dilution rate and pH in continuous cultures of Clostridium acetobutylicum in a fibrous bed bioreactor with high cell density and butyrate concentrations at pH 5.4 and 35°C. By feeding glucose and butyrate as co-substrates, the fermentation was maintained in the solventogenesis phase, and the optimal butanol productivity of 4.6 g L^{-1} h^{-1} and a yield of 0.42 g g^{-1} were obtained at a dilution rate of 0.9 h^{-1} and pH 4.3. Eight Clostridium acetobutylicum strains were examined for α-amylase and strains B-591, B-594 and P-262 had the highest activities. Defibered-sweet-potato-slurry containing starch supplemented with potassium phosphate, cysteine-HCl, and polypropylene glycol was used as continuous feedstock to a multistage bioreactor system. The system consisted of four columns (three vertical and one near horizontal) packed with beads containing immobilized cells of C. acetobutylicum P-262. The effluent contained 7.73 g solvents L^{-1} (1.56 g acetone; 0.65 ethanol; 5.52 g butanol) and no starch. Productivity of total solvents synthesized during continuous operation was 1.0 g L^{-1} h^{-1} and 19.5% yield compared to 0.12 g L^{-1} h^{-1} with 29% yield in the batch system [110]. Pierrot et al. introduced a hollow-fiber ultrafiltration to separate and recycle cells in continuous ABE fermentation. Under partial cell recycling and at a dilution rate of 0.5 h^{-1}, a cellular concentration of 20 g L^{-1} and a solvent productivity of 6.5 g L^{-1} h^{-1} is maintained for several days at a total solvent concentration of 13 g L^{-1} [111]. The device developed was sterilizable by steam and permitted drastic cleaning of the ultrafiltration membrane without interrupting continuous fermentation. With total recycle of biomass, a dry weight concentration of 125 g L^{-1} was attained, which greatly enhanced the volumetric solvent productivity averaging 4.5 g L^{-1} h^{-1} for significant periods of time (>70 h) and maintaining solvent concentration and yield at acceptable levels [112].

A stable continuous production system with nongrowing cells of C. acetobutylicum adsorbed to beechwood shavings was obtained by different types of adsorption procedures for production of ABE solvents by Foerberg and Haegsstroem [113]. The system was started with continuous flow of a complete nutrient medium. A thick cell layer was formed on the wood shavings during the 1st day but it disappeared rapidly. Under glucose limitation, a new cell layer developed during the following period (2-5 days). After this phase, a continuous flow of nongrowth medium with nutrient dosing (8 h dosing interval) was started. This led to a washout of most adsorbed cells and ~85% of suspended cells. Another cell layer was formed during this period and the system was controlled by the nutrient dosing technique. The system was stable with no cell leakage for weeks. The maximal productivity of butanol, acetone, and EtOH was 36 g L^{-1} d^{-1} with a product ratio of 6:3:1 [113].

A continuous ABE production system with high cell density obtained by cell-recycling of Clostridium Saccharoperbutylacetonicum N1-4 was also studied. In a conventional continuous ABE culture without cell-recycling, the cell concentration was below 5.2 g L^{-1} and the maximal ABE productivity was only 1.85 g L^{-1} h^{-1} at a dilution rate of 0.20 h^{-1}. To obtain a high cell density at a faster rate, we concentrated the solventogenic cells of the broth 10 times by membrane filtration and were able to obtain ~20 g L^{-1} of active cells after only 12 h of cultivation. Continuous culture with cell recycling was then started, and the cell concentration increased gradually through cultivation to a value greater than 100 g L^{-1}. The maximum ABE productivity of 11.0 g L^{-1} h^{-1} was obtained at a dilution rate of 0.85 h^{-1}. However, a cell concentration >100 g L^{-1} resulted in heavy bubbling and broth outflow, which made it impossible to carry out continuous culture. Therefore, to maintain a stable cell concentration, cell bleeding and cell recycling were performed. At dilution rates of 0.11 h^{-1} and above for cell bleeding, continuous culture with cell recycling could be operated for more than 200 h without strain degeneration and an overall volumetric ABE productivity of 7.55 g L^{-1} h^{-1} was achieved at an ABE concentration of 8.58 g L^{-1} [114].

Characteristics of the process	Yield	Content	Productivity	Ref.
	g g^{-1}	g L^{-1}	g $L^{-1}h^{-1}$	
Aspen hydrolysate (SO_2 and enzymatic), *Cl. Acetobutylicum P262,* extractive ferm., (dibutyl phthalate), cell recycling	0.36	17.7	0.73	[246]
Pine hydrolysate (SO_2 and enzymatic), *Cl. Acetobutylicum P262*, extractive ferm. (dibutyl phthalate), cell recycling	0.32	22.9	0.95	[246]
Corn stove hydrolysate (SO_2 and enzymatic), *Cl. Acetobutylicum P262*, extractive ferm., (dibutyl phthalate), cell recycling	0.34	25.7	1.07	[246]
Bagasse, alkali and enzymatic hydrolsis, C. saccharoperbutylacetonicum ATCC 27022 Simultaneous ferm., active C	0.33	18.1	0.30	[247]
Rice straw, alkali and enzymatic hydrolsis, C. saccharoperbutylacetonicum ATCC 27022 Simultaneous ferm., active C	0.28	13.0	0.15	[247]
Wheat straw, *Cl. Acetobutylicum IFP 921,* alkali-enzymatic hydrolysis and simultaneous fermentation	0.18	17.7	0.47	[248]
Corn fiber, sulphuric acid hydrolysis, XAD-4 resin purifn., *C. Beijerinckii BA101*	0.39	9.3	0.10	[249]
Corn fiber, enzymatic hydrolysis, *C. Beijerinckii BA101*	0.35	8.6	0.10	[249]
Wheat straw, *Cl. Beijerinckii P260*, simultaneous saccharification and fermentation, gas stripping	0.41	21.42	0.31	[250]
Rice straw, enzymatic simultanenous Hydrolysis and -fermentation, C. Acetobutylicum C375	0.30	12.8	0.21	[251]
Cornstalk stover, enzymetic hydrolysis, membrane reactor, steam exploding, C. Acetobutylicum ASI 132	0.21	-	0.31	[252]
Wheat straw, fed-batch, *Cl. Beijerinckii P 260*, simultaneous saccharification and fermentation, gas stripping	0.44	192.0	0.36	[253]

Table 3. Comparison of maximum solvent productivities, yields and concentrations with lignocellulose based sugar sources

6. Removal methods of ABE solvents from ferment liquors

The complexity and chemical interactions of the aqueous mixture of BuOH, 2-PrOH, acetone, and EtOH produced by the bacterial fermentation of various carbohydrate launched development of numerous innovative separation processes. Belafi-Bako et al. [115] reviewed with 246 reference results and developments on simultaneous product removal in ethanol and acetone/butanol/ethanol fermentation regarding thermal processes (e.g., evaporation, distillation), physical and chemical methods (e.g., extraction, adsorption as well as catalytic reactions), and different membrane separation techniques (e.g., perstraction, reversed osmosis, pervaporation, dialysis).

6.1. Removal by vacuum

The simplest recovering method is removal of ABE solvent during fermentation by using vacuum, because the relative volatility of ethanol, acetone or butanol is much higher than the volatility of water. The I.G.Farbenindustrie] [116] used this method periodically, and removed the butanol in vacuum before completion of the fermentation, fresh wort was added and the fermentation was continued. Dreyfus [117] removed the ABE solvents during fermentation by vaporization with the aid of cyclohexane forming an azeotropic mixture therewith and, passing an oxygen-free gas through the liquor. The cyclohexane may be added continuously or intermittently and may be carried as vapor by the gas stream. Removal of ABE solvents in vacuum-assisted in-situ pervaporation techniques at the temperature of fermentation is discussed in chapter 6.6.

6.2. Removal by gas stripping

Qureshi reviewed the ABE fermentation in various types of reactor systems and recovery by gas stripping with 13 references. Gas stripping is a simple technique which does not require expensive apparatus, does not harm the culture, does not remove nutrients and reaction intermediates and reduces butanol toxicity (inhibition). As a result of butanol removal by gas stripping, concentrated sugar solutions can be used to produce ABE solvents. Compared to sugar utilization of 30 g L^{-1} in a control batch reactor, sugar utilization of 199 g L^{-1} has been reported with 69.7 g L^{-1} solvent production. In fed-batch reactors concentrated sugar solutions (350 g L^{-1}) have been used. Additionally, the process of ABE production results in concentrated product streams containing 9.1-120 g L^{-1} ABE solvent. In the integrated ABE production and recovery systems, selectivity figures of 4-30.5 have been reported [118]. The effect of factors such as gas recycle rate, bubble size, presence of acetone, and ethanol in the solutions or broth were investigated in order to remove butanol from model solution or fermentation broth. Butanol stripping rate was found to be proportional to the gas recycle rate. In the bubble size range attempted (< 0.5 and 0.5-5.0 mm), the bubble size did not have any effect on butanol removal rate. In C. beijerinckii fermentation, ABE productivity was reduced from 0.47 g L^{-1} h^{-1} to 0.25 g L^{-1} h^{-1} when smaller (< 0.5 mm) bubble size and an excessive amount of antifoam (to inhibit production of foam caused by smaller bubbles) were used. This suggested that fermentation was negatively affected by antifoam [119].

Gas stripping can be performed by using fermentation gases (H_2 and CO_2) formed during fermentation. Concentrated sugar solutions (250-500 g/L) were used in continuous fermentation of Clostridium beijerinckii BA101, which operated for 21 d (505 h), producing 460 g acetone-BuOH/L [120]. In the integrated fed-batch fermentation and product recovery system, solvent productivities were improved to 400% of the control batch fermentation productivities. In a control batch reactor, the culture used 45.4 g glucose L^{-1} and produced 17.6 g total solvents L^{-1} (yield 0.39 g g^{-1}, productivity 0.29 g L^{-1} h^{-1}). Using integrated fermentation-gas stripping product recovery system with CO_2 and H_2 as carrier gases, the fed-batch reactor was operated for 201 h. At the end of fermentation, an unusually high concentration of total acids (8.5 g L^{-1}) was observed. A total of 500 g glucose was used to produce 232.8 g solvents (77.7 g acetone, 151.7 g butanol, 3.4 g ethanol) in 1 L culture broth. The average solvent yield and productivity were 0.47 g g^{-1} and 1.16 g L^{-1} h^{-1}, respectively [121]. Using a potential industrial substrate (liquefied corn starch, 60 g L^{-1}) in a batch process integrated with gas stripping resulted in the production of 18.4 g L^{-1} ABE solvents, with 92% utilization of sugars present in the feed. In a fed-batch reactor fed with saccharified liquefied corn starch, 81.3 g L^{-1} ABE was produced as compared to 18.6 g L^{-1} in the control. In this integrated system, 225.8 g L^{-1} corn starch sugar (487% of control) was consumed. In absence of product removal, it is not possible for C. beijerinckii BA101 to utilize more than 46 g L^{-1} glucose [122].

6.3. Removal by adsorption

Tomota and Fujiki [123] observed that the presence of a small amount of activated carbon promotes BuOH fermentation of corn. Oda [124] compared various activated carbons for the removal of BuOH in order to avoid its toxicity. The commercial supernorite proved to be the most effective with intermittent additions. Oda [125] studied the effect of pre-treatment of carbons on BuOH removing capacity, but similar beneficial results were obtained by adding commercial active C to the mash and the acid or alkali treated activated carbons. Yamazaki et al. [126-128] packed activated carbon into a column, and after saturation with ABE solvents it was heated at 150 °C, and steamed to recover solvents, when 98% of the BuOH and 99% of the Me_2CO could be recovered. Activated carbon could be used repeatedly without refreshing. The efficiency of carbon was a little reduced by repeated sorption with soft carbon but hardly reduced with hard carbon. Freundlich's adsorption isotherm by some commercial carbons was, respectively, for Me_2CO and BuOH at 37º, $x/m=0.151C^{0.52}$ and $x/m=0.275C^{0.57}$, where x was the amount of solvent (millimole L^{-1}) adsorbed by a mg of adsorbent, and C the concentration, (millimole L^{-1}) of solvent remaining after equilibrium was reached. The amount of BuOH absorbed by carbon was ≥4 times as large as that of Me_2CO, and this selective sorption was more marked with increasing concentration of solvents. The sorption of BuOH was slower than that of Me_2CO, and >48 h was necessary for reaching equilibrium. Smaller granules of carbon were more effective, and carbon packed in a bag suspended in fermenting mash was convenient. Fermentation experiments with 12-18 g sugar and 5-6 g C/100 ml proved to be the optimal. Carbon granules of the size 2-4 mm^3 were most adequate. Addition of carbon after the growth phase or the maximal acidity phase gave best results. A sugar mash (12 g/100 ml) was fermented with 6 g/100 ml. active C in 3 days by C. acetobutylicum to give a solvent yield of 36% (based on added sugar). The ratio of produced Me_2CO and BuOH was 1:2.

Urbas [129] developed a method for adsorption of ABE components from ferment mash produced by C. acetobutylicum on activated carbons with elution by a volatile solvent. Elution was carried out by feeding the solvent vapor to the carbon bed that is maintained at, or slightly less than, the solvent condensation temperature at a rate of ~1/2 bed vol h^{-1} until the volatile solvent is detected in the eluate and continuing until ~1/2 additional bed volume of eluate is collected. The 1st fraction was mainly water (up to ~96% of the initial amount) and the 2nd a concentrated aqueous solution of the organic compound in the volatile solvent. The solvent is distilled off. The concentration of the final aq. solution is ~30%. The volatile solvents were Me_2CO, 2-butanone, EtOAc, i-PrOH, MeOH, and Et_2O.

A series of adsorbents such as bone charcoal, activated charcoal, silicalite, polymeric resins (XAD series), bonopore, and polyvinylpyridine were tested in the separation of butanol from aqueous solutions and/or fermentation broth by adsorption. Usage of silicalite appeared to be the more attractive as it could be used to concentrate butanol from dilute solutions (5 to 790-810 g L^{-1}) and resulted in complete desorption of ABE solvents. In addition, silicalite could be regenerated by heat treatment. The energy requirement for butanol recovery by adsorption-desorption processes was 1,948 kcal kg^{-1} butanol as compared to 5,789 kcal kg^{-1} butanol during steam stripping distillation. Other techniques such as gas stripping and pervaporation required 5,220 and 3,295 kcal kg^{-1} butanol, respectively [130]. Milestone and Bibby [131] studied the usability of silicalite, which provided a possible economic route for the separation of alcohols from dilute solutions. Thus, EtOH was concentrated from a 2% (wt/vol) solution to 35% and BuOH from 0.5 to 98% (wt/vol) by adsorption on silicalite and subsequent thermal desorption. Maddox [132] found that 85 mg BuOH/g silicalite can be adsorbed from ferment liquors.

Polymeric resins with high n-butanol adsorption affinities were identified from a candidate pool of commercially available materials representing a wide array of physical and chemical properties. Resin hydrophobicity, which was dictated by the chemical structure of its constituent monomer units, most greatly influenced the resin-aqueous equilibrium partitioning of n-butanol, whereas ionic functionalization appeared to have no effect. In general, those materials derived from poly(styrene-co-divinylbenzene) possessed the greatest n-butanol affinity, while the adsorption potential of these resins was limited by their specific surface area. Resins were tested for their ability to serve as effective *in situ* product recovery devices in the n-butanol fermentation by C. acetobutylicum ATCC 824 [133]. In small-scale batch fermentation, addition of 0.05 kg L^{-1} Dowex Optipore SD-2 facilitated achievement of effective n-butanol titers as high as 2.22% (w/v), well above the inhibitory threshold of C. acetobutylicum ATCC 824, and nearly twice that of traditional, single-phase fermentation. Retrieval of n-butanol from resins via thermal treatment was demonstrated with high efficiency and predicted to be economically favorable [133]. Testing performed on four different polymeric resins in the fermentation by C. acetobutylicum showed that the pH increasing could prevent adsorption of intermediates such as acetic and butyric acids. Bonopore, the polymer giving the best adsorption pattern for butanol with no undesirable effects.The adsorption characteristic of butanol from aqueous fermentation broth were also determined on RA, GDX-105, and PVP resins. The adsorption order is GDX-105>RA>PVP and the isotherms could be represented by the Langmuir equation. The adsorption increases with increasing temperature excepting very low concentrations of butanol. The ΔG^0, ΔH^0 and ΔS^0 values for the butanol adsorption processes from aqueous solutions on GDX-105 showed that the enthalpy decreased and the entropy increased [134].

In butanol/isopropanol batch fermentation, adsorption of alcohols can increase the substrate conversion. The fouling of adsorbents by cell and medium components is severe, but this has no measurable effect on the adsorption capacity of butanol in at least three successive fermentations. With the addition of some adsorbents it was found that the fermentation was drawn towards production of butyric and acetic acids [135].

6.4. Basic considerations of solvent extraction

Solvent extraction techniques have the potential for tremendous energy savings in the recovery of fermentation products such as ABE solvents. Such savings will have a direct impact on the economics for the entire fermentation. In order to find the optimal conditions of extractive butanol recovery, however, numerous conditions and factors have to be taken into consideration. A special case of the extractive recovery is the so-called in-situ extractive fermentation (see in chapter 6.5), where extraction is performed during, and together with fermentation. In this case, not only the separation and recovering characteristics play key role in the process, but also the toxicity of the non-miscible solvents basically determines the applicability of the method. In presence of an extractant solvent, however, due to distribution equilibriums, the concentration of each component of the fermentation broth (acids used in decreasing the pH to initiate the solventogenic stage, substrates or intermediates such as glucose, acetate, butyrate) changes. Being aware of these concentration relationships is essential to be able to control the process. Mass and energy balances of side-stream and countercurrent extraction were compared with the appropriate parameters of a classic distillation procedure in recovery of ABE solvents from fermentation broth [136]

A general mathematical model for performance evaluation of acetone-butanol continuous flash extractive fermentation system was formulated in terms of productivity, energy requirement (energy utilization efficiency) and product purity. Simulation results based on experimental data showed that the most pronounced performance improvement could be achieved by using a highly concentrated substrate as feed and the increase in solvent dilution rate could only improve the total productivity at the expense of energy utilization efficiency. A two-vessel partial flash system, with the first vessel of two to three plates and the second vessel as a complete flash vessel, is required to ensure high product purity [137]. Extraction with solvents having distribution coefficients above one appears to have a more favourable energy balance than in case of distillation [136].

Distribution coefficients of ABE solvents between water and the selected extractant and biocompatibility of the extractant are crucial parameters. A solvent screening criterion was developed based on the maximum product concentartion attainable for the assessment of batch and semicontinuous multicomponent extractive fermentations [138]. Dadgar and Foutsch evaluated 47 solvents for the ability to recover Clostridium fermentatiuon products. Equilibrium distribution coefficients and separation factors from water for ethanol, butanol, and acetone were determined [139]. Griffith et al. [140] measured the organic/aqueous distribution coefficients of numerous potential BuOH extractants and simultaneously tested several in bacterial culture. The most effective appeared to be polyoxyalkylene ethers which had distribution coefficients in the range of 1.5-3 and showed little or no toxicity toward the fermentation. The esters and alcohols tested generally had better distribution coefficients but higher biotoxicity. Barton and Daugulis performed biocompatibility tests on 63 organic solvents, including alkanes, alcohols, aldehydes, acids, and esters. Thirty-one of these solvents were further tested to determine their partition coefficients for butanol in fermentation medium of C. acetobutylicum. The biocompatible solvent with the highest partition coefficient for BuOH (4.8) was poly(propylene glycol) 1200 which was selected for fermentation experiments F141G [141]. Thirty-six chemicals were tested for the distribution coefficients for BuOH, the selectivity of alcohol/water separation and the toxicity towards Clostridia. Convenient extractants were found in the group of esters with high molar mass. Liquid-liquid extraction was carried out in a stirred fermentor and a spray column. Formation of emulsions and fouling of the solvent in fermentation broth causes problems with the operation of this type of equipment [142].

Based on the solvent screening criterion and practical experience, one of the best solvents proved to be oleyl alcohol [143]. Oleyl alcohol was used in 40% that of the culture medium to extract BuOH and acetone from the fermentation broth produced from glucose by C. Acetobutylicum and fermentation of the raffinate was continued after the extraction [144]. With a known biocompatibility of extractants such as oleyl alcohol, 1-decanol, 1-octanol, 1-heptanol and ethyl acetate, considering economic viewpoints as well, a mixed extractant of oleyl-alkohol and decanol was chosen for extraction at phase rate of 1:5 [145].

Both butyric acid and butanol could readily be extracted from microbial fermentation broth with vinyl bromide. The vinyl bromide fraction was separated from the aqueous broth and evaporated to give substantially pure butyric acid and (or) BuOH. Three passes of broth through separation columns of vinyl bromide at 4° enabled to isolate ~65% of total butyric acid and ~60% of BuOH in the broth [146]. The methyl, ethyl, propyl and butyl esters of vegetable oils are effective extractants for butanol from aqueous solutions. The effect of four salts, three alcohols and a ketone could be expected to affect the extraction of BuOH from industrial fermentation systems were evaluated. Variations in NaCl, Na_2SO_4, Na_2SO_3 and KH_2PO_4 from 0 to 0.15 M on the extraction of 0.1-4.1% BuOH from aqueous solutions at 25, 40, and 55° gave small changes in distribution coefficients. Mild increases occurred with increasing temperature and increasing NaCl, Na_2SO_4, and KH_2PO_4. Mild decreases in BuOH extraction occurred with increasing Na_2SO_3. Variations in acetone, EtOH, and 2-PrOH concentration ranging between 0 and 4% at 25, 40, and 55° gave small changes in distribution coefficients at BuOH concentrations of 0.1-4.1%. A slight increase in BuOH extraction was observed with increasing 1-pentanol under similar conditions [147]. Extraction of ABE solvents with long-chain fatty acid esters/using the extracts without separation as diesel fuel is discussed in chapter 8.

Ionic liquids are novel green solvents that have the potential to be employed as extraction agents to remove butanol from aqueous fermentation media. An extraction procedure used 1-butyl-3-methylimidazolium-bis(trifluoromethylsulfonyl)imide or 1-butyl-3-methylimidazolium hexafluoro-phosphate ionic liquids was developed by Eom et al. [148]. Knowledge of phase behaviour of ionic-liquid-butanol-water systems is essential in selection the appropriate solvent [149,150]. The 1-hexyl-3-methylimidazolium bis(trifluoromethylsulfonyl)imide exhibits Type 2 liquid-liquid equilibrium behavior toward butanol-water system, thus this ionic liquid can easily separate 1-butanol from water [150].

6.5. *In situ* extractive fermentation

End product inhibition can be reduced by *in situ* removal of inhibitory fermentation products as they form. The first experiments were performed by Bekhtereva who studied the effect of BuOH on the ABE fermenting process and on the development of Clostridium acetobutylicum in concentrated mash. Experimental removal of neutral products from the substrate during fermentation was tested by continuous extraction with castor oil. This oil could extract acetone 13-60, EtOH 5-20 and BuOH 50-88% from the wort. By adding the oil to the medium in varying amounts depending on the carbohydrate content, it was possible to ferment corn mash of 3-5 times the usual concentrations. The yield of acetone was 20-37 g L^{-1} of wort, that of all neutral products 60-100 g L^{-1}. Their concentrations in the wort under the oil layer was usually lower than in control vessels, e. g., total products 1.4-2.3%, BuOH 0.4% against 1.2-1.3% in usual fermentation. The extraction was beneficial to the development of the bacteria [151]. Other starch-containing materials could also be fermented in the usual manner, and BuOH and Me_2CO were continuously removed by means of a solvent immiscible with H_2O, e.g. castor oil [152]. The extraction processes were coupled to batch, fed-batch, and continuous BuOH fermentation to affirm the applicability of recovery techniques in the actual process. In batch and fed-batch fermentation, a 3-fold increase in the substrate consumption, in continuous fermentation ~30% increase could be achieved. [142].

Toxicity and selectivity of 13 organic compounds were tested in extractive batch fermentations performed with C. acetobutylicum. Among them, oleyl alcohol and mixed alcohol (the mixture of oleyl alcohol and C_{18} alcohol) were the best for acetone-BuOH fermentation. The orthogonal-cross-test method with 3 elements and 3 levels was used to evaluate effects of fermentation temperature, inital glucose concentration, and solvent/water ratio on extractive batch ABE fermentation. Extractive batch ABE fermentation in a stirred fermentor was studied at different initial glucose concentrations at 41/35° and at solvent/water ratio 1:2. When initial glucose concentration was 110 g L^{-1}, at the end of extractive fermentation the BuOH concentration in the broth and in the solvent was 5.12 and 22.3 g L^{-1}, respectively. The total BuOH and ABE concentrations based on the broth volume were ~16.27 and 33.63 g L^{-1}, the conversion ratio of glucose was 98% and the total ABE yield was 0.312. *In situ* extractive fermentation could eliminate the inhibition of BuOH on microbial growth, increased the initial glucose concentration and reduced the wastewater amount, thus the consumption of energy could be reduced for the separation and purification of the products [153]. Roefler et al [154] studied the effect of six extractants in batch extractive fermentation: kerosene, 30 wt.% tetradecanol in kerosene, 50 wt.% dodecanol in kerosene, oleyl alcohol, 50 wt.% oleyl alcohol in a decane fraction, and 50 wt.% oleyl alcohol in benzyl benzoate. Best results were obtained with oleyl alcohol or a mixture of oleyl alcohol and benzyl benzoate. In normal batch fermentation of C.acetobutylicum, glucose consumption is limited to ~80 kg m^{-3} due to accumulation of BuOH in the broth. In extractive fermentation using oleyl alcohol or a mixture of oleyl alcohol and benzyl benzoate,

>100 kg m^{-3} of glucose can be fermented. Maximal volumetric BuOH productivity was increased by ~60% in extractive fermentation compared to batch fermentation. BuOH productivities obtained in extractive fermentation compare favorably with other *in situ* product removal fermentations [154].

A medium for ABE fermentation by C. acetobutylicum was mixed with 0.2-5.0% 1-octanol or 2-ethylhexanol and various parameters of fermentation were studied. Glucose consumption, cell growth, ABE formation, and acetate and butyrate formation were inhibited, especially at higher solvent concentrations. Octanol was more toxic than 2-ethylhexanol [155]. A mathematical model for simultaneous fermentation and extraction of the products was derived for ABE production by immobilized C.acetobutylicum cells in a microporous hollow fiber based tubular fermentor-extractor. The solvent, 2-ethyl-1-hexanol, is used for *in situ* dispersion-free extraction of products. Both predicted and experimental data follow the same trend. The experimentally observed value of total solvent productivity increased by >40% as a result of *in situ* solvent extraction [156]. Unfortunately, good extractants for BuOH, such as decanol, are toxic to C. acetobutylicum. The use of mixed extractants, namely, mixtures of toxic and nontoxic coextractants, was tested to circumvent this toxicity. Decanol appeared to inhibit BuOH formation by C. acetobutylicum when present in a mixed extractant that also contained oleyl alcohol, however, maintenance of the pH at 4.5 alleviated the inhibition of BuOH production and the consumption of butyrate during solventogenesis. A mixed extractant that contained 20% decanol in oleyl alcohol enhanced BuOH formation by 72% under pH-controlled conditions. A mechanism for the effects of decanol on product formation is proposed [157]. The same mixed extractant that contained 20% decanol in oleyl alcohol were combined by Wang et al. to carry out in-situ extractive acetone-butanol fermentation, resulting 19.21 g/L of butanol concentration. Butanol productivity could be 62.8% higher than that of control; meanwhile, total organic solvent productivity increases by 42.3% as compared to the control [145].

BuOH fermentation was carried out by contact with solvent containing C_{10-14} alcohols, as well. A seed culture of C. acetobutylicum IAM19013 was inoculated and mixed with tridecanol. The broth was anaerobically fermented, with stirring, at 37 °C for 60 h. The solvent layer at the top of the fermentor was circulated to the bottom. The concentrations of BuOH in the solvent and vapor were 41.6 g L^{-1} and 66%, respectively [158]. Higher alcohols, e.g. C_{16-18} unsaturated alcohols and C_{16-20} branched alcohols were also tested for continuous extraction of BuOH from the medium during the fermentation period. Extraction of the BuOH from the medium by using unsaturated or branched alcohols innoxious to the microorganism markedly increased BuOH yield. Thus, C. acetobutylicum was anaerobically cultivated at 37°C on a medium containing 10% glucose and, after 30 h, 40% oleyl alcohol was added to the broth to remove the BuOH from the aqueous phase and thereby reactivate the fermentation.This increased the total BuOH concentration to 2.5-fold in an additional 70 h [159]. Oleyl alcohol was found to be one of the best solvents for in-situ extractive ABE fermentation. Its butanol partition coefficents value was varied between 3.0 and 3.7 depending on the composition of the broth, nontoxic, nonmiscible and its boiling point is high as compared to ABE solvents. Batch and fed-batch extractive fermentation by C. acetobutylicum was studied with oleyl alcohol as extractant. Extractive fermentation could reduce the product inhibition, increase the initial glucose concentration and increase the fermentation rate. A mathematical model was suggested to describe batch fermentation processes. The proposed model could simulate the experimental data fairly well [160].

In situ removal of inhibitory products from C. acetobutylicum resulted in increased reactor productivity; volumetric butanol productivity increased from 0.58 kg m^{-3} h^{-1} in batch fermentation to 1.5 kg m^{-3} h^{-1} in fed-batch extractive fermentation using oleyl alcohol as the extraction solvent. The use of fed-batch operation allowed glucose solutions of up to 500 kg m^{-3} to be fermented, resulting in a 3.5-5-fold decrease in waste water vol. Butanol reached a concentration of 30-35 kg m^{-3} in the oleyl alcohol extractant at the end of fermentation, a concentration that is 2-3 times higher than is possible in regular batch or fed-batch fermentation. Butanol productivity and glucose conversion in fed-batch extractive fermentation was compared with continuous fermentation and *in situ* product removal fermentation [161].

In ABE fermentation using C. acetobutylicum IAM 19012, it was necessary not only to keep BuOH concentration below the toxic level (2 g L^{-1}), but also to control glucose concentration at <80 g L^{-1} and pH between 4.5 and 5.5. The amount of glucose consumed could approximatly be estimated as 4 times the volume of gas evolved, and BuOH was produced from glucose with an average yield of 0.173. It was thus possible to estimate the concentration of glucose and BuOH at any fermentation time using the volume of gas evolved as an indicator. As oleyl alcohol was an excellent extracting solvent for BuOH, a fed-batch culture system for the microorganism was developed, where withdrawing and feeding operations of the solvent were done automatically based on gas evolution [162]. Ohno combined the fermentation by C. beijerinckii ATCC 25752 which perfectly inhibited the process at the BuOH concentration of 12 kg m^{-3} with the extraction with oleyl alcohol and removing the butanol from its mixture with oleyl alcohol which was carried out by prevaporation with hollow fiber membrane. When the BuOH concentration in oleyl alcohol was 22 kg m^{-3}, the BuOH flux was 3.6′ 10^{-4} kg m^{-2} h^{-1} at 35 °C [163].

Extraction with the non-toxic immiscible solvent, oleyl alcohol was combined with fermentation performed with immobilized C. acetobutylicum to ferment glucose to ABE solvents in a fluidized-bed bioreactor. The extracting solvent had a distribution coefficient of near 3 for butanol. Nonfermenting system tests indicated that equilibrium between the phases could be reached in one pass through the column. Steady-state results are presented for the fermentation with and without extractive solvent addn. One run, with a continuous aqueous feedstream containing 40 g L^{-1} glucose, was operated for 23 d. Steady state was established with just the aqueous feedstream. About half of the glucose was consumed, and the pH fell from 6.5 to 4.5. Then, during multiple intervals, the flow of organic extractive solvent (oleyl alc.) began into the fermenting columnar reactor. A new apparent steady state was reached in about 4 h. The final aqueous butanol concentration was lowered by more than half. The total butanol production rate increased by 50-90% during the solvent extraction as the organic-to-aqueous ratio increased from 1 to 4, respectively. A maximal volumetric productivity of 1.8 g butanol h^{-1} L^{-1} was observed in this nonoptimized system. The butanol yield apparently improved because of the removal of the inhibition. More substrate is going to the desired product, butanol, and less to maintenance or acid production, resulting in a 10-20% increase in the ratio of butanol relative to all products [164].

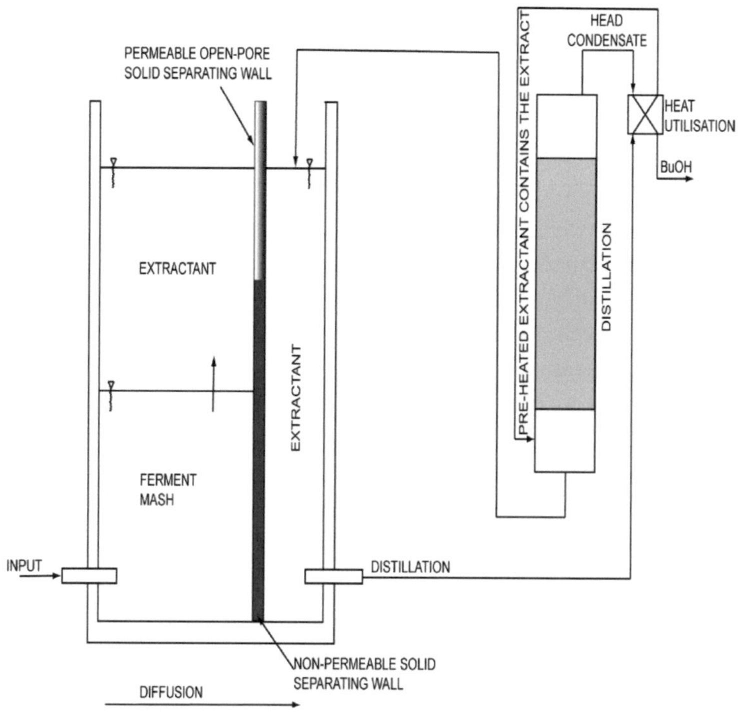

Figure 4. An in-situ extractive fermentor construction [166].

Whole broth containing viable cells of C. acetobutylicum was cycled to a Karr reciprocating plate extraction column in which acetone and butanol were extracted into oleyl alcohol flowing counter-currently through the column. A concentrated solution containing 300 g L^{-1} glucose was fermented at an overall butanol productivity of 1.0 g L^{-1} h^{-1}, 70% higher than productivity of normal batch fermentations. The continuous extraction process provides flexible operation and lends itself to process scale-up [165].

A new type of bioreactor containing a porous permeable wall to recover the biobutanol produced in anaerobic ABE fermentation processes was developed [166, 176]. The ferment liquor is contacted with a non-toxic organic solvent as oleyl alcohol and the butanol in the fermentation liquor distributes between the organic phase and the ferment liquor. The butanol containing solvent located at one side of the permeable wall is in diffusion equilibrium with a same kind of auxiliary solvent with lower butanol concentration located at the other side of the permeable wall. Due to concentration difference, butanol diffuses from one side of the wall to the other side. The concentration difference is kept to be constant by continuous removal of the butanol form the auxiliary solvent phase in which the butanol concentration is always lower than in the extractant phase but much higher than the butanol concentration in the ferment liquor phase. In this way, the primary extractant solvent contacting the ferment liquor is only a transmitting media between the ferment liquor and a small volume of the auxiliary solvent separated with the permeable wall. Energy demand of the distillation to remove the butanol from the auxiliary solvent is less than energy demand of the direct butanol recovery from the ferment liquor or from the extractant phase [166]. The porous composite membranes used as permeable walls for ABE production can be prepared by the method of Tamics et al. [167].

Not only simple alcohols but polyols can also be used in extractive fermentation systems for ABE production. Mattiasson et al. [168] produced acetone and BuOH by C. acetobutylicum in an aqueous two-phase system using 25 % polyethylene glycol 8000. Bacteria remained in the lower phase, and the partition coefficients of acetone and BuOH favoring the upper phase were 2.0 and 1.9, resp. Mean productivity was estimated at 0.24 g BuOH $L^{-1} h^{-1}$, producing 13 g BuOH L^{-1} in 50 h. Poly(propylene)glycol 1200 is the highest partition coefficient reported to date for a biocompatible ABE extracting solvents. Extractive fermentations using concentrated feeds produced ~58.6 g L^{-1} acetone and BuOH in 202 h, the equivalent of 3 control fermentations in a single run. Product yields (based on total solvent products and glucose consumed) of 0.234-0.311 g g^{-1} and within-run solvent productivities of 0.174-0.290 g $L^{-1} h^{-1}$ were consistent with conventional fermentation reported in the literature. The extended duration of fermentation resulted in an overall improvement in productivity by reducing the fraction of between-run down-time for fermentor cleaning and sterilization [141].

Two aqueous two-phased systems involving polyol-type extractants were investigated to determine their ability to reduce product inhibition in the acetone-BuOH-EtOH fermentation. An industrial-grade dextran (DEX) and a hydroxylpropyl starch polymer (Aquaphase PPT (APPT)) were tested as a copolymer with polyethylene glycol (PEG) to form a two-phased fermentation broth. Two-phase fermentation performances in the DEX-PEG and APPT-PEG 2-phase systems were compared to a single-phase conventional fermentation through a series of batch runs. Effects of the phase-forming polymers on C. acetobutylicum also were investigated. With a BuOH partition coefficient of 1.3, the BuOH yield with the two-phase system was increased by 27% over conventional fermentation [169].

Dibutyl phthalate is one of the ester-type extractants used in extractive fermentation of glucose, glucose-xylose mixtures and hydrolyzates of lignocellulosics to acetone-butanol solvents. Dibutyl phthalate has satisfactory physical properties, nontoxic and mildly stimulates the growth of the organism used, C. acetobutylicum P262. Sugar concentrations mainly in the range of 80-100 g L^{-1} resulted in solvent concentrations of 28-30 g L^{-1} in 24 h extractive fermentation compared to 18-20 g L^{-1} for nonextractive control fermentation. Conversion factors of 0.33-0.37 g solvents g^{-1} sugar consumed were obtained. Rapid fermentation was achieved by high cell concentrations and cell recycling from every 24 h fermentation to succeeding similar 24 h fermentation. Somewhat higher nutrients were also helpful. By this means, 255 L of acetone-butanol solvents were obtained per ton of aspen wood, 298 L per ton of pine, and 283 L per ton of corn stover. Such high product yields from inexpensive substrates offer the prospect of economic viability for the process [170].

Induction of flocculation of Clostridia led to a reduction of the specific solvent production rate. Cells adhering to sintered glass are better than flocculating cells for continuous BuOH-acetone fermentation. Due to low toxicity, in-situ application of paraffin, oleic alcohol or stearic acid butyl ester with the cells in the fermenter is possible. Solvent production by Clostridia can be considerably enhanced by the extractive process. Extraction may be directly integrated into a continuous fermentation. Separation of BuOH from oleic acid is easy due to the high boiling point of the extractant (260 °C) being far above the boiling point of BuOH (117 °C). Thus, BuOH can be obtained by normal distillation and the extractant can be recycled [171].

BuOH could be manufactured by cultivating BuOH-producing microorganisms such as C. acetobutylicum in a medium containing a fluorocarbon extractant. The generation time, the mean BuOH production rate, and the mean final BuOH concentration in the C. acetobutylicum culture medium containing Freon-11 (1 g L^{-1}) were increased by 29, 19, and 12%, respectively. Production of acetone and EtOH was not affected [172]. Continuous fermentation of a carbohydrate substrate with continuous extraction of the product by $CFCl_3$ took place in a cylindrical fermentor, with an inlet at the center and a filter membrane concentric with the outer wall, allowing the medium to diffuse outward and to retain microorganisms. The collected medium is pumped to an extractor, where it contacts $CFCl_3$ or another material with a high solvency for BuOH and a low solvency for H_2O and then separated into two phases. The extracted medium is recycled to a feed tank. The solvent is removed from BuOH in an evaporator, where BuOH is collected and the solvent pumped to a compressor and re-utilized [173].

Organic solvents having relatively high distribution coefficients for BuOH against water, often higher alcohols, esters, and organic acids, are very toxic to the microorganisms for BuOH fermentation. Most fermentation inhibition caused by solvent toxicity was eliminated by re-extracting the primary extractant solvent from the residual phase, to be recycled from the product extraction column to the fermentor by paraffin as an extractive fermentation process applied externally to product extraction. After selecting 2-octanol as the extractant from the standpoint of energy consumption in BuOH recovery, a two-stage-extraction BuOH extractive fermentation process having the possibility of reducing the production cost of BuOH was proposed [174]. Heptanal shows strong toxic effect towards C. Acetobutylicum R1 and T5 strains [175] but it has extremely high distribution coefficent (11.5) for butanol [175,176]. Ex-situ extraction with heptanal and recycling the residual broth into a new fermentation cycle proved to be unsuccesfull because the broth contained approx. two times higher heptanal concentration than the toxic limit. Diluting the recycled broth or extracting it with a secondary non-toxic apolar solvent such as hexane to remove the residual dissolved heptanal, inoculate the recycled broth with fresh bacteria in each cycle showed that 4-5 cycles of fermentation could be obtained without important decreasing in the ABE yields and productivity [175]. A multiple solvent extraction is described by Shi et al [177].

Mathematical formulation was made for the performance evaluation considering two types of solvent-supplying strategies. One is to add multiple solvents simultaneously and the product is removed at one time. Another is to add them one by one consecutively. Computer simulations were made for batch, fed-batch, and repeated fed-batch operation of acetone-BuOH fermentation to show the power of the approach. Significant improvement in terms of productivity and product concentration is expected when two extractants such as oleyl alcohol and benzyl benzoate are used, as compared to using only one solvent [177]. A two-stage-extraction butanol extractive fermentation process was developed and studied using a bench-scale extractive fermentation plant with a butanol production capacity of ~10 g h^{-1}. The production rate equation for extractive fermentation was simply expressed by a previously reported equation multiplied by an equation for the extraction raffinate recycling effect. A butanol production-cost calculation program for the two-stage-extraction process determined the optimum operational conditions to be when butanol concentration, residual sugar concentration and recycling ratio were 6 kg m^{-3}, 15 kg m^{-3} and 3, respectively. These optimal conditions were achieved in the bench-scale plant when it was operated with total sugar concentration, dilution rate and recycling ratio of 113 kg m^{-3}, 0.158 h^{-1} and 3, respectively [178].

A special kind of in-situ extractive fermentation is the so-called perstraction, where a selective membrane is located between the broth and the extractant phase. Both sides of the membrane contact with each phase and ensures a medium betwen two immiscible phases to exchange butanol content. Due to lack of direct contact between two phases, toxicity or other problems can be eliminated and a dispersion-free extraction is possible, leading to an easy operation of the equipment, but the mass transfer in the membrane becomes important. This extraction processes were coupled to batch, fed-batch, and continuous BuOH fermentation to affirm the applicability of the recovery techniques in the actual process. In batch and fed batch fermentation a 3-fold increase in the substrate consumption could be achieved, while in the continuous fermentation it increases by~30% [142]. Jeon and Lee [179] described a fed-batch operation for enhanced separation with a semipermeable silicon membrane which showed high specific permeability to BuOH and acetone. Among various solvents examined, oleyl alcohol and polypropylene glycol were the most suitable as extractants. In fed-batch operation of the membrane-assisted extractive BuOH fermentation system, significant improvements were found in comparison to a straight batch fermn. The total glucose uptake per run was raised to 10 times of the value normally found in batch fermentation. The solvent productivity increased by a factor of 2. The total solvent yield increased by 23% due to reduction of acid production and reuse of cells in the fed-batch operation [179]. A continuously operated membrane bioreactor was connected to a 4-stage mixer-settler cascade and Clostridium acetobutylicum was cultivated in this reactor. BuOH was selectively extracted with butyric acid-saturated decanol from the cell-free cultivation medium, and the BuOH-free medium was refed into the reactor. Due to high boiling point of decanol, recovery of BuOH from the decanol solution is easy. Both partition coefficient and selectivity of BuOH in the cultivation medium-decanol system are sufficiently high for removing it from the medium. Direct contact of cells with the decanol phase causes cell damage. However, decanol is practically insoluble in the fermentation medium, thus the contact of the cell-free medium with the solvent phase does not influence cell growth neither product formation. At a dilution rate of $D=0.1\ h^{-1}$, BuOH productivity was increased by a factor of 4 by removing BuOH from the medium [180].

6.6. Membrane techniques and other methods

Pervaporation is an energy-efficient alternative to distillation for removing volatile organic compounds from water, especially ABE solvents from their dilute solutions in a fermentation broth. Pervaporation is able to enrich acetone, BuOH, and EtOH with respect to water. The selectivity of this process is based mainly on superposition of the thermodinamical liquid-vapor selectivity, the chemical affinity selectivity, and the kinetic diffusional selectivity of the materials used. The liquids to be separated are not stressed in any chemical, thermal, or mechanical way. Gudernatsch et al. demonstrated the technical feasibility of the pervaporation process in continuous fermentation runs. Composite hollow fiber membranes with transmembrane fluxes in the range of 2 kg m^{-2} h^{-1} and sufficient selectivity were prepared and characterized [182]. El-Zanati et al. designed a special cell to separate the butanol from butanol/water solutions of different butanol concentrations between 6 and 50 g L^{-1}. The temperature of the mixture feed to the cell was 33 °C while the pressure of permeation side was about ~0 bar. Results revealed that butanol concentration changes non-linearly during the first 3 h, and then proceeds linearly. The percentage of butanol removal increases with increasing feed concentration [183]. A new type of pervaporation apparatus was designed and tested by Vrana et al. to develop an integrated fermentation and product recovery process for ABE fermentation. A cross-flow membrane module able to accommodate flat sheet hydrophobic membranes was used for the experiments. Permeate vapors were collected under vacuum and condensed in a dry ice/ethanol cold trap. The apparatus containing polytetrafluoroethylene membranes was tested using butanol-water and model solutions of ABE products. Parameters such as product concentration, component effect, temperature and permeate side pressure were examined [184].

Various kinds of polymeric, ceramic, and liquid membranes can be used for selective separation of solvent vapors at the temperature of fermentation. Polymeric and ceramic membranes have rather poor solvent selectivity compared to liquid membranes even though they achieve reasonable solvent mass fluxes. Liquid membranes have stability problems due to various losses. Groot et al. used silicon tubing membrane technology in the BuOH/iso-PrOH batch fermentation and the substrate conversion could be increased by simultaneous product recovery [185,186]. Geng and Park carried out fermentation by using a low acid producing C. acetobutylicum B18 and a pervaporation module with 0.17 m^2 of surface area was made of silicone membrane of 240 mm thickness. During batch and fed-batch fermentation, pervaporation at an air flow rate of 8 L min^{-1} removed butanol and acetone efficiently. Butanol concentration was maintained below 4.5 g L^{-1} even though C. acetobutylicum B18 produced butanol steadily. With pervaporation, glucose consumption rate increased as compared to that without pervaporation, and up to 160 g L^{-1} of glucose was consumed during 80 h [187]. Experiments using make-up solutions showed that BuOH and acetone fluxes increased linearly with their concentration in the aqueous phase. Fickian diffusion coefficients were constants for fixed air flow rates and increased at higher sweep air flow rates. During batch and fed-batch fermentation, pervaporation at an air flow rate of 8 L/min removed BuOH and acetone efficiently. BuOH concentration was maintained at <4.5 g/L even though C. acetobutylicum B18 produced BuOH steadily. Cell growth was not inhibited by possible salt accumulation or O_2 diffusion through the silicone tubing. The culture volume was maintained relatively constant during fed-batch operation because of offsetting effects of water and product removal by pervaporation and addition of nutrient supplements [188]. Fadeev et al evaluated poly[1-(trimethylsilyl)-1-propyne] (PTMSP) dense films for n-butanol recovery from ABE fermentation broth. Flux decline of a PTMSP film during pervaporation of 20 g L^{-1} BuOH/water mixture was linear. PTMSP films change their geometry when exposed to alcohol and alcohol/water mixtures and then dried. As a result of the relaxation process, polymer film becomes thicker and denser, effecting membrane performance. PTMSP films that were treated with 70% iso-propanol/water show linear flux decline vs. pervaporation time. Strong lipid adsorption seems to occur on the membrane surface when fermentation broth is used as a feed causing flux decline within short period of time [189].

Oya and Matsumoto used a hydrophobic polypropylene porous hollow fiber membrane of surface area 0.3 m^2, porosity 45%, and bubble point 12.5 kg/cm^2, under reduced pressure [190]. By Knapp et al, a vinyl-type norbornene polymer with average molar weight ~5000 was found to be useful as pervaporation membranes with separation factor of ~10 for separation of n-butanol and isobutanol [191]. Various membranes like styrene Butadiene Rubber (SBR), ethylene propylene diene rubber (EPDM), plain poly di-Me Siloxane (PDMS) and silicalite filled PDMS were studied for the removal of ABE solvents from binary aqueous mixtures and from a quaternary mixture. It was found that the overall performance of PDMS filled with 15% wt./wt. of silicalite was the best for removal of butanol in binary mixture study. SBR performance was best for the quaternary mixtures studied [192].

Composite membranes containing adsorbents such as silicalite or liquid extractants such as oleyl alcohol or other solvents proved to be effective materials in ABE solvent removal from fermentation broth. Thin-film silicalite-filled silicone composite membranes were fabricated by incorporating ultrafine silicalite-1 particles, 0.1-0.2 mm. It was found that with the increase of silicalite content in the top active layer, selectivity for n-butanol and n-butanol flux increased, while the total flux decreased. When the silicalite-1 content was over 60%, the active layer appeared to have defects, aggregation of silicalite-1 particles, which influenced the separation factor. By controlling the membrane thickness and silicalite-1 content, membranes with total flux of 600-700 g m^{-2} h^{-1} (n-butanol flux of 300 g m^{-2} h^{-1}) and selectivity of 90-100 at 70 °C using 10 g L^{-1} of n-butanol as feed solution were obtained. The effects of operation temperature and feed solution concentration on membrane performances were studied [193]. A membrane with a silicalite-1 (its adsorption capacity for a mixture of acetone, butanol and ethanol were 8-12, 85-90 and <5 mg g^{-1}, respectively, there was no apparent difference in absorption rate of butanol at 36° C and 79°C and desorption of butanol occurred efficiently at 78 °C and 1-3 Torr) to polymer ratio of 1.5:1 (g:g) (306 mm thick) had butanol selectivities of 100-108 and a flux of 89 g m^{-2} h^{-1} at feed butanol concnentrations ranging from 5 to 9 g L^{-1} and a retentive temperature of 78°C. A 170 mm silicone membrane under identical conditions had selectivity and flux of 30 and 84 g m^{-2} h^{-1}, respectively. A thin silicalite membrane offered low selectivity and high flux, while a thick membrane offered high selectivity and low flux. The effect of butanol concentration (0.37-78 g L^{-1}) on flux and selectivity was also studied [194].

Thongsukmak and Sirkar developed a new liquid membrane-based pervaporation technique to achieve high selectivity and avoid contamination of the fermentation broth. Trioctylamine as a liquid membrane was immobilized in the pores of a hydrophobic hollow fiber substrate having a nanoporous coating on the broth side. The coated hollow fibers demonstrated high selectivity and reasonable mass fluxes of solvents in pervaporation. The selectivities of butanol, acetone, and ethanol achieved were 275, 220, and 80, respectively, with 11.0, 5.0, and 1.2 g m^{-2} h^{-1} for the mass fluxes of butanol, acetone and ethanol, respectively, at a temperature of 54 °C for a feed solution containing 1.5 wt.% butanol, 0.8 wt.% acetone, and 0.5 wt.% ethanol. Mass fluxes were increased by as much as five times with similar selectivity of solvents when an ultrathin liquid membrane was used [195]. Other long-chain trialkylamines such as trilaurylamine or tri-decylamine could also be used as liquid membranes [196]. Acetic acid in the feed solution reduced selectivity of the solvents without reducing the solvent fluxes due to coextraction of water which increases the rate of water permeation to the vacuum side. The liquid membrane present throughout the pores of the coated substrate demonstrated excellent stability over many hours of experiment and essentially prevented the loss of liquid membrane to the feed solution and the latter's contamination by the liquid membrane [195].

In order to exclude toxic effect of the released liquid membrane ingredient, an oleyl alcohol based liquid membrane was developed. This liquid membrane was energy efficient and did not affect microorganism growth. Oleyl alcohol liquid membrane was proved to be useful for the separation of BuOH and isobutanol in a fermentation culture with immobilized Clostridium isopropylicum IAM 19239 [197].

An ionic liquid (IL)-polydimethylsiloxane (PDMS) ultrafiltration membrane (pore size 60 nm) guaranteed high stability and selectivity during ABE fermentation carried out at 37 °C. Overall solvent productivity of fermentation together with continuous product removal by pervaporation was 2.34 g L^{-1} h^{-1}. The supported ionic liquid membrane (SILM) was impregnated with 15 wt.% of a novel ionic liquid (tetrapropylammonium tetracyano-borate) and 85 wt. % of polydimethylsiloxane. Pervaporation, accomplished with the optimized SILM, led to stable and efficient removal of the solvents butan-1-ol and acetone out of a C. acetobutylicum culture [198].

Reverse osmosis for recovering water from broth can also be used to concentrate ABE fermentation products. Polyamide membranes exhibited BuOH rejection rates ≤85%. Optimum rejection of BuOH occurred at a pressure of 5.5-6.5 MPa and hydraulic recoveries of 50-70%. The flux range was 0.5-1.8 L m^{-1} [199]. Other membranes exhibited rejection rates as high as 98% and the optimal rejection of BuOH in the ferment liquor occurred at recoveries of 20-45% with flux ranging between 0.05-0.6 L m^{-2} min^{-1} [200].

Dialysis fermentation relieves BuOH toxicity with increased yield of product, and solvent extraction can be applied to the nongrowth side of the fermentor for concentration of the BuOH. C. acetobutylicum ATCC 824 and several other strains were studied for the fermentation of corn, potato, and glucose [201].

The ability of cyclodextrins to form crystalline insoluble complexes with organic components was explored as a selective separation of dilute ABE products from Clostridium fermentation systems. A product or a product mixture at a concentration of 0.150 mM each was treated with α-cyclodextrine or β-cyclodextrine in aqueous solutions or nutrient broth. In the acetone-butanol-ethanol system and in the butanol-isopropanol system, α-cyclodextrine selectively precipitated 48% and 46% butanol after 1 h agitation at 30°. However, β-CD was superior for the butyric acid-acetic acid system because it selectively precipitated 100% butyric acid under the same conditions. Cooling the three-product system with α-CD to 4° for 24 h significantly increased the precipitates but decreased the selectivity for either butanol or butyric acid [202].

Hypercrosslinked microporous ion-exchanger resins proved to be suitable agents to adsorb butanol into solid phase from fermentation broth. This ensures fermenting with a microorganism capable of producing butanol in a suitable fermentation medium and recovering butanol from the fermentation medium [203].

Integration of the abovementioned (Chapter 6) methods ensures new possibilities in the economic ABE solvent recovery. Some representative examples without demand of completeness are discussed here.

Mawasaki et al performed continuous extractive butanol fermentation with the microbe immobilized in gel beads and presented the recovery system of butanol from the solvent by pervaporation with hollow fiber membrane. This system was expected to be advantageous to prevent the fouling of membrane because butanol-oleyl alcohol mixtures obtained from extractive fermentation do not include solid particles [204]. Pervaporation method could also be used for *in situ* alcohol recovery in continuous iso-PrOH-BuOH-EtOH fermentation with immobilized cells. Fermentation was performed in a stirred tank and in a fluidized bed reactor as well. In the integrated process, the substrate consumption could be increased by a factor of 4 if compared to continuous fermentation without pervaporation product recovery. Experiments with a pilot plant plate-and-frame pervaporation module were described for the separation and dehydration of alcohols. This module was also coupled to continuous BuOH fermentation, however, sterilization of the module was troublesome, and it was frequently plugged by microbial cells [205]. ABE solvents were produced in an integrated fermentation-product recovery system using C. acetobutylicum and a silicalite-silicone composite membrane. Cells of C. acetobutylicum were removed from the cell culture using a 500,000 molecular weight cut-off ultrafiltration membrane and returned to the fed-batch fermentor. The ABE solvents were removed from the ultrafiltration permeate using a silicalite-silicone composite pervaporation membrane. The silicalite-silicone composite membrane (306 mm thick) flux was constant during pervaporation of fermentation broth at the same concentration of ABE solvents. Acetone butanol selectivity was also not affected by the fermentation broth, indicating that the membrane was not fouled by the ABE fermentation broth. The silicalite-silicone composite membrane was exposed to fermentation broth for 120 h. Acetic acid and ethanol did not diffuse through the silicalite-silicone composite membrane at low concentrations. The fed-batch reactor was operated for 870 h. Totally 154.97 g L^{-1} solvents was produced at solvent yield of 0.31-0.35 [206].

Application of membrane-assisted extraction to butanol fermentation was investigated as a means of product separation and also as a way of alleviating the problems concerning the end-product inhibition. The coupled reactor-separator system was stable enough to sustain continuous operation lasting several weeks. The data on continuous run reaffirmed most of the advantages found in a previous study on fed-batch system in that the reactor separator system rendered high productivity and yields due primarily to reduced product inhibition. Improvement in productivity was particularly notable, as a fourfold increase over straight batch operation was obtained. In normal continuous operation, spontaneous cell deactivation occurred after 200-400 h of operation despite the removal of inhibitory products. The presence of autolysin was one of the probable causes of cell deactivation. The cell viability, however, was prolonged significantly when the bioreactor was operated under glucose-limited conditions [207].

A calcium alginate-immobilized continuous culture was used in a novel gas-sparged reactor to strip the solvents from the aqueous phase and reduce their toxicity. A dilution rate of 0.07 h^{-1} was found to give maximal solvent productivity at 0.58 g dm^{-3} h^{-1}, although at 0.12 h^{-1} the productivity was slightly lower. In order to increase glucose uptake by the culture, feed glucose concentration was increased over time to attempt to acclimatize the culture. This resulted in productivity as high as 0.72 g dm^{-3} h^{-1} although this production rate was unstable [208].

An extractive acetone-BuOH fermentation process was developed by integrating bioproduction, ultrafiltration, and distillation, providing simultaneous retention of biomass, selective removal of inhibitors from permeate and separation and purification of acetone, BuOH, and EtOH. Successive batch fermentations were performed with normal pressure distillation (98°) which permitted prolonging and enhancing (by a factor of 3) solvent production with very few volume exchanges of medium (average dilution rate was 0.002 h^{-1}), and recovering the concentrated solvents online. Different operating conditions were also tested in order to study the presence of extracellular autolytic enzymes as inhibition factors. Extracellular autolytic activity was low most of the time, even without enzyme-inactivating heat treatment in the distillation boiler, and high-temperature distillation was deleterious to the culture medium. Improvements of the process were achieved, first, by managing continuous runs, providing a minimal renewal of the culture medium and, mainly, by decreasing the temperature and pressure of distillation. Solvent productivity reached 2.6 g L^{-1} h^{-1} for a 0.036 h^{-1} average dilution rate, corresponding to a feed concentration of 156 g L^{-1} glucose actually consumed [209].

Continuous extractive bioconversion processes were described for conversion of native starch granules to ABE solvent production using a selective adsorbent. In fermentation of carbohydrates with C. acetobutylicum selective synthetic zeolite or crosslinked divinylbenzene-styrene copolymer sorbents are integrated in the process to adsorb the products from the medium continuously [210]. The conversion of glucose to ABE solvents by C. acetobutylicum employing extractive fermentation by using a combination of membrane technology and solid adsorbents integrated into the fermentation process was studied. The adsorbent used was a nitrated divinylbenzene-styrene copolymer. Its ability to adsorb fermentation broth constituents was as follows: BuOH 82, EtOH 36, Me_2CO 51, butyric acid 99, and AcOH 21 mg/g sorbent. The polymer was then heat treated to release the bound solvents. In a long term experiment using an adsorption column, 400 g glucose was added successively to the column and fermentation allowed to proceed for 320 h. A total amount of 67 g of solvent was recovered by heating 930 g polymer [211]. It was found that the *in situ* adsorption process using polyvinylpyridine as the adsorbent enhanced the fermentation rates and the reactor productivity by C. acetobutylicum. In typical traditional acetone-butanol fermentation process only about 60 g/L of glucose could be used in a batch operation mode and thus, at maximum only 21 g L^{-1} of the total final products concentration could be achieved. In the adsorption-coupled system an initial glucose concentration of 94 g L^{-1} was fermented when a weight ratio of the adsorbent to the fermentation broth of 3/10 was used. An overall product concentration of 29.8 g L^{-1} and a productivity of 0.92 g L^{-1} h^{-1} were achieved in the adsorptive batch fermentation system. Compared with the controlled traditional batch acetone butanol fermentation, the integrated process increased the final product concentration by 54% and the productivity by 130% [212]. Integration of a repeated fed-batch fermentation (C. acetobutylicum) with continuous product removal (poly(vinylpyridine) adsorption) and cell recycling resulted in inhibitory product concentration reduction. Because of the reduced inhibition effect, a higher specific cell growth rate and thus a higher product formation rate were achieved. The cell recycle using membrane separation increased the total cell mass density and, therefore, enhanced the reactor productivity. The repeated fed-batch operation overcame the drawbacks typically associated with a batch operation such as down times, long lag period, and the limitation on the maximum initial substrate concentration allowed due to the substrate inhibition. Unlike a continuous operation, the repeated fed-batch operation could be maintained for a long time at a relatively higher substrate concentration without sacrificing the substrate loss in the effluent. As a result, the integrated process reached 47.2 g L^{-1} in the equivalent solvent concentration (including acetone, BuOH, and EtOH) and 1.69 g L^{-1} h^{-1} in the fermentor productivity, on average, over a 239.5-h period. Compared with controlled traditional batch acetone-BuOH fermentation, the equivalent solvent concentration and the fermentor productivity were increased by 140% and 320%, respectively [213].

Cells of C. acetobutylicum were immobilized by adsorption onto bonechar and used in a packed bed or fluidized bed reactor for the continuous production of ABE solvents from whey permeate. At dilution rates in of 0.35-1.0 h^{-1}, ABE solvent productivities of 3.0 to 4.0 g L^{-1} h^{-1} were observed, but lactose utilization values were poor. When operated in an integrated system with product removal by liquid-liquid extraction, there was a decrease in productivity, but lactose utilization was increased markedly. Of the three extractants tested, oleyl alcohol proved to be superior to both benzyl benzoate and dibutyl phthalate [214].

Shah and Lee studied simultaneous saccharification and extractive fermentation (SSEF) to produce ABE solvents from aspen tree. In SSEF employing cellulase enzymes and C. acetobutylicum, both glucan and xylan fractions of pretreated aspen are concurrently converted into acetone and butanol. Continuous removal of fermentation products from the bioreactor by extraction allowed long-term fed-batch operation. The use of membrane extraction prevented the problems of phase separation and extractant loss. Increase in substrate feeding as well as reduction of nutrient supply was found to be beneficial in suppressing the acid production, thereby improving the solvent yield. Because of prolonged low growth conditions prevalent in the fed-batch operation, the butanol-to-acetone ratio in the product was significantly higher at 2.6-2.8 compared to the typical value of two [215]. Integrated bioreactor-extractor was also tested in SSEF and production of ABE solvents from pretreated hardwood by C. acetobutylicum and cellulase enzymes. The SSEF system was constructed so that products of fermentation were extracted from the broth through a semipermeable membrane. *In situ* removal of inhibitory products was found to be beneficial in sustaining cell viability, thus allowing fed-batch operation of the bioreactor over a period of several weeks. Hardwood chips were pretreated by monoethanolamine in such a way that hemicellulose and cellulose were retained in high yield. The feed material thus prepd. was readily converted by SSEF. The ability of C. acetobutylicum to ferment both glucose and xylose was a major factor in simplifying the overall process into a single-stage operation [216].

7. Perspectives of butanol as biofuel

Biobutanol has excellent fuel properties compared to ethanol, thus it can be used directly as fuel or blending component for both diesel and gasoline powered internal combustion engines

[217-221]. Butanol has no corrosive properties and its miscibility with gasoline and water tolerance is higher than the appropriate properties of ethanol or methanol [222]. Butanol can also be used as hydrogen source for fuel cells [223] and proved to be useful as esterification alcohol in fatty acid ester type biodiesel production [229-233] or as raw material in the production of dibutyl ether [236] butoxylated butyl diesels [237] or can be converted into aromatic hydrocarbons on zeolite catalysts [238-241].

7.1. Butanol as fuel and blending component in fuel mixtures

Although ethanol as a gasoline extender has received a great deal of attention, this fluid has numerous problems, such as aggressive behaviour toward engine components and a relatively low energy content, the properties of butanol or butanol containing gasoline, diesel and biodiesel fuel compositions are more advantageous than the analogous properties of ethanol or ethanol containing fuels [222]. The performances of gasoline and diesel engines powered with gasoline contained 0-20% BuOH and diesel fuel contained 0-50% BuOH were evaluated. Tests showed that BuOH can be used as a gasoline or diesel fuel supplement in amounts of ≤20% and ≤40%, respectively, without significantly affecting unmodified engine performance. BuOH slightly decreased the octane rating of a blend of 20% BuOH in gasoline but in diesel fuel ≤40% BuOH had no detectable effect on the ignition of the fuel blend [217]. Diesel engines can be powered with 25-75% of a Bu-alcohol and 25-75% of vegetable oil mixtures which were normally liquids under operating conditions. A fuel mixture composed of 50% corn oil and 50% n-BuOH was used as the fuel for 2 tractors when the engine performance in both tractors and the behaviour of the fuel was entirely satisfactory, the engine running smoothly and evenly without significant smoke or odor, with quick acceleration and smooth idling. The above blend could be mixed in any proportion to no. 2 diesel oil without significant change in engine performance [218]. A diesel precombustion chamber engine powered with 70% BuOH-30% diesel fuel had, at an av. 5.9-bar pressure, an ignition delay of operation which was only 10% more than that when operated with diesel alone. The maximum pressure increase during the operation remained higher in both combustion chambers in operation with 70 vol.% BuOH than in operation with diesel alone. There is high potential of improvement of the exhaust gas quality with BuOH-diesel fuel mixtures, especially with regard to smoke value, particulate emissions, and nitrogen oxides. The engine performance under such conditions is similar to that with diesel fuel alone. The starting problem of the engine powered with diesel-BuOH mixture is avoided by using an electrically heated spark plug which maintains ~1000 ºC in the precombustion chamber. More than 200 h of satisfactory operation was attained in a BuOH-diesel mixture powered engine [219]. Substitute diesel fuel compositions consist of gas oil (b. 167-359 °C) 20-55, a 75:25 (wt.) mixture BuOH-Me_2CO 30-40, fatty acid esters 15-40 wt.%. Thus, substitute diesel fuel composition containing gas oil 20, BuOH-Me_2CO mixture 40, and gas oil and BuOH-Me2CO mixture 40 wt.% had cetane no. 40.6 and resulted in normal tractor operation for 50 h [220].

Coupled biodiesel and ABE production technology proceeds by extraction of the ABE containing broth with biodiesel oils forms a mixture which can directly be applied as fuel for diesel engines [224]. Using soybean-derived biodiesel as the extractant with an aqueous phase volume ratio of 1:1, butanol recovery ranged from 45 to 51% at initial butanol concentration of 150 and 225 mM, respectively. Using biodiesel-derived glycerol as feedstock for butanol production, the production of a biodiesel/butanol fuel blend could be a fully integrated process within a biodiesel facility [225]. The presence of surfactants had important influence on the amount of extracted butanol with biodiesel oil prepared from waste cooking oil [226]. This extraction was integrated into the fermentation process, when large quantity of gas (H_2 and CO_2), was released and the produced butanol and acetone were brought into extractant phase. Surfactants decreased the tension of gas-liquid interface and made the large bubble break down, therefore, the releasing gas passed through the extractant phase in form of small bubbles. The mass transfer rate of products from the aqueous phase to the extractant phase was enhanced and the balance time was shortened accordingly by addition of surfactants, consequently, the fermentation productivity was improved. Using waste cooking oil derived biodiesel as extractant the butanol concentration in the extractant phase was increased by 21.2% as compared to the control, while the concentration of surfactant (Tween-80) in culture medium was 0.140% (w/v). Under these conditions, gross solvent productivity was increased by 16.5% [226]. When the biodiesel derived from crude palm oil was used as extractant, the fuel properties of the biodiesel-ABE mixture were comparable to that of No.2 diesel, but its cetane no. and the boiling point of the 90% fraction were higher [227]. Biodiesels prepared from some waste oils proved to be somewhat toxic toward C. Acetobutylicum. Under this condition, the butanol concentration in the biodiesel phase also reached a level of 6.44 g L^{-1} [228].

7.2. Butyl–ester type biodiesels

Biodiesel is typically synthesized from triacylglycerides derived from vegetable oils and an alcohol with base catalysis, yielding the fatty acid ester type biodiesel. Wahlen et al. determined conditions that allowed rapid and high yield conversion of oil feedstocks containing significant concentrations of free fatty acids into biodiesel using an acid-catalyzed reaction with longer chain alcohols such as n-butanol at a slight molar excess. Biodiesel yields >98% were achieved in <40 min. Key properties of the resulting butyl-diesel were determined, including cetane number, pour point, and viscosity [229]. The batch and continuous-flow preparation of biodiesel derived from vegetable oil and 1-butanol using a microwave apparatus has been reported. The methodology allows for the reaction to be run under atmospheric conditions and in continuous-flow mode. It can be utilized with new or used vegetable oil with 1-butanol and a 1:6 molar ratio of oil to alcohol. Sulfuric acid or potassium hydroxide can be used as catalyst [230]. High conversion could be reached when the transesterification of triglycerides with 1-butanol was performed under near-critical or supercritical conditions with microwave heating [231].

Biodiesel synthesis by butanolysis of vegetable oils (soybean, sunflower, and rice bran) catalyzed by Lipozyme RM-IM), and the optimization of the enzyme stability over repeated batches has been described. The enzyme showed the highest activity at a 9:1 BuOH:oil molar ratio and in the 30-35 °C temperature range [232]. Transesterification reaction using sunflower oil and butanol catalyzed by immobilized lipases can be carried out without auxiliary solvent. Immobilized porcine pancreatic lipase (PPL) and Candida rugosa lipase (CRL) showed satisfactory activity in these reactions. Activities of immobilized lipases were highly increased in comparison with free lipases because its activity sites became more effective. Immobilized enzyme could be repeatedly used without difficult method of separation and the decrease in its activity was not largely observed [233].

7.3. Other types of butanol–based biofuels

Preparation of a fuel blending mixture characterized by viscosity breaking and clouding point decreasing abilities was carried out in the reaction of acetone (by-product of biobutanol production) and glycerol (by-product of biodiesel or butyldiesel production) in presence of acidic catalysts such as sulphuric acid, p-toluenesulfonic acid or strongly acidic cation exchangers. A mixture of 2,2-dialkoxy-propanes, 2,2-dimethyl-4-hydroxymethyl-1,3-dioxolane and 2,2-dimethyl-5-hydroxy-1,3-dioxane was formed [176]. Similar reaction of an oxidized ABE mixture consist of butyraldehyde, acetaldehyde and acetone was carried out with formation of a mixture contained 2,2-dialkoxy-propanes, 1,1-dialkoxyethanes, 1,1-dialkoxybutanes, and 2,2-dimethyl-, 2-methyl or 2-propyl derivatives of the appropriate 4-hydroxymethyl-1,3-dioxolane and 5-hydroxy-dioxane[176]. In this way, the by-products of the biodiesel or butyl-biodiesel production (glycerol) and the acetone from the biobutanol producing (or the oxidized ABE solvent mixtures) can completely be used as fuel components [176]. Fuel characteristics of a blended (15 %) biofuel prepared from oxidized ABE mixture and glycerol contains methanol can be seen in Table 4.

Parameter	Values measured by EU standardized methods [176, 243]		
	Commercial biodiesel	Experimental biodiesel	Commercial diesel No.2
Fatty acid methyl ester content, wt.%	100.00	85.00	4.00
Density at 15 °C, kg dm^{-3}	0.8879	0.8938	0.8495
Kinematic viscosity, 40 °C, mm^2 s^{-1}	2.98	4.23	5.65
Flash point, °C	179	56.5	76
Sulphur content, mg kg^{-1}	12	5.4	36
Conradson number, wt. %	0.08	0.02	0.16
Sulphate ash, wt. %	0.012	0.001	0.002
sClouding point, °C	-3	-15	-12

Table 4. Fuel parameters of a biodiesel oil contains 15 % acetal mixture prepared from mixture of glycerol, etanol, butanol and acetaldehyde, butyraldehyde and acetone [176].

Butanol and butyric acid prepared by optimized batch or fed-batch fermentation of wheat flour hydrolysate with selected strains of Clostridium strains, then butanol was recovered from the fermentation broth by distillation and butyric acid by solvent extraction. Esterification could be performed with a lipase in the solvent of extraction [234]. The butylbutyrate formed has a great value as novel biofuel [235]. D'amore at al. developed a catalytic process for making dibutyl ether as transportation fuel and diesel blending component from aq. butanol solutions

[236]. Butoxylation of the unsaturated fraction of biodiesel offers the potential benefit of reduced cloud point without compromising ignition quality or oxidation stability. Butyl biodiesel derived from canola oil was epoxidized via the *in situ* peroxyacetic acid method then the epoxy butyl biodiesel was butoxylated with n-butanol with sulfuric acid catalyst without use of solvents. Optimal conditions for the butoxylation of epoxy butyl biodiesel were 80 °C, 2% sulfuric acid, and a 40:1 molar ratio of n-butanol over a period of 1 h. Conversion of epoxy butyl biodiesel was 100%, and selectivity for butoxy biodiesel was 87.0%. Butoxy biodiesel is able to prevent an earlier onset of crystallization due to the decrease in unsaturated content, but only at lower concentrations [237]. One-step conversion reactions of the title products (6:3:1 volume BuOH-acetone-EtOH) with and without water to aromatic hydrocarbons over molecular shape-selective zeolite were carried out by Anunziata et al. The presence of water in the feed resulted in increased catalyst life. Deactivation reactions toward aromatic hydrocarbon synthesis with product-H_2O mixtures (50:50, 85:15, 99:1, vol.) shown the influence of secondary alkylation reactions leading to substituted aromatic hydrocarbons whose yields were related to the deactivation time of the catalyst [238]. Costa et al. studied the conversion of n-BuOH/Me_2CO mixtures to C_{1-10} hydrocarbons on ZSM-5-type zeolites with different Si-Al ratios. Best results were obtained with a HZSM-5 zeolite (Si/Al=36:1), using a 30 wt% Na montmorillonite binder. The formation of gaseous olefins and non-aromatic liquid hydrocarbons decreased with increasing reaction temperature or space velocity, whereas the amount of aromatic hydrocarbons and gaseous paraffins increased. The total yield of liquid hydrocarbons increased with pressure, although the aromatic content showed a smooth maximum at 1 atm. The yield of aromatic hydrocarbons decreased with increasing water content in the feed. A hydrocarbon distribution similar to that obtained from the anhydrous mixture can be obtained with water-containing feedstock, but lower space velocities were necessary [239]. Orio et al. described the conversion of low molecular-weight oxygenated compounds as ABE solvent mixture into gasoline components over HZSM-5 zeolites. Reagents were used in non-anhydrous form. Formation of C_{2-4} hydrocarbons decrease and aromatic hydrocarbons increase with increasing temperature, formation of C_{5-8} hydrocarbons increases to a maximum at ~300 °C and then decreases. The yields of aromatics from all reactants were ~60 to ~90%; the yields of C_{2-4} and C_{5-8} hydrocarbons were <30% and <10%, respectively. Highest production of aromatic hydrocarbons was attained with the fermentation products of starch (6:3:1 BuOH-acetone-EtOH) [240]. Butanol produced by the fermentation of starch can be presented as a key compd. to produce diesel and jet fuel. Butanol could be converted into Bu esters or into 1-butene which was catalytically oligomerized in a H_2 atmosphere into a hydrocarbon fuel [241].

8. Conclusion

Biobutanol proved to be a superior fuel substitute and blending component in gasolines or diesel fuels. It can be used as raw material in the preparation of so-called butyl-diesel (long-chain fatty acid butyl esters), in the butoxylation of unsaturated fatty acid esters and in the preparation of dibutoxy-acetals. Butanol can easily be transformed via butyraldehide into 2-butoxy-4-hydroxymethyl-1,3-dioxolane or 2-butoxy-5-hydroxy-1,3-dioxane fuel additives with using waste glycerol of biodiesel or butyldiesel production. The new fermentation techniques use renewable lignocellulosic raw materials, and integration with various recovering technologies, membrane techniques, together with new fermentor types and genetically engineered microorganisms make a solid base of a new generation of economic biobutanol production processes.

Author details

László Kótai[1,2], János Szépvölgyi[1], Mária Szilágyi[2], Li Zhibin[3], Chen Baiquan[3], Vinita Sharma[4] and Pradeep K. Sharma[4]

1 Institute of Materials and Environmental Chemistry, Research Centre for Natural Sciences, Hungarian Academy of Sciences, Budapest, Hungary

2 Deuton-X Ltd., Érd, Hungary

3 China New Energy Co. Ltd., Guangzhou, China

4 J.N.V University, Jodhpur, Rajasthan, India

References

[1] Pasteur L. New experiences and views on the nature of fermentation, Comptes Rendus, 1861; 52:1260-1264.

[2] Pasteur L. Studies on the fermentation of calcium lactate, Bulletin de la Societe Chimique de Paris, 1862; 52-53.

[3] Weizmann, C. Acetone and butyl alcohol, GB 1504845; 1915.

[4] Speakman HB. The production of acetone and butyl alcohol by a bacteriological process, Journal of the Society of Chemical Industry, London 1919; 38; 155-161.

[5] Beesch SC. Acetone-butanol fermentation of sugars, Journal of Industrial and Engineering Chemistry (Washington, D. C.) 1952; 44: 1677-1682.

[6] Chiao JS., Sun ZH. History of acetone-butanol-ethanol fermentation industry in China: development of continuous production technology, Journal of Molecular Microbiology and Biotechnology, 2007; 13(1-3): 12-14.

[7] Zverlov VV, Berezina O, Velikodvorskaya GA, Schwarz, WH. Bacterial acetone and butanol production by industrial fermentation in the Soviet Union: use of hydrolyzed agricultural waste for biorefinery, Applied Microbiology and Biotechnology, 2006; 71(5): 587-597.

[8] Bahl H, Duerre P, Clostridia – Biotechnology and Medical Applications, Wiley-VCH, Weinheim-New York-Chichester-Brisbane-Singapore-Toronto, 2001.

[9] Huesemann MHW, Papoutsakis ET. Solventogenesis in Clostridium acetobutylicum fermentations related to carboxylic acid and proton concentrations, Biotechnology and Bioengineering, 1988; 32(7): 843-52.

[10] Hartmanis, MGN, Klason T, Gatenbeck S. Uptake and activation of acetate and butyrate in Clostridium acetobutylicum, Applied Microbiology and Biotechnology 1984; 20(1): 66-71.

[11] Junelles AM, Janati-Idrissi R, Petitdemange H, Gay R. Iron effect on acetone-butanol fermentation, Curr. Microbiol., 1988; 7(5): 299-303.

[12] Terracciano JS, Kashket ER. Intracellular Conditions Required for Initiation of Solvent Production by Clostridium acetobutylicum, Appl Environ Microbiol. 1986; 52(1): 86-91

[13] Huang L, Forsberg CW, Gibbins LN. Influence of external pH and fermentation products on Clostridium acetobutylicum intracellular pH and cellular distribution of fermentation products. Applied and Environmental Microbiology, 1986; 51(6): 1230-4.

[14] Harris LH, Desai RP, Welker NE and Papoutsakis ET. Characterization of recombinant strains of the Clostridium acetobutylicum butyrate kinase inactivation mutant: need for new phenomenological models for solventogenesis and butanol inhibition, Biotechnology and Bioengineering, 2000; 67(1): 1-11.

[15] Zhao Y, Tomas CA, Rudolph FB, Papoutsakis ET, Bennett GN. Intracellular butyryl phosphate and acetyl phosphate concentrations in Clostridium acetobutylicum and their implications for solvent formation, Applied and Environmental Microbiology, 2005; 71(1): 530-537

[16] Costa JM. Solvent toxicity in the acetone-butanol fermentation, Proceedings of the Annual Biochemical Engineering Symposium, 1981; 11th: 83-90.

[17] Lin YL, Blaschek HP. Butanol production by a butanol-tolerant strain of Clostridium acetobutylicum in extruded corn broth, Applied and Environmental Microbiology, 1983; 45(3): 966-73.

[18] Moreira AR, Ulmer DC, Linden JC. Butanol toxicity in the butylic fermentation, Biotechnology and Bioengineering Symposium 1981; 11(Symp. Biotechnol. Energy Prod. Conserv., 3rd.): 567-79.

[19] Lepage C, Fayolle F, Hermann M, Vandecasteele JP. Changes in membrane lipid composition of Clostridium acetobutylicum during acetone-butanol fermentation: effects of solvents, growth temperature and pH, J. Gen. Microbiol., 1987; 133(1): 103-110.

[20] Compere AL, Griffith WL. Evaluation of substrates for butanol production, Developments in Industrial Microbiology Series, 1979(Volume Date 1978); 20: 509-17.

[21] Ounine K, Petitdemange H, Raval G, Gay R. Acetone-butanol production from pentoses by Clostridium acetobutylicum, Biotechnology Letters, 1983; 5(9): 605-610.

[22] Schoutens GH, Nieuwenhuizen MCH, Kossen NWF. Butanol from whey ultrafiltrate: batch experiments with Clostridium beyerinckii LMD 27.6, Applied Microbiology and Biotechnology 1984; 19(4): 203-206.

[23] Welsh FW, Veliky IA. Production of acetone-butanol from acid whey, Biotechnology Letters, 1984; 6(1): 61-64.

[24] Nakhmanovich BM, Shcheblykina NA. Investigation of hydrolysis products in the acetone-butanol fermentation of vegetable agricultural waste materials, Latvijas PSR Zinatnu Akademijas Vestis, 1960; No. 5: 125-128.

[25] Qureshi, N. Agricultural residues and energy crops as potentially economical and novel substrates for microbial production of butanol (a biofuel), CAB Reviews: Perspectives in agriculture, veterinary science, nutrition, and natural resources, 2010; 5: No.059.

[26] Qureshi N, Saha BC, Hector RE, Cotta MA. Removal of fermentation inhibitors from alkaline peroxide pretreated and enzymatically hydrolyzed wheat straw: Production of butanol from hydrolysate using Clostridium beijerinckii in batch reactors. Biomass and Bioenergy, 2008; 32(12): 1353-1358.

[27] Gu Y, Hu S, Chen J, Shao L, He H, Yang Y, Yang S, Jiang W. Ammonium acetate enhances solvent production by Clostridium acetobutylicum EA 2018 using cassava as a fermentation medium. Journal of Industrial Microbiology and Biotechnology, 2009; 36(9): 1225-1232.

[28] Harada R. Butanol-rich production in acetone-butanol fermentation of molasses. VIII. Isolation of butanol-rich producing bacteria. Hakko Kyokaishi, 1962; 20: 517-19.

[29] Harada R. Butanol-rich production in acetone-butanol fermentation of molasses. IX. Characteristics of the butanol-rich producing bacterium. Hakko Kyokaishi, 1962; 20: 520-4.

[30] Inoue N. The production of acetone and butyl alcohol by fermentation. IX. The effect of the time of inoculation and the mycological properties of the seed bacteria upon fermentation. Nippon Nogei Kagaku Kaishi , 1944; 20: 74-8.

[31] Harada R. Butanol-rich production in acetone-butanol fermentation of molasses. III. Influence of the incubation period of the seed culture. 1. Hakko Kyokaishi, 1962; 20: 157-61.

[32] Hermann M, Fayolle F, Marchal R, Podvin L, Sebald M and Vandecasteele JP. Isolation and characterization of butanol-resistant mutants of Clostridium acetobutylicum. Appl Environ Microbiol. 1985; 50(5): 1238-1243.

[33] Largier ST, Long S, Santangelo JD, Jones DT and Woods DR. Immobilized Clostridium acetobutylicum P262 Mutants for Solvent Production, Appl Environ Microbiol. 1985; 50(2): 477-481.

[34] Bergstrom SL, Foutch GL. Effect of cocultures on the production of butanol by Clostridium sp., Biotechnology and Bioengineering Symposium, 1984(Volume Date 1983); 13(Symp. Biotechnol. Fuels Chem., 5th, 1983): 251-6.

[35] Yasuda S. Acetone and butyl alcohol fermentation. JP 172304; 1946.

[36] Liu H, Zhang J, Liu Y, Cheng K, Zhou Y. Method for producing high-yield butanol producing Clostridium strain through irradiation with 60Co g ray. Faming Zhuanli Shenqing Gongkai Shuomingshu CN 101358187 A; 2009.

[37] Hermann M. Fayolle F. Marchal R. Production of mutants of Clostridium acetobutylicum with high productivity of butanol and acetone. Eur. Pat. Appl. EP 88656 A1; 1983.

[38] Nagashima K, Shirai K. Clostridium mutant for enhanced manufacture of butanol. Jpn. Kokai Tokkyo Koho JP 03058782 A2; 1991.

[39] Matta-el-Ammouri G, Janati-Idrissi R, Rambourg JM, Petitdemange H and Gay R. Acetone butanol fermentation by a Clostridium acetobutylicum mutant with high solvent productivity, Biomass, 1986; 10(2): 109-119.

[40] Lemmel SA. Mutagenesis in Clostridiumacetobutylicum, Biotechnol. Lett. 1985; 7(10): 711-16.

[41] Jiang Y, Xu C, Dong F, Yang Y, Jiang W, Yang S. Disruption of the acetoacetate decarboxylase gene in solvent-producing Clostridium acetobutylicum increases the butanol ratio. Metabolic Engineering, 2009; 11(4-5): 284-291.

[42] Larossa RA, Smulski DR. Metabolically engineered Escherichia coli for 1-butanol or 2-butanone production. U.S. Pat. Appl. Publ. US 20090162911 A1; 2009.

[43] Green EM, Bennett GN. Genetic manipulation of acid and solvent formation in Clostridium acetobutylicum ATCC 824. Biotechnology and Bioengineering, 1998; 58(2-3): 215-221

[44] Shen CR, Liao JC. Metabolic engineering of Escherichia coli for 1-butanol and 1-propanol production via the keto-acid pathways. Metabolic Engineering, 2008; 10(6): 312-320.

[45] Green E, Jenkinson E, Crow M. Fermentative production of butanol by genetically engineered Bacillaceae strains. PCT Int. Appl. WO 2009122192 A1; 2009.

[46] Young D, Lawrence A, Hudson M, Alsaker K, Vanessendelft J, Grenfell-Lee D. Microorganisms resistant to aliphatic alcohols and their use in the fermentation of lower alcohols. PCT Int. Appl. WO 2009143455 A2; 2009.

[47] Dijk EK, Raamsdonk LM. Process for the preparation of butanol and ethanol using a genetically engineered Saccharomyces cerevisiae. PCT Int. Appl. WO 2009013160 A2, 2009.

[48] Mueller UM, Wu L, Raamsdonk LM. Butanol production in genetically Saccharomyces cerevisiae. PCT Int. Appl. WO 2009013157 A1; 2009.

[49] McNeil B and Kristiahsen B. Effect of temperature upon growth rate and solvent production in batch cultures ofClostridiumacetobutylicum Biotechnol. Lett., 1985; 7(7): 499-502.

[50] Voget CE, Mignone CF, Ertola RJ. Influence of temperature on solvents production from whey. Biotechnology Letters, 1985; 7(8): 607-10.

[51] Oda M. Yamaguchi S. The effect of changing the temperature of fermentation. Nippon Nogei Kagaku Kaishi. 1948; 22: 3.

[52] Harada R. Butanol-rich production in acetone-butanol fermentation of molasses. II.Temperature. Hakko Kyokaishi, 1962; 20: 155-6.

[53] Fouad M, Abou-Zeil AA, Yassein M. The fermentative production of acetone-butanol by Clostridium acetobutylicum. Acta Biologica Academiae Scientiarum Hungaricae, 1976; 27(2): 107-17.

[54] Marchal R, Ropars M and Vandecasteele JP. Conversion into acetone and butanol of lignocellulosic substrates pretreated by steam explosion, Biotechnol. Lett. 1986; 8(5): 365-70.

[55] Fond O, Petitdemange E, Petitdemange H, Gay R. Effect of glucose flow on the acetone-butanol fermentation in fed-batch culture. Biotechnology Letters, 1984; 6(1): 13-18.

[56] Harada R. Butanol-rich production in acetone-butanol fermentation of molasses. VI. Nitrogen sources. Hakko Kyokaishi, 1962; 20: 341-4.

[57] Baghlaf AO, Abou-Zeid AA, Yassein M. Influence of nitrogen and phosphorus sources on the fermentative production of acetone and butanol by Clostridium acetobutylicum. Zentralblatt fuer Bakteriologie, Parasitenkunde, Infektionskrankheiten und Hygiene, Abteilung 2, Naturwissenschaftliche: Mikrobiologie der Landwirtschaft, der Technologie und des Umweltschutzes, 1980; 135(6): 515-22.

[58] Oda M. Butanol fermentation with sugar as the raw material. I. The effect of addition of various nitrogen sources. Nippon Nogei Kagaku Kaishi, 1946; 21: 15.

[59] Oda M. Butanol fermentation with sugar as the raw material. II. The effect of addition of calcium salts and stimulants. Nippon Nogei Kagaku Kaishi, 1946; 21: 16.

[60] Doi S, Sugama S, Yokoi M, Adachi I. Acetone-butanol fermentation. II. Amino acid requirements and growth-promoting substances in acid casein hydrolyzate, Nippon Nogei Kagaku Kaishi, 1958; 32: 548-52.

[61] Saji T, Nakano S, Sasama I. Acetone-butanol fermentation. I. The effects of acids on this fermentation. Nippon Nogei Kagaku Kaishi, 1943; 19: 409-20, 523-30.

[62] Kawano Y. Acetone-butanol fermentation. I. (1). Effect of acetate. Hakko Kogaku Zasshi, 1947; 25: 157-62.

[63] Nakhmanovich BM, Shcheblykina NA. Fermentation of calcium acetate and butyrate by Clostridium acetobutylicum. Mikrobiologiya, 1960; 29: 67-72.

[64] Tang B, Yu XB, Li LQ. Effect of acetic and butyric acids on fermentation production of acetone-butanol by Clostridium acetobutylicum. Shengwu Jiagong Guocheng, 2008; 6(3): 24-28.

[65] Gottschal JC, Morris JG. The induction of acetone and butanol production in cultures of Clostridium acetobutylicum by elevated concentrations of acetate and butyrate. FEMS Microbiology Letters, 1981; 12(4): 385-9.

[66] Martin JR, Petitdemange H, Ballongue J, Gay R. Effects of acetic and butyric acids on solvent production by Clostridium acetobutylicum. Biotechnology Letters, 1983; 5(2): 89-94.

[67] Fond O, Matta-Ammouri G, Petitdemange H and Engasser JM. The role of acids on the production of acetone and butanol by Clostridium acetobutylicum Journal Applied Microbiology and Biotechnology, 1985; 22(3): 195-200.

[68] Yu EKC, Saddler JN. Enhanced acetone-butanol fermentation by Clostridium acetobutylicum grown on D-xylose in the presence of acetic or butyric acid. FEMS Microbiology Letters, 1983; 18(1-2):_103-7.

[69] Holt RA, Stephens GM, Morris JG. Production of solvents by Clostridium acetobutylicum cultures maintained at neutral pH. Applied and Environmental Microbiology, 1984; 48(6): 1166-70.

[70] Bahl H, Andersch W, Gottschalk G. Continuous production of acetone and butanol by Clostridium acetobutylicum in a two-stage phosphate limited chemostat. European Journal of Applied Microbiology and Biotechnology 1982; 15(4): 201-5.

[71] Mollah AH, Stuckey DC. The influence of hydrogen, carbon dioxide and dilution rate on the continuous fermentation of acetone-butanol. Appl. Microbiol. Biotechnol., 1992; 37(5): 533-8.

[72] Klei HE, Sundstrom DW, Miller JE. Biotechnology and Bioengineering Symposium 1984; 14(Symp. Biotechnol. Fuels Chem., 6th): 353-64.

[73] Griffith WL, Compere AL, Googin JM. Novel neutral solvents fermentations. Developments in Industrial Microbiology Series, 1983; 24: 347-52.

[74] Yerushalmi, L.; Volesky, B.; Szczesny, T. Effect of increased hydrogen partial pressure on the acetone-butanol fermentation by Clostridium acetobutylicum. Applied Microbiology and Biotechnology (1985), 22(2), 103-7.

[75] Kim, Byung Hong; Bellows, Para; Datta, Rathin; Zeikus, J. G. Control of carbon and electron flow in Clostridium acetobutylicum fermentations: utilization of carbon monoxide to inhibit hydrogen production and to enhance butanol yields. Applied and Environmental Microbiology 1984; 48(4): 764-70.

[76] C. L. Meyer, J. K. McLaughlin and E. T. Papoutsakis, The effect of CO on growth and product formation in batch cultures of Clostridium acetobutylicum Biotechnol. Lett. 1985, 7(1), 37-42

[77] Charles L. Meyer, Joseph W. Roos and Eleftherios T. Papoutsakis, Carbon monoxide gasing leads to alcohol production and butyrate uptake without acetone formation in continuous cultures of Clostridium acetobutylicum Journal Applied Microbiology and Biotechnology, 1986, 24(2), 159-167

[78] Datta, Rathin; Zeikus, J. G. Modulation of acetone-butanol-ethanol fermentation by carbon monoxide and organic acids. Applied and Environmental Microbiology (1985), 49(3), 522-9.

[79] Wynne, A. M. Inhibition of the acetone-butyl alcohol fermentation by acids. Journal of Bacteriology (1931), 22 209-37.

[80] Fred, E. B.; Peterson, W. H.; Mulvania, M. The effect of lactic acid bacteria on the acetone-butyl alcohol fermentation. Journal of Bacteriology (1926), 11 323-43.

[81] Yota H. Acetone and butyl alcohol by fermentation. JP 172287 1946.

[82] Doremus, Michael G.; Linden, James C.; Moreira, Antonio R. Agitation and pressure effects on acetone-butanol fermentation. Biotechnology and Bioengineering (1985), 27(6), 852-60.

[83] S. Afschar, K. Schaller and K. Schügerl, Continuous production of acetone and butanol with shear-activated Clostridium acetobutylicum, Journal Applied Microbiology and Biotechnology, 1986; 23(5): 315-321.

[84] Korneeva, O. S.; Zherebtsov, N. A.; Lukina, G. P.Effect of the viscosity of the medium on an acetone-butylfermentation. Fermentn. Spirt. Prom-st., (1),34-7 (Russian) 1987.

[85] Nakhmanovich, B. M.; Kochkina, L. V. Effect of periodic introduction of oxygen on the oxidation-reduction potential and yield of butanol in the acetone-butanol fermentation. Tr. Tsentr. Nauchn. Issled. Inst. Spirt. i Likero-Vodochn. Prom. (1960), (No. 9), 147-53.

[86] Kim, Joong; Bajpai, Rakesh; Iannotti, Eugene L.. Redox potential in acetone-butanol fermentations. Appl. Biochem. Biotechnol., 18, 175-86 (English) 1988.

[87] Häggström L. and Molin N. Calcium alginate immobilized cells of clostridium acetobutylicum for solvent production. Biotechnol. Lett. 1980; 2(5): 241-246.

[88] Haeggstroem L. Immobilized cells of Clostridium acetobutylicum for butanol production. Adv. Biotechnol.,[Proc. Int. Ferment. Symp.], 1981 (Meeting Date 1980); 6th, 2: 79-83. Pergamon Toronto, Ont Editors: Moo-Young M and Robinson CW.

[89] Foerberg C, Enfors SO, Haeggstroem L. Control of immobilized, nongrowing cells for continuous production of metabolites. European Journal of Applied Microbiology and Biotechnology 1983; 17(3): 143-7.

[90] Allam RF, Shafei MS, Sallam LAR. Production of acetone-butanol by immobilized local strain of Clostridium acetobutylicum using different carriers and fermentation techniques. Egyptian Journal of Biomedical Sciences, 2001; 8: 104-113.

[91] Welsh FW, Williams RE, Veliky I. A.Solid carriers for a Clostridium acetobutylicum that produces acetone and butanol. Enzyme Microb. Technol., 1987; 9(8): 500-2.

[92] Xian YP. Application of Clostridium acetobutylicum method of immobilization in fermentation of acetone and butanol. Xibei Shifan Daxue Xuebao, Ziran Kexueban, 2001; 37(3): 70-73.

[93] Sun ZH, Wang S, Wu Y, Tang JQ, Jin M. Continuous production of acetone-butanol with immobilized Clostridium acetobutylicum. Gongye Weishengwu, 1988; 18(3): 12-16.

[94] Frick C, Schuegerl K. Continuous acetone-butanol production with free and immobilized Clostridium acetobutylicum. Appl. Microbiol. Biotechnol., 1986; 25(3): 186-93.

[95] Shamsudin S, Kalil MSH, Yusoff WMW. Production of acetone, butanol and ethanol (ABE) by Clostridium saccharoperbutylacetonicum N1-4 with different immobilization systems. Pakistan Journal of Biological Sciences, 2006; 9(10): 1923-1928.

[96] Schoutens GH, Nieuwenhuizen MCH, Kossen NWF. Butanol production by immobilized clostridia: achievable on a large scale ? Eur. Congr. Biotechnol., 1984; 3rd(Vol. 2): 139-43. Publisher: Verlag Chemie, Weinheim, Fed. Rep. Ger.

[97] Qureshi N, Maddox IS. Reactor design for the ABE fermentation using cells of Clostridium acetobutylicum immobilized by adsorption onto bonechar. Bioprocess Eng., 1988; 3(2): 69-72.

[98] Schoutens GH, Nieuwenhuizen, MCH, Kossen, NWF. Continuous butanol production from whey permeate with immobilized Clostridium beyerinckii LMD 27.6.Applied Microbiology and Biotechnology, 1985; 21(5): 282-6.

[99] Shafei MS, Allam RA, Sallam LAR. Physiological factors affecting the production of acetone and N-butanol by immobilized Clostridium acetobutylicum culture. Egyptian Journal of Biotechnology, 2002; 11: 206-216.

[100] Dyr J, Protiva J, Praus R. Formation of neutral solvents in continuous fermentation by means of Clostridium acetobutylicum. Continuous Cultivation of Microorganisms, Symposium, Prague, 1958; 210-26.

[101] Soni BK. Continuous acetone-butanol fermentation. The phenomenon of cell adhesion on a glass reactor. Appl. Biochem. Biotechnol., 1990; 24-25: 755-64.

[102] Nakhmanovich BM. Continuous acetone-butanol fermentation. Nepreryvnoe Brozhenie i Vyrashchivanie Mikroorganizmov, Materialy Soveshchaniya Provedennogo Inst. Mikrobiol., Akad. Nauk SSSR., 1960(Volume Date 1958); 110-13.

[103] Yarovenko VL. Principles of the continuous alcohol and butanol-acetone fermentation processes. Continuous Cultivation Microorganisms, 1964(Volume Date 1962), 2: 205-19.

[104] Yarovenko VL. Continuous alcohol and butanol-acetone fermentation process. Tr., Vses. Nauchn.-Issled. Inst. Fermentnoi i Spirt. Prom., 1964; No. 15: 105-24.

[105] KF Engineering KK. Batch method combined continuous production of acetone and butanol. Jpn. Kokai Tokkyo Koho JP 59088093 A; 1984.

[106] Meyer CL. and Papoutsakis ET. Continuous and biomass recycle fermentations of Clostridium acetobutylicum, Bioprocess Eng., 1989; 4(1): 1-10

[107] Vijjeswarapu WK, Chen WY, Foutch GL. Performance of a cell recycling continuous fermentation system for the production of n-butanol by Clostridium acetobutylicum. Biotechnology and Bioengineering Symposium, 1986; 15(Symp. Biotechnol. Fuels Chem., 7th, 1985): 471-8.

[108] Afschar AS, Biebl H, Schaller K, Schuegerl K. Production of acetone and butanol by Clostridium acetobutylicum in continuous culture with cell recycle. Applied Microbiology and Biotechnology, 1985; 22(6): 394-8.

[109] Huang WC, Ramey DE, Yang ST. Continuous production of butanol by Clostridium acetobutylicum immobilized in a fibrous bed bioreactor. Applied Biochemistry and Biotechnology 2004; 113-116: 887-898.

[110] Badr HR, Toledo R, Hamdy MK. Continuous acetone-ethanol-butanol fermentation by immobilized cells of clostridium acetobutylicum. Biomass and Bioenergy 2001; 20(2): 119-132.

[111] Pierrot P, Fick M, Engasser JM. Continuous acetone-butanol fermentation with high productivity by cell ultrafiltration and recycling. Biotechnology Letters, 1986; 8(4): 253-6.

[112] Ferras E, Minier M, Goma G. Acetonobutylic fermentation: improvement of performances by coupling continuous fermentation and ultrafiltration. Biotechnology and Bioengineering 1986; 28(4): 523-33.

[113] Foerberg C, Haeggstroem L. Adsorbed cell systems controlled by the nutrient dosing techniques. Eur. Congr. Biotechnol., 1984; 3rd(vol. 2): 115-20. Publisher: Verlag Chemie, Weinheim, Fed. Rep. Ger

[114] Tashiro Y, Takeda K, Kobayashi G, Sonomoto K, High production of acetone-butanol-ethanol with high cell density culture by cell recycling and bleeding. Journal of Biotechnology, 2005; 120(2): 197-206.

[115] Belafi-Bako K, Harasek M, Friedl A, Product removal in ethanol and ABE fermentations. Hung. J. Ind. Chem., 1995; 23(4): 309-19.

[116] . G. Farbenindustrie AG, Fermentation. FR 652407; 1928

[117] Dreyfus H. Acetone and other fermentation products. GB 424134; 1935.

[118] Qureshi N, Blaschek HP. Recovery of butanol from fermentation broth by gas stripping. Renewable Energy, 2001(Volume Date 2000); 22(4): 557-564

[119] Ezeji TC, Karcher PM, Qureshi N, Blaschek HP. Improving performance of a gas stripping-based recovery system to remove butanol from Clostridium beijerinckii fermentation. Bioprocess and Biosystems Engineering, 2005; 27(3): 207-214.

[120] Qureshi N,; Ezeji TC, Blaschek HP, Cotta MA. A novel biological process to convert renewable biomass to acetone and butanol. AIChE Annual Meeting, Conference Proceedings, Austin, TX, United States, Nov. 7-12, 2004, 029D/1-029D/2. American Institute of Chemical Engineers: New York, N. Y. ISBN: 0-8169-0965-2.

[121] Ezeji TC, Qureshi N, Blaschek HP. Acetone butanol ethanol (ABE) production from concentrated substrate: reduction in substrate inhibition by fed-batch technique and product inhibition by gas stripping. Applied Microbiology and Biotechnology, 2004; 63(6): 653-658.

[122] Ezeji TC, Qureshi N, Blaschek HP. Production of acetone butanol (AB) from liquefied corn starch, a commercial substrate, using Clostridium beijerinckii coupled with product recovery by gas stripping. Journal of Industrial Microbiology & Biotechnology, 2007; 34(12): 771-777.

[123] Tomota N, Fujiki K. Application of activated carbon in the butanol fermentation of concentrated "unrefined sake". Kogyo Kagaku Zasshi, 1945; 48: 17.

[124] Oda M. Butanol fermentation with sugar as the raw material. III. Increasing the concentration of the mash. Nippon Nogei Kagaku Kaishi, 1946; 21: 16-17.

[125] Oda M. Butanol fermentation with sugar as the raw material. IV. The effect of adding active carbon. Nippon Nogei Kagaku Kaishi, 1948; 22: 67.

[126] Yamazaki I, Hongo M, Akaboshi, K. Butanol fermentation. XXVII. Fermentation in high sugar concentration. 2.Sorption method by use of active carbon. 1. Sorption of each solvent by active carbon. Nippon Nogei Kagaku Kaishi, 1958, 32: 758-63.

[127] Yamazaki I, Hongo M, Butanol fermentation. XXVIII. Fermentation in high sugar concentrations. 2.Sorption method by use of active carbon. 2. Fermentation with addition of active carbon. Nippon Nogei Kagaku Kaishi, 1958; 32: 764-70.

[128] Yamazaki I, Hongo M, Akaboshi K, Nagata K. Butanol fermentation. XXIX. Fermentation in high sugar concentrations. 2.Sorption method by use of active carbon. 3. Recovery of solvents from carbon. Nippon Nogei Kagaku Kaishi, 1958, 32: 855-8.

[129] Urbas B. Concentration of organic chemicals from dilute aqueous solutions. Eur. Pat. Appl. EP 81211 A2; 1983.

[130] Qureshi N, Hughes S, Maddox IS, Cotta MA. Energy-efficient recovery of butanol from model solutions and fermentation broth by adsorption. Bioprocess and Biosystems Engineering, 2005; 27(4): 215-222.

[131] Milestone NB, Bibby DM. Concentration of alcohols by adsorption on silicalite. Journal of Chemical Technology and Biotechnology (1979-1982), 1981; 31(12): 732-6.

[132] Maddox IS. Use of silicalite for the adsorption of n-butanol from fermentation liquors. Biotechnology Letters, 1982; 4(11): 759-60.

[133] Nielsen DR, Prather KJ. *In situ* product recovery of n-butanol using polymeric resins. Biotechnology and Bioengineering, 2009; 102(3): 811-821.

[134] Nielsen L, Larsson M, Holst O, Mattiasson B. Adsorbents for extractive bioconversion applied to the acetone-butanol fermentation. Appl. Microbiol. Biotechnol., 1988; 28(4-5): 335-9.

[135] Groot WJ, Luyben KCAM. *In situ* product recovery by adsorption in the butanol/isopropanol batch fermentation. Appl. Microbiol. Biotechnol., 1986; 25(1): 29-31.

[136] Compere AL, Griffith WL, Googin JM. Integrated schemes for the production and separation of neutral solvents. Developments in Industrial Microbiology Series, 1984; 25: 791-4.

[137] Shi ZP, Zhang CG, Chen JX, Mao ZG. Performance evaluation of acetone-butanol continuous flash extractive fermentation process. Bioprocess and Biosystems Engineering, 2005; 27(3): 175-183.

[138] Shimizu K, Matsubara M. A solvent screening criterion for multicomponent extractive fermentation. Chem. Eng. Sci., 1987; 42(3): 499-504.

[139] Dadgar AM, Foutch GL. Evaluation of solvents for the recovery of Clostridium fermentation products by liquid-liquid extraction. Biotechnology and Bioengineering Symposium (1986), 15(Symp. Biotechnol. Fuels Chem., 7th, 1985), 611-20.

[140] Griffith WL, Compere AL, Googin JM. Novel neutral solvents fermentations. Developments in Industrial Microbiology Series, 1983; 24: 347-52.

[141] Barton WE, Daugulis AJ. Evaluation of solvents for extractive butanol fermentation with Clostridium acetobutylicum and the use of poly(propylene glycol) 1200. Appl. Microbiol. Biotechnol., 1992; 36(5): 632-9.

[142] Groot WJ, Soedjak HS, Donck PB, Van der Lans RGJM, Luyben KCAM, Timmer JMK. Butanol recovery from fermentations by liquid-liquid extraction and membrane solvent extraction. Bioprocess Eng., 1990; 5(5): 203-16.

[143] Schmidt A, Windsperger A, Friedl A. Process for the separation of oxygen containing organic compounds from aqueous media. Eur. Pat. Appl. EP 235117 A1; 1987.

[144] Taya M. Production of butanol by Clostridium acetobutylicum in extractive fermentation system. Kagaku Kogaku, 1988, 52(6): 408-10.

[145] Wang XX. Preliminary study on coupling between extraction and acetone-butanol fermentation to increase butanol production. Hebei Nongye Daxue Xuebao, 2008; 31(6): 62-64.

[146] Cenedella RJ. Vinyl bromide extraction of butyric acid and butanol from microbial fermentation broth. US 4628116 A; 1986.

[147] Compere AL, Googin JM, Griffith WL. Effect of possible interferences on the extraction of 1-butanol from aqueous solution by the ethyl esters of soybean oil fatty acids. Sep. Sci. Technol., 1987; 22(2-3): 373-8.

[148] Eom YS, Sang BI, Lee SM, Lee HJ, Seo YU, Seo DJ, Cho MO. Method for separating bio-butanol with ionic liquid. Repub. Korean Kongkae Taeho Kongbo KR2009003967; 2009.

[149] Davis SE, Morton SA., Investigation of ionic liquids for the separation of butanol and water. Separation Science and Technology, 2008; 43(9-10): 2460-2472.

[150] Chapeaux A, Simoni LD, Ronan TS, Stadtherr MA, Brennecke JF. Extraction of alcohols from water with 1-hexyl-3-methylimidazolium bis(trifluoromethylsulfonyl)imide. Green Chemistry, 2008; 10(12): 1301-1306.

[151] Bekhtereva MN. Acetone-butyl alcohol fermentation under continuous removal of the formed products by extraction. Mikrobiologiya, 1939; 8: 854-61.

[152] Shaposhnikov V N, Bekhtereva MN. Acetone and butyl alcohol. RU 57855; 19400831

[153] Yang LR, Cen PL, Zhu ZQ. Study on acetone/butanol fermentation with *in situ* solvent extraction. Zhejiang Daxue Xuebao, Ziran Kexueban, 1992; 26(4): 388-98, 387.

[154] Roffler SR, Blanch HW, Wilke CR. In-situ recovery of butanol during fermentation. Part 1: Batch extractive fermentation. Bioprocess Eng., 1987; 2(1): 1-12

[155] Shukla R, Kang W, Sirkar KK. Toxicity of organic solvents to Clostridium acetobutylicum for extractive ABE fermentation. Appl. Biochem. Biotechnol., 1988; 18: 315-24.

[156] Shukla R, Kang W, Sirkar KK. Acetone-butanol-ethanol (ABE) production in a novel hollow fiber fermentor-extractor. Biotechnol. Bioeng., 1989; 34(9): 1158-66.

[157] Evans PJ, Wang HY. Enhancement of butanol formation by Clostridium acetobutylicum in the presence of decanol-oleyl alcohol mixed extractants. Appl. Environ. Microbiol., 1988; 54(7): 1662-7.

[158] Chiyoda Chemical Engineering and Construction Co., Ltd. Butanol. Jpn. Kokai Tokkyo Koho JP 59216591 A; 1984.

[159] Kobayashi T, Taya M, Ishii S. Extracting butanol during fermentation. Jpn. Kokai Tokkyo Koho JP 60172290 A; 1985.

[160] Yang LR, Zhu ZQ, Cen PL. Study on in-situ acetone/butanol extractive fermentation. Gaoxiao Huaxue Gongcheng Xuebao, 1992; 6(4): 346-51.

[161] Roffler SR, Blanch HW, Wilke CR. *In situ* recovery of butanol during fermentation. Part 2: Fed-batch extractive fermentation. Bioprocess Eng., 1987; 2(4): 181-90.

[162] Taya M, Ishii S, Kobayashi T. Monitoring and control for extractive fermentation of Clostridium acetobutylicum. Journal of Fermentation Technology, 1985; 63(2): 181-7.

[163] Ohno T, Masawaki T, Taya M. Tone, S. Extractive butanol fermentation combined with membrane separation. Proc. Symp. Solvent Extr., 1991; 265-72.

[164] Davison BH, Thompson JE. Continuous direct solvent extraction of butanol in a fermenting fluidized-bed bioreactor with immobilized Clostridium acetobutylicum. Appl. Biochem. Biotechnol., 1993; 39-40: 415-26.

[165] Roffler SR, Blanch HW, Wilke CR. *In situ* extractive fermentation of acetone and butanol. Biotechnol. Bioeng., 1988; 31(2): 135-43.

[166] Kótai L. and Gács I. A porous permeable wall containing fermentor for in-situ extractive recovery of biobutanol from dilute ferment liquors, Eur. Chem. Bull., 2012; 1(1-2): 1-3.

[167] Tamics E, Bajnóczki Zs, Szilágyi M, Sharma PK. Porous composite materials for in-situ extractive recovery of biobutanol from dilute ferment liquors I. Composite materials manufactured from gypsum, J. Chem. Asia, 2012; 3(1-2): 1-15.

[168] Mattiasson B, Suominen M, Andersson E, Haggstrom L, Albertsson PA, Hahn-Hagerdal, B. Solvent production by Clostridium acetobutylicum in aqueous two-phase systems. Enzyme Engineering, 1982; 6: 153-5.

[169] Kim YJ, Weigand WA. Experimental analysis of a product inhibited fermentation in an aqueous two-phased system. Appl. Biochem. Biotechnol., 1992; 34-35: 419-30.

[170] Wayman M, Parekh R. Production of acetone-butanol by extractive fermentation using dibutylphthalate as extractant. J. Ferment. Technol., 1987; 65(3): 295-300.

[171] Hartmeier W, Buecker C, Wallrath J. Butanol-acetone fermentation with immobilized Clostridia and integrated solvent extraction. Biochem. Eng.-Stuttgart, [Proc. Int.

Symp.], 1991 (Meeting Date 1990); 2nd: 236-41. Edited by: Reuss, Matthias. Fischer: Stuttgart, Fed. Rep. Ger.

[172] Husted GR, Santangelo JD, Bostwick DW. Butanol fermentation and its enhancement with fluorocarbons. US 4777135 A; 1988.

[173] Levy S. Continuous process for producing N-butanol employing anaerobic fermentation. US 4424275 A; 1984.

[174] Araki K, Minami T, Sueki M, Kimura T. Elimination of fermentation inhibition caused by extractants in butanol extractive fermentation. Hakko Kogaku Kaishi, 1992; 70(4): 259-65.

[175] Kótai L, Szilagyi M, Li ZB, Chen BQ. Eur. Chem. Bull., 2012; submitted to publish.

[176] Kótai, L, Szépvölgyi J, Bozi J, Gács I, Bálint S, Gömöry Á, Angyal A, Balogh J, Li ZB, Chen MT, Wang C, Chen BQ. An integrated waste-free biomass utilization system for an increased productivity of biofuel and bioenergy, BIODIESEL, 2011; ISBN: 979-953-307-020-8.

[177] Shi ZP, Shimizu K, Iijima S, Morisue T, Kobayashi T. Theoretical development and performance evaluation for extractive fermentation using multiple extractants. Biotechnol. Bioeng., 1990; 36(5): 520-9.

[178] Araki K, Minami T, Sueki M, Kimura T. Optimum operational conditions of a 2-stage-extraction butanol extractive fermentation process. Seibutsu Kogaku Kaishi, 1993; 71(6): 389-95.

[179] Jeon YJ, Lee YY. Membrane-assisted extractive butanol fermentation. Ann. N. Y. Acad. Sci., 1987; 506(Biochem. Eng. 5): 536-42.

[180] Eckert G, Schugerl K. Continuous acetone-butanol production with direct product removal. Appl. Microbiol.Biotechnol., 1987; 27(3): 221-8.

[181] Grobben NG, Eggink G, Cuperus FP, Huizing HJ. Production of acetone, butanol and ethanol (ABE) from potato wastes: fermentation with integrated membrane extraction. Appl. Microbiol. Biotechnol., 1993; 39(4-5): 494-8.

[182] Gudernatsch W, Mucha H, Hofmann T, Strathmann H, Chmiel H. Use of pervaporation in the continuous fermentation of acetone, butanol, and ethanol. Ber. Bunsen-Ges. Phys. Chem., 1989; 93(9): 1032-3.

[183] El-Zanati E, Abdel-Hakim E, El-Ardi O, Fahmy M. Modeling and simulation of butanol separation from aqueous solutions using pervaporation. Journal of Membrane Science, 2006; 280(1+2): 278-283.

[184] Vrana DL, Meagher MM, Hutkins RW, Duffield, B. Pervaporation of model acetone-butanol-ethanol fermentation product solutions using polytetrafluoroethylene membranes. Sep. Sci. Technol., 1993; 28(13-14): 2167-78.

[185] Groot WJ, Van den Oever CE, Kossen, NWF. Pervaporation for simultaneous product recovery in the butanol/isopropanol batch fermentation. 1984, Biotechnology Letters, 1984; 6(11): 709-14.

[186] Groot WJ, Schoutens GH, Van Beelen PN, Van den Oever CE, Kossen NWF. Increase of substrate conversion by pervaporation in the continuous butanol fermentation. Biotechnology Letters, 1984; 6(12): 789-92.

[187] Geng QH, Park CH. Extractive butanol fermentation using pervaporation and a low acid producing Clostridium acetobutylicum B18. Nonmunjip - Sekye Hanminchok Kwahak Kisulcha Jonghap Haksul Taehoe: Kobunja, Chaelyo, Hwakong Punkwa, 1993; 1431-5.

[188] Geng QH, Park CH. Pervaporative butanol fermentation by Clostridium acetobutylicum B18. Biotechnol. Bioeng., 1994; 43(10): 978-86.

[189] Fadeev AG, Meagher MM, Kelley SS, Volkov VV. Fouling of poly[-1-(trimethylsilyl)-1-propyne] membranes in pervaporative recovery of butanol from aqueous solutions and ABE fermentation broth. Journal of Membrane Science, 2000; 173(1): 133-144.

[190] Oya H, Matsumoto K. Separating fermentation products from the broth with a porous membrane. Jpn. Kokai Tokkyo Koho JP 61199788 A; 1986.

[191] Knapp B, Elce E, Bedwell B, Langsdorf LJ, Wilks R. Norbornene polymer pervaporation membranes, preparation and use thereof. PCT Int. Appl. WO 2009097322 A1; 2009.

[192] Jitesh DK, Pangarkar VG, Niranjan K. Pervaporative stripping of acetone, butanol and ethanol to improve ABE fermentation. Bioseparation, 2000; 9(3): 145-154.

[193] Huang J, Meagher MM. Pervaporative recovery of n-butanol from aqueous solutions and ABE fermentation broth using thin-film silicalite-filled silicone composite membranes. Journal of Membrane Science, 2001; 192(1-2): 231-242.

[194] Qureshi N, Meagher MM, Hutkins RW. Recovery of butanol from model solutions and fermentation broth using a silicalite/silicone membrane. Journal of Membrane Science, 1999; 158(1-2): 115-125.

[195] Thongsukmak A, Sirkar KK. Pervaporation membranes highly selective for solvents present in fermentation broths. Journal of Membrane Science, 2007; 302(1+2): 45-58.

[196] Sirkar KK. Pervaporation membranes highly selective for volatile solvents present in fermentation broths.. U.S. Pat. Appl. Publ. US 20090114594 A1; 2009.

[197] Matsumura M, Kataoka H. Utilization of liquid membrane in butanol production by fermentation. Kemikaru Enjiniyaringu, 1991; 36(6): 501-8.

[198] Izak P, Schwarz K, Ruth W, Bahl H, Kragl U. Increased productivity of Clostridium acetobutylicum fermentation of acetone, butanol, and ethanol by pervaporation

through supported ionic liquid membrane. Applied Microbiology and Biotechnology, 2008; 78(4): 597-602.

[199] Garcia A, Iannotti EL, Fischer JR. Reverse osmosis application for butanol-acetone fermentation. Biotechnology and Bioengineering Symposium, 1984; 14(Symp. Biotechnol. Fuels Chem., 6th): 543-52.

[200] Garcia A, Iannotti EL, Fischer JL. Butanol fermentation liquor production and separation by reverse osmosis. Biotechnology and Bioengineering, 1986; 28(6): 785-91.

[201] Traxler RW, Wood EW, Mayer J, Wilson MP. Extractive fermentation for the production of butanol. Developments in Industrial Microbiology Series, 1985; 26: 519-25.

[202] Shity H, Bar R. New approach for selective separation of dilute products from simulated clostridial fermentation broths using cyclodextrins. Biotechnol. Bioeng., 1992; 39(4): 462-6.

[203] Dijk EK, Van de Graaf, MJ. Recovery of butanol using ion exchange. PCT Int. Appl. WO 2008095896 A1; 2008.

[204] Masawaki T, Ohno T, Taya M, Tone S. Separation of butanol from the extractive fermentation broth by pervaporation. Biochem. Eng. 2001, Proc. Asia-Pac. Biochem. Eng. Conf., 1992; 593-5. Edited by: Furusaki, Shintaro; Endo, Isao; Matsuno, Ryuichi. Springer: Tokyo, Japan.

[205] Groot WJ, Den Reyer MCH, Baart de la Faille T, Van der Lans RGJM, Luyben KCAM. Integration of pervaporation and continuous butanol fermentation with immobilized cells. I: Experimental results. Chem. Eng. J. (Lausanne), 1991; 46(1): B1-B10.

[206] Qureshi N, Meagher MM, Huang J, Hutkins RW. Acetone butanol ethanol (ABE) recovery by pervaporation using silicalite-silicone composite membrane from fed-batch reactor of Clostridium acetobutylicum. Journal of Membrane Science, 2001; 187(1-2): 93-102.

[207] Jeon YJ, Lee YY. *In situ* product separation in butanol fermentation by membrane-assisted extraction. Enzyme Microb. Technol., 1989; 11(9): 575-82.

[208] Mollah AH, Stuckey DCJ. Maximizing the production of acetone-butanol in an alginate bead fluidized bed reactor using Clostridium acetobutylicum. Chem. Technol. Biotechnol., 1993; 56(1): 83-9.

[209] Minier M, Grateloup R, Blanc-Ferras E, Goma G. Extractive acetonobutylic fermentation by coupling ultrafiltration and distillation. Biotechnol. Bioeng., 1990; 35(9): 861-9.

[210] Larsson M, Mattiasson B. Novel process technology for biotechnological solvent production. Chemistry & Industry (London, United Kingdom), 1984; No.12: 428-31.

[211] Larsson M, Holst O, Mattiasson B. Butanol fermentation using a selective adsorbent for product recovery. Eur. Congr. Biotechnol., 1984; 3rd(vol. 2): 313-16. Publisher: Verlag Chemie, Weinheim, Fed. Rep. Ger

[212] Yang XP, Tsai GJ, Tsao GT. Enhancement of *in situ* adsorption on the acetone-butanol fermentation by Clostridium acetobutylicum. Sep. Technol., 1994; 4(2): 81-92.

[213] Yang XP, Tsao GT. Enhanced acetone-butanol fermentation using repeated fed-batch operation coupled with cell recycle by membrane and simultaneous removal of inhibitory products by adsorption. Biotechnol. Bioeng., 1995; 47(4): 444-50.

[214] Qureshi N, Maddox IS. Continuous production of acetone-butanol-ethanol using immobilized cells of Clostridium acetobutylicum and integration with product removal by liquid-liquid extraction. J. Ferment.Bioeng., 1995; 80(2): 185-9.

[215] Shah MM, Lee YY. Process improvement in acetone-butanol production from hardwood by simultaneous saccharification and extractive fermentation. Appl. Biochem. Biotechnol., 1994; 45-46: 585-97.

[216] Shah MM, Lee YY. Simultaneous saccharification and extractive fermentation for acetone/butanol production from pretreated hardwood. Appl. Biochem. Biotechnol., 1992; 34-35: 557-68.

[217] Miller GL, Smith JL, Workman JP. Engine performance using butanol fuel blends. Transactions of the ASAE 1981; 24(3): 538-40.

[218] Elsea HR, Etheridge LL. Diesel engine fuel composition and its use for operating diesel engines. US 4359324 A; 1982.

[219] Pucher H, Sperling E. Butanol-diesel fuel mixture as an alternate fuel for diesel engines. Erdoel & Kohle, Erdgas, Petrochemie, 1986; 39(8): 353-6.

[220] Guibet JC, Born M, Vandecasteele JP. Combustible compositions containing a fatty acid ester and an alcoholic constituent based on n-butanol and useful as diesel fuel. Ger. Offen. DE 3150989 A1; 1982.

[221] Newton, DO. An evaluation of methanol, ethanol, the propanols, and the butanols as ship propulsion fuels. U. S. NTIS, Report 1976; AD-A033483, 13 pp.

[222] Bruno TJ, Wolk A, Naydich A. Composition-Explicit Distillation Curves for Mixtures of Gasoline with Four-Carbon Alcohols (Butanols). Energy &Fuels, 2009; 23(4): 2295-2306.

[223] Nakanishi H, Arikawa E. Jpn. Kokai Tokkyo Koho JP 2009295332 A; 2009.

[224] Zhang LY, Yang Y, Shi ZP Performance optimization of property-improved biodiesel manufacturing process coupled with butanol extractive fermentation. Shengwu Gongcheng Xuebao, 2008; 24(11): 1943-1948.

[225] Adhami L, Griggs B, Himebrook P, Taconi K. Liquid-Liquid Extraction of Butanol from Dilute Aqueous Solutions Using Soybean-Derived Biodiesel. Journal of the American Oil Chemists' Society, 2009; 86(11): 1123-1128.

[226] Yang Y, Zhang LY, Shi ZP Improvement of extractive fermentation performance of butanol by adding surfactants into fermentation medium. Shengwu Jiagong Guocheng, 2008; 6(4): 25-30.

[227] Crabbe E, Nolasco-Hipolito C, Kobayashi G, Sonomoto K, Ishizaki A. Biodiesel production from crude palm oil and evaluation of butanol extraction and fuel properties. Process Biochemistry (Oxford, United Kingdom), 2001; 37(1): 65-71.

[228] Hu CY, Du YP, Yang Y, Shi ZP, Duan ZY. Preliminary study on coupling between biodiesel oil and acetone-butanol fermentation. Shengwu Jiagong Guocheng, 2007; 5(1): 27-32.

[229] Wahlen BD, Barney BM, Seefeldt LC. Synthesis of Biodiesel from Mixed Feedstocks and Longer Chain Alcohols Using an Acid-Catalyzed Method. Energy & Fuels, 2008; 22(6): 4223-4228.

[230] Leadbeater NE, Barnard TM, Stencel LM. Batch and Continuous-Flow Preparation of Biodiesel Derived from Butanol and Facilitated by Microwave Heating. Energy & Fuels, 2008; 22(3), 2005-2008.

[231] Geuens J, Kremsner JM, Nebel BA, Schober S, Dommisse RA, Mittelbach M, Tavernier, S, Kappe CO, Maes BUW. Microwave-Assisted Catalyst-Free Transesterification of Triglycerides with 1-Butanol under Supercritical Conditions. Energy & Fuels, 2008; 22(1): 643-645.

[232] Rodrigues RC, Volpato G, Wada K, Ayub MAZ Enzymatic synthesis of biodiesel from transesterification reactions of vegetable oils and short chain alcohols. Journal of the American Oil Chemists' Society, 2008; 85(10): 925-930.

[233] Sagiroglu A. Conversion of Sunflower Oil to Biodiesel by Alcoholysis using Immobilized Lipase. Artificial Cells, Blood Substitutes, and Biotechnology, 2008; 36(2): 138-149.

[234] Fayolle F, Marchal R, Monot F, Béanchet D, Ballerini D. An example of production of natural esters: synthesis of butyl butrate from wheat flour. Enyzme Microb. Technol. 1991; 13: 215-220.

[235] Kang SY, Park CH, Yoon YS, Cho IH, Ahn HW, Park SR, Song JH, Lee SH, Sang BI, Suh, YW, Um YS, Lee SM. Method of extracting butyric acid from a fermented liquid and chemically converting butyric acid into biofuel. PCT Int. Appl. WO 2009120042 A2; 2009.

[236] D'Amore MB, Manzer LE, Miller ES, Dicosimo R, Knapp JP. Process for making dibutyl ethers from aqueous 1-butanol. PCT Int. Appl. WO 2007149400 A2; 2007

[237] Smith PC, O'Neill BK, Ngothai Y, Nguyen QD. Butoxylation of Butyl Biodiesel: Reaction Conditions and Cloud Point Impact. Energy & Fuels, 2009; 23(7): 3798-3803.

[238] Anunziata OA, Orio OA, Pierella LB, Aguirre MC, Perez CF. One-step conversion of butanol-acetone fermentation products with different water content to aromatic hy-

drocarbons over zeolite type HZSM-11. Revista Latinoamericana de Ingenieria Quimica y Quimica Aplicada, 1987; 17(3): 271-88.

[239] Costa E, Aguado J, Ovejero G, Canizares P. Conversion of n-butanol-acetone mixtures to C1-C10 hydrocarbons on HZSM-5 type zeolites. Ind. Eng. Chem. Res., 1992; 31(4): 1021-5.

[240] Orio OA, Anunziata OA, Herrero ER, Lopez AF, Perez CF, Suarez AR. Conversion of oxygenated compounds to hydrocarbons in one step over HZSM-5 zeolite. Revista Latinoamericana de Ingenieria Quimica y Quimica Aplicada, 1984; 14(1): 111-19.

[241] Wright ME, Harvey B, Quintana RL. Creating new navy fuels from biobutanol. Preprints of Symposia - American Chemical Society, Division of Fuel Chemistry, 2008; 53(1): 252-253.

[242] Cornillot E, Croux C and Soucaille, P. Physical and genetic map of the Clostridium acetobutylicum ATCC 824 chromosome. J. Bacteriol., 1997; 179: 7426-34.

[243] Kótai, L. et al., unpublished results, 2012.

[244] Leung, JCY, WAN DIC. Production of acetone-butanol ion continuous fermentation, World Congress of Chemical Engineering, Montreal, October 4-9, 1981, 2nd.

[245] Fick M, Pierrot P, Engasser JM. Optimal conditions for long-term stability of acetone-butanol production by continuous cultures of *Clostridium Acetobutylicum*, Biotechnology Letters, 1985; 7: 503-8.

[246] Parekh SR, Parekh RS. Ethanol and butanol production by fermentation of enzymatically saccharified SO_2-prehydrolysed lignocellulosics, entzyme Microbiol. Techn. 1988; 10: 660-666.

[247] Soni BK, Das K, Ghose TK, Bioconversion of agro-wastes into acetone-butanol. Biotechnology Letters, 1982; 4: 19-22.

[248] Marchal R, Rebeller M, Vandecasteele JP, Wayman M. Direct bioconversion of alkali-pretreated straw using simultaneous enzymatic hydrolysis and acetone-butanol fermentation, Biotechnology Letters, 1984; 6: 523-8.

[249] Qureshi N, Ezeji TC, Ebener J, Dien BS, Cotta MA, Blaschek HP. Butanol production by Clostridium beijerinckii. Part I: Use of acid and enzyme hydrolyzed corn fiber, Bioresource Technol., 2008; 99: 5915.

[250] Qureshi N, Saha BC, Hector RE, , Hughes SR, Cotta MA. Butanol production from wheat straw by simultaneous saccharification and fermentation using Clostridium beijerinckii: Part I—Batch fermentation, Biomass Bioenergy, 2008; 32: 168.

[251] Chen S, Ma X, Wang L, Zhao X. Acetone-butanol fermentation of rice straw enzymic hydrolyzate. Gongye Weishengwu 1998; 28(4): 30-34

[252] Li DM, Chen HZ. Fermentation of acetone and butanol coupled with enzymatic hydrolysis of steam exploded cornstalk stover in a membrane reactor. Guocheng Gongcheng Xuebao, 2007; 7(6): 1212-1216.

[253] Qureshi N, Saha BC, Cotta MA. Butanol production from wheat straw by simultaneous saccharification and fermentation using Clostridium beijerinckii: Part II-Fed-batch fermentation. Biomass and Bioenergy, 2008; 32(2): 176-183.

Chapter 8

Metabolic Engineering of Hydrocarbon Biosynthesis for Biofuel Production

Anne M. Ruffing

Additional information is available at the end of the chapter

http://dx.doi.org/10.5772/52050

1. Introduction

The world's supply of petroleum hydrocarbons, which serve as feedstock for the fuel and chemical industries, is rapidly diminishing to satisfy the global demand for energy and consumer goods. In response to this increasing demand and limited supply, the cost of crude oil has risen to over $100 per barrel in 2012, a 10-fold increase compared to prices in the late 1990s [1]. As fossil fuels are nonrenewable resources, the price of oil is only expected to increase in the future. This unavoidable reality necessitates the development of renewable energy sources in order to maintain the current standard of living. Among the alternative energy options under development, biofuels are anticipated to supplement and eventually replace the petroleum-based fuels that supply the transportation and chemical industries. Currently, first generation biofuels like corn-based ethanol are blended into conventional petroleum fuels, with biofuels supplying 2.7% of the world's transportation fuel in 2010 [2]. It appears that biofuels are on their way to becoming a viable renewable energy source, yet technological and biological advancements are necessary for sustainable and economical biofuel production at the scales necessary to support the world's energy needs.

The current practice of using food crops, like corn or soybean, as feedstocks for biofuel production is not a viable, long-term solution to the energy crisis. In fact, to replace our current petroleum usage with crop-based ethanol production, the entire surface area of land on Earth would be needed for corn production [3]. In addition to this shortcoming, first generation biofuels compete with food production for arable land, require significant nutrient resources (fertilizer and fresh water), and typically have low net energy yields due to the low energy density of the product fuel (i.e. ethanol) and the energy input required to harvest the feedstock and convert it into fuel [4]. Second and third generation biofuels address these limitations. Second generation biofuels use lignocellulosic biomass as the feedstock for fuel production.

Lignocellulose, the main component of plant biomass, is the most abundant form of renewable carbon on the Earth, making it an ideal feedstock for renewable hydrocarbon production. The cellulose and hemicellulose components of lignocellulose can be degraded into fermentable sugars to serve as the carbon source for microbial-based fuel production. The carbon feedstocks for both first and second generation biofuels are ultimately derived from carbon dioxide (CO_2) fixation through the process of photosynthesis. Third generation biofuels use photosynthetic microorganisms (i.e. microalgae) to directly convert CO_2 into fuel molecules or fuel precursors, eliminating the biomass intermediate (Figure 1). While both second and third generation biofuels require land, nutrients, and energy investment for harvesting and fuel production, the fuel production yields from these processes are predicted to be capable of meeting energy needs. However, these technologies have yet to be demonstrated at scale and still require further improvement before they can be economically competitive with fossil fuels.

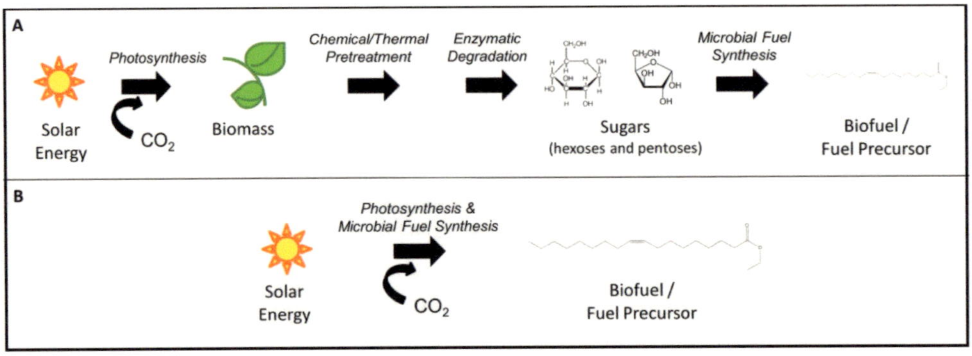

Figure 1. Process steps for (A) second (i.e. lignocellulosic feedstock) and (B) third (i.e. inorganic carbon feedstock) generation biofuels.

Both second and third generation biofuels rely on microbes to convert the carbon feedstock into the desired hydrocarbon fuels. Microorganisms have been identified that are capable of producing a range of fuel molecules and fuel precursors, yet the natural rates of microbial fuel synthesis are typically too low to support industrial-scale production. Metabolic engineering is a powerful tool to improve microbial fuel production, either through engineering the metabolic pathways within the native microorganism to encourage high fuel synthesis or though transferring the fuel production pathway into a model organism for optimization. This chapter will focus on the application of metabolic engineering to increase hydrocarbon fuel production. Within this chapter, hydrocarbon-based fuels are defined to include oxygen-containing fuel molecules with long hydrocarbon chains, such as fatty alcohols and fatty acid ethyl esters (FAEE), in addition to pure hydrocarbons like alkanes, alkenes, and isoprenoid-based molecules: hemiterpene (C5), monoterpenes (C10), and sesquiterpenes (C15). Hydrocarbon-based fuel precursors will also be considered, including free fatty acids (FFAs) and triacylglycerol (TAG). The structures of these hydrocarbon-based fuels and precursors are illustrated in Figure 2. Hydrocarbon-based fuels and precursors can be produced by both second and third generation biofuel processes. Therefore, the first section in this chapter will discuss the metabolic pathways for hydrocarbon fuel production and common metabolic engineering strategies for improving fuel synthesis. Because second and third generation biofuel processes rely on different carbon sources, sugars and CO_2 respectively, the remaining sections will focus on the use of organic carbon (heterotrophy) and inorganic carbon (autotrophy) as feedstocks for biofuel production. This division, based on carbon source, is important from both the biofuel production and metabolic engineering perspectives. The chapter will conclude with a discussion of the future outlook for microbial-based, hydrocarbon fuel synthesis.

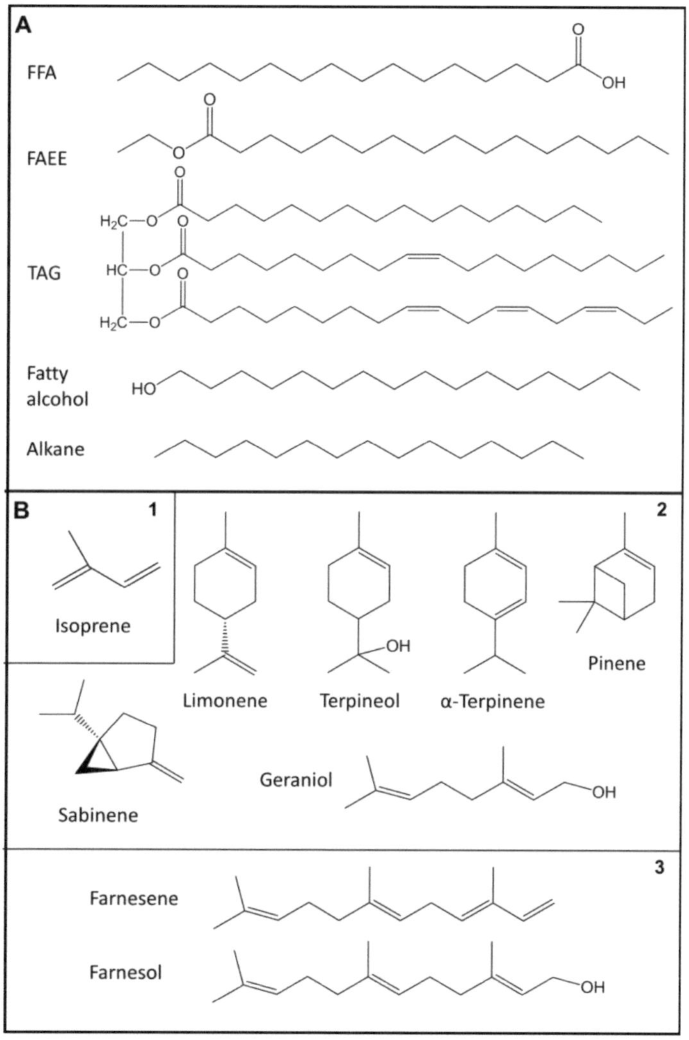

Figure 2. Chemical structures of hydrocarbon-based biofuels and fuel precursors. (A) Fuels derived from fatty acid biosynthesis and (B) fuels derived from isoprenoid biosynthesis, including (1) hemiterpene, (2) monoterpenes, and (3) sesquiterpenes.

2. Engineering hydrocarbon biosynthesis pathways

The hydrocarbon-based biofuels considered in this chapter (Figure 2) are all derived from two metabolites: fatty acids and isoprenoids. Thus, the two metabolic pathways commonly targeted by metabolic engineering strategies are the fatty acid biosynthesis pathway and the two pathways for isoprenoid production (Figure 3).

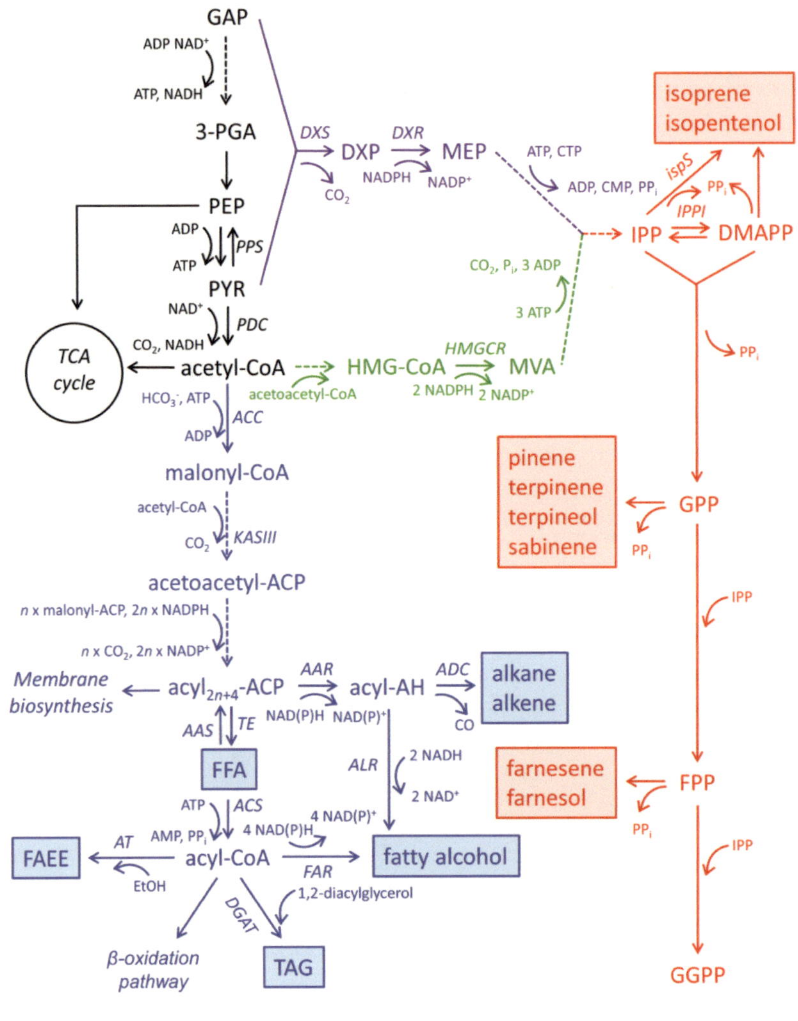

Figure 3. Hydrocarbon biosynthesis pathways for the production of biofuels, with the fatty acid biosynthesis pathway in blue, isoprenoid pathway in red, mevalonate pathway in green, and methylerythritol phosphate pathway in purple. Biofuels and biofuel precursors are highlighted in the colored boxes. Enyzmes are in italics. Solid arrows represent a single enzymatic step, while dashed arrows represent multiple enzymatic steps. Abbreviations for metabolites and enzymes are listed at the end of the chapter.

http://dx.doi.org/10.5772/52050

2.1. Fatty acid derived biofuels

As shown in Figure 3, fatty acid biosynthesis interfaces with the primary metabolism at the acetyl-CoA node. Fatty acid biosynthesis is initiated by the formation of acetoacetyl-ACP, the substrate for fatty acid chain elongation. The conversion of acetyl-CoA to acetoacetyl-ACP includes two key enzymatic steps: (1) the conversion of acetyl-CoA to malonyl-CoA, catalyzed by acetyl-CoA carboxylase (ACC) and (2) the conversion of malonyl-ACP to acetoacetyl-ACP via β-ketoacyl-ACP synthase III (KASIII). These two enzymes are common metabolic engineering targets for improving fatty acid biosynthesis. In fact, ACC has been shown to be a rate-limiting step of fatty acid synthesis in *Escherichia coli*, and overexpression of ACC has been shown to yield more than a 5-fold increase in FFA production [5]. Overexpression of KASIII in *E. coli* also improved FFA synthesis, increasing lipid production by 20-60% [6]. After acetoacetyl-ACP formation, fatty acid chain elongation proceeds by an iterative process, whereby the hydrocarbon chain is elongated in increments of 2 carbons. Once the elongation process terminates, the final acyl-ACP is divided among three possible paths: one leading to membrane biosynthesis, an essential pathway for cell growth, and the other two yielding hydrocarbon fuels or fuel precursors (Figure 3).

To produce biofuels with an even-numbered carbon chain, the acyl-ACP is cleaved by a thioesterase (TE), releasing the FFA. The TE is yet another key target for metabolic engineering. The final fuel properties, including viscosity, cloud point, flash point, oxidative stability, ignition delay, and combustion quality, are largely determined by the hydrocarbon chain length and degree of saturation [7]. Accordingly, numerous TEs have been cloned and characterized, predominantly from plant sources, to control the carbon chain length of the FFAs. Engineering strategies often exploit this collection of TEs to tailor the biofuel product. Favored TEs include a truncated TE (*'tesA*) from *E. coli* and acyl-ACP TEs from *Umbellularia californica* and *Cuphea hookeriana*, producing FFAs with carbon lengths of 16:0, 12:0, and 10:0 and 8:0, respectively [8-10]. The FFAs themselves can be extracted as fuel precursors and converted into biodiesel (FAMEs or FAEEs) using acid-catalyzed chemical processes [11]. To allow for FFA accumulation, the β-oxidation pathway and free fatty acid recycling are often eliminated by gene knockout of acyl-CoA synthetase (*acs*) and acyl-ACP synthetase (*aas*) [12]. An alternative strategy was recently demonstrated, whereby FFAs were synthesized through an engineered reversal of the β-oxidation cycle [13]. In this strategy, acetyl-CoA is used directly for fatty acid chain elongation, allowing for improved carbon and energy efficiency compared to the fatty acid biosynthesis pathway which requires activation of acetyl-CoA to malonyl-CoA. Engineering a reversed β-oxidation cycle required modification of multiple regulatory mechanisms, knockout of other fermentative pathways, expression of a TE or other fuel producing enzyme, and overexpression of key enzymes in the β-oxidation pathway [13]. While this strategy yielded the highest reported concentration of FFAs in *E. coli* (7 g/L), its application to other host organisms may be restricted by inadequate knowledge of the native regulatory mechanisms.

With an intact *acs*, FFAs can be converted into acyl-CoA, a precursor for other fuel products including the biodiesel precursor, TAG, and fuels such as FAEEs and fatty alcohols (Figure 3). The conversion of acyl-CoA to TAG requires the provision of 1,2-diacylglycerol and a diacylglycerol acyltransferase (DGAT) to catalyze transfer of the acyl chain. While DGAT has been overexpressed to improve TAG production in plants [14], the utility of this strategy still remains to be tested in microorganisms. Most metabolic engineering strategies for microbial TAG synthesis focus on improving the supply of the precursors: FFA and glycerol-3-phosphate (G3P) [15, 16]. Microbial production of FAEEs typically involves heterologous expression of both the pathway for ethanol production and an acyltransferase (AT) [17-19]. Selection of the two genes required for ethanol synthesis, pyruvate decarboxylase (*pdc*) and alcohol dehydrogenase (*adh*), will largely depend on the host organism, but generally, efforts involving prokaryotic hosts such as *E. coli* and cyanobacteria will use *pdc* and *adh* from *Zymomonas mobilis* due to their capacity for high ethanol production [20]. To date, only one AT has been heterologously expressed for FAEE production: the wax synthase gene (*aftA*) from *Acinetobacter baylyi* ADP1 [17-19]. A third biofuel product derived from acyl-CoA is fatty alcohols. The enzymatic conversion of acyl-CoA to a fatty alcohol is dependent upon whether the fatty acyl-CoA reductase (*far*) is of prokaryotic or eukaryotic origin. Most prokaryotic FARs reduce acyl-CoA to a fatty aldehyde, requiring another enzyme, fatty aldehyde reductase (ALR), for conversion to the fatty alcohol product. On the other hand, eukaryotic FARs catalyze the direct conversion of acyl-CoA to fatty alcohol without release of an aldehyde intermediate [21]. Metabolic engineering strategies for fatty alcohol production include: expression of a prokaryotic FAR, *acr1* from *Acinetobacter calcoaceticus* BD413, with reliance on native fatty aldehyde reductases for fatty alcohol synthesis [19]; expression of 5 different eukaryotic FAR homologs from the model plant organism *Arabidopsis thaliana* [22]; and expression of a eukaryotic FAR, *far1* from mouse [23]. The recent discovery of a prokaryotic FAR from *Marinobacter aquaeolei* VT8, capable of catalyzing the direct conversion of acyl-CoA to fatty alcohol, may be a beneficial alternative to the use of eukaryotic FARs for fatty alcohol production in prokaryotic hosts such as *E. coli* and cyanobacteria [24]. An alternative strategy used by Dellomonaco and colleagues identifies surrogates for *far* and *adh* in the native *E. coli* genome based on sequence homology [13]. With the numerous biofuel products derived from acyl-CoA and the natural enzymatic diversity for these conversions, we have only just begun to explore and develop the metabolic engineering tools essential to enable large-scale synthesis.

In addition to oxygen-containing biofuels, acyl-ACP can also be converted into pure hydrocarbon fuels in the form of alkanes and alkenes (Figure 3). In 2010, the discovery of an alkane synthesis pathway in cyanobacteria provided the genetic knowledge necessary for engineering microbial alkane production [25]. The pathway consists of two enzymatic steps: (1) reduction of acyl-ACP to a fatty aldehyde by means of an acyl-ACP reductase (AAR) and (2) decarbonylation of the aldehyde to an alkane or alkene, catalyzed by an aldehyde decarbonylase (ADC). Due to the recent discovery of this pathway, few metabolic engineering strategies have been applied for alkane production. Some strategies focus on improving supply of the acyl-ACP precursor, relying on the native cyanobacterial pathway for alkane synthesis [23], while others have simply transferred the alkane pathway (AAR and ADC) into another host organism [25-27]. With the rapidly growing database of genome sequence information, numerous homologs of AAR and ADC have been identified [26, 27], representing a diverse range of targets for metabolic engineering. Future optimization of the alkane biosynthesis pathway may result in the high alkane yields needed for biofuel production.

2.2. Isoprenoid-based biofuels

The chemical composition of petroleum-based fuels: gasoline, diesel, and jet fuel, includes linear, branched, and cyclic alkanes, aromatics, and chemical additives [28]. Isoprenoid-based biofuels have the structural diversity to mimic these petroleum compounds, with up to 50,000 known isoprenoid structures including branched and cyclic hydrocarbons with varying degrees of unsaturation [29, 30]. Isoprenoids reported to be potential fuel candidates include: the hemiterpene (C5) isoprene; monoterpenes (C10): terpinene, pinene, limonene, and sabinene; the sesquiterpene (C15) farnesene, and their associated alcohols: isopentenol, terpineol, geraniol, and farnesol [12, 31]. Two metabolic pathways are capable of producing the isoprenoid building blocks isopentenyl pyrophosphate (IPP) and dimethylallyl diphosphate (DMAPP): the mevalonate (MVA) pathway [32] and the methylerythritol phosphate (MEP) pathway, also known as the 1-deoxy-D-xylulose-5-phosphate (DXP) pathway and the non-mevalonate pathway (Figure 3) [33]. In general, the MVA pathway is found in eukaryotes and archaea while the MEP pathway is utilized by prokaryotes. In agreement with the proposed evolutionary origin of plants, they contain both isoprenoid pathways with the MEP pathway localized in the plastid and the MVA pathway in the cytosol [34]. The MVA and MEP pathways differ with respect to their requirement for carbon, energy, and reducing equivalents; this is illustrated by the net balances for IPP biosynthesis from glyceraldehyde-3-phosphate (GAP):

$$\text{MVA:} 3\ \text{GAP} + 3\ \text{ADP} + 4\ \text{NAD(P)}^{+} + 2\ \text{P}_{i} \rightarrow \text{IPP} + 4\ \text{CO}_{2} + 3\ \text{ATP} + 4\ \text{NAD(P)H} \quad (1)$$

$$\text{MEP:} 2\ \text{GAP} + \text{ADP} + \text{CTP} + \text{P}_{i} \rightarrow \text{IPP} + \text{CO}_{2} + \text{ATP} + \text{CMP} + \text{PP}_{i} \quad (2)$$

Based on these balances, IPP production via the MEP pathway is more efficient at carbon utilization, as only 2 GAPs are required and 1 CO_2 is emitted, compared to 3 GAPs and 4 CO_2 for the MVA pathway. On the other hand, IPP production via the MVA pathway is more energy efficient overall, resulting in ATP generation and yielding a net gain in reducing equivalents (NAD(P)H). These carbon, energy, and reducing equivalent requirements should be considered when designing a metabolic engineering strategy for isoprenoid biosynthesis.

The MVA pathway interfaces with the primary metabolism at the acetyl-CoA node (Figure 3), and it can be divided into two parts: the top, which involves 3 enzymatic steps to convert acetyl-CoA to MVA, and the 3 enzymatic conversions of the bottom portion to produce IPP from MVA. One novel metabolic engineering strategy compared the efficiencies of the top and bottom portions of the MVA pathway in *E. coli* using heterologously expressed pathways from 5 different eukaryotic sources. The most efficient top and bottom portions were combined to maximize the yield of isoprenoid building blocks [35]. Accumulation of an intermediate metabolite, 3-hydroxy-3-methyl-glutaryl-CoA (HMG-CoA), is a known bottleneck in the top MVA pathway, and HMG-CoA was also shown to inhibit cell growth in *E. coli* [36]. Thus, overexpression of the HMG-CoA reductase (HMGCR) increased MEV production and synthesis of subsequent FPP-derivatives in both *E. coli* and *S. cerevisiae* [36-38]. Whole pathway expression and elimination of the HMGCR bottleneck have proven to be successful techniques for enhancing the metabolic throughput of the MVA pathway.

The MEP pathway requires two primary metabolites as precursors: GAP and pyruvate (PYR) (Figure 3). Compared to the 6 enzymatic steps of the MVA pathway, the MEP pathway is comprised of 7 steps. Metabolic engineering strategies for the MEP pathway have primarily focused on the first two enzymatic steps. Overexpression of 1-deoxy-D-xylulose-5-phosphate synthase (*dxs*), catalyzing the conversion of GAP and PYR to 1-deoxy-D-xylulose-5-phosphate (DXP), resulted in 6-10-fold increases in the final isoprenoid product [39, 40]. Targeting the next enzymatic step through overexpression of DXP reductoisomerase (*dxr*) was shown to have little effect on isoprenoid production using the native gene; however, expression of *dxs* and *dxr* from *Bacillus subtilis* improved isoprenoid production 2.3-fold in *E. coli* [41]. The final step of the MEP pathway was also shown to be rate-limiting, as heterologous expression of IPP isomerases (IPPI) enhanced isoprenoid production in *E. coli* [42]. Based on its rate-limiting steps, the MEP pathway is a prime candidate for a push-pull metabolic engineering strategy, whereby overexpression of the first step 'pushes' carbon flux into the MEP pathway and overexpression of the final step 'pulls' the metabolic flux towards the end product. This strategy yielded nearly 2-fold improvements in isoprenoid production in *E. coli* [43, 44]. Lastly, overexpression of the entire MEP pathway can increase isoprenoid biosynthesis. In fact, Leonard and colleagues demonstrated that 5 additional copies of the MEP pathway genes yielded the highest production, while further increasing the gene copy number to 10 produced lower titers [45].

While targeted gene overexpression may alleviate pathway bottlenecks, the pathway is still subject to native regulatory mechanisms which may limit isoprenoid biosynthesis from either the MVA or MEP pathways. A highly successful strategy for overcoming regulatory limitations is overexpression of the non-native isoprenoid pathway. Expression of the MVA pathway from *Saccharomyces cerevisiae* in *E. coli* has enabled higher levels of isoprenoid synthesis compared to engineering the native MEP pathway as the sole isoprenoid pathway [46-50]. The success of this strategy has made it a favorite among metabolic engineers seeking to improve isoprenoid biosynthesis. Farmer and Liao presented a clever approach for regulating the carbon flux into an engineered MEP pathway in *E. coli* [51]. In this work, a native regulatory circuit was used to control the carbon flux into and through the MEP pathway by regulating expression of two key enzymes: phosphoenolpyruvate synthase (PPS) and isopentenyl diphosphate isomerase (IPPI). Under excess carbon flux, expression of *pps* and *idi* was activated using the regulatory circuit, redirecting carbon flux into and through the MEP pathway, yet when the carbon flux was growth limiting, expression of these genes was reduced. This strategy allows for high isoprenoid production without negatively impacting cell growth. As evidence, the regulated pathway improved isoprenoid titers by 50%, while simply placing *pps* and *ippi* under control of strong *tac* promoters resulted in growth inhibition [51]. Native regulatory mechanisms are often obstacles limiting isoprenoid biosynthesis, yet they can also be exploited to optimize the flux balance to support both cell growth and isoprenoid production.

Additional targets for improving isoprenoid-based fuel production include precursor supply, cofactor supply, and optimization of the downstream fuel synthesis pathway. Acetyl-CoA is the precursor for isoprenoid production via the MVA pathway. Overexpression of acetaldehyde dehydrogenase (ALDH) and acetyl-CoA synthetase (ACS), both of which produce acetyl-CoA, increased the acetyl-CoA supply and subsequently isoprenoid biosynthesis in *S. cerevisiae* [52]. On the other hand, the MEP pathway requires two precursors from the glycolysis pathway: PYR and GAP. The supply of these metabolites is complicated by the fact that PYR is derived from GAP, and consequently, the PYR/GAP balance is an important metabolic engineering target. The supply of GAP was shown to be limiting in *E. coli*, as modifying the conversion between PEP and PYR to redistribute the flux toward GAP synthesis increased isoprenoid production [53]. In addition to the carbon precursors, co-factors in the form of energy (ATP, CTP) and reducing equivalents (NADPH) are also required for isoprenoid synthesis. Co-factor supply is often overlooked in strategies for isoprenoid production, yet by improving the availability of NADPH in *S. cerevisiae*, isoprenoid synthesis through the MVA pathway increased by 85% [54]. This result emphasizes the importance of co-factor availability. Despite optimizing production of the isoprenoid building blocks, the downstream efficiency of assembling the final fuel product may still limit the overall yield. Successful strategies for improving downstream efficiency include overexpression of GPP and FPP synthases [47], overexpression and codon optimization of hemiterpene, monoterpene, and sesquiterpene synthases [41, 47, 48], fusion proteins to localize FPP synthesis and its conversion to sesquiterpene [47], and downregulation of competing products like squalene [37, 48]. The optimized production of isoprenoid-based fuels requires strategies to address limitations throughout the metabolic pathway, from precursor and co-factor supply to end product synthesis.

3. Influence of feedstock on hydrocarbon-based biofuel production

While hydrocarbon-based biofuel production relies on the biosynthetic pathways discussed in the previous section, the source of feedstock plays an important role in the overall production process. As discussed in the Introduction to this chapter, there are two main feedstocks for biofuel production: lignocellulosic biomass and gaseous CO_2, supporting the production of second and third generation biofuels, respectively (Figure 1). Both processes ultimately rely on CO_2 and sunlight as the carbon and energy source, but the microbial conversion processes are distinctly different between the two feedstocks. Lignocellulosic biomass deconstruction produces organic carbon, mostly in the form of hexoses and pentoses (C5 and C6 sugars); this feedstock requires heterotrophic microorganisms to convert the organic carbon into biofuel. Alternatively, the fixation of inorganic carbon feedstock (CO_2/HCO_3^-) into biofuel is reliant upon autotrophic microbes. The heterotroph vs. autotroph requirement of the respective feedstocks is an important distinction from both the metabolic engineering and biofuel production perspectives. Only a few model microorganisms are capable of both heterotrophy and autotrophy, resulting in different host candidates for second and third generation biofuel production. The feedstock will also influence the metabolic engineering targets, as heterotrophs utilize glycolysis and oxidative phosphorylation pathways for carbon consumption and energy production while oxygen-generating autotrophs utilize the Calvin-Benson-Bassham cycle and photosynthesis under light conditions (Figure 4). This section will discuss the host organisms, engineering strategies, and biofuel production processes specific to each carbon feedstock.

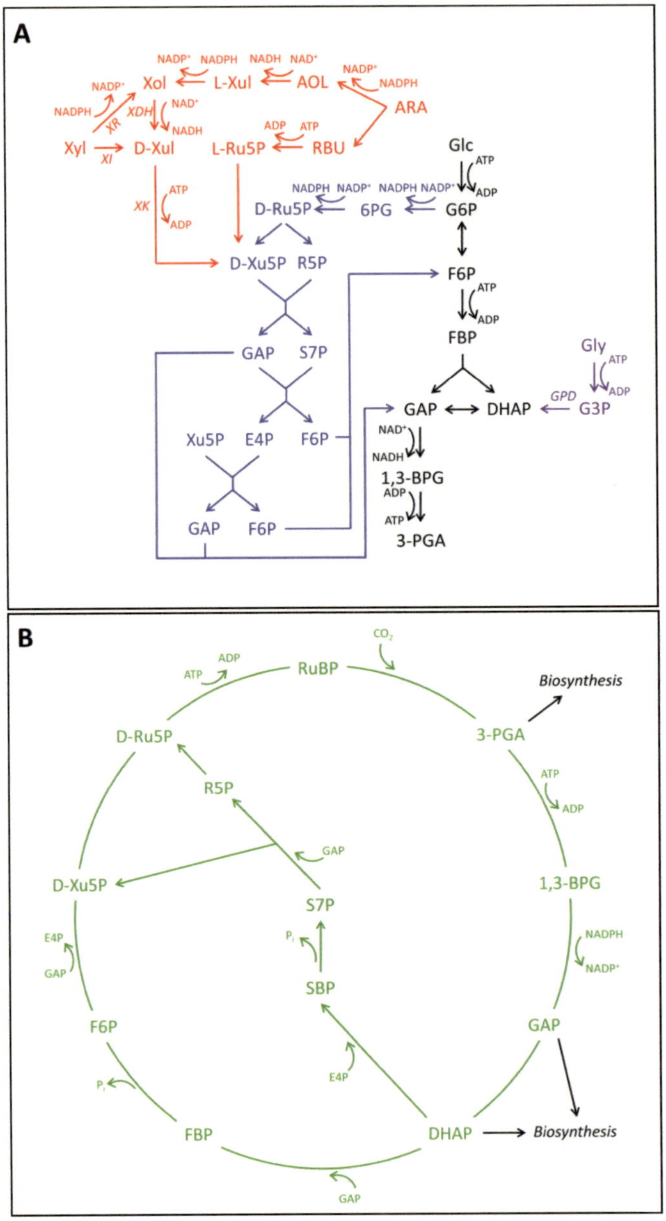

Figure 4. Heterotrophic (A) and autotrophic (B) pathways for carbon utilization, with the Embden-Meyerhof-Parnas (EMP) pathway (glycolysis) in black, the pentose phosphate pathway (PPP) in blue, pentose utilization pathways in red, glycerol metabolism in purple, and the Calvin-Benson-Bassham cycle in green. Abbreviations for metabolites and enzymes are listed at the end of the chapter.

3.1. Hydrocarbon biofuel production from organic carbon feedstocks

The release of C5 and C6 sugars from lignocellulosic biomass deconstruction supports the growth of heterotrophic microorganisms and the metabolic conversion of sugars into biofuel. Representative hydrocarbon-based fuel titers produced by engineered, heterotrophic hosts are listed in Table 1. The most common heterotrophic hosts for biofuel production are the model organisms *Escherichia coli* and *Saccharomyces cerevisiae*. These hosts are attractive candidates for fuel production due to their fast growth rates, well-known genetics and regulation, advanced molecular tools for genetic engineering, and established use in the industrial setting. Neither *E. coli* nor *S. cerevisiae* naturally produce significant amounts of hydrocarbon-based fuels, necessitating the application of metabolic engineering techniques. Heterotrophic organisms that naturally produce hydrocarbon-based fuels are also potential hosts for large-scale biofuel production. For example, *Bacillus subtilis* naturally produces higher concentrations of isoprene than other commonly known bacteria like *E. coli* [55]. *B. subtilis* is also a model organism for Gram-positive bacteria with established tools for genetic modification, advancing its appeal as a host for isoprene production. Similarly, heterotrophic algae can produce significant quantities of TAG. This has motivated some preliminary investigation into engineering the model green alga, *Chlamydomonas reinhardtii*, for TAG production [56-58]. While most metabolic engineering efforts have focused on these model heterotrophic hosts, genetic tools can be developed for other organisms with desirable fuel production traits.

Hydrocarbon Fuel/ Fuel Precursor	Concentration Range	Microbial Hosts	References
	Heterotrophic Production		
FFA	0.5 – 7 g/L	*Escherichia coli*	[5, 12, 13, 19, 59, 60]
	0.024 – 0.2 g/L	*Saccharomyces cerevisiae*	[61, 62]
TAG	20 - 32.6% dcw, 0.12 g/L	*Chlamydomonas reinhardtii*	[56-58]
	0.4 – 0.7 g/L	*Saccharomyces cerevisiae*	[63, 64]
FAEE	0.07 – 1.5 g/L	*Escherichia coli*	[18, 19, 65-67]
	N/A	*Saccharomyces cerevisiae*	[17]
Fatty alcohols	0.001 – 1.67 g/L	*Escherichia coli*	[13, 19, 22, 27, 59, 66, 68]
Alkanes/Alkenes	0.042 – 0.32 g/L	*Escherichia coli*	[25, 27]
Other Isoprenoids (lycopene, β-carotene, amorphadiene,	0.002 – 1 g/L	*Escherichia coli*	[35, 39, 42, 45, 50, 69]
	0.01 g/L	*Saccharomyces cerevisiae*	[37, 52]

Hydrocarbon Fuel/ Fuel Precursor	Concentration Range	Microbial Hosts	References
levopimaradiene, cubebol)			
Isoprene	0.31 – 0.53 g/L	*Escherichia coli*	[41, 49]
	0.002 g/L	*Bacillus subtilis*	[55]
Farnesol	N/A	*Escherichia coli*	[48]
	0.009 – 0.15 g/L	*Saccharomyces cerevisiae*	[37, 38, 70, 71]
Farnesene	0.38 – 1.1 g/L	*Escherichia coli*	[47, 72]
Autotrophic Production			
FFA	0.11 - 0.20 g/L	*Synechocystis* sp. PCC 6803	[73-75]
	0.015 - 0.06 g/L	*Synechococcus elongatus* PCC 7942	[73, 75, 76]
	0.051 g/L	*Synechococcus* sp. PCC 7002	[77]
TAG	28.5% dcw	*Chlamydomonas reinhardtii*	[57]
FAEE	0.077 – 0.086 g/L	*Synechococcus* sp. PCC 7002	[77]
Fatty alcohols	200 μg/L	*Synechocystis* sp. PCC 6803	[23]
Alkanes/Alkenes	150 μg/L/OD730	*Synechocystis* sp. PCC 6803	[23]
	0.05 g/L	*Synechococcus* sp. PCC 7002	[26]
	N/A	*Thermosynechococcus elongatus* BP-1	[26]
Isoprene	0.5 mg/L	*Synechocystis* sp. PCC 6803	[78]

Table 1. Hydrocarbon fuels and fuel precursors produced by genetically engineered microorganisms.

Most heterotrophic hosts for biofuel production utilize the Embden-Meyerhof-Parnas (EMP) pathway for sugar catabolism (Figure 4). The EMP pathway has evolved for efficient carbon utilization and is typically not rate-limiting for fuel production. As such, EMP pathway enzymes are not often targeted for genetic manipulation. However, the organic feedstock from lignocellulose deconstruction is comprised of a range of sugars, including hexoses: glucose, mannose, and galactose, and pentoses: xylose and arabinose [79]. A major concern in converting these sugars into fuel is the efficient utilization of all available hexoses and pentoses. While some organisms like *E. coli* can naturally metabolize these different forms of sugar, others, like *S. cerevisiae,* can only utilize specific forms [80]. *S. cerevisiae* does not naturally express pathways for catabolizing pentoses. There are two known pathways for xylose catabolism, both of which have been expressed in *S. cerevisiae* [81-83]. Xylose can be converted into xylulose-5-phosphate (Xu5P), an intermediate in the pentose phosphate pathway (PPP), through expression of a xylose isomerase (XI) and xylulose kinase (XK) [82]. Alternatively, the XI can be replaced by a xylose reductase (XR) and xylitol dehydrogenase (XDH) [81, 82]. Complications in these two xylose utilization pathways include the inhibition of XI by xylitol (Xol) and the reducing equivalents required by XR and XDH [80]. Successful strategies for engineering xylose utilization in *S. cerevisiae* include expression of a fungal XI from *Piromyces* sp. E2 along with overexpression of the non-oxidative PPP pathway [84] and expression of XR and XDH from the xylose-fermenting yeast *Pichia stipitis* [85]. Two pathways have also been expressed in *S. cerevisiae* for arabinose utilization [86, 87]. The bacterial pathway for arabinose catabolism consists of 3 enzymatic steps, while the fungal pathway involves 5 enzymatic steps, 4 of which require cofactors of NADPH or NAD^+ (Figure 4). Efficient arabinose utilization in *S. cerevisiae* has been achieved through heterologous expression of a bacterial arabinose catabolism pathway along with overexpression of the non-oxidative PPP and evolutionary engineering [88]. While most of these metabolic engineering examples focus on utilizing sugars for fermentation to ethanol, the strategies for engineering carbon utilization can also be applied for hydrocarbon-based fuel production.

Unlike *S. cerevisiae, E. coli* can utilize the hexoses and pentoses derived from lignocellulose; however, the carbon catabolite repression (CCR) system in *E. coli* leads to inefficient, diauxic growth [89]. Through CCR, *E. coli* sequentially consumes different sources of organic carbon based on substrate preference, leading to delayed and often incomplete utilization of unpreferred sugars like xylose and arabinose. This translates into lower productivities and yields along with downstream complications due to the presence of unmetabolized sugars [80]. As a result, CCR is often targeted by metabolic engineering to alleviate these undesired effects. A common engineering strategy is to use mutants of the transcriptional activator CRP (cyclic AMP receptor protein) which have been modified to eliminate the allosteric requirement for cAMP, thereby leading to expression of the pentose catabolizing pathways in the presence of the preferred substrate, glucose [90]. The phosphotransferase system (PTS), responsible for the preferential uptake of glucose, has also been deleted to encourage simultaneous utilization of mixed sugars [91]. Lastly, deletion of methylglyoxyal synthase was shown to improve the co-metabolism of sugars, ostensibly due to elimination of methylglyoxyal, an inhibitor of sugar metabolism [92]. Through modifying the components of CCR, *E. coli* can be engineered to efficiently utilize the organic carbon mixture resulting from lignocellulose degradation.

In addition to the hexoses and pentoses derived from lignocellulosic biomass, glycerol may soon become an inexpensive organic carbon source for fuel production. Glycerol is a byproduct of the conversion of TAG into biodiesel during algal biofuel processing, and thus, large quantities of glycerol may be available for use as an organic carbon source. The main pathway for aerobic glycerol utilization involves a two-step conversion to produce the glycolytic metabolite DHAP [93]. The glycerol utilization pathway is not a common target for metabolic engineering, yet glycerol has been reported as a supplementary carbon source for the production of isoprenoid-based fuels, farnesol and α-farnesene [47, 48]. Future metabolic engineering efforts may focus more on glycerol utilization as the availability of glycerol increases.

Second generation biofuel production still remains to be demonstrated at large scales, yet the overall process is easily integrated with current technologies. Equipment and practices used for agricultural harvesting can be directly applied to harvesting lignocellulosic biomass. In fact, some agricultural processes already produce biomass waste streams that can be utilized for feedstock, such as corn stover. Moreover, commercial fermenters can be employed as bioreactors for the microbial fuel conversion. The main technical difficulties in large-scale lignocellulosic fuel production center on provision of the carbon source. The quantities of biomass needed to support industrial-scale fuel production will require a significant investment of land and nutrient resources, and the supply will be subject to varying climate conditions. A supply chain infrastructure must also be constructed to harvest the biomass and transport it to the production facilities. A primary technical focus of current research on lignocellulosic-derived fuels is the deconstruction of biomass into useable sugars. The thermal, chemical, and enzymatic processes for biomass deconstruction have been a limiting factor for economical second generation biofuel production [94, 95]. As the cost of biomass deconstruction is reduced with new technology, the large-scale production of second generation biofuels will begin to contribute to the world's supply of renewable energy.

3.2. Hydrocarbon biofuel production from inorganic carbon feedstocks

The direct conversion of CO_2 into hydrocarbon-based fuels could greatly simplify the overall production process and reduce the cost of biofuel production (Figure 1). The search for autotrophic microorganisms capable of performing this CO_2-to-fuel conversion started in the late 1970's with the U.S. Department of Energy's Aquatic Species Program (ASP) [96]. The ASP isolated and screened over 3,000 species of microalgae from a diverse range of environmental habitats. The program focused mainly on eukaryotic algae, as they naturally produce significant amounts of TAG. During the course of the program, the recombinant DNA technology used in metabolic engineering was developed, yet due to the infancy of this technology, it was not applied to microalgae for fuel applications until near the end of the ASP [15]. With the development of recombinant DNA technology, prokaryotic microalgae (i.e. cyanobacteria, previously known as blue-green algae) were recognized as potential hosts for fuel production, and the successful engineering of cyanobacteria for ethanol production confirmed their potential [97]. Unfortunately, research funding for microalgal fuel production waned as crude oil prices fell in the 1990's. However, in the late 2000's, the cost of crude oil soared, spurring a resurgence of interest in microalgae for fuel production and in the application of metabolic engineering to enhance fuel yields. In general, both eukaryotic microalgae (referred to as algae in the subsequent text) and prokaryotic microalgae (referred to as cyanobacteria in the subsequent text) utilize photosynthesis for energy generation and the Calvin-Benson-Bassham cycle for CO_2 fixation (Figure 4). However, due to the cellular differences between algae and cyanobacteria, the strategies for engineering autotrophic fuel production will be discussed based on this host division.

3.2.1. Engineering algae for biofuel production

Algae are predicted to have first appeared approximately 1.5 billion years ago from an endosymbiotic event in which a eukaryotic cell engulfed a cyanobacterium [98]. The cyanobacterium evolved into the modern day chloroplast, the algal organelle responsible for photosynthesis and carbon fixation. Today, algae can be found in a wide-range of environmental habitats from freshwater lakes and oceans to deserts and even the snow of the Antarctic

[99]. Along with this diversity of habitat, algae have evolved diverse cellular physiologies and genetics, resulting in a wealth of potential hosts and genetic sources for engineering fuel production. Many types of algae are currently under consideration for fuel production due to their natural TAG synthesis, including diatoms, green algae, eustigmatophytes, prymnesiophytes, and red algae [100]. While many types of algae produce the fuel precursor TAG, few algal species have well-developed genetic tools available for engineering improved lipid production [101, 102]. Consequently, there are only a few reported examples of engineering algae for biofuel production.

To date, the only genetic mutation shown to improve lipid production in algae is the elimination of starch biosynthesis, a competing carbon sink. The generation of mutants with impaired starch synthesis using random mutagenesis techniques resulted in up to a 10-fold increase in cellular lipid production in *C. reinhardtii* [56-58, 103]. Other targeted metabolic engineering attempts, such as overexpression of ACC in the diatoms *Cyclotella cryptic* and *Navicula saprophila*, failed to improve TAG biosynthesis [15, 96]. In addition to targeting overall TAG production, metabolic engineering strategies have been applied to influence the chemical composition of the fatty acid side chains. By expressing two heterologous TEs, the diatom *Phaeodactylum tricornutum* produced TAG with increased levels of lauric acid (C12:0) and myristic acid (C14:0) [104]. These shorter chain length fatty acids are more desirable for fuel production, and this demonstrates the potential to control the chemical composition of the fuel product and its associated properties with metabolic engineering. While examples of engineering algal TAG production are sparse, many engineering strategies have proven successful at improving the fatty acid content in plants. These strategies include expression of ACC and KASIII involved in fatty acid biosynthesis, expression of G3P dehydrogenase (GPD) for production of the glycerol backbone of TAG, expression of ATs such as DGAT, expression of TEs to release FFAs, and deletion of desaturases to alter the fatty acid composition [105]. Similar strategies may also be successful at improving TAG production in algae.

The metabolic engineering of algae is complicated by several factors. Most algae have a rigid cell wall structure that makes transformation difficult. A common transformation technique uses glass beads (or silicon carbide whiskers) along with a cell wall-deficient algal strain [106]. The cell wall can be removed using enzymatic techniques or through genetic mutation. Alternatively, a microparticle bombardment technique has been applied successfully to transform many different algal species [107]. In this technique, the recombinant DNA is coated onto a metal microparticle and 'shot' into the algal cell using a helium-powered 'gun'. Other transformation methods include electroporation and the traditional plant transformation technique of *Agrobacterium tumefaciens* T-DNA-mediated transfer [107]. Once the recombinant DNA enters the cell, it must integrate into one of 3 algal genomes: nuclear, chloroplast, or mitochondrial (assuming the transformed DNA is not a stably maintained plasmid). DNA has been successfully integrated into the chloroplast genome via homologous recombination, whereby the recombinant gene and marker are flanked by homologous (i.e. matching) regions of the targeted chloroplast DNA, and the recombinant DNA replaces the matching region in the chloroplast. Unfortunately, homologous recombination does not occur in the nuclear genomes of many algae [108], and instead, the recombinant DNA is randomly integrated into the nuclear genome. This complicates metabolic engineering strategies due to the possibility of detrimental genetic effects resulting from the random integration and the lack of a technique for targeted gene knockout. Lastly, algal engineering attempts are often plagued by low gene expression. It has been discovered that many algae, like the model alga *C. reinhardtii*, employ RNA-mediated gene silencing [109]. Numerous strategies have been applied to combat the low gene expression brought about by gene silencing in algae, including codon optimization, the use of 5′ and 3′ untranslated regions which may participate in regulatory functions, and the inclusion of native intron sequences [108]. Knowledge of the gene silencing mechanisms in algae has led to the development of RNA interference (RNAi) technology for gene knock-down. RNAi exploits the native cellular machinery for gene silencing to reduce the expression of target genes [109]. As we continue to expand our knowledge of algal genetics, the list of engineered algae will rapidly increase. As evidence, the biofuel-relevant alga, *Nannochloropsis* sp., was recently shown to have a high efficiency of homologous recombination in the nuclear genome [110]. This will simplify future strategies for genetic engineering in *Nannochloropsis* sp. Another promising development is the construction of a plasmid for gene expression in *C. reinhardtii* that is now commercially available through Life Technologies [111]. The greater availability and standardization of tools for the genetic manipulation of algae will move algal engineering towards the advanced stages currently seen with other industrial organisms like *E. coli* and *S. cerevisiae*.

3.2.2. Engineering cyanobacteria for biofuel production

Cyanobacteria are predicted to be the first microorganisms to develop the capability of oxygenic photosynthesis, some 2.7 billion years ago [112]. Similar to algae, cyanobacteria have a great range of diverse morphologies, cellular functions, and genetics, presumably due to their long evolutionary history and their diverse habitats. As discussed previously, the ASP initially deemed cyanobacteria unfit for fuel production due to their lack of natural TAG accumulation. Since they are amenable to genetic manipulation, however, cyanobacteria can be engineered to produce a range of biofuel products (Table 1). As prokaryotes, cyanobacteria are subject to the traditional methods employed for engineering other well-developed bacterial hosts like *E. coli*. Some strains of cyanobacteria are even naturally transformable, uptaking exogenous DNA from their environment without the use of cell permeablization techniques [113]. As progenitors of the algal chloroplast, cyanobacteria also integrate DNA into their chromosomes using homologous recombination. Moreover, cyanobacteria do not possess the cellular components for gene silencing. The genetic tools for engineering some model strains of cyanobacteria are well developed and have been used to genetically modify cyanobacteria for several decades [113]. Another advantage of using cyanobacteria as the microbial host for hydrocarbon-based fuel production is that they have been shown to excrete potential fuel precursors such as FFAs [73]. Fuel excretion enables a continuous production process, eliminating the cost associated with harvesting the algal biomass and the time and nutrients needed to repeatedly grow new batches of algae for fuel production. The advantages of straightforward genetic manipulation and fuel excretion make cyanobacteria contenders for large-scale biofuel production despite the disadvantage of low natural lipid yields.

After the initial demonstration of engineering cyanobacteria for ethanol production [97], the production of hydrocarbon-based fuels in engineered cyanobacteria has expanded to include isoprene, FFAs, FAEEs, fatty alcohols, and alkanes/alkenes (Table 1). Isoprene biosynthesis was established in the model cyanobacterium, *Synechocystis* sp. PCC 6803, through expression of the isoprene synthase (*ispS*) from kudzu [78]. Codon optimization of *ispS* and the use of a strong promoter (*psbA2*) increased isoprene production. Engineering strategies targeting the upstream MEP pathway for isoprenoid biosynthesis, as described in Section 2.2 of this chapter, will likely further improve isoprene productivity. The remaining four hydrocarbon-based fuels are all derived from the fatty acid biosynthesis pathway. Common strategies for improving FFA production (see Section 2.1) have proven successful in cyanobacteria [74-76]. Eliminating non-essential, competing pathways such as polyhydroxybutyrate (PHB), cyanophycin, and acetate biosynthesis also improved FFA production [74]. Liu and colleagues engineered a more permeable peptidoglycan layer to improve FFA excretion in *Synechocystis* sp., yet this weakened cell membrane resulted in slower growth rates and may also make the engineered cyanobacterium more susceptible to external predators and toxins that may be present in large-scale cultivations. Initial engineering attempts for fatty alcohol and alkane/alkene production entail expression of a heterologous FAR and overexpression of AAR and ADC, respectively [23, 26]. Alkane/alkene synthesis was also observed with ACC overexpression and native AAR and ADC activities in cyanobacteria [23]. Despite being derived from fatty acids, the synthesis of fatty alcohols and alkanes/alkenes is up to 1000-fold lower than that observed with FFA production (Table 1), suggesting that the conversion of acyl-ACP to the final fuel product is rate limiting. These inaugural proof-of-concept reports illustrate the potential of cyanobacteria as hosts for autotrophic biofuel production, but additional metabolic engineering will be required to achieve the fuel titers necessary for large-scale synthesis.

3.3. Heterotrophic vs. autotrophic biofuel production

The selection of organic or inorganic carbon feedstock for biofuel production has downstream ramifications on host selection, product yields, and process requirements. Clearly, the feedstock choice will determine whether a heterotrophic or autotrophic host is required, and in turn, this will influence the metabolic engineering strategy. In general, heterotrophic hosts have generated higher fuel titers than autotrophic hosts, with more than 10-fold higher concentrations of FFAs, FAEEs, fatty alcohols, and alkanes/alkenes (Table 1). This does not imply that heterotrophic production is more advantageous than autotrophic production, for the entire production process must be considered (Figure 1). The sugars from lignocellulosic biomass deconstruction (heterotrophic feedstock) have a higher energy content compared to inorganic carbon (autotrophic feedstock). The overall balances for obtaining one molecule of GAP from heterotrophic and autotrophic metabolisms provide evidence for this:

$$\text{Heterotrophic: } \tfrac{1}{2}\,\text{Glc} + \text{ATP} \rightarrow \text{GAP} + \text{ADP} \tag{3}$$

$$\text{Autotrophic: } 3\,\text{CO}_2 + 9\,\text{ATP} + 6\,\text{NADPH} + 5\,\text{H}_2\text{O} \rightarrow \text{GAP} + 9\,\text{ADP} + 6\,\text{NADP}^+ + 8\,\text{P}_\text{i} \tag{4}$$

While autotrophic GAP generation requires a significant investment of energy (9 ATP) and reducing equivalents (6 NADPH), heterotrophic GAP production only requires one energy equivalent. However, if a life cycle perspective is considered, the carbon from lignocellulosic feedstocks is ultimately derived from photosynthesis, requiring the same energy and reducing equivalent input as autotrophic microorganisms. Overlooking this fact will bias a direct comparison between heterotrophic and autotrophic fuel production.

One major difference between heterotrophic and autotrophic fuel production is the design considerations for the bioreactor. Heterotrophic microbes, such as *E. coli* and *S. cerevisiae*, are traditional industrial microorganisms with well-established, large-scale cultivation practices and bioreactors. On the other hand, autotrophic hosts like algae and cyanobacteria require light as the energy source to drive photosynthesis and inorganic carbon fixation. This can have a dramatic effect on bioreactor design. Transparent materials can be used with traditional bioreactor designs to allow for light penetration. Light availability, however, will ultimately limit the cell densities of photosynthetic microalgae, and the surface area of light exposure with traditional bioreactor designs is not optimal. Some have proposed to use fiber-optics within the liquid culture to improve light availability [114], but a costly solution such as this is not feasible for a low-value, commodity product like fuel. A wide-range of photobioreactor (PBR) designs have been proposed [115], yet generally, PBRs are characterized by the use of transparent materials, high surface area to volume ratios, and a relatively short pathlength for light. Other PBR design factors include a mechanism for air/CO_2 delivery, dissipation of radiative heat, and removal of inhibitory O_2 [115]. Due to the low value of fuel products, PBRs for fuel synthesis favor low-tech designs and inexpensive materials to reduce both capital and operating costs. In fact, NASA has proposed to float plastic bags of algal cultures in wastewater to allow for nutrient exchange [116]. Alternatively, open pond systems, traditionally a raceway configuration with a paddle-wheel for mixing, have proven successful for cultivating microalgae at scale [117]. Unlike PBRs, ponds are open to the environment, allowing for evaporative water loss and pond crash due to contamination by predators and competitors. However, the low capital cost of an open pond system makes this design a contender for fuel production. Clearly, the large-scale cultivation techniques for autotrophic fuel production still require additional development and optimization compared to heterotrophic cultivation.

4. Other metabolic engineering strategies for industrial production of hydrocarbon fuels

In addition to improving hydrocarbon-based fuel synthesis, metabolic engineering strategies can also be applied to address other factors affecting large-scale production. Two main issues will be addressed in this section: product toxicity and industrial strain robustness.

Product toxicity was shown to be a limiting factor in the production of first generation biofuels like ethanol. Since the interest in hydrocarbon-based fuels has developed only during the past decade, the toxicities of these fuels have not been fully explored, particularly with respect to autotrophic hosts. Fortunately, interest in hydrocarbon inhibition of microbial growth dates back almost a century [118], and we can capitalize on this wealth of information to engineer improved product tolerance in microbial hosts. Most fatty acid derived fuel molecules have shown some antimicrobial activity. FFAs, with a diverse range of carbon chain lengths and degrees of unsaturation, impart inhibitory effects on organisms including algae, Gram-negative and Gram-positive bacteria, fungi, protozoans, and various cell types of multicellular organisms [119]. Medium chain fatty alcohols such as pentanol, hexanol, heptanol, and octanol inhibited the biological activity of several algal and cyanobacterial strains, including fuel-relevant hosts *C. reinhardtii* and *Dunaliella salina* [120]. Interestingly, long-chain fatty alcohols (>C14) did not exhibit inhibitory effects on yeasts, suggesting that targeting longer chain fatty alcohols may eliminate the toxicity concern [121]. Similarly, medium-length alkanes (hexane, heptane, and isooctane) were toxic to microalgae while long-chain alkanes (C12-C16) elicited no effect [120, 121]. Microbial TAG and FAEE toxicities have not been reported. However, the phospholipid membrane surrounding algal TAGs may mask potential inhibitory effects, and FAEE production has been linked to the toxic effects of alcohol consumption in humans [122]. Isoprenoid-based fuel molecules have also illustrated inhibitory effects. Cyclic terpenes, such as pinene and limonene (Figure 2), inhibited the growth of bacteria and *S. cerevisiae* [123, 124], while branched isoprenoids, such as farnesyl hexanoate and geranyl acetate, were shown to be toxic to *E. coli* [125]. In fact, *E. coli*'s tolerance to isoprenoid-derived biodiesels and bioaviation fuels only ranged from 0.025 – 1% (v/v) [125]. Based on these previous studies, product toxicity is a major limiting factor and should be integrated into the metabolic engineering strategy.

A variety of strategies can be adopted to address product toxicity. The easiest way to avoid complications from product toxicity is to select non-toxic fuel targets. Toxicity studies can be conducted for each potential host organism, and generally, fatty alcohols longer than C14, alkanes longer than C9, and alkenes longer than C12 have shown minimal microbial inhibition [120, 121]. Alternatively, metabolic engineering techniques can be applied to allow for a more diverse range of hydrocarbon fuel targets. Many cellular modifications have been shown to improve microbial solvent tolerance: changes in membrane lipid composition; altered enzymatic activities of membrane repair and energy transduction enzymes; solvent expulsion via efflux pump activity; and cellular stress responses including heat shock, phage shock, and general stress responses [118, 125, 126]. These natural mechanisms offer a range of engineering targets: expression of a cis-trans isomerase to alter lipid composition; overexpression of enzymes involved in membrane repair and energy transduction; expression of efflux pumps such as *tolC*, *mar*, *rob*, *soxS*, and *acrAB*; and overexpression of stress-induced enzymes such as phage shock protein, heat shock proteins, catalases, and superoxide dismutases [125, 126]. While few metabolic engineering efforts have focused on enhancing product tolerance, a recent study explored improving hydrocarbon-based fuel tolerance in *E. coli* by testing a library of 43 efflux pumps [127]. This work identified efflux pumps that improved tolerance to five potential isoprenoid derived fuels. This preliminary success at engineering solvent tolerance should inspire additional efforts to improve the microbial production of both fatty acid and isoprenoid derived fuels.

In addition to product tolerance, other host traits are desirable for industrial biofuel production, particularly for autotrophic microorganisms. As discussed in the previous section, light availability is often a growth limiting factor in microalgal cultures. Microalgae construct light harvesting complexes (LHC) to capture the available light for use in photosynthesis, and natural species actually absorb more light than is needed for photosynthesis under light intensities > 400 μmol photons m^{-2} s^{-1} [128]. As the sun can generate light intensities as high as 2,000 μmol photons m^{-2} s^{-1} during peak hours, it is estimated that as much as 80% of light absorbed by microalgae is 'wasted' as re-emitted fluorescence and heat [129]. In addition to this loss of energy, the excess energy can also cause cellular damage, known as photoinhibition [128]. In nature, this over-absorption of light will give the microalga a competitive advantage, but from a biofuel production perspective, this excess light harvesting will lead to lower culture cell densities and therefore lower biofuel productivities. Thus, there have been many attempts to engineer microalgae to absorb only the amount of light needed for photosynthesis. These efforts target genes of the light harvesting antenna complexes. Most LHC mutants were generated using random mutagenesis techniques including chemical, UV, and transposon mutagenesis [128, 130-134]. Many of these studies focus on the model alga *C. reinhardtii*, but other microalgal species, such as the diatom *Cyclotella* sp. and the cyanobacterium *Synechocystis* sp., have been mutated to reduce the size of their photosynthetic antennae [130, 133]. Several recent works have applied RNAi technology in *C. reinhardtii* to reduce the expression of targeted LHC genes in a more controlled manner [129]. In general, the antenna mutants have shown improved photosynthetic quantum yields, reduced photoinhibition, enhanced productivity under high light conditions, and increased light penetration within the culture [128, 129, 131-134]. While these results are promising, several questions remain to be addressed: Are the photosynthetic antenna mutants genetically stable, or will they revert back to their more competitive and less efficient forms over time? And are these mutants less fit and therefore more susceptible to predators and competitors in open pond systems?

Open pond systems are subject to a variety of changing environmental conditions, and as such, the optimal autotrophic host will have the necessary cellular mechanisms to adapt to these changing conditions. Desirable host traits may include temperature tolerance, salt tolerance, and resistance to predators. Open ponds are exposed to both daily and seasonal temperature fluctuations which often exceed the normal temperature ranges for optimal cell growth and may even cause cell death. Engineering efforts have successfully altered the temperature tolerance of cyanobacteria though either gene knockout or heterologous overexpression of desaturases which influence the viscosity of both the cell and photosynthetic membranes [135]. Alternatively, microalgae with different temperature optima can be rotated seasonally in the open ponds, similar to seasonal crop rotations in agricultural practices. As mentioned previously, open pond systems are complicated by evaporative water loss, particularly for the sunny, arid regions that are ideal for microalgal biofuel production. Evaporation can lead to fluctuations in the salt concentration within the pond culture, and many have proposed to utilize marine or brackish water sources to reduce the cost associated with freshwater systems. Moreover, high salt and saturated salt systems will have lower evaporative water loss compared to freshwater cultures. Naturally salt-tolerant microalgae, such as those isolated from marine or even hypersaline environments, may be selected as host for biofuel production, or efficient fuel-producing hosts can be engineered for increased salt tolerance. For example, the cyanobacterium *Synechococcus elongatus* PCC 7942, modified with expression of a Δ12 acyl-lipid desaturase (*desA*), showed improved resistance to salt and osmotic stress compared to the wildtype [136]. Lastly, pond crash due to microalgal predators like rotifers and chytrids is a major problem for open pond biofuel production systems. While there have not been any reported attempts at engineering predator-resistant microalgae, there have been reports of natural defense mechanisms such as palmelloid formation by *C. reinhardtii*, which produces non-motile cell aggregates that are simply too large to be consumed by grazing rotifers [137]. Once the genetic mechanism responsible for palmelloid formation is deciphered, it may be possible to transfer this resistance mechanism to other microalgae using genetic engineering techniques. When devising a metabolic engineering strategy for biofuel production, it is essential to consider the entire genomic landscape and the natural diversity of genetically-driven traits to design the optimal host for the specific industrial constraints.

5. Conclusions and future outlook

The microbial production of drop-in replacement fuels faces unprecedented challenges. The sheer quantity of hydrocarbon product required to meet the world's ever increasing demand for energy dwarfs the supply of any current microbially synthesized product. Moreover, both second (lignocellulosic feedstock) and third (microalgal feedstock) generation biofuels ultimately rely on sunlight and photosynthesis to supply the energy and carbon feedstocks necessary for production. This requires the development of new technology and infrastructure to facilitate the construction of this new supply chain. Finally, the low value of the final fuel product places additional financial restrictions on the development of large-scale biofuel production processes. For example, previous reports include the addition of exogenous metabolic precursors like mevalonate for isoprenoid production or FFA for FAEE biosynthesis [18, 50]. While these exogenous metabolites boost production of the desired hydrocarbon-based product, this practice is too expensive for large-scale biofuel applications. These challenges currently limit the industrial production of second and third generation biofuels.

Fortunately, new biological and technological tools are rapidly being developed and applied to overcome the obstacles in biofuel production. In addition to the metabolic engineering strategies previously described in this chapter, new global strategies are being applied to engineer microbes for biofuel production. With the affordability of next-generation DNA sequencing technologies, new microbial genomes are being reported at an unprecedented rate, and this information can be used to generate metabolic models for biofuel-producing hosts. In turn, these models can be leveraged to analyze proposed metabolic engineering strategies *in silico*, reducing the number of costly and time-intensive strain constructions and experiments. This technique was shown to be successful at increasing lycopene production, an isoprenoid derivative, in *E. coli* [69, 138]. The advancement of synthetic DNA technology enables new engineering approaches such as multiplex automated genome engineering (MAGE) [139]. In MAGE, synthetic oligomers, consisting of degenerate DNA sequences flanked by regions homologous to the target sequences, are simultaneously transformed into *E. coli,* and the modified strains are screened for improvements. MAGE was used to target ribosome binding sites, for optimization of protein translation, and to inactivate genes by inserting nonsense mutations; this technique can also be applied to target promoters for improved gene transcription and enzyme active sites for enhanced activities. The technique does have some limitations, however. MAGE will likely require modification of the host organism to allow for efficient integration of the single-stranded oligonucleotides, and a high-throughput screening method is essential for screening the billions of genetic variants that are generated with MAGE. Global or systems-level technologies can also be applied to advance our fundamental understanding of genetic and regulatory mechanisms within a microbial host; this is vital to host development of non-model organisms and newly isolated strains. Omics technologies including genomics, transcriptomics, metabolomics, and proteomics provide global insight at the cellular level, which can be compared across different conditions or time points to identify the native mechanisms that control the cell metabolism. Integration of omics data can identify bottlenecks at the transcriptional, translational, and protein levels, and as such, can be applied to inform the metabolic engineering strategy for biofuel production [34]. Systems-level tools for engineering microbial hosts, including metabolic modeling, MAGE, and omics technologies, will be integral to the successful development of hosts for biofuel production.

Commercial interest in the production of second and third generation biofuels has developed rapidly in the past decade. As evidence of this, there has been a flurry of activity in patent applications regarding microbial hydrocarbon production. Companies invested in heterotrophic hydrocarbon-based fuel production include LS9 [27, 59, 65, 66, 140, 141] and Amyris Biotechnologies [72, 142], which focus mainly on *E. coli* as the host, and Solazyme [143, 144], which initially focused on fuels derived from algae but has since moved toward more high-value markets, such as cosmetics and nutraceuticals. Most companies interested in algae and cyanobacteria are focused on autotrophically-produced hydrocarbon fuels. Notable companies in this industry include Sapphire Energy [145, 146], Joule Unlimited [26, 77, 147], and Synthetic Genomics [68, 75]. The hydrocarbon-based fuels targeted by these companies span the entire gamut of fatty acid and isoprenoid derived fuel products. Despite this commercial interest, hydrocarbon biofuel production still remains to be demonstrated at scale and in a sustainable manner.

This chapter has described the challenges in microbial hydrocarbon production and presented metabolic engineering strategies to resolve these issues. As is evident from this discussion, microbial-based fuel production is only in the initial stages of exploration, and additional research and innovation is necessary to enable large-scale biofuel production. New metabolic engineering tools and techniques are currently being developed for engineering untraditional hosts like eukaryotic algae and cyanobacteria, and as our understanding of these new hosts matures, significant improvement in hydrocarbon yields is anticipated.

Abbreviations

1,3-BPG	1,3-bisphosphoglycerate	GGPP	geranylgeranyl pyrophosphate
3-PGA	3-phosphoglycerate	Glc	glucose
AAR	acyl-ACP reductase	Gly	glycerol
AAS	acyl-ACP synthetase	GPD	glycerol-3-phosphate dehydrogease
ACC	acetyl-CoA carboxylase	GPP	geranyl pyrophosphate
ACP	acyl carrier protein	HCO_3^-	bicarbonate
ACS	acetyl-CoA synthetase	HMG-CoA	3-hydroxy-3-methyl- glutaryl-CoA
ADC	aldehyde decarbonylase	HMGCR	HMG-CoA reductase
ADH	alcohol dehydrogenase	IPP	isopentenyl Pyrophosphate
ADP	adenosine diphosphate	IPPI	isopentenyl diphosphate isomerase
AH	aldehyde	*ispS*	isoprene synthase
ALDH	acetaldehyde dehydrogenase	KASIII	β-ketoacyl-ACP synthase
ALR	aldehyde reductase	LHC	light harvesting complex
AMP	adenosine monophosphate	L-Ru5P	L-ribulose-5-phosphate
AOL	arabitol	L-Xu5P	L-xylulose-5-phosphate
ARA	arabinose	L-Xul	L-xylulose
ASP	aquatic species program	MEP	methylerythritol phosphate
AT	acyltransferase	MVA	mevalonate
ATP	adenosine triphosphate	NAD^+	nicotinamide adenine dinucleotide (oxidized)
cAMP	cyclic AMP	NADH	nicotinamide adenine dinucleotide (reduced)
CCR	carbon catabolite repression	$NADP^+$	nicotinamide adenine dinucleotide phosphate (oxidized)
CMP	cytosine monophosphate	NADPH	nicotinamide adenine dinucleotide phosphate (reduced)
CO_2	carbon dioxide	PBR	photobioreactor
CoA	coenzyme A	PDC	pyruvate decarboxylase
CRP	cyclic AMP receptor protein	PEP	phosphoenolpyruvate
CTP	cytosine triphosphate	P_i	phosphate
desA	Δ12 acyl-lipid desaturase	PP_i	pyrophosphate
DGAT	diacylglycerol acyltransferase	PPP	pentose phosphate pathway
DHAP	dihydroxyacetone phosphate	PPS	phosphoenolpyruvate synthase
DMAPP	dimethylallyl diphosphate	PTS	phosphotransferase system

D-Ru5P	D-ribulose-5-phosphate	PYR	pyruvate
DXP	1-deoxy-D-xylulose-5- phosphate	R5P	ribose-5-phosphate
DXR	1-deoxy-D-xylulose-5- phosphate reductoisomerase	RBU	ribulose
DXS	1-deoxy-D-xylulose-5- phosphate synthase	RNAi	ribonucleic acid interference
D-Xu5P	D-xylulose-5-phosphate	RuBP	ribulose-1,5-bisphosphate
D-Xul	D-xylulose	S7P	sedoheptulose-7- phosphate
E4P	erythrose-4-phosphate	SBP	sedoheptulose-1,7- bisphosphate
EMP	Embden-Meyerhof-Parnas	TAG	triacylglycerol
F6P	fructose-6-phosphate	TCA	tricarboxylic acid
FAEE	fatty acid ethyl ester	TE	thioesterase
FAR	fatty acyl-CoA reductase	XDH	xylitol dehydrogenase
FBP	fructose-1,6-bisphosphate	XI	xylose isomerase
FFA	free fatty acid	XK	xylulose kinase
FPP	farnesyl pyrophosphate	Xol	xylitol
G3P	glycerol-3-phosphate	XR	xylose reductase
G6P	glucose-6-phosphate	Xu5P	xylulose-5-phosphate
GAP	glyceraldehyde-3- phosphate	Xyl	xylose

Acknowledgements

This work was supported by the Harry S. Truman Fellowship in National Security Science and Engineering and the Laboratory Directed Research and Development program at Sandia National Laboratories. Sandia National Laboratories is a multi-program laboratory managed and operated by Sandia Corporation, a wholly owned subsidiary of Lockheed Martin Corporation, for the U.S. Department of Energy's National Nuclear Security Administration under contract DE-AC04-94AL85000.

Author details

Anne M. Ruffing

Sandia National Laboratories, Department of Bioenergy and Defense Technologies, Albuquerque, NM, USA

References

[1] Williams JL. (2011) Oil Price History and Analysis. WTRG Economics. Available: http://www.wtrg.com/prices.htm. Accessed 2012 April 16.

[2] Sawin JL, Martinot E, Barnes D, McCrone A, Roussell J, Sims R, et al. (2011) Renewables 2011 Global Status Report. Renewable Energy Policy Network for the 21st Century.

[3] Rittmann BE. (2008) Opportunities for Renewable Bioenergy Using Microorganisms. Biotechnol. Bioeng. 100(2):203-212.

[4] Christi Y. (2008) Biodiesel from Microalgae Beats Bioethanol. Trends Biotechnol. 26(3): 126-131.

[5] Davis MS, Solbiati J, Cronan JE. (2000) Overproduction of Acetyl-CoA Carboxylase Activity Increases the Rate of Fatty Acid Biosynthesis in *Escherichia coli*. J. Biol. Chem. 275(37):28593-28598.

[6] Lee S, Jeon E, Yun H, Lee J. (2011) Improvement of Fatty Acid Biosynthesis by Engineered Recombinant *Escherichia coli*. Biotech. Biopro. Eng. 16(4):706-713.

[7] Ramos MJ, Fernández CM, Casas A, Rodríguez L, Pérez Á. (2009) Influence of Fatty Acid Composition of Raw Materials on Biodiesel Properties. Bioresour. Technol. 100(1): 261-268.

[8] Dehesh K, Jones A, Knutzon DS, Voelker TA. (1996) Production of High Levels of 8:0 and 10:0 Fatty Acids in Transgenic Canola by Overexpression of Ch FatB2, a Thioesterase cDNA from *Cuphea hookeriana*. Plant J. 9(2):167-172.

[9] Jiang P, Cronan JE. (1994) Inhibition of Fatty Acid Synthesis in *Escherichia coli* in the Absence of Phospholipid Synthesis and Release of Inhibition by Thioesterase Action. J. Bacteriol. 176(10):2814-2821.

[10] Voelker TA, Davies HM. (1994) Alteration of the Specificity and Regulation of Fatty Acid Synthesis of *Escherichia coli* by Expression of a Plant Medium-Chain Acyl-Acyl Carrier Protein Thioesterase. J. Bacteriol. 176(23):7320-7327.

[11] Lotero E, Liu Y, Lopez DE, Suwannakarn K, Bruce DA, Goodwin JG. (2005) Synthesis of Biodiesel via Acid Catalysis. Ind. Eng. Chem. Res. 44(14):5353-5363.

[12] Liu T, Khosla C. (2010) Genetic Engineering of *Escherichia coli* for Biofuel Production. Annu. Rev. Genet. 44(1):53-69.

[13] Dellomonaco C, Clomburg JM, Miller EN, Gonzalez R. (2011) Engineered Reversal of the β-Oxidation Cycle for the Synthesis of Fuels and Chemicals. Nature. 476(7360): 355-359.

[14] Jako C, Kumar A, Wei Y, Zou J, Barton DL, Giblin EM, et al. (2001) Seed-Specific Over-Expression of an *Arabidopsis* cDNA Encoding a Diacylglycerol Acyltransferase Enhances Seed Oil Content and Seed Weight. Plant Physiol. 126(2):861-874.

[15] Dunahay T, Jarvis E, Dais S, Roessler P. (1996) Manipulation of Microalgal Lipid Production Using Genetic Engineering. Appl. Biochem. Biotechnol. 57-58(1):223-231.

[16] Nguyen HTT, Dieterich A, Athenstaedt K, Truong NH, Stahl U, Nevoigt E. (2004) Engineering of *Saccharomyces cerevisiae* for the Production of L-Glycerol 3-Phosphate. Metab. Eng. 6(2):155-163.

[17] Kalscheuer R, Luftmann H, Steinbüchel A. (2004) Synthesis of Novel Lipids in *Saccharomyces cerevisiae* by Heterologous Expression of an Unspecific Bacterial Acyl-transferase. Appl. Environ. Microbiol. 70(12):7119-7125.

[18] Kalscheuer R, Stölting T, Steinbüchel A. (2006) Microdiesel: *Escherichia coli* Engineered for Fuel Production. Microbiol. 152(9):2529-2536.

[19] Steen EJ, Kang Y, Bokinsky G, Hu Z, Schirmer A, McClure A, et al. (2010) Microbial Production of Fatty-Acid-Derived Fuels and Chemicals from Plant Biomass. Nature. 463(7280):559-562.

[20] Ohta K, Beall DS, Mejia JP, Shanmugam KT, Ingram LO. (1991) Genetic Improvement of *Escherichia coli* for Ethanol Production: Chromosomal Integration of *Zymomonas mobilis* Genes Encoding Pyruvate Decarboxylase and Alcohol Dehydrogenase II. Appl. Environ. Microbiol. 57(4):893-900.

[21] Hofvander P, Doan TTP, Hamberg M. (2011) A Prokaryotic Acyl-CoA Reductase Performing Reduction of Fatty Acyl-CoA to Fatty Alcohol. FEBS Lett. 585(22): 3538-3543.

[22] Doan TTP, Carlsson AS, Hamberg M, Bülow L, Stymne S, Olsson P. (2009) Functional Expression of Five *Arabidopsis* Fatty Acyl-CoA Reductase Genes in *Escherichia coli.* J. Plant Physiol. 166(8):787-796.

[23] Tan X, Yao L, Gao Q, Wang W, Qi F, Lu X. (2011) Photosynthesis Driven Conversion of Carbon Dioxide to Fatty Alcohols and Hydrocarbons in Cyanobacteria. Metab. Eng. 13(2):169-176.

[24] Willis RM, Wahlen BD, Seefeldt LC, Barney BM. (2011) Characterization of a Fatty Acyl-CoA Reductase from *Marinobacter aquaeolei* VT8: A Bacterial Enzyme Catalyzing the Reduction of Fatty Acyl-CoA to Fatty Alcohol. Biochem. 50(48):10550-10558.

[25] Schirmer A, Rude MA, Li X, Popova E, del Cardayre SB. (2010) Microbial Biosynthesis of Alkanes. Science. 329(5991):559-562.

[26] Reppas NB, Ridley CP, inventors; Joule Unlimited, Inc., assignee. (2010) Methods and Compositions for the Recombinant Biosynthesis of *n*-Alkanes. patent US 7794969.

[27] Schirmer A, Rude MA, Brubaker S, inventors; LS9, Inc., assignee. (2010) Methods and Compositions for Producing Hydrocarbons. patent US 2010/0249470.

[28] Lee SK, Chou H, Ham TS, Lee TS, Keasling JD. (2008) Metabolic Engineering of Microorganisms for Biofuels Production: From Bugs to Synthetic Biology to Fuels. Curr. Opin. Biotechnol. 19(6):556-563.

[29] Chandran SS, Kealey JT, Reeves CD. (2011) Microbial Production of Isoprenoids. Process Biochem. 46(9):1703-1710.

[30] Fortman JL, Chhabra S, Mukhopadhyay A, Chou H, Lee TS, Steen E, et al. (2008) Biofuel Alternatives to Ethanol: Pumping the Microbial Well. Trends Biotechnol. 26(7):375-381.

[31] Rude MA, Schirmer A. (2009) New Microbial Fuels: A Biotech Perspective. Curr. Opin. Microbiol. 12(3):274-281.

[32] Rilling H, K B. (1959) On the Mechanism of Sqalene Biogenesis from Mevalonic Acid. J. Biol. Chem. 234(6):1424-1432.

[33] Rohmer M, Knani M, Simonin P, Sutter B, Sahm H. (1993) Isoprenoid Biosynthesis in Bacteria: A Novel Pathway for the Early Steps Leading to Isopentenyl Diphosphate. Biochem. J. 295:517-524.

[34] Muntendam R, Melillo E, Ryden A, Kayser O. (2009) Perspectives and Limits of Engineering the Isoprenoid Metabolism in Heterologous Hosts. Appl. Microbiol. Biotechnol. 84(6):1003-1019.

[35] Yoon S-H, Lee S-H, Das A, Ryu H-K, Jang H-J, Kim J-Y, et al. (2009) Combinatorial Expression of Bacterial Whole Mevalonate Pathway for the Production of β-Carotene in *E. coli*. J. Biotechnol. 140(3–4):218-226.

[36] Pitera DJ, Paddon CJ, Newman JD, Keasling JD. (2007) Balancing a Heterologous Mevalonate Pathway for Improved Isoprenoid Production in *Escherichia coli*. Metab. Eng. 9(2):193-207.

[37] Asadollahi MA, Maury J, Schalk M, Clark A, Nielsen J. (2010) Enhancement of Farnesyl Diphosphate Pool as Direct Precursor of Sesquiterpenes Through Metabolic Engineering of the Mevalonate Pathway in *Saccharomyces cerevisiae*. Biotechnol. Bioeng. 106(1): 86-96.

[38] Ohto C, Muramatsu M, Obata S, Sakuradani E, Shimizu S. (2009) Overexpression of the Gene Encoding HMG-CoA Reductase in *Saccharomyces cerevisiae* for Production of Prenyl Alcohols. Appl. Microbiol. Biotechnol. 82(5):837-845.

[39] Kim S-W, Keasling JD. (2001) Metabolic Engineering of the Nonmevalonate Isopentenyl Diphosphate Synthesis Pathway in *Escherichia coli* Enhances Lycopene Production. Biotechnol. Bioeng. 72(4):408-415.

[40] Matthews PD, Wurtzel ET. (2000) Metabolic Engineering of Carotenoid Accumulation in *Escherichia coli* by Modulation of the Isoprenoid Precursor Pool with Expression of Deoxyxylulose Phosphate Synthase. Appl. Microbiol. Biotechnol. 53(4):396-400.

[41] Zhao Y, Yang J, Qin B, Li Y, Sun Y, Su S, et al. (2011) Biosynthesis of Isoprene in *Escherichia coli* via Methylerythritol Phosphate (MEP) Pathway. Appl. Microbiol. Biotechnol. 90(6):1915-1922.

[42] Kajiwara S, Fraser P, Kondo K, Misawa N. (1997) Expression of an Exogenous Isopentenyl Diphosphate Isomerase Gene Enhances Isoprenoid Biosynthesis in *Escherichia coli*. Biochem. J. 324:421-426.

[43] Ghimire GP, Lee HC, Sohng JK. (2009) Improved Squalene Production via Modulation of the Methylerythritol 4-Phosphate Pathway and Heterologous Expression of Genes from *Streptomyces peucetius* ATCC 27952 in *Escherichia coli*. Appl. Environ. Microbiol. 75(22):7291-7293.

[44] Morrone D, Lowry L, Determan M, Hershey D, Xu M, Peters R. (2010) Increasing Diterpene Yield with a Modular Metabolic Engineering System in *E. coli* Comparison of MEV and MEP Isoprenoid Precursor Pathway Engineering. Appl. Microbiol. Biotechnol. 85(6):1893-1906.

[45] Leonard E, Ajikumar PK, Thayer K, Xiao W-H, Mo JD, Tidor B, et al. (2010) Combining Metabolic and Protein Engineering of a Terpenoid Biosynthetic Pathway for Overproduction and Selectivity Control. Proceedings of the National Academy of Sciences.

[46] Martin VJJ, Pitera DJ, Withers ST, Newman JD, Keasling JD. (2003) Engineering a Mevalonate Pathway in *Escherichia coli* for Production of Terpenoids. Nat Biotech. 21(7): 796-802.

[47] Wang C, Yoon S-H, Jang H-J, Chung Y-R, Kim J-Y, Choi E-S, et al. (2011) Metabolic Engineering of *Escherichia coli* for α-Farnesene Production. Metab. Eng. 13(6):648-655.

[48] Wang C, Yoon S-H, Shah AA, Chung Y-R, Kim J-Y, Choi E-S, et al. (2010) Farnesol Production from *Escherichia coli* by Harnessing the Exogenous Mevalonate Pathway. Biotechnol. Bioeng. 107(3):421-429.

[49] Yang J, Zhao G, Sun Y, Zheng Y, Jiang X, Liu W, et al. (2012) Bio-Isoprene Production Using Exogenous MVA Pathway and Isoprene Synthase in *Escherichia coli*. Bioresour. Technol. 104(0):642-647.

[50] Yoon S-H, Lee Y-M, Kim J-E, Lee S-H, Lee J-H, Kim J-Y, et al. (2006) Enhanced Lycopene Production in *Escherichia coli* Engineered to Synthesize Isopentenyl Diphosphate and Dimethylallyl Diphosphate from Mevalonate. Biotechnol. Bioeng. 94(6):1025-1032.

[51] Farmer WR, Liao JC. (2000) Improving Lycopene Production in *Escherichia coli* by Engineering Metabolic Control. Nat Biotech. 18(5):533-537.

[52] Shiba Y, Paradise EM, Kirby J, Ro D-K, Keasling JD. (2007) Engineering of the Pyruvate Dehydrogenase Bypass in *Saccharomyces cerevisiae* for High-Level Production of Isoprenoids. Metab. Eng. 9(2):160-168.

[53] Farmer WR, Liao JC. (2001) Precursor Balancing for Metabolic Engineering of Lycopene Production in *Escherichia coli*. Biotechnol. Prog. 17(1):57-61.

[54] Asadollahi MA, Maury J, Patil KR, Schalk M, Clark A, Nielsen J. (2009) Enhancing Sesquiterpene Production in *Saccharomyces cerevisiae* Through in Silico Driven Metabolic Engineering. Metab. Eng. 11(6):328-334.

[55] Xue J, Ahring BK. (2011) Enhancing Isoprene Production by Genetic Modification of the 1-Deoxy-d-Xylulose-5-Phosphate Pathway in *Bacillus subtilis*. Appl. Environ. Microbiol. 77(7):2399-2405.

[56] Li Y, Han D, Hu G, Dauvillee D, Sommerfeld M, Ball S, et al. (2010) *Chlamydomonas* Starchless Mutant Defective in ADP-Glucose Pyrophosphorylase Hyper-Accumulates Triacylglycerol. Metab. Eng. 12(4):387-391.

[57] Li Y, Han D, Hu G, Sommerfeld M, Hu Q. (2010) Inhibition of Starch Synthesis Results in Overproduction of Lipids in *Chlamydomonas reinhardtii*. Biotechnol. Bioeng. 107(2): 258-268.

[58] Work VH, Radakovits R, Jinkerson RE, Meuser JE, Elliott LG, Vinyard DJ, et al. (2010) Increased Lipid Accumulation in the *Chlamydomonas reinhardtii* sta7-10 Starchless Isoamylase Mutant and Increased Carbohydrate Synthesis in Complemented Strains. Eukaryot. Cell. 9(8):1251-1261.

[59] Hu Z, Valle F, inventors; LS9, Inc, assignee. (2011) Enhanced Production of Fatty Acid Derivatives. patent US 2011/0256599.

[60] Lu X, Vora H, Khosla C. (2008) Overproduction of Free Fatty Acids in *E. coli*: Implications for Biodiesel Production. Metab. Eng. 10(6):333-339.

[61] Michinaka Y, Shimauchi T, Aki T, Nakajima T, Kawamoto S, Shigeta S, et al. (2003) Extracellular Secretion of Free Fatty Acids by Disruption of a Fatty Acyl-CoA Synthetase Gene in *Saccharomyces cerevisiae*. J. Biosci. Bioeng. 95(5):435-440.

[62] Scharnewski M, Pongdontri P, Mora G, Hoppert M, Fulda M. (2008) Mutants of *Saccharomyces cerevisiae* Deficient in Acyl-CoA Synthetases Secrete Fatty Acids due to Interrupted Fatty Acid Recycling. FEBS J. 275(11):2765-2778.

[63] Kamisaka Y, Tomita N, Kimura K, Kainou K, Uemura H. (2007) *DGA1* (Diacylglycerol Acyltransferase Gene) Overexpression and Leucine Biosynthesis Significantly Increase Lipid Accumulation in the Δ*snf2* disruptant of *Saccharomyces cerevisiae*. Biochem. J. 408:61-68.

[64] Nojima Y, Kibayashi A, Matsuzaki H, Hatano T, Fukui S. (1999) Isolation and Characterization of Triacylglycerol-Secreting Mutant Strain from Yeast, *Saccharomyces cerevisiae*. The Journal of General and Applied Microbiology. 45(1):1-6.

[65] del Cardayre SB, Milner Cockrem MC, inventors; LS9, Inc., assignee. (2009) Systems and Methods for the Production of Fatty Esters. patent WO/2009/009391.

[66] Keasling JD, Hu Z, Somerville C, Church G, Berry D, Friedman L, et al., inventors; LS9, Inc., assignee. (2007) Production of Fatty Acids and Derivatives Thereof. patent WO/2007/136762.

[67] Zhang F, Carothers JM, Keasling JD. (2012) Design of a Dynamic Sensor-Regulator System for Production of Chemicals and Fuels Derived from Fatty Acids. Nat Biotech. 30(4):354-359.

[68] Roessler PG, Watts K, Liu B, inventors; Synthetic Genomics, Inc., assignee. (2011) Microbial Production of Fatty Alcohols. patent US 2011/0195469.

[69] Alper H, Miyaoku K, Stephanopoulos G. (2005) Construction of Lycopene-Overproducing *E. coli* Strains by Combining Systematic and Combinatorial Gene Knockout Targets. Nat Biotech. 23(5):612-616.

[70] Muramatsu M, Ohto C, Obata S, Sakuradani E, Shimizu S. (2008) Various Oils and Detergents Enhance the Microbial Production of Farnesol and Related Prenyl Alcohols. J. Biosci. Bioeng. 106(3):263-267.

[71] Takahashi S, Yeo Y, Greenhagen BT, McMullin T, Song L, Maurina-Brunker J, et al. (2007) Metabolic Engineering of Sesquiterpene Metabolism in Yeast. Biotechnol. Bioeng. 97(1):170-181.

[72] Renninger NS, McPhee DJ, inventors; Amyris Biotechnologies, Inc., assignee. (2008) Fuel Compositions Comprising Farnesane and Farnesane Derivatives and Methods of Making and Using Same. patent WO/2008/045555.

[73] Kaczmarzyk D, Fulda M. (2010) Fatty Acid Activation in Cyanobacteria Mediated by Acyl-Acyl Carrier Protein Synthetase Enables Fatty Acid Recycling. Plant Physiol. 152(3):1598-1610.

[74] Liu X, Sheng J, Curtiss III R. (2011) Fatty Acid Production in Genetically Modified Cyanobacteria. Proceedings of the National Academy of Sciences. Epub April 11, 2011.

[75] Roessler PG, Chen Y, Liu B, Dodge CN, inventors; Synthetic Genomics, Inc., assignee. (2009) Secretion of Fatty Acids by Photosynthetic Microorganisms. patent US 2009/0298143.

[76] Ruffing AM, Jones HDT. (2012) Physiological Effects of Free Fatty Acid Production in Genetically Engineered *Synechococcus elongatus* PCC 7942. Biotechnol. Bioeng. 109(9): 2190-2199.

[77] Berry DA, Afeyan NB, Skraly FA, Ridley CP, Robertson DE, Wilpiszeski R, et al., inventors; Joule Unlimited, Inc., assignee. (2011) Methods and Compositions for the Recombinant Biosynthesis of Fatty Acids and Esters. patent US 2011/0111470.

[78] Lindberg P, Park S, Melis A. (2010) Engineering a Platform for Photosynthetic Isoprene Production in Cyanobacteria, Using *Synechocystis* as the Model Organism. Metab. Eng. 12(1):70-79.

[79] Hahn-Hägerdal B. (1996) Ethanolic Fermentation of Lignocellulose Hydrolysates. Appl. Biochem. Biotechnol. 57-58(1):195-199.

[80] Dellomonaco C, Fava F, Gonzalez R. (2010) The Path to Next Generation Biofuels: Successes and Challenges in the Era of Synthetic Biology. Microb. Cell Fact. 9(1):3.

[81] Jin Y, Lee T, Choi Y, Ryu Y, Seo J. (2000) Conversion of Xylose to Ethanol by Recombinant *Saccharomyces cerevisiae* Containing Genes for Xylose Reductase and Xylitol Dehydrogenase from *Pichia stipitis*. J. Microbiol. Biotech. 10(4):564-567.

[82] van Maris A, Winkler A, Kuyper M, de Laat W, van Dijken J, Pronk J. (2007) Development of Efficient Xylose Fermentation in *Saccharomyces cerevisiae:* Xylose Isomerase as a Key Component Biofuels. Adv Biochem Eng Biotechnol. 108:179-204.

[83] Walfridsson M, Anderlund M, Bao X, Hahn-Hägerdal B. (1997) Expression of Different Levels of Enzymes from the *Pichia stipitis* XYL1 and XYL2 Genes in *Saccharomyces cerevisiae* and its Effects on Product Formation During Xylose Utilisation. Appl. Microbiol. Biotechnol. 48(2):218-224.

[84] Kuyper M, Hartog MMP, Toirkens MJ, Almering MJH, Winkler AA, van Dijken JP, et al. (2005) Metabolic Engineering of a Xylose-Isomerase-Expressing *Saccharomyces cerevisiae* Strain for Rapid Anaerobic Xylose Fermentation. FEMS Yeast Res. 5(4-5): 399-409.

[85] Walfridsson M, Hallborn J, Penttilä M, Keränen S, Hahn-Hägerdal B. (1995) Xylose-Metabolizing *Saccharomyces cerevisiae* Strains Overexpressing the TKL1 and TAL1 Genes Encoding the Pentose Phosphate Pathway Enzymes Transketolase and Transaldolase. Appl. Environ. Microbiol. 61(12):4184-4190.

[86] Becker J, Boles E. (2003) A Modified *Saccharomyces cerevisiae* Strain That Consumes l-Arabinose and Produces Ethanol. Appl. Environ. Microbiol. 69(7):4144-4150.

[87] Richard P, Verho R, Putkonen M, Londesborough J, Penttilä M. (2003) Production of Ethanol from L-Arabinose by *Saccharomyces cerevisiae* Containing a Fungal L-Arabinose Pathway. FEMS Yeast Res. 3(2):185-189.

[88] Wisselink HW, Toirkens MJ, del Rosario Franco Berriel M, Winkler AA, van Dijken JP, Pronk JT, et al. (2007) Engineering of *Saccharomyces cerevisiae* for Efficient Anaerobic Alcoholic Fermentation of L-Arabinose. Appl. Environ. Microbiol. 73(15):4881-4891.

[89] Stülke J, Hillen W. (1999) Carbon Catabolite Repression in Bacteria. Curr. Opin. Microbiol. 2(2):195-201.

[90] Nichols NN, Dien BS, Bothast RJ. (2001) Use of Catabolite Repression Mutants for Fermentation of Sugar Mixtures to Ethanol. Appl. Microbiol. Biotechnol. 56(1):120-125.

[91] Hernández-Montalvo VH-M, Valle FV, Bolivar FB, Gosset GG. (2001) Characterization of Sugar Mixtures Utilization by an *Escherichia coli* Mutant Devoid of the Phosphotransferase System. Appl. Microbiol. Biotechnol. 57(1):186-191.

[92] Yomano L, York S, Shanmugam K, Ingram L. (2009) Deletion of Methylglyoxal Synthase Gene (*mgsA*) Increased Sugar Co-Metabolism in Ethanol-Producing *Escherichia coli*. Biotechnol. Lett. 31(9):1389-1398.

[93] da Silva GP, Mack M, Contiero J. (2009) Glycerol: A Promising and Abundant Carbon Source for Industrial Microbiology. Biotechnol. Adv. 27(1):30-39.

[94] Houghton J, Weatherwax S, Ferrell J. (2006) Breaking the Biological Barriers to Cellulosic Ethanol. Rockville, Maryland: Office of Energy Efficiency and Renewable Energy, Office of the Biomass Program, U.S. Department of Energy.

[95] NSF. (2008) Breaking the Chemical and Engineering Barriers to Lignocellulosic Biofuels: Next Generation Hydrocarbon Biorefineries. Washington, D.C.: University of Massachusetts Amerst. National Science Foundation. Chemical, Bioengineering, Environmental, and Transport Systems Division.

[96] Sheehan J, Dunahay T, Benemann J, Roessler P. (1998) A Look Back at the U.S. Department of Energy's Aquatic Species Program - Biodiesel from Algae. National Renewable Energy Laboratory.

[97] Deng M-D, Coleman JR. (1999) Ethanol Synthesis by Genetic Engineering in Cyanobacteria. Appl. Environ. Microbiol. 65(2):523-528.

[98] Falkowski PG, Katz ME, Knoll AH, Quigg A, Raven JA, Schofield O, et al. (2004) The Evolution of Modern Eukaryotic Phytoplankton. Science. 305(5682):354-360.

[99] Seckbach J, Oren A, Pattanaik B, Reisser W, Klaveness D, Lovhoiden F, et al. (2007) Algae and Cyanobacteria in Extreme Environments. Seckbach J, editor. The Netherlands: Springer.

[100] Singh J, Gu S. (2010) Commercialization Potential of Microalgae for Biofuels Production. Renew. Sustain. Ener. Rev. 14(9):2596-2610.

[101] León R, Fernández E. (2007) Nuclear Transformation of Eukaryotic Microalgae: Historical Overview, Achievements and Problems. In: León R, Galván A, Fernández E, editors.: Springer New York; p. 1-11.

[102] León-Bañares R, González-Ballester D, Galván A, Fernández E. (2004) Transgenic Microalgae as Green Cell-Factories. Trends Biotechnol. 22(1):45-52.

[103] Wang ZT, Ullrich N, Joo S, Waffenschmidt S, Goodenough U. (2009) Algal Lipid Bodies: Stress Induction, Purification, and Biochemical Characterization in Wild-Type and Starchless *Chlamydomonas reinhardtii*. Eukaryot. Cell. 8(12):1856-1868.

[104] Radakovits R, Eduafo PM, Posewitz MC. (2011) Genetic Engineering of Fatty Acid Chain Length in *Phaeodactylum tricornutum*. Metab. Eng. 13(1):89-95.

[105] Yu W-L, Ansari W, Schoepp N, Hannon M, Mayfield S, Burkart M. (2011) Modifications of the Metabolic Pathways of Lipid and Triacylglycerol Production in Microalgae. Microb. Cell Fact. 10(1):91.

[106] Stevens DR, Purton S. (1997) Genetic Engineering of Eukaryotic Algae: Progress and Prospects. J. Phycol. 33(5):713-722.

[107] Rosenberg JN, Oyler GA, Wilkinson L, Betenbaugh MJ. (2008) A Green Light for Engineered Algae: Redirecting Metabolism to Fuel a Biotechnology Revolution. Curr. Opin. Biotechnol. 19(5):430-436.

[108] Radakovits R, Jinkerson RE, Darzins A, Posewitz MC. (2010) Genetic Engineering of Algae for Enhanced Biofuel Production. Eukaryot. Cell. 9(4):486-501.

[109] Cerutti H, Ma X, Msanne J, Repas T. (2011) RNA-Mediated Silencing in Algae: Biological Roles and Tools for Analysis of Gene Function. Eukaryot. Cell. 10(9):1164-1172.

[110] Kilian O, Benemann CSE, Niyogi KK, Vick B. (2011) High-Efficiency Homologous Recombination in the Oil-Producing Alga *Nannochloropsis* sp. Proceedings of the National Academy of Sciences. 108(52):21265-21269.

[111] (2012) *Chlamydomonas* Engineering Kits. Life Technologies Corporation. Available: https://www.invitrogen.com/site/us/en/home/Products-and-Services/Applications/Protein-Expression-and-Analysis/Protein-Expression/algae-engineering-kits/chlamydomonas-engineering-kits.html. Accessed 2012 April 16.

[112] Rasmussen B, Fletcher IR, Brocks JJ, Kilburn MR. (2008) Reassessing the First Appearance of Eukaryotes and Cyanobacteria. Nature. 455(7216):1101-1104.

[113] Koksharova OA, Wolk CP. (2002) Genetic Tools for Cyanobacteria. Appl. Microbiol. Biotechnol. 58(2):123-137.

[114] Ogbonna JC, Soejima T, Tanaka H. (1999) An Integrated Solar and Artificial Light System for Internal Illumination of Photobioreactors. J. Biotechnol. 70(1–3):289-297.

[115] Kunjapur AM, Eldridge RB. (2010) Photobioreactor Design for Commercial Biofuel Production from Microalgae. Ind. Eng. Chem. Res. 49(8):3516-3526.

[116] Bullis K. (2012) NASA Wants to Launch Floating Algae Farms. Technology Review.

[117] Chaumont D. (1993) Biotechnology of Algal Biomass Production: A Review of Systems for Outdoor Mass Culture. J. Appl. Phycol. 5(6):593-604.

[118] Sikkema J, de Bont JA, Poolman B. (1995) Mechanisms of Membrane Toxicity of Hydrocarbons. Microbiol. Rev. 59(2):201-222.

[119] Desbois A, Smith V. (2010) Antibacterial Free Fatty Acids: Activities, Mechanisms of Action and Biotechnological Potential. Appl. Microbiol. Biotechnol. 85(6):1629-1642.

[120] León R, Garbayo I, Hernández R, Vigara J, Vilchez C. (2001) Organic Solvent Toxicity in Photoautotrophic Unicellular Microorganisms. Enzyme Microb. Technol. 29(2–3): 173-180.

[121] Gill C, Ratledge C. (1972) Toxicity of *n*-Alkanes, *n*-Alk-1-enes, *n*-Alkan-1-ols and *n*-Alkyl-1-bromides Towards Yeasts. J. Gen. Microbiol. 72(1):165-172.

[122] Best CA, Laposata M. (2003) Fatty Acid Ethyl Esters: Toxic Non-Oxidative Metabolites of Ethanol and Markers of Ethanol Intake. Front. Biosci. 8:202-217.

[123] Andrews RE, Parks LW, Spence KD. (1980) Some Effects of Douglas Fir Terpenes on Certain Microorganisms. Appl. Environ. Microbiol. 40(2):301-304.

[124] Uribe S, Ramirez J, Peña A. (1985) Effects of β-Pinene on Yeast Membrane Functions. J. Bacteriol. 161(3):1195-1200.

[125] Dunlop MJ. (2011) Engineering Microbes for Tolerance to Next-Generation Biofuels. Biotech. Biofuels. 4(32).

[126] Sardessai Y, Bhosle S. (2002) Tolerance of Bacteria to Organic Solvents. Res. Microbiol. 153(5):263-268.

[127] Dunlop MJ, Dossani ZY, Szmidt HL, Chu HC, Lee TS, Keasling JD, et al. (2011) Engineering Microbial Biofuel Tolerance and Export Using Efflux Pumps. Mol Syst Biol. 7.

[128] Melis A. (2009) Solar Energy Conversion Efficiencies in Photosynthesis: Minimizing the Chlorophyll Antennae to Maximize Efficiency. Plant Sci. 177(4):272-280.

[129] Mussgnug JH, Thomas-Hall S, Rupprecht J, Foo A, Klassen V, McDowall A, et al. (2007) Engineering Photosynthetic Light Capture: Impacts on Improved Solar Energy to Biomass Conversion. Plant Biotechnol. J. 5(6):802-814.

[130] Huesemann M, Hausmann T, Bartha R, Aksoy M, Weissman J, Benemann J. (2009) Biomass Productivities in Wild Type and Pigment Mutant of *Cyclotella* sp. (Diatom). Appl. Biochem. Biotechnol. 157(3):507-526.

[131] Lee J, Mets L, Greenbaum E. (2002) Improvement of Photosynthetic CO_2 Fixation at High Light Intensity Through Reduction of Chlorophyll Antenna Size. Appl. Biochem. Biotechnol. 98-100(1):37-48.

[132] Nakajima Y, Tsuzuki M, Ueda R. (2001) Improved Productivity by Reduction of the Content of Light-Harvesting Pigment in *Chlamydomonas perigranulata*. J. Appl. Phycol. 13(2):95-101.

[133] Nakajima Y, Ueda R. (1997) Improvement of Photosynthesis in Dense Microalgal Suspension by Reduction of Light Harvesting Pigments. J. Appl. Phycol. 9(6):503-510.

[134] Polle JEW, Kanakagiri S-D, Melis A. (2003) *tla1*; a DNA Insertional Transformant of the Green Alga *Chlamydomonas reinhardtii* with a Truncated Light-Harvesting Chlorophyll Antenna Size. Planta. 217(1):49-59.

[135] Gombos Z, Murata N. (2004) Genetic Engineering of the Unsaturation of Membrane Glycerolipid: Effects on the Ability of the Photosynthetic Machinery to Tolerate Temperature Stress Lipids in Photosynthesis: Structure, Function and Genetics. In: Paul-André S, Norio M, editors.: Springer Netherlands; p. 249-262.

[136] Allakhverdiev SI, Kinoshita M, Inaba M, Suzuki I, Murata N. (2001) Unsaturated Fatty Acids in Membrane Lipids Protect the Photosynthetic Machinery against Salt-Induced Damage in *Synechococcus*. Plant Physiol. 125(4):1842-1853.

[137] Lurling M, Beekman W. (2006) Palmelloids Formation in *Chlamydomonas reinhardtii*: Defence Against Rotifer Predators? Ann. Limnol. - Int. J. Limnol. 42:65-72.

[138] Jin Y-S, Stephanopoulos G. (2007) Multi-Dimensional Gene Target Search for Improving Lycopene Biosynthesis in *Escherichia coli*. Metab. Eng. 9(4):337-347.

[139] Wang HH, Isaacs FJ, Carr PA, Sun ZZ, Xu G, Forest CR, et al. (2009) Programming Cells by Multiplex Genome Engineering and Accelerated Evolution. Nature. 460(7257): 894-898.

[140] Friedman L, Rude MA, inventors; LS9, Inc., assignee. (2008) Process for Producing Low Molecular Weight Hydrocarbons from Renewable Resources. patent WO/2008/113041.

[141] Friedman L, da Costa B, inventors; LS9, Inc., assignee. (2008) Hydrocarbon-Producing Genes and Methods of Their Use. patent WO/2008/147781.

[142] Renninger NS, Ryder JA, Fisher KJ, inventors; Amyris Biotechnologies, Inc., assignee. (2008) Jet Fuel Compositions and Methods of Making and Using Same. patent WO/ 2008/140492.

[143] Trimbur DE, Im C-S, Dillon HF, Day AG, Franklin S, Coragliotti A, inventors; Solazyme, Inc., assignee. (2009) Lipid Pathway Modification in Oil-Bearing Microorganisms. patent US 2009/0061493.

[144] Trimbur D, Im C-S, Dillon HF, Day AG, Franklin S, Coragliotti A, inventors; Solazyme, Inc., assignee. (2009) Use of Cellulosic Materials for Cultivation of Microorganisms. patent US 2009/0011480.

[145] Mendez M, Fang S-C, Richard S, inventors; Sapphire Energy, Inc., assignee. (2010) Engineering Salt Tolerance in Photosynthetic Microorganisms. patent WO 2010/105095.

[146] Heaps NA, Behnke CA, Molina D, inventors; Sapphire Energy, Inc., assignee. (2010) Biofuel Production in Prokaryotes and Eukaryotes. patent WO/2010/104763.

[147] Berry DA, Robertson DE, Skraly FA, Green BD, Ridley CP, Kosuri S, et al., inventors; Joule Unlimited, Inc., assignee. (2011) Engineered CO_2 Fixing Microorganisms Producing Carbon-Based Products of Interest. patent US 2011/0008861.

Chapter 9

Catalytic Hydroprocessing of Liquid Biomass for Biofuels Production

Stella Bezergianni

Additional information is available at the end of the chapter

http://dx.doi.org/10.5772/52649

1. Introduction

The depletion of world petroleum reserves and the increased concern on climate change has stimulated the recent interest in biofuels. The most common biofuels are based on energy crops and their products, i.e. vegetable oil for Fatty Acid Methyl Esters (FAME) biodiesel [1] and sugars/starch for bioethanol. However these first generation biofuels and associated production technologies face several considerations related to their economic and social implications regarding energy crops cultivation, by-products disposal, necessity for large investments to ensure competitiveness and the "food versus fuel" debate.

As a result, **second generation biofuel technologies** have been developed to overcome the limitations of first generation biofuels production [2]. The goal of second generation biofuel processes is to extend biofuel production capacity by incorporating residual biomass while increasing sustainability. This residual biomass consists of the non-food parts of food crops (such as stems, leaves and husks) as well as other non-food crops (such as switch grass, jatropha, miscanthus and cereals that bear little grain). Furthermore the residual biomass potential is further augmented by industrial and municipal organic waste such as skins and pulp from fruit pressing, waste cooking oil etc. One such technology is **catalytic hydroprocessing**, which is an alternative conversion technology of liquid biomass to biofuels that is lately raising a lot of interest in both the academic and industrial world and is the proposed subject of this chapter.

Catalytic hydroprocessingis a key process in petrochemical industry for over a century enabling heteroatom (sulfur, nitrogen, oxygen, metals) removal, saturation of olefins and aromatics, as well as isomerization and cracking [3]. Due to the numerous applications of catalytic hydroprocessing, there are several catalytic hydroprocessing units in a typical refinery including distillate hydrotreaters and hydrocrackers (see Figure 1). As a result several refinery streams are treated with hydrogen in order to improve final product quality including straight-run naphtha, diesel, gas-oils etc. The catalytic hydroprocessing technology is evolving through the new catalytic materials that are being developed. Even though hydroprocessing catalysts development is well established [4], the growing demand of petroleum products and their specifications, which are continuously becoming stricter, have created new horizons in the catalyst development in order to convert heavier and lower quality feedstocks [5]. Furthermore the expansion of the technology to bio-based feedstocks has also broadened the R&D spam of catalytic hydrotreatment.

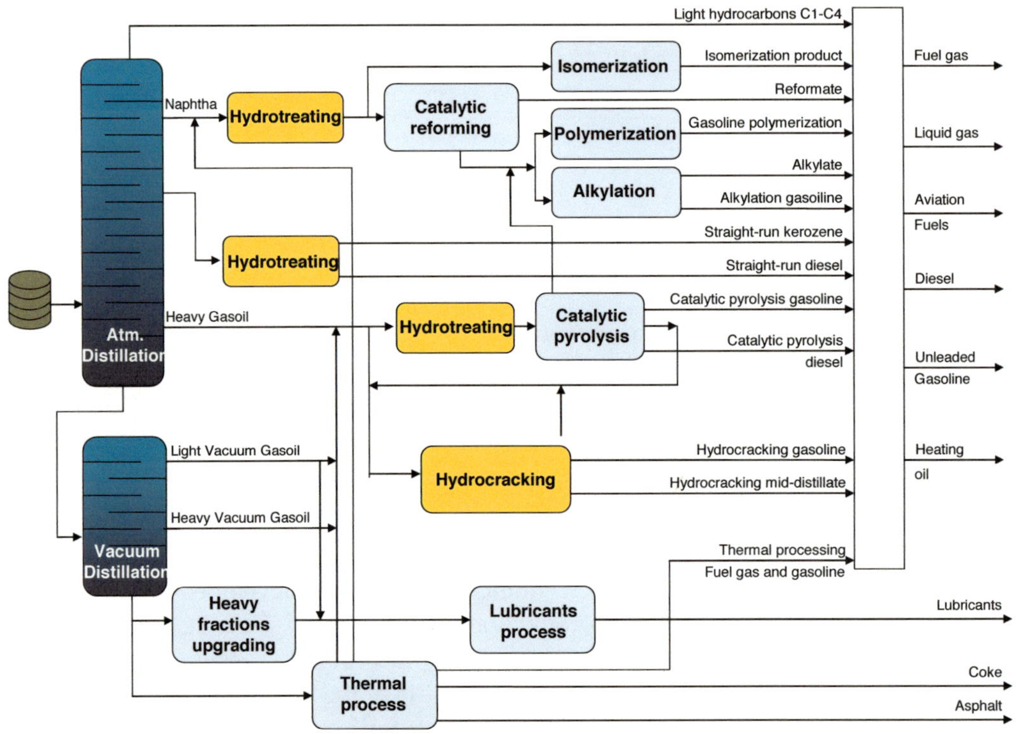

Figure 1. Catalytic hydroprocessing units within a refinery, including distillate hydrotreating and hydrocracking

Catalytic hydroprocessing of liquid biomass is a technology that offers great flexibility to the continuously increasing demands of the biofuels market, as it can convert a wide variety of liquid biomass including **raw vegetable oils**, **waste cooking oils**, **animal fats** as well as **algal oils** into biofuels with high conversion yields. In general this catalytic process technology allows the conversion of triglycerides and lipids into paraffins and iso-paraffins within the naphtha, kerosene and diesel ranges. The products of this technology have improved characteristics as compared to both their fossil counterparts and the conventional biofuels including high heating value and cetane number, increased oxidation stability, negligible acidity and increased saturation level. Besides the application of this catalytic technology for the production of high quality paraffinic fuels, catalytic hydroprocessing is also an effective technology for upgrading intermediate products of solid biomass conversion technologies such as **pyrolysis oils** and **Fischer-Tropsch wax** (Figure 2). The growing interest and investments of the petrochemical, automotive and aviation industries to the biomass catalytic hydroprocessing technology shows that this technology will play an important role in the biofuels field in the immediate future.

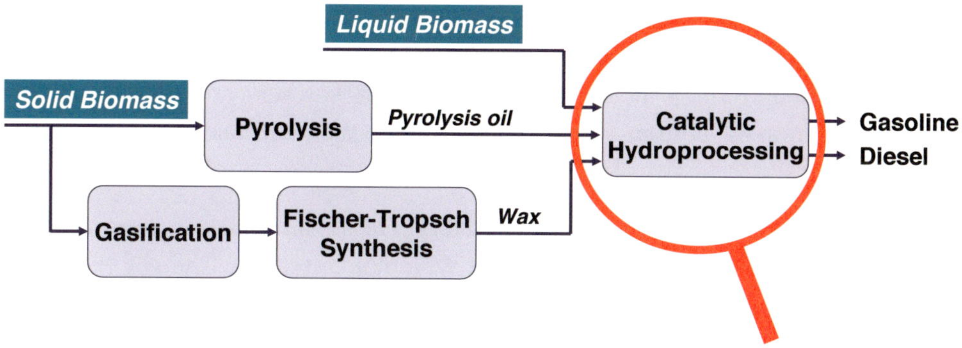

Figure 2. Catalytic hydroprocessing for biomass conversion and upgrading towards fuels production

In the sections that follow, the basic technical characteristics of catalytic hydrotreatment are presented including a description of the process, reactions, operating parameters and feedstock characteristics. Furthermore key applications of catalytic hydroprocessing of liquid biomass are outlined based on different feedstocks including raw vegetable oils, waste cooking oils, pyrolysis oils, Fischer-Tropsch wax and algal oil, and some successful demonstration activities are also presented.

2. Technical characteristics of catalytichydrotreatment

The catalytic hydrotreatment of liquid biomass converts the contained triglycerides/lipids into hydrocarbons at high temperatures and pressures over catalytic material under excess hydrogen atmosphere. The catalytic hydrotreatment of liquid biomass process is quite similar to the typical process applied to petroleum streams, as shown in Figure 3. A typical catalytic hydrotreatment unit consists of four basic sections: a) feed preparation, b) reaction, c) product separation and d) fractionation.

In the feed preparation section the liquid biomass feedstock is mixed with the high pressure hydrogen (mainly from gas recycle with some additional fresh make-up hydrogen) and is preheated before it enters the reactor section. The reactor section consists normally of two hydrotreating reactors, a first guard mild hydrotreating reactor and a second one where the main hydrotreating reactions take place. Each reactor contains two or more catalytic beds in order to maintain constant temperature profile throughout the reactor length. Within the reactor section all associated reactions take place, which will be presented in more detail at a later paragraph.

The reactor product then enters the separator section where, after it is cooled down, it enters the high pressure separator (HPS) flash drum in which the largest portion of the gas and liquid product molecules are separated. The gas product of the HPS includes the excess hydrogen that has not reacted within the reactor section as well as the side products of the reactions including CO, CO_2, H_2S, NH_3 and H_2O. The liquid product of the HPS is lead to a second flash drum, the low pressure separator (LPS), for removing any residual gas contained in the liquid product, and subsequently is fed to a fractionator section. The fractionator section provides the final product separation into the different boiling point fractions that yield the desired products including off-gas, naphtha, kerosene and diesel. The heaviest molecules return from the bottom of the fractionator into the reactor section as a liquid recycle stream.

In order to improve the overall efficiency, a liquid recycle stream is also incorporated, which in essence consists of the heavy molecules that were not converted. The gas product from the HPS and LPS, after being treated to remove the excess NH_3, H_2S, CO and CO_2, is compressed and fed back to the reactor section as a gas recycle stream in order to maintain a high pressure hydrogen atmosphere within the reactor section.

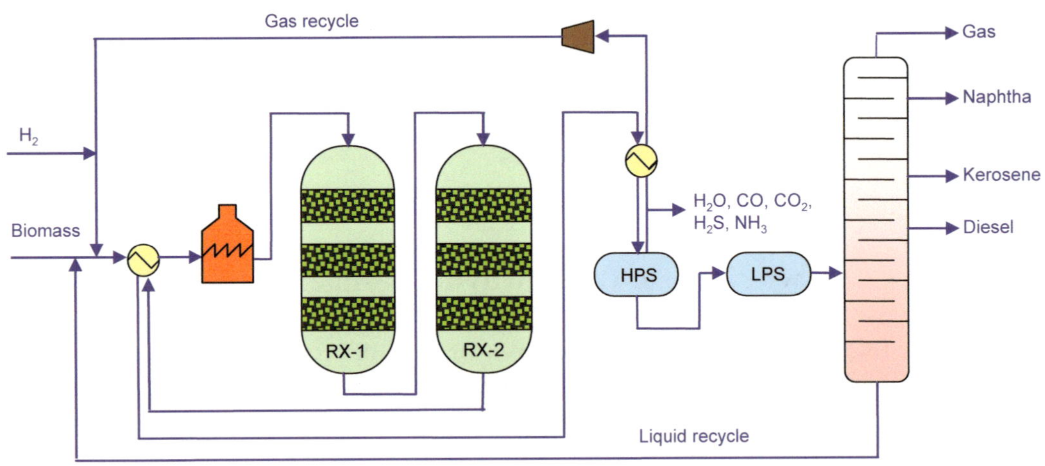

Figure 3. A typical process diagram of catalytic hydrotreatment of liquid biomass

2.1. Reaction mechanisms

Several types of reactions take place during catalytic hydrotreatment of liquid biomass, based on the type of biomass processed, operating conditions and catalyst employed. The types of reactions that liquid biomass undergoes during catalytic hydroprocessing include: a) cracking, b) saturation, c) heteroatom removal and d) isomerization, which are described in more detail in the following section.

http://dx.doi.org/10.5772/52649

2.1.1. Cracking

As the molecules included in the various types of liquid biomass can be relatively large and complicated, cracking reactions are desired to convert them into molecules of the size and boiling point range of conventional fuels, mainly gasoline, kerosene and diesel. A characteristic reaction that occurs during catalytic hydrotreating of oils / fats is the cracking of triglycerides into its consisting fatty acids (carboxylic acids) and propane as shown in Scheme 1 [5][6]. This reaction is critical as it converts the initial large triglycerides molecules of boiling point over 600°C into mid-distillate range molecules (naphtha, kerosene and diesel).

Other cracking reactions may take place however such as those described in Schemes 2 and 3, depending on the type of molecules present in the feedstock. For example Scheme 2 is a cracking reaction which may occur during catalytic hydrotreatment of pyrolysis oil which includes polyaromatic and aromatic compounds. Alternatively Scheme 3 may follow deoxygenation of carboxylic acids on the produced long chain paraffinic molecules, leading to smaller chain paraffins, during the upgrading of Fischer-Tropsch wax.

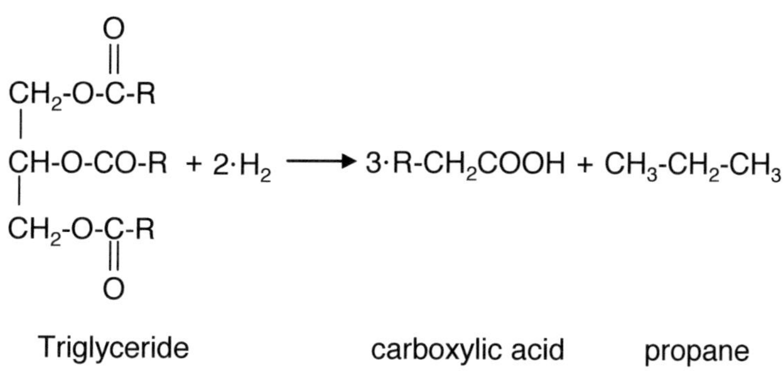

Scheme 1.

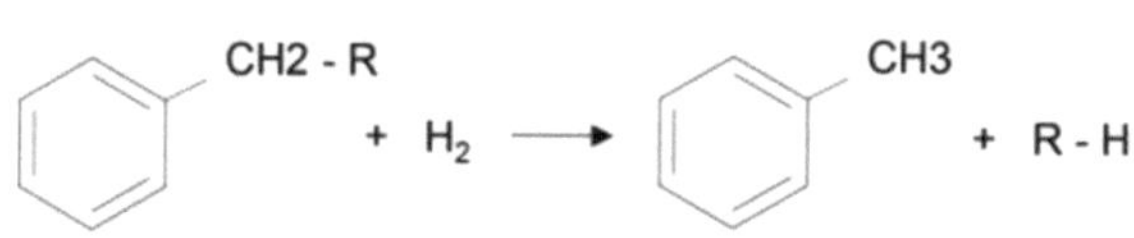

Scheme 2.

$$R - R' + H_2 \longrightarrow R - H + R' - H$$

Scheme 3.

2.1.2. Saturation

Saturation reactions are strongly associated with catalytic hydrotreating as the introduction of excess hydrogen allows the breakage of double C-C bonds and their conversion to single bonds, as shown in the following reactions. In particular the saturation of unsaturated carboxylic acids into saturated ones depicted in Scheme 4, is a key reaction occurring in lipid feedstocks. Furthermore other saturation reactions lead to the formation of naphthenes by converting unsaturated cyclic compounds and aromatic compounds as in Scheme 5 and 6, which are likely to occur during upgrading of pyrolysis oils.

$$RCH = CH\text{-}COOH + H_2 \longrightarrow RCH_2CH_2COOH$$

Scheme 4.

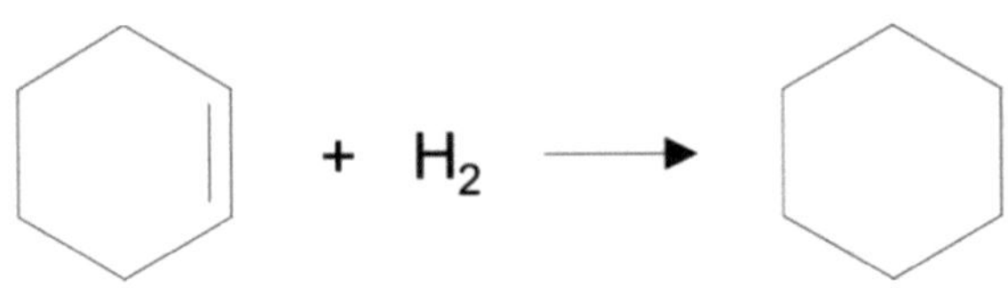

Scheme 5.

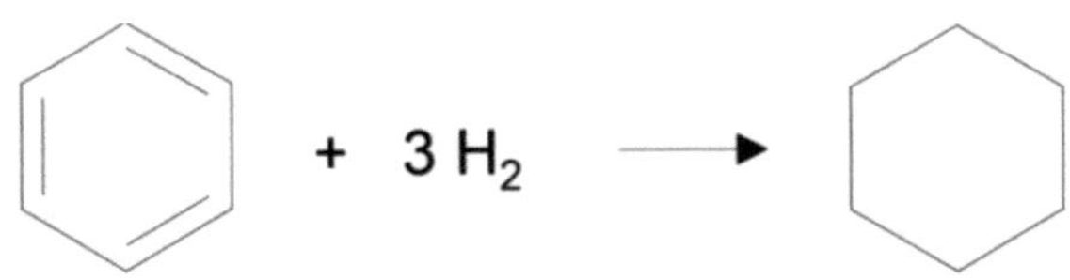

Scheme 6.

As a result of this reaction the produced saturated molecules are less active and less prone to polymerization and oxidation reactions, mitigating the sediment formation and corrosion phenomena appearing in engines.

2.1.3. Heteroatom removal

Heteroatoms are atoms other than carbon (C) and hydrogen (H) and are often encountered into bio- and fossil- based feedstocks. They include sulfur (S), nitrogen (N) and in the case of bio-based feedstocks oxygen (O). In particular oxygen removal is of outmost importance as the presence of oxygen reduces oxidation stability (due to carboxylic and carbonylic double bonds), increases acidity and corrosivity (due to the presence of water) and even reduces the heating value of the final biofuels. The main deoxygenation reactions that take place include deoxygenation, decarbonylation and decarboxylation presented in Schemes 7, 8 and 9 respectively [7]. The main products of deoxygenation reactions include n-paraffins, while H_2O, CO_2 and CO are also produced, but can be removed with the excess hydrogen within the flash drums of the product separation section. It should be noted however that these particular reactions give the paraffinic nature of the produced biofuels, and for this reason the hydrotreated products are often referred to as paraffinic fuels (e.g. paraffinic jet, paraffinic diesel etc)

$$R\text{-}CH_2COOH + 3 \cdot H_2 \longrightarrow R\text{-}CH_2CH_3 + 2 \cdot H_2O$$

Scheme 7.

$$R\text{-}CH_2COOH + H_2 \longrightarrow R\text{-}CH_3 + CO + H_2O$$

Scheme 8.

$$R\text{-}CH_2COOH + H_2 \longrightarrow R\text{-}CH_3 + CO_2$$

Scheme 9.

The other heteroatoms, i.e. S and N are removed according to the classic heteroatom removal mechanisms of the fossil fuels in the form of gaseous H_2S and NH_3 respectively.

2.1.4. Isomerization

The straight chain paraffinic molecules resulting from the aforementioned reactions, even though they offer increased cetane number, heating value and oxidation stability in the biofuels which contain them, they also degrade their cold flow properties. In order to improve the cold flow properties, isomerization reactions are also required, which normally take place during a second step/reactor as they require a different catalyst. Some examples of isomerization reactions are given in Schemes 10 and 11.

$$R\text{-}CH_2\text{-}CH_2\text{-}CH_3 \longrightarrow R\text{-}\underset{\substack{|\\CH_3}}{CH}\text{-}CH_3$$

Scheme 10.

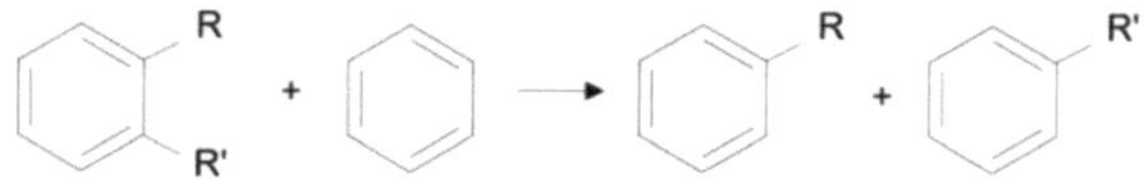

Scheme 11.

2.2. Hydroprocessing catalysts

Catalytic hydroprocessing of liquid biomass is a technology currently under developed and there is a lot of room for optimization. For example there are not many commercial catalysts specifically designed and developed for such applications, while conventional commercial catalysts, employed for catalytic hydroprocessing of refinery streams, are used instead. Common hydrotreating catalysts employed contain active metals on alumina substrate with increased surface area. The most known commercial catalysts employ Cobalt and Molybdenum (CoMo) or Nikel and Molybdenum (NiMo) in alumina substrate (Al_2O_3) as shown in Figure 4.

Figure 4. Typical hydrotreating catalysts (a) before use and (b) after use

Hydrotreating catalysts are dual action catalytic material, triggering both hydrogenation and cracking/isomerization reactions. On one hand hydrogenation takes place on the active metals (Mo, Ni, Co, Pd, Pt) which catalyze the feedstock molecules rendering them more active when subject to cracking and heteroatom removal, while limiting coke formation on the catalyst. Furthermore hydrogenation supports cracking by forming an active olefinic intermediate molecule via dehydrogenation. On the other hand both cracking and isomerization reactions take place in acidic environment such as amorphous oxides ($SiO_2 - Al_2O_3$) or crystalline zeolites (mainly z-zeolites) or mixtures of zeolites with amorphous oxides.

During the first contact of the feedstock molecules with the catalyst, a temperature increase is likely to develop due to the exothermic reactions that occur. However, during the continuous utilization of the catalyst and coke deposition, the catalyst activity eventually reduces from 1/3 to 1/2 of its initial one. The catalyst deactivation rate mainly depends on temperature and hydrogen partial pressure. Increased temperatures accelerate catalyst deactivation while high hydrogen partial pressure tends to mitigate catalyst deactivation rate. Most of the catalyst activity can be recovered by catalyst regeneration.

The selection of a suitable hydroprocessing catalyst is a critical step defining the hydroprocessing product yield and quality as well as the operating cycle time of the process in petroleum industry [5]. However the hydrotreating catalyst selection for biomass applications is particularly crucial and challenging for two reasons: a) catalyst activity varies significantly, as commercial catalysts are designed for different feedstocks, i.e. feedstocks with high sulfur concentration, heavy feedstocks (containing large molecules), feedstocks with high oxygen concentration etc, and b) there are currently no commercial hydroprocessing catalysts available for lipid feedstocks and other intermediate products of biomass conversion processes (e.g. pyrolysis biooil), and thus commercial hydrotreating catalysts need to be explored and evaluated as different catalyst have different yields (Figure 5) and different degradation rate [8]. Nevertheless, significant efforts have been directed towards developing special hydrotreating catalysts for converting/upgrading liquid biomass to biofuels [9-12].

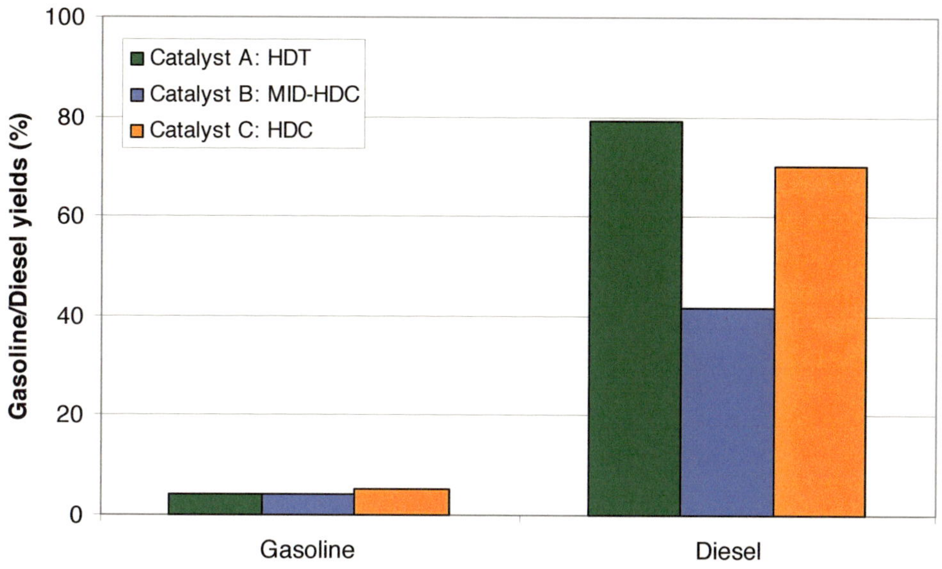

Figure 5. Catalyst comparison based on gasoline and diesel yields for WCO hydrotreating [8]. (Reprinted from Fuel, 93, S. Bezergianni, A. Kalogianni, A. Dimitriadis, Catalyst evaluation for waste cooking oil hydroprocessing, 638-641, 2012, with permission from Elsevier).

2.3. Operating parameters

As it was mentioned earlier, the choice of catalyst and operating parameters affect the reactions that take place within the hydroprocessing reactor. The key operating parameters of hydroprocessing include the reactor temperature, hydrogen partial pressure, liquid hourly space velocity and hydrogen feed-rate.

2.3.1. Temperature

Most catalytic hydrotreating and hydrocracking reactors operate between 290-450°C. The temperature range is selected according the type of catalyst and feedstock type to be processed. In the first stages of the catalyst life (after its loading in the reactor) the temperature is normally kept low as the catalyst activity is already very high. However as time progresses and the catalyst deactivates and cokes, the temperature is gradually increased to overcome the loss of catalyst activity and to maintain the desired product yield and quality.

2.3.2. Hydrogen partial pressure

Hydrogen partial pressure affects significantly the hydrotreating reactions as well as the catalyst deactivation. The catalyst deactivation rate is inverse proportional to the hydrogen partial pressure and to hydrogen feed-rate. However high hydrogen partial pressures correspond to high operational costs, which rise even higher for high olefinic feedstocks that exhibit higher hydrogen consumption due to the saturation reactions. Therefore hydrogen partial pressure should be balanced with the catalyst activity and catalyst life expectancy in order to optimize the overall process.

2.3.3. Liquid hourly space velocity

Liquid hourly space velocity (LHSV) is defined as the ratio of the liquid mass feed-rate (gr/h) over the catalyst mass (gr) and as a result is expressed in hr^{-1}. In fact the inverse of LHSV is proportional to the residence time of the liquid feed in the reactor. In essence the higher the liquid hourly space velocity, the less time is available for the contact of the feed molecules of the reaction mixture with the catalyst, thus the less the conversion. However, maintaining large LHSV imposes a faster degradation of the catalyst therefore in industrial applications the LHSV is maintained in as high values as it is practically possible.

2.3.4. Hydrogen feed-rate

The hydrogen feed-rate is another important parameter as it also defines hydrogen partial pressure depending on the hydrogen consumption of each application. It actually favours both heteroatom removal and saturation reaction rates. However, as hydrogen cost defines the overall unit operating cost, hydrogen feed-rate is normally optimized depending on the system requirements. Furthermore the use of renewable energy sources for hydrogen production is also envisioned as a potential cost improvement option.

3. Feed sand products

Even though liquid biomass is currently being exploited as a renewable feedstock for fuels production, its characteristics are far beyond suitable for its use as fuel. More specifically liquid biomass, just as other types of biomass, has a small H/C ratio and high oxygen content, lowering its heating value and increasing CO and CO_2 emissions during its combustion. Moreover liquid biomass contains water, which can cause corrosion in the downstream processing units if it's not completely removed, or even in the engine parts where its final products are utilized. In addition to the above, liquid biomass has an increased concentration in oxygenated compounds, mainly acids, aldehydes, ketones etc, which not only reduce the heating value, but also decrease the oxidation stability and increase the acidity of the produced biofuels. For all the aforementioned reasons it is imperative that liquid biomass should be upgraded and specifically that its H/C should be increased while the water and oxygen removed.

The effectiveness of catalytic hydroprocessing towards improving these problematic characteristics of liquid biomass is presented in Table 1, where the H/C ratio, the oxygen content and density before and after catalytic hydrotreatment of basic liquid biomass types are given. The H/C ratio exhibits a significant increase that exceeds 50% in all cases. This is due to the substitution of the heteroatoms by hydrogen atoms as well as in the saturation of double bonds that enriches the H/C analogy. The oxygen content (including the oxygen contained in the water) from over 15%wt can be decreased down to 5wppm. Actually the deep deoxygenation achieved via catalytic hydrotreatment is the most significant contribution of this biomass conversion technology, as it improves significantly the oxidation stability of the final biofuels. Furthermore significant improvement is also observed in the biomass density, which is never below 0.9 kg/l while after hydrotreatment it reduces to values less than 0.8 kg/l

	Liquid biomass (unprocessed)	Hydrotreated liquid biomass and produced biofuels
H/C ratio	0.08 – 0.1	0.13 – 0.18
Oxygen content (%wt)	15 - 40	10^{-4} – 3
Density (kg/l)	0.9 – 1.05	0.75 – 0.8

Table 1. Effect of catalytic hydrotreatment on the liquid biomass characteristics

Catalytic hydroprocessing has been proven as the most efficient technology for the upgrading of liquid biomass as it achieves to increase the H/C ratio and to remove oxygen and water. However the effectiveness of this technology is also shown in other parameters. For example the distillation curve of raw liquid biomass shows that over 90% of its molecules have boiling points exceeding 600°C and only 5% are within diesel range (220-360°C), while after catalytic hydrotreatment upgrading most of 90% of the product molecules are within diesel range [13].

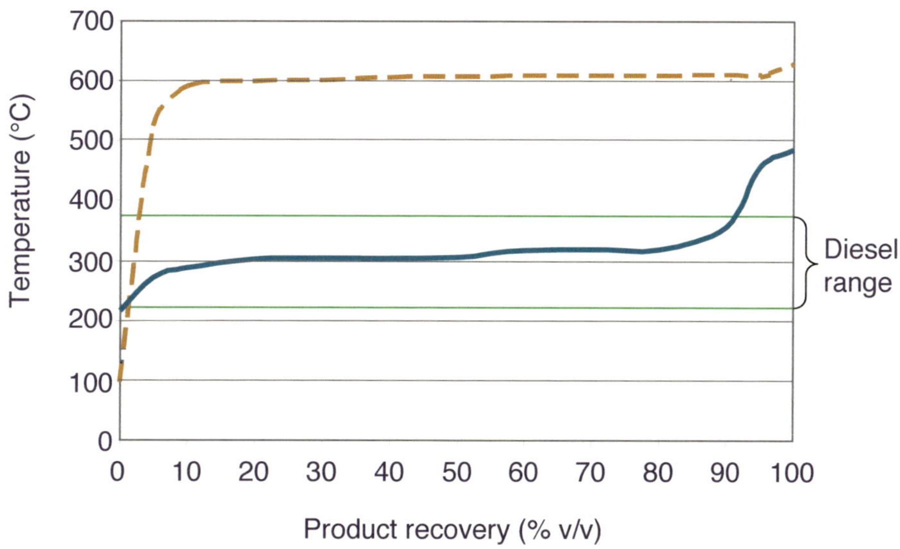

Figure 6. Distillation curves of untreated WCO (dashed) and catalytically hydrotreated WCO (solid)

In the following sections the basic types of liquid biomass and their corresponding products via catalytic hydrotreatment are presented.

3.1. Raw vegetable oils conversion to paraffinic biofuels

Vegetable oils are the main feedstock for the production of first generation biofuels, which can offer several CO_2 benefits and limit the consumption of fossil fuels. Raw vegetable oils consist of fatty acid triglycerides, the consistency of which depends on their origin (i.e. plant type) as shown in Table 2. Their production, however, is competing for the cultivated areas that were originally dedicated for the production of food and feed crops. As a result the production and utilization of vegetable oils for biofuels production has instigated the "food vs. fuel" debate. For this reason traditional energy crops (soy, cotton, etc) with low oil yield per hectare are being substituted by new energy crops (eg. jatropha, palm, castor etc).

		C8:0	C10:0	C12:0	C14:0	C16:0	C16:1	C18:0	C18:1	C18:2	C18:3	C20:0/ C22:0	C20:1/ C22:1
EU	**Rapeseed oil**	0.0	0.0	0.0	0.0	3.5	1.0	1.5	12.5	15.0	7.5	9.0	50.0
	Soybean oil	0.0	0.0	0.0	0.3	8.2	0.5	4.5	25.0	49.0	5.0	7.5	0.0
	Sunflower oil	0.0	0.0	0.0	0.0	6.0	0.0	4.2	18.8	69.3	0.3	1.4	0.0
	Corn oil	0.0	0.0	0.0	1.0	9.0	1.5	2.5	40.0	45.0	0.0	0.0	1.0
non-EU	**Palm oil**	0.0	0.0	0.0	3.5	39.5	0.0	3.5	47.0	6.5	0.0	0.0	0.0
	Peanut oil	0.0	0.0	0.0	0.5	8.0	1.5	3.5	51.5	27.5	0.0	7.5	0.0
	Canola oil	0.0	0.0	0.1	0.1	4.7	0.1	1.6	65.9	21.2	5.2	1.2	0.0
	Castor oil	0.0	0.1	0.2	10.6	1.4	9.5	29.7	29.7	41.3	3.3	3.8	0.0

Table 2. Fatty acid composition of most common vegetable oils [14][15]

Catalytic hydrotreatment was explored for conversion of vegetable oils in the early 90's. The investigation of the hydrogenolysis of various vegetable oils such as maracuja, buritimtucha and babassu oils over a Ni–Mo/γ-Al_2O_3 catalyst as well as the effect of temperature and pressure on its effectiveness was firstly investigated [16][17]. The reaction products included a gas product rich in the excess hydrogen, carbon monoxide, carbon dioxide and light hydrocarbons as well as a liquid organic product of paraffinic nature. In more detail these studies showed the conversion of triglycerides into carboxyl oxides and then to high quality hydrocarbons via decarboxylation and decarbonylation reactions. Rapeseed oil hydroprocessing was also studied in lab-scale reactor for temperatures 310° and 360°C and hydrogen pressures of 7 and 15 MPa using three different Ni–Mo/alumina catalysts [18]. These products contained mostly n-heptadecane and n-octadecane accompanied by low concentrations of other n-alkanes and i-alkanes [19].

3.2. Waste cooking oils conversion to paraffinic biofuels

Even though vegetable oils are the main feedstock for the production of first generation biofuels, soon their production has troubled the public opinion due to their abated sustainability and to their association with the food vs. fuel debate. As a result the technology has shifted towards the exploitation of both **solid and liquid residual biomass**. Waste Cooking Oils (WCOs) is a type of residual biomass resulting from frying with typical vegetable frying oils (e.g. soybean-oil, corn-oil, olive-oil, sesame-oil etc). WCOs have particular problems regarding their disposal. In particular grease may result in coating of pipelines within the residential sewage system and is one of the most common causes of clogs and sewage spills. Furthermore, in the cases that sewage leaks into the environment, WCOs can cause human and environmental health problems because of the pathogens contained. It has been estimated that by disposing 1 lit of WCO, over 1,000,000 of liters of water can be contaminated, which is estimated as the average demand of a single person for 14 years.

Catalytic hydroprocessing of WCO was studied as an alternative approach of producing 2nd generation biofuels [20-24]. Initially **catalytic hydrocracking** was investigated over commercial hydrocracking catalysts leading not only to biodiesel but also to lighter products such as biogasoline [20], employing a continuous-flow catalytic hydroprocessing pilot-plant with a fixed-bed reactor. During this study several parameters were considered including hydrocracking temperature (350-390°C) and liquid hourly space velocity or LHSV (0.5-2.5 hr^{-1}) under high pressure (140 bar), revealing that the conversion is favoured by high reaction temperature and low LHSV. Lower and medium temperatures, however, were more suitable for biodiesel production while higher temperatures offered better selectivity for biogasoline production. Furthermore, heteroatom removal (S, N and particularly O) was increased while saturation of double bonds was decreased with increasing hydrocracking temperature, indicating the necessity of a pre-treatment step.

However **catalytic hydrotreatment** was later examined in more detail as a more promising technology particularly for paraffinic biodiesel production (Figure 1). The same team has studied the effect of temperature (330-398°C) on the product yields and heteroatom removal [21]. The study was conducted in the same pilot plant utilizing a commercial $NiMo/Al_2O_3$ hydrotreating catalyst over lower pressure (80 bar). According to this study, the hydrotreating temperature is the key operating parameter which defines the catalyst effectiveness and life. In fact lower temperatures (330°C) favour diesel production and selectivity. Sulfur and nitrogen removal were equally effective at all temperatures, while oxygen removal and saturation of double bonds were favoured by hydrotreating temperature. The same team also studied the effect of the other three operating parameters i.e. pressure, LHSV and H_2/WCO ratio [22]. Moreover they also studied the hydrocarbon content of the products [23] qualitatively via two-dimensional chromatography and quantitatively via Gas Chromatography with Flame Ionization Detector (GC-FID), which indicated the presence of C15-C18 paraffins. Interestingly this study showed that as hydrotreating temperature increases, the content of normal paraffins decreases while of iso-paraffins increases, revealing that isomerization reactions are favoured by temperature.

Figure 7. Catalytic hydrotreatment of WCO to 2nd generation biodiesel

The total liquid product of WCO catalytic hydrotreatment was further investigated in terms of its percentage that contains paraffins within the diesel boiling point range (220-360°C) [24]. The properties of WCO, hydrotreated WCO (total liquid product) and the diesel fraction of the hydrotreated WCO are presented in Table 3. Based on this study the overall yield of the WCO catalytic hydrotreatment technology was estimated over 92%v/v. The properties of the new 2nd generation paraffinic diesel product indicated a high-quality diesel with high heating value (49MJ/kg) and high cetane index (77) which is double of the one of fossil diesel. An additional advantage of the new biodiesel is its oxidation stability (exceeding 22hrs) and negligible acidity, rendering it as a safe biofuel, suitable for use in all engines. The properties and potential of the new biodiesel were further studied [25], for evaluating different fractions of the total liquid product and their suitability as an alternative diesel fuel.

		WCO	Hydrotreated WCO	Final biodiesel
Density	gr/cm³	0.896	0.7562	0.7869
C	wt%	76.74	84.59	86.67
H	wt%	11.61	15.02	14.74
S	wppm	38	11.80	1.54
N	wppm	47.42	0.77	1.37
O	wt%	14.57	0.38	0
Recovery 0%	°C	431.6	195.6	234.1
Recovery 10%	°C	556.4	287.4	294.1
Recovery 30%	°C	599	304.0	296.8
Recovery 50%	°C	603.2	314.4	298.3
Recovery 70%	°C	609	319.0	300
Recovery 90%	°C	612.4	320.4	298.3
Recovery 100%	°C	727.2	475.4	306.2

Table 3. Basic properties of waste cooking oil, hydrotreated waste cooking oil and final biodiesel

3.3. Pyrolysis oil upgrading

Pyrolysis oil is the product of fast pyrolysis of biomass, a process that allows the decomposition of large organic compounds of biomass such as lignin at medium temperatures in the presence of oxygen. Pyrolysis, that is in essence thermal cracking of biomass, is a well established process for producing bio-oil, the quality of which however is far too poor for direct use as transportation fuel. The product yields and chemical composition of pyrolysis oils depend on the biomass type and size as well as on the operating parameters of the fast pyrolysis. However, a major distinction between pyrolysis oils is based on whether catalyst is employed for the fast pyrolysis reactions or not. Non-catalytic pyrolysis oils have a higher water content than catalytic pyrolysis oils, rendering the downstream upgrading process a more challenging one for the case of the non-catalytic pyrolysis oils.

Untreated pyrolysis oil is a dark brown, free-flowing liquid with about 20-30% water that cannot be easily separated. It is a complex mixture of oxygenated compounds including water solubles (acids, alcohols, ethers) and water insolubles (n-hexane, di-chloor-methane), which is unstable in long-term storage and is not miscible with conventional hydrocarbon-based fuels. It should be noted that due to its nature pyrolysis oil can be employed for the production of a wide range of chemicals and solvents. However, if pyrolysis oil is to be used as a fuel for heating or transportation, it requires upgrading leading to its stabilization and conversion to a conventional hydrocarbon fuel by removing the oxygen through catalytic hydrotreating. For this reason, a lot of research effort is focused on catalytic hydrotreating of pyrolysis oil, as it is a process enabling oxygen removal and conversion of the highly corrosive oxygen compounds into aromatic and paraffinic hydrocarbons.

For non-catalytic pyrolysis oils, the catalytic hydrotreating upgrading process involves contact of pyrolysis oil molecules with hydrogen under pressure and at moderate temperatures (<400°C) over fixed bed catalytic reactors. Single-stage hydrotreating has proved to be difficult, producing a heavy, tar-like product. Dual-stage processing, where mild hydrotreating is followed by more severe hydrotreating has been found to overcome the reactivity of the bio-oil. Overall, the pyrolysis oil is almost completely deoxygenated by a combination of hydro deoxygenation and decarboxylation. In fact less than 2% oxygen remains in the treated, stable oil, while water and off-gas are also produced as byproducts. The water phase contains some dissolved organics, while the off-gas contains light hydrocarbons, excess hydrogen, and carbon dioxide. Once the stabilized oil is produced it can be further processed into conventional fuels or sent to a refinery. Table 1 shows the properties of some common catalytic pyrolysis oils according to literature.

Catalytic pyrolysis oils have been reported to getting upgraded via single step hydroprocessing, most of the times utilizing conventional CoMo and NiMo catalysts. During the single step hydroprocessing, the catalytic pyrolysis oil feedstock is pumped to high pressure, then mixed with compressed hydrogen and enters the hydroprocessing reactor. In Table 5 the typical operating parameters for single stage hydroprocessing and associated deoxygenation achievements are given according to literature [29;33-38].

		Types of Pyrolysis Biooils						
Properties	Test Methods	[26]	[27]	[28]	[29]	[30]	[31]	[32]
H_2O content (%wt)	KarlFisher	20	23.9	30	20-30	29.85		
pH	pHmeter	2.2		2.5	2~3		2.5	
Density 15C (Kg/L)	ASTM D4052	1.207		1.2	1.15-1.2	1.192	1.2	1.19
HHV (MJ/Kg)	DIN51900	17.57						
LHV (MJ/Kg)	DIN51900	15.83						
Solids Content (%wt)	Insolubles in Ethanol	0.06						
Ash content (%wt)	ASTM D482	0.0034	<0.1		0.1	0.15	0-0.2	
Pour point	ASTM D97	-30				-30		
Flash point	ASTM D93	48		40-65	40-65	51		
Viscosity (cP) @ 40C				40	40-100		40-100	43-1510
Viscosity 20°C 9mm2/s)	ASTM D445	47.18						
Viscosity 50°C (mm2/s)	ASTM D445	9.726						
Carbon (%wt)	ASTM D5291	42.64	40.1	51.1	~52	39.17	54-58	39.4-46.7
Hydrogen (wt%)	ASTM D5291	5.83	7.6	7.3	~6.4	8.04	5.5-7	7.2-7.9
Nitrogen (wt%)	ASTM D5291	0.1	0.1		~0.2	0.05	0-0.2	0.2
Sulphur (%wt)	ASTM	0.01						0.032
Clorine (%wt)	ASTM	0.012						
AlkaliMetals (%wt)	ICP	<0.003						
Oxygen (wt%)		52.1		41.6	~40	52.74	35-40	45.7-52.7

Table 4. Properties of different pyrolysis oils according to literature

Catalyst	CoMo [29][33][34][35][36], NiMo [34][35][36], others [37][38]
Temperature (°C)	350-420
Pressure (psig)	1450-2900
LHSV (Hr^{-1})	0.1-1.2
Deoxygenation (wt%)	78-99.9
Density (kg/l)	0.9-1.03

Table 5. Single-stage pyrolysis oil hydroprocessing operating parameters

However, in the case of non-catalytic pyrolysis oils or for achieving better quality products, multiple-stage hydroprocessing can be employed for upgrading pyrolysis oils. Multiple-stage hydroprocessing utilizes at least two different stages of hydroprocessing, which may include hydrotreating or hydrotreating and hydrocracking reactions. In the first stage the catalytic hydrotreatment reactor stabilizes the pyrolysis oil by mild hydrotreatment over Co-Mo or NiMo hydrotreating catalyst [32;40-42]. The first stage product is then further processed in the second-stage hydrotreater, which operates at higher temperatures and lower space velocities than the first stage hydrotreater, employing also CoMo or NiMo catalysts within the reactor. The 2nd stage product is separated into an organic-phase product, waste-water, and off-gas streams. In the literature [41], even a 3rdstage hydroprocessing has been used for the heavy fraction (which boils above 350°C) of the 2ndstage product, where hydrocracking reactions take place for converting the heavy product molecules into gasoline and diesel blend components.

Feed	1st stage	2nd stage	3rd stage
Catalyst	CoMo[32][40],NiMo[32][42], others[39]	CoMo[32][40]NiMo[32][42], others [39]	CoMo[4141]
Temperature (C°)	150-240	225-370	350-427
Pressure (psig)	1000-2000	2015	1280
LHSV (hr^{-1})	0.28-1	0.05-0.14	
Deoxygenation (wt%)		60-98.6	

Table 6. Multiple-step pyrolysis oil hydroprocessing operating parameters

3.4. Fischer-Tropsch wax upgrading

Biofuels production via the Fischer-Tropsch technology is a conversion process of solid biomass into liquid fuels (Biomass-To-Liquid or BTL) as it is depicted in Figure 2. More specifically the solid biomass is gasified in the presence of air and the produced biogas rich in CO and H_2 (synthesis gas), after being pretreated to remove coke residues and sulfur compounds, enters the Fischer-Tropsch reactor. The Fischer-Tropsch reactions allow the catalytic conversion of the synthesis gas into a mixture of paraffinic hydrocarbons consisting of light (C_1-C_4), naphtha (C_5-C_{11}), diesel (C_{12}-C_{20}) and heavier hydrocarbons (>C_{20}). Even though the Fischer-Tropsch reactions yields depend on the catalyst and operating parameters employed [43-45], the liquid product (naphtha, diesel and heavier hydrocarbons) yield is high (~95%). The produced synthetic naphtha and diesel fuels can be used similarly to their fossil counterparts. The heavier product however, which is called as Fischer-Tropsch wax, due to its waxy/paraffinic nature should get upgraded via catalytic hydrocracking to get converted to mid-distillate fuels (naphtha and diesel).

The conversion of Fischer-Tropsch wax into mainly diesel was studied in virtue of the European Project RENEW [46]. During this project Fischer-Tropsch wax with high paraffinic content of C_{20}-C_{45} was converted into a total liquid product consisting of naphtha, kerosene and diesel fractions via catalytic hydrocracking. However the total liquid product content of diesel molecules was the highest and the diesel fraction was further separated and characterized having density of 0.78gr/ml and cetane index of 76 [47]. The schematic of the BTL process with actual images of the feedstock, Fischer-Tropsch wax and synthetic diesel are given in Figure 8.

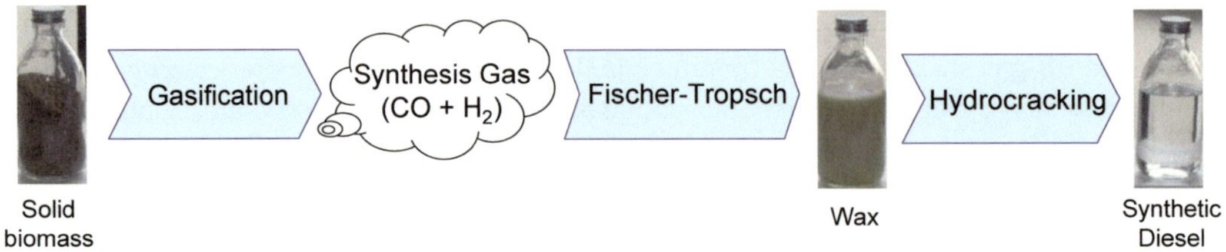

Figure 8. Biomass-to-Liquid production of synthetic diesel

3.5. Micro-algal oil conversion to biofuels

The rapid development of the biofuels production technologies from different biomass types has given rise to the biomass and food markets as it was aforementioned. Besides the use of residual biomass, research and in particular biotechnology has moved forward into seeking alternative biomass production technologies that will offer higher yields per hectare as well as lipids and carbohydrates, which are not part of the human and animal food-chain, avoiding competition between food/feed and energy crops. Targeted research efforts have offered a promising solution by the selection of unicellular microorganisms for the production of biofuels [48][49]. **Micro-algae** are photosynthetic microorganisms that can produce lipids, proteins and carbohydrates in large amounts over short periods of time.

Micro-algae are currently considered a prominent source of fatty acids, which offers large yields per hectare with various fatty acid foot-prints from each strain. In fact, there are certain strains that offer fatty acids of increased saturation (small content of unsaturated fatty acids) and of smaller carbon-chain length such as *Dunaliellasalina, Chlorella minutissima, Spirulina maxima, Synechococcus sp.*[50] etc. Another advantage of algal oils is that their fatty acid content can be directed to small carbon-chain molecules either genetically or by manipulating the aquaculture conditions such as light source and intensity [51], nitrogen starvation period [52], nutrients and CO_2 feeding profiles [53].

Micro-algae and their products formulated the so called 3rd generation biofuels, as they incorporate various characteristics, which render them superior over other biofuels and biomass types. Micro-algae can also be produced in sea water [54] or even waste water, while they are biodegradable and relatively harmless during an eventual spill. Furthermore, their yield per hectare can reach 3785-5678lit, which is 20-700 higher over the conventional energy crops yield (soy, rape and palm). The lipids contained in most micro-algal oils have a similar synthesis with that of soy-bean oil, while they also contain some poly-saturated fatty acids with four double bonds. As a result catalytic hydrotreating of micro-algal oil is the most promising technology for converting it into biofuels.

3.6. Co-hydroprocessing

The effectiveness of catalytic hydroprocessing was also explored for co-processing of lipid feedstocks with petroleum fractions as catalytic hydroprocessing units are available in almost all refineries. The first co-processing study involved experiments of catalytic hydrotreating of sunflower oil mixtures with heavy petroleum fractions aiming to produce high quality diesel [55]. The experiments were conducted in a continuous fixed-bed reactor over a wide range of temperatures 300-450°C employing a typical $NiMo/Al_2O_3$ hydrotreating catalyst. The study was focused on the hydrogenation of double C-C bonds and the subsequent paraffin formation via the three different reactions routes: decarbonylation, decarboxylation and deoxygenation. Furthermore the large carbon-chain paraffins can also undergo isomerization and cracking leading to the formation of smaller paraffins. This study concluded that the selectivity of products on decarboxylation and decarbonylation is increasing as the temperature and vegetable oil content in the feedstock increase [55].

In a similar study catalytic hydrocracking over sunflower oil and heavy vacuum gas oil mixtures was investigated [56]. The experiments were conducted in a continuous-flow hydroprocessing pilot-plant over a range of temperatures (350-390°) and pressures (70-140bar). Three different hydrocracking catalysts were compared under the same conditions and four different feedstocks were employed, incorporating for 10% and 30%v/v of lipid bio-based feedstock and considering non-pretreated and pretreated sunflower oil as a bio-based feedstock. The results indicated that a prior mild hydrogenation step of sunflower oil is necessary before hydrocracking. Furthermore, conversion was increased with increasing sunflower oil ratio in the feedstock and increasing temperature, while the later decreased diesel selectivity.

The effect of the process parameters and the vegetable oil content of the feedstocks on the yield, physical properties, chemical properties and application properties during co-hydrotreating of sunflower oil and gas-oil mixtures utilizing a typical $NiMo/Al_2O_3$hydrotreating catalyst was also studied [57]. The experimental results of this study indicated that catalytic co-hydrogenation of gas oil containing sunflower oil in different percentages allowed both vegetable oil conversion reactions (saturation, deoxygenation) and the gas oil quality improvement reactions (hetero atom removal, aromatic reduction). The optimal operating conditions (360-380°C, P=80 bar, LHSV=$1.0h^{-1}$, H_2/oil=600 $Nm^3/m^3$and 15% sunflower oil content of feed) resulted in a final diesel product with favorable properties (e.g. less than 10 wppm sulfur, ~20% aromatics) but poor cold flow properties (CFPP=3°C). The study also indicated that for sunflower content in the feedstock higher than 15% reduced the desulfurization efficiency. Furthermore, the authors also concluded that the presence of sunflower oil in the feedstock has augmented the normal and iso-paraffins content of the final product and as a result has increased the cetane number but degraded the cold flow properties, indicating that an isomerization step is required as an additional step.

The issue of catalyst development suitable for co-hydrotreating and co-hydrocracking of gas-oil and vegetable oil mixtures was recently addressed [10], as there are no commercial hydroprocessing catalysts available for lipid feedstocks. New sulfided $Ni–W/SiO_2–Al_2O_3$ and sulfided $Ni–Mo/Al_2O_3$ catalysts were tested for hydrocracking and hydrotreating of gas-oil and vegetable oil mixtures respectively. The results indicated that the hydrocracking catalyst was more selective for the kerosene hydrocarbons (140–250°C), while the less acidic hydrotreating catalyst was more selective for the diesel hydrocarbons (250–380°C). The study additionally showed that the deoxygenation reactions are more favored over the hydrotreating catalyst, while the decarboxylation and decarbonylation reactions are favored over the hydrocracking catalyst.

4. Demo and industrial applications

As catalytic hydrotreating of liquid biomass has given promising results, the industrial world has given enough confidence to apply it in pilot and industrial scale. The NesteOil Corporation has developed the NExBTL technology for converting vegetable oil (primarily palm oil) into a renewable diesel also known as "green" diesel (Figure 9). Based on this technology the first catalytic hydrotreatment of vegetable oils unit was constructed in Finland in 2007, within the existing Poorvo refinery of NesteOil, with a capacity of 170 kton/hr. The primary feedstock is palm oil, while it can also process rapeseed oil and even waste cooking oil. The same company has constructed a second unit within the same refinery while it has also planned to construct two new units, one in Singapore and one in Rotterdam, with the capacity of 800 kton/yr each.

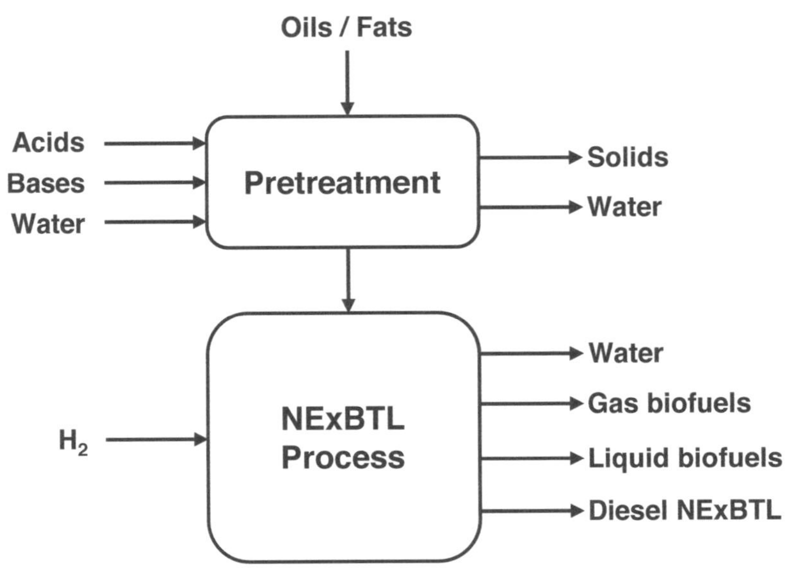

Figure 9. NExBTL catalytic hydrotreating of oils/fats technology for biodiesel production [58]

The catalytic hydrotreatment technology of 100% waste cooking oil for biodiesel production was developed in the Centre for Research and Technology Hellas (CERTH) in Thessaloniki, Greece [21-24] and later demonstrated via the BIOFUELS-2G project [59], which was co-

http://dx.doi.org/10.5772/52649

funded by the European Program LIFE". In this project WCO was collected from associated restaurants and the produced 2nd generation bio-diesel, to be called "white diesel" was employed. For the demonstration of the new technology, 2 tons of "white diesel' were produced via catalytic hydrotreatment of WCO based on the large-scale pilot units available in CERTH. The production process simplified diagram is given in Figure 10. The new fuel will be applied to a garbage truck in a 50-50 mixture with conventional diesel in August 2012, aiming to promote the new technology as it exhibits overall yields exceeding 92% v/v.

In the USA the Dynamic Fuels company [60] has constructed in Baton Rouge a catalytic hydrotreating unit dedicated to oils and animal fats with 285 Mlit capacity. The unit employs the Syntroleum technology based on Fischer-Tropsch for the production of synthetic 2nd generation Biodiesel while it also produces bio-naphtha and bio-LPG. The Bio-Synfining technology of Syntroleum converts the triglycerides of fats and oils into n- and iso-paraffins via catalytic hydrogenation, thermal cracking and isomerization as it is applied in the Fischer-Tropsch wax upgrading to renewable diesel (R-2) and renewable jet (R-8) fuel.

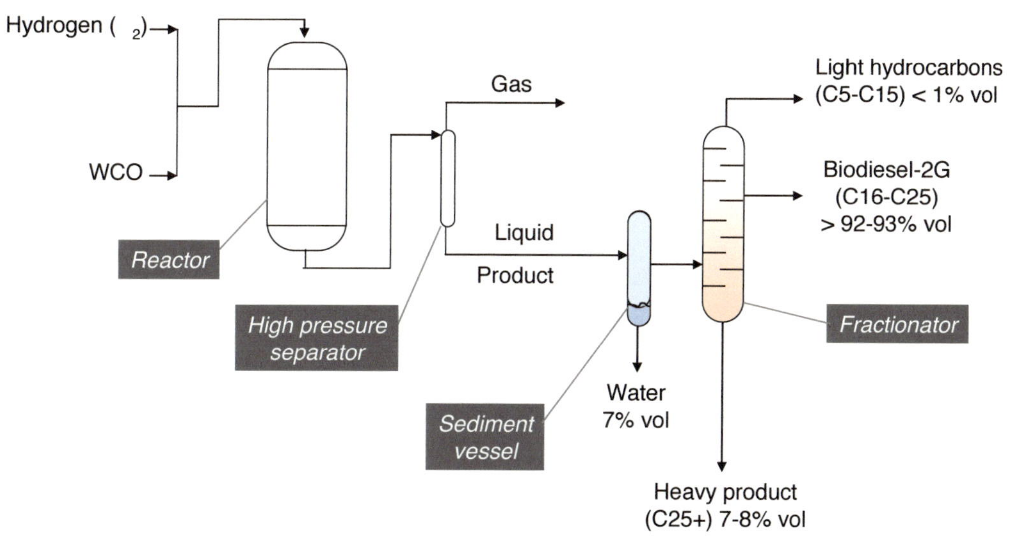

Figure 10. Pilot-scale demo application of WCO catalytic hydrotreatment during the BIOFUELS-2G project [59]

5. Future perspectives

Catalytic hydrotreating of liquid biomass is continuously gaining ground as the most effective technology for liquid biomass conversion to both ground- and air-transportation fuels. The UOP company of Honeywell, via the technology it has developed for catalytic hydrotreating of liquid biomass (Figure 11), has announced imminent collaboration with oil and airline companies such as Petrochina, Air China and Boeing for the demonstration of the sustainable air-transport in China. This initiative will lead a strategic collaboration between the National Energy Agency of china with the Commerce and Development Agency of USA leading to the development of the new biofuels market in China.

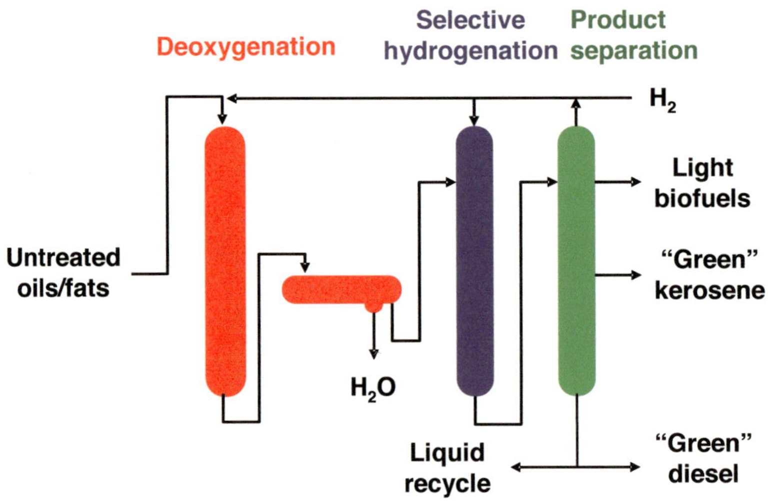

Figure 11. Vegetable oil and animal fats conversion technology to renewable fuels of UOP [61]

In the EU airline companies collaborate with universities, research centers and biofuels companies in order to confront their extensive contribution to CO_2 emissions. Since 2008 most airline companies promote the use of biofuels in selected flights as shown in Table 7 [62]. As it is obvious most pilot flights have taken place with Hydrotreated Renewable Jet (HRJ), which is kerosene/jet produced via catalytic hydrotreatment of liquid biomass. Moreover, Lufthansa has also completed a 6-month exploration program of employing HRJ in a 50/50 mixture with fossil kerosene in one of the 4 cylinders of a plane employed for the flight between Hamburg-Frankfurt-Hamburg with excellent results [63].

Besides the future applications for air-transportation, the automotive industry is also exhibiting increased interest for the broad use of biofuels resulting from catalytic hydrotreatment of liquid biomass. In fact these paraffinic biofuels can be employed in higher than 7%v/v blending ratio (which is the maximum limit for FAME) as they exhibit high cetane number and have significant oxidation stability [64]

http://dx.doi.org/10.5772/52649

Airline	Aircraft	Partners	Biofuel (lipid sources)	Blend*
Virgin Atlantic	B747-400	Boeing, GE Aviation	FAME (coconut & palm)	20%
Air New Zealand	B747-400	Boeing, Rolls-Royce	HRJ (Jatropha)	50%
Contintental Airlines	B737-800	Boeing, GE Aviation, CFM, Honeywell UOP	HRJ (Jatropha&algea)	50%
JAL	B747-300	Boeing, Pratt&Whitney, Honeywell UOP	HRJ (Camelina, Jatropha& algae)	50%
KLM	B747-400	GE, Honeywell UOP	HRJ (Camelina)	50%
TAM	A320	Airbus, CFM	HRJ (Jatropha)	50%

Table 7. Pilot flights with biofuels [62]

The highest interest is exhibited by oil companies around the catalytic hydrotreatment of liquid biomass technology for the production of biofuels and particularly to its application to oil from micro-algae. ExxonMobil has invested 600M$ in the Synthetic Genomics company of the pioneer scientist Craig Ventner aiming to research of converting micro-algae to biofuels with minimal cost. BP has also invested 10M$ for collaboration with Martek for the production of biofuels from micro-algae for air-, train-, ground- and marine transportation applications.

6. Conclusion

Catalytic hydrotreatment of liquid biomass is the only proven technology that can overcome its limitations as a feedstock for fuel production (low H/C ratio, high oxygen and water content). Even though it has recently started to be investigated as an alternative technology for biofuels production, it fastly gains ground due to the encouraging experimental results and successful pilot/demo and industrial applications. Catalytic hydrotreatment of liquid biomass leads to a wide range of new alterative fuels including bio-naphtha, bio-jet and biodiesel, are paraffinic in nature and as a result exhibiting high heating values, increased oxidation stability and negligible acidity and corrosivity. As a result it is not over-optimistic to claim that this technology will broaden the biofuels market into scales capable to actually mitigate the climate change problems.

Acknowledgements

The author would like to thank Ms Iva Simcic and InTech Europe for enabling her to publish this book chapter, while she is grateful to Mr Athanasios Dimitriadis who provided support, offered comments, proofreading and design. Finally she would like to express her appreciation for the financial support provided by the EU project BIOFUELS-2G which is co-financed by the European Program LIFE+.

Author details

Stella Bezergianni*

Address all correspondence to: sbezerg@cperi.certh.gr

Chemical Processes & Energy Resources Institute (CPERI), Centre for Research & Technology Hellas (CERTH), Thermi-Thessaloniki, Greece

References

[1] Meher LC, Vidya SD, Naik SN. Technical aspects of biodiesel production by transesterification—a review. Renewable and Sustainable Energy Reviews 2006;10 248–268.

[2] Naik SN., Goud VV, Rout OJ, Dalai AK. Production of first and second generation biofuels: A comprehensive review. Renewable and sustainable Energy Reviews 2010;14 578-597.

[3] Gary JH, Handwerk GE. Petroleum Refining, Technology and Economics, 4th ed. New York: Marcel Dekker; 2001.

[4] Furimsky E. Selection of catalysts and reactors for hydroprocessing. Applied Catalysis A: General 1998;171(2) 177-206.

[5] Birchem T. Latest Improvements in ACETM Catalysts Technology for ULSD Production & Deep Cetane Increase. 5th ERTC Annual Meeting: conference proceedings, November 2010, Istanbul, Turkey.

[6] Donnis B, Egeberg RG, Blom P, Knudsen KG. Hydroprocessing of bio-oils and oxygenates to hydrocarbons. Understanding the reaction routes. Top. Catal. 2009; 52 229–240.

[7] Boda L, Onyestyák G, Solt H, Lónyi F, Valyon J, Thernesz A. Catalytic hydroconversion of tricaprylin and caprylic acid as model reactionfor biofuel production from triglycerides. Applied Catalysis A: General 2010;374 158–169.

[8] Bezergianni S, Kalogianni A, Dimitriadis A, Catalyst Evaluation for Waste Cooking Oil Hydroprocessing, Fuel 2012;93 638-647.

[9] Funimsky E. Catalytic Hydrodeoxygenation. Applied Catalysis A: General 2000;199147-190.

[10] Tiwari R, Rana BS, Kumar R, Verma D, Kumar R, Joshi RK, Garg MO, Sinha AK. Hydrotreating and hydrocracking catalysts for processing of waste soya-oil and refinery-oil mixtures. Catalysis Communications 2011;12 559-562.

[11] Lavrenov AV, Bogdanets EN, Chumachenko YA, Likholobov VA. Catalytic processes for the production of hydrocarbon biofuels from oil and fatty raw materials: Contemporary approaches. Catalysis in Industry 2011;3(3) 250-259.

[12] Bulushev DA, Ross JRH. Catalysis for conversion of biomass to fuels via pyrolysis and gasification: A review. Catalysis Today 2011; 171(1) 1-13.

[13] Bezergianni S, Dimitriadis A, Kalogianni A., Pilavachi PA. Hydrotreating of waste cooking oil for biodiesel production. Part I: Effect of temperature on product yields and heteroatom removal. Bioresource Technology 2010;101 6651–6656.

[14] Tyson KS. Biodiesel Handling and Use Guidelines. National Renewable Energy Laboratory 2001, NREL/TP-580-30004

[15] Winayanuwattikun P, Kaewpiboon C, Piriyakananon K, Tantong S, Thakernkarnkit W, Chulalaksananukul W, Yongvanich T. Potential plant oil feedstock for lipase-catalyzed biodiesel production in Thailand. Biomass & Bioenergy 2008;321279–1286.

[16] Da Rocha Filho GN, Brodzki D, Djega-Mariadassou G. Formation of alkanes alkylcykloalkanes and alkylbenzenes during the catalytic hydrocracking of vegetable oils. Fuel 1993;72 543–549.

[17] Gusmao J, Brodzki D, Djega-Mariadassou G, Frety R. Utilization of vegetable oils as an alternative source for diesel-type fuel: Hydrocracking on reduced Ni/ SiO2 and sulphided Ni–Mo/c-Al2O3. Cat Today 1989;5 533–544.

[18] Simacek P, Kubicka D, Sebor G, Pospisil M. Fuel properties of hydroprocessed rapeseed oil. Fuel 2010;89611–615.

[19] Simacek P, Kubicka D, Sebor G, Pospisil M. Hydroprocessed rapeseed oil as a source of hydrocarbon-based Biodiesel. Fuel 2009;88456–60.

[20] Bezergianni S, Kalogianni A. Hydrocracking of used cooking oil for biofuels production. Bioresource Technology 2009;100(17) 3927-3932.

[21] Bezergianni S, Dimitriadis A, Kalogianni A, Pilavachi PA. Hydrotreating of waste cooking oil for biodiesel production. Part I: Effect of temperature on product yields and heteroatom removal. Bioresource Technology 2010;101 6651-6656.

[22] Bezergianni S, Dimitriadis A, Kalogianni A, Knudsen KG. Toward Hydrotreating of Waste Cooking Oil for Biodiesel Production. Effect of Pressure, H2/Oil Ratio, and

Liquid Hourly Space Velocity. Industrial Engineering Chemistry Research 2011;50(7) 3874-3879.

[23] Bezergianni S, Dimitriadis A, Sfetsas T, Kalogianni A. Hydrotreating of waste cooking oil for biodiesel production. Part II: Effect of temperature on hydrocarbon composition. Bioresource Technology 2010;101 7658-7660.

[24] Bezergianni S, Dimitriadis A, Voutetakis S. Catalytic Hydrotreating of Waste Cooking Oil for White Diesel Production. Proceedings of First International Congress on Catalysis for Biorefineries (CatBior), October 2-5,Torremolinos-Málaga, Spain; 2011

[25] Karonis D, Bezergianni S, Chilari D, Kelesidis E. Maximizing the yield of biodiesel from waste cooking oil hydroprocessing by cut-point optimization. Proceedings Am. Chem. Soc., Div. Fuel Chem. 2010; 55(1)

[26] DynamotiveBioOil Information Booklet 2009, http://ingenieria.read.net/blog/PIB-Bio-Oil.pdf (last visited August 6, 2012)

[27] Venderbosch RH, Ardiyanti AR, Wildschut J, Oasmaa A, Heeres HJ. Stabilization of biomass-derived pyrolysis oils. J Chem.Technol.Bioetchnol. 2010; 85 674-686.

[28] Wildschut J, Mahfud FH, Venderbosch RH, Heeres HJ. Hydrotreatment of fast pyrolysis Oil Using Heterogeneous Noble-Metal Catalysts. Ind. Eng. Chem. Res. 2009;48 10324-10334.

[29] Wildschut J. Pyrolysis Oil Upgrading to Transportation Fuels by Catalytic Hydrotreatment. PhD Thesis. Rijksuniversiteit Groningen; 2009.

[30] Baldauf W, Balfanz U, Rupp M. Upgrading of Flash Pyrolysis Oil and Utilization in Refineries. Biomass and Bioenergy 1994;7(1-6) 237-244.

[31] Czernik S, Bridgwater AV. Overview of Applications of Biomass Fast Pyrolysis Oil. Energy& Fuels 2004;18 590-598.

[32] Elliott DC, Neuenschwander GG. Liquid fuels by low severity hydrotreatment of biocrude. Developments in Thermochemical Biomass Conversion, Vol. 1, pp. 611-621, A. V. Bridgwater and D. G. B. Boocock, eds., Blackie Academic & Professional, London: 1996

[33] Baker EG, Elliott DC. Catalytic Hydrotreating of Biomass-Derived Oils. Pacific Northwest Laboratory 1988.

[34] BorjeGevert S, Anderson PV, Sandqvist SP. Hydroprocessing of directly liquefied biomass with large-pore catalysts. Energy & Fuels 1990;4 78-81.

[35] Elliott DC, Baker EG. Upgrading Biomass Liquefaction Products through hydrodeoxygenation. Biotechnol.Bioeng.Symp. 1984, suppl. 14, 159.

[36] Samolada MC, Baldauf W, Vasalos IA. Production of a Bio-Gasoline by Ugrading Biomass Flash Pyrolysis Liquids via Hydrogen Processing and Catalytic Cracking. Fuel 1998;77 1667.

[37] de Miguel Mercader F, Groeneveld MJ, Kersten SRA, Way NWY, Schaverien CJ, Hogendoorn JA. Production of advanced biofuels: Co-processing of upgraded pyrolysis oil instandard refinery units. Applied Catalysis B: Environmental 2010;96 57–66.

[38] Oasmaa A, Kuoppala E, Ardiyanti A, Venderbosch RH, Heeres HJ. Characterization of Hydrotreated Fast Pyrolysis Liquids. Energy Fuels 2010;24 5264–5272.

[39] Venderbosch RH, Heeres HJ. Pyrolysis Oil Stabilisation by Catalytic Hydrotreatment. Biofuel's Engineering Process Technology.

[40] Jones SB, Holladay JE, Valkenburg C, Stevens DJ, Walton C, Kinchin C, Elliott DC, Czernik S. Production of Gasoline and Diesel from Biomass via Fast Pyrolysis, Hydrotreating and Hydrocracking: A Design Case. Prepared for the U.S. Department of Energyunder Contract DE-AC05-76RL01830. February 2009.

[41] Pindoria RV, Megaritis A, Herod AA, Kandiyoti R. A two-stage fixed-bed reactor for direct hydrotreatment of volatiles from the hydropyrolysis of biomass: effect of catalyst temperature, pressure and catalyst ageing timeon product characteristics. Fuel 1998;77(15) 1715–1726.

[42] Conti L, Scano G, Boufala J, Mascia S. Experiments of Bio-oil Hydrotreating in a Continuous Bench-Scale Plant. In: Bridgwater, A.V.; Hogan, E.N., editors. Bio-Oil Production and Utilization. CPL Press, Newbury Berks, 1996, 198

[43] Sharma RK, Bakhshi NN. Catalytic Upgrading of Pyrolysis Oil. Energy & Fuels 1993;7 306-314.

[44] Subiranas M, Schaub A. Combining Fischer-Tropsch (FT) and hydrocarbon reactions under FT reaction conditions - Catalyst and reactor studies with Co or Fe and Pt/ZSM-5, International Journal of Chemical Reactor Engineering 2007; 5(A78).

[45] Rosyadi E, Priyanto U, Suprapto, Roesyadi A, Nurunnabi M, Hanaoka T, Miyazawa T, Sakanishi K, Biofuel production by hydrocracking of biomass FT wax over NiMo/Al2O3-SiO2 catalyst, Journal of the Japan Institute of Energy 2011;90(12) 1171-1176.

[46] www.renew-fuel.com (accessed 8 August 2012)

[47] Lappas AA, Voutetakis SS, Drakaki N, Papapetrou M., Vasalos IA. Production of Transportation Biofuels through Mild-Hydrocracking and catalytic cracking of waxes produced from Biomass to Liquids (BTL) Process. Proceedings of the 14thEuropean Biomass Conference, Paris France; 2005

[48] Posten C, Schaub G. Microalgae and terrestrial biomass as source for fuels—a process view. J.Biotechnol. 2009;142 64–69.

[49] Schenk PM, Thomas-Hall SR, Stephens E, Marx UC, Mussgnug JH, Posten C, Kruse O, Hankamer B. Second Generation Biofuels: High-Efficiency Microalgaefor Biodiesel Production. Bioenerg Res 2008;120-43

[50] Patil V, Källqvist T, Olsen E, Vogt G, Gislerød HR. Fatty acid composition of 12 microalgae for possible usein aquaculture feed. Aquaculture International 2007;151-9.

[51] Kachroo D, Jolly SMS, Ramamurthy V. Modulation of unsaturated fatty acids content in algae Spirulinaplatensis and Chlorella minutissima in response to herbicide SAN 9785. Electronic Journal of Biotechnology 2006;9(4)386-390.

[52] Chen M, Tang H, Ma H, Holland TC, Ng KYS, Salley SO. Effect of nutrients on growth and lipid accumulation in the green algae Dunaliellatertiolecta, Bioresource Technology 2011;1021649-1655.

[53] Yoo C, Jun SY, Lee JY, Ahn CY, Oh HM. Selection of Microalgae for Lipid Production under High Levels Carbon Dioxide. Bioresource Technology 2010;101 S71-S74.

[54] Popovich CA, Damiani C, Constenla D, Martínez AM, Freije H, GiovanardiM, Pancaldi S, Leonardi PI. Neochlorisoleoabundans grown in enriched natural seawater for Biodiesel feedstock: Evaluation of its growth and biochemical composition. Bioresource Technology 2012;114 287–293.

[55] Huber GW, O'Connor P, Corma A. Processing biomass in conventional oil refineries: production of high quality diesel by hydrotreating vegetable oils in heavy vacuum oil mixtures. Applied Catalysis A: General 2007;329120–129.

[56] Bezergianni S, Kalogianni A, Vasalos IA. Hydrocracking of vacuum gas oil-vegetable oil mixtures for biofuels production. Bioresource Technology 2009;1003036–3042.

[57] Tóth C, Baladincz P, Kovács S, Hancsók J. Producing Diesel Fuel by Co-Hydrogenation of Vegetable Oil with Gas Oil. Chemical Engineering Transactions 2010;21 1219-1224.

[58] Neste Oil Corporation.NexBTL diesel. http://www.nesteoil.com/default.asp?path=1,41,11991,12243,12335 (accessed 15 July 2012).

[59] BIOFUELS-2G project. http://www.biofuels2g.gr(accessed 20 July 2012).

[60] Dynamic Fuels. http://www.dynamicfuelsllc.com/ (accessed 18 July 2012).

[61] Honneywell's UOP. http://www.uop.com (accessed 5 July 2012).

[62] Steele P. Powering the Future of Flight. In Proceedings of the World Biofuels Markets 2011, March 22-24, Rotterdam, The Netherlands.

[63] Buse J. Commercialization of Aviation Biofuels. In Proceedings of the World Biofuels Markets 2011, March 22-24, Rotterdam, The Netherlands.

[64] Lahaussois D, Desaeger M. 10% Renewable Energy in transport by 2020: a car manufacturer's perspective. In Proceedings of the World Biofuels Markets 2011, March 22-24, Rotterdam, The Netherlands.

Chapter 10

Hydrotreating Catalytic Processes for Oxygen Removal in the Upgrading of Bio-Oils and Bio-Chemicals

Iñaki Gandarias and Pedro Luis Arias

Additional information is available at the end of the chapter

http://dx.doi.org/10.5772/52581

1. Introduction

In a future sustainable scenario a progressive transition by the chemical and energy industries towards renewable feedstock will become compulsory. Energy demand is expected to grow by more than 50% by 2035 [1], with most of this increase in demand emerging from developing nations. Clearly, increasing demand from finite petroleum resources cannot be a satisfactory policy for the long term. The transition to a more renewable production system is now underway; however, this transition needs more research and investment in new technologies to be feasible.

Biomass appears as the only renewable source for liquid fuels and most commodity chemicals [2]. This is the reason why, in the near future, bio-refineries in which biomass is catalytically converted to pharmaceuticals, agricultural chemicals, plastics and transportation fuels will take the place of petrochemical plants [3]. Indeed, biomass represents 77.4% of global renewable energy supply [4]. Current technologies to produce liquid fuels from biomass are typically multistep and energy-intensive processes, including the production of ethanol by fermentation of biomass derived glucose [5],bio-oils by fast pyrolysis or high pressure liquefaction of biomass [6,7], polyols and alkanes from hydrogenolysis of biomass derived sorbitol [8],and biodiesel from vegetable oils [9].Biomass can also be gasified to produce CO and H_2(synthesis gas), which can be further processed to produce methanol or liquid alkanes through Fischer–Tropsch synthesis [10].

The so-called "First Generation" biofuels, such as sugarcane ethanol in Brazil, corn ethanol in US, oilseed rape biodiesel in Germany, and palm oil biodiesel in Malaysia,already present mature commercial markets and well developed technologies. Nonetheless, there is a worldwide increasing awareness against the use of edible oils and seeds to generate transportation fuels, and critical voices have aroused questioning the actual sustainability of these

"First Generation" biofuels. In fact, nowadays 95 % of biodiesel is made from edible oil [9]. This means that possible food resources are being used as automotive fuels when some part of the World's population is suffering from hunger. Therefore, large-scale production of biodiesel from edible oils may bring about a global imbalance in the food supply market. Another significant concern of using "First Generation" technologies is the deforestation and the destruction of ecosystems. Indeed, the expansion of oil-crop plantations for biofuel production on a large scale has caused deforestation in countries such as Malaysia, Indonesia and Brazil because more and more forest has been cleared for plantation purposes. In addition to this, in developing countries energy crops are powerful competitors for scarce water resources [11].

Being the non-edible portion of the plant and the most abundant source of biomass, lignocellulosic biomass materials are attracting growing attention as sustainable and renewable energy sources. The so-called "Second Generation" technologies for the production of fuels and chemicals can use a wide range of lignocellulosic biomass residues such as agricultural, industrial, and forest wastes, and also energy crops (willow, switchgrass) that do not compete with food crops for available land. The average composition of lignocellulosic material is as follows: 50% cellulose, 25% hemicellulose, and 20% lignin [12]. Cellulose is a linear polysaccharide with β-1,4 linkages of D-glucopyranose monomers (Figure 1). Hemicellulose is a more complex polymer containing five different sugar monomers: five carbon sugars (xylose and arabinose) and six carbon sugars (galactose, glucose, and mannose). Lignin is a highly branched aromatic polymer, that consists of an irregular array of variously bonded "hydroxy-" and "methoxy-" substitutedphenylpropane units. Lignin is mainly found in woody biomass. Lignocellulosic materials can be converted into liquid fuels by three primary routes, including (i) syngas production by gasification, (ii) bio-oil production by pyrolysis or liquefaction, and (iii) acid hydrolysis reactions [13].

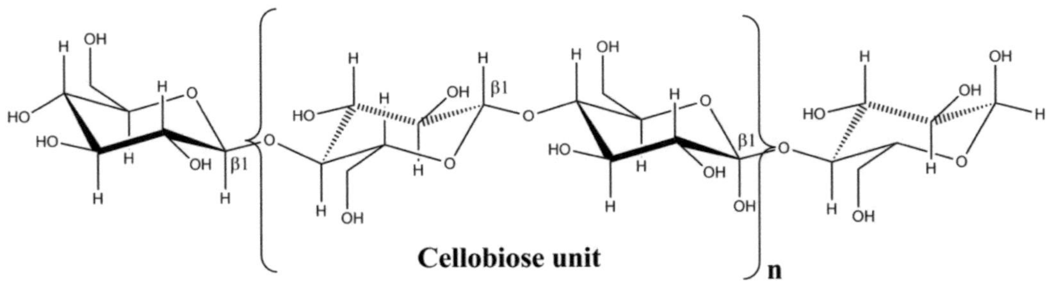

Figure 1. Chemical structure of cellulose.

In the pyrolysis process, biomass feedstock is heated in the absence of oxygen, forming a gaseous product, which after cooling condenses. Depending on the operating conditions that are used, pyrolysis processes are known as slow or fast pyrolysis. Fast pyrolysis processes are characterized by high rates of particle heating (heating rate > 1000ºC/min) to temperatures around 500ºC, and rapid cooling of the produced vapors to condense them (vapor residence time 0.5-5s). In order to obtain that fast heating rates, it is essential to use reactors that provide high external heat transfer (such as fluidized bed reactors) and to guarantee an efficient heat transfer through the biomass particle, using biomass particle size of less than 5 mm [7]. Fast pyrolysis produce 60-75 wt% of liquid bio-oil, 15-25 wt% of solid char, and 10-20 wt% of non condensable gases, depending on the feedstock. In slow pyrolysis biomass is heated to around 500ºC at much lower heating rates than those used in fast pyrolysis. The vapor residence times are much longer; they vary from 5 min to 30 min. As a consequence of the lower heating rate and of the longer vapor residence time, lower yields to pyrolysis oils and higher yields to char and gas products are obtained (Figure 2). As a result of all this, for bio-oil production from biomass, fast pyrolysis processes are preferred.

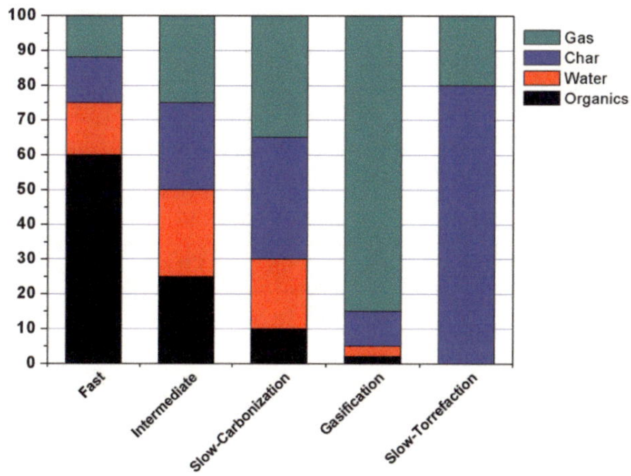

Figure 2. Product spectrum from pyrolysis. Data from [14].

Bio-oils are dark-red brown color liquids. They are also known as pyrolysis oils, bio-crude oil, wood oil or liquid wood. Bio-oils usually have higher density, viscosity and oxygen content compared to fuel-oil. While the sulfur and nitrogen content is usually smaller (Table 1). The high oxygen content of bio-oils generates some negative characteristics like low heating value (HV), immiscibility with conventional fuels and high viscosity. A serious problem of bio-oils is their instability during storage, as their viscosity, HV and density are affected. This is because some of the organic compounds present in bio-oils are highly reactive. For instance, ketones, aldehydes and organic acids react to form ethers, acetals and hemiacetals respectively [15]. Therefore, bio-oils need to be upgraded to reduce their oxygen content in order to increase their stability, to be miscible with conventional oil, and to increase their H/C ratio. This upgrading can be carried out through three different routes: (i) catalytic hydrotreating, usually known as hydrodeoxygenation (HDO), which consists mainly on decarboxylation, hydrocracking, hydrogenolysis and hydrogenation reactions, (ii) zeolite upgrading or (iii) through esterification reactions. Zeolite upgrading is carried out without external hydrogen sources, and therefore the resulting oil has lower HV and H/C than conventional fuels. Esterification can significantly increase the chemical and physical properties of bio-oil, however it requires using high amounts of alcohols, which are highly demanded. Catalytic hydrotreating appears to have the greatest potential to obtain high grade oils which are compatible with the already available infrastructure for fossil fuels.

Property	Pyrolysis Oil	Heavy Oil
Moisture Content, wt %	15-30	0.1
pH	2.5	
Elemental Composition, wt %		
Carbon	54-58	85
Hydrogen	5.5-7.0	11
Oxygen	35-40	1.0
Nitrogen	0-0.2	0.3
Ash	0-0.2	0.1
Higher Heating Value, MJ/kg	16-19	40
Viscosity (50°C), cP	40-100	180
Solids (wt%)	0.2-1.0	1

Table 1. Typical Properties of Wood Pyrolysis Bio-Oil, and Heavy Fuel Oil [13].

Not only fuels, but also commodity chemicals are nowadays derived from petroleum-based resources. Commodity chemicals are involved in the production of a wide variety of products and thus are an essential and integral part of the modern societies. Hence, in the search for a sustainable scenario, it is crucial to also look towards alternative biorenewable sources for these chemicals. In the case of platform chemicals coming from biomass, such as glucose, levulinic acid, 5-(hydroxyl-methyl furfural), sorbitol, or glycerol,they usually have higher O/C ratio than most commodity chemicals. Therefore, the conversion of these platform chemicals into value-added chemicals usually requires O removal reactions.

This book chapter summarizes the main aspects involved in the catalytic hydrotreating processes for the oxygen removal from bio-oils and from biomass based platform chemicals.

2. Hydrotreating catalytic processes in bio-oil upgrading

As it has been stated in the introduction, a general characteristic of bio-oils coming from the pyrolysis of biomass is their high oxygen content (35-40 wt%). More than 300 compounds have been identified in bio-oil, most of them containing oxygen atoms. The exact composition of the bio-oil depends on the type of biomass fed. These compounds can be classified in five broad categories: (i) hydroxyaldehydes, (ii) hydroxyketones, (iii) sugars and dehydrosugars, (iv) carboxylic acids, and (v) phenolic compounds [16]. Hydroprocessing of biomass-derived oils differs from processing petroleum because of the importance of deoxygenation as compared to nitrogen or sulfur removal. Bio-oil hydrodeoxygenation (HDO) process implies complex reaction networks that includes cracking, decarbonylation, decarboxylation, hydrocracking, hydrogenolysis, hydrogenation and polymerization. The upgrading process should yield a product with lower amount of water and oxygen, decreased acidity and viscosity, and higher HV. The complexity of the reactions and the high variety of oxygenated compounds make the evaluation of bio-oil upgrading difficult and has brought the use of model compounds such as phenol, guaicol, 2-ethylphenol, methyl heptanoate or benzofuran to test different catalysts and to understand the main characteristics of the HDO process. Elliot [17] has reported the HDO reactivity of different organic compounds that are typically present in bio-oils (see Figure 3). Olefins, aldehydes and ketones can easily be reduced by H_2 at temperatures as low as 150–200 °C. Alcohols react at 250–300 °C by hydrogenation and thermal dehydration to form olefins. Carboxylic and phenolic ethers react at around 300 °C. Regarding the operating pressures, due to the low solubility of hydrogen in organic and aqueous solutions, high pressures are required to guarantee high availability of hydrogen in the vicinity of the catalyst (80-300 bar of H_2 pressure) [15].

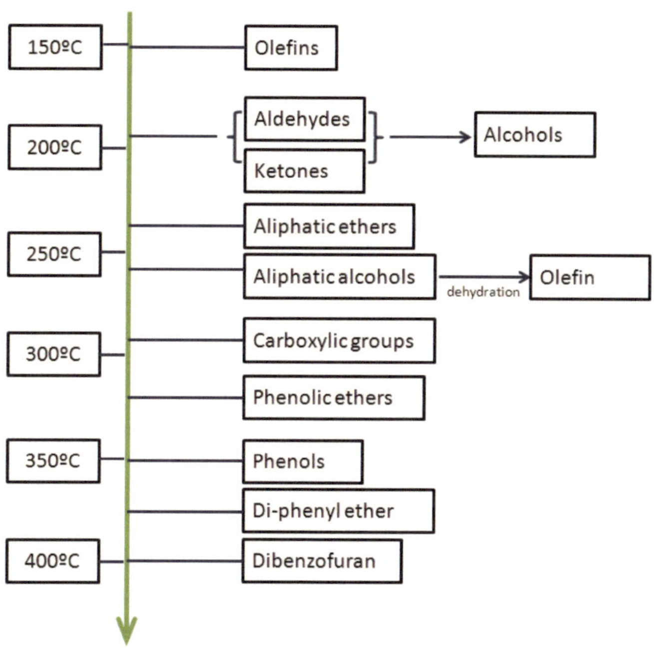

Figure 3. Reactivity scale of organic components under HDO conditions. Adapted from [17].

2.1. Catalysts and reaction mechanisms

HDO is a process closely related to hydrodesulphurization (HDS), which is highly developed in the oil-refinery industry. In both processes, hydrogen is used to remove the heteroatom in the form of H_2O and H_2S respectively. This is the reason why several works on bio-oil HDO use catalytic systems already used in HDS processes, such as Co-Mo or Ni-Mo based catalysts. These catalysts are active in their sulphide form, so they need to be pretreated with H_2S before operation to obtain Co-MoS_2 or Ni-MoS_2 active sites. Romero et al. [18] using Co-MoS_2 type catalysts for the HDO of 2-ethylphenol at 340ºC and 7 MPa of hydrogen pressure proposed the reaction mechanism described in Figure 4. It is suggested that the oxygen from the molecule adsorbs on a vacancy of a MoS_2 matrix. At the same time, the H_2 from the feed dissociatively adsorbs on the catalyst surface forming S-H species. The addition of a proton to the adsorbed oxygenated molecule leads to an adsorbed carbocation. This intermediate can directly undergo a C–O bond cleavage and the aromatic ring is regenerated leading to ethylbenzene. The vacancy is afterwards recovered by elimination of water.

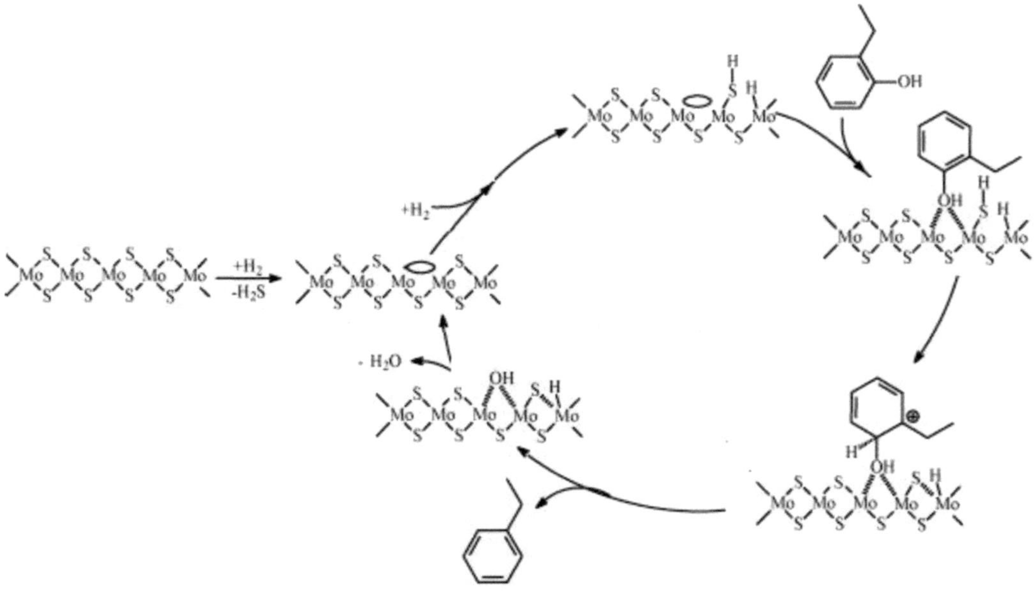

Figure 4. Proposed mechanism of HDO of 2-ethylphenol over a schematic Co-MoS_2 catalyst Adapted from [18].

The problem of using MoS_2 type catalysts for HDO of bio-oils is that during prolonged operation sulfur stripping and oxidation of the surface of the catalyst occurs, causing deactivation of the catalyst. The reason is that as compared to conventional oil, the sulfur content of bio-oil is very low (less than 0.1 wt % [19]). One alternative to avoid this problem is the co-feeding of H_2S to the system, in order to regenerate the sulfide sites. For instance, in the HDO of alyphatic esters over a $CoMoS_2/Al_2O_3$ and $NiMoS_2/Al_2O_3$ catalysts a promoting effect was observed in the activity of the catalyst when co-feeding H_2S, however this co-feeding did not prevent from catalyst deactivation. This promoting effect was related to the increase in Brönsted acidity in the presence of H_2S [20]. Nonetheless, the use of H_2S has also some drawbacks. In the HDO of phenol over a Ni-MoS_2-Al_2O_3 catalyst, it was observed an

http://dx.doi.org/10.5772/52581

inhibitory effect of H_2S, leading to a decrease in phenol conversion and not preventing catalyst deactivation. This was ascribed to the competitive adsorption between phenol and H_2S [21]. Moreover, the formation of sulfur-containing compounds such as dimethyl sulfide, diheptyl sulfide, hexanethiol and heptanethiol was observed in the HDO of aliphatic oxygenates over Co-MoS_2 catalysts, even in the absence of sulfiding agents [22]. Therefore, the use of MoS_2 type catalysts in bio-oil HDO seems challenging, becouse sulfur free bio-oil can be contaminated by sulfur, and because wood-based bio-oils contain high amounts of phenolic compounds that would compete with H_2S for the active sites of the catalyst.

Another alternative is the use of bi-functional catalysts formed by the combination of transition metals and oxophilic metals, such as MoO_3, Cr_2O_3,WO_3 or ZrO_2. In this case, the oxophilic metal acts as a Lewis acid site. The oxygen ion pair of the target molecule is attracted by the unsaturated oxophilic metal. The second step of the mechanism is hydrogen donation. In this case, the hydrogen molecule is dissociatively adsorbed and activated on the transition metal. Finally, the activated hydrogen is transferred to the adsorbed molecule.

Regarding the support, γ-Al_2O_3 is the most commonly used one. Nonetheless, it has to be taken into account the structural changes that γ-Al_2O_3 might suffer under the typical operating conditions in HDO. In contact with hot water (T > 350ºC), γ-Al_2O_3 is converted into a hydrated boehmite (AlOOH) phase with a significant decrease in the acidity and surface area [23]. Moreover, the relatively high surface acidity of Al_2O_3 is thought to promote the formation of coke precursors. In fact, coke formation is one the main factors affecting the stability of the catalyst. Therefore, the use of less acidic or neutral support like active carbon or SiO_2 is an interesting alternative [24]. For instance, Echeandia et al. [25] using Ni-WO_3 on active carbon for the HDO of 1 wt% phenol in n-octane at 150-300ºC and 15 bar observed lower coke formation on the surface of the active carbon with respect to alumina support. Based on product analysis, they also concluded that HDO of phenol occurs via two separate pathways: one leading to aromatics through a direct hydrogenolysis route, and the other one to cyclohexane, through a hydrogenation-hydrogenolysis route (see Figure 5). In terms of obtaining a final product with high octane number and reducing the consumption of hydrogen, direct hydrogenolysis reaction is preferred. Nonetheless, aromatics are harmful to human health and its content in transportation fuels is limited by legislation. Therefore, it is important to understand which sites are responsible of each route, in order to obtain an upgraded product with the desired aromatic content. CeO_2 and ZrO_2 supports have also shown to give good results in the HDO of different molecules. ZrO_2-supported noble metal catalysts (Rh, Pd and Pt) [26] were compared with the conventional sulfided CoMo/Al_2O_3 catalyst in the HDO of Guaiacol in the presence of H_2 at 300 °C. Sulfided CoMo/Al_2O_3 deactivated due to carbon deposition, and the products were contaminated with sulfur, however, neither problem was observed with the ZrO_2-supported noble metal catalysts. As a conclusion, a good support for HDO should provide high affinity for the oxygen-containing molecule while presenting moderate acidity in order to minimize the formation of coke deposits.

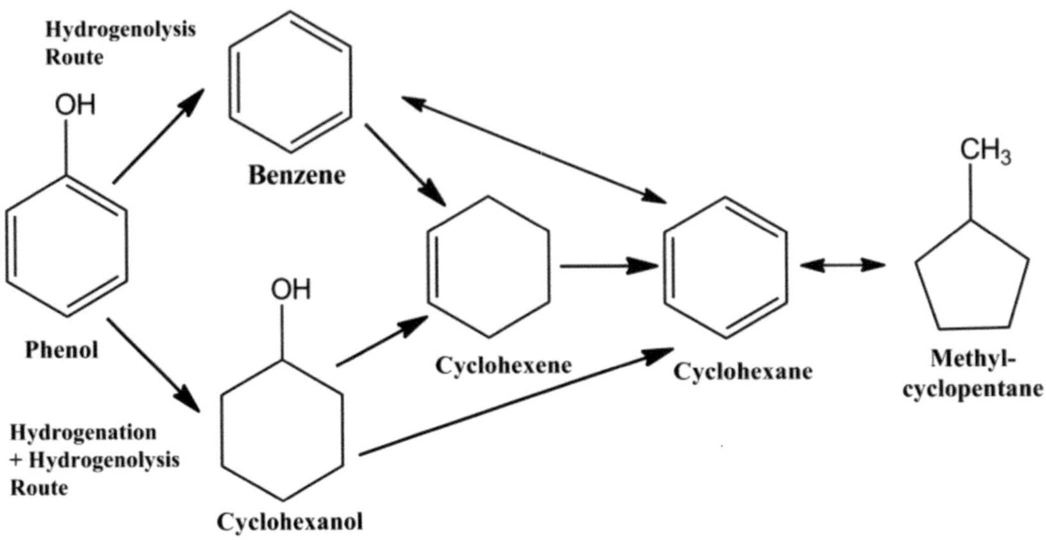

Figure 5. Scheme of phenol HDO. Adapted from [25].

2.2. Upgrading of real bio-oils

An important aspect in the HDO of bio-oils is the required degree of deoxygenation. It is assumed that the upgraded oil should contain less than 5 wt% oxygen so that the viscosity is decreased to that required for fuel applications [17]. However, during the hydrotreating, not only the oxygen is removed in the form of water, but also the saturation of double bounds occurs. This saturation has two significant negative effects. The first one is related to the quality of the upgraded oil, because the saturation of the aromatic components has a highly detrimental effect in the octane number. For instance, the octane number of toluene (119) decreases to 73 when the aromatic ring is hydrogenated [10]. The second negative effect is related to the consumption of hydrogen. According to Venderbosh et al. [27] in order to achieve 50% of deoxygenation 16 g H_2/Kg of bio-oil is required, which is close to the expected stoichiometry value. Nonetheless, if the aim is to obtain the total removal of oxygen, the H_2 consumption increases to 50 g H_2/Kg of bio-oil; which means that the H_2 consumption is 56% higher than the stoichiometry value. Some other studies suggest even higher H_2 consumption requirements, 62 g H_2/Kg of bio-oil [28]. This deviation of the H_2 consumption from the stoichiometry value is explained on the basis of the different reactivity of the oxygenated compounds present in the bio-oil. High reactive compounds, such as ketones, are easily converted with low hydrogen consumption. However, more complex molecules, such as phenols, might suffer the hydrogenation/saturation of the molecule and therefore the hydrogen consumption exceeds the stoichiometric prediction at the high degree of deoxygenation.

In order to obtain high degrees of HDO but minimizing the hydrogenation of aromatics in bio-oil, two step hydrogenating processes have been developed. In the first stage, high reactive and unstable compounds are transformed into more stable ones at low temperature (270ºC, 136 atm H_2) and without a catalyst. In the second step, a deeper HDO is carried out at higher temperatures (400ºC, 136 atm H_2) and using hydrotreating catalysts. The two-step hydrotreatment allows 13% reduction in hydrogen consumption for equivalent oil yield. Nonetheless, the reported octane number of the upgraded bio-oil, 72, is still lower than that of gasoline [17].

Environmental aspects should also be taken into account. Aromatic compounds have on one hand high octane number; however, they are also harmful to health. Indeed, environmental standards for aromatics in transportation fuels are becoming more restrictive. Thus, it seems challenging to achieve an agreement between obtaining oils with high octane number while fulfilling aromatic content policies.

3. Hydrogenolysis reactions in the valorization of platform chemicals

Biomass components have a great potential as building block intermediates. Indeed, sugars, vegetable oils and terpenes can be employed for synthesizing products with a high added value, such as chemicals and fine chemicals. There are hundreds of different processes to obtain chemicals from biomass origin building blocks. This chapter deals with those processes involving hydrotreating for the removal of oxygen. In the first part of this section, some examples of significant hydrogenolysis reactions in the valorization of platform chemicals will be given, while the last part will be focused on one of the most studied hydrogenolysis proccesses; the conversion of glycerol into propanediols (PDO).

As it has been previously stated, platform chemicals coming from biomass usually contain higher O/C ratio than most commodity chemicals; thus main valorization processes require the removal of oxygen. One widely used process to remove oxygen is hydrogenolysis. Hydrogenolysis is a type of reduction that involves chemical bond dissociation in an organic substrate and simultaneous addition of hydrogen to the resulting molecular fragments [33]. Therefore, reaction for oxygen removal involves the cleavage of the C-O bond and the addition of hydrogen (oxygen is removed in the form of H_2O). This is a significant aspect, because, in those processes where the starting and target molecule have the same number of carbons it is important to use catalytic systems that present high activity in C-O bond hydrogenolysis while low activity in C-C bond hydrogenolysis.

3.1. Hydrogenolysis of sugars

Two types of sugars are present in biomass: hexoses (six-carbon sugars), of which glucose is the most common one, and pentoses (five-carbon sugars), of which xylose is the most common one. Glucose and xylose can be easily hydrogenated to yield sorbitol [29] and xylitol [30] respectively. These two molecules can undergo C-C and C-O hydrogenolysis in the presence of hydrogenation catalysts, leading mainly to a mixture of ethyleneglycol, glycerol, and 1,2-propanediol. Other products such as butanediols, lactic acid, methanol, ethanol, and propanol can also be formed (Figure 6). Ni is known to show high hydrogenolysis activity towards C-C and C-O bond hydrogenolysis, this is the reason why, the use of Ni on different acid supports seems an interesting alternative for this process. For instance, Ni supported on NaY zeolite gave 68% sorbitol conversion with 75% combined selectivity to 1,2-PDO and glycerol at 220ºC and 60 bar H_2 pressure after 6 h [8]. The addition of Pt to the catalyst did not influence its activity and selectivity significantly. However, in the case of 20 wt% Ni/Al_2O_3 prepared by coprecipitation, the addition of 0.5 wt% of Ce significantly increased sorbitol conversion (from 41% to 91%) and the stability of the catalyst [31]. It seems that the addition of Ce considerably reduces Ni leaching, and hence improves the stability of the catalyst. Other catalytic systems have also been reported besides the Ni acid-support ones. For instance, Ru supported on carbon nanofiber and graphite felt composite catalysts gave 68% sorbitol conversion and 79% propylene glycol selectivity at 220ºC and 8.0 MPa hydrogen pressure [32].

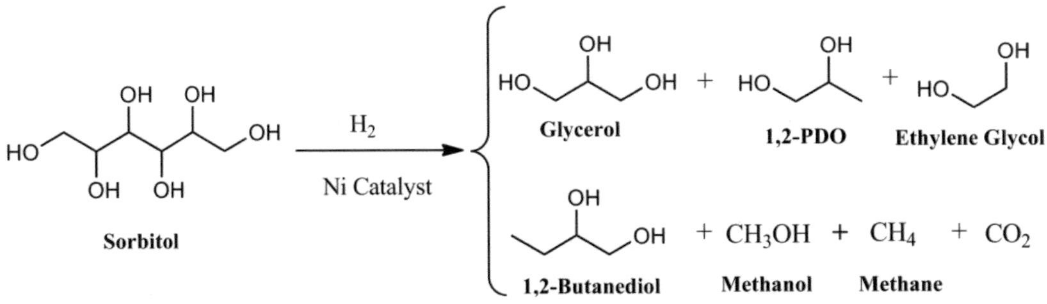

Figure 6. Reaction products of catalytic hydrogenolysis of sorbitol over supported Ni catalyst in the aqueous phase. Adapted from [31].

3.2. Hydrogenolysis of 5-Hydroxymethyl-Furfural (HMF)

5-Hydroxymethyl-furfural (HMF) can be obtained in a biphasic reactor from the acid-catalyzed dehydration of hexoses[33]. HMF by itself cannot be used as motor fuel due to its high boiling point (283ºC). However, it can be transformed to 2,5-dimethylfuran (DMF) through a two consecutive hydrogenolysis reactions (see Figure 7). DMF not only decreases the boiling point to a value suitable for liquid fuels, but also attains the lowest water solubility and the highest octane number (RON) of the mono-oxygenated C_6 compounds, while preserving a high energy density 30 kJ cm^{-3}, which is 40% higher that the energy density of bio-ethanol and comparable to the one of gasoline (35 KJ cm^{-3}) [34]. Roman-Leshkov et al. [34] used CuRu/C catalysts (prepared by incipient wetness impregnation) in a flow reactor using 5 wt % HMF in a 1-butanol solution at 220 ºC and 6.8 bar H_2 pressure. Yields to DMF of 71% were measured. An important aspect in their process is that the catalyst should be chloride-resistant, because, NaCl was used in the dehydration step of hexoses to HMF to increase their solubility in water. Very recently, Luijkx et al. [35] reported the production of 2,5-DMF by the hydrogenolysis of 5-HMF over a Pd/C catalyst in 1-propanol. Due to simultaneous alcoholysis, significant amount of ethers products were formed.

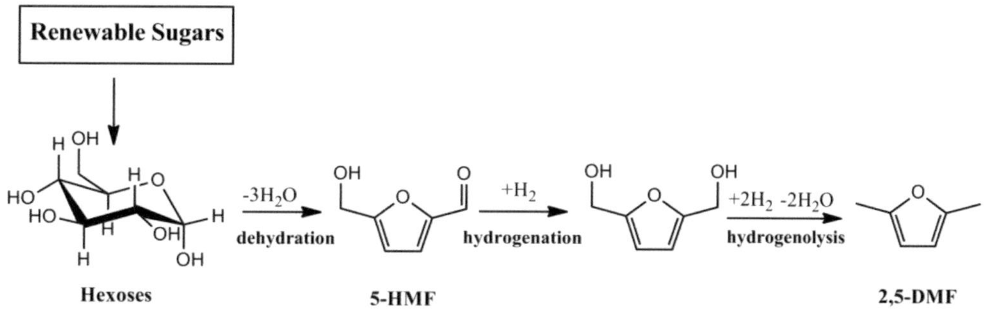

Figure 7. Reaction scheme for the conversion of sugars into 2,5-dimethylfuran. Adapted from [36]

3.3. Hydrotreating of vegetable oils and hydrogenolysis of fatty acids

Biodiesel is currently obtained from the transesterification reaction of vegetable oils. A possible drawback of this technology is that large investment is required to build up new biodiesel plants. An interesting alternative is to directly feed the vegetable oil into the hydrotreating unit of a petroleum refinery, for instance, vegetable oil can be co-fed with heavy vacuum oil HVO. Under typical hydrotreating conditions (300-450ºC, 50 bar H_2 pressure, sulfidedNiMo/Al_2O_3 catalyst), vegetable oils are transformed into alkanes through three different pathways: decarboxylation, decarbonylation and HDO. The straight chain alkanes can undergo isomerization and cracking to produce lighter and isomerized alkanes (Figure 8) [37]. It was reported that mixing the sunflower oil with HVO does not decrease the rate of desulfurization. Moreover, the rate of vegetable oil hydrotreating is faster that the rate of HVO desulfurization. For industrial application, corrosion problems should be taken into account and the formation of waxes should be minimized, as they can plug the reactor.

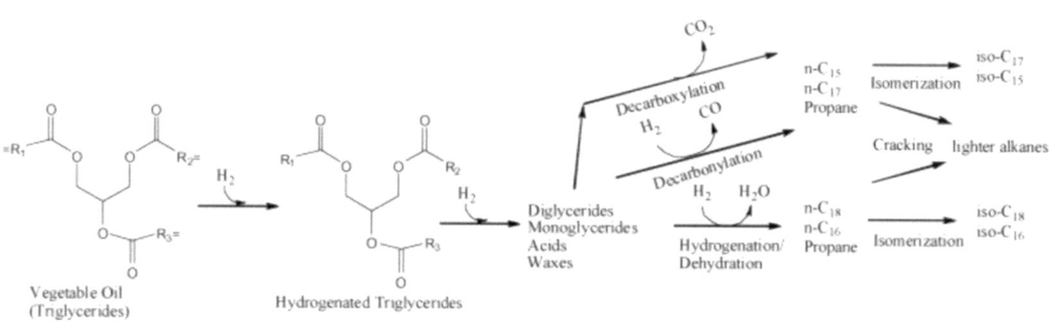

Figure 8. Reaction pathway for conversion of tri-glycerides to alkanes [37].

Fatty alcohols can be obtained by catalytic hydrogenolysis of fatty acid methyl esters. Small-chain fatty alcohols are used in cosmetics and food and as industrial solvents or plasticizers, while the large-chain fatty alcohols are important as biofuels and as nonionic surfactants or emulsifiers. Fatty alcohols are produced by hydrogenolysis, in the presence of Cu based heterogeneous hydrogenation catalysts, operating under H_2 pressures between 20 and 30 bar and temperatures in the range of 97-197ºC [38]. High hydrogen pressures are required to increase the solubility of hydrogen in the reaction mixture, in order to boost the availability of H_2 at the catalyst surface and to reduce mass transport limitations [39].The stoichiometry of the reaction is presented below:

$$R\text{-}COOCH_3 + 2H_2 \rightarrow R\text{-}CH_2OH + CH_3OH$$

3.4. Hydrogenolysis of glycerol

In the last years, much attention has been devoted to the valorization of glycerol. Glycerol is obtained as byproduct in the transesterification reaction of fatty acids to produce biodiesel. With the significant increase of worldwide biodiesel production, there is also an important increase in glycerol availability. Due to the increments in biodiesel manufacture, important amounts of glycerol have been placed in the market, and glycerol has become a waste difficult to handle. The volumes of glycerol remaining unsold in recent years are a clear example of wasted energy and material resources. This is the reason why intense research activity has started worldwide in order to find an exit to the big amounts of glycerol produced. Glycerol price has experimented constant reduction during the last years. Low glycerol prices allow new interesting applications like the production of high added value chemicals. Effective valorization of glycerol will enable to make more cost effective biodiesel production and to replace fossil fuels as the raw material for the production of commodity chemicals.

Among the different possible transformations of glycerol, the hydrogenolysis to propanediols (PDO) presents special interest due to the big number of applications of both 1,2 and 1,3-propanediol (PDO). 1,3-PDO has traditionally been considered a specialty chemical; it has been used in the synthesis of polymers and other organic chemicals, but its market has been quite small. However, over the past years this situation has changed significantly. 1,3-PDO is a starting material in the production of polyesters. It is used together with terephthalic acid to produce polytrimethylene terephthalate (PTT), which is in turn used for the manufacture of fibers and resins. This polymer is currently manufactured by Shell Chemical (Corterra polymers) and DuPont (Sorona 3GT).1,2-PDO is a major commodity chemical traditionally derived from propylene oxide, and hence also based on fossil feedstock. It is a widely used commodity chemical that plays a significant role in the manufacture of a broad array of industrial and consumer products, including unsaturated polyester resins, plasticizers and thermoset plastics, antifreeze products, heat-transfer and coolant fluids, aircraft and runway deicing products, solvents, hydraulic fluids, liquid detergents, paints, lubricants, cosmetics and other personal care products. Today, the industry estimates a global demand for 1,2- PDO between 2.6 and 3.5 billion lb/yr [48]. One of the future main markets for 1,2-PDO shall be the substitution of ethylene glycol (EG) in cooling water systems to prevent freezing, as ethylene glycol is harmful to health.

3.4.1. Reaction mechanisms

Glycerol hydrogenolysis to PDOs consists of hydrogen addition and removal of one oxygen atom in the form of H_2O. In order to design efficient catalysts, it is fundamental to understand the mechanism of this reaction. Three main reaction mechanisms have been proposed in the literature, depending on whether the reaction runs on acid or basic catalytic sites and with or without the formation of intermediate compounds:

i. dehydrogenation – dehydration – hydrogenation (glyceraldehyde route),

ii. dehydration– hydrogenation,

iii. direct glycerol hydrogenolysis.

Below, the main features of each mechanism will be discussed

i. Glyceraldehyde route

One of the first studies related to glycerol hydrogenolysis was developed by Montassier et al. [40] in the late 1980s. They suggested that over Ru/C catalyst glycerol is first dehydrogenated to glyceraldehyde on the metal sites. Next, a dehydroxylation reaction takes place by a nucleophilic reaction of glyceraldehyde with water or with adsorbed -OH species. Finally, hydrogenation of the intermediate yields 1,2-PDO (Figure 9). The main controversial point of this mechanism is the initial dehydrogenation step, which is thermodynamically unfavored due to the high hydrogen pressures used [41]. Therefore, in order to shift the equilibrium, glyceraldehyde dehydration should be faster than glycerol dehydrogenation. Otherwise glyceraldehyde would be hydrogenated back to glycerol on the metal sites. Several authors observed that the addition of a base notably increased glycerol conversion, and this was related to the fact that bases enhance glyceraldehyde dehydration [42-44]. It is interesting to point out that when glycerol hydrogenolysis is carried out under alkaline conditions, marginal 1,3-PDO selectivities are measured.

Apart from 1,2-PDO, other products stemming from C-C bond cleavage were also reported when glycerol hydrogenolysis is conducted under alkaline conditions; mainly, ethylene glycol (EG), methanol and methane. It is suggested that glyceraldehyde can either undergo dehydration or retro-aldolization reactions.The so formed intermediates are hydrogenated in the last step to yield the products of C-C bond cleavage. Because both the glyceraldehyde dehydration and glyceraldehyde retro-aldol reaction are catalyzed by OH^-, the addition of a base increases the glycerol reaction rate but does not improve the selectivity to 1,2-PDO [45].

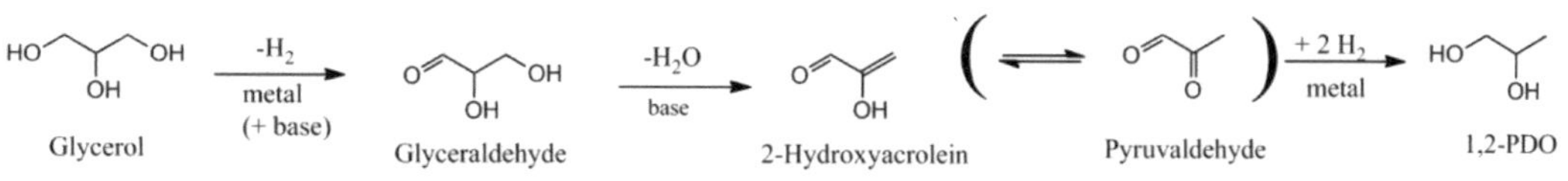

Figure 9. PDO formation from glycerol under alkaline conditions.

ii. Dehydration-hydrogenation route

Dasari et al. [46] observed the formation of acetol (hydroxyacetone) together with 1,2-PDO using copper-chromite catalyst at 473 K and 15 bar hydrogen pressure. Moreover, glycerol hydrogenolysis to 1,2-PDO occurred even in the absence of water. Since the copper-chromite catalyst was reduced in a stream of hydrogen prior to the reaction, no surface hydroxyl species were present to take part in the reaction. Therefore, the mechanism suggested by Montassier et al. (Figure 9) was not able to explain these results. Dasari et al. proposed a new mechanism in which glycerol is first dehydrated to acetol, which is further hydrogenated to 1,2-PDO (Figure 10). Based on their findings, a two step process was developed [47]. In the first step, acetol is generated from glycerol dehydration by a reactive distillation process, operating at 513 K, slight vacuum and using copper-chromite catalyst. The acetol obtained is then hydrogenated at 15 bar H_2 pressure using the same catalyst. The process was patented in the USA in 2005 [48].

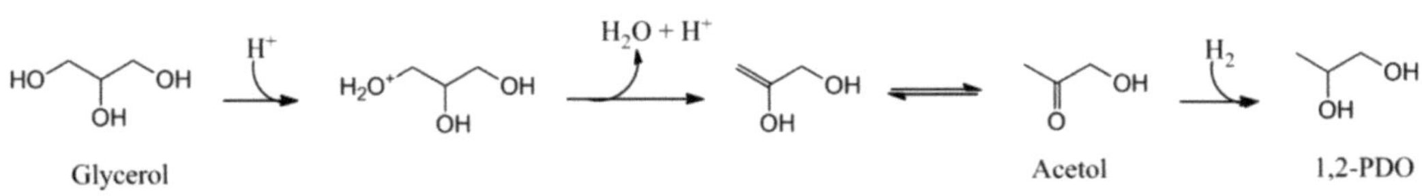

Figure 10. PDO formation via the dehydration-hydrogenation route.

According to Schlaf, acid-catalyzed hydrogenolytic cleavage of -OH group occurs through an initial protonation of the hydroxyl group that leads to the formation of a carbocation and water [49]. Thermodynamically, the formation of a secondary carbocation is more favored than the formation of a primary carbocation. Therefore, operating under acid conditions should bring about higher selectivity to 1,3-PDO. The fact that product distribution is usually shifted towards 1,2-PDO seems to be a complex function of operating conditions, catalyst and starting materials. Ethylene glycol, ethanol, methanol and methane are usually reported as degradation products. Ethylene glycol and methanol are formed from the C-C bond cleavage reaction of glycerol, while ethanol stems from the further hydrogenolysis of ethylene glycol.

iii. Direct glycerol hydrogenolysis

A direct glycerol hydrogenolysis mechanism was recently proposed by Yoshinao et al. [50]. The experiments were carried out using Rh-ReO_x/SiO_2 and Ir-ReO_x/SiO_2 catalysts at 393 K and 80 bar H_2 pressure. The low reaction temperature implies that the dehydration-hydrogenation route was not further possible, due to the endothermic character of glycerol dehydration and the required activation energy, and suggests the energetically more favored direct hydrogenolysis reaction [51]. They suggested a direct hydride ▪ proton mechanism. The selected catalysts are able to activate hydrogen easily and to form hydride species. It is proposed that glycerol is adsorbed on the surface of ReO_x clusters to form alkoxide species. Glycerol can form two adsorbed alkoxides: 2,3-dihydroxypropoxide and 1,3-dihydroxyisopropoxide; it is suggested that the formation of 2,3-dihydroxypropoxide is preferred as it requires a smaller adsorption cross-section than 1,3-dihydroxyisopropoxide [52]. Next, the hydride attack to the 2-position of 2,3-dihydroxypropoxide gives 1,3-PDO, while the hydride attack to the 3-position of 2,3-dihydroxyisopropoxide yields 1,2-PDO. The higher selectivity to 1,3-PDO obtained (1,3-PDO/1,2-PDO ratio = 2.7) is explained on the basis of the higher stability of the six membered-ring transition state that leads to the formation of 1,3-PDO as compared to the stability of the seven membered-ring transition state that leads to the formation of 1,2-PDO (Figure 11).

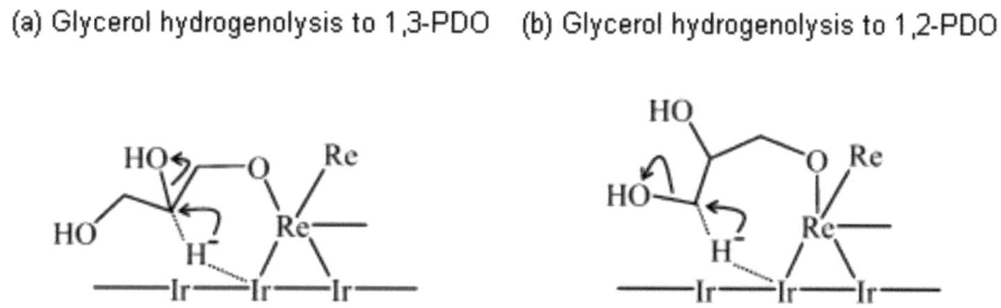

Figure 11. Model structures of the transition states of the hydride attack to the adsorbed substrate in the glycerol hydrogenolysis [52].

A different direct glycerol hydrogenolysis mechanism was established by Chia et al. [53] trying to explain the hydrogenolysis of different polyols and cyclic ethers over a Rh-ReO_x/C catalyst. They concluded from DFT calculations that the -OH groups on Re associated with Rh are acidic. The acidic nature of ReO_x was also reported before [54]. Such acidic Re sites can donate a proton to the reactant molecule and form carbenium ion transition states. In the case of glycerol hydrogenolysis, the first step involves the formation of a carbocation by protonation-dehydration reaction. This carbocation is stabilized by the formation of a more stable oxocarbenium ion intermediate resulting from the hydride transfer from the primary -CH_2OH group. Final hydride transfer step leads to 1,2-PDO or 1,3-PDO [53]. The authors also reported that the secondary carbocation is more stable than the primary carbocation. Nevertheless, higher selectivity to 1,2-PDO was obtained (1,3-PDO/1,2-PDO ratio = 0.65).

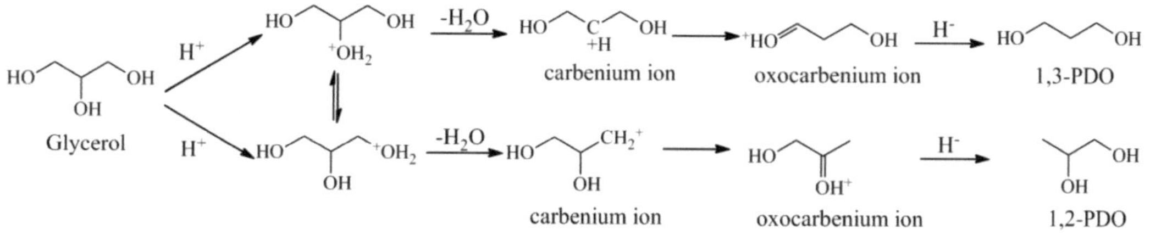

Figure 12. Reaction mechanism for direct glycerol dehydrogenation. Adapted from[55].

3.4.2. *Catalytic systems*

i. Noble metals

Hydrogenolysis reactions involve the addition of hydrogen to an organic molecule. Therefore, hydrogenolysis catalysts must be able to activate hydrogen molecules. Noble metals are known to be active for the dissociation of hydrogen molecules and are widely used in hydrogenation reactions. The first studies on glycerol hydrogenolysis were carried out using Ru based catalysts [56]. Feng et al. [57] studied the effect of different supports (TiO_2, SiO_2, NaY, γ-Al_2O_3) on Ru based catalysts. The TiO_2 supported catalyst exhibited the highest activity giving a glycerol conversion of 90.1%; however, it also favored the production of ethylene glycol over 1,2-PDO. In contrast, Ru/SiO_2 showed the lowest activity, but resulted in much higher selectivity to 1,2-PDO. They also performed blank reactions with the supports, achieving no significant conversions; which indicated that the supports cannot catalyze the reaction independently. Ru particle size was affected by the type of support, and a correlation was established between the size of the Ru particle and the activity of the catalyst, being higher with decreasing Ru particle size.

Apart from Ru, other noble metals have also been studied. For instance, Furikado et al. [58] compared the activity of various supported noble-metal catalysts (Rh, Ru, Pt and Pd over C, SiO_2 and Al_2O_3). Among all the catalysts, the best results in terms of 1,2-PDO selectivity were achieved with Rh/SiO_2 at low reacting temperature and low glycerol conversions (7.2). Nevertheless, the selectivities to 1,2-PDO obtained were rather low, due to the over-hydrogenolysis of 1,2- and 1,3-PDO to 1 and 2-PO.

The use of noble metal-base bifunctional catalytic systems has also been reported. As it was previously described in the glyceraldehyde based mechanism, the dehydration of glycerol to glyceraldehyde, and further dehydration of glyceraldehyde to pyruvaldehyde are both thought to be catalyzed by adsorbed hydroxyls.The effect of different base additives on the performance of Ru/$TiO_2$was reported [45]. The addition of Li or Na hydroxides dramatically increased the glycerol hydrogenolysis activity of Ru/$TiO_2$and the selectivity to 1,2-PDO. The highest conversion of glycerol (89.6%) and the highest selectivity to 1,2-PDO (86.8%) were observed with LiOH. The selectivity to 1,2-PDO was similar with all the bases added, which showed that the selectivity to 1,2-PDO is independent of base concentration within a certain range. However, the selectivity to ethylene glycol decreased no matter which base was added. Almost no reaction was observed in the absence of Ru/TiO_2, indicating that the presence of metal is required in order to take place glycerol hydrogenolysis. The lower selectivity to ethylene glycol with increasing base addition to the reacting solution was explained by the fact that ethylene glycol presented higher affinity to adsorb in the surface of the catalyst and to suffer the attack of hydroxyl groups, whose concentration was higher at elevated pH values [59].

Noble metal-acid catalytic systems have also been used. According to the mechanism in Figure 10, glycerol is firstly dehydrated to acetol, which is then hydrogenated to 1,2-PDO. The first dehydration step is supposed to be catalyzed by acid sites while the second one by metal sites. Therefore, one interesting option to increase the selectivity to target product, 1,2-

PDO, is the use of bifunctional noble metal-acid catalysts. Different Bronsted acids like sulfonated zirconia, zeolites, homogeneous H_2SO_4 and Amberlyst 15 were tested together with Ru/C [60,61]. Acid-type cation-exchange resin Amberlyst 15 was the most effective co-catalyst. Nevertheless, a weak point in the system of Ru/C with Amberlyst 15 is that the reaction temperature is limited to 393 K. At higher temperatures sulfur compounds such as SO_2 and H_2S, which are formed by the thermal decomposition of the sulphonic groups of the resins, poison the catalyst. Using Amberlyst 70 the reacting temperature can be increased to 453 K before observing thermal decomposition [62].

Catalyst	H_2 (bar)	Temp. (°C)	Glyc. Conc. (wt.%)	mg_{cat}/g_{glyc}	Time (h)	Conv. (%)	Product Selectivity (%)	Ref
Ru/TiO_2, 5wt%	50	180	20	96	12	90.1	1,2-PDO (21), EG (41)	[57]
Pt/C, 3wt% + CaO 0.8 M	40	200	1	233	5	40	1,2-PDO (71), lacticacid (19), EG (9)	[42]
Ru/C, 5wt% + Amberlyst 15	80	120	20	112.5	10	79.3	1,2-PDO (75), 1-PO(8), 2-PO (2), EG (7)	[63]
Ru/C, 5wt% + Amberlyst 70	80	180	20	12.2	10	48.8	1,2-PDO (70), 1,3-PDO (1.3), 1-PO (7.1), EG (8.3)	[62]
Cu/Al_2O_3, 60wt%	1	120 -200	30	-	0.066 $h^{-1\,a}$	100	1,2-PDO (96.9), acetol (1.4)	[64]
Cu/SiO_2, 30wt%	90	180	80	62.5	12	32.7	1,2.PDO (98), EG (1)	[65]
$Cu_{0.4}/Mg_{5.6}Al_2O_9$ + NaOH	30	180	75	166	20	91	1,2-PDO (96), EG (3)	[44]
$Pd_{0.04}Cu_{0.4}Mg_{5.6}Al_2(OH)$	20	180	75	166	10	77	1,2-PDO (98), EG (1.6)	[66]
$Ir–ReO_x/SiO_2$, 4wt% (Re/Ir = 1)	80	120	20	37.5	36	81.0	1,2-PDO (4.2), 1,3-PDO (46.3), 1-PO (41.2)	[72]

[a] WHSV (weight hour space velocity)

Table 2. Selected examples of hydrogenolysis of aqueous glycerol over heterogeneous catalysts. PDO: Propanediol, PO: Propanol, EG: Ethylene Glycol.

The use of more stable inorganic salts can avoid the temperature problems related to ion-exchange resins. Balaraju et al. [67] used the combination of Ru/C catalyst with different inorganic salts such as niobia, zirconia-supported 12-tungstophosphoric acid or acid caesium 12-tungstophosphate in glycerol hydrogenolysis at 453 K. The best results were achieved with those co-catalysts presenting a high number of medium strength acid sites. Particularly, with niobia as co-catalyst 62.8% glycerol conversion and 66.5% 1,2-PDO selectivity were reported. Another option is the use of a noble metal on acid supports. Vasiliadou et al. [68] investigated glycerol hydrogenolysis on Ru-based (γ-Al_2O_3, SiO_2, ZrO_2) catalysts at 513 K and 80 bar. The nature of the oxidic support was found to influence the ability of the catalyst to both activate the glycerol substrate and selectively convert it to propanediol. The characterization of the catalytic materials revealed a correlation between catalytic activity for the hydrogenolysis reaction and total acidity, as the yield to hydrogenolysis products increased with the concentration of the acid sites. However, increased acidity was also responsible for the promotion of the excessive hydrogenolysis of the desired 1,2-propanediol to propanols.

ii. Cu based catalysts

Cu has been extensively investigated in the glycerol hydrogenolysis reactions. Although its hydrogenation activity is generally lower than that of noble metals, its much lower price and its ability to catalyze C-O bond but not C-C bond hydrogenolysis make Cu catalysts attractive for this process. There are some works in the literature that report the use of other transition metals like Ni or Co, however, Cu based catalysts are predominant. Vapor phase glycerol dehydration reaction was studied by Sato et al. [69] over different copper catalysts at 513 K and atmospheric N_2 pressure. They observed that basic MgO, CeO_2, and ZnO supports showed low acetol selectivity, while acidic supports, such as Al_2O_3, ZrO_2, Fe_2O_3, and SiO_2, effectively promoted acetol formation. The best results were obtained with Cu/Al_2O_3 catalyst. Increments in copper content lead to increments in acetol selectivity. Moreover, the activity of the Al_2O_3 support alone was rather low, which indicates that copper metal sites play a significant role in glycerol dehydration. Continuing with vapor phase processes, Akiyama et al. [64,70] studied glycerol hydrogenolysis in a fixed-bed down-flow glass reactor at temperatures between 340 and 473 K, atmospheric hydrogen pressure, and using Cu/Al_2O_3 catalysts. In the two step reaction they observed that glycerol dehydration to acetol was favored at relatively high temperatures. However, acetol hydrogenation to 1,2-PDO was favored at lower temperatures, because it is an exothermic reaction and the dehydrogenation of 1,2-PDO occurs preferentially at high temperatures. Based on these findings, they developed a reactor with gradient temperatures, at the top of the reactor glycerol dehydration reaction occurred at 453 K while at the bottom of the reactor acetol was hydrogenated to 1,2-PDO at 418 K. Really high 1,2-PDO yields (94.9%) were reported.

Some of the best results in terms of glycerol conversion and 1,2-PDO selectivity were recently reported using Cu on base supports. For instance, Yuan et al. [44] developed a Cu based solid catalyst ($Cu_{0.4}/Mg_{5.6}Al_2O_{8.6}$)via thermal decomposition of the as-synthesized $Cu_{0.4}Mg_{5.6}Al_2(OH)_{16}CO_3$ layered double hydroxides. This bifunctional highly dispersed Cu-solid base catalyst was effective for hydrogenolysis of aqueous glycerol. The measured conversion of glycerol reached 80.0% with a 98.2% selectivity of 1,2-propanediol at 180 °C, 30 bar H_2 and 20 h. The addition of Pd to the same catalytic system notably increased the activity of the catalyst [71]. It was suggested that the hydrogen spill over from Pd to Cu favored glycerol hydrogenolysis to 1,2-PDO.

iii. Metal oxide modified-noble metal

As stated above, the use of acid or base as a co-catalyst gives 1,2-PDO as a main product. To obtain more valuable 1,3-PDO, the most effective approach has shown to be the use of noble metal (Ir, Rh or Pt) combined with oxophilic metals. Shinmi et al. [52] modified Rh/SiO_2 catalyst with Re, W and Mo. Re addition showed the largest enhancing effect on catalytic activity and also increased the selectivity to 1,3-PDO. The Rh–ReOx/SiO_2 (Re/Rh = 0.5) exhibited 22 times higher glycerol conversion (79%) and 37 times higher 1,3-PD yield (11%) than

Rh/SiO_2. In a more recent work, an Ir–ReO_x/SiO_2 (Re/Ir = 1) catalyst prepared by a similar method to that for Rh–ReO_x/SiO_2, catalyzed the hydrogenolysis of glycerol to 1,3-PDO in a more effectively way (1,3-PDO/1,2-PDO ratio = 11) [72]. Based on characterization results, the authors suggested that oxidized low-valence Re clusters are attached to the Ir or Rh metal particles. Glycerol is adsorbed on the surface of MO_x species (M = Mo, Re and W) at the OH group to form alkoxide. Hydrogen is activated on the noble-metal (Rh or Ir) surface. The alkoxide located on the interface between MO_x and the noble-metal surface is attacked by the activated hydrogen species, and the C–O bond neighboring to the C–O–M group is dissociated. The hydrolysis of the resulting alkoxide releases the product (see Figure 11). One of the weak point of these catalytic systems is that they are also active in the further hydrogenolysis of both 1,2 and 1,3-PDO to 1-PO.

In summary, Cu based catalysts are active and selective for the production of 1,2-PDO from glycerol. However, if the aim is to produce the more valuable 1,3-PDO, different approaches are required. The used of noble metals combined with low-valence metal oxide seems to be a promising alternative. Nonetheless, there is still room for improvement; both in catalyst design and in process engineering, as PDOs further hydrogenolysis significantly affect the final yields to target products.

4. Main alternatives to the use of molecular hydrogen

In the previous sections the significance that hydrogenolysis reactions have and will have in the future bio-refineries has been highlighted. In fact, they will be essential in fuel and chemical manufacturing. Hydrogenolysis involves chemical bond dissociation in an organic substrate and simultaneous addition of hydrogen. Therefore, hydrogen is required as reactant in all hydrogenolysis reactions. This is the reason why, most of the literature works referred to hydrogenolysis report experiments conducted under molecular hydrogen (H_2) atmosphere. Nevertheless, the use of molecular hydrogen has some important drawbacks:

i. Liquid phase processes are preferred to gas phase processes as they are more energy efficient. However, H_2 presents really low solubility on aqueous or organic solutions. As a consequence, when operating in liquid phase it is necessary to operate at elevated hydrogen pressures to obtain significant hydrogen concentrations near the catalysts. This, on one hand, notably increases the cost of design and building of the future plants, and on the other hand, increases the operating cost related to safety measures, as hydrogen is easily ignited and shows high diffusivity.

ii. Most of the nowadays available hydrogen gas is produced from fossil fuels by energy intensive processes. Therefore, if sustainability is the goal it is a contradiction that the main reactant in most of the biorefinery processes is based on fossil resources.

iii. The low density and high diffusivity of hydrogen make problematic and expensive its transportation and storage. This problem is more relevant for small size biomass conversion facilities.

Hydrogen from non fossil origin will surely be a reality in the oncoming years, as reforming processes from various renewable compounds (like biomethane, glycerol or ethanol) and water splitting processes using solar light are being intensively developed. Nonetheless, the problems of transportation, storage and low solubility in liquid solutions will remain. One interesting option that could solve the problems associated to the use of molecular hydrogen is to directly generate the required hydrogen in the active sites of the catalyst.

4.1. Bio-oil upgrading using hydrogen donating solvents

One interesting approach to reduce the consumption of molecular hydrogen during the HDO of bio-oils is to use hydrogen donating solvents. For instance, Elliott has reported that when the bio-oil upgrading is carried out in the presence of a hydrogen donor solvent (tetralin, 1-1 ratio with bio-oil feedstock) the oxygen removal increases from 70 to 85% and less deactivation of the catalyst was observed. Some of the components already present in the bio-oil, such as alcohols or acids, may also provide hydrogen for the deoxygenation reactions [10]. Traditional catalysts active in hydrogen transfer reactions, such as Pd, Ni or Cu should be used in this process [73].

Another attractive option is to use hydrogen donating solvents during the hydrotreating of biomass. The idea is to obtain a bio-oil with a lower oxygen content, and therefore, easier to upgrade. This concept has been mainly applied in the pyrolysis of lignin. If a hydrogen donor molecule is added during the pyrolysis, both depolymerization and hydrogenation occur simultaneously. Remarkable results have been obtained using hydrogen-donating solvents, such as tetralin or 9,10-dihydroanthracene [74]. However, a major drawback is the need for large quantities of these solvents. At this point, formic acid appears to be a promising donor molecule, as it can be obtained together with levulinic acid from the hydrolysis of biomass. On heating, formic acid decomposes completely into CO_2 and two active hydrogen atoms, which are efficient scavengers of any radical species formed in the lignin. By successive homolytic cleavage of the covalent linkages of the lignin, including aromatic rings, most of the oxygen is removed as water and hydrocarbons are formed (Figure 13). When pyrolysis is carried out with formic acid, lignin can be converted into hydrogen-rich, oxygen depleted products with no added catalyst [75].

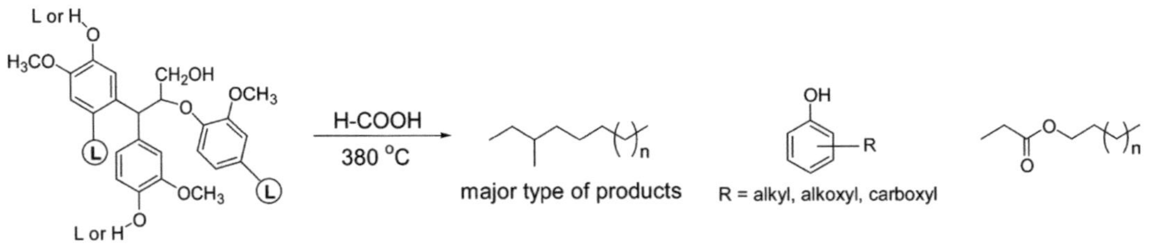

Figure 13. Schematic picture of the products formed upon the pyrolysis of lignin in thepresence of formic acid [75].

4.2. Hydrogenolysis with in-situ generation of hydrogen

4.2.1. Aqueous Phase Reforming (APR)

One interesting option to in situ generate the required hydrogen for hydrogenolysis reactions is through aqueous phase reforming (APR). APR is a quite well known process in which a polyol is converted to hydrogen and CO_2 in the presence of water. The hydrogen generated can be further used in the hydrogenolysis reaction. The specific case for combined glycerol APR and hydrogenolysis to 1,2-PDO is shown in Figure 14. If the process is perfectly balanced, glycerol is fully converted into 1,2-PDO, being CO_2 and H_2O the only byproduct. Tailored metal-acid bifunctional catalysts or combination of catalysts are required to obtain high yields to 1,2-PDO. Indeed, there must be a proper balance between the C-C bond cleavage reactions that lead to the production of hydrogen, and the C-O bond cleavage reactions that lead to the formation of PDOs [76]. While Pt is known to be active in C-C bond cleavage, its combination with other metals active in C-O bond hydrogenolysis, like Ni, Sn or Ru, over acidic supports appears as promising formulations to obtain high yields to 1,2-PDO [77]. However, glycerol APR itself runs at elevated pressure and therefore the advantage over conventional hydrogenolysis at high hydrogen pressure is marginal with regard to equipment and safety costs.

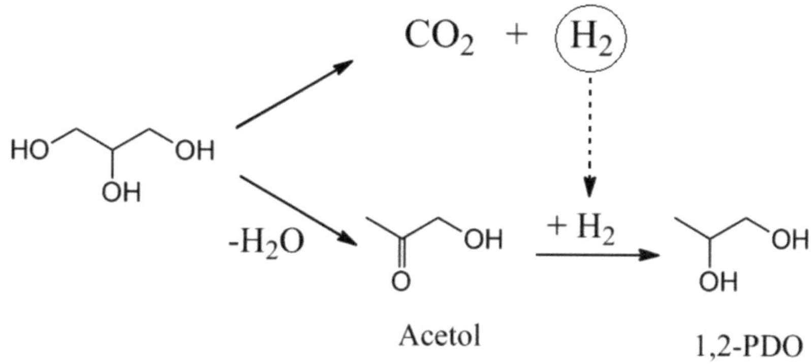

Figure 14. Combined glycerol APR and hydrogenolysis to 1,2-PDO.

The same benefits that have been previously addressed for the use of *in situ* generated hydrogen in glycerol hydrogenolysis can be applied to the conversion of other higher polyols, like sorbitol or xylitol. However, the considerable research effort that has been directed to the conversion of glycerol yet has not been paid to other biomass based polyols. Therefore, the amount of works related to high polyol hydrogenolysis with *in situ* generation of the required hydrogen is quite scarce. As a consequence of this, it is a really interesting and open research field.

Huber et al. [78] studied the production of renewable alkanes (C1-C6) from the aqueous phase reforming of sorbitol using a Pt/SiO_2-Al_2O_3catalyst.They suggested a multistep bifunctional reaction pathway. The first step involves the formation of CO_2 and H_2 on the Pt sites, and the dehydration of sorbitol on the acid sites of the silica-alumina support. These initial steps are followed by hydrogenation of the dehydrated reaction intermediates on the metal catalyst (Scheme 9). 64 % alkane selectivity at 92% sorbitol conversion were recorded at 498 K and 39.6 bar. When hydrogen was co-fed, alkane selectivity significantly increased up to 91%. Glucose showed to be less active than sorbitol over a Pt/Al_2O_3 catalyst at 538 K and 52.4 bar of N_2pressure, achieving moderate alkane selectivities (49.5%) [79]. Therefore, it seems that initial hydrogenation of glucose to sorbitol and subsequent aqueous phase reforming of the sugar is more effective than direct aqueous phase reforming of glucose.

4.2.2. Catalytic Transfer Hydrogenation

Catalytic transfer hydrogenation (CTH) is a process in which hydrogen is transferred from a hydrogen donor molecule to an acceptor [80]. CTH reactions can be of industrial importance as the renewable production, transportation and storage of hydrogen donors can be cheaper than those for molecular hydrogen. For CTH, it has been reported that adjacent sites may be necessary for donor and acceptor molecules [73]. Therefore, the first criterion to be fulfilled by the selected hydrogen donor molecules is to be soluble in the compound to be hydrotreated. Moreover, in order to improve the yield of desired products, reactions other than dehydrogenation of the donor should be minimized under the operating conditions. The best hydrogen donors for heterogeneous CTH include simple molecules like cyclohexene, hydrazine, formic acid and formates [81]. Alcohols like 2-propanol (2-PO) or methanol can also be used as hydrogen donors; primary alcohols are generally less active than the corresponding secondary alcohols due to the smaller electron-releasing inductive effect of one alkyl group as against two [82]. The most active catalysts for heterogeneous transfer reduction are based on palladium metal. Other noble metals such as Pt and Rh are also widely utilized. Sometimes, other transition metals such as Ni and Cu have also been reported but for operation at higher temperature [73].

In this area, the most studied process has been the conversion of glycerol into 1,2-PDO. Musolino et al. [83] studied glycerol hydrogenolysis by transfer hydrogenation under 5 bar inert atmosphere, using ethanol and 2-PO as solvents and hydrogen donor molecules over 10PdFe_2O_3 catalyst at 453 K. They observed that complete glycerol conversion and high selectivities to 1,2-PDO could be obtained when the hydrogen came from the dehydrogenation of the solvent. Formic acid has also been used as a hydrogen donor molecule in the glycerol hydrogenolysis process using Ni-Cu/Al_2O_3 catalysts [84]. Under the operating conditions used, formic acid was readily converted into CO_2 and H_2, therefore, a semi-continuous setup was used to continuously pump formic acid to the glycerol water solution, in order to ensure a constant supply of hydrogen at an appropriate rate [85]. For a constant metal content of 35 wt-% (Ni+Cu), increasing Ni proportion caused an increase in glycerol conversion but also an increase in C-C bond cleavage reactions. Cu is known to be active in the C-O bond cleavage but not in the C-C bond cleavage. The presence of Cu and the creation of a

Ni-Cu alloy notably reduced formation of products <C_3. This was related to the fact that C-C bond cleavage reactions are ensemble size sensitive and that the formation of a Cu-Ni alloy causes a decrease in the Ni ensemble size. Therefore, the presence of both metals is required for obtaining high 1,2-PDO yields: Ni to provide high hydrogenolysis activity and Cu to shift the selectivity towards C-O bond cleavage. It was also observed that above a certain metal content, further increments led to a decrease in glycerol conversion. This was correlated to the total acidity of the catalyst that also decreased with increasing metal content. A direct glycerol hydrogenolysis mechanism was also proposed (Figure 15).

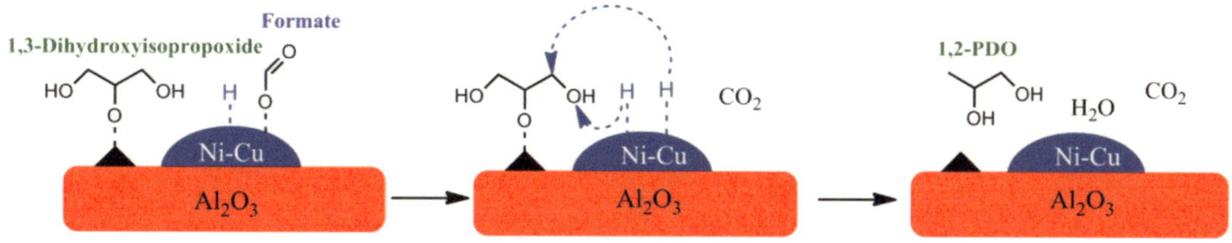

Figure 15. Proposed mechanism for glycerol hydrogenolysis by CTH using formic acid as hydrogen donor molecule [84].

5. Conclusions

Bio-oils coming from the pyrolysis of biomass feedstocks and biomass based platform chemicals present a common limiting feature: their high oxygen content. This oxygen can be removed by catalytic hydrotreating in the form of H_2O. Intensive research is required in this field in order to develop catalytic systems active and stable under the hard operating conditions used: high temperatures and pressures, and high concentrations of sub-critical water. The required bifunctional catalysts must have Brönsted acidity to catalyze dehydration reactions or/and Lewis acid sites to attract the oxygen ion pair of the target molecule; but also metal sites that show the ability to activate hydrogen molecules. In this sense, the combination of oxophilic metals (Re, Mo or W) with Ni or noble metals has shown to be a promising approach. In the case of bio-oil upgrading, the developed catalysts should promote hydrodeoxygenation reactions against hydrogenation reactions that lead to higher hydrogen consumption and reduction in the octane number of the oil. In order to avoid coke formation under the hard operating conditions used, neutral supports appear as an interesting option. In the case of catalysts for platform chemical valorization, C-C bond cleavage reactions should be avoided. Therefore, for some applications, like glycerol hydrogenolysis to 1,2-PDO, Cu based catalysts have to be considered due to the high selectivity of Cu for C-O bond cleavage reactions.

Hydrogenolysis processes for oxygen removal require the use of large amounts of hydrogen, which is commonly supply by operating under high molecular hydrogen pressures. Nonetheless, this might be a problem because nowadays, most technologies to obtain hydrogen are energy intensive and non-renewable. An interesting alternative might be to in-situ generate the required hydrogen. Among all the alternatives, the use of hydrogen donor molecules that can be obtained from biomass in a renewable way, such as formic acid, appears as a promising approach.

Author details

Iñaki Gandarias* and Pedro Luis Arias

*Address all correspondence to: inaki_gandarias@ehu.es

Department of Chemical and Environmental Engineering, University of the Basque Country (UPV/EHU) Alameda Urquijo s/n, Bilbao, Spain

References

[1] USA Energy Information Administration (EIA): International Energy Outlook. http://www.eia.gov/forecasts/ieo/index.cfm (accessed 08 August 2012).

[2] Bull TE. Biomass in the Energy Picture. Science 1999; 285 (5431) 1209-1209.

[3] Ragauskas AJ, Williams CK, Davison BH, Britovsek G, Cairney J, Eckert CA, Frederick WJ, Hallett JP, Leak DJ, Liotta CL, Mielenz JR, Murphy R, Templer R, Tschaplinski T. The Path Forward for Biofuels and Biomaterials. Science 2006; 311 (5760) 484-489.

[4] Carlos RM, Ba Khang D. Characterization of biomass energy projects in Southeast Asia. Biomass and Bioenergy 2008; 32 (6) 525-532.

[5] Sarkar N, Ghosh SK, Bannerjee S, Aikat K. Bioethanol production from agricultural wastes: An overview. Renewable Energy 2012; 37 (1) 19-27.

[6] Mohan D, Pittman CU, Steele PH. Pyrolysis of Wood/Biomass for Bio-oil: A Critical Review. Energy Fuels 2006; 20 (3) 848-889.

[7] Venderbosch R, Prins W. Fast pyrolysis technology development. Biofuels, Bioproducts and Biorefining 2010; 4 (2) 178-208.

[8] Banu M, Sivasanker S, Sankaranarayanan TM, Venuvanalingam P. Hydrogenolysis of sorbitol over Ni and Pt loaded on NaY. Catalysis Communications 2011; 12 (7) 673-677.

[9] Yusuf NNAN, Kamarudin SK, Yaakub Z. Overview on the current trends in biodiesel production. Energy Conversion and Management 2011; 52 (7) 2741-2751.

[10] Bulushev DA, Ross JRH. Catalysis for conversion of biomass to fuels via pyrolysis and gasification: A review. Catalysis Today 2011; 171 (1) 1-13.

[11] International Energy Agency. IEA: Sustainable Production of SECOND -Generation Biofuels.http://www.iea.org/papers/2010/second_generation_biofuels.pdf(accessed 25 July 2012).

[12] Bobleter O. Hydrothermal degradation of polymers derived from plants. Progress in Polymer Science 1994; 19 (5) 797-841.

[13] Huber,G.W., Iborra,S., Corma,A. Synthesis of Transportation Fuels from Biomass: Chemistry, Catalysts, and Engineering. Chemical Reviews. 2006; 106 (9) 4044-4098.

[14] Bridgwater AV. Review of fast pyrolysis of biomass and product upgrading. Biomass and Bioenergy 2012; 38 (1) 68-94.

[15] Mortensen PM, Grunwaldt J-, Jensen PA, Knudsen KG, Jensen AD. A review of catalytic upgrading of bio-oil to engine fuels. Applied Catalysis A: General 2011; 407 (1–2) 1-19.

[16] Piskorz J., Scott DS.,Radlien D. Composition of oils obtained by fast pyrolysis of different woods. In: Pyrolysis Oils from Biomass: Producing Analyzing and Upgrading. Washington, DC: American Chemical Society; 1988.

[17] Elliott DC. Historical Developments in Hydroprocessing Bio-oils. Energy Fuels 2007; 21 (3) 1792-1815.

[18] Romero Y, Richard F, Brunet S. Hydrodeoxygenation of 2-ethylphenol as a model compound of bio-crude over sulfided Mo-based catalysts: Promoting effect and reaction mechanism. Applied Catalysis B: Environmental 2010; 98 (3–4) 213-223.

[19] Ba T, Chaala A, Garcia-Perez M, Roy C. Colloidal Properties of Bio-Oils Obtained by Vacuum Pyrolysis of Softwood Bark. Storage Stability. Energy Fuels 2004; 18 (1) 188-201.

[20] Şenol Oİ, Viljava T-, Krause AOI. Effect of sulphiding agents on the hydrodeoxygenation of aliphatic esters on sulphided catalysts. Applied Catalysis A: General 2007; 326 (2) 236-244.

[21] Gandarias I, Barrio VL, Requies J, Arias PL, Cambra JF, Güemez MB. From biomass to fuels: Hydrotreating of oxygenated compounds. International Journal of Hydrogen Energy 2008; 33 (13) 3485-3488.

[22] Şenol Oİ, Ryymin E-, Viljava T-, Krause AOI. Reactions of methyl heptanoatehydrodeoxygenation on sulphided catalysts. Journal of Molecular Catalysis A: Chemical 2007; 268 (1–2) 1-8.

[23] Ravenelle RM, Copeland JR, Kim W, Crittenden JC, Sievers C. Structural Changes of Î³-Al2O3-Supported Catalysts in Hot Liquid Water. ACS Catalysis 2011; 1 (5) 552-561.

[24] Centeno A, Laurent E, Delmon B. Influence of the Support of CoMo Sulfide Catalysts and of the Addition of Potassium and Platinum on the Catalytic Performances for the Hydrodeoxygenation of Carbonyl, Carboxyl, and Guaiacol-Type Molecules. Journal of Catalysis 1995; 154 (2) 288-298.

[25] Echeandia S, Arias PL, Barrio VL, Pawelec B, Fierro JLG. Synergy effect in the HDO of phenol over Ni–W catalysts supported on active carbon: Effect of tungsten precursors. Applied Catalysis B: Environmental 2010; 101 (1–2) 1-12.

[26] Gutierrez A, Kaila RK, Honkela ML, Slioor R, Krause AOI. Hydrodeoxygenation of guaiacol on noble metal catalysts. Catalysis Today 2009; 147 (3–4) 239-246.

[27] Venderbosch RH, Ardiyanti AR, Wildschut J, Oasmaa A, Heeres HJ. Stabilization of biomass-derived pyrolysis oils. Journal of Chemical Technology & Biotechnology 2010; 85 (5) 674-686.

[28] Bridgwater AV. Production of high grade fuels and chemicals from catalytic pyrolysis of biomass. Catalysis Today 1996; 29 (1–4) 285-295.

[29] Gallezot P, Nicolaus N, Flèche G, Fuertes P, Perrard A. Glucose Hydrogenation on Ruthenium Catalysts in a Trickle-Bed Reactor. Journal of Catalysis 1998; 180 (1) 51-55.

[30] Yadav M, Mishra DK, Hwang J. Catalytic hydrogenation of xylose to xylitol using ruthenium catalyst on NiO modified TiO_2 support. Applied Catalysis A: General 2012; 425–426 (1) 110-116.

[31] Ye L, Duan X, Lin H, Yuan Y. Improved performance of magnetically recoverable Ce-promoted Ni/Al2O3 catalysts for aqueous-phase hydrogenolysis of sorbitol to glycols. Catalysis Today 2012; 183 (1) 65-71.

[32] Zhao L, Zhou JH, Sui ZJ, Zhou XG. Hydrogenolysis of sorbitol to glycols over carbon nanofiber supported ruthenium catalyst. Chemical Engineering Science 2010; 65 (1) 30-35.

[33] Nikolla E, Romaìn-Leshkov Y, Moliner M, Davis ME. One-Pot Synthesis of 5-(Hydroxymethyl)furfural from Carbohydrates using Tin-Beta Zeolite. ACS Catalysis 2011; 1 (4) 408-410.

[34] Roman-Leshkov Y, Barrett CJ, Liu ZY, Dumesic JA. Production of dimethylfuran for liquid fuels from biomass-derived carbohydrates. Nature 2007; 447 (7147) 982-985.

[35] Luijkx GCA, Huck NPM, van Rantwijk F, Maat L, van Bekkum H. Ether Formation in the Hydrogenolysis of Hydroxymethylfurfural over Palladium Catalysts in Alcoholic Solution. ChemInform 2009; 40 (28).

[36] Tong X, Ma Y, Li Y. Biomass into chemicals: Conversion of sugars to furan derivatives by catalytic processes. Applied Catalysis A: General 2010; 385 (1–2) 1-13.

[37] Huber GW, O'Connor P, Corma A. Processing biomass in conventional oil refineries: Production of high quality diesel by hydrotreating vegetable oils in heavy vacuum oil mixtures. Applied Catalysis A: General 2007; 329 (1) 120-129.

[38] Rovetto LJ, Bottini SB, Brignole EA, Peters CJ. Supercritical hydrogenolysis of fatty acid methyl esters: Phase equilibrium measurements on selected binary and ternary systems. The Journal of Supercritical Fluids 2005; 35 (3) 182-196.

[39] Veldsink JW, Bouma MJ, Schöön NH, Beenackers AACM. Heterogeneous Hydrogenation of Vegetable Oils: A Literature Review. Catalysis Reviews 1997; 39 (3) 253-318.

[40] Montassier C, Ménézo JC, Moukol J, Naja J, Hoang LC, Barbier J. Polyol conversions into furanic derivates on bimetallic catalysts: Cu-Ru, Cu-Pt and Ru-Cu. Journal of Molecular Catalysis 1991; 70 65-84.

[41] Ruppert AM, Weinberg K, Palkovits R. Hydrogenolysis Goes Bio: From Carbohydrates and Sugar Alcohols to Platform Chemicals. AngewandteChemie International Edition 2012; 51 (11) 2513-2513.

[42] Maris EP, Davis RJ. Hydrogenolysis of glycerol over carbon-supported Ru and Pt catalysts. Journal of Catalalysis 2007; 249 (2) 328-337.

[43] Lahr DG, Shanks BH. Effect of sulfur and temperature on ruthenium-catalyzed glycerol hydrogenolysis to glycols. Journal of Catalysis 2005; 232 (2) 386-394.

[44] Yuan Z, Wang L, Wang J, Xia S, Chen P, Hou Z, Zheng X. Hydrogenolysis of glycerol over homogenously dispersed copper on solid base catalysts. Applied Catalysis B: Environmental 2011; 101 (3–4) 431-440.

[45] Feng J, Wang J, Zhou Y, Fu H, Chen H, Li X. Effect of Base Additives on the Selective Hydrogenolysis of Glycerol over Ru/TiO_2 Catalyst. Chemistry Letters 2007; 36 1274-1275.

[46] Dasari MA, Kiatsimkul P, Sutterlin WR, Suppes GJ. Low-pressure hydrogenolysis of glycerol to propylene glycol. Applied Catalysis A: General 2005; 281 (1-2) 225-231.

[47] Chuang-Wei C, Dasari MA, Suppes GJ. Dehydration of Glycerol to Acetol via Catalytic Reactive Distillation. AICHE Journal 2006; 52 3543-3548.

[48] Suppes,G.J., Sutterling,W.R., Dasari,M.A., 2005; WO2005095536.

[49] Schlaf M. Selective deoxygenation of sugar polyols to α,ω-diols and other oxygen content reduced materials-a new challenge to homogeneous ionic hydrogenation and hydrogenolysis catalysis. Dalton Transactions 2006; (39) 4645-4653.

[50] Nakagawa Y, Shinmi Y, Koso S, Tomishige K. Direct hydrogenolysis of glycerol into 1,3-propanediol over rhenium-modified iridium catalyst. Journal of Catalysis 2010; 272 (2) 191-194.

[51] Chaminand J, Djakovitch L, Gallezot P, Marion P, Pinel C, Rosier C. Glycerol hydrogenolysis on heterogeneous catalysts. Green Chemistry 2004; 6 (8) 359-361.

[52] Shinmi Y, Koso S, Kubota T, Nakagawa Y, Tomishige K. Modification of Rh/SiO2 catalyst for the hydrogenolysis of glycerol in water. Applied Catalysis B: Environmental 2010; 94 (3-4) 318-326.

[53] Chia M, Pagan-Torres YJ, Hibbitts D, Tan Q, Pham HN, Datye AK, Neurock M, Davis RJ, Dumesic JA. Selective Hydrogenolysis of Polyols and Cyclic Ethers over Bi-

functional Surface Sites on Rhodium-Rhenium Catalysts. Journal of the American Chemical Society 2011; 133 (32) 12675-12689.

[54] King DL, Zhang L, Xia G, Karim AM, Heldebrant DJ, Wang X, Peterson T, Wang Y. Aqueous phase reforming of glycerol for hydrogen production over Pt–Re supported on carbon. Applied Catalysis B: Environmental 2010; 99 (1-2) 206-213.

[55] Qin L, Song M, Chen C. Aqueous-phase deoxygenation of glycerol to 1,3-propanediol over Pt/WO3/ZrO2 catalysts in a fixed-bed reactor. Green Chemistry 2010; 12 (8) 1466-1472.

[56] Montassier C, Ménézo JC, Hoang LC, Renaud C, Barbier J. Aqueous polyol conversions on ruthenium and on sulphur-modified ruthenium. Journal of Molecular Catalysis 1991; 70 (1) 99-110.

[57] Feng J, Fu H, Wang J, Li R, Chen H, Li X. Hydrogenolysis of glycerol to glycols over ruthenium catalysts: Effect of support and catalyst reduction temperature. Catalysis Communications 2008; 9 (6) 1458-1464.

[58] Furikado I, Miyazawa T, Koso S, Kunimori K, Tomishige K. Catalytic performance of Rh/SiO_2 in glycerol reaction under hydrogen. Green Chemistry 2007; 9 (6) 582-588.

[59] Lahr DG, Shanks BH. Kinetic Analysis of the Hydrogenolysis of Lower Polyhydric Alcohols: Glycerol to Glycols. Industrial & Engineering Chemistry Research 2003; 42 (22) 5467-5472.

[60] Kusunoki Y, Miyazawa T, Kunimori K, Tomishige K. Highly active metal-acid bifunctional catalyst system for hydrogenolysis of glycerol under mild reaction conditions. Catalysis Communications 2005; 6 (10) 645-649.

[61] Miyazawa T, Kusunoki Y, Kunimori K, Tomishige K. Glycerol conversion in the aqueous solution under hydrogen over Ru/C + an ion-exchange resin and its reaction mechanism. Journal of Catalysis 2006; 240 (2) 213-221.

[62] Miyazawa T, Koso S, Kunimori K, Tomishige K. Glycerol hydrogenolysis to 1,2-propanediol catalyzed by a heat-resistant ion-exchange resin combined with Ru/C. Applied Catalysis A: General 2007; 329 30-35.

[63] Miyazawa T, Koso S, Kunimori K, Tomishige K. Development of a Ru/C catalyst for glycerol hydrogenolysis in combination with an ion-exchange resin. Applied Catalysis A: General 2007; 318 (1) 244-251.

[64] Akiyama M, Sato S, Takahashi R, Inui K, Yokota M. Dehydration–hydrogenation of glycerol into 1,2-propanediol at ambient hydrogen pressure. Applied Catalysis A: General 2009; 371 (1-2) 60-66.

[65] Huang Z, Cui F, Kang H, Chen J, Xia C. Characterization and catalytic properties of the CuO/SiO_2 catalysts prepared by precipitation-gel method in the hydrogenolysis of glycerol to 1,2-propanediol: Effect of residual sodium. Applied Catalysis A: General 2009; 366 (2) 288-298.

[66] Xia S, Yuan Z, Wang L, Chen P, Hou Z. Catalytic production of 1,2-propanediol from glycerol in bio-ethanol solvent. Bioresource technology 2012; 104 (1) 814-817.

[67] Balaraju M, Rekha V, Prasad PSS, Devi BLAP, Prasad RBN, Lingaiah N. Influence of solid acids as co-catalysts on glycerol hydrogenolysis to propylene glycol over Ru/C catalysts. Applied Catalysis A: General 2009; 354 (1-2) 82-87.

[68] Vasiliadou ES, Heracleous E, Vasalos IA, Lemonidou AA. Ru-based catalysts for glycerol hydrogenolysis—Effect of support and metal precursor. Applied Catalysis B: Environmental 2009; 92 (1–2) 90-99.

[69] Sato S, Akiyama M, Takahashi R, Hara T, Inui K, Yokota M. Vapor-phase reaction of polyols over copper catalysts. Applied Catalysis A: General 2008; 347 (2) 186-191.

[70] Sato S, Akiyama M, Inui K, Yokota M. Selective Conversion of Glycerol into 1,2-Propanediol at Ambient Hydrogen Pressure. Chemistry Letters 2009; 38 (6) 560-561.

[71] Xia S, Yuan Z, Wang L, Chen P, Hou Z. Hydrogenolysis of glycerol on bimetallic Pd-Cu/solid-base catalysts prepared via layered double hydroxides precursors. Applied Catalysis A: General 2011; 403 (1–2) 173-182.

[72] Amada Y, Shinmi Y, Koso S, Kubota T, Nakagawa Y, Tomishige K. Reaction mechanism of the glycerol hydrogenolysis to 1,3-propanediol over Ir-ReOx/SiO_2 catalyst. Applied Catalysis B: Environmental 2011; 105 (1-2) 117-127.

[73] Johnstone RAW, Wilby AH, Entwistle ID. Heterogenous Catalytic Transfer Hydrogenation and Its Relation to Other Methods for Reduction of Organic Compounds. Chemical Reviews 1985; 85 (2) 129-170.

[74] Thring RW, Chornet E, Overend RP. Thermolysis of Glycol Lignin in the Presence of Tetralin. Canadian Journal of Chemical Engineering 1993; 71 (1) 107-115.

[75] Kleinert M, Barth T. Towards a lignocellulosic Biorefinery: Direct One-Step Conversion of Lignin to Hydrogen-Enriched Biofuel. Energy Fuels 2008; 22 (2) 1371-1379.

[76] D′Hont E, Van de Vyver S, Sels BF, Jacobs PA. Catalytic glycerol conversion into 1,2-propanediol in absence of added hydrogen. Chemical Communications 2008; (45) 6011-6012.

[77] Van de Vyver S, D'Hondt E, Sels BF, Jacobs PA. Preparation of Pt on NaY zeolite catalysts for conversion of glycerol into 1,2-propanediol. Studies in Surface Science and Catalysis 2010; 175 (1) 771-774.

[78] Huber GW, Cortright RD, Dumesic JA. Renewable alkanes by aqueous-phase reforming of biomass-derived oxygenates. AngewandteChemie-International Edition 2004; 43 (12) 1549-1551.

[79] Davda RR, Shabaker JW, Huber GW, Cortright RD, Dumesic JA. Aqueous-phase reforming of ethylene glycol on silica-supported metal catalysts. Applied Catalysis B: Environmental 2003; 43 (1) 13-26.

[80] Wolfson A, Dlugy C, Shotland Y, Tavor D. Glycerol as solvent and hydrogen donor in transfer hydrogenation-dehydrogenation reactions. Tetrahedron letters 2009; 50 (43) 5951-5953.

[81] Bulushev DA, Ross JRH. Vapour phase hydrogenation of olefins by formic acid over a Pd/C catalyst. Catalysis Today 2011; 163 (1) 42-46.

[82] Aramendía MA, Borau V, Jiménez C, Marinas JM, Ruiz JR, Urbano FJ. Liquid-phase heterogeneous catalytic transfer hydrogenation of citral on basic catalysts. Journal of Molecular Catalysis A: Chemical 2001; 171 (1-2) 153-158.

[83] Musolino MG, Scarpino LA, Mauriello F, Pietropaolo R. Selective transfer hydrogenolysis of glycerol promoted by palladium catalysts in absence of hydrogen. Green Chemistry 2009; 11 (10) 1511-1513.

[84] Gandarias I, Requies J, Arias PL, Armbruster U, Martin A. Liquid-phase glycerol hydrogenolysis by formic acid over Ni–Cu/Al_2O_3 catalysts. Journal of Catalysis 2012; 290 (1) 79-89.

[85] Gandarias I, Arias PL, Fernández SG, Requies J, El Doukkali M, Güemez MB. Hydrogenolysis through catalytic transfer hydrogenation: glycerol conversion to 1,2-propanediol. Catalysis Today 2012; http://dx.doi.org/10.1016/j.cattod.2012.03.067.

Chapter 11

Synthesis of Biomass-Derived Gasoline Fuel Oxygenates by Microwave Irradiation

Armando T. Quitain, Shunsaku Katoh and
Motonobu Goto

Additional information is available at the end of the chapter

http://dx.doi.org/10.5772/52539

1. Introduction

Recent concerns about climate change and problems associated with the use of fossil-derived fuels and nuclear energy have inspired researchers to seriously explore environmentally benign and economically viable renewable energy and fuels. As potential solution to reduce fossil-derived carbon dioxide (CO_2) emissions from gasoline-run automobiles, addition of biomass-derived oxygenates was proposed. Bioethanol has been considered, however, ether oxygenates such as ethyl *tert*-butyl ether (hereby referred to as ETBE), has gained popularity over ethanol (EtOH) due to its superior properties which blend well with gasoline [1]. ETBE also outranks MTBE as an octane enhancer due to its low blending Reid vapor pressure. Moreover, ETBE is a better option because it is derived from EtOH which can be obtained from biomass. ETBE is produced from the reaction of isobutene (IB) and EtOH, however, the current supply of IB, which is mostly derived from non-renewable crude oil, may not be sufficient to cope up with the expected high demand in the future. For this reason, alternative routes for its synthesis are also currently being explored. *tert*-Butyl alcohol (TBA), which can also be derived from biomass can be employed instead of IB [2]. Research for the development of efficient and energy-saving methods for the production of these gasoline oxygenates had gained significant momentum over the past few years. The application of microwave technology was proposed for the synthesis of the above mentioned gasoline oxygenate.

Microwave technology relies on the use of electromagnetic waves to generate heat by the oscillation of molecules upon microwave absorption. Unlike the conventional heating, the heat is generated within the material, thus rapid heating occurs. Other than the advantages of rapid heating, microwave effects on reaction likely occur, thus obtaining dramatic increase in the yield even at low temperatures. The benefits of using microwave irradiation to the synthesis of the abovementioned biofuels include energy efficiency, development of a compact process, rapid heating and instant on-off process (instant heating-cooling process), among others.

In this chapter, syntheses of ether oxygenates by microwave irradiation will be discussed and summarized, focusing on our recent studies on microwave-assisted heterogeneously catalyzed processes.

2. Common gasoline oxygenates and their production processes

Gasoline oxygenates such as methyl *tert*-butyl ether (MTBE), ethyl *tert*-butyl ether (ETBE), *tert*-amyl methyl ether (TAME) and *tert*-amyl ethyl ether (TAEE) are produced commercially by the reaction of olefins (IB and isoamylenes) or C4-C7 hydrocarbons with MeOH or EtOH, in the presence of homogeneous catalysts (e.g. H_2SO_4) or a heterogeneous catalysts (e. g. ion-exchange resin) as shown in Figure 1, in case of ETBE.

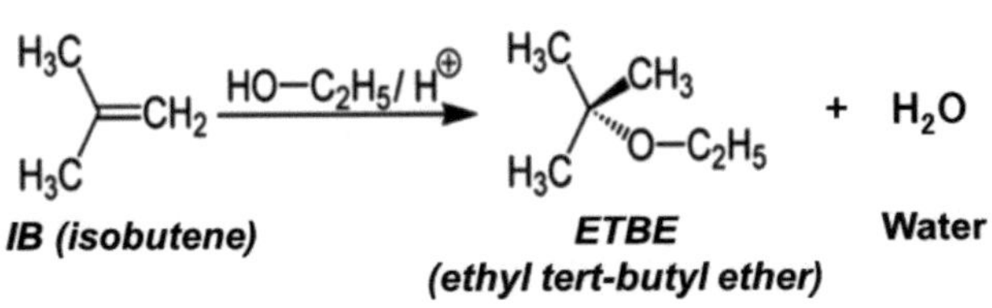

Figure 1. Reaction scheme for ETBE synthesis using IB as a reactant

The differences of blending characteristics of the ether oxygenates are summarized in Table 1.

Property	MTBE	TAME	ETBE	EtOH
Blending Rvp (psi)	8.0	2.5	4.4	18.0
Octane blending	110	105	112	115
Boiling point (K)	328	358	345	351
Oxygen content (wt%)	18.2	15.7	15.7	34.7
Solubility in water	4.3	1.15	1.20	Infinite
Fungibility in gasoline				
distribution system	High	High	High	Low

Table 1. Comparison of blending characteristics of various gasoline oxygenates

Brockwell et al. [3] discussed the process schemes for the production of these ethers. The schemes generally consist of primary reactor, distillation column and additional column to purify the products (e. g. extractor). Due to the upsurge of the demand for gasoline oxygenates, the MTBE process having a world production capacity of 20 million tons per year in 1994 [4] is the most established one. With legislations banning MTBE, it is assumed that the currently existing processes for the production of MTBE can be converted for ETBE production. Similarly, the processes can be integrated for MTBE and TAME production, since these ethers are both produced from MeOH [3].

Some of the patented processes for the production of ether oxygenates are reported in "Refining 1996" [5]. The main features of each process are summarized here.

2.1. CDTECH process

The CDTECH process utilizes C4s and alcohol as feed to a fixed-bed downflow adiabatic reactor. The equilibrium-converted reactor effluent is introduced to the reactive distillation column where the reaction continues. Concurrently, ether is separated from unreacted C4s as the bottom product.

In case of MTBE, the reactive distillation column overhead is washed in an extractor with a countercurrent H_2O stream to extract MeOH. The H_2O extract stream is sent to a MeOH recovery column to recover both MeOH and H_2O for recycle.

This scheme can provide overall IB conversions of up to 99.99% for MTBE process. Conversion is slightly less for ETBE than MTBE. For TAME and TAEE, isoamylene conversions of 95%+ are achievable.

2.2. ETHERMAX process (by Huls AG and UOP)

This process which uses reactive distillation technology is developed by combined expertise of Huls AG and UOP. The feed consists of MeOH or EtOH and hydrocarbon streams containing reactive tertiary olefins such as isoamylene and IB. Reaction takes place over an acidic ion exchange resin at mild temperature and moderate pressure.

In the MTBE case, feed first passes through an optional water wash system to remove the resin contaminants. The majority of the reaction is carried out in a simple fixed-bed reactor. The reactor effluent feeds the reactive distillation column containing a proprietary packing where simultaneous reaction of the remaining IB and distillation occur.

Overhead from the reactive distillation column is routed to MeOH recovery, a simple countercurrent extraction column using H_2O, and a MeOH-H_2O distillation column. The recovered MeOH is recycled to the reactor section. Hydrocarbon raffinate is typically sent to a downstream alkylation or oligomerization unit.

2.3. NExTAME and NExETHERS process (by Bechtel and Neste)

In the NExTAME process, the feed is typically light fluid catalytic cracker (FCC) gasoline and/or a light pyrolysis gasoline fraction from which the diolefins are removed by selective hydrogenation. It is an integrated process consisting of reactors and a distillation tower. Etherification is carried out close to the thermodynamic equilibrium in the reactors from where the reaction product is directed into a distillation tower. A side stream containing unreacted tertiary olefins and alcohol is recycled back into the reactors. This proprietary technology leads to a high conversion of a broad range of olefins and alcohol. The process does not require a separate alcohol recovery section.

Similar to NExTAME, the NExETHERS process consists of fixed bed reactors and two distillation towers. In fixed bed reactors, the etherification synthesis is completed to the thermodynamic equilibrium in the reactors. The reaction product is then directed into the main fractionator. The bottom product of the main fractionator comprises the ether product. A side stream containing unreacted tertiary olefins and unreacted alcohol is recycled back into the etherification section. The second tower is used to recycle the rest of the excess alcohol and to remove oxygenates from C4 cut which can be directed for alkylation without any further oxygenate removal or drying.

2.4. Institutfrancais du petrole process

The process includes alcohol purification, hydrocarbon purification and a main reaction that uses acid resins. The reactants are converted at temperatures lower than 90 °C and pressures lower than 2 MPa. Then, the main effluents are purified for further applications or recycle.

The reactor column uses CATACOL technology that combines catalysis (in a well-controlled liquid phase) and distillation in separated sections.

2.5. Phillips Etherification process (by Philips Petroleum Co.)

This process uses olefins (i. e. isoamylene and IB) to react with EtOH or MeOH over acidic ion-exchange resin. Mixed olefins from a fluid catalytic cracking unit (FCCU) or steam cracker, along with fresh alcohol are fed to the reactor section. The reactor operation is liquid phase at mild temperature and pressure.

In case of MTBE, high purity MTBE is removed as a bottom product from the fractionator and all the unreacted MeOH is taken overhead. The overhead product is then stripped of MeOH in an extractor using H_2O. The extract is sent to the fractionator, while the denuded H2O are returned to the MeOH extractor.

2.6. Snamprogetti process (by Snamprogetti SpA)

Similar to Philips Etherification Process, ethers are produced by the addition of alcohol to reactive olefins in the presence of an ion exchange resin at mild temperature and pressure.

The feed passes through two reactors in series – an isothermal tubular reactor and an adiabatic drum reactor. The second reactor effluent goes to the product fractionation tower where the ether product leaves the bottom stream and hydrocarbon is recovered overhead.

In the MTBE process, MeOH in the overhead stream is extracted with H_2O in the MeOH removal tower. The extract from the bottom enters the MeOH-H_2O fractionator, while the MeOH overhead is recycled to reactor feed.

3. Catalysts for synthesis of ether oxygenates

In commercial practice, cation-exchange resins (*e. g.* Bowex 50w, Amberlyst 15 (A15), Lewatit SPC 118 or Nacite) which are sulphonated copolymers of styrene and divinylbenzene (DVB), the cross-linking agent, are used as fixed-bed catalysts for etherification reaction [4].

Other catalysts such as HPA [6] and zeolites [7] are also being considered. Le Van Mao et al. [8] studied the synthesis of MTBE over triflic acid loaded Y-type zeolites. Ahmed et al. [9] developed MFI-type zeolites which were synthesized by the rapid crystallization method for production of MTBE. Collignon et al. [10] evaluated several acid zeolites, including H-Beta (ZB25, ZB75, ZBF, ZBSC), US-Y (CBV760) and ion-exchange resin (Amberlyst 15) for the liquid phase synthesis of MTBE. From all of the works, it was found that all zeolites appear to be as active as Amberlyst 15 but the zeolite catalysts produce less by-products and are more thermally stable than the resin catalyst.

4. Direct synthesis of ETBE from TBA and EtOH

As a replacement for IB, TBA was first used in the synthesis of ETBE over 80 years ago utilizing concentrated H_2SO_4 as a catalyst as shown in reaction scheme in Figure 2 [11]. Habenicht et al. [12] suggested that TBA was preferred over IB as a reactant for ETBE synthesis at elevated pressures and temperatures. The reason for this is that the protonated IB (the key component in ETBE formation) forms only from TBA (not IB) under the conditions employed. Yin et al. [13] also studied liquid-phase synthesis of ETBE from TBA and EtOH catalyzed by ion-exchange resin and heteropoly acid (HPA) at mild pressures and temperatures. Knifton et al. [7] also investigated different types of zeolites catalysts for direct synthesis of ETBE from TBA and EtOH. At temperature ranges of 40-140 °C and pressure ranges of 0.1-7 MPa, liquid-phase synthesis of ETBE resulted to a 40-70 % yield and 65-95 % selectivity.

In our previous work, reactive distillation, a configuration in which the reactive section was located inside the column, was employed to continuously synthesize ETBE from bioethanol and TBA using Amberlyst 15 in pellet form as a catalyst. Results under standard operating conditions indicated that ETBE at about 60 mol% could be obtained in the distillate, and almost pure water in the residue. The conversion of TBA and the selectivity of ETBE were 99.9 and 35.9 %, respectively. The effects of operating conditions on conversion and selectivity were also investigated. Further purification of the distillate using the residue results in 95 mol% ETBE. Simulation of the process was also carried out using ASPEN PLUS simulator, and results showed good agreement with the obtained experimental results as shown in Figure 3 [14].

$$TBA + EtOH \Leftrightarrow ETBE + H_2O$$
$$TBA \Leftrightarrow IB + H_2O$$
$$IB + EtOH \Leftrightarrow ETBE$$

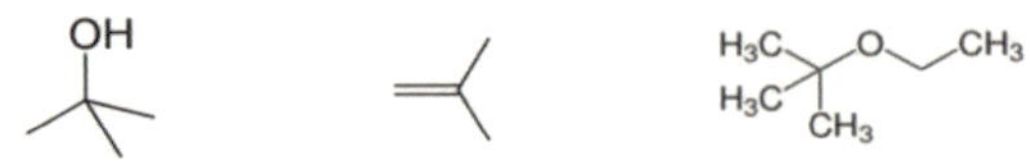

TBA ***(tert-butyl alcohol)*** | ***IB*** ***(isobutene)*** | ***ETBE*** ***(ethyl tert-butyl ether)***

Figure 2. Reaction scheme of ETBE synthesis utilizing TBA instead of IB as a reactant

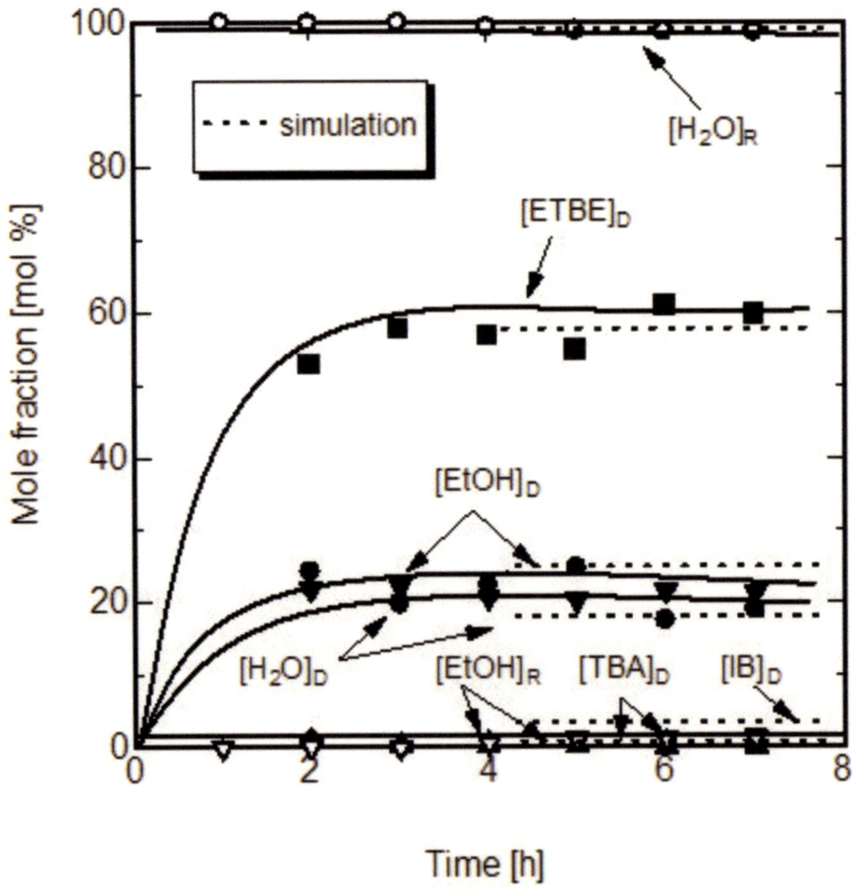

Figure 3. Comparison of concentration profiles of distillate and residue at standard operating conditions (Total feed molar flowrate = 4.13x10-3 mol/s, Reflux ratio = 7.0, Catalyst = 0.1 kg, Feed molar ratio = 1:1:38 (TBA:EtOH:H2O)

5. Fundamentals of microwave technology and its benefits

Microwave technology utilizes electromagnetic waves to generate heat by the oscillation of molecules upon microwave absorption. The electromagnetic spectrum for microwaves is in between infrared radiation and radiofrequencies of 30 GHz to 300 MHz, respectively, corresponding to wavelengths of 1cm to 1 m. Domestic and industrial microwave systems are required to operate at either 12.2 cm (2.45 GHz) or 33.3 cm (900 MHz) in order not to interfere with the wavelength ranges being utilized for RADAR transmissions and telecommunications [15].

In microwave-assisted heating, unlike the conventional methods, the heat is generated within the material, thus rapid heating occurs. As a result of this rapid heating, many microwave-assisted organic reactions are accelerated, incomparable with those obtained using the conventional methods. Thus, higher yields and selectivity of target compounds can be obtained at shorter reaction times. In addition, many reactions not possible using the conventional heating methods, had been reported to occur under microwave heating. Some very useful information on the fundamentals of microwave-enhanced chemistry, its sample preparation and applications are well presented in the book edited by Kingston and Haswell [16].

Other than the above mentioned advantages of rapid, internal and selective heating, microwave non-thermal effects on reaction likely occur, obtaining dramatic increase in the yield even at milder conditions. The microwave non-thermal effect is defined as the system response to electromagnetic energy not attributed to temperature variation [17]. Although doubts are cast on the true existence of non-thermal effects, some evidences had been reported and postulates had also been made by several researchers. These were summarized in a review article published by de la Hoz et al [18] comparing them with the thermal effects. The review of Jacob et al [19] on thermal and non-thermal interaction of microwaves with materials attributed some interesting results on specific microwave effects. Evidences on reaction rate enhancement due to some reasons other than the thermal effects such as "hotspots" or localized heating, molecular agitation, improved transport properties were discussed. They suggesed that due to the interaction of microwave with the materials, heating cannot be simply treated as that similar to the conventional methods as there are a lot of possible mechanisms of activation of materials that might possibly occur.

6. Advantages of microwave technology as applied to the solid-catalyzed synthesis of oxygenates from biomass-derived alcohols

Most alcohols, especially the ones used as raw materials for the synthesis of ether oxygenates, have very high microwave absorptivity. Table 2 shows attained temperatures if 50 ml of typical alcohols at room temperature is heated for 1 min at microwave power of 560 W and frequency of 2.45 GHz [15]. The data indicate the benefits of using microwave irradiation, such as effective and efficient use of energy,to the synthesis of ether oxygenates. Reaction relying on microwave can also be easily terminated by turning the supply of microwave irradiation off, thus further reaction or decomposition of the target compounds can be avoided resulting into higher selectivity. Also, reaction could reach completion in shorter time due to rapid heating, thus development of a compact process for a more efficient energy utilization could be possible.

Solvent	T (°C)	Boiling point (°C)
Methanol	81	100
Ethanol	78	78
1-Propanol	97	97
1-Butanol	109	117
1-Pentanol	106	137
1-Hexanol	92	158

Table 2. Attained temperatures of 50 ml of several solvents when heated from room temperature for 1 min at 560 W and 2.45 GHz

Using solid catalysts, instead of homogeneous ones, rapid heating on the surface of the catalysts likely occur upon microwave irradiation. Due this localized heating, the actual temperature at which reaction takes place may be higher than the measured bulk temperature as depicted in Figure 4, thus significantly increasing reaction rates compared to those with the conventional heating.

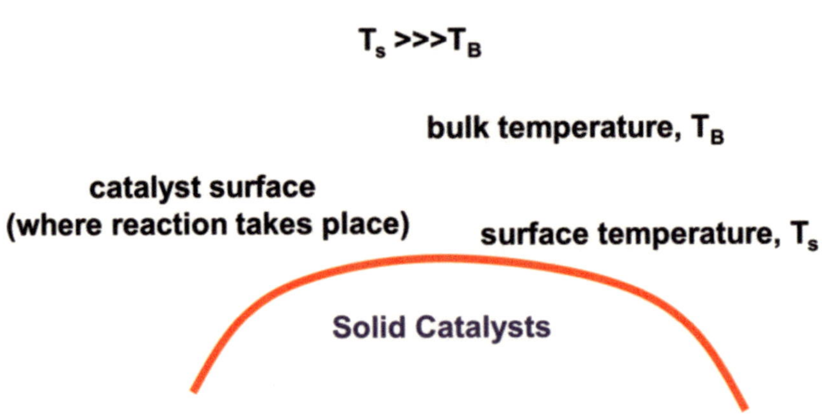

Figure 4. Phenomenon for microwave-assisted solid catalyzed reaction

http://dx.doi.org/10.5772/52539

7. Recent works on microwave-assisted synthesis of oxygenates

Due to the previously mentioned advantages and benefits, we applied microwave irradiation technique to the synthesis of ETBE from EtOH and TBA. Microwave-assisted experiments were performed using various microwave apparatuses working at 2.45 GHz frequency, with a power programmable from 0 to 1000W.

7.1. Batch experiments under atmospheric conditions

At first, experiments were carried out in batch mode under atmospheric pressure using the apparatus, shown as an actual image in Figure 5.

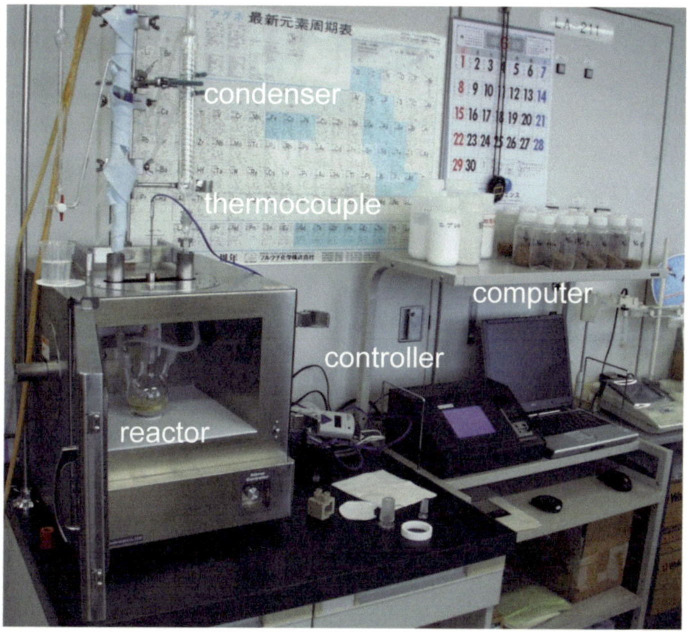

Figure 5. Microwave apparatus for batch synthesis of ETBE under atmospheric conditions (Shikoku Instrumentation Co., Ltd.)

In a typical run, about 0.25mol of EtOH and TBA, and 20 g of catalyst were placed in a reactor vessel, and heated using a microwave apparatus described above. Amberlyst 15, an ion exchange resin in H+ form was used as catalyst, unless otherwise specified. GC-FID apparatus equipped with a CP-Sil 8CB-MS (60mx0.25mm, df=0.25) column for component separation was used for the analyses of the products. Isopropanol was used as an internal standard.

Figure 6 shows typical experimental results obtained under atmospheric conditions. Experiments at atmospheric pressures using a batch reactor showed that the yield hardly increased above the 20% level. Similar results were obtained by other researchers [20,21], and the lower yield was likely due to the selective dehydration of TBA to IB, a highly volatile compound that easily escaped from the reaction zone. The maximum attained temperature was 80 °C, corresponding to the boiling point of the mixture.

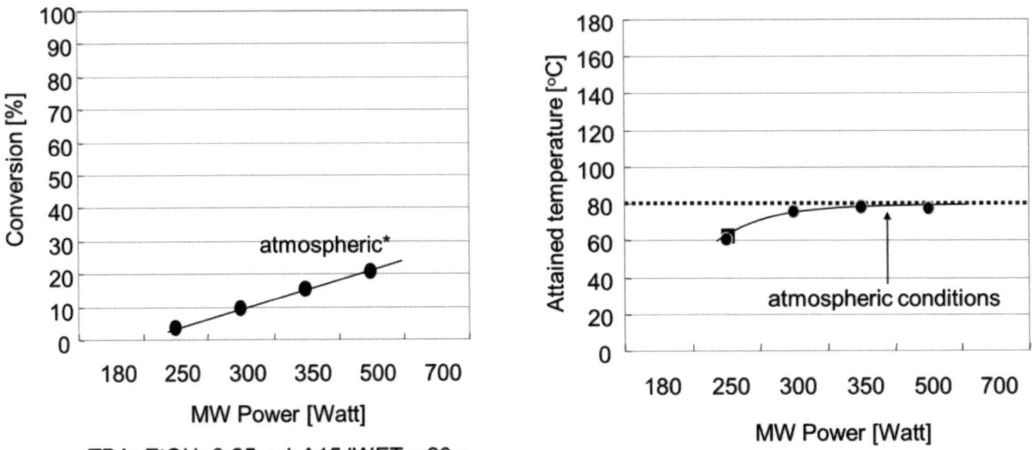

TBA=EtOH=0.25mol, A15JWET = 20g

Figure 6. Typical experimental results of batch experiments carried out under atmospheric conditions

7.2. Continuous-flow experiments under atmospheric conditions

Experiments under atmospheric conditions were extended to a continuous flow system using the apparatus shown in Figure 7. A Masterflex digital pump was used to deliver the reactants into the glass reactor filled with about 50g Amberlyst 15 catalyst. The temperature was set at 70 °C. Flowrates were varied to study the effect of residence times. The residence time of the reactants inside the reactor was calculated based on the reactor void space volume and the flowrates. Products were collected continuously after certain time has elapsed, and until the system reached equilibrium.

Results in Figure 8 show that the yield increases with increasing residence time, getting maximum yield of about 30% at a microwave duty of 13%. Increasing the duty to 20% did not have any significant effect on the yield. The maximum yield of 30% obtained using this method was in agreement with our previous studies on reactive distillation [14] and the results obtained by other researchers [20]. This low yield of ETBE was likely due to the selective dehydration of TBA to IB, a highly volatile compound that easily escaped from the reaction zone. If IB could be allowed to further react with EtOH to produce ETBE, better yield could be obtained. Thus performing the experiments in a sealed reactor vessel was thought to be effective in overcoming this limitation under atmospheric conditions.

http://dx.doi.org/10.5772/52539

Figure 7. Microwave apparatus for continuous synthesis of ETBE (Shikoku Instrumentation Co., Ltd.)

Catalyst：Amberlyst A15JWET = 50g;
EtOH：TBA = 1:1 (mol)

Set Temperature: 70 °C; No Stirring
Pump: Masterflex Digital Pump

	Run 1	Run 2	Run 3	Run 4	Run 5	Run 6	Run 7	Run 8
Flowrate (ml/min)：	100	100	50	50	20	20	10	10
Residence Time(min)：	1	1	2	2	5	5	10	10
Product Collection(min)：	1	1	2	2	5	5	10	10
Microwave duty:	13%	20%	13%	20%	13%	20%	13%	20%
Temperature(℃)：	~28	66–68	68-71	68-73	69-71	68-73	69-71	69-73

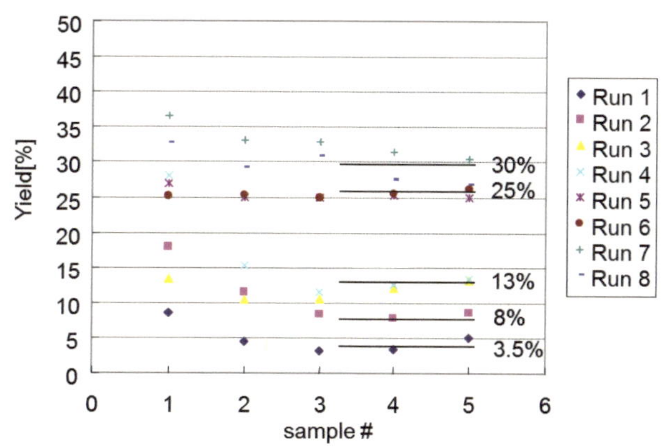

Figure 8. Typical results of continuous-flow synthesis of ETBE

7.3. Microwave-assisted pressurized synthesis of ETBE

Microwave apparatus (Microwave Accelerated Reaction System (MARS 5), CEM Corporation) shown in Figure 9 was used to study the effect of carrying out experiments in a sealed reactor under high pressure and high temperature. The microwave apparatus operates at

2.45 GHz frequency, while the microwave output can be manipulated up to maximum power of 1,200 Watts. This apparatus also consists of a fluoropolymer-coated microwave cavity, a cavity exhaust fan and tubing to vent fumes and a digital computer programmable for 100 programs consisting of up to five stages each. Inside the cavity is an alternating turntable system which can hold up to 13 reactor vessels, thus performing simultaneous experiments on 13 samples is possible.

The sealed Teflon-reactor vessel can handle pressures up to 5MPa and temperatures up to 250 °C. The reactor can be connected to the pressure and temperature control mechanisms of the MARS 5, for online monitoring and for operational safety. Some of the experiments were also carried out in a similar apparatus (Ethos) manufactured by Milestone General Co, Ltd.

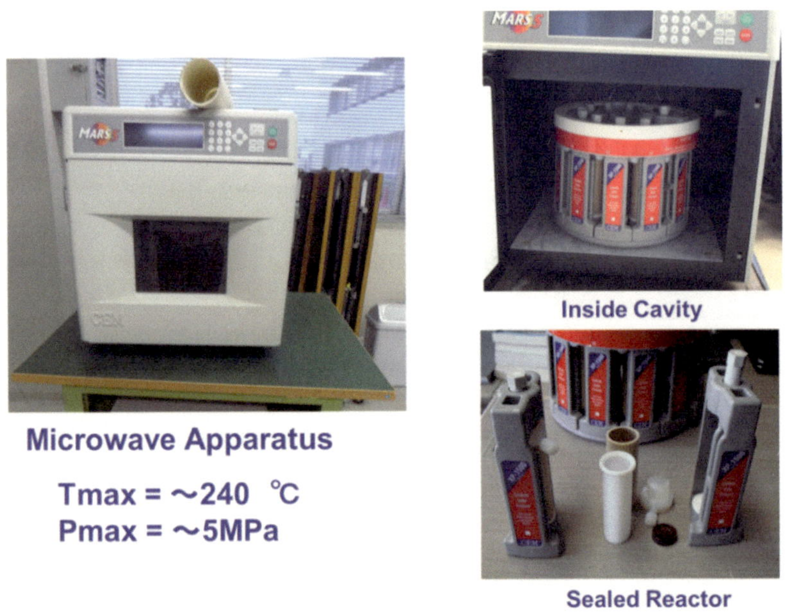

Figure 9. Microwave apparatus for pressurized synthesis of ETBE in a sealed reactor vessel (MARS5, CEM Corporation, Japan)

In a typical experiment, about 0.25 mol each of the reactants (TBA and ETOH) were placed inside the vessel, and mixed with 20g of Amberlyst 15 catalysts. The reactor was sealed and connected to temperature and pressure sensors, then microwave was irradiated until the set irradiation time has elapsed. After cooling the vessel to reach a temperature below 50 °C, the reactor was opened and an aliquot part of the products was taken for analysis.

Figure 10 shows the yields of ETBE using MW at various power at irradiation time of 1 min. A maximum yield of about 87% was obtained at MW power of 350W. At this condition, the attained temperature was around 87°C as shown in Figure 11, higher than the boiling points of the two alcohols. The yield was also found to be dependent on the amount of catalysts, reaction time and microwave power.

http://dx.doi.org/10.5772/52539

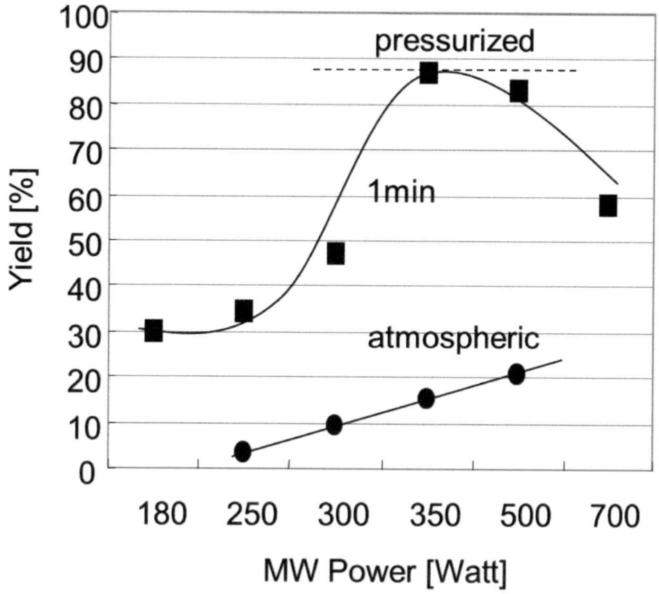

Figure 10. Yields obtained using a sealed reactor compared to those at atmospheric conditions (TBA=EtOH=0.25mol, A15 = 20g)

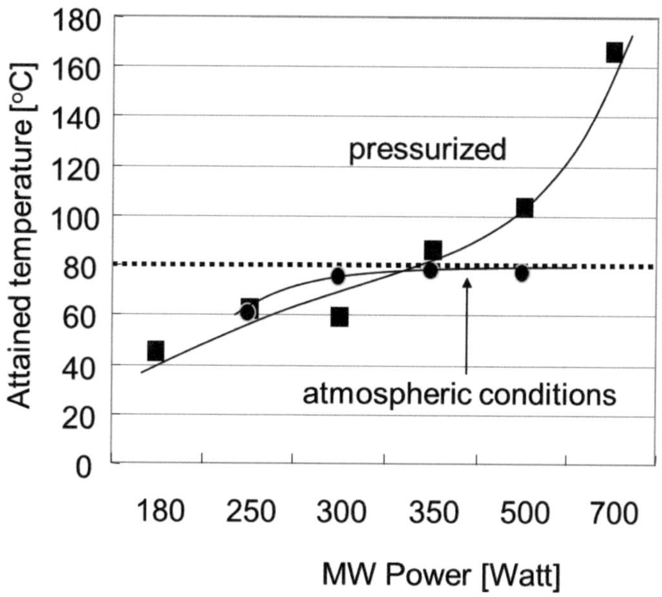

Figure 11. Attained temperature inside the pressurized vessel

7.4. Preliminary studies on combined reaction and separation inside microwave cavity

In an equilibrium reaction, simultaneous separation of obtained products will shift the reaction forward, thus further increasing the yield. In case of ETBE synthesis, simultaneous removal of the product (ETBE) and byproduct (H_2O) could increase the yield of the target compound. Preliminary studies were carried out using the apparatus shown in Figure 12. In this set-up, the catalysts were placed at the middle of the column. Initially, an equimolar mixture of TBA and EtOH was placed in a round bottom flask in B, then microwave irradiated. The products (distillates) were condensed and collected outside the cavity. After 25 min, the collected products in the distillates (D) and bottoms (B) were analyzed of its composition.

Results in Figure 13, show that the bottom consisted of mostly H_2O and unreacted EtOH, while the distillates consisted of ETBE, unreacted TBA, unreacted EtOH and H_2O indicating possibility of simultaneous separation of the products. The future will look at the control and optimization of its operation to obtain better yield of the target compound.

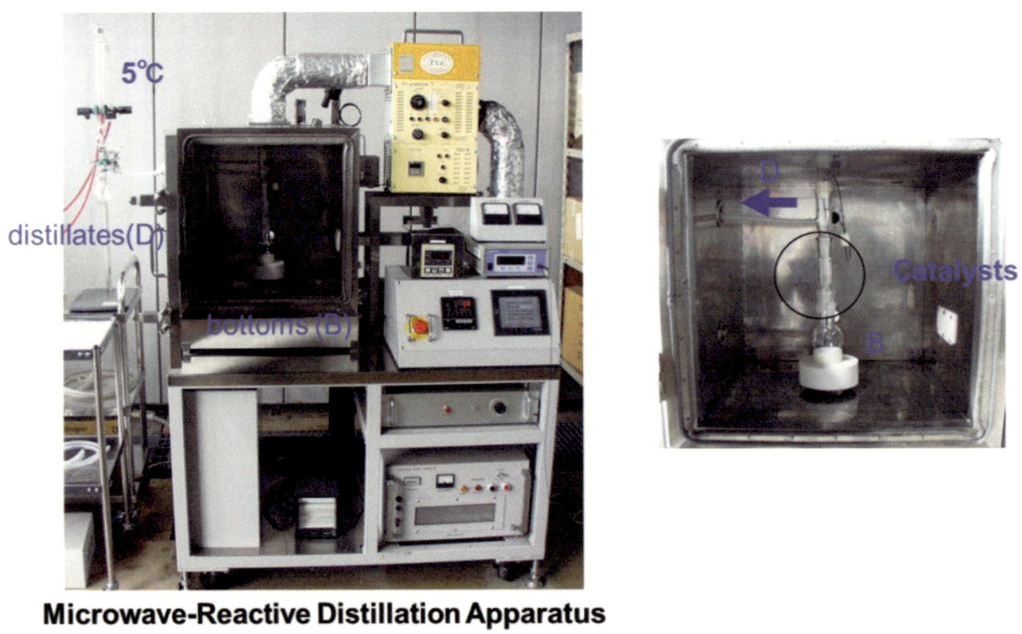

Figure 12. Microwave apparatus for preliminary studies on reactive separation for ETBE synthesis

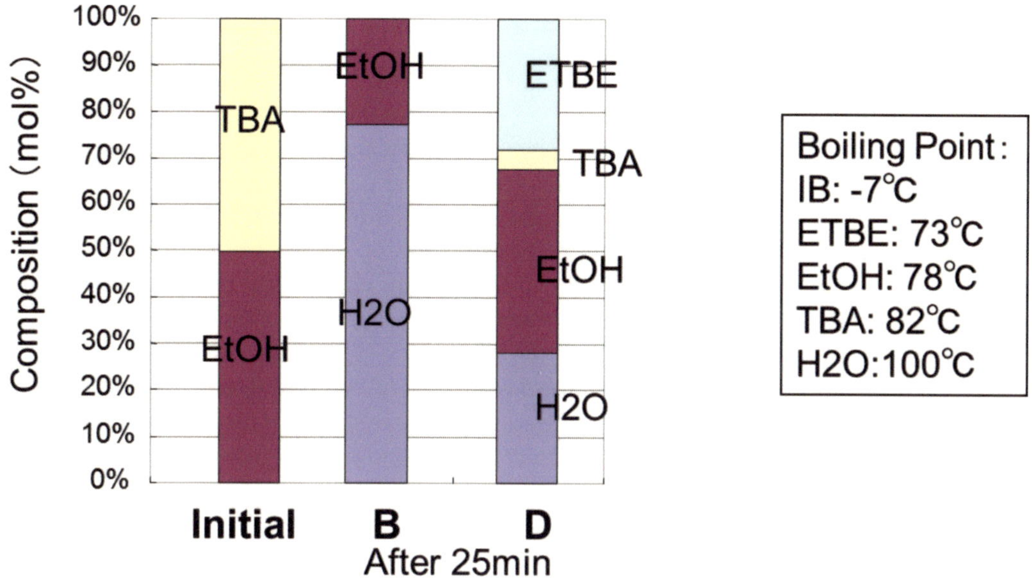

Figure 13. Results of preliminary studies on reactive separation for ETBE synthesis

8. Conclusion and future directions

Microwave irradiation was applied to syntheses of ETBE from biomass-deriveable alcohols (TBA and EtOH). The maximum yield obtained under atmospheric conditions was low at around 30%, which agreed with our previous studies on reactive distillation and the results obtained by other researchers. Performing the experiments in a sealed reactor at a microwave power of 350W, and irradiation time of 1 min, the mixture reached pressurized conditions obtaining EtOH conversion closed to 90%. The conversion and yield were found to be also dependent on operating parameters such as temperature, microwave irradiation power, time and amount of catalysts. Preliminary studies on combined reaction and separation process inside the cavity showed promising results but need further investigation for control and optimization of its operation.

The application of microwave irradiation to the synthesis of this promising biofuel offer benefits including energy efficiency, development of a compact process, rapid heating and instant on-off process (instant heating-cooling process), among others. Unlike the conventional heating, the heat is generated within the material, thus rapid heating occurs. Besides, microwave effects on reaction also likely occur, thus obtaining dramatic increase in the yield even at low bulk temperatures.

The benefits have been indicated by the above mentioned results. However, there are some drawbacks including the problems with non-homogeneous heating that would require thorough investigation prior to its commercialization. Although the field is in its infancy, the outlook is bright for the proposed methods due to foreseen high global demands for biofuels. The next few years should see development of continuous compact process, along with cheap, effective and stable solid catalysts.

As the demand for biofuels continue to increase in the near future, and while the search for an efficient and low-cost production process continues, the global outlook is positive for the use of microwave irradiation to the synthesis of ETBE. To overcome the limitations for scaling up microwave-assisted technology for ETBE production, development of a compact continuous process is suggested, but still poses several challenges that require detailed investigation. The future also calls for the development of cheap, effective and stable solid catalysts for the synthesis of the abovementioned fuels. While the use of microwave irradiation offers great benefits with regards to rapid reaction or synthesis, safety is a big factor to consider in designing a large scale production plant. However, this can be avoided if multilayered compact reactors operating under microwave irradiation can be developed instead.

Acknowledgments

Most of the research works were supported financially by the Japan Science and Technology carried out at the Research Institute for Solvothermal Technology, Kagawa Industry Support Fundation. The MARS 5 apparatus was donated by AIST-Shikoku through the kindness of Dr. Akinari Sonoda. This work was also partly supported by Kumamoto University Global COE Program "Global Initiative Center for Pulsed Power Engineering".

Author details

Armando T. Quitain[1], Shunsaku Katoh[2] and Motonobu Goto[2]

1 Graduate School of Science and Technology, Kumamoto University, Japan

2 RIST Kagawa, Kagawa Industry Support Foundation, Japan

Department of Chemical Engineering, Nagoya University, Japan

References

[1] Cunnill F, Villa M, Izquierdo JF, Iborra M, Tejero J. Effect of Water Presence on Methyl tert-Butyl Ether and Ethyl tert-Butyl Ether Liquid Phase Syntheses. Ind. Eng. Chem. Res. 1993; 32, 564-569.

[2] Rihko LK, Outi A, Krause L. Etherification of FCC Light Gasoline with Methanol. Ind. Eng. Chem. Res. 1996; 35, 2500-2507.

[3] Brockwell HL, Sarathy PR and Trotta R. Hydrocarbon Processing, 9, 133-141 (1990)

[4] Hutchings GJ, Nicolaides CP, Scurrell MS. Developments in the Production of Methyl tert-Butyl Ether. Catalysis Today 1992; 15, 23-49.

[5] "Refining 1996," Hydrocarbon Processing, 75, 113-116(1996)

[6] Matouq M, Tagawa T, Goto, S. Combined Process for Production of Methyl tert-Butyl Ether from tert-Butyl Alcohol and Methanol. J. Chem. Eng. Japan 1994; 27, 302-306.

[7] Knifton, J. F. and P. E. Dai; US Patent 5,387,722 (1995)

[8] Le Van Mao R, Carli R, Ahlafi H and Ragaini V. Synthesis of Methyl tert-Butyl Ether (MTBE) Over Triflic Acid Loaded ZSM-5 and Y Zeolites, Catalysis Letters 1990; 6, 321-330.

[9] Ahmed S, El-FaerMZ, Abdillahi MM, Shirokoff J, Siddiqui MAB, Barri SAI. Production of Methyl tert-butyl Ether (MTBE) Over MFI-type Zeolites Synthesized by the Rapid Crystallization Method and Modified by Varying Si/Al Ratio and Steaming. Applied Catalysis A 1997; 161, 47-58.

[10] Collignon F, Londers R, Martens JA, Jacobs PA, Poncelet G. Liquid Phase Synthesis of MTBE from Methanol and Isobutene Over Acid Zeolites and Amberlyst-15. J. of Catalysis 1996; 182, 302-312.

[11] Norris JF, Rigby GW. The Reactivity of Atoms and Groups in Organic Compounds. J. Am. Chem. Soc. 1932, 54, 2088-2100.

[12] Habenicht CL, Kam C, Wilschut MJ, Antal J. Homegeneous Catalysis of Ethyl tert-Butyl Ether Formation from tert-Butyl Alcohol in Hot, Compressed Liquid Ethanol. Ind. Eng. Chem. Res. 1995; 34, 3784-3792.

[13] Yin, X., B. Yang and S. Goto; Kinetics of Liquid-Phase Synthesis of Ethyl tert-Butyl Ether from tert-Butyl Alcoholand Ethanol Catalyzed by Ion-Exchange Resin and Heteropoly Acid. Int. J. Chem. Kinetics. 1995; 271065-1074.

[14] Quitain AT, Itoh H, Goto S. Reactive Distillation for Syntheizing Ethyl tert-Butyl Ether from Bioethanol. J ChemEng Japan 1999; 32 (3) 280-287

[15] Mingos DMP, Baghurst DR. Applications of microwave dielectric heating effects to synthetic problems in chemistry, In: Microwave-Enhanced Chemistry, H. M. (Skip) Kingston & S. J. Haswell, (Ed.), pp. 3-50, American Chemical Society, ISBN 0-8412-3375-6, Washington DC, USA, 1997.

[16] Kingston HM (Skip), Haswell SJ. Microwave-enhanced chemistry: fundamentals, sample preparation and applications, American Chemical Society, ISBN 0-8412-3375-6, Washington DC, USA, 1996.

[17] Banik S, Bandyopadhyay S, Ganguly S. Bioeffects of microwave—a brief review. Bioresource Technology 2003; 87,155–159.

[18] De la Hoz A, Diaz-Ortiz A, Moreno A. Critical review: microwaves in organic synthesis. thermal and non-thermal mi-crowave effects. ChemSoc Rev 2005; 34,164-178.

[19] Jacob J, Chia LHL, Boey FYC. Review: thermal and non-thermal interaction of microwave irradiation with materials. Journal of Material Science 1995; 30, 5321-5327.

[20] Japanese Patent Application: JP2007-126450. ETBE Synthesis Using Microwave.

[21] Japanese Patent Application: JP2008-133250. Production of Alkyl Ether by Microwave.

Section 2

Gases and Other Products

Chapter 12

Generation of Biohydrogen by Anaerobic Fermentation of Organics Wastes in Colombia

Edilson León Moreno Cárdenas,
Deisy Juliana Cano Quintero and
Cortés Marín Elkin Alonso

Additional information is available at the end of the chapter

http://dx.doi.org/10.5772/53351

1. Introduction

1.1. The trouble of organics solids wastes

In the protection of environment, the adequate handling of solids wastes occupy a main place, the integral handling of wastes is a term applied to all activities associated with the wastes management in the society. The main aim is the administration of wastes associated with the environment and public health. The handling of solid wastes is one of the main environmental problems in the cities due to its generations increase simultaneously with the growth of the cities, its industrialization and the increase of population. In addition, the actual life style carries out a high demand of consumption of goods that generally are thrown out in a short time; this generates more production of wastes and therefor having to search for solutions to the final disposition.

A solution for the trouble of the urban solids wastes is the implementation of process of reusing and giving value to the different materials that form what is known as "garbage", with the purpose of obtaining products or sub products that can be to introduce into new economic cycles. The maximization of reusing and giving value to solids wastes carry out benefits as: less consumption of natural sources, reduction of energy consumption, less environmental pollution, better use of the location where the garbage is placed and economic benefits from recovered materials. So the changes of consumption patron and the sustainable production are essential for the reduction of wastes production.

Is very difficult to stop the production of solids wastes, the idea is consider the solids wastes as a source of material reusable, raw matter, organics nutrients, biofuels and energetics fuel. The set of process to recover and treatment the wastes are known as valorization of solids wastes. This production of wastes is due to origin, social context and production activities [1]. During the valorization and reusing of wastes, is necessary take account aspect as recollection and transport, with this is possible to obtain highs benefices by the transformation. Additionally is necessary to include applications of new concepts related to the financial services, decentralized management, community contribution and the options of transformation, valorization and incorporation to economic cycles [2].

1.2. Source of wastes

At the whole world, the solids wastes from different sources are generating negative environmental impact to the nature, the biodiversity and life in the planet. This is caused by the inappropriate disposition of wastes, the increase of population, the processes of industrial transformation, agroindustrial and life habits of people [3]. At the present time, one characteristic of the society is the increase unbridled of the production and accumulation of solids wastes, which are generated without a solution to its final disposition. In the most of cases, this produced an inappropriate final disposition, an increase in the environment deterioration (air, surface water and groundwater, soil, landscape), problems in the public health and personal security [2].

The characteristics of solids wastes changed in function of the main activity (industry, trade, tourism and others), the habits of the population, type of fed, consumption models, environment conditions and others. The solids wastes can be classified according to: the source (domestic activities, institutional, commercial, industry, farming, municipal services and construction); the constitution (recyclable material and non-recyclable) and grade of danger (commons and dangerous).

The present chapter shows the energetic potential of the solid organic wastes generated in Colombia and its capacity to produce biohydrogen by anaerobic fermentation; additionally is presented a research carried out at the Laboratory of Agricultural Mechanization of the National University of Colombia in Medellín between the years 2009 and 2012, which main aim was determinate the initial feasibility to generate biohydrogen from urban organics wastes and to establish some conditions to operate a bioreactor type batch.

2. Generation of solids wastes in Colombia

The quantity of wastes produced depend of factors as: the number of inhabitant in the city, urbanization rate, consumption habits, cultural practices to handle of wastes, the income, the application of technology and industrial development. According to the information reported by the "Superintendencia de Servicios Públicos Domiciliarios" by 2008, see [4], in Colombia were generated daily 25.079 tons of urban solids wastes, 10 million of tons/year, which 77% were domestics (19.310,8 ton); 15% Industrials (3.761,9) and 8% others (2.006,3 ton). In the country, the management of wastes is focused to the final disposal in landfill; only 2,4% is dedicated to recycle and valorization [1].

Disposing of wastes	tons/day	Participation (%)	Municipalities
Landfill	22.204	88,5	653
Open dumpsite	2.185	8,7	297
Treatment facility	615	2,4	98
Buried	75	0,3	19
Discharged into rivers		<0,1	10
Incineration		<0,1	11
Total	25.076	100	1.088

Source: [4].

Table 1. Disposing of wastes in Colombia

In the country, the solid wastes are mainly composed of organic material (65%), followed by the plastics (14%), paper and cardboard (5%), glass (4%), other components with minor participation.

Type of waste	Percentage %
Organic	65
Paper and cardboard	5
Plastic	14
Glass	4
Rubber	1
Metals	1
Textiles	3
Dangerous and pathogens	2
Others	5

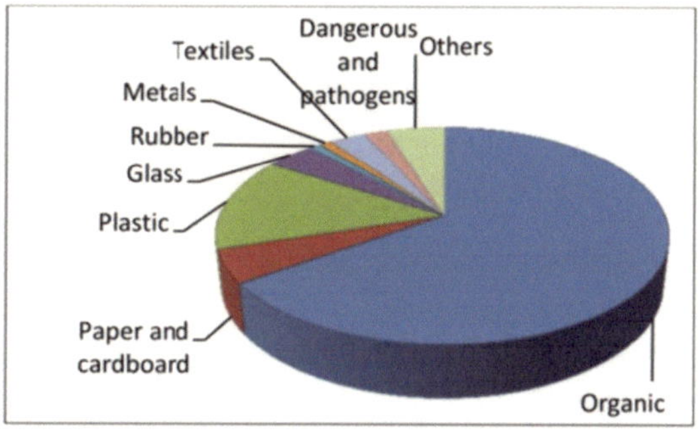

Source: [2].

Table 2. Composition of solids wastes in Colombia

In Colombia the major quantity of solid wastes generated are collected and treated by municipal companies (waste from domestic activities, commercials and industrials); however in some regions the problem of wastes solids is very important as the final disposition is made with little control, generating environmental pollution. The production of wastes (kg/habitant/day) is approximately 0,5 kg/habitant/day, oscillating between 1 kg/habitant/day for the big cities until 0,2 kg/habitant/day in the small towns [5]. The "Superintendencia de Servicios Públicos Domiciliarios" published by 2002 a study about the final disposition of the solids wastes in 1.086 cities. The technologies more frequent are: dumpsite and open incineration (52%), then landfill (30 %), and finally the use of composting, incineration and others (18%), [6].

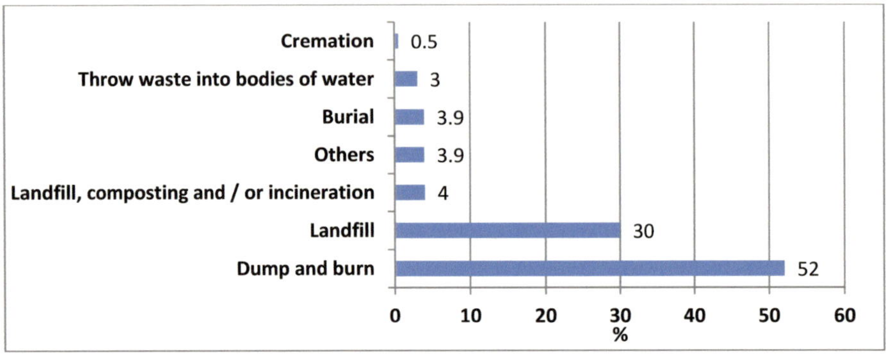

Source: [6].

Figure 1. Final disposition of solids wastes in Colombian for 1.086 municipalities, 2002.

There are two options to solve the problems generated by urban solid wastes which can be applied simultaneously to reach an optimum result:

- The first option according to the National Politics of Solid Wastes is give priority to integral management of solids wastes, focusing the operations management in the reuse and valorization of different materials that integrate the urban solid wastes.

- The second option is to take the wastes and give them an adequate final disposal in landfill operated technically.

The biomass in Colombia has calorific values between 4,384 kcal / kg for stems of coffee and 1,800 kcal / kg for banana rachis [7]. These values are comparable with reports from other countries as China where biomass from agricultural and forest activities have values between 3,827 to 4,784 kcal / kg [8]. In Argentina the lignocellulose biomass has values between 3,000 – 3,500 Kcal / kg and the municipal wastes between 2,000 and 2,500 Kcal / kg [9].

2.1. Colombian normativity about solid wastes

The Colombian normativity related to management of organics solids wastes began with the code of renewable natural sources (decree 2811 of 1974) and were implemented the followed norms:

- Decree 2104 of 1983: management of solids wastes.

- Resolution 2309 of 1986: special solids wastes.

- Law 142 of 1994: Law of public services.

http://dx.doi.org/10.5772/53351

- Decree 605 of 1996: Indications for an adequate cleaning service, from the generation, storage, collection, transport, to final disposition.
- Committee Technical ICONTEC 000019 about environmental management of solids wastes.
- Decree 1716 of August 2002 of "Ministerio de Desarrollo Económico" (In English: Economic Development Ministry) by mean the law 142 of 1994, law 632 of 2000 and the law 689 of 2001, related to the cleaning public service, the law 2811 of 1974 and the law 99 of 1993. The article 8 related to the program for the integral management of solids wastes, which should be realized by the cities in a maximum time of 2 years [10].

2.2. Organic solids wastes and its energetic potential in Colombia

The increase of energy demand in recent decades driven in particular by developed countries and countries with economic growth as Colombia, is leading to rapid depletion of nonrenewable energy resources, increasing pollution and global warming. The alternatives energetics sources emerge as a great option to reduce the adverse effects of this development. The biomass is considered as the alternative energetic source the most potential, according to reports from the World Energy Council [11] it is estimated that energy from biomass will account for 25,4% of global consumption by 2030 and 80% by 2080. Biomass is very varied due to its production and origin, a particular type are the wastes of natural processes, industrial or agroindustrial. It is estimated that Colombia has an energy potential from residual biomass of 449.485 TJ / year, also has a land area of 114,174,800 hectares, of which 44,77% are engaged to agricultural activities, this places to sector as the main source of wastes (with an energy potential of 331.645 TJ / year, mainly from annual and permanent crops). At the second place are the wastes from livestock activities (with an energy potential of 117.747 TJ / year), then the urban organic wastes (wastes from food and homes with an energy potential of 91 TJ / year) and finally the wastes from agroindustrial activities [12].

Among the methods to profit energetically the residual biomass, the anaerobic fermentation is a way of great interest, with this bioprocess is possible to generate a gas with high energy characteristics such as hydrogen and sludge that could be employed as fertilizer on crop. The generation of biohydrogen by anaerobic fermentation of wastes has generated great interest in the last decades. Hydrogen is a promising option as energy source [13, 14], it is a clean renewable resource because its combustion produces only water as emissions, in addition has the highest energy content per unit mass, with a value of 122 kJ / g [13]. The biological production of hydrogen can be seen as a promising option [15], two types of bacteria are involved in the process: acidogenic bacteria which initially to reduce the substrate in H_2 (biohydrogen), acetic acid and CO_2 and the methanogenic bacteria that converted these elements in methane gas. If the purpose is to produce biohydrogen, favorable conditions for the growth of the first type of bacteria (acidogenic) should be provided, inhibiting or eliminating the population of methanogenic bacteria [16]. Currently there are two methods to inhibit this type of bacteria: thermal shock and acidification [17, 18].

The residual biomass in Colombia has a high potential as alternative energetic source, only in wastes of sugar cane, rice husk, coco fiber, coffee pulp, oil palm, bean seed and barley, the potential is 12.000 MW/year approximately. The wastes are produced in different regions of the country and during all year. The country has a potential for generation of biomass of 331'638.720 ton/year, if all agricultural and urban wastes were treated by fermentation anaerobic, could be generated 28'825.609 m^3 of biohydrogen, this might give a energetic potential of 144 GW, upper value to country potential in wind energy (21 GW), tidal energetic potential (30 GW with two coasts) and geothermic energetic potential (1 GW).

In Colombia this quantity of biohydrogen could replace all diesel requested by the diesel electrical plants installed in the country. This has a great important especially in regions without connection to national electrical grid. In the country approximately the 66% of the territory are not connection to national electrical grid, this is 1,4 millions of people, namely the 4% of the population. The country has an installed electric capacity at the region without connection to national electrical grid of 102 MW of which 97 MW are produced by diesel plants, this quantity could be generated, using only the 40% of the urban organic wastes generated at the country. Colombia produces 250.000 tons/year of banana wastes with a potential to generated 100.000 m^3 of biohydrogen by anaerobic fermentation, this represent 500 MW of energy per year, quantity enough to supply the electric energy demand of 200.000 people during a year.

3. Generation of biohydrogen in Colombia

A research in order to determine the initial feasibility to generate biohydrogen from urban organics wastes and then established some conditions to operate a batch bioreactor was developed in Colombia. This section presents the results of this research and analysis the potential use of urban wastes as sources to generate hydrogen.

3.1. Localization

The research was performance between the years 2009 and 2012, at the Laboratory of Agricultural Mechanization of the National University of Colombia in Medellín, localized in 6°13′55″N and 75°34′05″W, with average annual temperature of 24°C, relative humidity of 88% and average annual precipitation of 1571mm.

3.2. Methods

Two stages were established to develop the research, the first had five phases.

3.2.1. First stage

Phase 1. Identification of organic wastes generated at the Central Wholesaler of Antioquia

The Central Wholesaler of Antioquia is the main company dedicated to trade food in the city of Medellín (fruits, vegetable and some grains). At the first phase historical information related to organic wastes production during two year was supplied by Central Wholesaler of Antioquia and was made a photographic register of solids wastes generated. The photographs were taken twice per day at the morning and afternoon.

Phase 2. Selection of wastes with greater production

According to the information collected and the photographic register from the first phase, the wastes with greater production were selected to be introduced into a batch bioreactor.

Phase 3. Elemental Composition and chemical composition analysis

The quantity of volatile solids, total solids and elemental composition on both wet and dry basis (coal, nitrogen and hydrogen) were obtained for each wastes. Were taken samples of 5 grams and the analysis method applied was the Wendee method (the analysis was made at the chemical analysis laboratory of National University in Medellin). With that information was calculated the quantity of wastes to use. Six samples of 3 grams in each wastes were taken in order to obtain the elemental analysis, in this case the method applied was burn of sample and the equipment employed was an elemental analyzer CE – 440 (Figure 2a). The samples were triturated with a precision crusher – IKA WERNE with sieve of 0,5 mm (Figure 2b) and then were dried in a lyophilizer LABCONCO Freezone 12L (Figure 2c). In order to determine the quantity of wastes and water to be employed, 6 grams of volatile solids per liter-day were used as organics load [19], additionally was employed a concentration on volatile solids of 5% [20].

Figure 2. Elemental analyzer CE – 440 (a), Crusher MF Basic- IKA WERKE (b), Lyophilizer LABCONCO - Freezone 12L (c).

Phase 4. Installation of bioreactor

A batch bioreactor of 2000 liters was installed, the wastes were triturated to facilitate its access into bioreactor and its process by the bacteria. The quantity of gas generated was registered with a gas flow meter Metrex G 2,5 with accurate of 0,040 m^3/h; maximum pressure of 40 kPa, additionally was employed a gel of silica to remove the wet of gas. The load of bioreactor was made during four days, each day was used the same quantity until to complete the total load.

Phase 5. Principal variables to register

The relativity humidity and environment temperature were registered daily, was used a thermohygrometer with rank in temperature until 120°C and 100% in relativity humidity (Figure 4). The pH into the bioreactor was registered daily too, in this case was employed a digital pH-meter Hanna Instruments, with accurate of ± 0,2 (reference temperature of 20°C).

Figure 3. Installation of bioreactor and equipment to trituration

Figure 4. Thermohygrometer and pH-meter

The organics load was determined at the beginning and end of bioprocess; in this case the total suspended solids (TSS), total solids (TS), volatile fatty acids (VFAs), chemical oxygen demand (COD) and biochemical oxygen demand (BOD) were determined. The analytics method employed were Standard Method by water and residual water of the APHA-AWWA-WPCF, edition 19 of 1995.

The production of gas was registered daily, samples were collected in Tedlar bags (with capacity of 1 liter, Figure 5) and then were analyzed in a chromatographic gas (Perkin Elmer) to establish its composition (percentage of CO_2, O_2, H_2, CH_4 and N_2). During the tests, the wastes were subjected to an acid pretreatment to eliminate the methanogenic bacteria, after several days, agricultural lime was added to increase the pH until to obtain a value most adequate to the acidogenic bacteria.

3.2.2. *Second stage*

With the information from the first stage was elaborated an experiment with three treatments and three repetitions, were used three bioreactors of 2000 liters each them. The treatments were integrated by three values of duration to acid pretreatment (3, 7 and 10 days) and three values of pH to operation of the bioprocess (4,5 – 5,0; 5,1 – 5,5 and 5,6 – 6,0). The materials used were wastes of different fruits and vegetal from Central Wholesaler of Antioquia. The wastes were

Figure 5. Gas flow meter and Tedlar bag

triturated and mixed with water in a relation of 1:2,5. In each test were taken samples of wastes and sludge to determinate the organic load, in addition was recorded daily the pH and the gas production. When the pretreatment of acidification ended, agricultural lime was added like at the first stage. The methodology employed to obtain the quantity of gas generated was the same of the first stage, was used a gas flow meter Metrex G2,5 with accurate of 0,040 m^3/h and maximum pressure of 40 kPa. Samples of gas were collected in Tedlar bags and then were analyzed in a chromatographic gas (Perkin Elmer) to determinate its composition.

The organics load of wastes was obtained at the beginning and end of bioprocess; this included the total suspended solids (TSS), total solids (TS), volatile fatty acids (VFAs), chemical oxygen demand (COD) and biochemical oxygen demand (BOD). The analytic method employed was the Standard Method by water and residual water like the first stage. Was calculated the production of gas (liters/day), hydrogen percentage (% de H_2) and yield of biohydrogen (liters of H_2/day).

3.3. Results

3.3.1. First stage

First and second phase: The quantity and percentage of wastes generated at the Central Wholesaler of Antioquia during the year 2011 are show at the Table 3 and Figure 6. The highest production of wastes was associated to cabbage and lettuce leaves then wastes of citrics (orange and lemon) and finally wastes of mango, guava and others tropical fruits. The Figure 7 shows some pictures of wastes in the storage containers at the Central Wholesaler of Antioquia. With the information of production were selected wastes of cabbage and lettuce leaves, orange, mango, papaya and guava to be employed at the bioprocess.

Third phase: The elemental analysis of wastes selected is show at the Table 4. To orange wastes, the relation C/N obtained was less than values reported in others research. In the others cases the results were close to values reported for wastes with similar characteristics. A relation C/N close to 30 is considered appropriate to growth of anaerobic bacteria [22].

Organic wastes	Volumen average month (m^3)
Lettuce and cabbage leaves	360
Orange and lemon	38
Pimento, cucumber	26
Mango	23
Tomato	23
Papaya and guava	22
Total	492
Source: [21].	

Table 3. Organics wastes generated at the Central Wholesaler of Antioquia

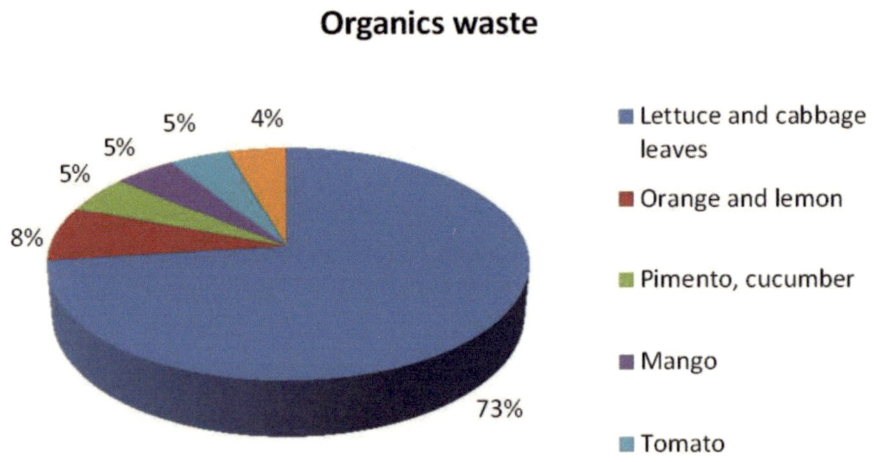

Figure 6. Percentage of wastes generated at the Central Wholesaler of Antioquia, year 2011

Wastes	C	H	N	C/N	C/N (Literature)
Mango	37.6	6.0	1.5	25.9	34.8
Orange	40.6	5.5	1.3	31.5	75.6
Guava	40.9	5.7	1.4	29.2	34.8
Papaya	36.7	5.7	1.3	27.6	34.8
Lettuce and cabbage leaves	37.6	5.3	1.5	25.1	18.0

Table 4. Result of elemental analysis on dry basis (%), (Coil laboratory, National University of Colombia)

Figure 7. Pictures of wastes in the storage containers at the Central Wholesaler of Antioquia

The chemical composition analysis of wastes showed that the highest values of volatile solids were found in the tropical fruits (mango, orange, guava and papaya). The volatile solids are the proportion of the raw material that bacteria using to generate biogas and have an outstanding role during the anaerobic fermentation process.

Waste	ST (%)	SV (%ST)	SV (%)
Mango	97,4	15,1	14,71
Orange	96,6	14,3	13,81
Guava	96,7	15,3	14,79
Papaya	97,1	12,7	12,33
Lettuce and cabbage leaves	86,5	8	6,92

Table 5. Chemical composition analysis, (Chemical composition analysis laboratory, National University of Colombia)

In order to determinate the quantity of wastes to be used was obtained the density of each wastes, to this were taken samples and then were triturated, weighed and finally was calculated the volume to employ. The bioreactor was loaded with 422 kilograms of wastes and 1110 kilograms of water, this provided an average relation (wastes: water) of 1:2,5.

Wastes	Density (kg/l)	So (g SV)/l	Organic load (g SV)/day [19]	Wastes to use (l)	Wastes to use (kg)	Concentration of volatile solid (% SV)/day [20]	Relation (wastes: water)	Water to use (l)
Mango	0,8820	129,7	6,0	22	19	5	1 : 2.94	64
Orange	0,9639	133,1	6,0	22	21	5	1 : 2.76	61
Guava	1,1655	172,4	6,0	29	33	5	1 : 2.96	85
Papaya	1,1907	146,8	6,0	24	29	5	1 : 2.47	60
Lettuce and cabbage leaves	0,4579	31,7	6,0	5	2	5	1 : 1.38	7
Total				102	105			278
Total to 4 days				409	422			1110

Table 6. Quantity of wastes and water to the fermentation process

Fourth phase:

Each waste was triturated and mixed with water during three minutes until to reach an average size of 2 centimeters. In order to reduce the quantity of methanogenic bacteria, the wastes were submitted to acidic conditions during three months with a value of pH close to 3,5. Afterwards was added during three days agricultural lime until to reach a pH of 6,2; in that moment the production of biohydrogen started. The quantity of agricultural lime added was 7 kilograms (Figure 9).

Figure 8. Bioreactor used by the first stage and wastes triturated

Fifth phase:

The organics load showed an important reduction during the process, the total suspend solids were reduced in 83%, the chemical oxygen demand was reduced in 65% and the biochemical

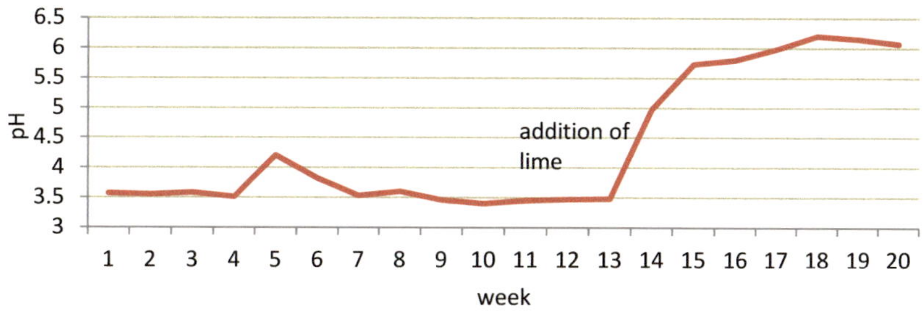

Figure 9. Behavior of pH during the first stage

oxygen demand was reduced in 63,6%. The environment temperature was between 21,8 y 31 °C, this mean that the biohydrogen production was developed under mesophilic conditions. The average relative humidity was between 38 y 73%.

Analysis	Beginning	End
SST (mg/l)	1920	325
STV(mg/l)	54815	8296
ST(mg/l)	62395	9893
COD(mg/lO_2)	54000	19133
BOD(mg/lO_2)	37633	13713

Table 7. Organic load of wastes at the first stage (Laboratory of Sanitary Engineering, National University of Colombia)

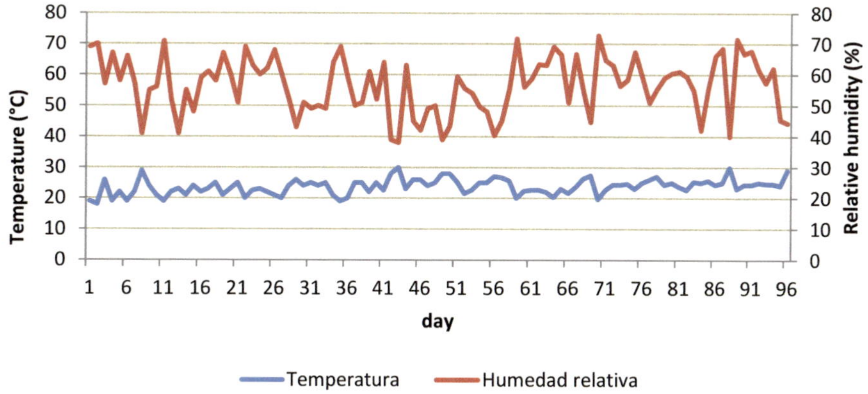

Figure 10. Behavior of temperature and relative humidity average

The gas production started three days after application of agricultural lime and continued for 22 days more. The hydrogen (biohydrogen) percentage found in gas ranged between 6,37 y 17,26; with a percentage of hydrogen less than 13,3; there was carbon dioxide and nitrogen in the biogas, however when the percentage of hydrogen was greater than 13,3; the gas composition was only hydrogen and carbon dioxide. The greater value of methane was 1,25% and less was 0%, this mean that the pretreatment to reduce the methanogenic bacteria was satisfactory.

Sample	CO_2 (%)	H_2 (%)	N_2 (%)	O_2 (%)	CH_4 (%)
1	31,79	6,72	48,19	13,06	0
2	70,99	13,31	2,63	0,42	1,25
3	75,67	17,26	0,65	0,096	0,73
4	80,98	13,51	ND	ND	0,6
5	32,80	6,37	48,13	13,16	0,24

ND: not detected

Table 8. Composition of gas generated (Coil laboratory, National University of Colombia)

The total production of hydrogen was 177 liters in 22 days, with a maximum value of 14,5 liters, an average of 7,4 liters of H_2/day and maximum yield of 83 liters of H_2/m^3 of bioreactor. The maximum value of generation of hydrogen was registered 7 days after from started the gas production and the maximum rate of hydrogen generation was obtained between first and seventh days. The Figure 12 shows a several pictures of biohydrogen generated, the color blue is from silica gel used to remove the wet of the gas. The quantity of organic load removed was 26.400 mg/liter of O_2, (COD).

Source: Information personal from research

Figure 11. Pictures of biohydrogen generated

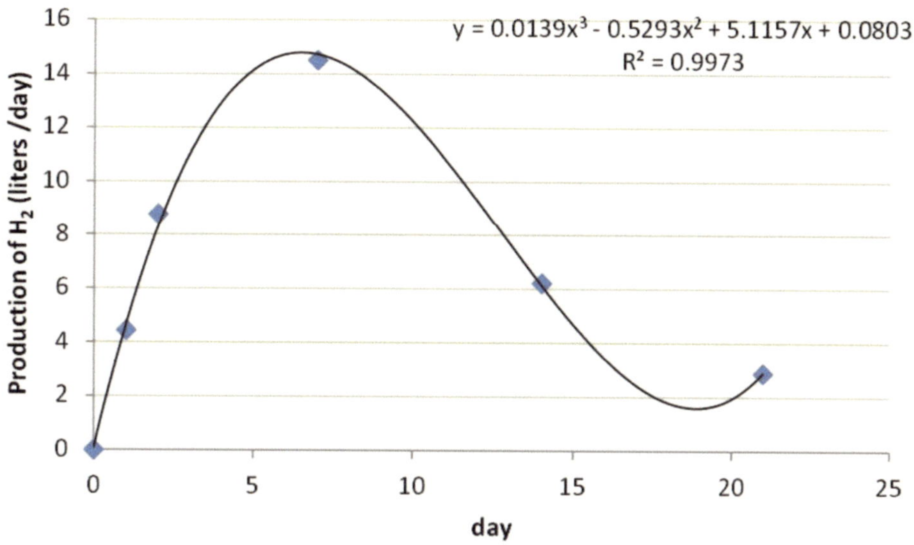

Figure 12. Production daily of hydrogen

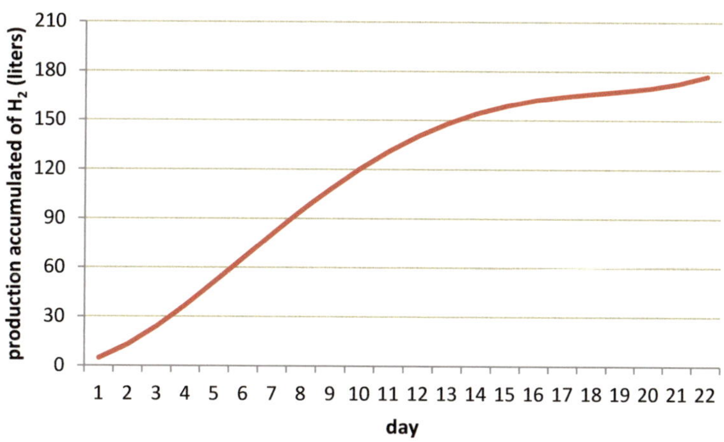

Figure 13. Production accumulated of hydrogen

3.3.2. Second stage

Installation of bioreactors and wastes to use

At the second stage were used the same wastes of first stage, but additionally were employed wastes of tomato, onion, garlic and husk of cape gooseberry (Table 10). The wastes were triturated and mixed with water during three minutes, the relation of wastes: water was 1:2,5 like at first stage. The quantity of wastes employed in each treatment was similar to quantity used at first stage. The volume of work in each bioreactor was 70% and 30% was dedicated to storage the gas generated.

Figure 14. Set of bioreactors employed at the second stage

Repetition	Wastes	Quantity of wastes (kg)		
		T1	T2	T3
1	Lettuce and cabbage leaves, tomato, onion, garlic, pimento, orange, lemon, mango, guava and papaya	506	514	453
2	Lettuce and cabbage leaves, tomato, onion, garlic and husk of cape gooseberry	450	450	450
3	Lettuce and cabbage leaves, orange, mango, guava and papaya	500	500	400
Average		485,3	488	434,3

Table 9. Wastes used in each repetition

The highest values of chemical oxygen demand (COD) were obtained in the treatment 2 during the repetition 1 and the treatments 1 and 3 of the repetition 3 respectively, namely in these cases there were more quantity of food available to the microorganisms.

Analysis	Repetition 1			Repetition 2	Repetition 3		
	T1	T2	T3	T1, T2 y T3	T1	T2	T3
ST (mg/l)	5322	8198	4700	11280	16290	9830	10500
COD (mg/ lO_2)	8667	18000	8000	12000	27140	17940	23340
BOD (mg/lO_2)	7617	10983	6200	5840	24415	3775	14455

Table 10. Organic composition of wastes employed

http://dx.doi.org/10.5772/53351

Behavior of pH during the pretreatment and operation of bioreactors

Due to type of wastes employed in the repetition 2, was necessary to add muriatic acid into all bioreactors to achieve the pH of acidification, but there were not response (pH between 3,5 and 4,5). However, during the repetitions 1 and 3, in all treatments was used wastes of orange and lemon, this allowed to apply the pretreatment of acidification, afterwards was added agricultural lime and was reached a pH between 5 and 6, values adequate to generate biohydrogen.

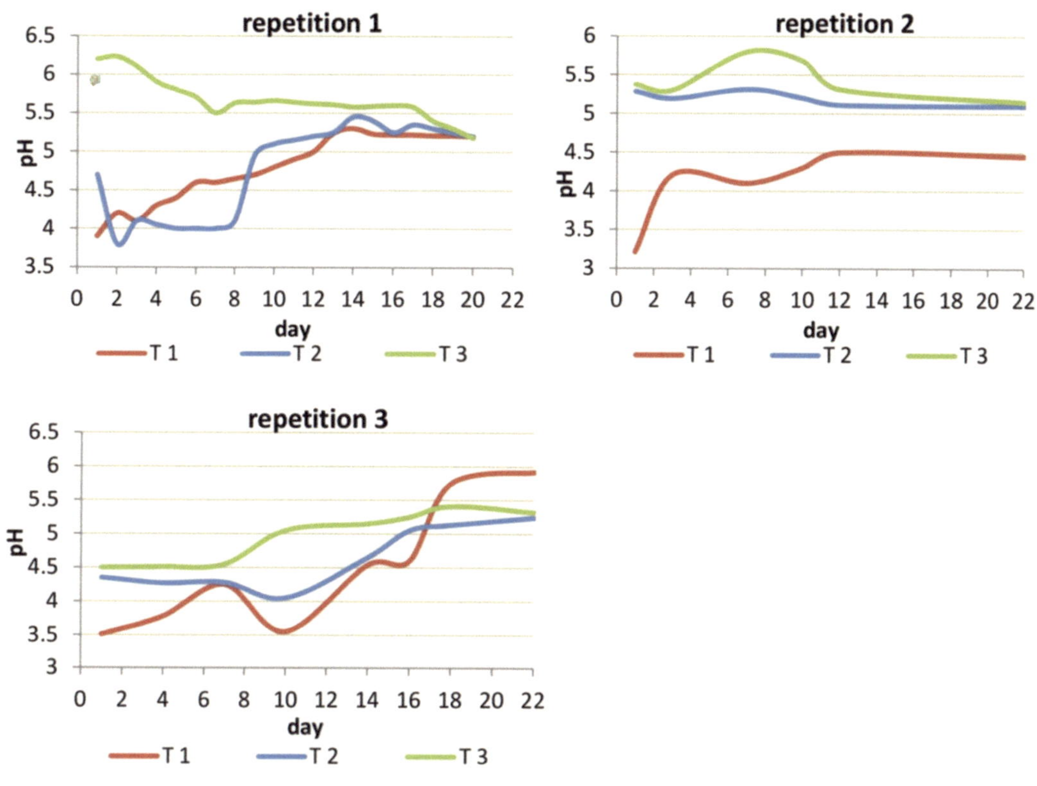

Figure 15. Behavior of pH in each treatment and repetition

Production of biohydrogen

The gas generated in all treatments was compound of hydrogen, carbon dioxide, nitrogen and oxygen. The greater value of methane was 3,7% and the less was 0%, in many times the methane was not detected (ND), this mean that the pretreatment to reduce the methanogenic bacteria was satisfactory. The percentage of hydrogen in gas was between 5 and 18,08; this was the highest value in the research and was obtained in the treatment 3 during the repetition 3 when the wastes used were Lettuce and cabbage leaves, orange, lemon and papaya. In the repetition 2 in all treatments, there were not generation of hydrogen (NG). The oxygen content in some repetitions show maybe that some air entered to bioreactor when the samples were taken.

Treatment	Sample	CO_2 (%)	H_2 (%)	N_2 (%)	O_2 (%)	CH_4 (%)
1	1	35,1	6,7	53,9	3,0	1,4
	2	16,2	6,4	64,8	13,6	ND
	3	18,2	0,02	67,2	13,9	0,8
2	1	40,0	10,5	45,7	0,9	0,8
	2	46,5	7,5	43,7	1,8	ND
	3	34,1	0,02	54,3	7,0	3,7
3	1	40,2	7,7	45,0	2,5	0,4
	2	43,9	7,0	47,0	2,1	ND
	3	4,0	0,02	77,3	18,1	ND

Table 11. Composition of gas generated, first repetition

Treatment	Sample	CO_2 (%)	H_2 (%)	N_2 (%)	O_2 (%)	CH_4 (%)
1	1	NG	NG	NG	NG	NG
	2	NG	NG	NG	NG	NG
	3	NG	NG	NG	NG	NG
2	1	11,4	5,5	41	13,4	0,2
	2	37,8	5,5	36,2	5,4	0,5
	3	20,1	5,7	43,5	12	3,3
3	1	NG	NG	NG	NG	NG
	2	NG	NG	NG	NG	NG
	3	NG	NG	NG	NG	NG

Table 12. Composition of gas generated, second repetition

Treatment	Sample	CO_2 (%)	H_2 (%)	N_2 (%)	O_2 (%)	CH_4 (%)
1	1	29,08	7,5	50,3	3,0	0,2
	2	23,88	6,5	41,7	10,3	2,6
	3	42,14	8,2	36,1	7,1	1,4
2	1	28	7,0	52,2	4,4	0,1
	2	45,29	5,8	38,1	3,0	0,1
	3	40,69	5,3	40,8	3,2	0,1
3	1	8,96	18,0	51,3	12,1	0,2
	2	43,32	6,3	36,5	7,1	0,7
	3	52,85	5,0	30,5	5,1	0,3

ND: not detected
NG: not generated

Table 13. Composition of gas generated, third repetition

http://dx.doi.org/10.5772/53351

The maximum production of biohydrogen per day obtained at the first stage was 15 liters, however at the second stage the maximum production was 38 liters, this mean that the production was duplicated during the second stage. In the repetition 3, in the treatment 3 were generated 32 liters of biohydrogen, meanwhile in the treatment 1 were generated 15 liters. The wastes employed in both treatments were Lettuce and cabbage leaves, orange, lemon and tropical fruits as mango, guava and papaya, the initial pH was lesser to 4,5 during 7 and 8 days, and the pH of bioreactors operation was between 5 and 5,5.

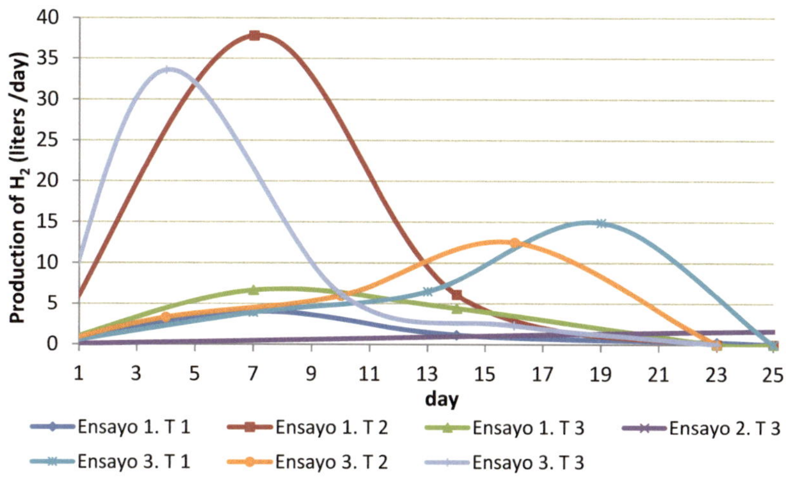

Figure 16. Production of hydrogen

The greater accumulated production was reached in the treatment 2 and the repetition 1, followed by the treatment 3 in the repetition 3. In both cases were used vegetal wastes and tropical fruits in same proportions in addition, initially the wastes were subjected to acid conditions with a pH less to 4,0 during 8 days and a pH for operation of bioreactor between 5 and 5,5. Under those conditions the hydrogen content into gas ranged between 7,5 and 10,5%. The total production of hydrogen in the treatment 2 and the repetition 1 was 317,8 liters in 22 days, with 14,44 liters of H_2/day (twice the result from the first stage), and maximum yield of 159 liters of H_2/m^3 of bioreactor. Other outstanding result was reached when were employed the same wastes, at beginning were applied acid conditions during 7 days under a pH less to 4,5; and then was used a pH for the operation of bioreactor between 5 and 5,5. In this case the percentage of hydrogen into gas ranged between 5 and 18,08% (the last value was the maximum content reached in the research). The production of hydrogen was 231,1 liters of H_2 in 22 days with 10,5 liters of H_2/day (duplicated the value reached from the first stage) and a maximum yield of 115,6 liters of H_2/m^3 of bioreactor.

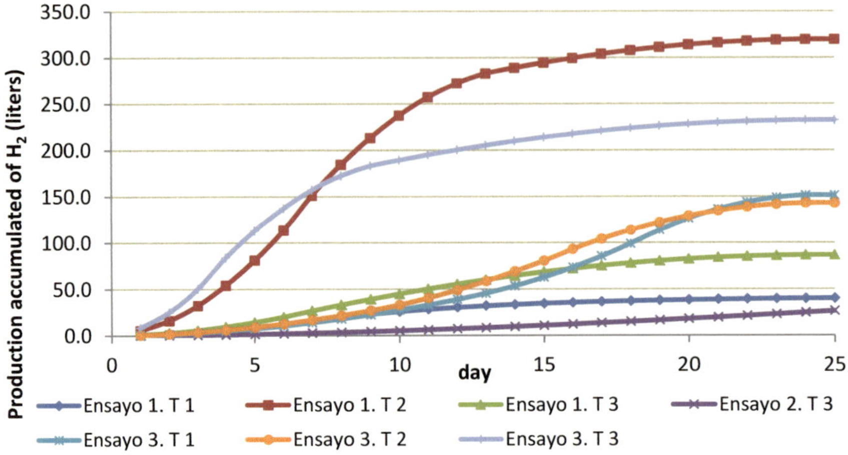

Figure 17. Production accumulated of hydrogen

4. Final analysis

Two treatments showed the highest production of biohydrogen, the treatment 2 in the repetition 1 and the treatment 3 in the repetition 3, the maximum value was obtained with the treatment 2 in the repetition 1 in which were used wastes from lettuce and cabbage leaves, tomato, onion, garlic, pimento, orange, lemon, mango, guava and papaya. The acid conditions were implemented 8 days with value of pH near to 4, the operation of bioreactor was between 5 and 5,5. In the treatment 3 in the repetition 3 were used the same wastes, the acid condition was applied during 7 days with value of pH near to 4,5; the pH of bioreactor operation was between 5 and 5,5. Although was generated more quantity of hydrogen in the treatment 2 during the repetition 1, was in the treatment 3 in the repetition 3 where was obtained the greater hydrogen content in the gas (18,04%) and greater rate of generation of hydrogen.

The maximum production of hydrogen was obtained at the second stage when the pretreatment of acidification was applied during 8 days with a value for the pH of 4, a pH of reactor operation between 5 and 5,5; and a value of chemical oxygen demand (COD) near to 20.000 mg/liter of O_2. At the first stage when was used a quantity of wastes from tropical fruits greater than wastes of lettuce and cabbage leaves, the chemical oxygen demand (COD) initial was 54.000 mg/liter of O_2, however the hydrogen production was significantly less respect to second stage. This indicates that a high value of chemical oxygen demand could inhibit the hydrogen generation; this result is according to reports of different authors [13, 23-29]. When were used vegetal wastes (without wastes of tropical fruits) as lettuce and cabbage leaves, tomato, onion, garlic and husk of cape gooseberry, there were no acid conditions at beginning the process and was necessary to add acid, however there was no response in the biosystems and the pH always was upper than 4,5. Under these conditions there was no production of hydrogen. In addition the chemical oxygen demand was low (12.000 mg/liter of O_2).

The results shows that is feasible to produce biohydrogen (hydrogen) when are employed organic wastes from the Central Wholesaler of Antioquia. The wastes should be submitted to a pretreatment acid with a pH between 3,5 y 4,0; during 7 days (or less), then the operation pH should be increased until a value between 5 and 5,5. The chemical oxygen demand (COD) should be between 20.000 and 54.000 mg/liter of O_2, this is possible to reach when in the bioprocess is employed a proportion similar of tropical fruit waste and vegetal waste.

5. Conclusions

- It was possible to generate hydrogen from organic wastes of Central Wholesaler of Antioquia and to improve the bioprocess.
- The chemical oxygen demand (COD) promoted the biohydrogen production, the best results were obtained to values between 20.000 and 54.000 mg/liter of O_2. These values were achieved with a heterogeneous mix of fruits and vegetal wastes.
- There was not generation of biohydrogen when the bioprocess started with a pH upper than 4. This ratifies that to generate biohydrogen by anaerobic fermentation is necessary to apply a pretreatment, in this research, a pretreatment under acid conditions (pH between 3,5 and 4,0) was successful.
- Colombia has a high potential to generate hydrogen by anaerobic fermentation due to organic wastes available, these wastes could generate until 28'825.609 m^3 of biohydrogen and supply an energetic potential of 144 GW, value upper than the installed potential (13,5 GW).
- The results show that is possible to produce biohydrogen by anaerobic fermentation of organic wastes and providing new sources energetic.

Acknowledgements

The authors acknowledge to National University of Colombia in Medellín and the Central Wholesaler of Antioquia by the financial support to research.

Author details

Edilson León Moreno Cárdenas*, Deisy Juliana Cano Quintero and Cortés Marín Elkin Alonso

*Address all correspondence to: elmorenoc@unal.edu.co

Engineering Agricultural Department, National University of Colombia, Medellín, Colombia

References

[1] Contreras C. Manejo integral de aspectos ambientales – residuos sólidos. 2006. Diplomado gestión ambiental empresarial para funcionarios de ETB. Universidad Pontificia Javeriana, Bogotá, Colombia.

[2] Ministerio de Ambiente, Vivienda y Desarrollo Territorial. Viceministerio de Ambiente. Construcción de criterios técnicos para el aprovechamiento y valorización de residuos sólidos orgánicos con alta tasa de biodegradación, plásticos, vidrio, papel y cartón. Manual 1: generalidades. Bogotá: epam s.a. e.s.p; 2008.

[3] Acurio G, Rossin A, Teixeira F.P, and Zepeda F. Diagnóstico de la situación de residuos sólidos municipales en América Latina y el Caribe. Banco Interamericano de Desarrollo y la Organización Panamericana; 1997. http://www. Bvsde.paho.org/acrobat/diagnost.pdf (accessed 12 June 2012).

[4] Superintendencia de Servicio Públicos Domiciliarios. Estadísticas 2008. Bogotá; 2008. http://www.superservicios.gov.co. (accessed 21 April 2012).

[5] Puerta E. S. M. Los residuos sólidos municipales como acondicionadores de suelos. Revista lasallista de investigación. 2004; 1(1) 56 - 65.

[6] Superintendencia de Servicio Públicos Domiciliarios. Estadísticas 2002. Bogotá; 2002. http://www.superservicios.gov.co. (accessed 09 July 2012).

[7] Ministerio de Minas y Energía. Estadísticas: MME. http://www.minminas.gov.co/minminas (accessed 14 May 2012).

[8] Cuiping L, Chuangzhi W, Yanyongjie, Haitao H. Chemical Elemental Characteristics of Biomass Fuels in China. Biomasss and Bioenergy 2004; 27 (2) 119-130.

[9] Secretaria Argentina de Energía: Biocombustibles. http://energia3.mecon.goc.ar/home (accessed 13 August 2012).

[10] Ministerio de Desarrollo Económico. MCIT. http://www.mincomercio.gov.co/ (accessed 10 March 2012).

[11] World Energy Council. WEC. http://www.worldenergy.org/activities/knowledge_networks/energy_efficiency/default.asp (accessed 12 June 2012).

[12] Unidad de Planeación Minero Energética. UPME. http://www1.upme.gov.co (accessed 27 March 2012).

[13] Hallenbeck C. P. Fermentative Hydrogen Production: Principles, progress and prognosis. International Journal of Hydrogen Energy 2009; 34(17) 7379-7389.

[14] Lee Z. K, Shiue L. L, Jian S. L, Yu H. W. Effect of pH in fermentation of vegetable kitchen wastes on hydrogen production under a thermophilic condition. International Journal of Hydrogen Energy 2008; 33(19) 5234-5241.

[15] Vázquez D. G, Navarro C. C. B, Colunga R. L. M, Rodríguez A. L, Flores R. E. Continuous biohydrogen production using cheese whey: Improving the hydrogen production rate. International Journal of Hydrogen Energy 2009; 34(10) 4296-4304.

[16] Lee Y.W, Chung J. Bioproduction of hydrogen from food wastes by pilot-scale combined hydrogen/methane fermentation. International Journal of Hydrogen Energy 2010; 35(21) 11746-11755.

[17] Cano Q.D.Y, Moreno C.E.L. Viabilidad del Aprovechamiento de Coproductos como fuente de combustible para una Celda de hidrógeno en la central mayorista de Antioquia. Tesis de Grado en Ingeniería Agrícola. Universidad Nacional de Colombia Sede Medellín, 2010.

[18] Ozmihci S, Kargi F. Dark fermentative bio-hydrogen production from wastes wheat starch using co-culture with periodic feeding: Effects of substrate loading rate. International Journal of Hydrogen Energy 2011; 36(12) 7089-7093.

[19] Shin H. S, Youn J. H, Kim S. H. Hydrogen production from food waste in anaerobic mesophilic and thermophilic acidogenesis. International Journal of Hydrogen Energy 2004; 29(13) 1355-1363.

[20] Kim D.G, Kim S.H, Shin H.S. Hydrogen fermentation of food wastes without inoculum addition. Enzyme and Microbial Technology 2009; 45(3)181-187.

[21] Central Wholesaler of Antioquia. Wastes production, statistics 2011. Document interno. 20011.

[22] Piedrahita V. D. R. Elementos para una tecnología sobre la producción de biogas. Medellín: Universidad Nacional de Colombia; 2000.

[23] Bouallagui H. O, Haouari Y, Touhami R, Ben C, Marauani L, Hamdi M. Effect of Temperature on the Performance of an Anaerobic Tubular Reactor Treating Fruit and Vegetal Waste. Process Biochemistry 2004; 39 (12) 2143-2178.

[24] Chenlin L, Fang H. Fermentative hydrogen production from wasteswater and solid wastes by mixed cultures. Critical Reviews in Environmental Science and Technology 2007; 37(1) 1-39.

[25] Das D, Veziroglu T.N. Advances in biological hydrogen production processes. International Journal of Hydrogen Energy 2008; 33(21) 6046-6057.

[26] Das D, Veziroglu T.N. Hydrogen production by biological processes: a survey of literature. International Journal of Hydrogen Energy 2001; 26(16) 13-28.

[27] Hawkes F.R, Hussy I, Kyazze G, Dinsdale R, Hawkes D.L. Continuous dark fermentative hydrogen production by mesophilic microflora: Principles and progress. International Journal of Hydrogen Energy 2007; 32(2) 172-184.

[28] Hwanga J.J, Chang W.R. Life-cycle analysis of greenhouse gas emission and energy efficiency of hydrogen fuel cell scooters. International Journal of Hydrogen Energy 2010; 35(21) 11947-11956.

[29] Nishio N, Nakashimada Y. High rate production of hydrogen/methane from various substrates and wastes. Advances in Biochemical Engineering/Biotechnology 2004; 90 63-87.

Chapter 13

Hydrogen Conversion in DC and Impulse Plasma-Liquid Systems

Valeriy Chernyak, Oleg Nedybaliuk, Sergei Sidoruk, Vitalij Yukhymenko, Eugen Martysh, Olena Solomenko, Yulia Veremij, Dmitry Levko, Alexandr Tsimbaliuk, Leonid Simonchik, Andrej Kirilov, Oleg Fedorovich, Anatolij Liptuga, Valentina Demchina and Semen Dragnev

Additional information is available at the end of the chapter

http://dx.doi.org/10.5772/53764

1. Introduction

It is well known [1] that hydrogen (H_2) as the environmentally friendly fuel is considered to be one of the future most promising energy sources. Recently, interest in hydrogen energy has increased significantly, mainly due to the energy consumption increase in the world, and recent advances in the fuel cell technology. According to the prognosis, in the next decades, global energy consumption will be increased by 59%, and still most of this energy will be extracted from the fossil fuels. Because of the traditional fossil fuels depletion, today there's a growing interest in renewable energy sources (f.e. – bioethanol, biodiesel). Bioethanol can be obtained from the renewable biomass, also it can be easily and safely transported due to its low toxicity, but it's not a very good fuel. Modern biodiesel production technologies are characterized by a high percentage of waste (bioglycerol) which is hard to recycle.

It is common knowledge [2] that addition of the syn-gas to the fuel (H_2 and CO) improves the combustion efficiency: less burning time, rapid propagation of the combustion wave, burning stabilization, more complete mixture combustion and reduction of dangerous emissions (NO_x). Besides, the synthesis gas is an important stuff raw for the various materials and synthetic fuels synthesizing. There are many methods of synthesis gas (including hydrogen) production, for example – steam reforming and partial liquid hydrocarbons oxidation. Also, there is an alternative approach – biomass reforming with low-temperature plasma assistance. Plasma is a very powerful source of active particles (electrons, ions, radicals, etc.), and therewith it can be catalyst for the various chemical processes activation. However, a major disadvantage of chemical processes plasma catalysis is weak processes control.

There is a bundle of electrical discharges that generate both equilibrium and non equilibrium plasma. For plasma conversion – arc, corona, spark, microwave, radio frequency, barrier and other discharges are used. One of the most effective discharges for the liquid hydrocarbons plasma treatment is the "tornado" type reverse vortical gas flow plasma-liquid system with a liquid electrode ("TORNADO-LE") [3]. The main advantages of plasma-liquid systems are – high chemical plasma activity and good plasma-chemical conversions selectivity. It may guarantee high performance and conversion efficiency at the relatively low power consumption. Moreover, those are systems of atmospheric pressure and above, and this increases their technological advantages.

Also, syn-gas ratio – hydrogen and carbon monoxide concentration ratio should be mentioned. As well, it should be taken into consideration that for efficient combustion (in terms of energy) of the synthesis gas it should contain more hydrogen, and in the case of the synthesis materials – they should contain more CO.

Relatively new possible solution to this problem – carbon dioxide recycling. Many modern energy projects have difficulties with the large amount of CO_2 storing and disposing. And it is also known that the addition of CO_2 to plasma during the hydrocarbons reforming may help to control plasma-chemical processes [4]. That is why the objective of the research is to study the influence of different amounts of CO_2 in the working gas on the plasma-chemical processes during the hydrocarbons conversion.

This research deals with hydrocarbons (bioethanol, bioglycerol) reforming by means of the combined system, which includes plasma processing and pyrolysis chamber. As a plasma source the "tornado" type reverse vortical gas flow plasma-liquid system with liquid electrode has been used [5].

Qualitatively new challenge is connected with a selectivity of the plasma chemistry strengthening by the transition of the chemical industry to "green chemistry". The last is a transition from the traditional concept of evaluating the effectiveness of the chemical yield to the concept that evaluates the cost-effectiveness as the exclusion of hazardous waste and non-toxic and/or hazardous substances [6].

A quantitative measure of the environmental acceptability of chemical technology is the ecology factor, which is defined as the ratio of the mass of waste (waste) to the mass of principal product. Waste is all that is not the principal product.

By the way, the most promising approaches in green chemistry is the implementation of processes in supercritical liquids (water, carbon dioxide) [7].

Water in supercritical condition unlimitedly mixes with oxygen, hydrogen and hydrocarbons, facilitating their interaction with each other - oxidation reactions are very fast in scH_2O (supercritical water). One particularly interesting application of this water - efficient destruction of chemical warfare agents. When mixed with other substances scH_2O can be used not only for oxidation but also in the reactions of hydrolysis, hydration, the formation and destruction of carbon-carbon bonds, hydrogenation, and others.

Besides, the use of pulsed electrical discharges in the liquid brings up new related factors: strong ultraviolet emission and acoustic or shock waves. In literature it can be found that systems with energies more than 1 kJ/pulse, that have negative influence on the lifetime of such systems. Reasonable from this perspective is the usage of pulsed systems with relatively low pulse energy and focusing of acoustic waves. In addition, the acoustic oscillations in such systems can be used as an additional mechanism of influence on chemical transformations.

In using of acoustic oscillations for chemical reactions the most attention is paid to systems with strong convergent waves. However, the processes during the collapse of the powerful convergent waves are studied unsufficiently. In the literature the systems of cylindrical, spherical or parabolic surfaces used in the focusing of shock waves for technological needs are known [8]. However, among their disadvantages should be noted that partial usage of the energy of acoustic wave and the problem of it's peripheral sources synchronization, which leads to distortion of the shock wave front ideality and reduces the focusing effectiveness.

Probably, more perspective method of using acoustic waves is their generation by single axial pulse electric discharge with further reflection from an ideal cylindrical surface. This approach can provide better symmetry of compression by convergent acoustic wave both in the gas and in the liquid. Probably, such mechanism can be exploited for scH_2O production

In addition, the re-ignition of electrical discharge at the moment of collapse convergent acoustic waves can lead to the plasma temperature increasing due to compression of the discharge channel, as well as the appropriate amplification of acoustic waves after the collapse.

It's clear that plasma-liquid systems (PLS) mentioned above have some sharp differences. Therefore, the first section of this article presents the results of our research on the addition of CO_2 to the "TORNADO-LE". And the second section of the article is devoted to investigation of double-impulse system in underwater electric discharge.

2. Bioethanol and bioglycerol conversion in "tornado" type plasma-liquid system with the addition of CO_2

2.1. Experimental set up

The experimental setting is shown in Fig. 1. Its base is a cylindrical quartz chamber (1) with diameter of 90 mm and height of 50 mm. Top (2) and bottom (3) it is hermetically closed with metal flanges. Camera is filled with fluid (4), the level of which has been maintained by the injection pump through the hole (5). Bottom flange is made of stainless steel. The stainless steel T-shaped cylindrical electrode (6), cooled with water, immerses in the liquid through the central hole in the bottom flange. There is a 5 mm thick metal washer on its surface (7) in the middle of which there is a hole in diameter of 10 mm. Sharp corners are rounded. This washer is used for reducing the waves (which have been moving to the quartz wall) amplitude on the liquid surface.

The top flange, made from duralumin, contains copper sleeve (13) with a diameter of 20 mm is placed in the center (2), and plays the role of the second electrode. The nozzle with diameter of 4 mm and length of 6 mm is located in the center of the copper sleeve (8). Gas is introduced into the flange (2) through the aperture (9). Gas flow changes the direction at 90 degrees inside the flange and injects tangentially into the channel (10). (10) The gas is rotated in the circular channel. Rotating gas (11) lands on the surface liquid and moves to the central axis of the system, where fells into the quartz cell through the nozzle (14), forming a plasma torch (12). Camera (14), in its turn, plays a role of pyrolytic chamber. Flow rate reaches the maximum value near the nozzle. Due to this, the zone of lower pressure is formed in the center of the gas layer, compared to the periphery. The conical structure appears over the liquid's surface near the system axis (Fig. 1). External static pressure is 1 atm. and internal - 1.2 atm (during discharge burning). Gas from quartz chamber (14) gets into the refrigerator (15), which is cooled with water at room temperature.

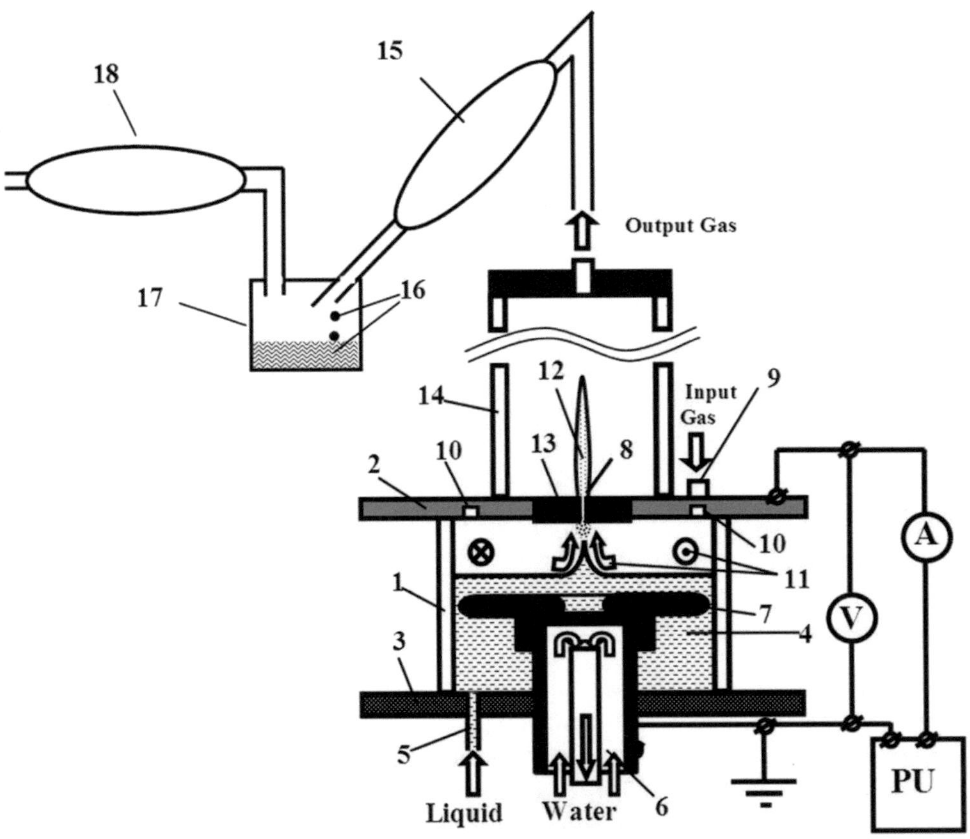

Figure 1. Schematic set up of the "TORNADO-LE".

Condensed matter (16) together with the gas from the refrigerator gets to the chamber (17). At the chamber exit (17) there's a flask (18), where gas is gathered for its composition diagnostics by means of mass spectrometry and gas chromatography. Study of plasma parameters is performed by emission spectrometry. The emission spectra registration procedure uses the system which consists of optical fiber, the spectral unit S-150-2-3648 USB, and the computer. Fiber is focusing on the sight line in the middle between the top flange (2) and the surface of the liquid (4).

The spectrometer works in the wavelength range from 200 to 1100 nm. The computer is used in both control measurements process and data processing, received from the spectrometer.

The voltage between the top flange and electrode, immersed in the liquid, is supplied by the power unit "PU". DC voltage provided is up to 7 kV. Two modes of operation have been considered:

1. "liquid" cathode (LC) – electrode immersed in the liquid has "minus" and the top flange has "plus";
2. solid" cathode (SC) - with the opposite polarity.

Electrode which has "plus" is grounded. Breakdown conditions are controlled by three parameters: the fluid level, the gas flow value and the voltage magnitude between the electrodes. The several modes of operation have been studied:

1. Various air flow and CO_2 ratio;
2. Discharge voltage varied within $U_d = 2.2 \div 2.4$ kV;
3. Discharge current varied within $I_d = 220 \div 340$ mA (ballast resistance hasn't been used).

At first, for the analysis of the plasma-chemical processes kinetics the distilled water (working fluid), and ethanol (ethyl alcohol solution in distilled water with a molar ratio C_2H_5OH/H_2O = 1/9.5), as a hydrocarbon model have been used. As the working gas mixture of air with CO_2, in a wide range of air flow and CO_2 ratios has been used. The ratio between air and CO_2 in the working gas changes in the ranges: CO_2/Air = 1/20 ÷ 1/3 for the working fluid C_2H_5OH/H_2O (1/9.5) and CO_2/Air = 0/1 ÷ 1/0 (by pure air to pure CO_2) - for distilled water.

Plasma component composition and population temperature of the excited electron (T_e^*), vibration (T_v^*) and rotational (T_r^*) levels of plasma components and relative concentrations of these components have been determined by the emission spectra. For the temperature population determination of the excited oxygen atoms electron levels the Boltzmann diagram method has been used [5]. T_e^* of oxygen atoms has been determined for the three most intense lines (777.2 nm, 844 nm, 926 nm). Temperature population of excited hydrogen atoms electron levels has been determined by the two lines of 656.3 nm and 486 nm relative intensities.

The effect of the presence of CO_2 in the system on the initial gas products has been investigated by means of "TORNADO-LE" current-voltage characteristics with changes in the working gas composition. T_v^* and T_r^* have been determined by comparing the experimentally measured emission spectra with the molecules spectra simulated in the SPECAIR program [9]. With help of this program and measured spectra, relative component concentrations in plasma have been determined. Also, the concentration of atomic components has been obtained by calculating the amount of oxygen that fell into a working system with the working gas flow. The hydrogen amount has been received from the electrolysis calculations. The output gas in reforming ethanol has been analyzed by gas chromatography and infrared absorption.

2.2. Results

The process of discharge ignition occurred as follows: the chamber is filled with liquid to a fixed level (5 mm above the washer). At the next stage a certain amount of gas flow forms the stationary cone from liquid; the voltage applied between the top flange and electrode immersed in a liquid starts gradually increase. When the voltage reaches a break-out value - U_b, a streamer appears for the first time. After that, burning discharge starts in split second, and then voltage decreases and current increases. After a second or two it is stabilized. During this time – static pressure rises inside the chamber from 1 to 1.2 atm. If you maintain the liquid level fixed, then the discharge is quite steady.

Liquid layer thickness of 5 mm has been chosen because that is the minimum liquid thickness in which the discharge burns between the liquid surface and the top flange. If the thickness is smaller plasma pushes the water toward the electrode immersed in the liquid and the discharge starts burning between two metal electrodes. Discharge goes into the arc regime. When the thickness of the distilled water layer above the washer is 5 mm (in the case of air flow only) break voltage reaches 4.5 kV and for a CO_2 flow - 6 kV. It is known [10], this increase in break-out voltage derives from the appearance of an additional loss channel of electrons – due to their sticking onto CO_2 molecules. This sticking has dissociative character and it is accompanied by the energy expense.

For example, the threshold reaction with CO_2 is 3.85 eV. Therefore CVC in pure CO_2 is decreased (Fig. 2). When the thickness of the C_2H_5OH/H_2O (1/9.5) solution layer above the washer is 5 mm (in the case of air flow only) the break voltage is 5.5 kV, and for the air flow mixture with CO_2 (CO_2/Air = 1/3) – 6.5 kV. Adding CO_2 to the air leads to the increase in the break-out voltage value. Adding ethanol to distilled water (C_2H_5OH/H_2O = 1/9.5) results in the increase of break voltage on 1 kV. Power supply unit provides maximum voltage of 7 kV. Increasing the thickness of the fluid layer above the washer (> 5 mm) leads to the increase of the break-out voltage value. There is no discharge ignition with a break-out voltage value of more than 7 kV. Therefore, 5 mm thickness of the liquid layer above the surface immersed in a liquid metal electrode (washer) has been chosen as the optimum one.

The current-voltage characteristics of the discharge are shown for the SC mode (Fig. 2 a; 2b). The cell has been filled with distilled water (Fig. 2a) or bioethanol (Fig. 2b).

The "tornado" type reverse vortex gas flow is formed by gas flow, which is a mixture of air with CO_2 in varying proportions. Ratio of CO_2/Air is changed in the range from 1/20 to 1/3, and in the case of ethanol and 1/0 in the case of water. Current varied in the range from 230 to 400 mA. The initial level of the working liquid is the same in all cases.

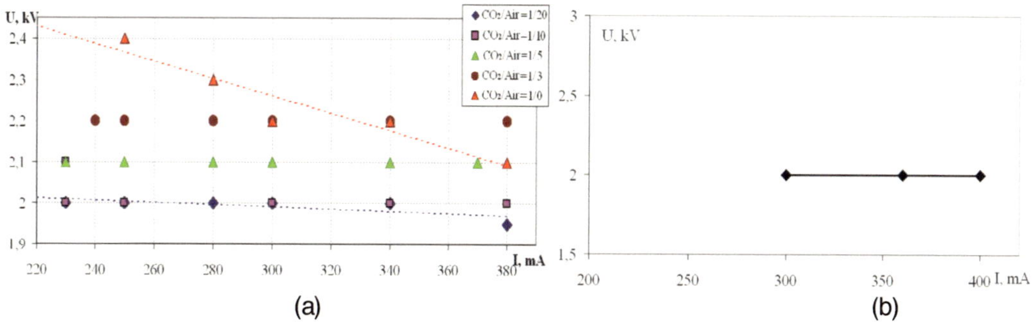

Figure 2. a) Current-voltage characteristics of the discharge at different ratios of CO_2/Air in the working gas. Working liquid - distilled water. Airflow - 55 and 82.5 cm^3/s, the flow of CO_2 - 4.25, 8.5 and 17 cm^3 /sec. b) Current-voltage characteristics of the discharge at the ratio CO_2/Air = 1/5 in the working gas. Working liquid – C_2H_5OH/H_2O (1/9.5) solution. Airflow - 82.5 cm^3/s, the flow of CO_2 - 17 cm^3/s.

The current-voltage characteristics show that adding a small amount of CO_2 (near 20%) to the working gas has no effect on the discharge type in various studied working liquids. In the range of flow ratios CO_2/Air from 1/20 to 1/5 characteristics are straight lines. It was observed that the increasing of CO_2 share in working gas causes discharge voltage supply rise.

Typical emission spectra of the plasma are shown in Fig. 3a and Fig. 3b for the cases with the distilled water as the working liquid and the solution of C_2H_5OH/H_2O (1/9.5).

The emission spectra show that when the working liquid is distilled water, plasma contains the following components: atoms H, O, and hydroxyl OH. In case when the working liquid is C_2H_5OH/H_2O solution (1/9.5), plasma has the following components: atoms – N, O, C, Fe, Cr, molecules – OH, CN, NH. The emission spectra shows that the replacement of the working liquid with distilled water with ethanol CN and lower electrode material made of stainless steel (anode) occur in plasma. Occasionally, during discharge burning breakdown may occur in C_2H_5OH/H_2O layer solution (1/9.5).

Those breakdowns may occur due to the fact that during the discharge burning, thickness of liquid layer, when the working fluid has a significant share of C_2H_5OH, a current channel is formed through the liquid layer to the metal electrode. And in the case of distilled water - plasma channel discharge ends near the surface of the liquid. It may indicate the presence of large liquid surface charge.

It was observed that the increase of CO_2 in the working gas (CO_2/air > 0.3) leads to an increase in the intensity of hydrogen and oxygen radiation lines (H and O) at the time when the intensity of the molecular component (OH) radiation, within the error, is stable (I = 300 mA, U = 1.9-2.4 kV, air flow 0 - 82.5 cm^3/sec, the flow of CO_2 - 4.25 - 85 cm^3/sec).

Fig. 4 shows the ratio of the hydrogen (Hα λ = 656.3 nm) and oxygen (O, λ = 777.2 nm) radiation intensity to the highest point of band hydroxyl (OH, λ = 282.2 nm) small intensity at different ratios of CO_2/air (I = 300 mA, U = 1.9 – 2.4 kV, air flow - 27.5, 55 and 82.5 cm^3/s, the flow of CO_2 - 4.25, 8.5, 17, 42.5 and 85 cm^3/s). High intensity bands haven't been used in the calculations because of the possible reabsorption. (I = 300 mA, U = 2 – 2.2 kV). In the case of distilled water

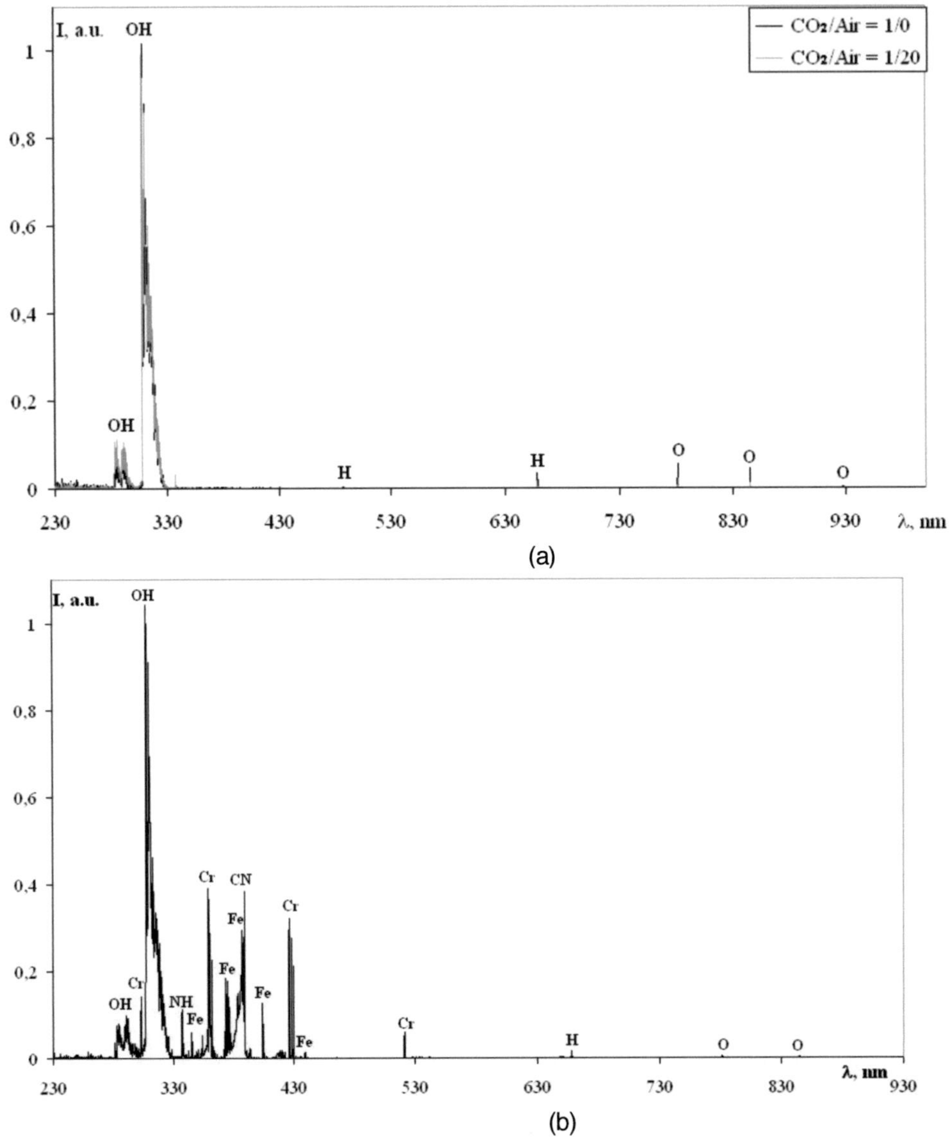

Figure 3. a) Emission spectrum of the plasma in TORNADO-LE plasma-liquid system, where the working liquid is distilled water. Working gas - a mixture of CO_2/air = 1/0, Id = 300 mA, U = 2.2 kV, the flow of CO_2 - 85 cm^3/s and CO_2/air = 1/20, Id = 300 mA, U = 1.9 -2.0 kV, air flow - 82.5 cm^3/s, the flow of CO_2 - 4.25 cm^3/s. b) Emission spectrum of the plasma in the TORNADO-LE plasma-liquid system, where the working fluid is bioethanol. Working gas - a mixture of CO2/air = 1/20, Id = 300 mA, U = 2 kV, air flow - 82.5 cm3/sec, the flow of CO2 - 4.25 cm3/sec.

(Fig. 4a), results are presented for the three air flows - 27.5, 55 and 82.5 cm^3/s and five CO_2 streams - 4.25, 8.5, 17, 42.5 and 85 cm^3/s (I = 300 mA , U = 1.9 – 2.4 kV). Air and CO_2 flows are variated so that the total flow compiles similar values and achieves ratios of CO_2/air in a wide range from 1/20 to 1/0.

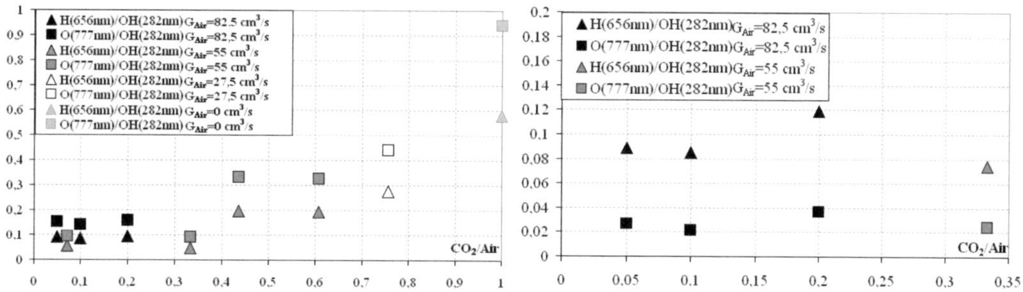

(a) (b)

Figure 4. The ratio of the radiation intensity of hydrogen (H_α λ = 656.3 nm) and oxygen (O, λ = 777.2 nm) to the peak of the band hydroxyl (OH λ = 282,2 nm) at different ratio CO_2/Air in the working gas. Working liquid - distilled water (a) I = 300 mA, U = 1.9 – 2.4 kV and bioethanol (b) I = 300 mA, U = 2 -2.2 kV.

In calculating the relative concentration ratio of hydrogen to oxygen from the emission spectra, it was observed that the hydrogen concentration is two times as much of the oxygen concentration for the case of distilled water as the working liquid - (I = 300 mA, U = 1.9 – 2.4 kV, airflow - 27.5, 55 and 82.5 cm^3/sec, the flow of CO_2 - 4.25, 8.5, 17, 42.5 and 85 cm^3/sec), and ten times as much when the working liquid is C_2H_5OH/H_2O solution (1/9.5) (I = 300 mA, U = 2 - 2.2 kV, air flow - 55 and 82.5 cm^3/sec, the flow of CO_2 - 4.25, 8.5 and 17 cm^3/sec). However, according to the calculations, these components production by means of electrolysis and their extraction from the working gas, the oxygen concentration exceeds the average hydrogen concentration in three orders of magnitude, unless the case when the pure CO_2 is used as a working gas.

It should be noted that the addition of CO_2 reduces the discharge stability, especially in the case of bioethanol. In determination of the temperature population excited electron levels of plasma atomic component the most intense lines (spectra with the smallest possible accumulation in the experiment measurement of 500 ms) have been used, according to the discharge burning particularity. Also, it affects the parameters determination accuracy.

Temperature of excited hydrogen electron population levels is determined by the relative intensities (two lines of 656 nm and 486 nm). For the case where the working liquid is distilled water – T_e^*(H) = 5500 ± 700 K (I = 300 mA, U = 1.9-2.4 kV, air flow - 27.5, 55 and 82.5 cm^3/s, the flow of CO_2 - 4.25, 8.5, 17, 42.5 and 85 cm^3/s) and as for the bioethanol T_e^*(H) = 6000 ± 500 K - (I = 300 mA, U = 2 – 2.2 kV, air flow - 55 and 82.5 cm^3/s, the flow of CO_2 - 4.25, 8.5 and 17 cm^3/s). Also, the oxygen T_e^*(O) has been defined by the Boltzmann diagrams method, in the case of distilled water. The three most intense lines (777.2 nm, 844 nm, 926 nm) are used in this method. So, we have T_e^*(O) = 4700 ± 700 K.

Temperatures of OH excited vibrate and rotational population levels have been determined by comparing the experimentally measured emission spectra with the molecular spectra modeled in The SPECAIR program. In the case when the working liquid is distilled water, appropriate temperatures are: T_r^*(OH) = 3000 ± 1000 K, T_v^*(OH) = 4000 ± 1000 K (I = 300 mA,

U = 1.9-2.4 kV, air flow - 27.5, 55 and 82.5 cm3 / s, the flow of CO_2 - 4.25, 8.5, 17, 42.5 and 85 cm³/s). Also, population temperatures of vibration and rotational levels for OH and CN have been determined in case of C_2N_5ON/H_2O (1/9.5) solution as the working liquid, they are: T_r^* (OH) = 3500 ± 500 K, T_v^*(OH) = 4000 ± 500 K, T_r^*(CN) = 4000 ± 500 K, T_v^*(CN) = 4500 ± 500 K) (I = 300 mA, U = 2 - 2.2 kV, air currents - 55 and 82.5 cm³/s, the flow of CO_2 - 4.25, 8.5 and 17 cm³/s). Temperatures for other molecular components haven't been determined because of their bands low intensity.

During the study, it turned out that the addition of CO_2 weakly affects the population temperature of excited electron, vibration and rotational levels of plasma components (Fig. 5) (I = 300 mA, U = 1.9 - 2.4 kV, air flow - 27.5, 55 and 82.5 cm³/s, the flow of CO_2 - 4.25, 8.5, 17, 42.5 and 85 cm³/s). Weak tendency to temperature decrease has been observed, but these changes do not exceed the error.

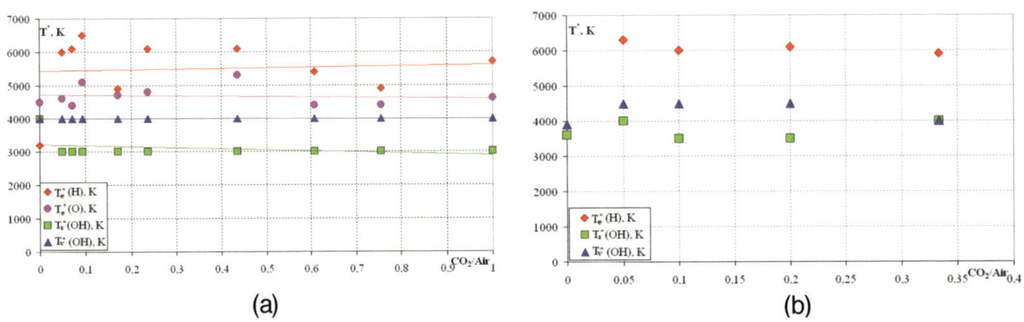

Figure 5. Population temperatures of excited electron, vibration and rotational levels of plasma components at different ratio of CO_2/Air in the working gas. Working liquids - distilled water (a) and ethanol (b)

Fig. 6–7. shows the results of gas chromatography bioethanol conversion output products. Results are presented for two air streams 55 and 82.5 cm³/s + three CO_2 streams: 4.25, 8.5 and 17 cm³/s (I = 300 mA, U = 2 – 2.2 kV). CO_2/Air ratio in the range from 1/20 to 1/3 has been changing exactly this way. Selection of gas into the flask has been taken place at the refrigerator output. The flask has been previously pumped by the water-jet pump to the pressure of saturated water vapor (23 mm Hg).

Fig. 6 shows the gas chromatography comparison of bioethanol conversion output products with and without the addition of CO_2. The air flow is constant – 55 cm³/s, in case of CO_2/Air = 1/3 – 17 cm³/s of CO_2 has been added to the air (the total flow has been increased, which may explain the decrease in the percentage of nitrogen at a constant air flow; I = 300 mA, U = 2 - 2.2 kV). This histogram shows that adding of carbon dioxide leads to a significant increase of the H_2 component percentage, CO (syn-gas) and CH_4 in the output gas. This may indicate that the addition of CO_2 during the ethanol reforming increases the conversion efficiency, because CO_2 plays a burning retarder role.

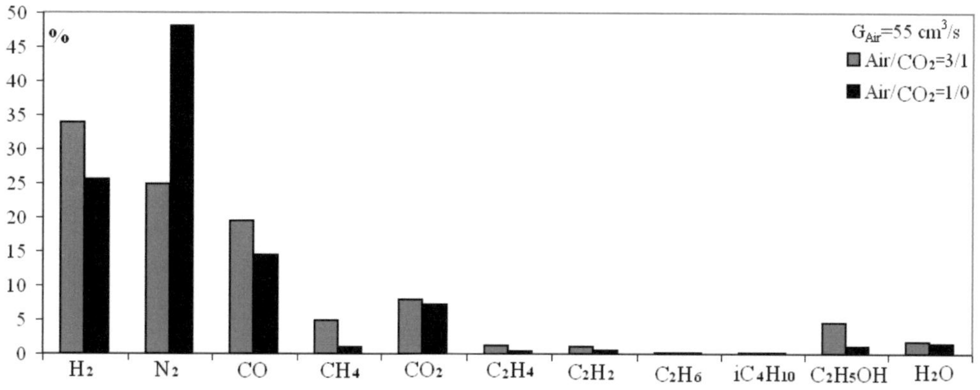

Figure 6. Gas chromatography comparison of bioethanol conversion output products with and without the addition of CO_2.

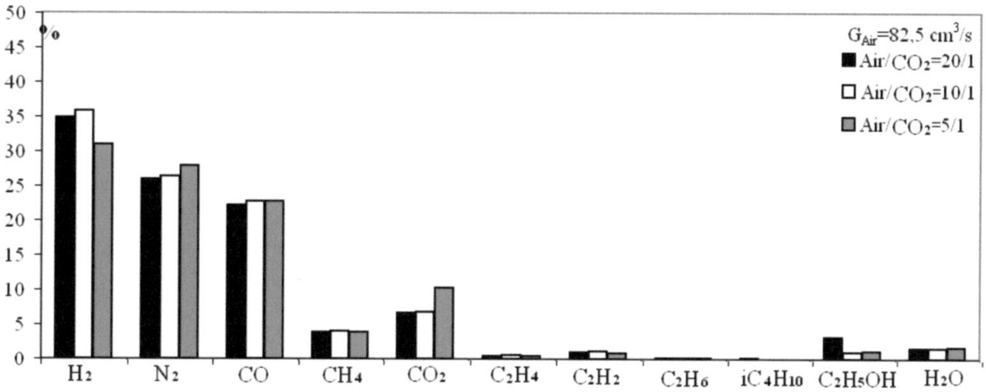

Figure 7. Gas chromatography comparison of bioethanol conversion output products by adding different amounts of CO_2.

The ethanol solution consumption for the SC mode with current of 300 mA and air flow of 55 cm^3/ equals 6 ml/min, and for the air flow of 82.5 cm^3/s and CO_2 of 17 cm^3/s mixture - 10 ml/min.

According to the gas chromatography, in the studied correlations range of CO_2/Air, syn-gas ratio ($[H_2]/[CO]$), changes slightly – look at Fig. 8. Measurements were made by two air streams of 55 and 82.5 cm^3/s and three CO_2 streams of – 4.25, 8.5 and 17 cm^3/s; I = 300 mA, U = 2 – 2.2 kV.

Besides the gas chromatography, the output gas composition has been studied by means of infrared spectrophotometry (IRS). Fig. 9 shows a typical IRS spectrum of the output gas. In the SC mode (current 300 mA, voltage 2 kV) the working liquid is ethyl alcohol and distilled water mixture (C_2H_5OH/H_2O = 1/9.5), and the working gas – air (82.5 cm^3/s) and CO_2 (4.25 cm^3/s) mixture. Research has been carried out in a ditch with a length of 10 cm and a diameter of 4 cm. Pressure inside the ditch has been 1 atm. The ditch walls have been made of BaF_2.

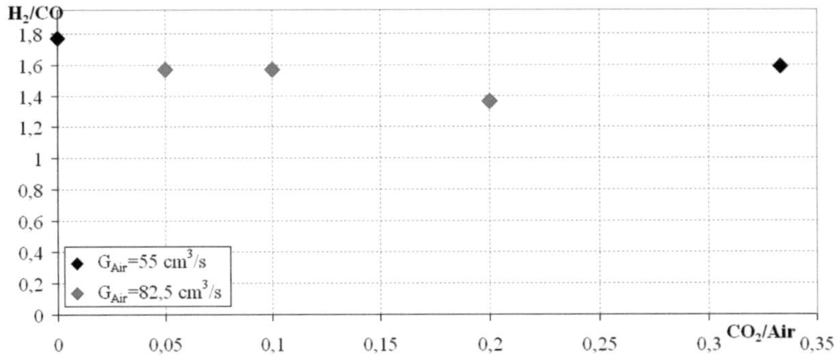

Figure 8. Syn-gas ratio of bioethanol conversion output products for various ratios of CO_2/Air in the range between 0/1 - 1/3.

Fig. 10 shows the dependence of the CO transmission standardized maximum intensity peaks (2000 - 2250 cm^{-1}) in the syn-gas, depending on the CO concentration according to gas chromatography results. Standardization has been conducted for the maximum intensity value of the CO transmission peak bandwidth at the SC mode with the current of 300 mA, voltage - 2 kV, the mixture of ethyl alcohol and distilled water (C_2H_5OH/H_2O = 1/9.5) as the working liquid, and the mixture of air (82.5 cm^3/s) and CO_2 (4.25 cm^3/s), as the working gas.

According to IR spectrophotometry CO fraction in the synthesis gas is practically the same. According to gas chromatography CO fraction in the synthesis gas in the different operation modes stays on the same level as well (the changes are in the range of 1%). So IR spectrophotometry can be used determine the composition of synthesis gas under ethanol reforming.

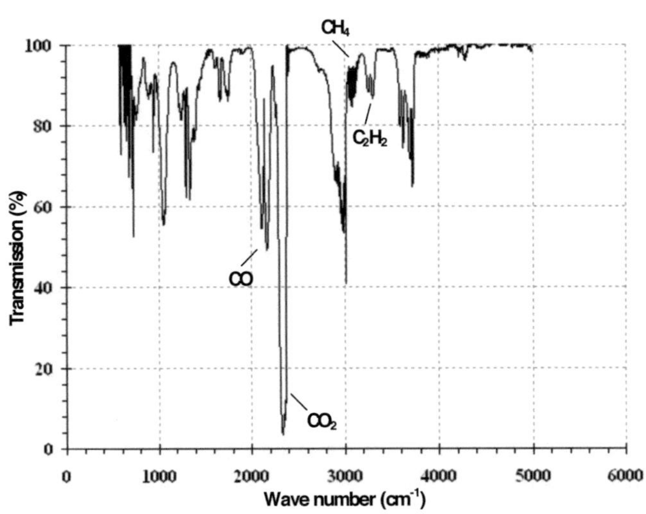

Figure 9. The SC mode (current 300 mA, voltage 2 kV) the working liquid is ethyl alcohol and distilled water mixture (C_2H_5OH/H_2O = 1/9.5), the working gas – air (82.5 cm^3/s) and CO_2 (4.25 cm^3/s) mixture

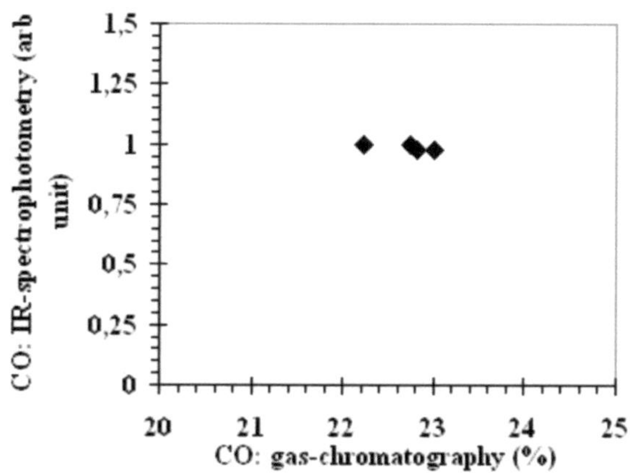

Figure 10. Dependence of the normalized maximum intensity peaks (2000-2250 cm^{-1}) transmission of CO in the syn-gas, depending on the concentration of CO according to gas chromatography data.

Plasma provides gas generation, which contains a certain amount of the syn-gas. The energy needed for this plasma support (Q_p) has been calculated by the following formula:

$$Q_p = P_d t, \tag{1}$$

where P_d - power that has been embedded into the discharge, t – production time of gas volume unit during the reforming process. Electrical energy transformation coefficient α has been calculated by the formula:

$$\alpha = \frac{Q_s}{Q_p} \tag{2}$$

where Q_s - energy that is released during the complete combustion of syngas (obtained in the reforming process).

Electrical energy transformation coefficient α has value of 0.81 for the "TORNADO-LE" with an ethanol solution and pure air flow 55 cm^3/s. And the CO_2 addition (the ratio of CO_2/air = 1/3) gives the value of α = 1,01. System electrical parameters are as follows: I = 300 mA, U = 2 – 2.2 kV.

2.3. Model and calculations

In the model of calculations was assumed that the discharge is homogeneous over the entire volume. It is justified at zero approximation, because the time of gas mixing in the radial direction is less than the times of characteristic chemical reactions. Also we neglect the processes in the transitive zone between the discharge to post-discharge. Thus, the time of gas pumping through the transition region is too short for the chemical reactions to have a sufficient influence on the concentration of neutral components.

The total time of calculation is divided into two time intervals: the first one is the calculation of the kinetic processes of fast generation of active atoms and radicals in the discharge region. Those components accelerate the formation of molecular hydrogen, carbon oxides and production of other hydrocarbons. The second time interval is the oxidation of the gas mixture in the post-discharge region as a result of the high gas temperature and the presence of O and OH. These components remain in the mixture after the dissociation of water and oxygen molecules by electron impacts in the plasma. The oxidation of generated hydrocarbons has a noticeable influence on kinetics in the investigated mixture due to the high gas temperature.

Under the aforementioned conditions, the characteristic time of oxidation is approximately equal to the air pumping time through the discharge region (~10^{-3}–10^{-2} s). The following system of kinetic equations is used in order to account for the constant air pumping through the system:

$$\frac{dN_i}{dt} = S_{ei} + \sum_j k_{ij} N_j + \sum_{j,l} k_{ijl} N_j N_l + \cdots + K_i - \frac{G}{V} N_i - kN_i \tag{3}$$

N_i, N_j, N_l in the equation (3) are the concentrations of molecules and radicals; k_{ij}, k_{iml} are the rate constants of the processes for the i-th component. The rates of electron–molecule reactions S_{ei} are connected with discharge power and discharge volume. The last three terms in equation (1) describe the constant inflow and outflow of gas from the discharge region. The term K_i is the inflow of molecules of the primary components (nitrogen, oxygen, carbon dioxide, water and ethanol) into the plasma, G/VN_i and kN_i are the gas outflow as the result of air pumping and the pressure difference between the discharge region and the atmosphere. In order to define the initial conditions, the ethanol/water solution is assumed to be an ideal solution. Therefore, the vapor concentrations are linear functions of the ethanol-to-water ratio in the liquid. The evaporation rates K_i of C_2H_5OH and H_2O are calculated from the measured liquids' consumption. The inflow rates K_i of nitrogen and oxygen are calculated by the rate of air pumping through the discharge region:

$$K_i = \frac{G}{V} N_i^0 \tag{4}$$

where N^0_i correspond to $[N_2]$ and $[O_2]$ in the atmospheric pressure air flow.

The gas temperature in the discharge region is taken to be constant in the model. In reality, the gas temperature T is dependent on the gas pumping rate and the heat exchange with the environment. Therefore, in order to take into account those influences, T is varied in the interval 800–2500K (similarly to the experimentally obtained temperature spread). After ~10^{-2} s, the balance between the generation and decomposition of the components leads to saturation of concentrations of all species. This allows us to stop the calculations in the discharge region and to investigate the kinetics in the post-discharge region. System (3) is solved without accounting for the last three terms on the time interval without the plasma. The calculations are terminated when the molecular oxygen concentration reaches zero level.

The full mechanism developed for this experimental work is composed of 30 components and 130 chemical reactions between them and its closed to [11]. The charged particles (electrons and ions) are ignored in the mechanism, because of low degree of ionization of the gas (~ 10^{-6} – 10^{-5}). Nitrogen acts as the third body in the recombination and thermal dissociation reactions. In the non-equilibrium plasma almost the entire energy is deposited into the electron component. The active species, generated in the electron–molecular processes, lead to chain reactions with ethanol molecules.

Numerical simulation of kinetics showed that the main channels of H_2 generation in the plasma were ethanol abstraction for the first 10–100μs, and hydrocarbon abstraction afterwards. Additionally, the conditions when the reaction between H_2O and hydrogen atoms was the main channel of H_2 production were found. A kinetic mechanism, which adequately described the chemistry of the main components, was proposed. The model did not account for nitrogen-containing species, and nitrogen was considered only as a third body in recombination and dissociation reactions. The comparison between experiments and calculations showed that the mechanism can adequately describe the concentrations of the main components (H_2, CO, CO_2, CH_4, C_2H_4, C_2H_6, and C_2H_2).

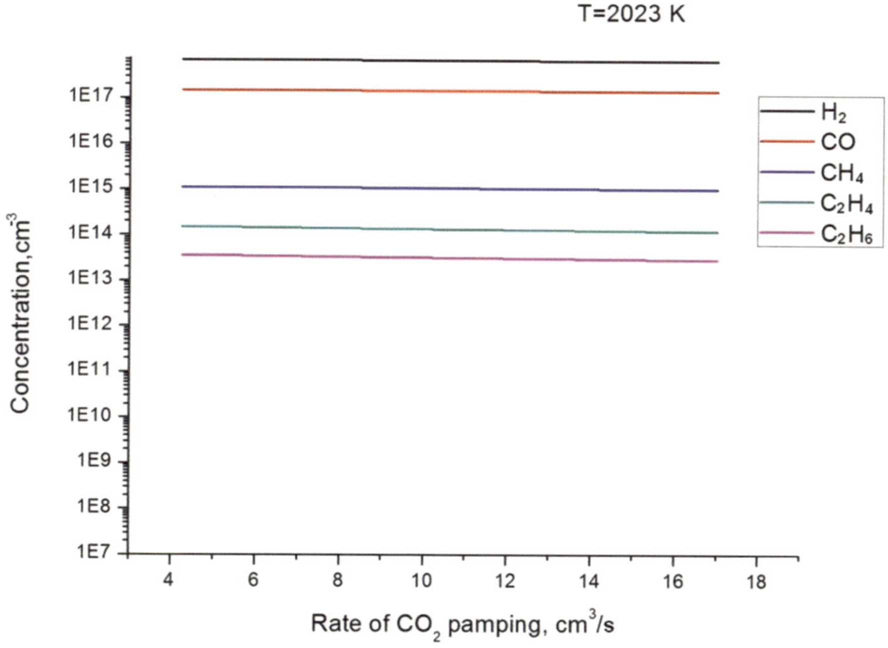

Figure 11. The dependence of the reaction main products of the flow rate of CO_2 (inside discharge), T = 2023 K

However, it should be noted that with the increase in temperature to 2523 K leads to the fact that the output of the reactor is not observed almost no light hydrocarbons. They simply "fall apart" and burned. That leaves the most stable elements such as H_2O, N_2, CO_2. This suggests that the increase in temperature up to these values is not advisable because of the decrease in the yield of useful products (see Fig. 11 and Fig. 12a,b).

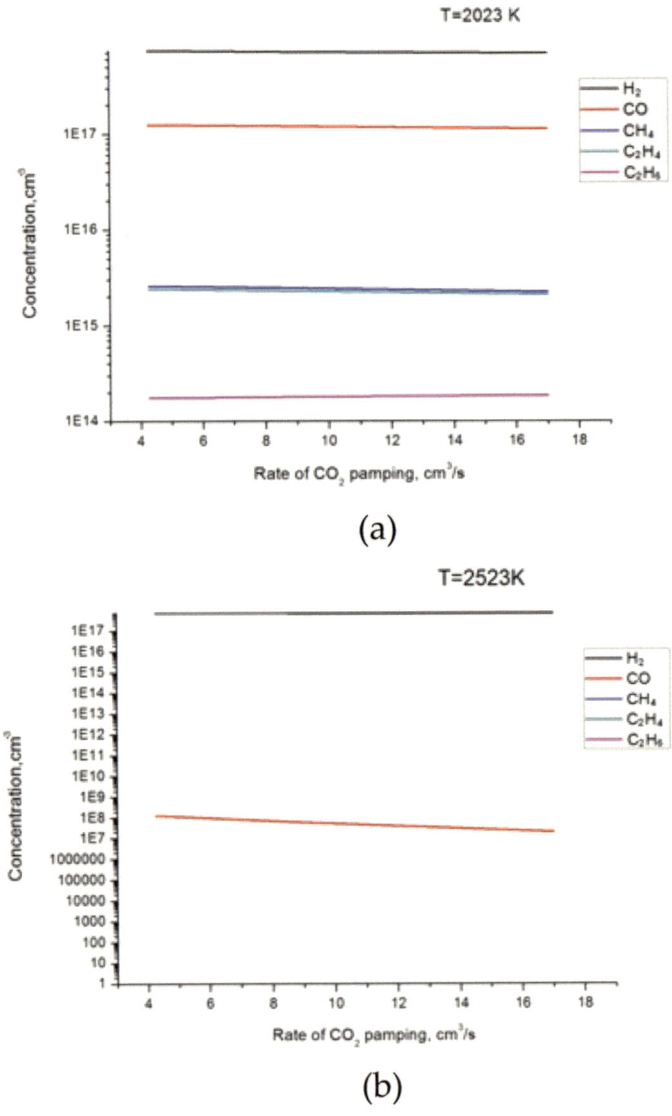

Figure 12. a). The dependence of the reaction main products of the flow rate of CO_2 (after discharge), T = 2023 K. b). The dependence of the reaction main products of the flow rate of CO_2 (after discharge), T = 2023 K

These calculations are based in good correspondence with the experimental data (see Fig. 8).

Other model hydrocarbon is bioglycerol (crude glycerol) which is a byproduct of the biodiesel manufacture. Biodiesel is a popular alternative fuel. It is carbon neutral, has emissions equivalent or below diesel, is biodegradable, non-toxic, and is significantly cheaper to manufacture than its petroleum equivalent. However there is one significant drawback: for every 10 gallons of biodiesel produced, roughly 1 gallon of bioglycerol is created as a byproduct.

Biodiesel is produced by mixing vegetable oil and potassium hydroxide KOH. Therefore, the large-scale production of environmentally friendly and renewable fuel may lead to possible bioglycerol accumulation in large quantities, which, in turn, can cause environmental problems, as it is comparably bad fuel. In addition, it has a rather large viscosity of 1.49 Pa•s, which is larger for almost three orders of magnitude than ethanol and water viscosity. The solution to this problem would be "TORNADO-LE" usage for bioglycerol reforming. Pure glycerol chemical formula is $C_3H_5(OH)_3$. However, bioglycerol contains various impurities (including a set of alkali).

Fig. 13 shows a photograph of burning discharge, where the working liquid is bioglycerol and working gas - air. Research is conducted by the SC polarity, because this mode has lowest liquid consumption.

Figure 13. Photo of the combustion discharge in which the working liquid is bioglycerol and working gas - air.

Fig. 14 shows the typical emission spectrum of the plasma discharge in a "TORNADO-LE" where the working liquid is bioglycerol doped with alkali. It is registered at a current of 300 mA, voltage – 2 kV, air flow – 110 cm^3/s. Optical fiber is oriented on the sight line, parallel to the liquid surface in the middle of the discharge gap. The distance from the liquid surface to the top flange equals 10 mm.

Emission spectrum (Fig. 14) is normalized to the maximum Na doublet (588.99 nm, 589.59 nm). It contains K (404.41 nm, 404.72 nm, 766.49 nm, 769.89 nm), Na (588.99 nm, 589.59 nm), Ca (422.6 nm) lines, and a part of continuous spectrum, which indicates that the there's a soot in the discharge. Temperature, which is defined by the plasma continuous emission spectrum is 2700 ± 100 K.

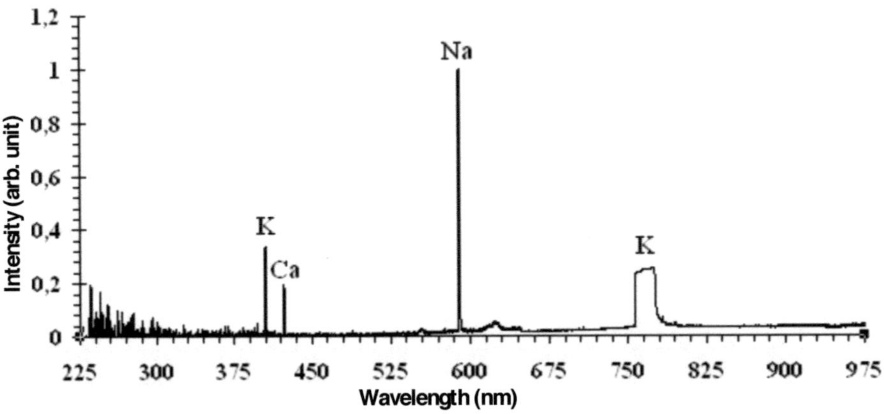

Figure 14. Typical emission spectrum of the plasma discharge, which burns in a mixture of air and bioglycerol / alkali.

The K, Na, Ca elements presence in the discharge gap complicates the plasma kinetics numeric modeling of the bioglycerol reform process. The gas flow rate at the system outlet is 190 cm^3/s, i.e. by 80 cm^3/s larger than the initial (110 cm^3/s), which indicates bioglycerol reforming to the syn-gas. Liquid flow is 5 ml/min. Change of the CO_2 share in the working gas weakly affects the spectrum appearance.

Based on the continuous nature of the plasma emission spectra, we compared the experimental results with the calculated spectra of the blackbody radiation. Calculations have been performed by using Planck's formula.

Fig. 15 shows the computational grid with step of 200-300 K in the temperature range from 2500 K to 3500 K and the plasma emission spectrum in the case of bioglycerol, as a working fluid (air flow - 82.5 cm^3/s, the flow of CO_2 - 17 cm^3/s, CO_2/Air = 1/5, I_d = 300 mA, U = 600 V). All spectra are normalized to the intensity, which is located at a wavelength of 710 nm.

The data in Fig. 15 show that the plasma emission spectrum coincides with the calculated by the Planck formula for the temperature T = 2800 ± 200 K. Since bioglycerol contains alkali metals, which represent an aggressive environment, the gas chromatography can't be used. Therefore, in order to determine the gas composition, formed the bioglycerol reformation IR and mass spectrometry have been used.

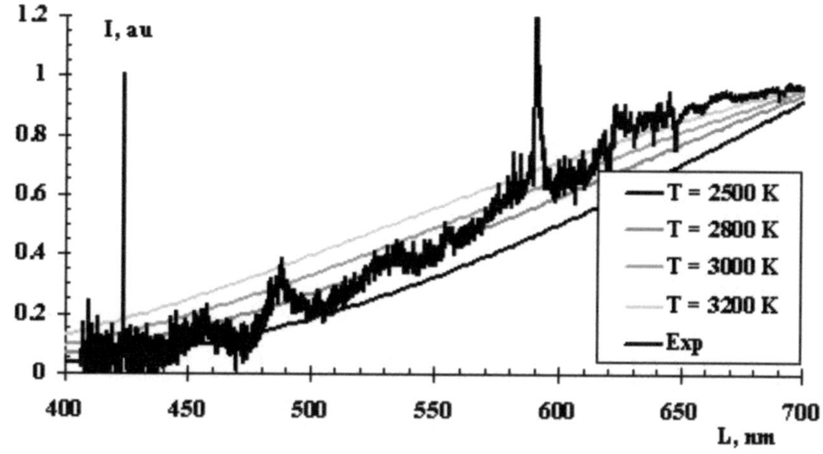

Figure 15. Plasma emission spectrum in the case when the working gas is a mixture SO_2/Air = 1/5 (air flow - 82.5 cm^3/s, the CO_2 flow - 17 cm^3/s), I_d = 300 mA, U = 600 V and calculated spectra of blackbody radiation)

With infrared transmission spectra one can see that the transition to bioglycerol increases the amount of such components as CO_2 (2250-2400 cm^{-1}), CO (2000-2250 cm^{-1}), CH_4 (3025-3200 cm^{-1}), C_2H_2 (3200-3350 cm^{-1}).

2.4. Discussions

Electrical energy is added to the "TONADO-LE" plasma-liquid system in the form of plasma power. Plasma acts as a catalyst and thus this power should be controlled. In addition to electric energy for plasma we incorporate hydrocarbon (ethanol or bioglycerol) as an input to the system. These hydrocarbons are raw material for syn-gas generation but they are also a fuel which has some energy associated with it. So, we input some energy to the system (hydrocarbon + electricity) and we get syn-gas, which is potentially a source of energy as well.

Carbon dioxide adding leads to a significant increase the percentage of H_2 + CO (syn-gas) and CH_4 components in the exhaust. This may indicate that the CO_2 addition under the ethanol reforming increases the conversion efficiency, because CO_2 plays a role of the retarder in the system by reducing the intensity of the conversion components combustion.

The transmission spectra of infrared radiation indicate that the exhaust gas obtained by ethanol solution conversion, contains such components as CO, CO_2, CH_4, C_2H_2. It was found that CO_2 adding reduces the CH_4 and C_2H_2 amount, but does not affect the amount of producted CO.

The possibility of hydrocarbons reforming, which have considerable viscosity (bioglycerol) in the "TORNADO-LE" is shown. This gives a possibility to avoid environmental problems due to the bioglycerol accumulation during biodiesel production.

The α coefficient [see (2)] in bioglycerol reforming is higher than ethanol reforming at the same ratios of CO_2/Air in the input gas. This may be connected with the lower power consumption on the plasma generation in case of bioglycerol reforming. Bioglycerol contains alkaline dash, which increases the bioglycerol conductivity. Bioglycerol reforming products contain mainly CO and hydrocarbons CH_4, C_2H_2, which also gives some contribution to energy yield.

3. Dynamic impulse plasma–liquid systems

3.1. Experimental set up

The experimental setting is shown in Fig. 16. The main part of the system is cylinder with height H = 10 mm, and radius R = 135 mm. Its lateral surface made of stainless steel with a thickness of 5 cm. This cylinder is filled with liquid for experimental operations. The electrodes are placed perpendicular to the cylinder axis. They have the diameter of 10 mm, made of brass, and their ends are shaped hemispheres with a radius of curvature of 5 mm. The discharge (2) is ignited between the rounded ends of the electrodes. At a distance of 40 mm from the lateral surface of the cylinder is piezo-ceramic pressure sensor (3), which records acoustic vibrations in the fluid, caused by electric discharge under water. The distance between the sensor head and the system axis = L.

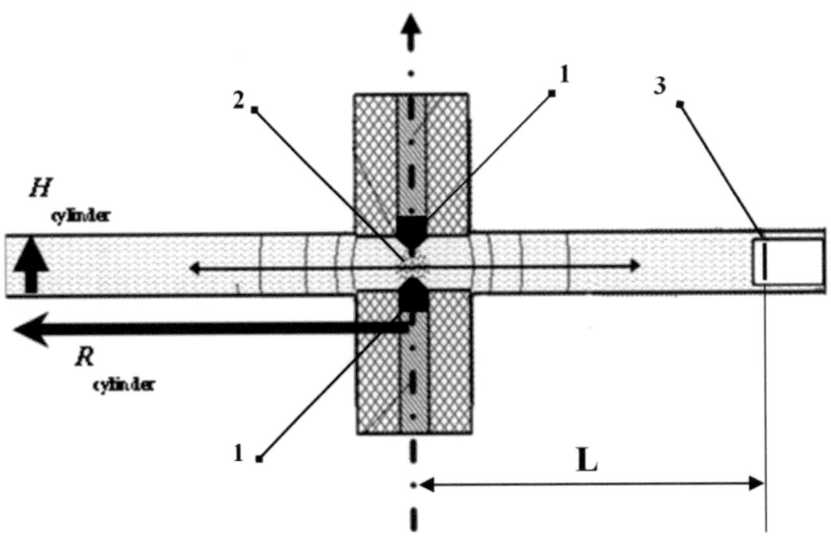

Figure 16. Schematic diagram of plasma-liquid system with a pulsed discharge, 1 - electrodes with brass tips, 2 – plasma, 3 – piezo-ceramic pressure sensor.

The cylindrical system could be located in a horizontal position (Fig. 17a) or vertical one (Fig. 17b). The full volume (0.5 l) of system is fluid-filled. The fluid in the system can be processed as in static mode (no flow), and dynamic one (with flow ~ 15 cm^3/s). Additional supply of gas may be realized in the system also (airflow ~ 4 cm^3/s), which is injected through a spray nozzle (source diameter 8 mm) located near the inner wall of the cylinder at a distance of 130 mm from the discharge gap (Fig. 16). The working fluids are: the tap water (with and without flow), distillate and ethanol (96%, no flow).

The main feature of electrical scheme for pulsed power feeding of discharge in a liquid is usage of two independent capacitors which are supplied two independent sources of power (1 kW). Pulsed discharge realized in two modes: single and double pulses. In the single pulse mode only one capacitor is discharged with a frequency of 0 - 100 Hz.

Double pulse mode is realized as follows: one capacitor discharges in the interelectrode gap through air spark gap; the clock signal from the Rogowski belt after first breakdown is applied to the thyratron circuit and second capacitor discharges through it. This set of events leads to the second breakdown of the discharge gap and second discharge appearance.

Delay of the second discharge ignition may be changed in range of 50 - 300 microseconds. The following parameters are measured: discharge current and the signal from the pressure sensor. The Rogowski belt has the sensitivity 125 A/V, and its signal is recorded with an oscilloscope. Capacity for the first discharge (C_1) = 0.105 μF and it is charged to U_1 = 15 kV (energy E_1 = 12 J), capacity for the second discharge C_2 = 0.105 μF and it is charged to U_2 = 18 kV (energy E_2 = 17 J).

A distance between electrodes can be changed in the range of 0.25 - 1 mm. The second discharge can be ignited at the moment (according to the delay tuning) when the reflected acoustic wave, created by the first electric discharge in liquid, returns to the center of the system (the time of its collapse ~ 180 ms).

(a) (b)

Figure 17. Photograph of the cylinder from the outside: a) horizontal position, b) vertical position.

The composition of ethanol and bioethanol reforming products is studied with gas chromatography, in case of bioglycerol reforming - mass spectrometry and infrared spectrophotometry.

3.2. Results

Oscillograms of current and acoustic signal for different distances between electrodes (0.5 and 1 mm) are presented in Fig. 18. These oscillograms show the presence of electrolysis phase before breakdown, while duration of electrolysis increases with interelectrode distance.

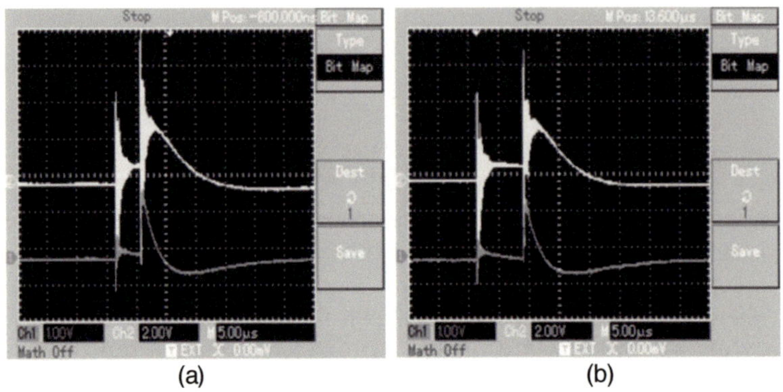

Figure 18. Oscillograms of the discharge current (top curve) and signal piezo-ceramic pressure sensor (lower curve): a) d = 0.5 mm, b) d = 1 mm. Tap water flow = 15 cm³/sec, without the input gas stream, C = 0.18 uF, U = 13.5 kV; ballast resistor in the discharge circle: R_b = 20 Ohm, the cylinder is in horizontal position.

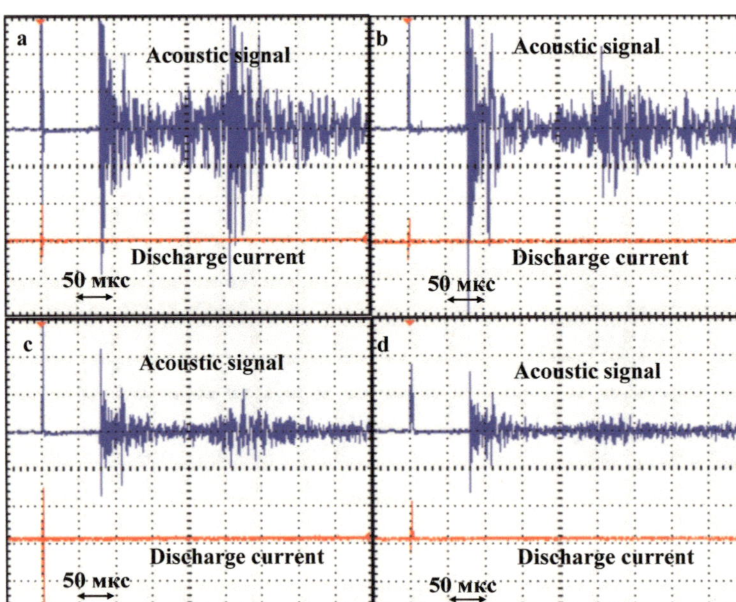

Figure 19. Oscillograms of current and acoustic signal in the single pulse mode at the different discharge ballast resistor: R_b: a) - 0 Ohm, b) - 10 Ohm; c) - 20 Ohm, d) - 50 Ohm. Tap water flow 15 cm³/s, without the input gas stream, d = 0.5 mm, C = 0.015 μF, U = 19.5 kV, the cylinder is in horizontal position.

http://dx.doi.org/10.5772/53764

Fig. 19 shows the acoustic signal dependence from ballast resistor in the discharge electric circuit. The acoustic signal has two splashes: №1 - the first diverging acoustic wave, and №2 - the second diverging acoustic wave. When the ballast resistor is increased, first and second acoustic signal splashes are decreased. This may be due to the fact: we increase the ballast resistor and set measures to the discharge current, as a result the injected into the discharge gap energy is diminished.

Also, there is a signal immediately behind the front of the first splash, which is founded in all cases at 110 microseconds interim from the beginning of the discharge current. The acoustic wave passes the way near 17 cm during this time. The pressure sensor is located at the distance of 2 cm from the lateral surface, so the acoustic signal passes the way near 12 cm to the sensor. Thus, there is a second stable signal after the first splash through time ~ 29 μs, which corresponds to the path ~ 4.4 cm, so the signal can be the convergent acoustic waves reflected from the wall.

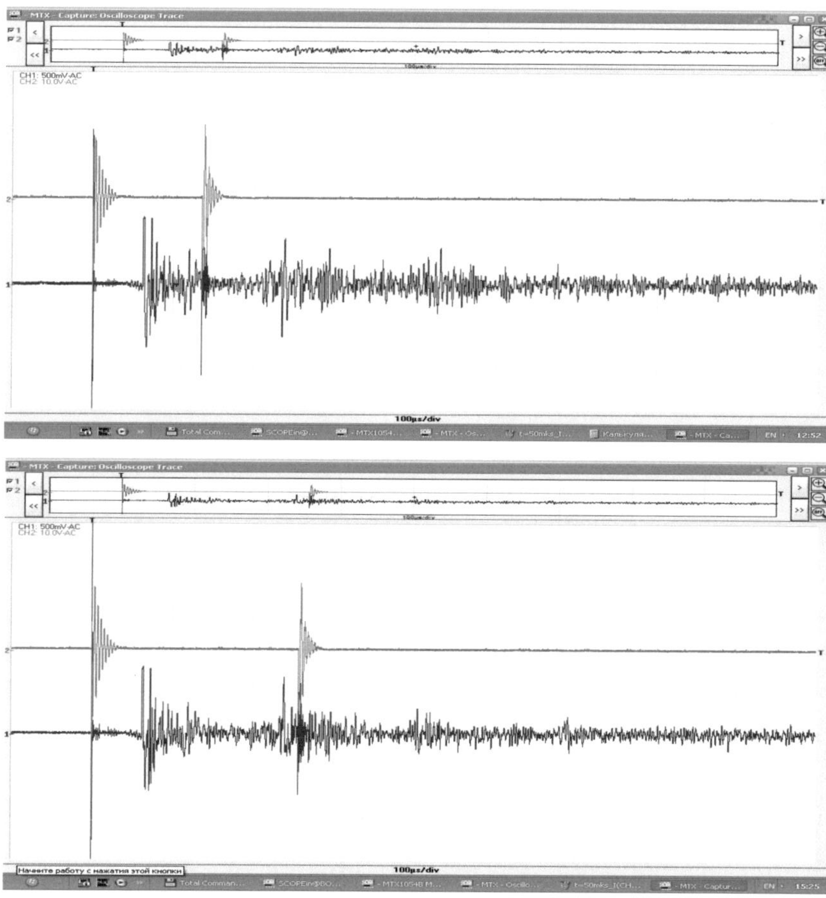

Figure 20. Oscillograms of the discharge current (top oscillogram) and acoustic signal (lower oscillogram) at different delays of the second discharge pulse.

There is the third acoustic signal splash in the experiment, but it does not affect the second discharge pulse delay in relation to the first. In addition, there is no acoustic signal from to the second discharge pulse in the double pulse mode, although the single pulse signal is present in the single pulse mode (Fig. 20).

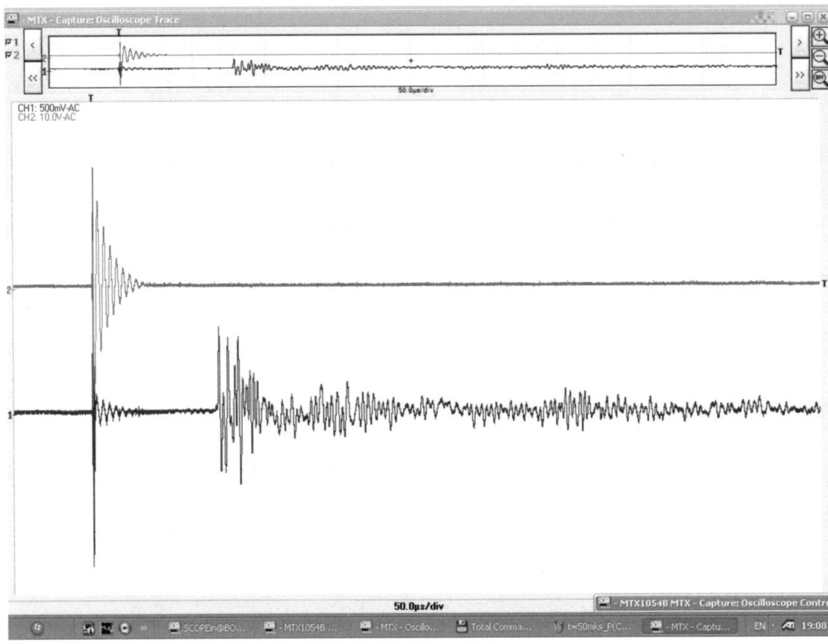

Figure 21. Oscillograms of the discharge current (top oscillogram) and the acoustic signal (lower oscillogram) in the single pulses mode. Working fluid - ethanol, d = 0.25 mm, C_1 = 0.105 μF, U_1 = 15 kV, the cylinder in the vertical position

Fig. 21 shows clearly that the duration and amplitude parameters for the first current pulse in the ethanol are virtually indistinguishable from the first current pulse in distilled water at any cylinder orientations. The ratio of the second acoustic signal amplitude to the first acoustic signal amplitude in the ethanol is noticeably less than in the tap water and distillate.

The results of oscillographic studies of the discharge current and acoustic signals in double pulses mode demonstrate that the first discharge in double pulses mode takes place in the narrow gas channel with a radius comparable to the size of the plasma channel, and the second discharge takes place in the wide channel with radius larger than the plasma channel.

Next, we present the results of ethanol reforming studies in the impulse plasma-liquid system with double pulses mode and their comparison with the results obtained for "TORNADO-LE".

The mass spectrometer studies of ethanol reforming in the impulse PLS of cylindrical geometry were carried out in the following modes: single pulse mode (C = 0.105 μF, U = 15 kV, f = 15 Hz, power 180 W) and double-pulse mode (C_1 = C_2 = 0.105 μF, U_1 = 15 kV, U_2 = 15 kV, f = 15 Hz, second pulse delay = 170 μs, this time is less on 10 μsec than collapse time, the power is 435

Wt), the interelectrode distance - 0.25 mm, working liquid - ethanol (96%), the input airflow is 4 cm³/s.

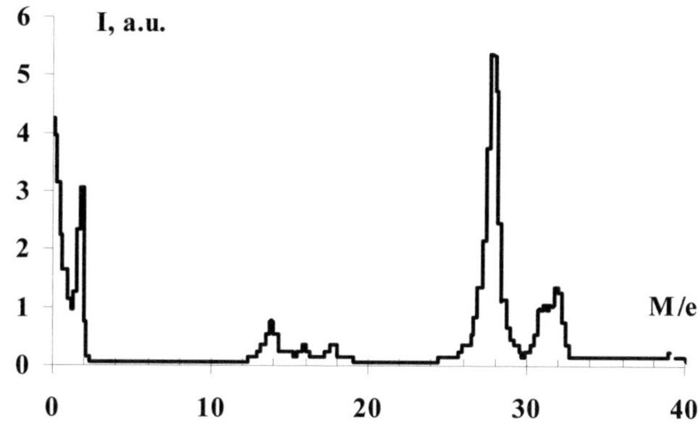

Figure 22. Mass spectrum for double pulse mode. Ethanol is without flow, inlet gas stream - 4 cm³/s, d = 0.25 mm, $C_1 = C_2 = 0.105$ μF, $U_1 = 15$ kV, $U_2 = 18$ kV, the cylinder is in the vertical position, f = 15 Hz.

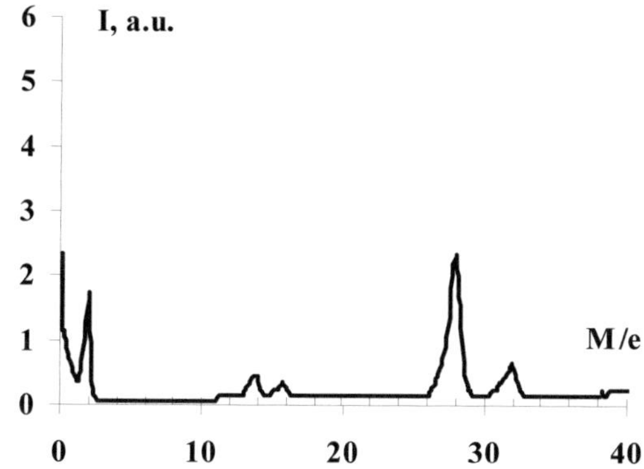

Figure 23. Mass spectrum for the single pulse mode. Ethanol without flow, inlet air flow - 4 cm³/s, d = 0.25 mm, $C_1 = 0.105$ μF, U1 = 15 kV, the cylinder is in the vertical position, f = 15 Hz.

The mass spectrometric studies show that the main components of the output fuel mixture are: hydrogen, carbon dioxide, and molecular nitrogen. The values of these components in the mixture: H_2 - 29%, CO - 17% for double pulse mode and H_2 - 35%, CO - 7% for single pulse mode. That is, with the same molecular hydrogen output, the carbon dioxide yield is significantly increased in double pulses mode.

The typical mass spectrum (Fig. 24) of the ethanol reforming (ethanol aqueous solution ethanol with concentrations 3.5, 13 and 26 percents) in the "TORNADO-LE". The power is 640 Wt. It is injected in the plasma for its generation, and inlet air flow is 55 cm³/s.

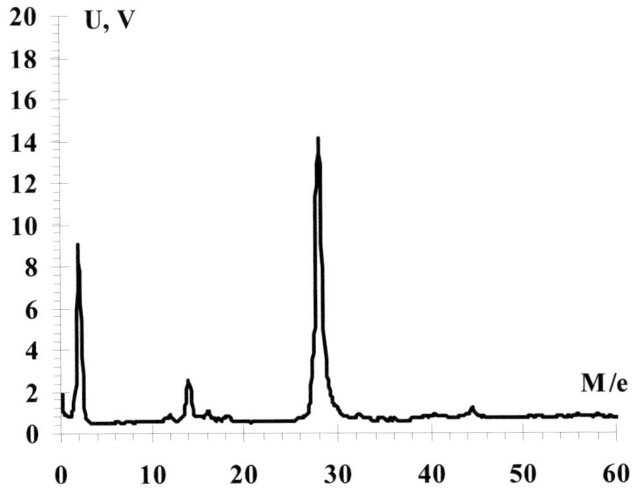

Figure 24. Mass spectrum of the output mixture in the ethanol reforming (ethanol - 26%) in "TORNADO-LE" PLS

The following Tab.1 shows the values ratio generating the volume unit of (H_2 + CO) mixture per unit of electrical power, which is injected into the plasma under reforming process in the impulse PLS of cylindrical geometry with double pulses mode, and in the "TORNADO-LE":

	Impulse PLS of cylindrical geometry with double pulses mode	**"TORNADO-LE"**
Single pulse	0.027 cm^3/Wt	
Double pulses	0.0082 cm^3/Wt	
Bioethanol 6,5%		0.0024 cm^3/Wt
Bioethanol 13%		0.0079 cm^3/Wt
Bioethanol 26%		0.0615 cm^3/Wt

Table 1. The volume unit of (H_2 + CO) mixture per unit of electrical power in various PLS

The H_2 and CO components yield increases with increasing of the ethanol aqueous solution concentration. This concentration has maximum value 26%, and H_2 - 26%, CO - 14%. The results of these systems studies show, that the pressure, in region collapse of converging shock waves (with pulse energy > 10 J), exceeds critical (Tab. 2). So, the additional increase chemical activity due to supercritical processes inclusion can be achieved in this situation.

http://dx.doi.org/10.5772/53764

Solvent	Molecular mass	Critical temperature, T_{crit}	Critical pressure, P_{crit}	Critical density, ρ_{crit}
	g/mol	K	MPa (bar)	g/sm³
CO_2	44.01	303.9	7.38 (72.8)	0.468
H_2O	18.015	647.096	22.064 (217.755)	0.322
ethanol	46.07	513.9	6.14 (60.6)	0.276

Table 2. Critical parameters of different solvents

4. Discussion

The presence of electrolysis phase preceding electrical breakdown of heterophase environment demonstrates that the discharge development in the liquid perform with microbubbles. This result confirms the theory of "bubble" breakdown proposed by Mark Kushner [12].

The formation of convergent acoustic wave after reflection from the ideal solid cylindrical surface was investigated. It is shown that acoustic waves may be effectively focused during these waves passage inside the system.

The research of ethanol reforming in pulse plasma-liquid system has shown that transition from single pulse mode to double pulse mode is accompanied by reduction syn-gase ratio ($[H_2]/[CO]$).

When the working fluid is bioglycerol the K, Na, Ca lines are presented in emission spectra and there is a solid continuous spectrum, which indicates that microparticles are present in the discharge. Its temperature is $T = 2800 \pm 200$ K.

5. General conclusions

On the base of our results in bioethanol and bioglycerol CO_2-reforming by "TORNADO-LE" plasma-liquid system, we can say that:

1. This process has special features, connected with CO_2 retarding role in the conversion components combustion;
2. In this system there is the possibility of reforming of hydrocarbons with significant viscosity (such as bioglycerol);
3. All the diagnostic methods, used in the "TORNADO-LE" plasma-liquid system, indicate that there're no NO_x compounds in the bioethanol and bioglycerol reforming products.

The investigations of bioethanol and bioglycerol in pulse plasma-liquid system have shown:

1. The main components of the output fuel mixture in this case are: hydrogen, carbon dioxide, and molecular nitrogen, but the carbon dioxide yield is significantly increased in double pulses mode;
2. The formation of supercritical water in such system and its possible applications for recycling of organic waste and for nanocrystalline particles (in particular, oxide catalysts and other nanocrystalline materials, such as nanotubes) productions needs for additional researches.

Author details

Valeriy Chernyak[1*], Oleg Nedybaliuk[1], Sergei Sidoruk[1], Vitalij Yukhymenko[1], Eugen Martysh[1], Olena Solomenko[1], Yulia Veremij[1], Dmitry Levko[2,3], Alexandr Tsimbaliuk[2], Leonid Simonchik[4], Andrej Kirilov[4], Oleg Fedorovich[5], Anatolij Liptuga[6], Valentina Demchina[7] and Semen Dragnev[8]

*Address all correspondence to: chernyak_v@ukr.net

1 Taras Shevchenko National University of Kyiv, Ukraine

2 Institute of Physics, National Academy of Sciences of Ukraine, Kyiv, Ukraine

3 Physics Department, Technion, 32000, Haifa, Israel

4 B.I. Stepanov Institute of Physics, National Academy of Sciences, Minsk, Belorus

5 Institute of Nuclear Research, National Academy of Sciences of Ukraine, Kyiv, Ukraine

6 V.E.Lashkaryov Institute of Semiconductor Physics, National Academy of Science of Ukraine, Kyiv, Ukraine

7 The Gas Institute, National Academy of Science of Ukraine, Kyiv, Ukraine

8 National University of Life and Environmental Sciences of Ukraine, Kyiv, Ukraine

References

[1] AEO2011 Early Release Overview, available on the sight: www.eia.gov/forecast/aeo/pdf/0383er(2011).pdf

[2] Sharvin E.A., Aristova Ye. Yu., Syn-gas generator for internal combustion engines, "Alternative energy and ecology" (in Russain), 8 (88) 2010, c.31-37

[3] Nedybaliuk O.A., Chernyak V.Ya., Olszewskij S.V., Plasma-liquid system with reverse vortex flow of "tornado" type (Tornado-LE) //Problems of atomic science and technology, № 6. Series: Plasma Physics (16), p. 135-137. (2010).

[4] Xumei Tao, Meigui Bai, Xiang Li, e.a., CH_4-CO_2 reforming by plasma – challenges and opportunities // Progr. in Energy and Combustion Science 37, №2, pp. 113-124, 2011

[5] V. Chernyak, Eu. Martysh, S. Olszewski, D. Levko. e.a., Ethanol Reforming in the Dynamic Plasma - Liquid Systems, Biofuel Production-Recent Developments and Prospects, Marco Aurélio dos Santos Bernardes (Ed.), (2011). ISBN: 978-953-307-478-8, InTech, P.101-136.

[6] Lukes, P.; Sunka, P.; Hoffer, P.; Stelmashuk, V.; Benes, J.; e.a.,Book of Abstracts: NATO Science Advanced Research Workshop on Plasma for bio decontamination, medicine and food security, Jasná, Slovakia, March 15–18, 2011.

[7] Sheldon R. A. C. Catalytic conversions in water and supercritical carbon dioxide from the standpoint of sustainable development (in Russian) // Rus. Chem. J., 48(2004), 74-83.

[8] N.A.Popov, V.A. Shcherbakov, e.a. Thermonuclear fusion by exploding a spherical charge (gas-dynamical thermonuclear fusion problem) // Uspekhi, 10 (2008), 1087-1094.

[9] Laux, C.O. Optical diagnostics of atmospheric pressure air plasma SPECAIR / C.O. Laux, T.G. Spence, C.H. Kruger, and R.N.Zare // Plasma Source Sci. Technol. – 2003. - Vol. 12, No. 2. - P. 125-138.

[10] Raizer Yu.P. Gas discharge physics (Springer, 1991)

[11] D. Levko, A. Shchedrin, V. Chernyak e.a., Plasma kinetics in ethanol/water/air mixture in a 'tornado'-type electrical discharge // J. Phys. D: Appl. Phys. 44 (2011) 145206 (13pp)

[12] Kushner, M.J.; Babaeva, N.Yu. Plasma production in liquids: bubble and electronic mechanism, Bulletin of the APS GES10, Paris, France, October 4–8, 2010.

Chapter 14

Biofuels from Algae

Robert Diltz and Pratap Pullammanappallil

Additional information is available at the end of the chapter

http://dx.doi.org/10.5772/53827

1. Introduction

1.1. Current sources of biofuels

The United States, as well as numerous other countries throughout the world, is seeing a rapid rise in the amount of power and fuel required to maintain the current and future lifestyles of its citizens. With the rapid increase in global consumerism and travel seen over the recent decades due to improvements in technology and the increase in international interactions, the demand for fuel is rapidly growing, as can be seen in Figure 1. Due to the worldwide demand for fuel, which currently is primarily fossil-derived, supplies are being strained and costs are rapidly rising. In order to satiate this rapid increase in demand and stem the shrinking supply, new alternative sources of fuel must be brought to the market that can be used to replace standard petroleum based fuels.

Currently, there are several sources of alternative fuels that can be used to replace or supplement traditional petroleum based fuels. Some of these sources include alternative fossil-derived sources such as coal, natural gas, and hydrogen derived from hydrocracking, while other sources come from more renewable sources such as biomass. Biomass has several advantages when it comes to fuels in that there are numerous sources such as terrestrially grown starch based or cellulosic material, waste derived material, or aquatic and marine based organisms, each of which has unique components and characteristics useful for fuel production.

Due to the structural variability of the various types of biomass available, a wide range of technologies can be used to convert the organic molecules into a useable form of fuel. As food substrates (such as carbon dioxide in autotrophic organisms or sugars in heterotrophic organisms) are metabolized, a range of cellular components are assembled to perform numerous duties to keep organisms alive and reproducing. Starches and celluloses are assembled from carbohydrates to provide rigid structural support in many woody biomasses as well as acting as a sugar storage method for quick conversion to a food source in times of famine. Proteins and amino acids are the building blocks of DNA structures and additional biomass. Lipids provide a highly energy dense storage system while also serving as a transport mechanism for several nutrients vital to metabolic activity. However, when broken down to the most basic levels, these organic compounds all contain energy which can be extracted through several methods. Table 1 shows a breakdown of some common algal biomass cellular components.

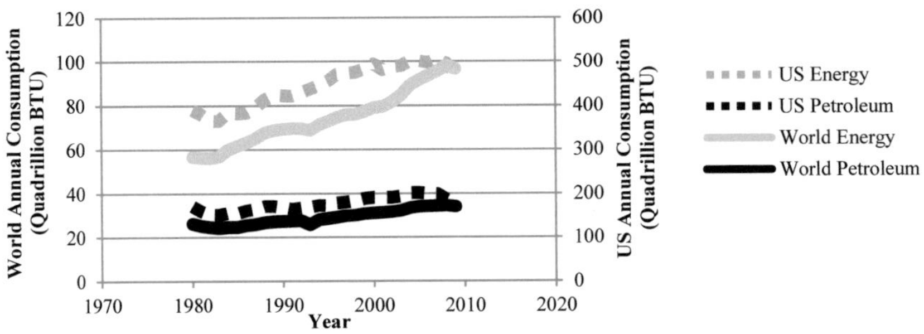

Figure 1. Annual Consumption of Total Energy and Petroleum in the United States and the World [1]

Species	Protein (%)	Carbohydrate (%)	Lipid (%)
Ankistrodesmus	36	24	31
Nitzschia	36	14	22
Chlorella	55	24	21
C. protothecoides	38	52	11
C. emersonii	32	41	29
C. vulgaris	29	51	18

Table 1. Variations in the chemical composition of selected algal species [2, 3]

Sources of biofuel currently being produced range in production rate from the laboratory scale through full scale implementation. Technologies to break down starches and cellulosic materials into sugars for subsequent conversion to bioalcohols has been extensively developed and scaled to produce billions of gallons per year to add into petroleum derived gasoline. Other structural components such as lipids have a high energy content to them and have characteristics that closely mimic petroleum diesel and kerosene, and thus, only require simple chemical reaction (i.e., transesterification) for use as a biofuel, and have been developed up to a quasi-large scale of volumetric output that can be seen in some regional market places, as well as in home production for personal use.

1.2. Aquatic biomass

In order to produce the vast amounts of fuel needed by the United States, and the rest of the world, there will be a demand for massive quantities of biomass to be grown. This could be problematic when using terrestrial biomass, since in most cases, growing plants would require a switch from using land for food sources to energy sources. An alternative source of biomass, however, is available in the form of aquatic and marine species of biomass such as kelps, algae, and other types of water borne plants or bacteria. Aquatic and marine biomass (excluding bacteria) are typically plant-like in that they are autotrophic organisms that contain photosynthetic pigmentation, can utilize inorganic carbon for biomass development, and express molecular oxygen as a byproduct. However, these organisms do not suffer from the inherent liability of requiring fertile soil to grow, minimizing competition with the food supply chain. Also, as microscopic organisms, they do not require abundance of land to develop root systems and large floral brush in order to absorb sunlight and nutrients, and therefore, a much more effective utilization of space. With rapid growth rates that can typically double in concentration in less than a day, it is possible to have daily harvests, creating a steady and abundant supply of biomass for harvesting. As such, marine and aquatic biomass can be a useful alternative source of biomass that can be used to produce a wide range of biofuels for commercial use, while avoiding several of the more common pitfalls associated with more traditional sources of terrestrial biomass, and thus, will be the biomass focus of discussion for the remainder of this article.

Primarily, growth of algae for the production of oils and energy conversion has focused on microalgae, including species of diatoms and cyanobacteria (as opposed to macroalgae, such as seaweed), although some bacterial species (such as *Clostridium sp.)* have been demonstrated for production of biologically derived hydrogen and methane [4]. To date, there have been numerous studies of algae and other water based biomass in order to identify strong candidates for biomass accumulation rates as well as lipid content for production of biodiesel. Some strains are summarized for these characteristics in Table 2. There is also a wealth of microbial biomass resources available as a by-product of industrial activities such as sewage treatment, brewing industries and food processing that could provide biomass or nutrients for further microbial biomass growth [5, 6]. With this concept, it is feasible to use algae as a means for tertiary wastewater treatment in order to utilize trace nutrients such as phosphorous- and nitrogen-containing compounds, or can be used at industrial processes as a way to absorb carbon dioxide by entraining algal cultures to gaseous exhaust streams.

Growth of aquatic and marine biomass is not without challenges though. Maximum growth rates of the microorganisms typically occur under very specific conditions, and any variance on these conditions can cause substantial delays in biomass development. Also, open pond algal systems (which are common for algae production due to their ease of construction and inexpense) are susceptible to contamination from various airborne microorganisms that can decrease overall productivity. And of prime concern, is the ability to separate algae from water, which due to their very dilute nature, can be expensive and inefficient. Several methods are used to do this, such as flocculation with chemicals (such as hydroxides or alum) or electric fields, filtration, centrifugation, or thermal drying, but each of these methods is not without bulky equipment, expensive materials, or long processing times.

2. Lipids and biodiesel

The diesel engine, created by Rudolph Diesel in 1893 as an alternative to steam engines, has seen a marked rise in use over the past decades as newer engines coming to market have become such cleaner combustors. Since the engines are so efficient, they are ideal for use in heavy transport such as rail and ship, but as technology and advances in fuel make the engine emissions cleaner, more and more small engine vehicles are coming to market in light trucks and passenger cars in the US and Europe as well as the rest of the world.

Species	Biomass Productivity (g/L/D)	Growth Rate (d^{-1})	Biomass Conc. (g/L)	Lipid Content (% by dry weight)	Reference
Chlorella lutereorividis		0.55		28.5	[7]
Chlorella protothecides		1.32		31.2	[7]
Chlorella regularis		3		44.4	[7]
Chlorella vulgaris	1.9		1.9	53	[8]
Scenedesmus bijuga		6.1		35.2	[7]
Scenedesmus dimorphus		5.9		43.1	[7]
Scenedesmus obliquus		5.4		42.6	[7]
Dunaliella salina	0.3			35	[9]
Spirulina platensis	0.1				[9]
Tetraselmus chui	1			23.5	[10]
Botryoccocus braunii	10.8			25-75	[11]
Nannochloropsis sp	72			31-68	[11]
Nannochloropsis oculata	2.4			22.8	[10]
Phaeodactylum tricornutum		0.003		2.5	[12]

Table 2. Productivity of Selected Algal Species

Diesel engines have the ability to run on various sources of fuel. Originally the engine was tested using pure peanut oil and vegetable oil, though today, the engine is commonly run on fossil fuel based diesel fuel, a type of kerosene. To reduce the amount of petroleum based diesel being used in today's market several alternative types of fuel have been introduced that are compatible with these engines. Among the alternatives, generally seen are the lipid based straight vegetable oils and the modified biodiesels. Straight vegetable oil will burn

http://dx.doi.org/10.5772/53827

without problem in diesel engines; however, preheating of the fuel is required in order to reduce viscosity to pumpable levels. Biodiesel fuels, which are generally from the same source of lipids as straight vegetable oils or algal oils, are a much better suited fuel because they match several of the same characteristics as modern diesel fuel, and thus, require little to no engine modifications or fuel pretreatment modifications.

2.1. Sources of lipids

Lipids are a general set of cellular components that are grouped together by the common trait that they are soluble in non-polar solvents. Throughout living organisms, there are several sources of lipids that play various roles in biochemical processes including energy storage and water insoluble nutrient transport across cell membranes that include neutral lipids, phospholipids, steroids, waxes, and carotenoids. Since lipids have a generally low oxygen and high carbon and hydrogen content, they are very energy dense molecules. This characteristic, along with their natural abundance and similarities with petroleum based fuels, make them ready targets for processing and use as a blend or replacement to traditional fuels.

Neutral lipids (commonly referred to as "fats"), which are widely regarded as one of the most common sources of lipids, and which has the highest potential for use as an alternative fuel, can be found in various forms throughout different organisms, and will be the primary topic of focus for this discussion. Most marine and aquatic biomass can store lipids within the cell that can range from a small fraction to upwards of 80% of the cellular weight. Due to this trait, research and production scale operations have been centered on utilizing aquatic biomass for lipid production and conversion to fuel with the remaining cellular components being recycled for mineral content or discarded.

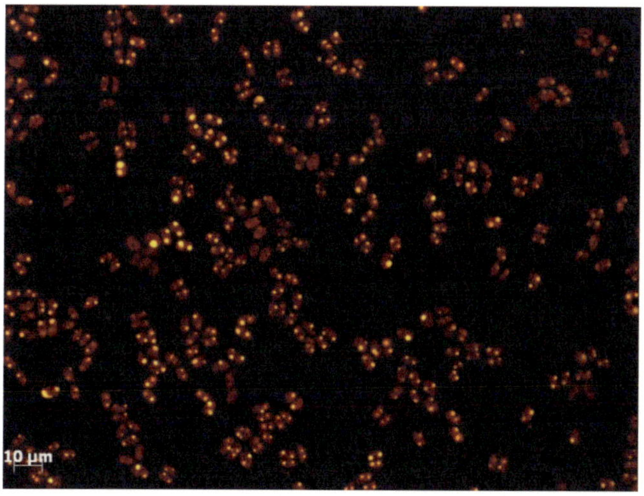

Figure 2. Nile Red Fluorescence Image of *Nitzchia sp.*

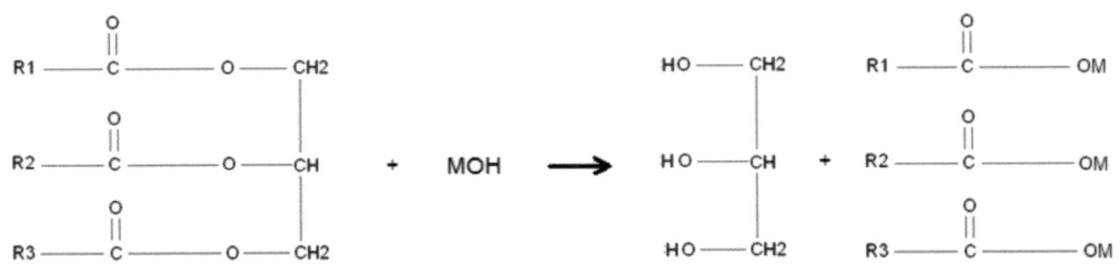

Scheme 1. Transesterification reaction schematic

Figure 2 shows an example of a marine diatom *Niztchia sp.* stained with Nile Red fluorescence stain (red color shows chlorophyll and yellow shows lipid fluorescence).

Neutral lipids consist of a glycerol molecule (a three carbon alcohol) and one to three fatty acids (referred to as mono-, di-, or tri- acylglycerols depending on number of fatty acids present) with the fatty acids being various carbon chain lengths and having various levels of unsaturation (unsaturated, mono-unsaturated, poly-unsaturated, etc.). Fatty tissues in animals serve as both an energy storage mechanism as well as a means of insulation against temperature extremes. Algae primarily store fats in the cell membrane to serve as an energy storage medium as well as a nutrient transport system to shuttle metabolites into and out of the cell. Several studies have been conducted to attempt to identify the distribution of fatty acids in algae and other aquatic biomass [13-15].

2.2. Ideal lipid characteristics for biodiesel

Biodiesel is produced primarily through the transesterification reaction of triglycerides and alcohol usually in the presence of a metal catalyst and can be visualized by the chemical reaction equation found in Scheme 1. where "R" groups are functional carbon chains varying in length and level of saturation and "M" is a metal, usually referring to sodium or potassium. The resultant glycerol that is produced is generally treated as a by-product and either sold for commodities use or burned to provide heating if necessary. This process is dependent on water content and pH, which dictates pre-processing demands in order to minimize the formation of soaps and maximize the production of wanted fatty acid ester compounds.

During this reaction the fatty acids tails are removed from the glycerol backbone leaving a glycerol molecule and one to three fatty acid esters (almost always either ethyl or methyl alcohol yielding a methyl or ethyl ester). These fatty acid methyl esters (FAME) or ethyl esters (FAEE) will vary in characteristics as a fuel based on carbon chain length as well as degree of unsaturation and location of unsaturated bonds. Some of the characteristics of biodiesel that are affected by fatty acid chemistry are viscosity, cloud point, and freezing point, among other factors important to engine performance. In general, there are several tradeoffs that must be made with regards to saturation of fatty acids, branching of the fatty acid chain, and the carbon chain length, as each will have positive and negative attributes affecting fuel performance.

As the length of the molecule increases, the cetane number, and thus the heat of combustion, increases, this in turn decreases NO_x emissions. However, as the length of the fatty acid chain increases, the resultant biodiesel has increased viscosity leading to a pre-heating requirement. Also, as fatty acids become more branched there is a benefit of the gel point (the temperature at which the fuel becomes gel-like and has complications flowing through fuel lines) decreasing. The negative to higher branching is that the cetane number will decrease due to a more difficult combustion. As saturation of the fatty acid chain increases, there is a decrease in NO_x emissions and an improvement in fuel stability. As saturation increases, there is an increase in melting point and viscosity, both undesirable traits in a fuel.

Since there are so many trade-offs in the production of biodiesel, it is very difficult, if not impossible, to pick one ideal source of fatty acid for conversion to fuel. The multitude of climates across the globe will necessitate various traits in fuel such as the gel point, melting/ freezing point, and oxidative stability. This leads to the argument of localized production of specific biomass sources that can be tailored to produce the types of lipids most suited to fuel that specific region, which will keep transportation costs down, as well as provide for the local economy. In following this method, there will be ample biomass produced to meet the specific needs of each climate, reducing environmental stresses that can occur due to overproduction for large scale purposes.

2.3. Enhancement of lipid production

Due to the various conditions that microorganisms grow and the constant flux of nutrients that can persist in nature, there are numerous types of lipids found that can change in concentration as the local environments evolve through typical ebbs and flows of materials. In response to these changes, microorganisms will change their cellular structures (i.e., lipid accumulation) by storing energy in various forms in order to utilize existing nutrients and energy to prepare for leaner conditions that may occur. In practical terms, this concept can be leveraged in order to produce high concentrations of intracellular lipids in marine and aquatic biomass in order to maximize the amount of lipids that can be harvested. Several studies have been conducted to determine what conditions affect the lipid composition and concentration of microorganisms. The more common techniques applied to increase the production of lipids from algae have through genetic manipulation [16], where genetic markers are manipulated that allow for increased lipid production to occur in the cell under normal conditions, by alteration of the cultivation conditions[17, 18], or by addition and manipulation of nutrients and chemicals added to the media [19]. By utilizing methods such as these, algal lipids can be increased by a substantial amount without increasing the footprint of required reactor space, nor greatly increasing the amount of time between harvests.

3. Synthetic fuels from biomass

3.1. Synthetic fuels

Unlike biofuels, which transform biological molecules into petroleum substitutes, synthetic fuels take a raw biological material, and through chemical processing, create compounds identical to petroleum fuel. This has a very distinct advantage over common biofuels in that there are no compatibility issues between the traditional fuels nor is there a need for any engine or fuel line modifications required. Synthetic fuels are usually made by utilizing a complex biological molecule and through thermal processing, break down the material into simple chemical building blocks (i.e., methane, carbon monoxide, hydrogen, etc.) and reform them into target chemicals. There are limitations with synthetic fuels production, especially when pertaining to production from aquatic and marine biomass where the water content is naturally higher than 99% by weight in its natural state, since initial breaking down of the products is usually through thermal processing that require dry or near dry conditions. However, since algae and aquatic biomass has such diverse characteristics and high cellular energy density, there is benefit for using either algae where the lipids have been extracted or whole algal cells as feedstock for these thermal synthetic fuel processes and thus can be considered as an option for production of synthetic fuels.

3.2. Methods for synthetic fuel production

There are three common methods for producing molecular precursors for synthetic fuels from biomass, and several variants of each method, dependent on the specific feedstock characteristic. These three methods are gasification, pyrolysis, and liquefaction. Pyrolysis and liquefaction will both produce of form of bio-oil that can be processed along with petroleum oil stocks and made into useful fuel products, while gasification will produce gaseous products such as carbon monoxide, methane, and hydrogen (commonly called syngas or synthesis gas in this process), and can be further refined directly to produce specific fuel molecules.

3.3. Gasification

Gasification is a process in which carbonaceous materials are exposed to heat and a sub-stoichiometric concentration of air to produce partially oxidized gaseous products that still have a high heating value with relatively lower concentrations of carbon dioxide due to limited oxygen [20]. Syngas can be catalytically reformed into a liquid fuel through the Fischer–Tropsch process, which converts carbon monoxide and hydrogen into long-chain hydrocarbons. By-products of the process include ash (formed from alkali-metal promoters present in the original reaction), char and tars that are created due to inefficiencies in mixing and heat distribution. This can be problematic when using water based biomass as the feedstock, since there will either be very high costs (in both energy and cost) to dry, or numerous unwanted products formed through side reactions. Three main types of gasification reactor are commonly used in industry: fixed bed, fluidized bed and moving bed. Each process has inherent advantages and drawbacks based on the complexity of the reactors, operating costs and product quality for use in the combustion of biomass. A more in-depth discussion of the design criteria and problems associated with using biomass as a fuel source for gasification reactors can be found in recent review articles [20, 21].

3.4. Liquefaction

Liquefaction is a process of converting biomass into a bio-oil in the presence of a solvent—usually water, an alcohol, or acetone—and a catalyst [22]. Liquefaction operates at milder temperatures than gasification, but requires higher pressures. Liquefaction can be indirect, wherein biomass is converted into gas and thence into liquid, or direct, in which biomass is converted directly into liquid fuel [23]. Bio-oils produced in direct liquefaction processes usually produce heavy oils with high heating values and value-added chemicals as by-products. Direct liquefaction also produces relatively little char compared to other thermochemical processes that do not utilize solvents. In addition, liquefaction has the advantage that the method is not hindered by the water content of the biomass, giving credence to utilizing this method for water based biomass. The use of water as a solvent can significantly reduce operating costs, and recent studies with sub-and super-critical water have demonstrated increased process productivity by overcoming heat-transfer limitations [24, 25]. Operating parameters and feed quality significantly influence the overall quality of the oil produced by these processes. A recent review presented an exhaustive comparison of the operational variables that affect the liquefaction of biomass and concluded that a well-defined temperature range is the most influential parameter for optimizing bio-oil yield and biomass conversion [22]. Similarly, catalyst choice can alter the heating value of the final liquefaction product and reduce the quantity of solid residue [25].

3.5. Pyrolysis

Pyrolysis is a process in which organic matter is exposed to heat and pressure in the absence of oxygen. The primary components of this process are syngas molecules like those found in gasification, as well as bio-oils and charred solid residues [26]. Pyrolysis methods are defined by the rate of heating, which directly affects the residence time of the reaction [27]. In slow pyrolysis, for example, the material is exposed to reactor conditions for five minutes; in fast pyrolysis, residence time is reduced to one to two minutes and in flash pyrolysis to less than five seconds. The residence time of the pyrolysis reaction greatly influences the composition of oils, gases and chars that are formed [28-30]. Several studies have been performed to identify the effect of operational variables— reactor conditions and variations in feedstock material —on the quality of the pyrolysis oils, gases, and chars [27, 30]. The oils typically produced during pyrolysis reactions are high in moisture content, and corrosive due to low pH. Pyrolysis of biomass is typically constrained by the high water content of the raw material, and current pyrolysis methods for biomass conversion have not reached the stage of commercial development. Ongoing research, however, aims at maximizing energy potential from biomass and optimizing conversion methods to achieve commercialization at marketable levels [31, 32].

4. Ethanol

Several species of cyanobacteria, including *Chlamydomonnas reinhardtii, Oscillatoria limosa, Microcystis PCC7806, Cyanothece PCC7822, Microcystis aeruginosa PCC7806 and Spirulina platensis* produce ethanol via an intracellular photosynthetic process. After selecting strains for ethanol, salt and pH tolerance, ethanol production can be enhanced through genetic modification [33]. These strains are long-lived and can be grown in closed photobioreactors to produce an ethanol containing algae slurry. This process for ethanol production from algae is currently being demonstrated by Algenol Biofuels [34-36]. The cyanobacteria are grown in flexible-film, polyethylene-based closed photobioreactors containing seawater or brackish water as medium. Industrial (or other waste) CO_2 is sparged into the bags to enhance growth of the microorganisms. Nutrients (primarily nitrogen and phosphorus) are supplied to sustain growth. At maturity, the microorganisms produce ethanol. The ethanol in the liquid phase will maintain an equilibrium with the ethanol-water in the vapor phase. The ethanol-water in vapor phase condenses along the walls of headspace which is collected by gravity for ethanol recovery. Algenol aims to produce 56,000 L of ethanol per hectare per year using 430 polyethylene bags established over a one hectare footprint each containing 4500 L of culture medium with a cyanobacteria concentration of 0.5 g/L. Unlike other algae derived biofuel processes, the algae are retained in the bags while the ethanol water condensate is removed for ethanol recovery. It is expected that the photobioreactors will be emptied once a year to replace the seawater, growth media and cyanobacteria.

The ethanol concentration in the algal cultures is expected to range between 0.5 and 5 % (w/w) depending on the ethanol tolerance levels of the strain and that of the condensate between 0.5 and 2% [36]. Since the maximum ethanol concentration is expected to be only 2 %, conventional distillation for ethanol recovery will not be energy efficient. A vapor compression steam stripping (VCSS) process is being developed to concentrate the ethanol to 5-30 % (w/w) range. VCSS is a highly heat integrated process that offers the potential for energy efficient separation even at low ethanol concentrations. This is then followed by a vapor compression distillation process to concentrate ethanol to an azeotropic 94% concentration. Life cycle energy requirements and greenhouse gas emissions for the process are dependent on the ethanol content of the condensate from the photobioreactors. Detailed analysis using process simulation software have shown that net life cycle energy consumption (excluding photosynthesis) is 0.55 down to 0.2 $MJ/MJ_{ethanol}$ and net life cycle greenhouse gas emissions is 29.8 down to 12.3 g $CO_2e/MJ_{ethanol}$ for ethanol concentrations ranging from 0.5 to 5% by weight [36]. Compared to gasoline these values represent a 67% and 87% reduction in the carbon footprint on an energy equivalent basis [36].

One of the technological challenges for this approach appears to be developing genetically engineered cyanobacterial strains that can tolerate high concentrations of ethanol. The ethanol concentration in the growth medium will affect the vapor phase ethanol content which in turn will affect the content of the condensate recovered from the photobioreactor. There is a dramatic increase in energy consumption in a conventional distillation process as ethanol

http://dx.doi.org/10.5772/53827

content decreases below 7.5% (by volume). Energy required almost doubles when ethanol content decreases from 12% down to 5% (by volume).

Another challenge would be the economical disposal of spent algal cultures. Sterilization and inactivation of large volumes of biomass can involve extremely energy intensive unit operations like heating, or expensive processes like ultra violet treatment or chlorination.

5. Anaerobic digestion

Biogasification (or anaerobic digestion) is a biochemical process that converts organic matter to biogas (a mixture of methane, 50-70%, and balance carbon dioxide) under anaerobic conditions. Biogas can be used as a replacement for natural gas or it can be converted to electricity. The process is mediated by a mixed, undefined culture of microorganisms at near ambient conditions. Several terrestrial biomass feedstocks (agricultural residues, urban organic wastes, animal wastes and biofuel crops) have been anaerobically digested and commercial scale digesters exist for the biogasification of such feedstocks.

Anaerobic digestion offers several advantages over other biofuel production processes like ethanol fermentation or thermochemical conversion. The microbial consortia in an anaerobic digester are able to naturally secrete hydrolytic enzymes for the solubilization of macromolecules like carbohydrates, proteins and fats. Therefore, unlike in ethanol fermentation process there is no need to incorporate a pretreatment step to solubilize the macromolecules prior to fermentation. In addition, since the process is mediated by a mixed undefined culture, issues of maintaining inoculum (or culture) purity does not arise. Being a microbial process, there is no need to dewater the feedstock prior to processing unlike in thermochemical conversion where the feedstock is dried, to improve net energy yield. This is advantageous when it comes to processing aquatic biomass as these can be processed without dewatering. The anaerobic digestion process will also mineralize organic nitrogen and phosphorous, and these nutrients can be recycled for algae growth [37].

The process primarily takes place in four steps. A mixed undefined culture of mciroorganisms mediates hydrolysis, fermentation, acetogenensis and methanogenesis of the organic substrates as shown in Figure 3. During hydrolysis, the complex organic compounds are broken down into simpler, soluble compounds like sugars, amino acids and fatty acids. These soluble compounds are fermented to a mixture of volatile organic acids (VOA). The higher chain VOAs like propionic, butyric, and valeric acids are then converted to acetic acid in the acetogenesis step. Acetic acid is converted to methane during methanogenesis. Hydrogen and carbon dioxide are also liberated during fermentation and acetogenesis. A different group of methanogens converts hydrogen and carbon dioxide to methane. This mixed microbial culture thrives in the pH range of 6-8. Digestion can be performed either at mesophilic conditions (30 - 38ºC) or thermophilic conditions (49 - 57ºC).

Aquatic biomass – macrophytes [38], micro and macro algae, have all been tested as feedstock for biogasification. Microalgae have proportions of proteins (6–52%), lipids (7–23%)

and carbohydrates (5–23%) that are strongly dependent on the species and environmental conditions [39-41]. Compared with terrestrial plants microalgae have a higher proportion of proteins, which is characterized by a low carbon to nitrogen (C/N) ratio. The average C/N for freshwater microalgae is around 10.2 while it is 36 for terrestrial plants [40]. Usually the digestion of terrestrial plants is limited by nitrogen availability; however for microalgae this situation does not arise. Besides carbon, nitrogen and phosphorus, which are major components in microalgae composition, oligo nutrients such as iron, cobalt, zinc are also found [42]. These characteristics of microalgae make it a good feedstock for anaerobic digestion.

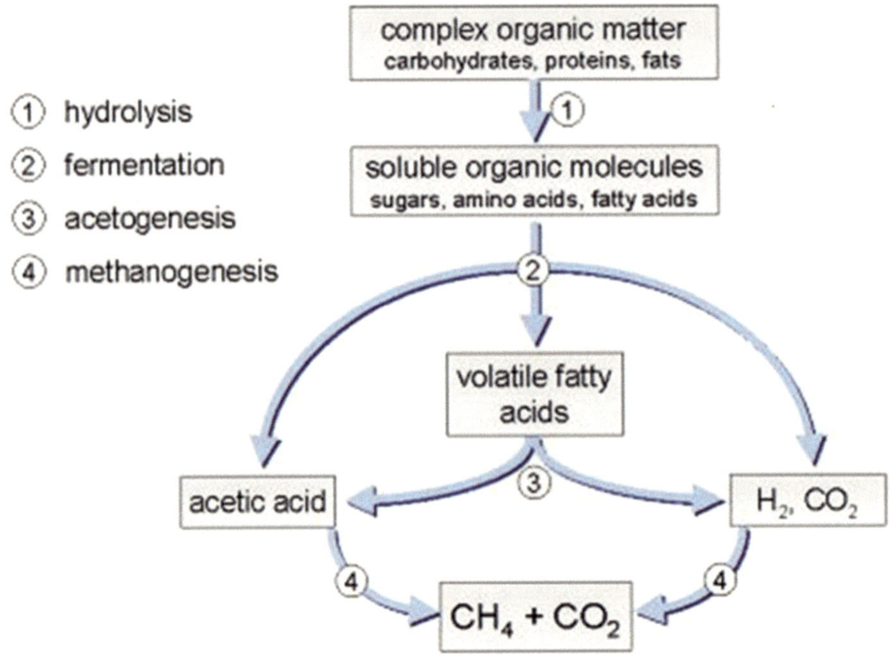

Figure 3. Pathways for mineralization of organic matter to biogas in an anaerobic digestion process

Previous studies have shown that macro algae like *Ulva lactuca, Gracillaria vermiculophylla, Saccharina latissima* etc. can be anaerobically digested producing methane at yields ranging from 0.1-0.3 LCH_4/g volatile solids (VS) [43]. Methane yields of microalgae like *Spirulina platensis* (fresh water), and *Scenedesmus* spp. and *Chlorella* spp. (fresh water) ranged between 0.2 and 0.3 L CH_4/g VS [44, 45] when these were codigested with other feedstocks like dairy manure and waste paper sludge, whereas other microalgae like *Tetraselmis sp* (marine), *Chlorella vulgaris* (fresh water), *Scendesmus obliquess* (fresh water) and *Phaeodactylum tricornutum* (fresh water) produced an average methane yield ranging from 0.17 to 0.28 L CH_4/g VS [45-47] when digested as sole feedstock. Table 3 summarizes microalgae digestion studies reported in the literature. The Table also lists the methane yield of cellulose powder as a benchmark to compare the methane potentials of microalgae feedstocks. Depending on the type of microalgae, the methane potentials range from 5 to 78% of methane potential of cellulose. Choice of microalgae has an impact on the methane yield.

More recently when *Nannochloropsis oculata* was biogasified [48] in laboratory scale digesters at thermophilic temperature, the methane yield obtained was 0.20 L at STP/g VS. *N. oculata* was chosen because it can be grown easily in brackish or seawater, has a satisfactory growth rate and can tolerate a wide range of pH (7-10) and temperature (17 – 27º C). *N. oculata* is not rich in lipids but contains predominantly cellulose and other carbohydrates, which makes it a good feedstock for anaerobic digestion instead of biodiesel production. On a % (w/w dry matter) basis, the composition of *N. oculata* is: 7.8% carbohydrate, 35% protein and 18% lipid. Rest of the components are amino acids, fatty acids, omega-3, unsaturated alcohols, ascorbic acid [49]. About 88% of the carbohydrate is polysaccharide. Of the polysaccharides, 68.2% is glucose, and the rest are fucose, galactose, mannose, rhamnose, ribose and xylose.

Based on *N. oculata* growth observed in the pilot raceways and the methane yield from digestion of this alga, an analysis was carried out to estimate energy production and land requirements. Currently the algae harvesting rate from the raceways are 9.64 g ash free dry weight (afdw)/m^2/d. Note that afdw (ash free dry weight) is the same as volatile solids content. An often cited study for algae growth has yielded a much higher productivity of 50 g afdw/m^2/d for *Platyomonas sp* [50]. The algae biomass yield obtained in this study was only about 20% of the productivity potentially attainable. Optimization of growth conditions for *N. oculata* may improve its productivity. Using the methane yield value of 204 L/kg VS for anaerobic digestion of *N. oculata*, the annual energy output from a facility that grows the algae and subsequently digests it would be 27 MJ/m^2/year. The area occupied (or footprint) of the digester(s) would be far less than the land area required for growing the algae. If the methane produced from this facility is converted to electricity, the electrical energy output would be 2.25 kWH_e/m^2/year assuming that the efficiency of converting thermal energy to electrical energy is 30%. The household electrical energy and natural gas consumption in the US for the year 2010 was 11,496 kWH/year and 2070 m^3/year respectively. If the algae biogasification facility were to supply the entire electrical energy requirements for a household, the land area required would be 5108 m^2 (1.26 acres). If in addition, the facility were to supply the natural gas needs, then an additional 2900 m^2 (0.77 acres) would be needed. In other words ~2 acres of land could supply all the energy needs of a household in US. If the algae productivities were improved then land requirement could be further reduced. At 50 g afdw/m^2/d algae productivity, the land requirement would only be about 0.4 acres.

Despite useful methane production potential from biogasification and the ability to process dilute algal slurries in a digester, there are challenges to be overcome to commercialize this approach for producing bioenergy from microalgae. One bottleneck is that some feedstock characteristics can adversely affect anaerobic digestion. Unlike defined cultures used for production of biofuels like ethanol or butanol, the microbial consortia in an anaerobic digester is capable of secreting extracellular enzymes to hydrolyze and solubilize macromolecules like cellulose, hemicellulose, proteins and fats. This characteristic has enabled several terrestrial biomass feedstocks like sugarbeets, sugarbeet tailings, napier grass, sorghum and aquatic biomass like water hyacinth and giant kelp to be successfully digested using practical retention times. However, degradability of feedstocks containing high fraction of lignin (for example sugarcane bagasse, switchgrass, miscanthus and woody biomass like pine, eucalyptus) is poor in an anaerobic digester. The refractoriness of these feedstocks has been attributed to low moisture, crystalline nature of the cellulose, and complex association of the component carbohydrates within lignin [51]. As seen from Table 3, the digestibility of microalgae varies. Species with no cell wall or cell encapsulation composed of proteins like *Chlorella vulgaris* and *Phaeodactylum tricornutum*, has a higher yield of methane. *Dunaliella tertiolecta* has very low methane yield of 0.018 L/kg VS due to the presence of a cell wall consisting of cellulose fibers distributed within an organic matrix. So depending on the type of microalgae used it may be necessary to carry out some form of pretreatment of algae to improve methane yield and rate of methane production. The type of pretreatment may depend on algae type.

Strain	Source	Pretreatment	Digester operating conditions	Methane Yield L/kg VS	Reference
Chlorella vulgaris *	Fresh	None	No co-digestion Digestion at 30±5° C	0.22	[47]
Tetraselmis sp.	Marine	None	No co-digestion Digestion at 35°C	0.25	[46]
Scendesmus obliquus	Fresh	None	Hybrid flow through at 33±2°C and 54±2°C	0.17	[48]
Phaeodactylum tricornutum	Marine	None	Hybrid flow through at 33±2°C and 54±2°C	0.28	[48]
Dunaliella tertiolecta *	Marine	None	Serum bottle at 37°C	0.018	[47]

*Sample dried and then frozen at -24°C

Table 3. Summary of microalgae anaerobic digestion studies

6. Conclusion

Aqueous and marine biomass can be processed into a variety of sources of energy. Due to the extreme dilution in water, non-thermal processes such as anaerobic digestion, fermentation to bioalcohols, and lipid extraction are logical and useful methods to utilize key components of microorganisms to produce biofuels for the replacement or supplementing of traditional fossil fuels. However, thermal methods such as gasification of wet biomass may play a role in producing specialty fuels such as jet fuel that require a specific ratio of higher hydrocarbons that would prove otherwise difficult to manufacture, even given the requirement of intense drying.

In order for biofuels sourced from aqueous and marine biomass to secure a market share in the world, research and development needs to further nature's ability to produce higher concentrations of biomass with targeted characteristics and reduced footprints, while better utilizing available nutrients. This will allow for an ample supply of biomass to be produced without competition with the human food chain, that can be used renewably produce fuel that can power the world's mobile fleet.

Author details

Robert Diltz[1] and Pratap Pullammanappallil[2]

*Address all correspondence to: Robert.diltz@us.af.mil

1 Air Force Research Laboratory, Tyndall AFB, FL, USA

2 Department of Agricultural and Biological Engineering, University of Florida, Gainesville, FL, USA

References

[1] U.S. Energy Information Administration. International Energy Statistics. http://www.eia.gov/cfapps/ipdbproject/IEDIndex3.cfm?tid=5&pid=5&aid=2 (accessed August 6, 2012).

[2] Illman A, Scragg A, Shales S. Increase in Chlorella strains calorific values when grown in low nitrogen medium. *Enzyme and Microbial Technology* 2000; 27 631-635.

[3] Pan P, Hu C, Yang W, Li Y, Dong L, Zhu L, Tong D, Qing R, Fan Y. The direct pyrolysis and catalytic pyrolysis of *Nannochloropsis sp* residue for renewable bio-oils. *Bioresource Technology* 2010; 101 4593-4599.

[4] Collet C, Adler N, Schwitzguebel J, Peringer P. Hydrogen production by *Clostridium thermolacticum* during continuous fermentation of lactose. *International Journal of Hydrogen Energy* 2004; 29 1479-1485.

[5] Iakovou E, Karagiannidis A, Vlachos D, Toka A, Malamakis A. Waste biomass-to-energy supply chain management: A critical synthesis. *Waste Management* 2010; 30 1860-1870.

[6] Virmond E, Schacker R, Albrecht W, Althoff C, Souza M, Moreira R, Jose H. Organic solid waste originating from the meat processing industry as an alternative energy source. *Energy* 2010; 36 3897-3906.

[7] Liu A, Chen W, Song L. Identification of high-lipid producers for biodiesel production from forty-three green algal isolates in China. *Progress in Natural Science: Materials International* 2011; 21 269-276.

[8] Mujtaba G, Choi W, Lee C-G, Lee K. Lipid production by Chlorella vulgaris after a shift from nutrient-rich to nitrogen starvation conditions. *Bioresource Technology* 2012; doi: http://dx.doi.org/10.1016/j.biortech.2012.07.057.

[9] Griffiths M, Harrison T. Lipid productivity as a key characterisitic for choosing algal species for biodiesel production. *J Appl Phycol* 2009; 21 493-507.

[10] Araujo G, Matos L, Goncalves L, Fernandes F, Farias W. Bioprospecting for oil producing microalgal strains; Evaluation of oil and biomass production for tenn microalgal strains. *Bioresource Technology* 2011; 102 5248-5250.

[11] Borugadda V, Goud V. Biodiesel production from renewable feedstocks: Status and opportunities. *Renewable and Sustainable Energy Reviews* 2012; 4763-4784.

[12] Miron A, Garcia M, Camacho F, Grima E, Chisti Y. Growth and biochemical characterization of microalgal biomass produced in bubble column and airlift photobioreactors; studies in fed-batch culture. *Enzymy and Microbial Technology* 2002; 31 1015-1023.

[13] Khotimchenko S, Vaskovsky V, Titlyanova T. Fatty acids of marine algae from the Pacific Coast of North Carolina. *Botanica Marina* 2002; 45 17-22.

[14] Li X, Fan X, Han L, Lou Q. Fatty acids of some algae from the Bohai Sea. *Phytochemistry* 2002; 59 157-161.

[15] Gressler V, Yokoya N, Fujii M, Colepicolo P, Filho J, Torres R, Pinto E. Lipid, fatty acid, protein, amino acid and ash contents in four Brazilian red algae species. *Food Chemistry* 2010; 120 585-590.

[16] Courchesne N, Parisien A, Wang B, Lan C. Enhancement of lipid production using biochemical, genetic and transcription factor engineering approaches. *Journal of Biotechnology* 2009; 141 31-41.

[17] Lv J, Cheng L, Xu X, Zhang L, Chen H. Enhanced lipid production of *Chlorella vulgaris* by adjustment of cultivation conditions. *Bioresource Technology* 2010; 101 6797-6804.

[18] Zhao G, Y J, Jiang F, Zhang X, Tan T. The effect of different trophic modes on lipid accumulation of *Scenedismus quadricauda*. *Bioresource Technology* 2012; 114 466-471.

[19] Devi M, Subhash G, Mohan S. Heterotrophic cultivation of mixed microalgae for lipid accumulation and wastewater treatment during sequential growth and starvation phases: Effect of nutrient supplementation. *Renewable Energy* 2012; 43 276-283.

[20] Balat M, Balat M, Kirtay E. Main routes for the thermo-conversion of biomass into fuels and chemicals. Part 2: Gasification systems. *Energy Conversion and Management* 2009; 50 3158-3168.

[21] Wang L, Weller C, Jones D, Hanna M. Contemporary issues in thermal gasification of biomass and its application to electricity and fuel production. *Biomass and Bioenergy* 2008; 32 573-581.

[22] Akhtar J, Amin N. A review on process conditions for optimum bio-oil yield in hydroghermal liquefaction of biomass. *Renewable and Sustainable Energy Reviews* 2011; 15 (3) 1615-1624.

[23] Rustamov V, Abdullayev K,Samedov E. Biomass conversion to liquid fuel by two-stage thermochemical cycle. *Energy Conversion and Management* 1998; 39 (9) 869-875.

[24] Xu C, Etcheverry T. Hydro-liquefaction of woody biomass in sub- and super-critical ethanol with iron-based catalysts. *Fuel* 2008; 87 (3) 335-345.

[25] Sun P, Heng M, Sun S, Chen J. Direct liquefaction of paulownia in hot compressed water: Influence of catalysts. *Energy* 2010; 35 5421-5429.

[26] Balat M, Balat M, Kirtay E. Main routes for the thermo-conversion of biomass into fuels and chemicals. Part 1: Pyrolysis systems. *Energy Conversion and Management* 2009; 50 3147-3157.

[27] Onay O, Kockar O. Slow, fast and flash pyrolysis of rapeseed. *Renewable Energy* 2003; 28 2417-2433.

[28] Hajaligol M, Waymack B, Kellogg D. Low temperature formation of aromatic hydrocarbon from pyrolysis of cellulosic materials. *Fuel* 2001; 80 1799-1807.

[29] M.F. Demirbas, M. Balat, and H. Balat, *Energy Conversion and Management* 50, 1746 (2009).

[30] Smets K, Adriaensens P, Reggers G, Schreurs S, Carleer R, Yperman J. Flash pyrolysis of rapeseed cake: Influence of temperature on the yield and the characteristics of the pyrolysis liquid. *Journal of Analytical and Applied Pyrolysis* 2011; 90 118-125.

[31] Fowler P, Krajacic G, Loncar D, Duic N. Modeling the energy potential of biomass-H_2RES. *Hydrogen Energy* 2009; 34 7027-7040.

[32] Kelly-Yong T, Lee K, Mohamed A, Bhatia S. Potential of hydrogen from oil palm biomass as a source of renewable energy worldwide. *Energy Policy* 2007; 35 5692-5701.

[33] Baier K, Dühring I, Enke H, Gründel M, Lockau W, Smith C, Woods P, Ziegler K, Oesterhelt C, Coleman J,Kramer D. Genetically modified photoautotrophic ethanol producing host cells, methods for producing the host cells, constructs for the transformation of the host cells, method for testing a photoautotrophic strain for desired growth property and method for producing ethanol using the host cells. International Application No. PCT/EP 2009/000892 (2009).

[34] Woods, P. Algenol Biofuels' Direct to Ethanol Technology AIChE, Nashville, TN (2009).

[35] Woods R, Legere E, Moll B, Unamunzanga C, Mantecon E. Closed photobioreactor system for continued daily in situ production, separation, collection and removal of

ethanol from genetically enhanced photosynthetic organisms. US Patent 7,682,821 (2007).

[36] Luo D, Hu Z, Choi D, Thomas V, Realff M, Chance R. Life cycle energy and greenhouse gas emissions for an ethanol production process based on blue-green algae. *Environ. Sci. Technol.* 2010; 44 8670-8677.

[37] Sialve B, Bernet N, Bernard O. Anaerobic digestion of microalgae as a necessary step to make microalgal biodiesel sustainable. *Biotechnology Advances* 2009; 27 409–416.

[38] Singhal V, Rai J. Biogas production from water hyacinth and channel grass used for phytoremediation of industrial effluents. *Bioresource Technology* 2003; 86 221–225.

[39] Brown M. The amino-acid and sugar composition of 16 species of microalgae used in mariculture. *Journal of Experimental Marine Biology and Ecology* 1991; 145 79–99.

[40] Elser J, Fagan W, Denno R, Dobberfuhl D, Folarin A, Huberty A, Interlandi S, Kilham S, McCauley E, Schulz K, Siemann E, Sterner R. Nutritional constraints in terrestrial and freshwater food webs. *Nature* 2000; 408 578–580.

[41] Droop M. 25 Years of Algal Growth Kinetics A Personal View. *Botanica Marina* 1983; 26 99–112.

[42] Grobbelaar J. Algal Nutrition - Mineral Nutrition. *Handbook of Microalgal Culture: Biotechnology and Applied Phycology* 2007; 95–115.

[43] Nielsen H, Heiske S. Anaerobic digestion of macroalgae: methane potentials, pretreatment, inhibition and co-digestion. *Water Science & Technology* 2011; 64 1723-1729.

[44] Sánchez Hernández E, Travieso Córdoba L. Anaerobic digestion of *Chlorella vulgaris* for energy production. *Resources, Conservation and Recycling* 1993; 9 127–132.

[45] Marzano C, Legros A, Naveau H, Nyns E. Biomethanation of the Marine Algae *Tetraselmis*. *International Journal Of Solar Energy* 1982; 1 263–272.

[46] Lakaniemi A, Hulatt C, Thomas D, Tuovinen O, Puhakka J. Biogenic hydrogen and methane production from *Chlorella vulgaris* and *Dunaliella tertiolecta* biomass. *Biotechnology for biofuels* 2011; 34 (4).

[47] Zamalloa C, Boon N, Verstraete W. Anaerobic digestibility of *Scenedesmus obliquus* and *Phaeodactylum tricornutum* under mesophilic and thermophilic conditions. *Applied Energy* 2012; 92 733-738.

[48] Buxy S, Diltz R, Pullammanappallil P. Biogasification of marine algae *Nannochloropsis oculata*. Accepted for publication in Proceedings of Material Challenges in Alternative and Renewable Energy. Trans of American Ceramic Society (2012).

[49] Foree E, McCarty P. Anaerobic decomposition of algae. *Environ. Sci. Technol.* 1970; 4 842–849.

[50] Sheehan J, Dunahay T, Benemann J, Roessle P. A Look Back at the U.S. Department of Energy's Aquatic Species Program—Biodiesel from Algae. National Renewable Energy Laboratory report, July 1998.

[51] Chynoweth D, Jerger D. Anaerobic digestion of woody biomass. *Developments in Industrial Microbiology* 1985; 26 235-246.

Chapter 15

Biofuel: Sources, Extraction and Determination

Emad A. Shalaby

Additional information is available at the end of the chapter

http://dx.doi.org/10.5772/51943

1. Introduction

Biofuel is a type of fuel whose energy is derived from biological carbon fixation. Biofuels include fuels derived from biomass conversion (Figure 1, JICA, Okinawa, Japan), as well as solid biomass, liquid fuels and various biogases. Although fossil fuels have their origin in ancient carbon fixation, they are not considered biofuels by the generally accepted definition because they contain carbon that has been "out" of the carbon cycle for a very long time. Biofuels are gaining increased public and scientific attention, driven by factors such as oil price hikes, the need for increased energy security, concern over greenhouse gas emissions from fossil fuels, and support from government subsidies. Biofuel is considered carbon neutral, as the biomass absorbs roughly the same amount of carbon dioxide during growth, as when burnt. The chemical composition of different kinds of biomass was shown in Table 1.

Biodiesel as one from important biofuel types is made from vegetable oils and animal fats. Biodiesel can be used as a fuel for vehicles in its pure form, but it is usually used as a diesel additive to reduce levels of particulates, carbon monoxide, and hydrocarbons from diesel-powered vehicles. Biodiesel is produced from oils or fats using transesterification and is the most common biofuel in Europe.

Bioethanol is an alcohol made by fermentation, mostly from carbohydrates produced in sugar or starch crops such as corn or sugarcane. Cellulosic biomass, derived from non-food sources such as trees and grasses, is also being developed as a feedstock for ethanol production. Ethanol can be used as a fuel for vehicles in its pure form, but it is usually used as a gasoline additive to increase octane and improve vehicle emissions. Bioethanol is widely used in the USA and in Brazil. Current plant design does not provide for converting the lignin portion of plant raw materials to fuel components by fermentation.

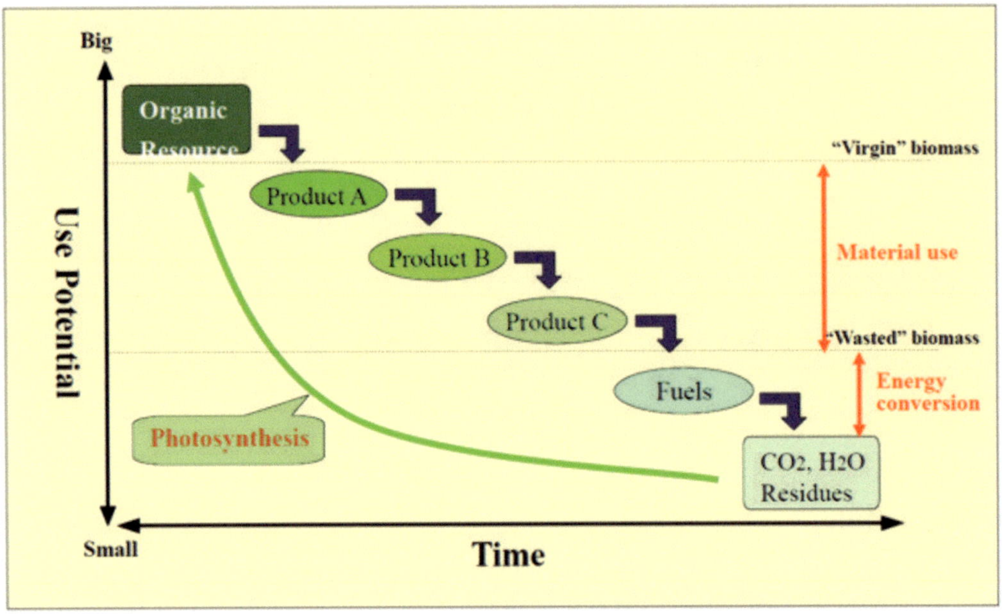

Figure 1. Cascade use of biomass

Kinds of bimass			Characteristics								
			Moisture content (%)	Solid substance(%)				Minerals(dry matter%)			
				Organic matter			Minerals	Carbon (T-C)	Nitrogen (T-N)	Phosphrus (T-P)	Potasium(T-K)
				Bacteria decomposi-tion	Hard decomposi -tion	Total					
Livestock waste	Beef cow dung		78.0	5.3	12.3	17.6	4.4	35.1	2.5	1.0	1.5
	Dairy cow	Dung	85.0	3.6	8.4	12.0	3.0	35.1	2.5	1.0	1.5
		Mixed	90.0	2.4	5.6	8.0	2.0	35.1	2.5	1.0	1.5
	Pig	Dung	72.0	10.5	10.5	21.0	7.0	35.1	3.5	2.5	1.5
		Mixed	90.0	3.8	3.8	7.6	2.5	35.1	3.5	2.5	1.5
	Poultry		70.0	15.8	5.2	21.0	9.0	35.1	5.5	3.0	3.0
Food residue	Tofu grounds		80.0	17.4	1.6	19.0	1.0	44.2	4.5	0.5	1.0
Garbage	Garbage		80.0	14.2	2.5	16.7	3.3	44.2	2.5	2.4	3.4
	Rubbish		50.0	29.4	12.6	42.0	8.0	55.0	1.0	–	–
	Used cooking oil		–	–	–	–	–	71.4	–	–	–
Slurry	Human waste		97.8	0.7	1.0	1.7	0.5	38.4	4.5	2.1	0.2
	Septic tank		99.0	0.3	0.5	0.8	0.2	38.4	4.5	1.1	0.1
	Sewage		78.0	5.8	8.6	14.4	7.6	38.4	4.5	2.3	0.3
Crop-wood residue	Wood residue		35.0	12.8	51.2	64.0	1.0	52.0	0.4	0.02	0.1
	Crop residue		12.0	32.0	38.7	70.7	17.3	40.9	0.6	0.09	0.4

Table 1. Average properties of biomass

In 2010 worldwide biofuel production reached 105 billion liters (28 billion gallons US), up 17% from 2009, and biofuels provided 2.7% of the world's fuels for road transport, a contribution largely made up of ethanol and biodiesel. Global ethanol fuel production reached 86 billion liters (23 billion gallons US) in 2010, with the United States and Brazil as the world's top producers, accounting together for 90% of global production. The world's largest biodiesel producer is the European Union, accounting for 53% of all biodiesel production in 2010. As of 2011, mandates for blending biofuels exist in 31 countries at the national level and in 29 states/provinces. According to the International Energy Agency, biofuels have the potential to meet more than a quarter of world demand for transportation fuels by 2050.

2. Different sources of biofuel

Here are 4 biofuel sources, with some of their application in developmental stages, some actually implemented:

2.1. Algae

Algae come from stagnant ponds in the natural world, and more recently in algae farms, which produce the plant for the specific purpose of creating biofuel. *Advantage of algae focude on the followings:* No CO_2 back into the air, self-generating biomass, Algae can produce up to 300 times more oil per acre than conventional crops. Among other uses, algae have been used experimentally as a new form of green jet fuel designed for commercial travel. At the moment, the upfront costs of producing biofuel from algae on a mass scale are in process, but are not yet commercially viable (Figure 2)

2.2. Carbohydrate (sugars) rich biomaterial

It comes from the fermentation of starches derived from agricultural products like corn, sugar cane, wheat, beets, and other existing food crops, or from inedible cellulose from the same. Produced from existing crops, can be used in an existing gasoline engine, making it a logical transition from petroleum. It used in Auto industry, heating buildings ("flueless fireplaces") At present, the transportation costs required to transport grains from harvesting to processing, and then out to vendors results in a very small net gain in the sustainability stakes.

2.3. Oils rich biomaterial

It comes from existing food crops like rapeseed (aka Canola), sunflower, corn, and others, after it has been used for other purposes, i.e food preparation ("waste vegetable oil", or WVO), or even in first use form ("straight vegetable oil", or SVO). Not susceptible to microbial degradation, high availability, re-used material. It is used in the creation of biodiesel fuel for automobiles, home heating, and experimentally as a pure fuel itself. At present, WVO or SVO is not recognized as a mainstream fuel for automobiles. Also, WVO and SVO are susceptible to low temperatures, making them unusable in colder climates.

Figure 2. Dense algal growth in four pilot-scale tank bioreactors fed by treated wastewater from the Lawrence, Kansas (USA) wastewater treatment plant (photo by B. Sturm). Each fiberglass bioreactor has an operating volume of ten cubic meters of water, and is operated as an air-mixed, flow-through vessel. Nutrientrich wastewater inflows are pumped in through the clear plastic hose (blue clamp), and water outflow occurs through the white plastic pipe shown at the waterline. These bioreactors are intended to be operated year-round, as the temperature of the inflowing wastewater is consistently ca. 10 - 8°C.

2.4. Agriculture wastes (organic and inorganic sources)

It comes from agricultural waste which is concentrated into charcoal-like biomass by heating it. Very little processing required, low-tech, naturally holds CO_2 rather than releasing it into the air. Primarily, biochar has been used as a means to enrich soil by keeping CO_2 in it, and not into the air. As fuel, the off-gasses have been used in home heating. There is controversy surrounding the amount of acreage it would take to make fuel production based on biochar viable on a meaningful scale. Furthermore, use of agriculture wastes which rich with inorganic elements (NPK----) as compost (fertilizer) in agriculture.

http://dx.doi.org/10.5772/51943

3. Comparison between different extraction methods of bio-diesel, bio-ethanol, biogas (bio-methane)

3.1. Biodiesel

3.1.1. Biodiesel extraction

Biodiesel is a clean-burning diesel fuel produced from vegetable oils, animal fats, or grease. Its chemical structure is that of fatty acid alkyl esters (FAAE). Biodiesel as a fuel gives much lower toxic air emissions than fossil diesel. In addition, it gives cleaner burning and has less sulfur content, and thus reducing emissions. Because of its origin from renewable resources, it is more likely that it competes with petroleum products in the future. To use biodiesel as a fuel, it should be mixedwith petroleum diesel fuel to create a biodiesel-blended fuel. Biodiesel refers to the pure fuel before blending. Commercially, biodiesel is produced by transesterification of triglycerides which are the main ingredients of biological origin oils in the presence of an alcohol (e.g. methanol, ethanol) and a catalyst (e.g. alkali, acid, enzyme) with glycerine as a major by-product [Ma and Hanna, 1999 ; Dube et al., 2007]. After the reaction, the glycerine is separated by settling or centrifuging and the layer obtained is purified prior to using it for its traditional applications (pharmaceutical, cosmetics and food industries) or for the recently developed applications (animal feed, carbon feedstock in fermentations, polymers, surfactants, intermediates and lubricants) [Vicente et al., 2007].

However, one of the most serious obstacles to use biodiesel as an alternative fuel is the complicated and costly purification processes involved in its production. Therefore, biodiesel must be purified before being used as a fuel in order to fulfil the EN 14214 and ASTM D6751 standard specifications listed in Table 2; otherwise the methyl esters formed cannot be classified as biodiesel. Removing glycerine from biodiesel is important since the glycerine content is one of the most significant precursors for the biodiesel quality. Biodiesel content of glycerine can be in the form of free glycerine or bound glycerine in the form of glycerides. In this work we refer to the total glycerine, which is the sum of free glycerine and bound glycerine. Severe consequences may result due to the high content of free and total glycerine, such as buildup in fuel tanks, clogged fuel systems, injector fouling and valve deposits (Hayyan et al., 2010).

3.1.2. Biodiesel extraction methods:

3.1.2.1. One step transesterification

For the synthesis of biodiesel, the following materials were used: oil sample (FFM Sdn Bhd), methanol (Merck 99%), and potassium hydroxide (KOH) as a catalyst (HMGM Chemicals >98%). Methanol and potassium hydroxide were pre-mixed to prepare potassium methoxide, and then added to oil in the reactor with a mixing speed of 400 rpm for 2 h at 50 °C. The molar ratio of oil to methanol was 1:10. Finally, the mixture was left overnight to settle forming two layers, namely: biodiesel phase (upper layer) and the glycerin-rich phase (Figure 3).

Property	EN 14214		ASTM D 6751	
	Test method	Limits	Test method	Limits
Ester content	EN 14103	96.5% (mol mol^{-1}) min	-	-
Linolenic acid content	EN 14103	12.0% (mol mol^{-1}) max	-	-
Content of FAME[a] with ≥4 double bonds	-	1.0% (mol mol^{-1}) max	-	-
MAG[b] content	EN 14105	0.80% (mol mol^{-1}) max	-	-
DAG[c] content	EN 14105	0.20% (mol mol^{-1}) max	-	-
TAG[d] content	EN 14105	0.20% (mol mol^{-1}) max	-	-
Free glycerine	EN 14105	0.02% (mol mol^{-1}) max	ASTM D 6584	0.020% (w/w) max
Total glycerine	EN 14105	0.25% (mol mol^{-1}) max	ASTM D 6584	0.240% (w/w) max
Water and sediment or water content	EN ISO 12937	500 mg kg^{-1} max	ASTM D 2709	0.050% (v/v) max
Methanol content	EN 14110	0.20% (mol mol^{-1}) max	-	-
(Na + K) content	EN 14108	5.0 mg kg^{-1}max	UOP 391	5.0 mg kg^{-1}max
(Ca + Mg) content	prEN 14538	5.0 mg kg^{-1} max	-	-
P content	EN 14107	10.0 mg kg^{-1} max	ASTM D 4951	0.001% (w/w) max
Oxidative stability (110 °C)	EN 14112	6 h min	-	-
Density (15 °C)	EN ISO 3675	860–900 kgm^{-3}	-	-
Kinematic viscosity or viscosity (40 °C)	EN ISO 3104	3.5–5.0 mm^2s^{-1}	ASTM D 445	1.9–6.0 mm^2s^{-1}
Flash point	EN ISO 3679	120 °C min	ASTM D 93	130 °C min
Cloud point	-	-	ASTM D 2500	Not specified
Sulfur content	EN ISO 20864	10.0 mg kg^{-1} max	ASTM D 5453	0.05% (w/w) max
Carbon residue	EN ISO 10370	0.30% (mol mol^{-1}) max	ASTM D 4530	0.050% (w/w) max
Cetane number	EN ISO 5165	51 min	ASTM D 613	47 min
Sulphated ash	ISO 3987	0.02% (mol mol^{-1}) max	ASTM D 874	0.020% (w/w) max
Total contamination	EN 12662	24 mg kg^{-1} max	-	-
Copper strip corrosion (3 h, 50 °C)	EN ISO 2160	1 (degree of corrosion	ASTM D 130	No. 3 max
Acid number or acid value	EN 14104	0.50 mg KOH g^{-1} max	ASTM D 664	0.50 mg KOH g^{-1}max
Iodine value	EN 14111	120 g I2·100 g^{-1} max	-	-
Distillation temperature (90% recovered)	-	-	ASTM D 1160	360 °C max

[a] FAME = fatty acid methyl esters.
[b] MAG = monoacylglycerines.
[c] DAG = diacylglycerines.
[d] TAG = triacylglycerine.

Table 2. Biodiesel specifications according to EN 14214, and ASTM D6751 standards.

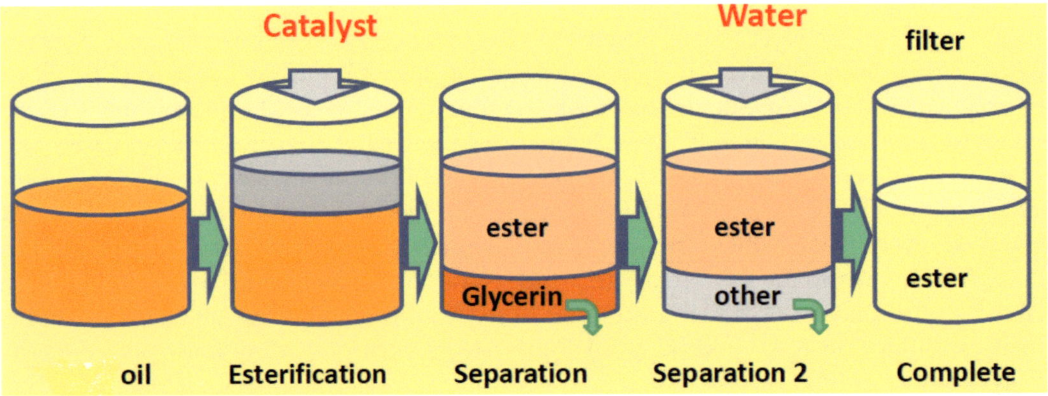

Figure 3. The biodiesel extraction process (steps).

3.1.2.2. Second step transterification

The production methodology followed in this study was according to Tomosevic and Siler-Marinkovic [2003] with some modification, where the alkali-catalyzed transesterification was applied. Basically, methanol was the alcohol of choice and KOH was used as the catalyst. Potassium methoxide solution (PMS) was prepared freshly by mixing a predetermined amount of methanol (≈ 12 wt % of oil) with KOH (≈ 1.0 wt % of oil) in a container until all the catalyst dissolved. The PMS was then added to 200 g of oil and stirred vigorously for 30 min at 30°C. Then after, the mixture was carefully transferred to a separating funnel and allowed to stand for 4 h. the lower layer (glycerol, methanol and most of the catalysts) was drained out. The upper layer (methyl esters MEs, some methanol and traces of the catalyst) was transferred into another flask containing freshly prepared PMS mixed at 60 rpm under reflux at 60°C for 30 min. afterwards; the mixture was carefully transferred to a separating funnel and allowed to stand there over night. The glycerol was removed by gravity settling, whereas the obtained crude esters layer was transferred into water bath to remove excess methanol at 65°C and 20 kPa. The obtained crude methyl esters were then cleaned thoroughly by washing with warm (50°C) deionized water, dried over anhydrous Na_2SO_4, weighted and applied for further analysis (Shalaby and Nour, 2012; Shalaby, 2011).

3.1.2.3. Qualitative analysis of glycerol

The Borax/phth test is special test for detection on the compound contain two neighboring hydroxyl group as in glycerol organic compound as the following:

1 ml glycerol layer mix with 1 ml of Borax/phth (red color) if the red color disappear in cold and appearing after heating (direct) this positive control.

3.1.2.4. Fourier transforms infrared spectroscopy (FTIR) analysis

FTIR analysis was performed using instrument, Perkin Elmer, model spectrum one, for detection of transesterification efficiency of oil by determination of the active groups produced from these process.

The results obtained by Shalaby and Nour (2012) found that, two step transterification of oil led to 100 % disappearance of hydroxyl group but this was less than 100 % in case of one step transterification as shown in Figure (4).

3.2. Bioethanol

3.2.1. Bioethanol extraction

Bioethanol is one of the most important renewable fuels due to the economic and environmental benefits of its use. The use of bioethanol as an alternative motor fuel has been steadily increasing around the world for the number of reasons. 1) Fossil fuel resources are declining, but biomass has been recognized as a major reasons World renewable energy source. 2) Greenhouse gas emissions is one of the most important challenges in this century because of fossil fuel consumption, biofuels can be a good solution for this problem. 3) Price of petroleum in global market has raising trend. 4) Petroleum reserves are limited and it is monopoly of some oil-importing countries and rest of the world depends on them. 5) Also known petroleum reserves are estimated to be depleted in less than 50 years at the present rate of consumption. At present, in compare to fossil fuels, bioethanol is not produced economically, but according to scientific predictions, it will be economical about 2030.

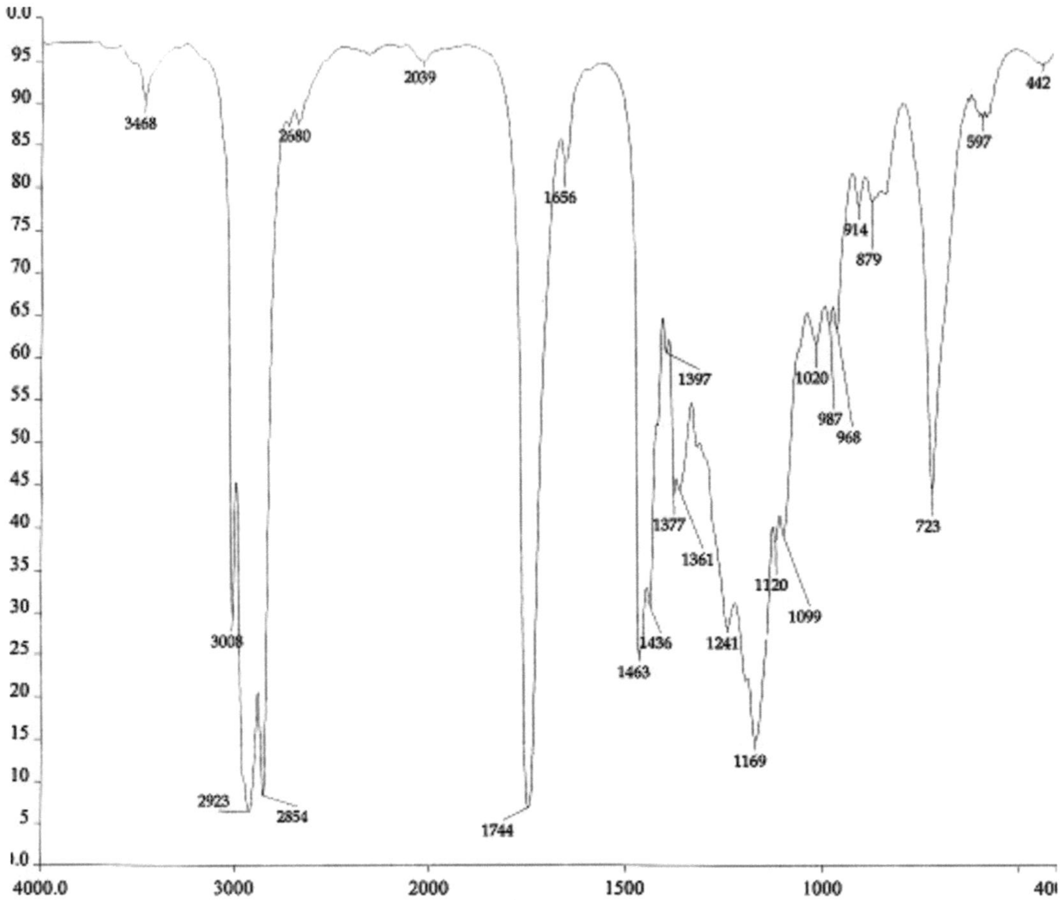

Figure 4. The IR spectrum of oil after two step transterification (produced biodiesel) process

Biomass commonly gathers from agricultural, industrial and urban residues. The wastes used for bioethanol production are classified in three groups according to pretreatment process in sugary, starchy and lignocellulosic biomasses. Lignocellulosic biomass, including forestry residue, agricultural residue, yard waste, wood products, animal and human wastes, etc., is a renewable resource that stores energy from sunlight in its chemical bonds. Lignocellulosic biomass typically contains 50%-80% (dry basis) carbohydrates that are polymers of 5C and 6C sugar units. Lignocellulosic biomasses such as waste wood are the most promising feedstock for producing bioethanol.

Bioconversion of lignocellulosic biomass to ethanol is significantly hindered by the structural and chemical complexity of biomass, which makes these materials a challenge to be used as feedstock for cellulosic ethanol production. Cellulose and hemicellulose, when hydrolyzed into their component sugars, can be converted into ethanol through well-established fermentation technologies. However, sugars necessary for fermentation are trapped inside the crosslinking structure of the lignocellulose.

http://dx.doi.org/10.5772/51943

Conventional methods for bioethanol production from lignocellulosic biomasses take three steps: *pretreatment* (commonly acid or enzyme hydrolyses), *fermentation, distillation*. Pretreatment is the chemical reaction that converts the complex polysaccharides to simple sugar. pretreatment of biomass is always necessary to remove and/or modify the surrounding matrix of lignin and hemicellulose prior to the enzymatic hydrolysis of the polysaccharides (cellulose and hemicellulose) in the biomass. Pretreatment refers to a process that converts lignocellulosic biomass from its native form. In general, pretreatment methods can be classified into three categories, including physical, chemical, and biological pretreatment. In this step, biomass structure is broken to fermentable sugars. This project focused on chemically and biologically pretreatment. For example: this project shows the effect of sulfuric acid, hydrochloric acid and acetic acid with different concentration by different conditions also shows the effect of cellulase enzyme by different techniques. Then fermentation step in which there are a series of chemical or enzymatic reactions that converted sugar into ethanol. The fermentation reaction is caused by yeast or bacteria, which feed on the sugar such as *Saccharomyces cerevisae*. After that, distillation step in which the pure ethanol is separated from the mixture using distiller which boil the mixture by heater and evaporate the mixture to be condensate at the top of the apparatus to produce the ethanol from joined tube.

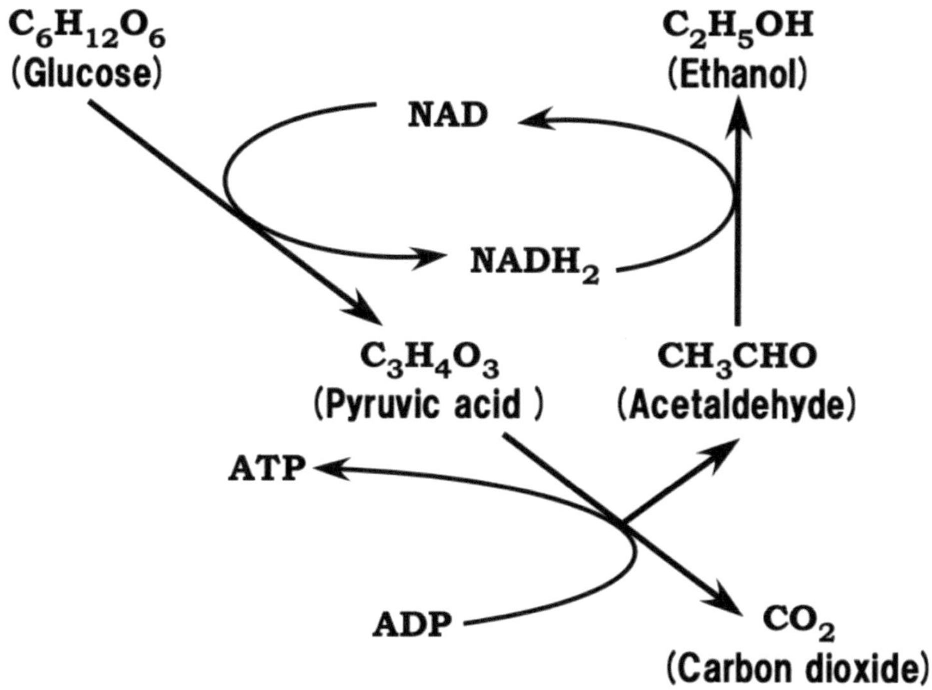

Figure 5. Ethanolic fermentation metabolism chart

The way to manufacture bioethanol is basically the same as that of liquor. Generally, saccharinity material such as sugar and starchy material such as rice and corn are saccharified (Figure 5-7), fermented and distilled till absolute ethanol whose alcoholicity is over 99.5%. It is technically possible to manufacture ethanol from cellulosic material such as rice straw or wood remains.

3.2.2. How to produce bio-ethanol:

- Materials

Sugarcane stems 5kg

Dry yeast, 15g

- Items

Brix meter, 5L flask, Dimroth condenser, Liebig condenser, Stick, Beaker

Cloth filter

1. Fermentation method
2. Mill juice out of Sugarcane stems. (about 3L of juice)
3. The juice is filtered out impurities.
4. Measurement Brix of juice.
5. Dry yeast is added to juice, the rate of 6g/L.
6. It keeps in the flask which sealed except the vent.
7. A cover is opened one day and once, then juice and dry yeast mixes so that air may enter with stick.
8. It continues until Brix becomes fixed.
9. Distillation method (Fig. 8)
10. Fermented juice is filtered out sediment.
11. It heats to boiling point in distiller.
12. Dimroth condenser is kept warm (about 70 degree) with hot water which is made to circulate by a pump.
13. Allihn condenser cools with tap water (about 20 degree).
14. Bio-ethanol which falls from the point of a allihn condenser is caught with beaker on ice.

3.2.3. Qualitative analysis for ethanol

Iodoform test on cold is special test for ethanol as the following: I ml ethanol layer mix with iodide and sodium hydroxide after that, the presence of yellow crystal and iodoform odor produced, this meaning presence of ethanol.

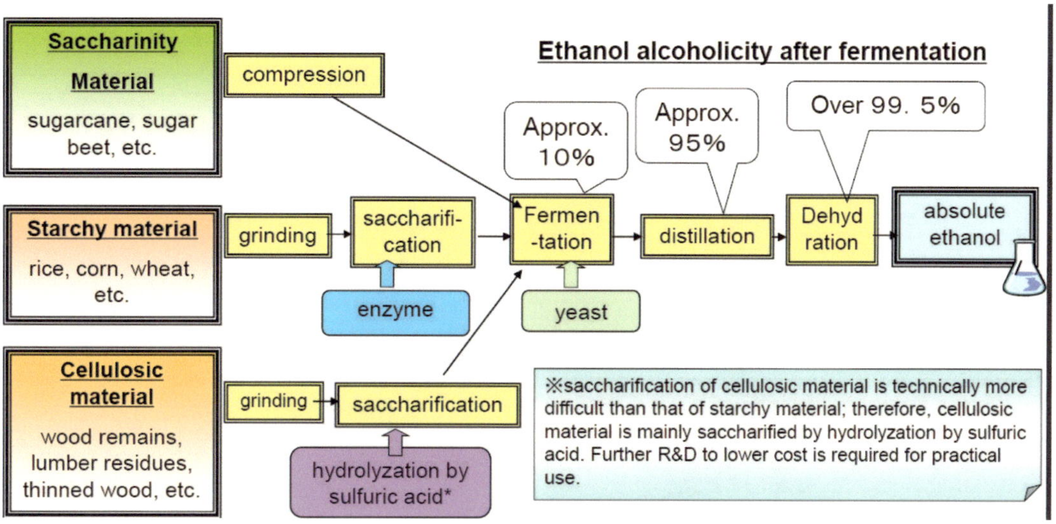

Figure 6. Production of absolute ethanol from Saccharinity, Starch and Cellulosic materials

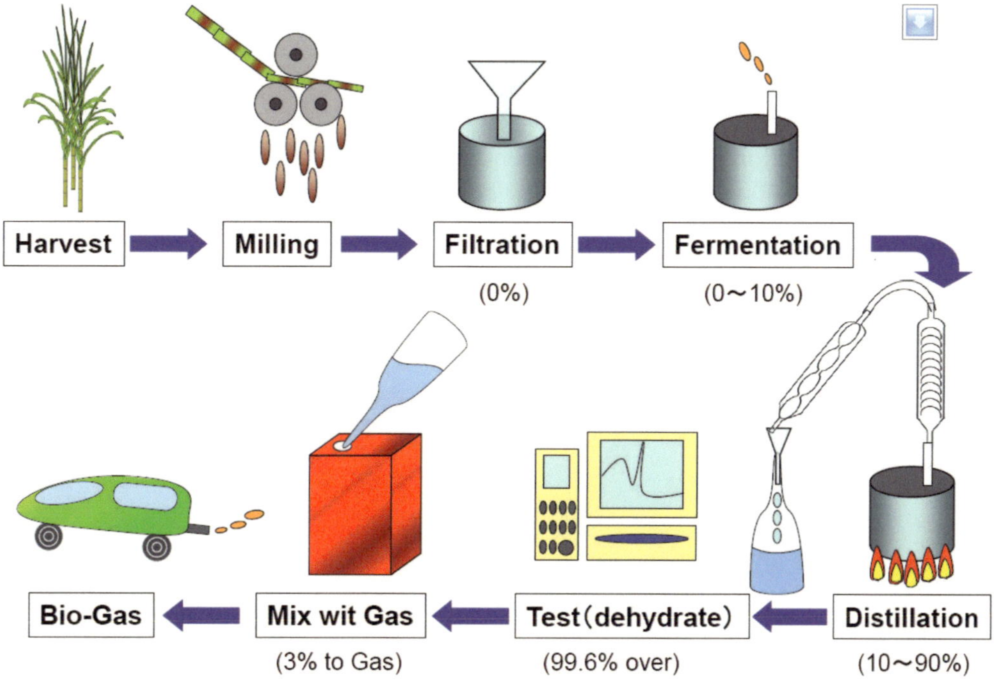

Figure 7. The main steps of bioethanol production from Starchy and cellulosic materials (Masami YASUNAKA / JIR-CAS)

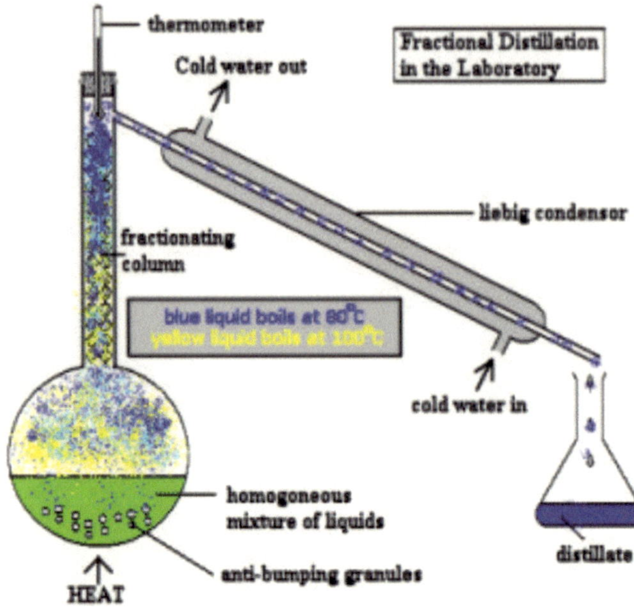

Figure 8. The distillation process for ethanol production.

3.2.4. Quantitative ethanol determination

3.2.4.1. Direct injected GC method

Beverage sample solution (0.5 mL) was dispensed into an l-mL caped sample vial, and then 5 mL of 1% internal standard solution (equivalent to 50 mg) was added. After mixing, 0.1 μL of the sample solution was injected directly into a GC or GC/MS (Figure 9) with syringe (Anonymous. 1992; Collins et al., 1997).

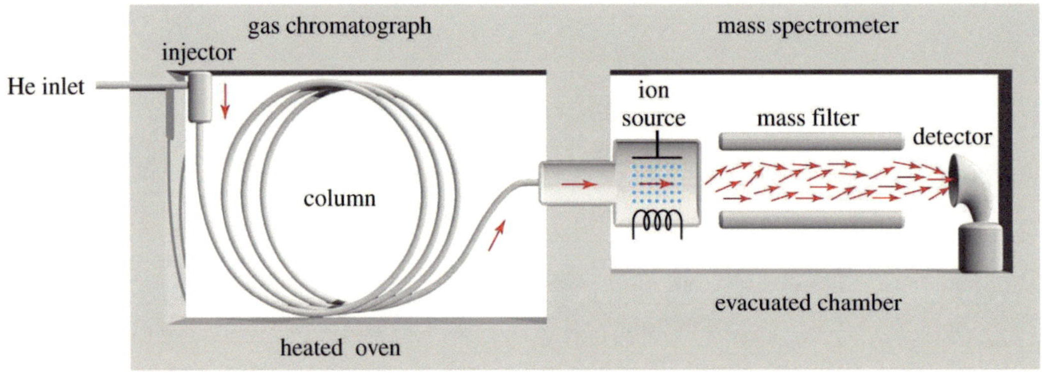

Figure 9. The GC/MS used for determination of ethanol.

3.2.4.2. Dichromate oxidation method

Beverage sample solution (1~5 mL) was steam distillated to obtain alcoholic eluate (> 50 mL), and then oxidized with acidified dichromate. The excessive potassium dichromate was then titrated with ferric oxide. The ethanol content in beverage sample could be obtained by calculating the volume difference of potassium dichromate consumption between sample solution and control solution (Anonymous. 1992; Collins et al., 1997).

3.2.4.3. Distillation-hydrometric method

Alcoholic volatile compounds in beverage samples were separated by distillation, and the gravity of the distillate was measured by hydrometer. The ethanol content was then converted (Anonymous. 1992; Collins et al., 1997).

3.3. Biogas (bio-methane) extraction

Methane fermentation is a versatile biotechnology capable of converting almost all types of polymeric materials to methane and carbon dioxide under anaerobic conditions. This is achieved as a result of the consecutive biochemical breakdown of polymers to methane and carbon dioxide in an environment in which varieties of microorganisms which include fermentative microbes (acidogens); hydrogen-producing, acetate-forming microbes (acetogens); and methane-producing microbes (methanogens) harmoniously grow and produce reduced end-products (Fig. 10-11). Anaerobes play important roles in establishing a stable environment at various stages of methane fermentation.

Methane fermentation offers an effective means of pollution reduction, superior to that achieved via conventional aerobic processes. Although practiced for decades, interest in anaerobic fermentation has only recently focused on its use in the economic recovery of fuel gas from industrial and agricultural surpluses.

The biochemistry and microbiology of the anaerobic breakdown of polymeric materials to methane and the roles of the various microorganisms involved are discussed here. Recent progress in the molecular biology of methanogens is reviewed, new digesters are described and improvements in the operation of various types of bioreactors are also discussed.

Methane fermentation is the consequence of a series of metabolic interactions among various groups of microorganisms. A description of microorganisms involved in methane fermentation, based on an analysis of bacteria isolated from sewage sludge digesters and from the rumen of some animals,. The first group of microorganisms secretes enzymes which hydrolyze polymeric materials to monomers such as glucose and amino acids, which are subsequently converted to higher volatile fatty acids, H_2 and acetic acid (Fig. 10). In the second stage, hydrogen-producing acetogenic bacteria convert the higher volatile fatty acids *e.g.*, propionic and butyric acids, produced, to H_2, CO_2, and acetic acid. Finally, the third group, methanogenic bacteria convert H_2, CO_2, and acetate, to CH_4 and CO_2 (Nagai et al., 1986).

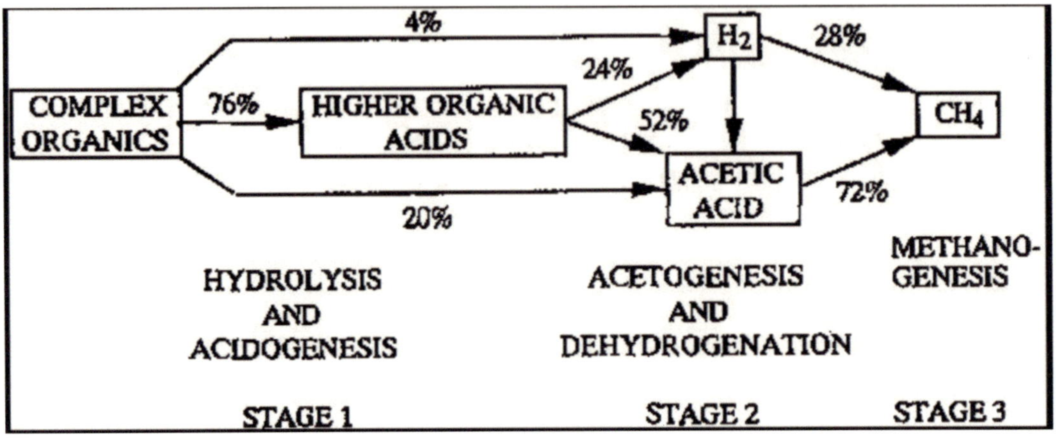

Figure 10. The main steps for production of methane gas

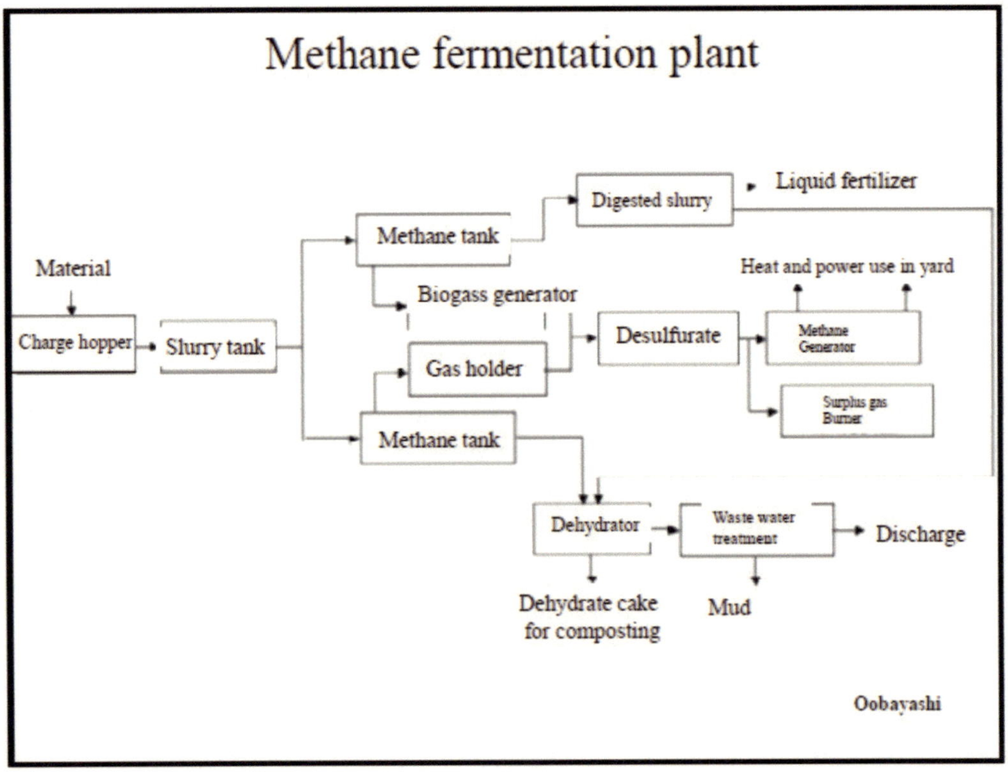

Figure 11. The principles methods for biomethane production

3.4. Determination of methane concentration

Methane will be measured on the gas chromatogram (Figure 9)using a FID (flame ionization) detector.

Note, unless you want smelly hands, it is recommended that you wear gloves. A lab coat is recommended for similar reasons.

- Using a 20 ml syringe connected to a 2-way stopcock, collect a little more than 5 ml of water from a port on your Winogradsky column.
- With the syringe pointing up, remove any air (tapping the sides of the syringe) and expel any extra water so that the final liquid volume in the syringe is 5 ml. Do this over a sink.
- Now, draw in 15 ml of air into the syringe so that the total air+water volume in the syringe is 20 ml. Close the stopcock.
- Shake the syringe to equilibrate the methane between the air and water.
- With the syringe pointing down, eject all the water from the syringe into the sink and close the stopcock. Try to get all the water out, but leave at least 10 ml of gas in the syringe
- We will now move to the GC lab in Starr 332 to measure methane.
- Repeat the above procedure for each of the ports on your Winogradsky column.

3.5. Calculations

To assist in plotting up results, measure the distance from the top of the sediment-water interface to each of the ports on the Winogradsky column, with distance to the ports in the sediment as positive and those in the water column negative. Also, measure the distance from the sediment-water interface to the surface of the water and the bottom of the sediments.

3.6. Methane concentration calculation

- From the standards, determine the concentration of methane in ppmv. Use the ideal gas law to determine the number of moles of methane in the 15 ml gas volume:

$$n=\frac{PV}{RT}=\frac{\frac{ppm}{10^6}\frac{15}{1000}}{(0.08205)(293)} \tag{1}$$

4. Physico-chemical parameters of extracted biofuel

4.1. Biodiesel

Most of the physical and chemical properties of the obtained methyl esters were determined by methods listed in JUS EN 14214:2004 standard [JUS EN 14214:2004] equivalent to EN 14214:2003, which defines requirements and test methods for fatty acid methyl esters (FAME) to be used in diesel engine. It must be emphasized that the characterization of crude methyl esters (i.e. those obtained before the purification) was not performed as it is well known fact that such raw products represent mixtures that were not in compliance with the strict restrictions for alternative diesel fuels, as it contains glycerol, alcohol, catalyst, mono- and diglycerides besides fatty acid esters. Measurements of the density at 15 _C by hydrometer method and of the kinematic viscosity at 40 _C were carried out according to JUS EN ISO 3675:1988 and JUS ISO 3104:2003, respectively. The acid value (Av) was determined by titration in accordance to EN 14104:2003; the iodine value was obtained by Hannus method (EN 14111:2003) this property has been also previously used for the biodiesel characterization [Karaosmanog et al., 1996; Šiler-Marinkovic et al., 1998]. The method for the cetane index (CI) estimation based on the saponification (Sv) and iodine (Iv) values was previously described [Krisnangkura, 1986] as simpler and more convenient than experimental procedure for the cetane number determination utilizing a cetane engine (EN ISO 5165:1998). The Krisnangkura's equation [Krisnangkura, 1986] used for CI calculation was as follows: CI = 46.3 + 5458/Sv _0.225 Iv. The cloud polint of MEs was determined according to ASTM D-2500 and Total sulfur content according to ASTM D-4294, Copper strip corrosion at 100 C according to ASTM D-130.The methyl ester composition was obtained by gas chromatograph equipped with DB-WAX 52 column (Supelco) and flame ionization detector. All the properties of frying oils as example were analyzed in two replicates and the final results given below were obtained as the average values (Table 3).

4.1.1. Density at 15 ⁰C

It is known that biodiesel density mainly depends on its methyl esters content and the remained quantity of methanol (up to 0.2% m/m according to JUS EN 14214 [JUS EN 14214:2004]); hence this property is influenced primarily by the choice of vegetable oil [Mittelbach, 1996], and in some extent by the applied purification steps. the mean density value of produced biodiesel was 0.90 g/cm3, while this value was more than Egyptian diesel (0.82-0.87g/cm3). but met the density value specified by JUS EN 14214 [JUS EN 14214:2004] to be in the range 0.860–0.900 g/cm3 at 15 ⁰C. This property is important mainly in airless combustion systems because it influences the efficiency of atomization of the fuel [Felizardo et al., 2006].

4.1.2. Kinematic viscosity at 40 ⁰C

Even more than density, kinematic viscosity at 40 ⁰C is an important property regarding fuel atomization and distribution. With regard to the kinematic viscosities that were in the range from 32.20 to 48.47 mm2/s, the feedstocks differed among themselves significantly. The viscosities of MEs were much lower than their respective oils (about 10 times) and they met the required values that must be between 3.5 and 5.0 mm2/s [JUS EN 14214:2004]. Comparing our MEs, the increase of the viscosities was observed more than Egyptian diesel, EN14214 and D-6751 (14.3, 7, 5 and 6 respectively) as shown in Table (3). However, the kinematic viscosity at 100 ⁰C of MEs produced from frying oil was met the viscosity range of Egyptian diesel, EN14214 and D-6751 (4.3, 7, 5 and 6 respectively). Predojevic (2008).

4.1.3. *Acid value*

The acid value measures the content of free acids in the sample, which have influence on fuel aging. It is measured in terms of the quantity of KOH required to neutralize sample. The base catalyzed reaction is reported to be very sensitive to the content of free fatty acids, which should not exceed a certain limit recommended to avoid deactivation of catalyst, formation of soaps and emulsion [Sharma et al., 2008, Meher et al., 2004]. The feedstock acid values obtained in this study differed significantly ranging from 1.86 to 3.31 mg KOH/g oil. Thus, in the light of the previous discussion on the requirements for the feedstock acid values, it could be concluded that frying oil had the values above the recommended 2 mg KOH/g. However, these values did not turn out to be limiting for the efficiency of the applied two-stage process, as it will be discussed along to the obtained product yields and purity later on. Acid values of MEs were less than 0.5 mg KOH/g specified as the maximum value according to JUS EN14214 (Table 4), Sharma et al. (2008) reviewed the literature and found that acid value of the feedstock for alkaline transesterification has to be reduced to less than 2 mg KOH/g (i.e. 1%), while only few examples of transesterification with feedstock acid value of up to 4.0 mg KOH/g (i.e. 2%) were found. They also reported that when waste cooking oil is used as feedstock, the limit of free fatty acids is a bit relaxed and the value a little beyond 1% (i.e. 2 mg KOH/g) did not have any effect on the methyl ester conversion. Acid values of MEs produced from frying oil was 1.16 mgKOH/g when compared with 0.5 mg KOH/g specified as the maximum value according to JUS EN14214 [JUS EN 14214:2004].

4.1.4. *Iodine value*

The iodine value of the feedstocks used in this study, which is a measure of unsaturation degree, was in the range of 70-78 mg I_2/100 g. According to JUS EN 14214 [JUS EN 14214:2004], MEs used as diesel fuel must have an iodine value less than 120 g I2 per 100 g of sample. Methyl esters obtained in this study had iodine value in the range 72-80g I2/100 g and this finding is in accordance to the fatty acid composition, i.e. the calculated total unsaturation degree of MEs (see Table 4). Iodine value depends on the feedstock origin and greatly influences fuel oxidation tendency. Consequently, in order to avoid oxidation.

4.1.5. *Saponification value*

The saponification value represents milligrams of potassium hydroxide required to saponify one gram of fat or oil. The obtained results indicated that in general, esters had higher saponification values than the corresponding oils. Saponification values of the feedstocks and products analyzed here, ranged from 199 to 207 mg KOH/g oil. However, knowing that a triglyceride has 3 fatty acid chains associated and each triglyceride will give 3 methyl esters, stoichiometrically it may be expected that the same amount of fatty acid carbon chain in neat feedstock oil and the biodiesel will react with the same amount of KOH giving the soaps, i.e. their saponification values will be the same. But, could this assumption be also applied on the waste frying oils knowing that their properties differ significantly from the neat oils as a consequence of cyclization, polymerization and degradation of fatty acids.

4.1.6. Cetane index

Krisnagkura [1986] proposed the equation for the estimation of cetane index (CI) based on the saponification and iodine values, recommending not to be used for oils, only for methyl esters. Namely, it has been previously documented that despite the fact that triglycerides and fatty acid methyl esters have similar saponification and iodine values, like it was obtained in this study too, cetane indexes of oils are generally much lower than those of methyl ester derivates. Thus, discussion on CI of frying oil will not be made. In this work, the CI value was 38 and this value less than the CI of Egyptian diesel, EN 14214 and D-6751 (55, 51 and 47 respectively). Šiler-Marinkovic′ and Tomaševic [1998] also used CI for the characterization of methyl esters produced from crude frying oils, and the estimated values were from 49.7 to 50.9. As an alternative to cetane number, cetane index is also an indicator of ignition quality of the fuel and is related to the time that passes between injection of the fuel into the cylinder and onset of ignition [Knothe, 2005].

Test	Produced Biodiesel	Egyptian Diesel oil	Biodiesel (EN14214)	Biodiesel D-6751
Flash point °C	202	"/> 55	"/> 101	"/> 130
Density g/cm^3 @ 15.56 °C	0.9055	0.82-0.87	0.86-0.9	----
Kinematic Viscosity cSt @ 40 °C	8.38	1.6-7	3.5-5	1.9-6
Kinematic Viscosity cSt @ 100 °C	4.34	---	----	----
Total acid number (mg KOH/g)	0.48	Nil	< 0.5	< 0.8
Cloud point °C	3	---	- 4	----
Pour point °C	0	4.5-15	----	----
Initial boiling point IBP °C	229	----	----	
Cetane number	63.8	Min. 55	"/> 51	"/> 47
Calorific value MJ/Kg	38.54	Min. 44.3	32.9	----
Total S wt%	0.12	Max. 1.2	< 0.01	< 0.05
Ash content wt%	0.002	Max. 0.01	0.02	< 0.02
Carbon residue wt%	0.63	Max.0.1	< 0.03	< 0.05
Copper strip corrosion @ 100°C	1a	1a	Class 1	No. 3 Max.
Water content wt%	0.08	Max. 0.15	< 0.05	0.05
Iodine number mg I_2/100 g	60	---	120	---

Table 3. Physicochemical properties for produced biodiesel compared to the Egyptian standards of petro-diesel fuel and two international biodiesel standards

http://dx.doi.org/10.5772/51943

4.1.7. Fatty acid composition

As can be observed from Table 5, regardless of the fatty acid profiles were observed in the biodiesel produced from frying oil, consisting mainly of methyl esters of oleic (C 18:1), palmitic (C 16:0), and stearic (C 18:0) acids (30.60, 3.0 and 66.40 % respectively) and 2.8 % unknown fatty acid. these results are in agreement with the results obtained by Predojvic (2008) who reported that, fatty acid profiles were observed in the biodiesels produced from sun flower oil consisting mainly of methyl esters of oleic (C 18:1), palmitic (C 16:0), linoleic (C 18:2) and stearic (C 18:0) acids.

Parameters	Feedstock	Produced biodiesel
Acid value mg KOH/g	5.1	0.48
Iodine value mg I_2/g	62.0	60.0
Saponification value mg KOH/g	199.5	207.0

Table 4. Some chemical properties of waste cooking oil (WCO) used as feedstock for methyl esters preparation and produced biodiesel

Fatty acid ester	Carbon number chain	Wt%	Molecular formula
Palmetic	16	3.00	$C_{16}H_{32}O_2$
Stearic	18	66.40	$C_{18}H_{36}O_2$
Oleic	18	30.60	$C_{18}H_{34}O_2$

Table 5. Composition of biodiesel obtained by transesterification of WCO using GC

4.2. Bioethanol

4.2.1. Property of ethanol

Melting point: -114.15

Boiling point: 78.3

Molecular formula: C_2H_5OH

Molecular weight: 46.07

Specific gravity: 0.789

Toxicity: Get intoxicated

4.3. Biomethane

4.3.1. Gas properties

4.3.1.1. Molecular weight

- Molecular weight : 16.043 g/mol

4.3.1.2. Solid phase

- Melting point : -182.5 °C
- Latent heat of fusion (1,013 bar, at triple point) : 58.68 kJ/kg

4.3.1.3. Liquid phase

- Liquid density (1.013 bar at boiling point) : 422.62 kg/m^3
- Liquid/gas equivalent (1.013 bar and 15 °C (59 °F)) : 630 vol/vol
- Boiling point (1.013 bar) : -161.6 °C
- Latent heat of vaporization (1.013 bar at boiling point) : 510 kJ/kg

4.3.1.4. Critical point

- Critical temperature : -82.7 °C
- Critical pressure : 45.96 bar

4.3.1.5. Gaseous phase

- Gas density (1.013 bar at boiling point) : 1.819 kg/m^3
- Gas density (1.013 bar and 15 °C (59 °F)) : 0.68 kg/m^3
- Compressibility Factor (Z) (1.013 bar and 15 °C (59 °F)) : 0.998
- Specific gravity (air = 1) (1.013 bar and 21 °C (70 °F)) : 0.55
- Specific volume (1.013 bar and 21 °C (70 °F)) : 1.48 m^3/kg
- Heat capacity at constant pressure (Cp) (1 bar and 25 °C (77 °F)) : 0.035 kJ/(mol.K)
- Heat capacity at constant volume (Cv) (1 bar and 25 °C (77 °F)) : 0.027 kJ/(mol.K)
- Ratio of specific heats (Gamma:Cp/Cv) (1 bar and 25 °C (77 °F)) : 1.305454
- Viscosity (1.013 bar and 0 °C (32 °F)) : 0.0001027 Poise
- Thermal conductivity (1.013 bar and 0 °C (32 °F)) : 32.81 mW/(m.K)

4.3.2. Miscellaneous

- Solubility in water (1.013 bar and 2 °C (35.6 °F)) : 0.054 vol/vol
- Autoignition temperature : 595 °C

5. Biofuel blending

It is important that when you are purchasing fuel you make sure it is high quality by meeting all ASTM specifications. Fuel that is off specification on just one of the ASTM standards can not only cause serious engine problems, but it can void engine warranties if it is determined that the fuel caused damage. This can cause unnecessary costly repairs for vehicles/equipment. To review specifications for diesel fuel, biodiesel and biodiesel blends, see the specifications in the Appendix. In an effort ensure that producers and marketers operate in a manner consistent with proper specifications, the National Biodiesel Accreditation Commission created the BQ-9000 program in 2005. This voluntary program establishes quality systems for producers and marketers of biodiesel in the areas of storage, sampling, testing, blending, shipping, distribution and fuel management practices. If purchasing B100 or a biodiesel blend, ask if the biodiesel is from a BQ-9000 biodiesel producer/marketer. If you are unable to get fuel from a BQ-9000 producer/marketer, the next best thing is to verify with your supplier that the fuel meets all ASTM specifications.

In most cases the blending process takes place right at the terminal rack by a process called in-line blending. This is the preferred method because it ensures complete blending. In-line blending occurs when warm biodiesel is added to a stream of diesel fuel as it travels through a pipe or hose in such a way that the biodiesel and diesel fuel become thoroughly mixed by the turbulent movement. This product is sold directly to customers, petroleum jobbers or a distribution company for sale to customers.

The blend level (percentage of biodiesel in the biodieseldiesel mixture) determines many important characteristics of the blended fuel. A higher-than-specified level of biodiesel may exceed the engine manufacturer's recommended limitation, compromising the engine performance. A lower blend level of biodiesel may reduce the expected benefits, such asfuel lubricity and tail pipe emission. In addition, cloud point and pour point of biodiesel are usually higher than that of diesel fuel, and a higher blend level makes the fuel unsuitable or difficult to use in cold weather conditions. Engine injection timing can be adjusted based on the blend level in order to improve the engine emission and performance (Tat and Van Gerpen, 2003).

It has been reported that the actual biodiesel content of blended biodiesel fuel sold at gas stations can be significantly different from the nominal blend level. A 2% nominal blend has been found to actually contain anywhere from 0% to 8% biodiesel (Ritz and Croudace, 2005). There are several reasons why the actual blend level may differ from the specified level. For instance, if biodiesel is blended at a temperature less than 10°F above its cloud point, it will not mix well with diesel, causing a rich mixture in one portion of the tank and a lean mixture in another portion (NBB, 2005). Other reasons for the discrepancy may include profit-driven fraud and involuntary mixing of diesel into the blend to lower the overall blend level of biodiesel. Biodiesel is usually sold at a higher price than diesel fuel; therefore, the price of the fuel is dependent on the blend level. Knothe (2001) has shown that near-infrared (NIR) spectroscopy and nuclear magnetic resonance (NMR) can be used to detect biodiesel blend levels. However, the NMR method depends on the biodiesel fatty acid profile; hence, knowledge of the biodiesel feedstock is required before this method can be used. In addition, using NMR only to detect blend level may not be cost effective. For NIR spectroscopy, Knothe suggested using wavelengths around 1665 nm or 2083 to 2174_nm. Since aromatic compounds produce strong and sharp infrared bands due to their relatively rigid molecular structure and diesel fuels have varying amounts of aromatics between 20% and 35% (Song et al., 2000), the absorbance of a blend may not directly correlate to the percentage of biodiesel. The absorbance is defined as the logarithm of the radiation intensities ratio, that is, before and after being absorbed by a sample.

Diesel fuel is distilled from crude petroleum, which is composed primarily of hydrocarbons of the paraffinic, naphthenic, and aromatic classes. Each class contains a very broad range of molecular weights. One of the features of diesel fuel is the presence of 20% to 35% aromatic compounds by weight. Aromatics are a class of hydrocarbons that are characterized by a stable chemical ring structure. They are determined primarily by the composition of the crude oil feed, which is usually selected based on considerations of availability and cost (Chevron, 2006). On the other hand, biodiesel is a mixture of fatty acid esters. Fatty acids with 16 to 22 carbon chain lengths are predominant in oils and fats. The resulting mixture of fatty acid esters depends on the kind of feedstock used. Neat biodiesel contains essentially no aromatic compounds.

The presence of aromatics in diesel and their absence in biodiesel creates the possibility of distinguishing these two fuels using ultraviolet spectroscopy. Benzene, the simplest aromatic compound, has maximum absorption at 278 nm (Zawadzki et al., 2007). Biodiesel, which is esters of long-chain fatty acids when adequately diluted in *n*-heptane, has negligible absorbance compared to the aromatics at the same frequency. Hence, differences in biodiesel feedstocks will have a minimal impact on absorbance at this wavelength. The ultraviolet (UV) range between 200 and 380 nm is also referred to as near-UV. In general, light sources, filters, and detectors are less expensive for this vicinity of the spectrum than for IR at 8621 nm, as used by the CETANE 2000. Hence, near-UV spectroscopy may present a low-cost alternative method for biodiesel blend level sensing (Figure 12 and 13).

6. Material balance of biofuel product

Biomass conversion plant has many components which are connected each other. Material and energy flow among the components, therefore we should grasp the detail of the balance (Figures 13-16). If there is a choke point, the flow stagnation causes to the troubles of operation and low efficiency of the performance. (Masami UENO, University of Ruyku, Faculty of Agriculture, Okinawa, Japan).

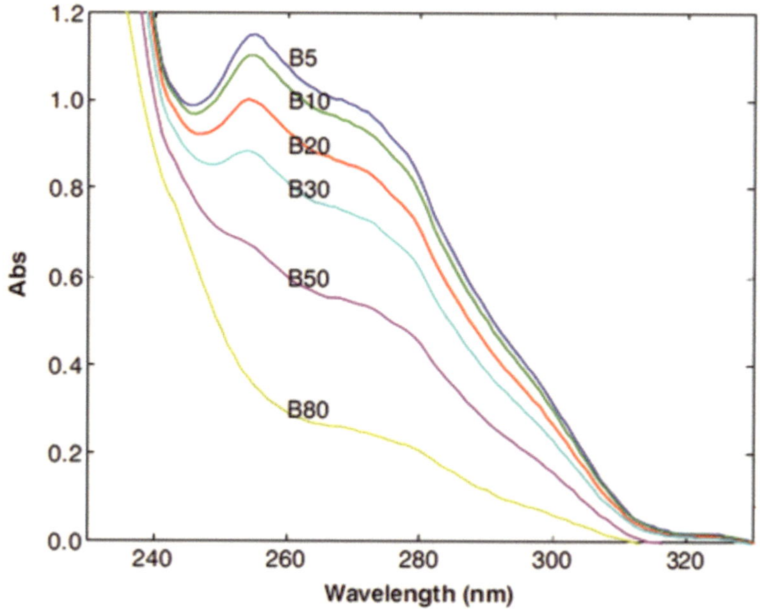

Figure 12. UV absorbance spectra of soy methyl ester and No. 2 diesel blend diluted 1:2915 in *n*-heptane.

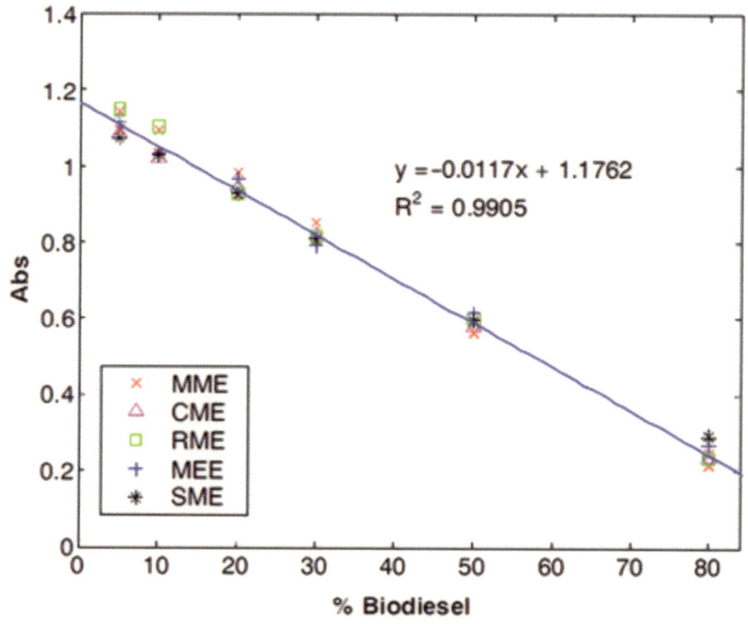

Figure 13. Absorbance of diluted biodiesel-diesel blends from different feedstocks at 260 nm wavelength (MME = mustard methyl esters, CME_=canola methyl esters, RME = rapeseed methyl esters, MEE = mustard ethyl esters, and SME = soybean methyl esters).

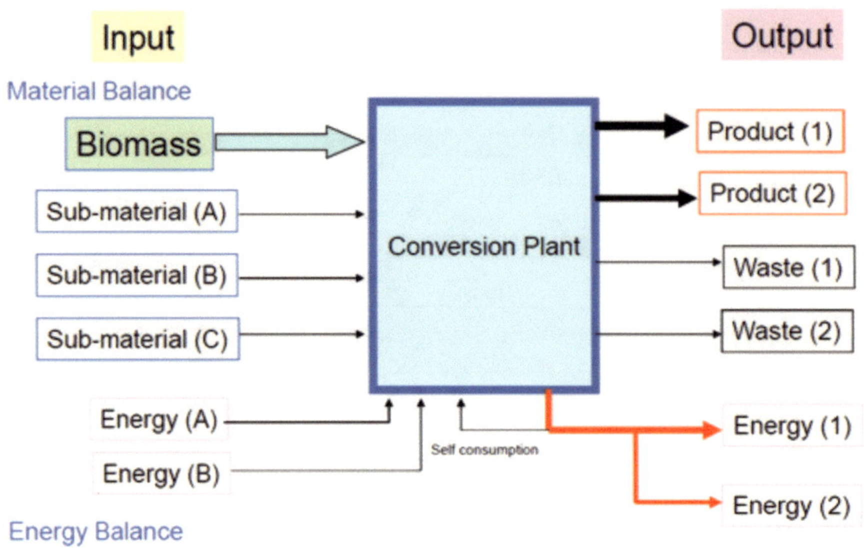

Figure 14. Material balance and energy balance.

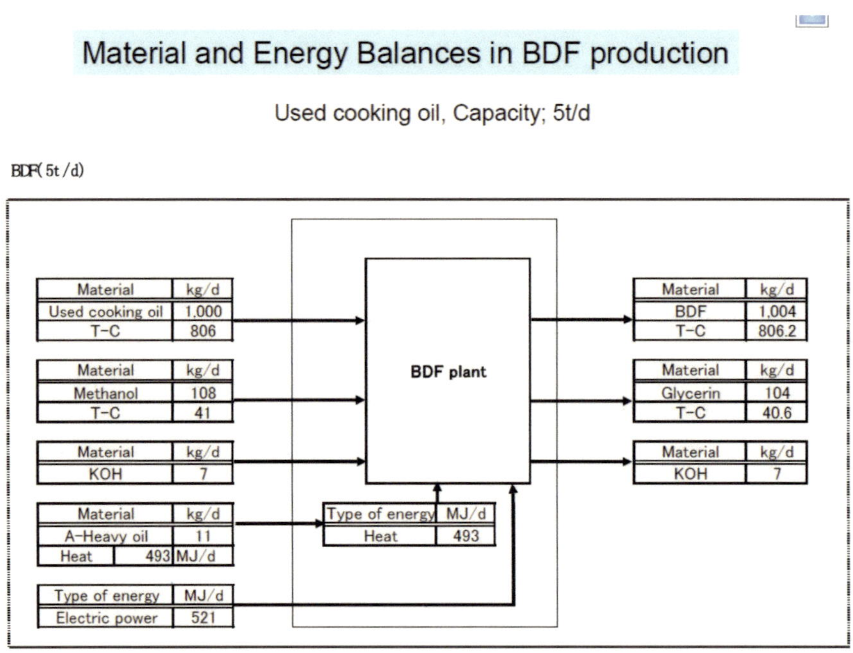

Figure 15. Material and energy balance in biodiesl fuel production

http://dx.doi.org/10.5772/51943

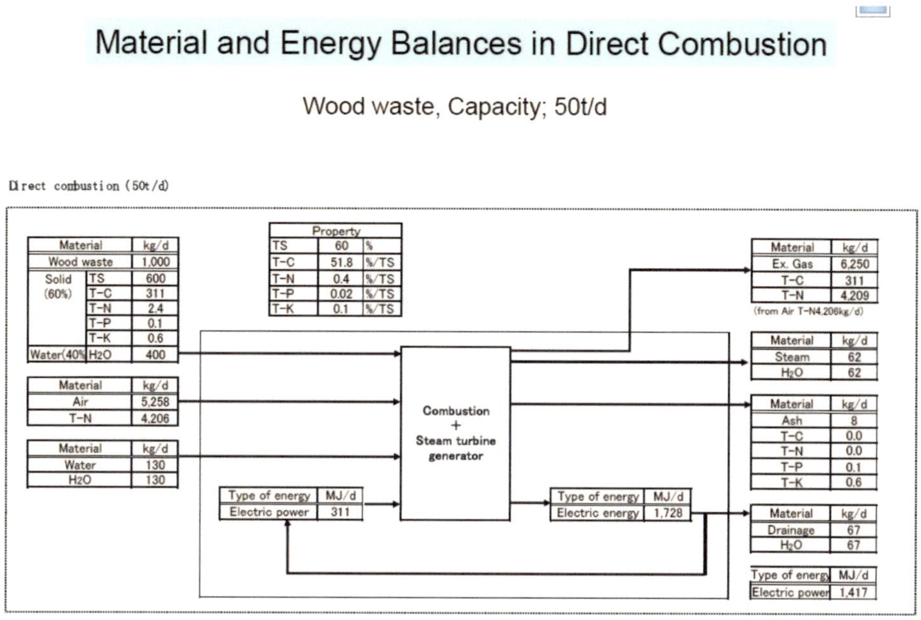

Figure 16. Material and energy balance in direct combustion

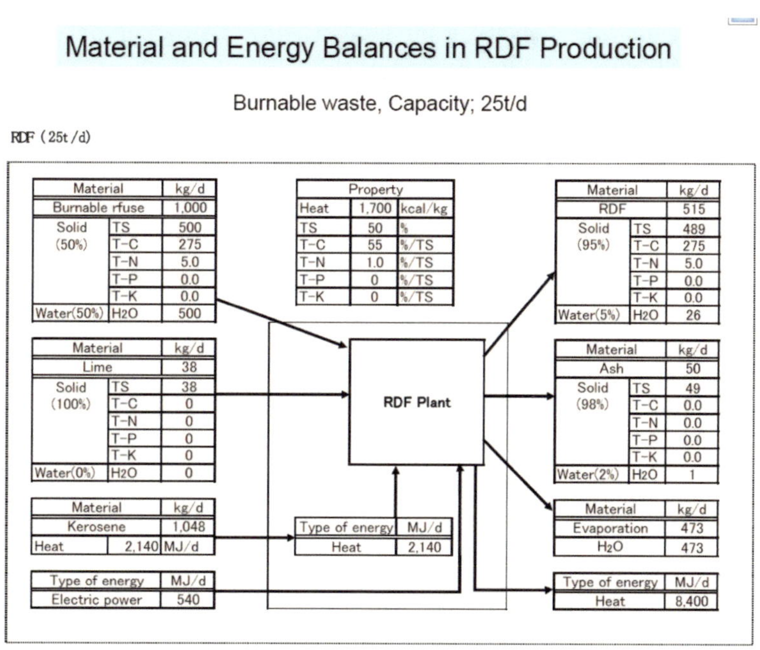

Figure 17. Material and energy balance in RDF production

7. Conclusion

The different kind of biomass considered as main source for biofuel (diesel-methane, ethanol, compost –etc). The cost of extraction and blending is very effective point for use of biomass in addition to ability for use of all part from biomass as multipurpose.

Author details

Emad A. Shalaby

Biochemistry Dept., Facult. of Agriculture, University of Ruyku, Cairo University, Egypt

References

[1] Anonymous. 1992. Chapter 4-Alcohols. In Wine Analysis. Research Institute for Wines, Taiwan Tobacco & Wine Monopoly Bureau. pp. 1-7.

[2] Chevron. 2006. Diesel fuels technical review. San Ramon, Cal.: Chevron Corp.

[3] Collins, T. S., Miller, C. A., Altria, K. D. Waterhouse, A. L. 1997. Development of a rapid method for the analysis of ethanol in wines using capillary electrophoresis. Am. J. Enol. Vitic. 48: 280-284.

[4] Dube′MA.; A.Y. Tremblay, J. Liu, Biodiesel production using a membrane reactor, Bioresour. Technol. 98 (2007) 639–647.

[5] Felizardo P, Neiva Correia MJ, Raposo I, Mendes JF, Berkemeier R, Bordado JM. Production of biodiesel from waste frying oils. Waste Manage 2006;26:487–94.

[6] Hayyan , M., Mjalli , FS., Hashim, MA.; , lNashef, IA. (2010). A novel technique for separating glycerine from palm oil-based biodiesel using ionic liquids. Fuel Processing Technology 91 :116–120.

[7] JUS EN 14214:2004. Automotive fuels. Fatty acid methyl esters (FAME) for diesel engines-requirements and test methods. Standardization Institute,Belgrade, Serbia.

[8] Karaosmanog˘lu F, Cig˘izog˘lu KB, Tuter M, Ertekin S. Investigation of the refining step of biodiesel production. Energ Fuel 1996;10:890–5.

[9] Krisnangkura K. A simple method for estimation of cetane index of vegetable oil methyl esters. JAOCS 1986;3:552–3.

[10] Knothe, G. 2001. Determining the blend level of mixtures of biodiesel with conventional diesel fuel by fiber-optic near-infrared spectroscopy and H-1 nuclear magnetic resonance spectroscopy. J. American Oil Chem. Soc. 78(10): 1025-1028.

[11] Ma F.; M.A. Hanna, Biodiesel production: a review, Bioresour. Technol. 70 (1999) 1–5.

[12] Meher LC, Sagar DV, Naik SN. Techical aspects of biodiesel production by transesterification-a review. Renew Sust Energ Rev 2004:1–21.

[13] Mittelbach M. Diesel fuel derived from vegetable oils, VI: specifications and quality control of biodisel. Bioresour Technol 1996;56:7–11.

[14] Nagai, S. and Nishio, N., In "Handbook of Heat and Mass Transfer, vol. 3 Catalysis, Kinetics and Reactor Engineering" Ed. Cheremisinoff, N.P., 701-752 (1986) Gulf Publishing Corn., Houston, London, Paris, Tokyo.

[15] NBB. 2005. Biodiesel cold weather blending study. Jefferson City, Mo.: National Biodiesel Board.

[16] Predojevic, ZJ. (2008).The production of biodiesel from waste frying oils: A comparison of different purification steps. Fuel 87 : 522–3528.

[17] Ritz, G. P., and M. C. Croudace. 2005. Biodiesel or FAME (fatty acid methyl ester): Mid-infrared determination of ester concentration in diesel fuel. Houston, Tex.: Petroleum Analyzer Company (PAC).

[18] Shalaby, EA (2011). Algal Biomass and Biodiesel Production, Biodiesel - Feedstocks and Processing Technologies, Margarita Stoytcheva and Gisela Montero (Ed.), ISBN: 978-953-307-713-0, InTech, Available from: http://www.intechopen.com/articles/show/title/algal-biomass-and-biodiesel-production

[19] Shalaby, EA and El-Gendy. NS. (2012). Two steps alkaline transesterification of waste cooking oil and quality assessment of produced biodiesel. International Journal of Chemical and Biochemical Sciences. IJCBS, 1(2012): 30-35.

[20] Sharma YC, Singh B, Upadhyay SN. Advancements in development and characterization of biodiesel: a review. Fuel 2008;87:2355–73.

[21] Šiler-Marinkovic′ SS, Tomaševic′ VA. Transesterification of sunflower oil in situ. Fuel 1998;77:1389–91.

[22] Song, C., C. S. Hsu, and I. Mochida. 2000. Introduction to chemistry of diesel fuel. In Chemistry of Diesel Fuels, 1-60. New York, N.Y.: Taylor and Francis.

[23] Tat, M. E., and J. H. Van Gerpen. 2003. Biodiesel blend detection with a fuel composition sensor. Applied Eng. in Agric. 19(2): 125-131.

[24] Tomasevic V. A. and Siler-Marinkovic S. S. "Methanolysis of used frying oil" Fuel Process Technol. 81: 1-6 (2003).

[25] Vicente G.; M. Martínez, J. Aracil, Optimisation of integrated biodiesel production. Part I. A study of the biodiesel purity and yield, Bioresour. Technol. 98 (2007). 1724–1733.

[26] Zawadzki, A.; Shrestha, D. S.; He, B. (2007). Biodiesel Blend Level Detection Using Ultraviolet Absorption Spectra. American Society of Agricultural and Biological Engineers, 50(4): 1349-1353.

Chapter 16

Conversion of Oil Palm Empty Fruit Bunch to Biofuels

Anli Geng

Additional information is available at the end of the chapter

http://dx.doi.org/10.5772/53043

1. Introduction

Crude palm oil production is reaching 48.99 million metric tonnes per year globally in 2011 and Southeast Asia is the main contributor, with Indonesia accounting for 48.79%, Malaysia 36.75%, and Thailand 2.96% (Palm Oil Refiners Association of Malaysia, 2011). Oil palm is a multi-purpose plantation and it is also an intensive producer of biomass. Accompanying the production of one kg of palm oil, approximately 4 kg of dry biomass are produced. One third of the oil palm biomass is oil palm empty fruit bunch (OPEFB) and the other two thirds are oil palm trunks and fronds [1-3].

Figure 1. Oil palm and oil palm empty fruit bunch.

The supply of oil palm biomass and its processing by-products are found to be seven times that of natural timber [4]. Besides producing oils and fats, there are continuous interests in using oil palm biomass as the source of renewable energy. Among the oil palm biomass, OPEFB is the most often investigated biomass for biofuel production. Traditionally, OPEFB is used for power and steam utilization in the palm oil mills, and is used for composting and soil mulch. Direct burning of OPEFB causes environmental problems due the incomplete combustion and the release of very fine particles of ash. The conversion of OPEFB to biofuels, such as syngas, ethanol, butanol, bio-oil, hydrogen and biogas etc., might be a good alternative and have less environmental footprint. The properties of OPEFB is listed in Table 1 [5].

	Literature values % (w/w)	Measured % (w/w)	Method
Components			
Cellulose	59.7	na	na
Hemicellulose	22.1	na	na
Lignin	18.1	na	na
Eelemental analysis			
Carbon	48.9	49.07	Combustion analysis
Hydrogen	6.3	6.48	
Nitrogen	0.7	0.7	
Sulphur	0.2	<0.10	
Oxygen	36.7	38.29	By difference
K	2.24	2.00	Spectrometry
K_2O	3.08–3.65	na	na
Proximate analysis			
Moisture	na	7.95	ASTM E871
Volatiles	75.7	83.86	ASTM E872
Ash	4.3	5.36	NREL LAP005
Fixed carbon	17	10.78	By difference
HHV (MJ/kg)	19.0	19.35	Bomb calorimeter
LHV (MJ/kg)	17.2	na	na

Notes: na - not available.

Table 1. Properties of oil palm empty fruit bunch

While all the OPEFB components can be converted to biofuels, such as bio-oil and syngas through thermo-chemical conversion, cellulose and hemicellulose can be hydrolysed to sugars and subsequently be fermented to biofuels such as ethanol, butanol, and biogas etc. Although many scientists around the world are developing technologies to generate biofuels from OPEFB, to-date, none of such technologies has been commercialized. This is largely due to the recalcitrance of the OPEFB and therefore the complexity of the conversion technologies making biofuels from OPEFB less competitive than the fossil-based fuels. Continual efforts in R&D are still necessary in order to bring such technology to commercialization. The aim of this paper is to review the progress and challenges of the OPEFB conversion technologies so as to help expedite the OPEFB conversion technology development.

2. Pretreatment

Similar to all other lignocellulosic biomass, OPEFB are composed of cellulose, hemicellulose and lignin. Among the three components, lignin has the most complex structure, making it recalcitrant to both chemical and biological conversion. Pretreatment of OPEFB is therefore necessary to open its structure and increase its digestibility and subsequently the degree of conversion. Pretreatment of OPEFB can be classified as biological pretreatment, physical pretreatment, chemical pretreatment, and physical-chemical pretreatment.

For biological pretreatment, oxidizing enzymes and white-rot fungi were used to degrade the lignin content in OPEFB. For example, enzymes such as lignin peroxidase (LiP) and manganese peroxidase (MnP) was used to pretreat OPEFB for fast pyrolysis and the bio-oil yield was improved from 20% to 30% [6]. Syafwina et al. used white-rot fungi to pretreat OPEFB and the saccharification efficiency was improved by 150% compared to that of the untreated OPEFB [7].

Among all the pretreatment methods, chemical pretreatment is most often reported for OPEFB. Two-stage dilute acid hydrolysis [8], alkali pretreatment [9], sequential dilute acid and alkali pretreatment [10], alkali and hydrogen peroxide pretreatment [11], sequential alkali and phosphoric acid pretreatment [10], aqueous ammonia [12], and solvent digestion [5] were used to increase the digestibility of OPEFB. Among all the chemical methods investigated, alkali pretreatment seemed to be the most effective. Umikalsom et al. autoclaved the milled OPEFB in the presence of 2% NaOH and 85% hydrolysis yield was obtained [13]. Han and his colleagues investigated NaOH pretreatment of OPEFB for bioethanol production [9]. The optimal conditions were found to be 127.64°C, 22.08 min, and 2.89 mol/L NaOH. With a cellulase loading of 50 FPU /g cellulose a total glucose conversion rate (TGCR) of 86.37% was obtained using the Changhae Ethanol Multi Explosion (CHEMEX) facility. The effectiveness of alkali pretreatment might be attributed to its capability in lignin degradation. Mission et al. investigated the alkali treatment followed H_2O_2 treatment and found that almost 100% lignin degradation was obtained when OPEFB was firstly treated with dilute NaOH and subsequently with H_2O_2 [11]. This confirmed the lignin degradation by NaOH and its enhancement by the addition of H_2O_2.

Besides alkali pretreatment, physical-chemical pretreatment such as ammonium fibre explosion (AFEX) [14] and superheated steam [15] were also shown to be effective in the increase of OPEFB digestibility. Hydrolysis efficiency of 90% and 66% were obtained, respectively.

3. Thermo-chemical conversion

Thermo-chemical conversion is one of the important routes to obtain fuels from lignocellulosic biomass. Thermo-chemical conversion of biomass involves heating the biomass materials in the absence of oxygen to produce a mixture of gas, liquid and solid. Such products can be used as fuels after further conversion or upgrading. Generally, thermo-chemical processes have lower reaction time required (a few seconds or minutes) and the superior ability to destroy most of the organic compounds. These mainly include biomass pyrolysis and biomass gasification. Recently, thermo-chemical pretreatment of biomass, such as torrefaction was introduced to upgrade biomass for more efficient biofuel production [16-17].

3.1. OPEFB pyrolysis

Pyrolysis is defined as the thermal degradation of the biomass materials in the absence of oxygen. It is normally conducted at moderate temperature (400 – 600°C) over a short period of retention time. Its products comprise of liquids (water, oil/tars), solids (charcoal) and gases (methane, hydrogen, carbon monoxide and carbon dioxide). The efficiency of pyrolysis and the amount of solid, liquid, and gaseous fractions formed largely depend on the process parameters such as pretreatment condition, temperature, retention time and type of reactors.

Misson et al investigated the effects of alkaline pretreatment using NaOH, $Ca(OH)_2$, in conjunction with H_2O_2, on the catalytic pyrolysis of OPEFB [11]. They proved that consecutive addition of NaOH and H_2O_2 decomposed almost 100% of OPEFB lignin compared to 44% for the $Ca(OH)_2$ and H_2O_2 system, while the exclusive use of NaOH and $Ca(OH)_2$ could not alter lignin composition much. In addition, the pretreated OPEFB was catalytically pyrolysed more efficiently than the untreated OPEFB samples under the same conditions.

Fast pyrolysis represents a potential route to upgrade the OPEFB waste to value-added fuels and renewable chemicals. For woody feedstock, temperatures around 400-600°C together with short vapour residence times (0.5-2 s) are used to obtain bio-oil yields of around 70%, along with char and gas yields of around 15% each. Sulaiman and Abdullah investigated fast pyrolysis of OPEFB using and bench top fluidized bed reactor with a nominal capacity of 150 g/L [18]. After extensive feeding trials, it was found that only particles between 250 and 355 ▪ m were easily fed. The maximum liquid and organics yields (55% total liquids) were obtained at 450°C. Higher temperature was more favourable for gas production and water content was almost constant in the range of temperature investigated. The maximum liquids yield and the minimum char yield were obtained at a residence time of 1.03 s. The pyrolysis liquids produced separated into two phases; a phase predominated by tarry organic compounds (60%) and an aqueous phase (40%). The phase separated liquid product would represent a challenging fuel for boilers and engines, due to the high viscosity of the organics phase and the high water content of the aqueous phase. These could be overcome by upgrading. However, the by-product, charcoal, has been commercialized for quite some time. It is worth noting that the first pilot bio-oil plant by Genting Bio-oil has already started operation in Malaysia [19].

3.2. OPEFB gasification

Gasification process is an extension of the pyrolysis process except that it is conducted at elevated temperature range of 800–1300 °C so that it is more favourable for gas production [20]. The gas stream is mainly composed of methane, hydrogen, carbon monoxide, and carbon dioxide. Biomass gasification offers several advantages, such as reduced CO_2 emissions, compact equipment requirements with a relatively small footprint, accurate combustion control, and high thermal efficiency. The main challenge in gasification is enabling the pyrolysis and gas reforming reactions to take place using the minimum amount of energy and gasifier design is therefore important [21].

Ogi et al. used an entrained-flow gasifier for OPEFB gasification at 900°C [22]. During gasification with H_2O alone, the carbon conversion rate was greater than 95% (C-equivalent), and hydrogen-rich gas with a composition suitable for liquid fuel synthesis ($[H_2]/[CO]$ = 1.8–3.9) was obtained. The gasification rate was improved to be greater than 99% when O_2 was added to H_2O; however, under these conditions, the gas composition was less suitable for liquid fuel synthesis due to the increase of CO_2 amount. Thermogravimetric (TG) analysis suggested that OPEFB decomposed easily, especially in the presence of H_2O and/or O_2, suggesting that OPEFB is an ideal candidate for biomass gasification. Lahijani and Zainal investigated OPEFB gasification in a pilot-scale air-blown fluidized bed reactor [23]. The effect of bed temperature (650–1050°C) on gasification performance was studied and the gasification results were compared to that of sawdust. Results showed that at 1050°C, OPEFB had almost equivalent gas yield and cold gas efficiency compared with saw dust, however, with low maximal heating values and higher carbon conversion. In addition, it was realized that agglomeration was the major issue in OPEFB gasification at high temperatures. This can be overcome by lowering the temperature to 770 ± 20 °C. Mohammed et al. studied OPEFB gasification in a bench scale fluidized-bed reactor for hydrogen-rich gas production [24]. The total gas yield was enhanced greatly with the increase of temperature and it reached the maximum value (~92 wt.%) at 1000 °C with big portions of H_2 (38.02 vol.%) and CO (36.36 vol.%). The feedstock particle size of 0.3–0.5 mm, was found to obtain a higher H_2 yield (33.93 vol.%), and higher LHV of gas product (15.26 MJ/m^3). The optimum equivalence ratio (ER) (0.25) was found to attain a higher H_2 yield (27.31 vol.%) at 850 °C. Due to the low efficiency of bench scale gasification unit the system needs to be scaling-up. The cost analysis for scale-up EFB gasification unit showed that the hydrogen supply cost is $2.11/kg OPEFB. Recently, a characterization and kinetic analysis was done by Mohammed et al. and it was found that a high content of volatiles (>82%) increased the reactivity of OPEFB, and more than 90% decomposed at 700 °C; however, a high content of moisture (>50%) and oxygen (>45%) resulted in a low calorific value [25]. The fuel characteristics of OPEFB are comparable to those of other biomasses and it can be considered a good candidate for gasification.

3.3. OPEFB torrefaction

Torrefaction is a thermal conversion method of biomass in the low temperature range of 200-300 °C. Biomass is pretreated to produce a high quality solid biofuel that can be used for combustion and gasification [16-17]. It is based on the removal of oxygen from biomass to produce a fuel with increased energy density. Different reaction conditions (temperature, inert gas, reaction time) and biomass resources lead to the differences in solid, liquid and gaseous products.

Uemura et al. [16] studied the effect of torrefaction on the basic characteristics of oil palm empty fruit bunches (EFB), mesocarp fibre and kernel shell as a potential source of solid fuel. It was found that mesocarp fibre and kernel shell exhibited excellent energy yield values higher than 95%, whereas OPEFB, on the other hand, exhibited a rather poor yield of 56%. Torrefaction can also be done in the presence of oxygen. Uemura and his colleagues [17] carried out OPEFB torrefaction in a fixed-bed tubular reactor in the presence of oxygen at varied oxygen concentration. The mass yield decreased with increasing temperature and oxygen concentration, but was unaffected by biomass particle size. The energy yield decreased with increasing oxygen concentrations, however, was still between 85% and 95%. It was found that the oxidative torrefaction process occurred in two successive steps or via two parallel reactions, where one reaction is ordinary torrefaction, and the other is oxidation.

3.4. Summary

The analysis of thermo-chemical conversion of OPEFB suggests that gasification is the most suitable thermo-chemical route for OPEFB conversion to biofuels. It has the highest carbon conversion (>90%) and biofuel yield. Due to the high viscosity and high water content of pyrolysis products, application of bio-oil as a biofuel is still very challenging. Compared to other oil palm residues, such as oil palm kernel, due to its high water content, OPEFB may not be a good candidate for solid fuels even after torrefaction pretreatment.

4. Bioconversion

Bioconversion of lignocellulosic biomass to fuels involves three major steps: 1) pretreatment- to effectively broken the biomass structure and release the biomass components i.e. cellulose, hemicellulose, and lignin, and therefore increase the digestibility of the biomass; 2) enzymatic hydrolysis – to hydrolyse cellulose and hemicellulose and produce fermentable sugar, such as glucose, xylose etc.; 3) fermentation – to convert the biomass hydrolysate sugars to the desired products. OPEFB was intensively investigated as a potential substrate for the production of biofuels, such as ethanol, butanol, and biogas etc. Among the biofuels produced through bioconversion of OPEFB, cellulosic ethanol is the most intensively studied.

Two stage dilute acid hydrolysis was applied for OPEFB bioconversion to ethanol, 135.94 g xylose/kg OPEFB and 62.70 g glucose/kg OPEFB were produced in the first stage and 2nd stage, respectively [8]. They were then fermented to ethanol using *Mucor indicus* and *Saccharomyces cerevisiae*, respectively, and the corresponding ethanol yields were 0.45 and 0.46 g ethanol/g sugar.

Alkali is the most often used pretreatment chemical for cellulosic ethanol production from OPEFB. Kassim et al. pretreated OPEFB using 1% NaOH followed by mild acid (0.7% H_2SO_4) hydrolysis and enzymatic saccharification [26]. A total of 16.4 g/L of glucose and 3.85 g/L of xylose were obtained during enzymatic saccharification. The OPEFB hydrolysate was fermented with *Saccharomyces cerevisiae* and an ethanol yield of 0.51 g/g yield was obtained, suggesting that OPEFB is a potential substrate for cellulosic ethanol production. Han and his colleagues investigated ethanol production through pilot scale alkali pretreatment and fermentation [9]. The best pretreatment condition was 127.64 °C, 22.08 min, and 2.89 mol/L

NaOH. Enzyme loading of 50 FPU/g cellulose resulted in 86.37% glucose conversion in their Changhae Ethanol Multi Explosion (CHEMEX) facility. An ethanol concentration of 48.54 g/L was obtained at 20% (w/v) pretreated biomass loading, along with simultaneous saccharification and fermentation (SSF) processes. This is so far the highest reported ethanol titre from OPEFB. Overall, 410.48 g of ethanol were produced from 3 kg of raw OPEFB in a single run, using the CHEMEX 50 L reactor.

Jung and his colleagues tried aqueous ammonia soaking for the pretreatment of OPEFB and its conversion to ethanol [12]. Pretreated OPEFB at 60°C, 12 h, and 21% (w/w) aqueous ammonia, showed 19.5% and 41.4% glucose yields after 96h enzymatic hydrolysis using 15 and 60 FPU of cellulase per gram of OPEFB, respectively. An ethanol concentration of 18.6 g/L and a productivity of 0.11 g/L/h were obtained with the ethanol yield of 0.33 g ethanol/ glucose.

Lau et al. successfully applied ammonia fibre expansion (AFEX) pretreatment for cellulosic ethanol production from OPEFB [14]. The sugar yield was close to 90% after enzyme formulation optimization. Post-AFEX size reduction is required to enhance the sugar yield possibly due to the high tensile strength (248 MPa) and toughness (2,000 MPa) of palm fibre compared to most cellulosic feedstock. Interestingly, the water extract from AFEX-pretreated OPEFB at 9% solids loading is highly fermentable and up to 65 g/L glucose can be fermented to ethanol within 24 h without the supplement of nutrients.

OPEFB was also used for butanol production. Noomtim and Cheirsilp (2011) studied butanol production from OPEFB using *Clostridium acetobutylicum* [27]. Again, the pretreatment by alkali was found to be the most suitable method to prepare OPEFB for enzymatic hydrolysis. 1.262 g/L ABE (acetone, butanol and ethanol) was obtained in RCM medium containing 20 g/L sugar obtained from cellulase hydrolysed OPEFB. Ibrahim et al also investigated OPEFB as the potential substrate for ABE production [28]. Higher ABE yield was obtained from treated OPEFB when compared to using a glucose-based medium using *Clostridium butyricum* EB6. A higher ABE level was obtained at pH 6.0 with a concentration of 3.47 g/L. The accumulated acid (5 to 13 g/L) had inhibitory effects on cell growth.

Nieves et al. investigated biogas production using OPEFB. OPEFB was pre-treated using NaOH and phosphoric acid [29]. When 8% NaOH (60 min) was used for the pretreatment, 100% improvement in the yield of methane production was observed and 97% of the theoretical value of methane production was achieved under such pretreatment condition. The results showed that the carbohydrate content of OPEFB could be efficiently converted to methane under the anaerobic digestion process. O-Thong et al. investigated the effect of pretreatment methods for improved biodegradability and biogas production of oil palm empty fruit bunches (EFB) and its co-digestion with palm oil mill effluent (POME) [30]. The maximum methane potential of OPEFB was 202 mL CH_4/g VS-added corresponding to 79.1 m^3 CH_4/ton OPEFB with 38% biodegradability. Co-digestion of treated OPEFB by NaOH presoaking and hydrothermal treatment with POME resulted in 98% improvement in methane yield comparing with co-digesting untreated OPEFB. The maximum methane production of co-digestion treated OPEFB with POME was 82.7 m^3 CH_4/ton of mixed treated OPEFB and POME (6.8:1), corresponding to methane yield of 392 mL CH_4/g VS-added. The study showed that there was a great potential to co-digestion treated OPEFB with POME for bioenergy production.

In summary, OPEFB has been frequently investigated as a substrate for biofuel production through bioconversion. Cellulosic ethanol production was most intensively investigated and the highest ethanol titre of 48.54 g/L was obtained through alkali pretreatment in a pilot scale reactor [9]. Although not much research has been done for ABE and biogas production, the few reports summarized in this paper suggest that OPEFB is also potential substrate for butanol and biogas production. Throughout the reports reviewed, alkali-based pretreatment methods, such as NaOH alone, NaOH followed by acid, and ammonium fibre expansion (AFEX) pretreatment are the most effective in enhancing OPEFB digestibility.

5. Conclusion

In conclusion, OPEFB is the most potential renewable resource for biofuel production in Southeast Asia. It can be converted to biofuels through thermo-chemical or biological conversion. Pretreatment of OPEFB is necessary for both routes of conversion and alkali pretreatment is the most effective. A summary of OPEFB conversion technology is shown in Fig. 2.

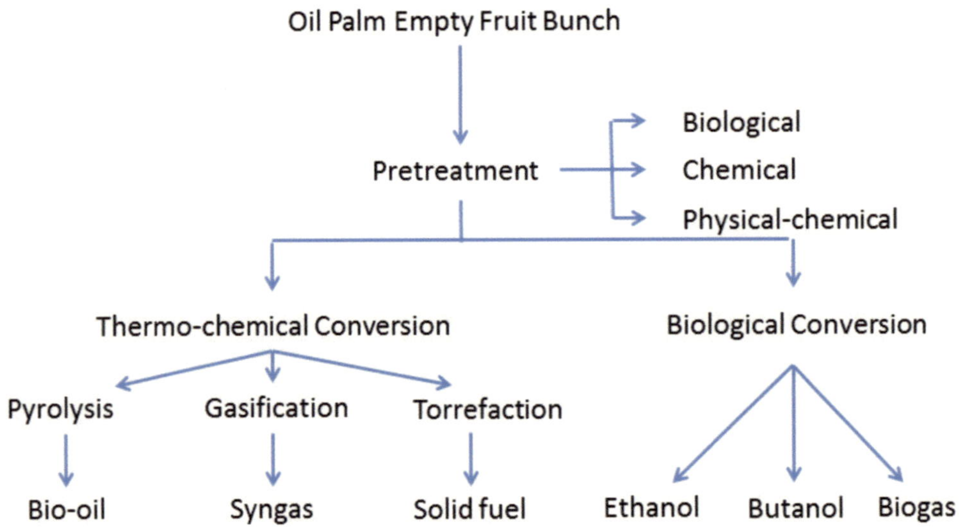

Figure 2. Biofuel production from OPEFB.

Among the studies on OPEFB thermo-chemical conversion, it seems that gasification is the most suitable approach to obtain bioenergy from OPEFB and has potential in commercialization. Pyrolysis, on the other hand, produced very complex bio-oil with high viscosity and water content, making it challenging for commercialization. However, charcoal from OPEFB pyrolysis can be a potential commercial product. Compared to other palm oil residues, such as oil palm kernel, OPEFB may not be a good candidate for solid fuel production, even after torrefaction pretreatment due to its high water content and low energy capacity.

Biological conversion of OPEFB is another route to obtain biofuels from OPEFB. Cellulosic ethanol production was most intensively studied and around 50 g/L titre was obtained with 20% (w/v) biomass loading through NaOH pretreatment. AFEX also showed potential in OPEFB pretreatment and a glucose yield of 90% was obtained with 9% biomass loading. The water extract of the AFEX pretreated OPEFB was highly fermentable. OPEFB also showed some promising preliminary results in ABE (acetone, butanol and ethanol) and biogas production; however, further investigation is necessary to enhance OPEFB conversion potentials in these areas.

For both thermo-chemical and biological conversion of OPEFB, pretreatment technology is the key for the process cost. Although alkali pretreatment is effective, scaling-up the process requires huge amount of acid to neutralize the base in the pretreatment solution. In addition, before alkali pretreatment, OPEFB should be milled to reduce its size, which is energy-consuming. Steam explosion is effective for a lot of lignocellulosic biomass, however not much research was found on its pretreatment of OPEFB. A cost-effective pretreatment is the key for the successful commercialization of OPEFB conversion technologies for biofuel production.

Acknowledgement

The. authors are grateful for the financial support to the research on cellulosic ethanol by the Science and Engineering Research Council of the Agency for Science, Technology and Research (A*STAR), Singapore.

Author details

Anli Geng

Address all correspondence to: gan2@np.edu.sg

School of Life Sciences and Chemical Technology, Ngee Ann Polytechnic, Singapore

References

[1] Husin M, Ramli R, Mokhtar A, Hassan WHW, Hassan K, Mamat R. Research and development of oil palm biomass utilization in wood-based industries. Palm Oil Development 2002; 36: 1-5.

[2] Wahid MB, Abdullah SNA, Henson IE. Oil palm achievements and potential. New Directions for a diverse planet. In: Proceedings of the fourth international crop science congress, 26 Sep - 1 Oct 2004. Brisbane, Australia.

[3] Yusoff S. Renewable energy from palm oil e innovation on effective utilization of waste. Journal of Cleaner Production 2006; 14: 87-93.

[4] Basiron Y, Chan KW. The oil palm and its sustainability. Journal of Oil Palm Research 2004; 16: 1-10.

[5] Abdullah1 N, Sulaiman1 F, Gerhauser H. Characterisation of oil palm empty fruit bunches for fuel application. Journal of Physical Science, 2011; 22: 1–24.

[6] Amin NAS, Ya'aini N, Misson M, Haron R, Mohamed M. Enzyme pretreatment empty palm fruit bunch for biofuel production. Journal of Applied Science 2010; 10: 1181-1186 .

[7] Syafwina, Honda Y, Watanabe T, Kuwahara M. Pre-treatment of oil palm empty fruit bunch by white-rot fungi for enzymatic saccharification. Wood Research 2002; 89: 19-20.

[8] Millati R, Wikandari R, Trihandayani ET, Cahyanto MN, Taherzadeh MJ, Niklasson C. Ethanol from oil palm empty fruit bunch via dilute-acid hydrolysis and fermentation by Mucor indicus and Saccharomyces cerevisiae. Agriculture Journal 2011; 6: 54-59.

[9] Han MH, Kim Y, Kim SW, Choi G-W. High efficiency bioethanol production from OPEFB using pilot pretreatment reactor. Journal of Chemical Technology and Biotechnology 2011; 86: 1527-1534.

[10] Kim S, Park JM, Seo J-W, Kim CH. Sequential acid-/alkali-pretreatment of empty palm fruit bunch fibre. Bioresource Technolog 2012; 109: 229-233.

[11] Misson M, Haron R, Kamaroddin MFA, Amin NAS. Pretreatment of empty palm fruit bunch for production of chemicals via catalytic pyrolysis. Bioresource Technology 2009; 100: 2867 – 2873.

[12] Jung YH, Kim IJ, Han J-I, Choi I-G, Kim KH. Aqueous ammonia pretreatment of oil palm empty fruit bunches for ethanol production. Bioresource Technology 2011; 102: 9806-9809.

[13] Umikalsom MdS, Ariff AB, Karim MIA. Saccharification of pretreated oil palm empty fruit bunch fibre using cellulase of Chaetomium globosum. Journal of Agriculture and Food Chemistry 1998; 46: 3359-3364.

[14] Lau MJ, Lau MW, Gunawan C, and Dale BE. Ammonia fibre expansion (AFEX) pretreatment, enzymatic hydrolysis, and fermentation on empty palm fruit Bunch fibre (EPFBF) for cellulosic ethanol production. Applied Biochemistry and Biotechnology 2010; 162: 1847-1857.

[15] Bahrin EK, Baharuddin AS, Ibrahim MF, Razak MNA, Sulaiman A, Abd-Aziz S, Hassan MA, Shirai Y, Nishida H. Physicochemical property changes and enzymatic hydrolysis enhancement of oil palm empty fruit bunches treated with superheated steam. BioResource 2012; 7: 1784-1801.

[16] Uemura Y, Omar WN, Tsutsui T, Bt Yusup S. Torrefaction of oil palm wastes. Fuel 2011; 90: 2585-2591.

[17] Uemura Y, Omar W, Othman NA, Yusup S, Tsutsui T. Torrefaction of oil palm EFB in the presence of oxygen. Fuel, 2011, doi 10.1016/j.fuel.2011.11.018.

[18] Sulaiman F, Abdullah N. Optimization conditions for maximising pyrolysis liquids of oil palm empty fruit bunches. Energy 2011; 36: 2352-2359.

[19] Genting Group Unveils Malaysia's First Commercially Produced Bio-oil using Breakthrough Technology. Report by Genting Group, Press Release; 21 August, 2005.

[20] Balat M, Kırtay E, Balat H. Main routes for the thermo-conversion of biomass into fuels and chemicals. Part 2. Gasification systems. Energy Conversion and Management 2009; 50: 3158–3168.

[21] Panwar NL, Kothari R, Tyagi VV. Thermo chemical conversion of biomass – Eco friendly energy routes. Renewable and Sustainable Energy Reviews 2012; 16: 1801-1816.

[22] Ogi T, Nakanishi M, Fukuda Y, Matsumoto K. Gasification of oil palm residues (empty fruit bunch) in an entrained-flow gasifier. Fuel, 2010. DOI: 10.1016/j.fuel. 2010.08.028.

[23] Lahijani P, zainal ZA. Gasification of palm empty fruit bunch in a bubbling fluidized bed: A performance and agglomeration study. Bioresource Technology 2011; 102: 2068-2076.

[24] Mohammed MAA, Salmiaton A, Wan Azlina WAKG, Mohammad Amran MS, Fakhru'l-Razi A. Air gasification of empty fruit bunch for hydrogen-rich gas production in a fluidized-bed reactor. Energy Conversion and Management 2011; 52: 1555-1561.

[25] Mohammed MAA, Salmiaton A, Wan Azlina WAKG, Mohamad Amran MS. Gasification of oil palm empty fruit bunches: A characterization and kinetic study. Bioresource Technology 2011; 110: 628-636.

[26] Kassim MA, Kheang LS, Bakar NA, Aziz AA, Som RM. Bioethanol production from enzymatically saccharified empty fruit bunches hydrolysate using Saccharomyces cerevisiae. Research Journal of Environmental Sciences 2011; 5: 573-586.

[27] Noomtim P, Cheirsilp B. Production of butanol from palm empty fruit bunches hydrolysate by Clostridium acetobutylicum. Energy Procedia 2011; 9: 140-146.

[28] Ibrahim MF, Aziz SA, Razak MNA, Phang LY, Hassan MA. Oil palm empty fruit bunch as alternative substrate for acetone-butanol-ethanol production by Clos-tridium butyricum EB6. Applied Biochemistry and Biotechnology 2012; 166: 1615-1625.

[29] Nieves DC, Karimi K, Horváth IS. Improvement of biogas production from oil palm empty fruit bunches (OPEFB). Industrial Crops and Products 2011; 34: 1097-1101.

[30] O-Thong S, Boe K, Angelidaki. Thermo[philic anaerobic co-digestion of oil palm empty fruit bunches with palm oil mill effluent for efficient biogas production. Applied Energy 2012; 93: 648-654.h

Coproducts of Biofuel Industries in Value-Added Biomaterials Uses: A Move Towards a Sustainable Bioeconomy

S. Vivekanandhan, N. Zarrinbakhsh, M. Misra and A. K. Mohanty

Additional information is available at the end of the chapter

http://dx.doi.org/10.5772/55382

1. Introduction

World population is expected to grow nearly 9 billion in 2040 and eventually increases the global energy demand by 30% compared to current conception [1]. The issues related to increasing trend of crude oil cost, depleting source of fossil fuels and emerging threat on greenhouse gas emissions are leading the global energy sector to undergo a fundamental transformation towards renewable energy sources [1-2]. As the result, a main focus is motivated on renewable energy technologies that are based on solar, wind and biofuels. In transportation point of view, biofuels receive extensive attention due to their versatility in storage and refilling. Both bioethanol and biodiesel come together as biofuel currently produced from renewable resources through two different pathways. In some countries like Brazil, biofuels are produced and marketed at competitive cost compared to petroleum-based fuels employing existing technology [3-4]. They also carry following advantages comparing to petro fuels; (i) create significantly less pollutants (SO_x and NO_x), which also mitigates CO_2 emission, (ii) biodegradable nature lead to the less environmental leak risk and (iii) provides better lubricant effect, which enhances the engine life [5]. In addition, these emerging biofuel technologies will be expected to create more economic benefits to agriculture sectors and new rural job opportunities. Moreover, biofuels are attractive options for future energy demand since they can be produced domestically by many countries while the respective retail and consumer infrastructure needs minimum modification; so does the existing engine and fueling technology [6].

However, biofuel foresees a challenging journey to benefit from its highest potentials and to guarantee a viable future. Primarily, it needs policy support and commercialization. At the same time, research and development is crucial to conquer the challenges and bring sustainability to biorefinery facilities [7]. Major motivation for biofuels usage arises from the execution of biofuel policies by many countries, which mandates the incorporation of bio-counterpart into traditional fuels. United Kingdom introduced the Renewable Transport Fuels Obligation (RTFO) and encouraged the oil suppliers to incorporate biofuel into transport fuel between 2.5 and 5% during 2008-2010. RTFO's ultimate aim is to increase this 5% upto 10% by 2020, which will reflect in the demand of minimum 5 million tonnes [8]. Renewable fuel blending mandates in Canada was implemented through Canadian Environmental Protection Act, which recommends 5% ethanol with gasoline (in 2010) and 2% biodiesel with diesel (in 2012) [9]. In South Africa, the National Biofuels Industrial Strategy was introduced by the government in 2007, which recommends the implementation of 2% biofuels into liquid road transport fuels by 2013 [10]. Currently India's ~80% crude oil demand is satisfied by foreign suppliers, which is projected to rise 90% in 2025. In order to reduce this foreign dependency, India has announced the target of ethanol blending with gasoline 20% by 2017 [11]. In biofuel production, China has clear production goals to meet emerging demand in near future. China's integrated biofuel polices (rural welfare, improved energy security, reduced fossil fuel dependence, and CO_2 emissions) aimed to meet ~ 15% of the total transportation fuel demand by 2020 [12]. In Malaysia, the National Biofuel Policy initially planned to proceed with 5% biodiesel blend with 95% petroleum diesel, which is similar to Europe's B5, which has been started from 2009. This will be implemented through short, medium and long term strategies aiming to reduce their petroleum imports [13]. In addition to that, many countries have already designed various incentive programs for the effective promotion of biofuel production including bioethanol and biodiesel. This implements 5-20% biofuel supplement into traditional fuels [14].

Such blending mandates of biofuels adopted by the E.U. and U.S. created a dispute of increased food prices. Besides, the contribution of corn bioethanol in addressing the global warming issues is very modest while having a small positive net energy balance; i.e. the energy return on investment (EROI) of corn bioethanol is low (=1.2-1.6) compared to oil (=9) [6]. The emerging challenges for 1st generation of biofuel industries that utilizes corn and soybean as a major feed stock for biofuel production motivated the search for non-food and more efficient energy feedstocks like jatropha, lignocellulosic biomass and algae. Among them, cellulosic matter will be the major feedstock for second generation biofuel, since it exhibits much higher yield per hectare in comparison with sugar or starch crops [6]. As a result, cellulosic biomass can potentially yield higher land fuel (135 GJ/ha) than corn kernel (85 GJ/ha) and soy (18 GJ/ha) [15]. Moreover, significantly higher carbon sequestration is another advantage of the use of cellulosic biomass in biofuel production compared to the first generation biofuel crops [6].

This biomass-biofuel conversion can be performed under three major classes and they are (i) conversion of renewable polysaccharides into sugar molecules and their effective fermentation into ethanol, (ii) syngas production and their bio/chemical conversion into alcohols and (iii) production of bio-oil though fast-pyrolysis and their upgrade into transportation fuels. Considering the lignocellulosic feedstock as the biofuel precursor, it is crucial to create the necessary infrastructure in many levels from biomass to biofuel production; agriculture–technology–policy. The new utilization of biomass would largely affect the agriculture sector and necessitates effective actions to ease the adaptation process. Biofuel production uses land which keeps it from food production and environmental preservation. Other issues might be considered; soil erosion may worsen by expanding the biomass production, reduction of environmental land affects biodiversity and more pesticides and fertilizers may be used. Thus, the sustainability of biofuel is not achieved solely by a positive net energy balance [6]. In spite of all raised issues, it is important to bear in mind that biofuel still offers its advantages even if it has a small contribution compared to fossil fuels. In this regard, the two main challenges in biomass production can be (a) developing crops with suitable physical and chemical traits for biofuel production and (b) increasing biomass yields (double or more) [7]. How to put these two different strands into an integrated production strategy is important and brings new research topics into the whole agriculture picture. The outcome of such productivity-enhancing innovations, research and development motivated by biofuel can be such that by 2050, the whole world population could be supplied enough diet while less cropland is used than today [6]. The respective biorefinery operation also needs improvements so that sugars can be produced from cellulosic biomass and fermented economically feasible and able to compete with production from corn and sugar. These include improvement in lignocellulosic pretreatment, reduction in enzyme (cellulase) cost, both cellulase production and ethanol fermentation by using modified microorganisms [7].

In general, the growth of biofuel industry consists of (i) increased production capacity and (ii) successful transformation of industrial technology from discrete batch method (small-scale) into continuous flow method (larger-scale) [16]. In most of the small scale manufacturing, the industries do not have the practice of collecting coproducts, thus they run with increased operating costs. Hence, larger-scale industries are keen in capturing their coproducts in order to reuse them in the production process, which results in the reduction of operating cost significantly. Thus, value-added processing may serve as a viable alternative that not only reduces the impact on the environment, but also generates additional revenue source for biofuel plants [17]. Sustainable bioeconomy road map that integrates renewable resources, biofuel production/ utilization and the value-addition to the respective coproducts is shown schematically in Figure 1. In recent years, biofuel coproducts have been utilised for the fabrication of various chemicals for diversified applications and used as the filler/reinforcement for polymer blends as well as composites. The emerging opportunities for the biofuel coproducts in biomaterials (polymers/ composites) applications make successive transformation of coproducts to renewable feedstock with economic benefits. Capitalizing this transformation enhances the economic viability and also the sustainability of biofuel industries. Thus, this chapter summarizes the various aspects in biomaterial applications of the biofuel coproducts and their role in sustainable bioeconomy.

2. Current status of coproducts from biofuel industries

2.1. The global biofuel industry status

The biofuel industry has been growing rapidly during recent years and continues to expand for the next decade. Such expansion is basically driven by renewable energy goals and different policy supports as for example use mandates, tax relief, fuel quality specifications and investment capacities in leading producing countries [18]. Based on the projection reported by

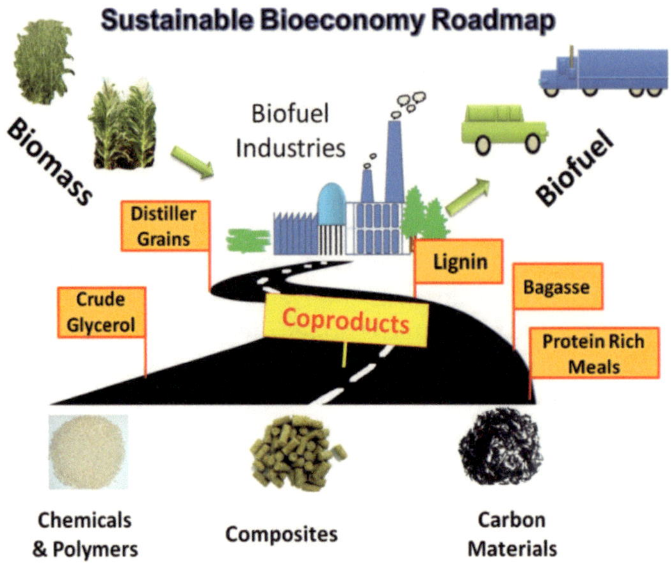

Figure 1. Sustainable bioeconomy roadmap.

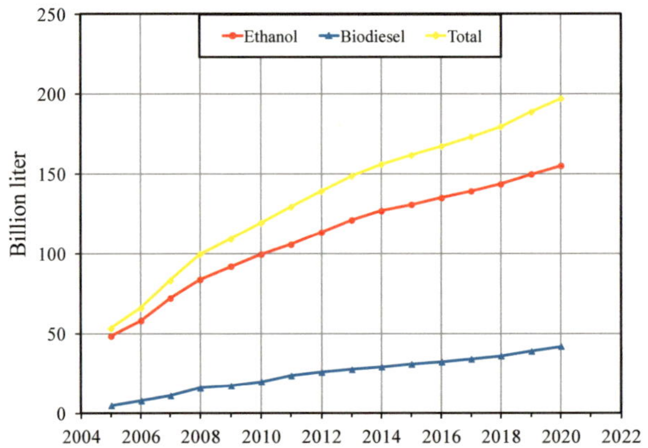

Figure 2. Development of global biofuel industry (drawn from data reported in [19, 21]).

the OECD (organization for economic co-operation and development) – FAO (food and agriculture organization of the united nations) Agricultural Outlook for the 2011-2020 period, the global bioethanol industry will be growing almost 68% from an average of 92 billion liter in the 2008-2010 period to 155 billion liter in 2020 (Figure 2) [19]. In this regard, coarse grains and sugarcane are going to remain the major precursors in bioethanol production and in 2020 they are expected to account for 78% of bioethanol feedstock (Figure 3) although this value was 81% on average during the 2008-2010 period [20]. The large scale production of cellulosic ethanol is still not achieved and under research and development. Therefore, it is expected to expand in the latter projected years reaching up to more than 4 billion liter in 2020. This is far less than the respective value for the first generation ethanol. The rest feedstocks include wheat (3.9%), molasse (3.2%), non-agricultural feedstock (2.6%), sugar beet (2%) and other (5.8%) [20].

Similar trend has been presented for biodiesel as illustrated in Figure 2 [21]. The growth in this industry in 2020 is projected to be almost 138% compared to 2008-2010 period on average; an increase from 17.6 to 41.9 billion liter. Vegetable oils will contribute more than 78% as the main feedstocks for biodiesel production. The application of non-food oils such as jatropha in biodiesel production still remains very less as compared with the contribution of vegetable oils such as soybean and palm oil. Feedstocks other than edible oils in biodiesel production include non-agriculture feeds (12.3%), biomass-based (6%) and jatropha oil (3.2%) (Figure 3) [22]. The huge impact of such expansion in biofuel industry on the respective coproducts is incontestable. Based on the feedstock share in the biofuel production by 2020, the major coproducts of different sectors of the biofuel industry can be listed as dried distillers' grains with solubles (DDGS) from dry mill corn ethanol, corn gluten meal and corn gluten feed from wet mill corn ethanol, bagasse from sugarcane ethanol, lignin from second generation lignocellulosic ethanol, and soy meal and crude glycerol from biodiesel. More focus on these coproducts will be dedicated in the following sections.

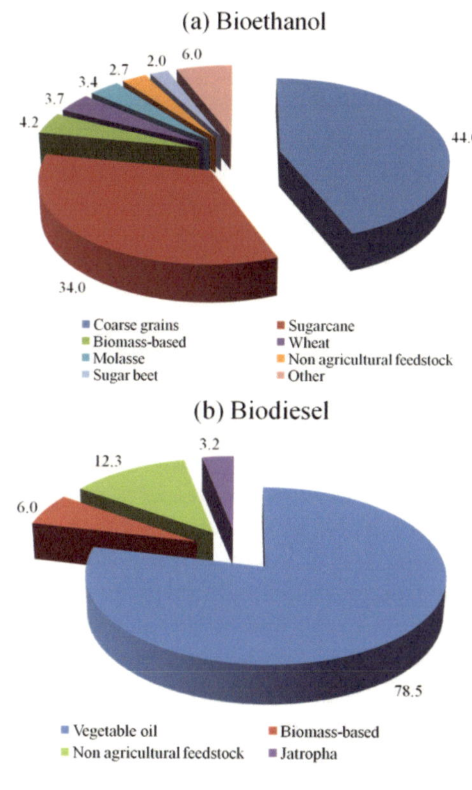

Figure 3. The global biofuel production by feedstocks contribution (%) in 2020 (redrawn from data reported in [20, 22]).

2.2. Biofuel coproducts

2.2.1. Corn bioethanol

Starch-based ethanol can be obtained from corn, wheat, barley, sorghum or any other starchy grain by fermentation. However, due to highly fermentable starch content, corn is the main feedstock for ethanol production by fermentation and accounts for 98% of all starch-based ethanol feedstocks [1]. Bioethanol from corn is produced in both dry mill and wet mill plants each of which producing specific coproducts as described below.

2.2.1.1. Dry mill

In a dry mill, ethanol is produced from corn after several steps including grinding, slurrying, cooking, liquefaction, saccharification, fermentation and distillation. Further steps are implemented to separate coproducts such as centrifugation, evaporation and drying. From the original corn mass before processing, approximately one third results in carbon dioxide during fermentation, one third is converted into ethanol and the residue are nonfermentable components in the form of different coproducts namely dried distillers' grains with solubles (DDGS), dried distillers' grains (DDG), wet distillers' grains (WDG) and condensed distillers' solubles (CDS). The coproducts are mainly dried and sold as dried distillers' grains with solubles (DDGS). This way, it is possible to store the coproduct for a longer time or ship it to far distances with less probability of fungi attack. A smaller part of the coproducts are shipped wet locally for immediate usage [23-24]. Distillers' grains have been traditionally using as animal feed due to its nutritious value as shown in Table 1 [25-31]. At the end of ethanol production process, when most of the grain's starch portion is fermented, there is an increase of 3 to 4 times in other components of the grain including protein, lipid and fibre over that contained in the unconverted whole grains [27].

Several attempts and studies have been published on distillers' grains application as animal feed in many different species such as dairy cattle [32], beef cattle [33], swine [34], broiler [35], laying hen [36], turkey [37], lamb [38], catfish [39], tilapia [40], trout [41] and prawn [42]. However, four major livestock species to which distillers' grains is practically fed are beef cattle, dairy cattle, swine and poultry [43-45]. Renewable Fuel Association (RFA) reports the distillers' grains consumption in 2009 in different species at approximately 39% for dairy cattle, 38% for beef cattle, 15% for swine, 7% for poultry and 1% for other species [24]. The important question here is whether the increasing supply of distillers' grains can be totally consumed by animals or the supply far exceeds its demand as feed. According to Hoffman and Baker [44] and Tokgoz et al. [46] the potential domestic and export use of distillers' grains in U.S. exceeds its production and the U.S. beef sector is the dominant user of distillers' grains. However, such opinions need precise consideration with respect to the fact that incorporation of distillers' grains within animal diets exhibits some limitations. Since distillers' grains are highly concentrated in terms of nutritious content, it should be included as a part of animal feed. In this regard, Canadian Food Inspection Agency (CFIA) has set out the policy for the maximum inclusion rates of distillers' grains in the feed of different species [47]. For example, the inclusion rates of distillers' grains in the diet of beef cattle and swine must not exceed 50% on a dry basis. This suggests that continuing the use of distillers' grains in animal diet in order to keep the track with its increasing supply from ethanol production can only come true if the number of consumer animals is also increasing. In other words, finding new value-added usages for distillers' grains within feed sector should also be considered in the future.

	DDGS [25-27]	CGM [28]	CGF [28]	SM [29]	CM [29-30]	JM [31]
Dry Matter (%)	88.8-91.1	90	87-90	NA	91.5	NA
Protein (% DM)	24.7-32.8	60	18-22	53.5-54.1	38.3	55.7-63.8
Fat (% DM)	11.0-16.3	2.5	2-5	1.4-2.3	3.6	0.8-1.5
Acid Detergent Fiber (ADF) (% DM)	12.4-15.2	5	13	7.2-10.2	17.5	5.6-7.0
Neutral Detergent Fiber (NDF) (% DM)	46.1-51.6	NA	35	9.6-13.8	21.5	8.1-9.1
Ash (% DM)	4.2-12.0	1.8	6.5-7.5	7.2-8.1	8.1-8.6	9.6-10.4

DDGS: dried distillers' grains with solubles, CGM: corn gluten meal, CGF: corn gluten feed, SM: soybean meal, CM: canola meal, JM: jatropha meal, NA: not available

Table 1. Composition of different biofuel coproducts

It is worth to note that the U.S. dried distillers' grains with solubles (DDGS) exports already doubled in 2009 compared to 2008 and U.S. has managed to increase its export of DDGS in 2010 by 60% compared to 2009 [48]. This may suggest that there is an excess of DDGS supply over its consumption in animal feed sector in the United States. Moreover, it should be carefully examined how the revenue from distillers' grains sale as feed, returning to the biofuel industry will economically help the ethanol industry. For corn biofuel industry to stay viable, the applications of its coproduct, distillers' grains, need to be expanded [23]. Consequently, the new outlets of distillers' grains may add value to it and create revenue for the corn ethanol biofuel. Such new usages can be value-added animal [23, 49] and human food [50], burning [51-52], extraction of zein [53], cellulose [54] and oil [55-56] from distillers' grains, and biobased filler for polymer composites, which is going to be discussed more later on.

2.2.1.2. Wet mill

Wet milling is a corn processing process in which the produced corn starch can be fermented into ethanol. Thus, ethanol is not the only product of a wet mill. In the beginning, the feedstock goes through a steeping step that soaks it in warm water containing small quantities of dissolved sulfur dioxide for almost 40 hours. This step facilitates the separation of the grain components. The different processes will be then applied such as grinding, screening, germ separation, oil refining, starch-gluten separation, drying, fermentation and syrup refining. Other products of a wet mill plant produced along with ethanol in these processes include starch, corn oil, high fructose corn syrup (HFCS) and glucose/dextrose. The coproducts of different steps are corn gluten feed, corn gluten meal, corn germ meal and corn steep liquor [28]. Corn gluten feed and corn gluten meal are the major feeds for livestock produced in a wet mill. As compared to the produced ethanol in a wet mill, corn gluten feed and corn gluten meal are produced almost as much as 70% and 17% of the mass of the produced ethanol, respectively [44]. The composition of the wet mill coproducts are presented in Table 1. Similar to distillers' grains, these coproducts account for a good source of nutritious components such as protein and fiber for feed applications. The value of these coproducts for animal feed has been realized for many years now and they are used as the feed for a wide range of animals including beef cattle [57], calf and lamb [58], dairy cattle [59], poultry [60], swine [61], pet [62] and fish [63]. Also, it has been reported that corn gluten meal can be used as the pre-emergence weed control (to control weeds before the weed seeds germinate) [64] and has been regulated by US Environmental Protection Agency (EPA) [65].

2.2.1.3. Corn ethanol coproducts production

Although there is a lack of comprehensive statistics on global production of corn bioethanol coproducts, a general insight can be obtained considering the production of these coproducts in the United States during last 10 years. This may be reasonable since U.S. is the largest ethanol producer globally with more than 50% contribution in 2007 and 2008 [8]. As reported by the Renewable Fuels Association (RFA) (Figure 4), the production of distillers' grains, corn gluten feed and corn gluten meal in the U.S. have shifted totally more than 10 times from 3.1 million tonnes in 2001 to 32.5 million tonnes in 2010 [24].

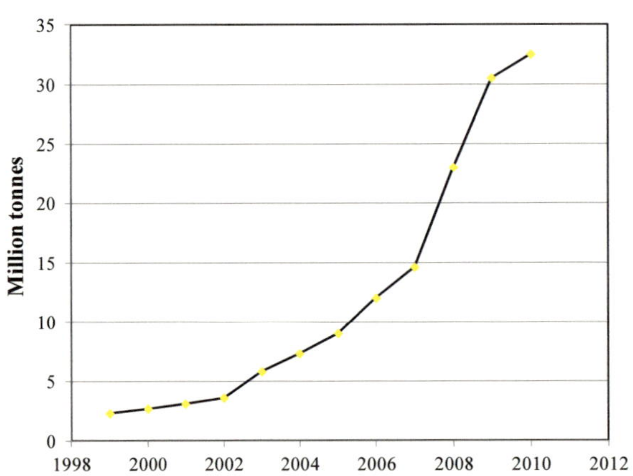

Figure 4. Production of ethanol biorefineries coproducts in US including distillers grains, corn gluten feed and corn gluten meal (drawn from data reported in [24]).

2.2.2. Sugarcane bioethanol

In a sugar mill, the crushed sugarcane is washed to go for juice extraction. The resulted juice can be used for sugar as well as ethanol production. Bagasse is the lignocellulosic coproduct after the sugarcane is crushed for juice extraction [66]. Approximately, it consists of 50%

cellulose, 25% hemicellulose and 25% lignin [67]. Bagasse has been widely used as the fuel for generating electricity. One metric ton of bagasse containing 50% moisture will produce heat equivalent to that from 0.333 tons of fuel oil [68]. This coproduct has been considered for such purpose in different countries such as Zimbabwe [69], Nicaragua [70] and Brazil [71]. Another large utilization of bagasse is in paper and pulp industry. This was patented in 1981 [72] and found huge application in many places such as India as early as 1990 [73]. The particleboard production is another industrial utilization of this biomass [74]. Bagasse has been also used in composting to a limited extent [68] and use of fungal strains on bagasse has been reported to produce compost with low pH and high soluble phosphorus [75]. Fermentation of bagasse using mold cultures was also considered to produce animal feed [76].

As a lignocellulosic material, bagasse has the potential of a feedstock for biofuel production either by gasification or hydrolysis method. In this context, still the biofuel production from bagasse via gasification has not been reported. However, as an alternative method competitive to the direct combustion of bagasse, gasification using a two-stage reactor has been proposed to be economically viable and more efficient [77]. Also, studies have been conducted in order to improve the bagasse gasification as far as retention and separation of alkali compound is concerned during the process. Considering the lignocellulosic ethanol production from bagasse, several investigations has been published on liquid hot water, steam pretreatment [78] and acid hydrolysis of it [79] as well as simultaneous saccharification and co-fermentation (SSCF) method [80]. As published in 2004, about 180 MT of dry sugarcane bagasse is produced globally and can be utilized to produce about 51 GL of bioethanol [81].

The expansion in production of the sugar-based ethanol is one of the key factors affecting the bagasse production. The sugarcane harvest of Brazil, the global leader of sugar-based ethanol, has shifted upward approximately 45% during recent five years from 425.4 MT in 2006-07 to 620.4 MT in 2010-11 (Figure 5) [82]. The ethanol production in Brazil generally shows a similar trend. To produce one liter of ethanol, 12.5 Kg of sugarcane is required. The weight of the produced bagasse is about 30% of the weight of sugarcane used for sugar or ethanol production [66]. Therefore, the bagasse production of Brazil in 2010-11 can be estimated as more than 180 MT, almost equal to global bagasse production before 2004.

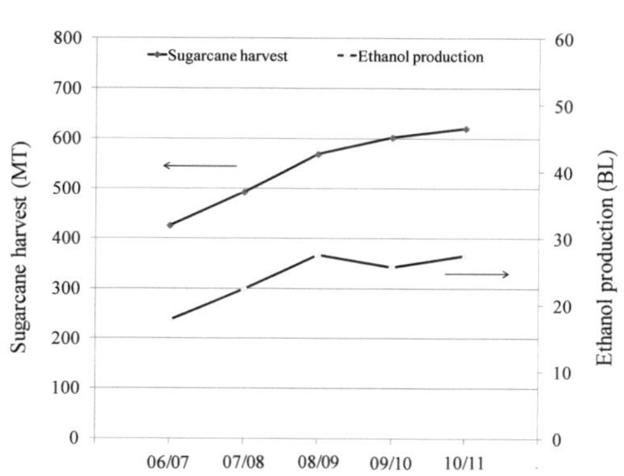

Figure 5. Brazilian sugarcane harvest and ethanol production (drawn from data reported in [82]).

2.2.3. *Lignocellulosic bioethanol*

In a second generation bioethanol plant, lignocellulosic polysaccharides (cellulose and hemicellulose) are broken down into monosaccharides (hexose and pentose) to be further fermented into ethanol. This includes a sequence of processes such as pretreatment, hydrolysis (enzymatically or chemically), fermentation and purification. At the end, the residual from the original lignocellulosic biomass is the coproduct mainly in the form of lignin [83]. The amount and quality of the produced lignin depends on the original lignocellulosic matter and the process. Typically, lignocelluloses contain 10–30 % lignin, which depends on various factors such as nature of biomasses, growth as well as isolation process [84]. As we know, lignin is a polymer that exists in the cell walls of plants, which is available in nature next to cellulose. The role of lignin in plant is to save them from compression, impact and bending. In addition to that, the major role of lignin extends to prevent the plant tissues from various kinds of naturally occurring microorganisms [85].

Chemically the polymeric structure of lignin is highly complicated and consists of three different monomer units (Figure 6) and they are called as p-coumaryl alcohol, coniferyl alcohol and sinapyl alcohol [86-87]. In addition to plant varieties, the lignin extraction process (known to be delignification) play a major role in the determination of local structure. Thus, the chemical structure of lignin extracted from plant biomass is never similar as exist in plant. There is a possibility for the alteration of monomer arrangement in the lignin structure. Hence, the lignin in plants, called as "natural lignin", is termed as "technical lignin" after isolation. These technical lignins can be further classified into three classes based on the domination of monomer units and they are [88]: (i) Softwood lignin: dominated with coniferyl monomer units, (ii) Hard wood lignin: combinations of equal quantities of guaiacyl and syringyl monomer units and (iii) Grass lignin: equally formulated with all three monomers of coniferyl, sinapyl and p-coumaryl.

The sources for lignin production can be divided into two major categories and they are (i) paper and (ii) bioethanol industries. They use different types of de-lignification process thus the lignin from paper industries and the lignin from bioethanol industries are not similar. Paper industries adopt Kraft, Sulphite, and Soda pulping processes to remove lignin from biomass, where as lignocellulosic ethanol industries, prefer to go with organosolv, steam explosion, dilute acid as well as ammonia fibre explosion for the removal of lignin [89]. Due to the tough food-versus-fuel concerns, non-food crops such as switchgrass, miscanthus and sugarcane bagasse become an effective feedstock for ethanol production and those related biofuel are named as second generation biofuels [90]. As the global demand for biofuel continues to grow, there will be emerging opportunities for lignocellulosic ethanol industries, which is expected to create a huge amount of lignin and it is predicted to be ~225 million tons by 2030 [85]. The challenge is to dispose them effectively that includes using them as a feedstock for energy products as well as for the fabrication of various chemicals and materials [91]. In common practice, lignin has been used for energy fuel, which is also the easiest way of disposal. In other hand, value-added uses of lignin can give economic return to the lignocellulosic ethanol industries and can improve their sustainability.

Only 2% of the lignin, which is produced from various sources, has been used as the feedstock for various chemicals including phenol, terephthalic acid, benzene, xylene, toluene, etc. [89] In addition to these, lignin has also been used in fertilizer, wood adhesives, surfactants, and also some kind of coloring agents [89]. Recently, lignin has been included as filler/reinforcing agent for blends and composites in both thermoplastic as well as thermoset platforms [85, 92-94]. In addition to that, lignin is found to be a suitable renewable carbon source for the synthesis of carbon materials [95]. Lignin has been widely exploited for the fabrication of activated carbon for various purposes including hydrogen storage, waste water removal and energy storage/conversion. Especially synthesizing nanostructured carbon materials from renewable resource-based lignin receives a great scientific interest due to the unique morphology as well as their physicochemical properties. As the global demand grows for the carbon fibre composites, there is a huge demand for the low cost carbon fibres. Lignin-based carbon fibres can substitute polyacrylonitrile (PAN)-based carbon fibre, hence the opportunity for lignin as a successful feedstock for carbon fibre is in near future [96]. This will be possible by understanding the basics of lignin chemistry and their application for fibre fabrication.

2.2.4. Biodiesel

2.2.4.1. Proteineous meal

The major biodiesel feedstocks are vegetable oils which are generally produced by crushing oil seeds, leaving significant quantity of proteineous meals as coproducts. The global consumption of proteineous meal in 2011/12 as reported by USDA Economic Research Services [97] is depicted in Figure 7 indicating soybean as the predominant crop producing proteineous meal with 67 % contribution. According to the FAOSTAT database [98], US, Brazil and Argentina were the global premiers of soybean production in 2010 with 35, 26 and 20% contribution, respectively. The next largest producer is China with almost 6% share in the global soybean production. In this context, biodiesel production of the US, as for example, has increased, during 2006-11 period, 340% from 250 to 1100 millions of gallons [99], which promoted the proteineous meal production. Soybean meal is traditionally used as a filler in animal feed including poultry, swine, beef, dairy, pet and other animals due to its concentrated protein content (Table 1).

Other examples of plant-based feedstock potentially suitable for oil extraction and biodiesel production can be listed as canola and linseed [100], palm [101], karanja [102] and jatropha [103]. Jatropha is a non-edible seed from a large shrub commonly found throughout most of the tropical and subtropical regions of the world. As shown previously in section 2.1, jatropha is projected to have a 3.2 % contribution in biodiesel production by 2020. The growing utilization of plant-based feedstocks other than soybean meal in biodiesel production also brings new streams of proteineous meal as coproducts.

2.2.4.2. Crude glycerol

Biodiesel is chemically known as methyl esters, which is produced through transestrification reaction by reacting a vegetable oil or animal fat with an alcohol under a strong base catalysis environment [104]. Along with biodiesel, such transestrification reaction produces significant

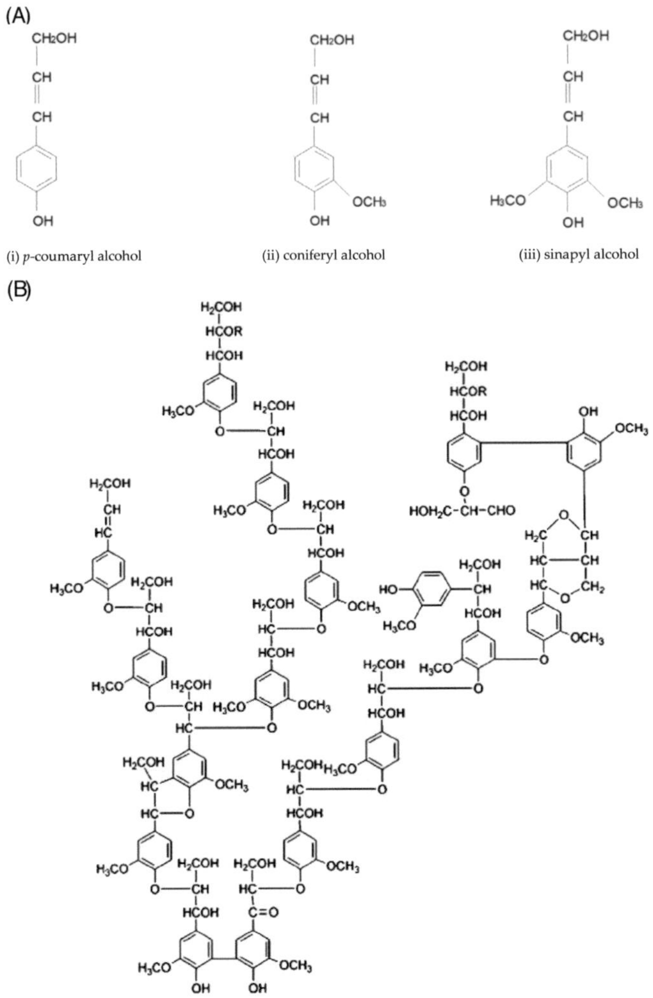

Figure 6. (A) Phenolic precursors that form the lignin and (B) Chemical structure of lignin (reprinted the figure with permission) [87].

quantity of glycerol (also called as glycerin), which is normally collected with other ingredients such as catalysis, water and unreacted alcohol and it is termed as crude glycerol [105]. Normally biodiesel industries utilize excess amount of methanol as required for the completion of reaction, which leaves unreacted methanol to the glycerol after the reaction. With every 3 gallons of biodiesel, 1kg of crude glycerol is produced and it shows very low value because of its impurity [106]. As the global biodiesel production increases exponentially, the resulting crude glycerol is extensively high and become issues due to their disposal or effective utilization. On the other hand, pure glycerol has found a wide range of applications that includes food, cosmetics, and drugs. In order to upgrade the crude glycerol to those high end applications, it should undergo various purification stages such as bleaching, deodoring, and ion exchange. Normally, this is not affordable nor economically feasible for most of the small/

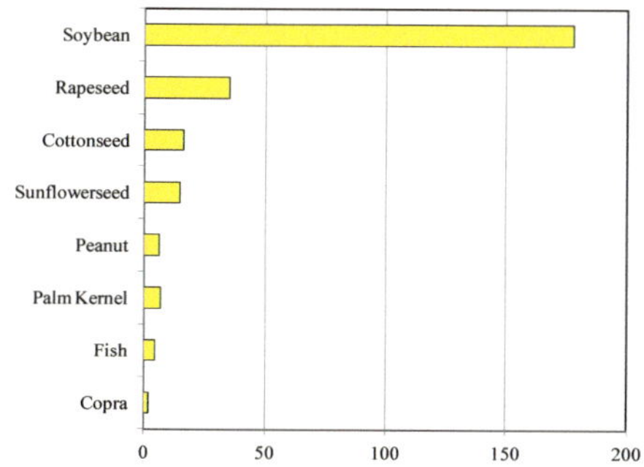

Figure 7. World protein meal consumption (million tonnes) in 2011 (drawn from data reported in [97]).

medium ranged industries. Hence, it is necessary to investigate the value-added uses of crude glycerol in various applications.

Prior to that, it is necessary to understand the relationship between the oil feedstock and the crude glycerol. Thompson and He [105] performed a research on the characterization of crude glycerol samples from various feedstocks. Their research shows that the compositions of different crude glycerol are highly varying with their feedstocks. This creates the challenge to adopt a universal protocol to fabricate value-added products from crude glycerol from various feedstocks. Crude glycerol has been used to produce various products including 1,3/1,2-propanediol, dihydroxyacetones, polyesters and hydrogen [107]. Mu et al. [108] reported the synthesis of 1,3-propanediol using crude glycerol produced during biodiesel preparation through fermentation process using Klebsiella pneumonia. They used the crude glycerol obtained during soybean oil-based biodiesel production employing alkali catalysis. They ultimately compared the product of 1,3-propanediol obtained from pure glycerol and found that they are similar to each other [108]. Soares et al. [109] demonstrated the generation of synthesis gas or syngas (hydrogen and carbon monoxide) from glycerol at very low temperature between 225-300 °C employing a Pd-based catalyst. Further it can be converted into fuels/chemicals by Fischer–Tropsch methanol synthesis. They also suggest this process for the effective utilization of various crude glycerol feedstocks for the fabrication of high value fuels/chemicals. Mothes et al. [110] reported the synthesis of poly (3-hydroxybutyrate), PHB, using crude glycerol (rape seed oil-based) as the feedstock via biotechnological process employing Paracoccus denitrificans and Cupriavidus necator microbes. They compared the properties of the synthesized PHBs from two different feedstocks and found that the properties are very similar. Zhou et al. [111] reviewed the chemo-selective oxidation of crude glycerol into various products such as glyceric acid, hydroxypyruvic acid and mesooxalic acid which can be used as precursor for various fine chemicals and polymeric materials. These reports indicate the emerging opportunities for crude glycerol for various applications including chemical, fuel and materials.

2.3. Sustainability through value addition

Due to the uncertainty in long term availability of fossil fuels and their continuous threat to environment through greenhouse gas emission, there is a drive across the globe towards the exploration of various biorefinery systems. Production of biofuel creates impact in utilization of biomass, replacement of possible extend of gasoline, reduction of greenhouse gas emission and the creation of significant amount of coproducts [112]. Biofuel production from biomass tends to strengthen the entire value chain (farming community-biofuel industries-consumers) and claims as the probable sustainable alternate for the conventional fossil fuel systems. Biomass is the sustainable feedstock for biofuel industries and biofuel provides ecological safety towards sustainable transportation, however the emerging concern is about the co/by-products. If they create challenging environmental issues in disposal, sustainability of this technology is challenged. Thus, value addition to these biofuel coproducts plays key role for the sustainability of biofuel technology in long term perspective [113]. With this understanding, the ultimate aim of biorefinery is focused to satisfy the conceptual "triple bottom line" of sustainability that includes (i) economic development; commercial value for biomasses, biofuels and coproducts, (ii) social development; appearance of new manufacturing sectors as well as creation of rural job opportunities, and (iii) environmental/resource sustainability; greenhouse gas reduction and eco-friendly green products [114]. In order to ensure the sustainability of biofuel technology it is essential to address various issues including (i) "food vs. fuel" due to the usage of edible resource for biofuel production, (ii) resource availability/management; effective utilization of land/water resources, (iii) environmental impact; issues related to land/water quality retention, conversion of grasslands/forests to agricultural fields and the efficient disposal/utilization of biofuel coproducts and (iv) validated measures: policy making and certification/ standardization [115]. Scale-up activities of biofuel production is essential due to the increasing demand for the substitution of fossil fuel. However, it is significantly controlled by various factors such as effective land usage for the larger biomass generation, water availability for agricultural forming, retention of soil quality, environmental impact of biofuel coproducts and labor market shift towards biorefinery [116-117]. The successive transformation of biofuel production from conventional to second generation effectively addresses the issues related to water consumption. Lignocellulosic ethanol industries utilise perennial crops such as miscanthus and switchgrass, which grows on marginal land and consume very less water. The challenge is towards biofuel coproducts. Failure of handling these large quantity coproducts will ultimately create serious environmental issues. These emerging technologies related to the effective utilization of biofuel coproducts that holds significant quantity of renewable content significantly substitute/ replace petroleum-based products and helps in reduction of greenhouse gas emission.

3. Value-added biomaterials from biofuel coproducts

3.1. Distillers' grains

3.1.1. As biofiller in producing polymeric biocomposites

The low cost of distillers' grains (DDG and DDGS) is a key incentive for researchers to utilize them as biobased fillers in manufacturing polymer composites. Also, addition of DDG(S) to the polymer matrix can result in improved stiffness as long as proper treatments and processing aspects are taken into account. It is less than ten years that DDGS-containing biocomposites has been reported in the literature. In this regard, the very first produced DDGS-containing composites exhibited low mechanical properties so that the utilization of DDG(S) as a biofiller in composite materials seemed to be not worthy in the beginning. However, recent works project a better future for DDG(S)-based biocomposites.

3.1.1.1. DDGS-polyolefin biocomposites

Polyolefins such as polypropylene (PP) and polyethylene (PE) were the first polymers compounded with DDGS up to 30 wt% [118]. The mentioned work included a comparison of four types of biofiller such as big blue stem (BBS) grass, soybean hull, pinewood and DDGS in terms of the mechanical performance of the their biocomposites with PP and PE. The composite processing was performed in a twin screw extruder and mechanical properties tests were conducted. Generally, the studied biofillers increased the flexural and tensile moduli. However, DDGS increased the modulus not significantly in comparison with neat PE and PP. Moreover, the tensile and flexural strengths decreased drastically as a result of compounding with DDGS. In general, the authors of [118] came to a conclusion that DDGS is not a suitable biofiller because of the decreased mechanical properties of the studied DDGS/polyolefin composites. In another work, the composite of high density PE with DDGS has been produced via extrusion and injection molding technique with DDGS content of 25 wt% [119]. The effect of maleated polyethylene (MAPE) as the compatibilizer was studied. Moreover, DDGS was solvent-treated to remove its oil and polar extractables. It has been reported that the application of the MAPP with solvent-treated DDGS resulted in better tensile and flexural properties of the composite compared to the respective properties of the neat HDPE.

3.1.1.2. DDGS-polyurethane biocomposites

This is another attempt of biomaterial application of DDGS. The overall idea of such work was to utilize a tough binder between rigid particles of DDGS to create an acceptable flexible material. Thus, polyurethane prepolymer (PUP) from castor oil was used as a binder with different compositions of DDGS and the biocomposites were prepared in a two-step process; PUP and DDGS were premixed in a micro-extruder and then compression molded to the shape of sheet. The mechanical and dynamic mechanical properties characterizations showed that the produced PUP/DDGS sheet was more flexible compared to the brittle DDGS material, thus polyurethane enhanced the properties of DDGS [120].

3.1.1.3. DDGS-phenolic resin biocomposites

Tatara et al. [121-122] bonded DDGS particles, from 0 to 90 wt%, with phenolic resin via compression molding process. Mechanical properties of the blends showed a reduction in modulus, tensile strength and elongation at yield by increasing the DDGS amount. However, the researchers believed that the cost saving resulted from the addition of low-cost DDGS filler in a reasonable quantity may offset the reduction in property performance; i.e. a cost-performance balance is achieved. In this context, inclusion of 25 and 50 wt% of DDGS maintained the mechanical strength to an acceptable value. In another work, the effect of DDGS particle size and content (25 and 50%) was studied when phenolic resin-based glue and wood glue were used to produce DDGS composites [123]. Overall, the DDGS composite with resin glue showed better mechanical properties and curing uniformity compared to wood glue/DDGS composites. Also, DDGS enhanced the flexural properties such as modulus and maximum stress. Also, composites of DDGS with smaller particle size (0.7 mm) had higher mechanical properties compared to those with higher particle size (0.34 mm).

3.1.1.4. DDGS biopolymer biocomposites

During recent years, biopolymer thermoplastics such as poly(lactic acid), PLA [124], poly(butylene succinate), PBS [125], polyhydroxy(butyrate-co-valerate)/poly(butylene succinate), PHBV/PBS, blend [126] and poly(butylene adipate-co-terephatalate), PBAT [127] have been utilized to produce DDGS composites. The influence of DDGS amount from 20 to 50 wt% as well as compatibilizer in PLA/DDGS composites were investigated [124]. Drastic decrease in tensile modulus and strength was observed by increasing the wt% of DDGS when no compatibilizer was used. On the other hand, after using isocyanate type of compatibilizer, a huge improvement in tensile modulus and strength was observed in the PLA/20% DDGS composite (Figure 8). In comparison with pure PLA, the compatibilized formulation showed higher modulus and almost equal strength. In another work, the thermal degradation of DDGS was studied with considerations for biocomposite processing and it was reported that water-washing of DDGS improved the thermal stability of DDGS to the extent that its thermal decomposition was highly prevented at typical temperatures of polymer melt processing. Such improvement in thermal stability of DDGS resulted in better strength and modulus of the PBS/DDGS biocomposite with 30 wt% DDGS [125]. The effect of compatibilizer was studied in a composite of 30 wt% DDGS and PHBV/PBS blend processed in a micro-extruder/micro-injection molding machine [126]. The DDGS used in this biocomposite had a water-washing step prior to compounding with bioplastics. Using a compatibilizer (isocyanate type), the interfacial adhesion was enhanced. The optimized biopolymer/DDGS composite exhibited improved tensile modulus compared to the biopolymer matrix while having almost equal strength. The influence of DDGS on the biodegradability properties of PBAT/DDGS biocomposites has been evaluated [127]. It was observed that PBAT/DDGS biocomposite was found to be more bio-susceptible material compared to virgin PBAT and was totally biodegraded. During the biodegradation experiment DDGS domains were preferentially attacked by microorganisms and influenced the biodegradability of the PBAT matrix. The produced biocomposite showed a degree of biodegradation similar to the biodegradation rate of natural materials such as DDGS and cellulose.

3.1.2. Other biomaterial applications

DDGS as a source of protein, fiber and fat has been used to isolate these components. Xu et al. [53] implemented a novel acidic method to extract zein from DDGS which is the main protein in corn and corn coproducts such as DDGS. The resulted zein has the potential for uses in fibers, films, binders and paints applications. Their method also isolated DDGS oil during protein extraction. Other researchers have also investigated the extraction of oil from DDGS

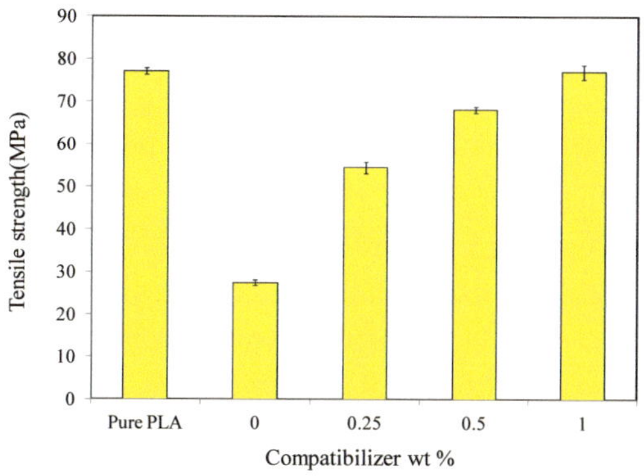

Figure 8. The effect of compatibilizer on tensile strength of the biocomposite PLA-20wt% DDGS (drawn from the data table with permission) [124].

[55, 128]. The obtained oil can be used as precursor for biodiesel production. It has been also tried to extract fiber from DDGS with physical methods such as sieving and elutriation [129] or extract cellulose chemically by sodium hydroxide solution [54]. The obtained cellulose had properties suitable for films and absorbents. The bioadhesive formulation obtained from DDGS is another biomaterial application of this coproduct [130-131]. This glue is prepared by reaction with an aqueous base solution. Urea can be also included with the base. The obtained bioadhesive is particularly useful as boxboard glue.

Recently, it has been tried to produce thermoplastics from DDGS by chemical methods such as acetylation [132] or cyanoethylation [133] with almost similar approaches. Hu et al. [133] were successful in producing highly flexible thermoplastic films from DDGS. The oil and zein protein of the DDGS were extracted first and the resultant underwent cyanoethylation using acrylonitrile. A compression molding machine was implemented to produce oil-and-zein-free DDGS films. It was observed that the produced films had much higher strength even at high elongations compared to films developed from other various biopolymers. Therefore, cyanoethylation could be a viable approach to develop bio-thermoplastics from biopolymers for applications such as packing films, extrudates and resins for composites. It has been also tried to find a novel use of dried distillers' grains (DDG) as a feedstock for bio-polyurethane preparation [134]. The procedure consist of, first, liquefaction of DDG in acidic conditions at atmospheric pressure and then reaction of hydroxyl-rich biopolyols in the liquefied DDG with methylene diphenyl diisocyanate (MDI) to form networks of cross-linked polyurethane. Thus, DDG-based bio-polyols were the precursor in this way to synthesize flexible and rigid polyurethane foams. The biodegradation tests showed that the degradation of these polyurethane foams in a 10-month period was about 12.6% most probably because of natural extracts such as proteins and fats in DDG and partially cross-linked or uncross-linked residue in the foam [134].

3.2. Bagasse

Bagasse has been used in biomaterial applications since a very long time ago. It has been used for interior panels and particleboard production. The first bagasse composition panel plant in Americas was built by Celotex, Louisiana, in 1920. Since then, more than 20 bagasse particleboard plants have been built throughout the world [135]. However, recent characterization of bagasse fiber for its chemical, physical and mechanical properties indicates that the potential of this coproduct of sugar and biofuel industries is much more than its applications in interior and structural components [136]. Bagasse is mostly burnt to generate energy for the sugar industry itself. Considering the fact that for such purpose almost 50% of the bagasse's production is enough [137], it is necessary to develop new uses for these fibers to implement the rest 50 % and reduce their environmental impact. Moreover, the burning of bagasse fiber is also a matter of concern as far as atmospheric pollution because of smoke, soot and ash is concerned [138]. Chemical composition as well as physical and mechanical properties of bagasse fiber are presented in Table 2 [136, 139-142]. Bagasse fiber consists of structural components such as cellulose and hemicellulose that can provide stiffness and rigidity to the polymers and enhance their engineering applications. Besides, bagasse exhibits a porous cellular structure with a hollow cavity called lumen existing in unit cell of the fibers. Therefore, the bulk density of bagasse fiber is lower than other natural fibers and bagasse fibers can act more effectively as thermal and acoustic insulators [142]. For example, the densities of kenaf and banana fibers are 749 kg m^{-3} [140] and 1350-1500 kg m^{-3} [139], respectively, which are higher than that of bagasse (344-492 kg m^{-3} [139-140]). Also, cellulosic fibers such as bagasse with low Young's modulus can act as useful crack growth inhibitors [143].

Properties			Ref.
	Cellulose (%)	32.0-55.2	[139]
	Hemicellulose (%)	16.8-32.0	[139]
Chemical composition	Lignin (%)	19.0-25.3	[139]
	Ash (%)	1.1-4.3	[139]
	Extracts (%)	0.7-3.5	[139]
	Density (kg m^{-3})	344-492	[139-140]
Physical properties	Diameter (μm)	394-490	[136, 140-141]
	Moisture content (%)	52.2	[136]
	Water absorption (%)	235	[136]
	Tensile strength (MPa)	29.6-96.2	[136, 141-142]
Mechanical properties	Tensile modulus (GPa)	4.5-6.4	[140, 142]
	Elongation at break (%)	4.0	[142]

Table 2. Chemical composition, physical and mechanical properties of bagasse fiber

3.2.1. Bagasse particleboards

Bagasse particleboards generally consist of bagasse fibers bound together with either an organic or inorganic binder. The organic binders are mostly a phenolic or polyester thermoset resins and the board is produced by compression molding under high pressure and temperature. Different inorganic binders such as cement, gypsum and calcined magnesite can also be used to produce bagasse boards [144-146]. Besides, binderless bagasse particleboards have been produced and patented in 1986 which can simplify the manufacturing process and reduce production cost since the blending operation and equipment are eliminated [147]. In this regards, different processing techniques such as hot pressing [148] and steam-injection pressing [74] have been conducted.

3.2.2. Bagasse-thermoset biocomposites

Phenolic resins are the major thermosets used for bagasse particleboards and several studies have been published on using resol [137], Novolac [149], lignophenolic [150] and other phenolic resins [143, 151] with bagasse fiber. Zárate et al. [137] studied the effect of fiber volume fraction on the density and flexural properties of composites from resol and several fibers including bagasse. They compared the efficiency criterion for mechanical performance, which relates the strength and stiffness with density, of the composites with those of typical structural materials including aluminum, magnesium, polyethylene and steel. Based on this comparison, it was concluded that the stiff composite materials produced from bagasse fibers and resol matrix are better compared to typical structural materials such as steel [137]. The effect of maleic anhydride (MA) treatment of bagasse fiber on properties of its composite with Novolac has been studied [149]. It has been reported that the composites with MA treated fibers had a hardness of 2–3 times more than that of the untreated bagasse composite and MA treatment reduced water and steam absorption of the fibers. Paiva and Frollini [150] extracted lignin from sugarcane bagasse by the organosolv process and used it as a partial substitute of phenol in resole phenolic matrices to produce bagasse-lignophenolic composite by compression molding. They observed improvement in the impact strength when sugarcane bagasse was used, but no improvement was found as a result of fiber treatments such as mercerization and esterification.

Unsaturated polyesters are another family of thermoset resins used for bagasse-based composite purposes. The effect of fiber size, its surface quality and the compression molding parameters on the flexural properties of composites from polyester and chopped bagasse fiber has been investigated. It was found that composites produced with bagasse particle size of less than 2 mm, and pre-treated for the extraction of sugar and alcohol exhibited the highest mechanical performance [138]. The effect of chemical treatments using sodium hydroxide and acrylic acid on the properties of bagasse-polyester composites has been studied. The treatments resulted in the better interaction between fiber and matrix as well as lower water absorption than composites with untreated fiber [142].

3.2.3. Bagasse-thermoplastic biocomposites

Bagasse has been used as reinforcing filler in different thermoplastic matrices such as poly(ethylene-co-vinyl acetate) (or EVA) [152-153], polyolefins [154-155] and starch-based biodegradable polyester [140-141, 156-157]. The effect of cultivar type and surface cleaning of the bagasse fiber on the tensile properties of the bagasse-EVA composites have been investigated [152]. The results suggested that blends of bagasse from various cultivars can be used for commercial applications of these composites. Also, the surface cleaning of the bagasse obtained from sugar mill was good enough to use the bagasse without further surface treatment. Another study on the impact behavior of the bagasse-EVA composites showed that the mechanical performance of this type of composites could be tailored by varying the bagasse volume fraction in order to reproduce the behavior of wood-based particleboards [153]. Luz et al. [154] explored the efficiency of two different processing methods, injection molding and compression molding, to produce bagasse-polypropylene (PP) composite. They found that the injection molding under vacuum process was more efficient and created homogeneous distribution of fibers without blisters. It was observed that bagasse incorporation into PP improved the flexural modulus. High density polyethylene (HDPE) was used as the matrix for incorporation of cellulose obtained from bagasse [155]. It has been reported that modification of bagasse cellulose with zirconium oxychloride helped in improving the tensile strength of the biocomposite.

Bagasse fibers have been used to produce biocomposites from bagasse and biodegradable corn starch-based polyester which is reported as a blend of starch and polycaprolactone (PCL). The effects of volume fraction and fiber length were investigated and an optimum value for both factor were reported beyond which the decrease in mechanical performance was observed [140]. Also, it was reported that after alkali treatment of the bagasse fibers the improvement in fibre–matrix adhesion occurred that resulted in enhancement of mechanical properties [141]. Moreover, incorporation of bagasse fiber into the polyester matrix improved tensile as well as impact strength. Acetylated starch has been reinforced with bagasse fiber [156-157]. The matrix in that case was a blend of starch, PCL and glycerol. It was observed that incorporation of alkali-treated bagasse fiber up to 15 wt% increased the tensile strength while it decreased when bagasse content was more than this value. Also, the water absorption of the composite was improved as the bagasse content increased due to hydrophobic nature of bagasse compared to acetylated starch.

3.2.4. Other biomaterial applications

It has been also tried to convert bagasse fibers into a thermo-formable material through esterification [158-159]. This has been done without any solvent using succinic anhydride followed by hot-pressing to produce the test samples. The thermoplasticization of the esterified fibers was proven to occur by scanning electron microscopy. By studying the effect of pressing parameters, it has been found that that de-esterification and hemicellulose degradation could occur on certain pressing conditions. It has been claimed that the mechanical properties of the produced composite could be superior to the standard properties of conventional high density wood particleboards.

3.3. Lignin

Biomaterials applications of lignin for the fabrication of polymers, blends and their reinforced composites are highly motivated due to the following reasons (i) abundance/occurrence in nature, (ii) phenolic chemical structure and the possibility of chemical modification and (iii) eco friendliness and reduced carbon footprint [160]. The challenge to use lignin as the materials feedstock is their complicated chemical structure as well as molecular weight, which are highly dependent on the lignin extracting process and also their sources [161]. The biomaterials application of lignin is vast and it has been blended/ reinforced with a wide range of polymeric systems such as thermoplastics, thermosets and elastomers as a renewable low cost filler [162]. Utilization of lignin as the raw material for the fabrication carbonaceous materials such as activated carbon and carbon fibres has a long history, however the creation of nanostructured materials such as carbon nanoparticles/nanofibres are quite new, which has a huge commercial potential [95, 163]. The emerging opportunities for lignin in these areas are summarized in this section.

3.3.1. Polymeric blends and composites from lignin

3.3.1.1. Lignin in thermoplastics

In polymers, lignins have been used as low-cost fillers aiming to retain their mechanical properties. Nitz et al. [164] reported the influence of various types of lignin reinforcement with the thermoplastics on their mechanical properties. Their results indicate that they are able to incorporate ~40 wt% lignin in to polyamide 11 (PA11), polyester (Ecoflex®) and polyesteramide (BAK®) systems without impairing their mechanical properties [164]. Generally lignin shows high cross-linking/intramolecular interactions, which limits their application in solid material systems. This can be overcome through polymer blending; however, achieving miscibility is very essential to develop a material system with superior properties [165]. This is possible in lignin-based blends by manipulating the chemistry of hydrogen bonding between the OH groups and interacting sites of polymers, either polar or semi polar [165]. Moreover, the hydrogen bond with a polymer varies with lignin to lignin since the monomer combinations of the lignins are unique [166].

Lignin–thermoplastic blends can be classified into two categories and they are (i) lignin – petro-based polymer blends and (ii) lignin–renewable resources based polymer blends. Blending the lignin with polyethylene and polypropylene is well known [166-169]. Alexy et al. [166] reported the effect of lignin concentration in the fabrication of polymeric blend with PP and PE. They measured the tensile strength as the measure of mechanical properties over the various lignin compositions. For both the polymer systems they identified that the mechanical properties decrease with increasing lignin content [166]. In addition to mechanical properties, Canetti et al. [169] and Mikulášová et al. [170] reported the fabrication of lignin/PP blends and investigated their thermal and biodegradable properties respectively. Poly(vinyl chloride) (PVC) is the next popular thermoplastic, which has been produced globally and exhibits a wide range of applications [171]. Raghi and coworkers [171] reported the fabrication of lignin/PVC blend and studied their mechanical/weathering properties. Their research investigation confirmed that the addition of lignin to PVC enhanced their tensile strength and not influenced their weathering behavior. Banu et al. [172] reported the fabrication of PVC/lignin blends and investigated the effect of plasticizer in their formulations. They concluded that the specific thermal and mechanical properties are feasible in some formulations with the addition of plasticizer. In addition to that, lignin/ poly(vinyl alcohol) (PVA) and lignin/ poly(ethylene oxide) (PEO) blend systems with various types of lignins are also investigated for the effective electrospinning performance [173-176]. Sahoo et al., [92] reported the fabrication of polybutylene succinate (PBS) reinforced with renewable resource-based lignin employing a melt extrusion process. They found that lignin reinforcement in PBS enhances their properties synergistically and also achieved the incorporation of high fraction of lignin of about 65%. In addition to that, they also reported the fabrication of PBS-based composite materials with the hybrid reinforcement of lignin and other natural fibre [93]. They found that the hybrid reinforcement is more beneficial over individual reinforcement for the better flexural strength.

The research on lignin-based polymer blends with renewable resource-based biopolymer is very limited. Only few publications are available in this content. Camargo et al. [177] reported the melt processing of poly(3-hydroxybutyrate-co-hydroxyvalerate) (PHBV) with lignin, in which they used the lignin isolated from sugarcane bagasse. They found that the addition of lignin to PHBV caused a reduction in their mechanical properties, which is due to the zero integration of lignin and PHBV [177]. Mousavioun et al. [178] performed the processing of poly(hydroxybutyrate) (PHB)-soda lignin blend and studied its thermal behavior. They found that the addition of soda lignin formed the miscible blend and improved their overall thermal stability. However, they have not reported their mechanical properties [178]. Vengal et al. [179] investigated the blending effect of lignin with starch and gelatin for the fabrication of biodegradable polymeric films. They found that the addition of lignin into starch can create better film with the composition of 90:10 (lignin: starch) and further increment of lignin content decreases their properties. Casetta et al. [180] fabricated the PLA and lignin blend and investigated their flame retardant behavior. They observed that the addition of lignin to PLA enhanced their flame retardant property comported to virgin PLA.

3.3.1.2. Lignin in thermosets

The lignins of different resources have been incorporated into various thermoset resins. Amorphous/heterogeneous nature and the complex structural composition of lignin result their behavior as either like a filler or like a reactive macromonomer in epoxy resin systems. Mansouri et al. [181] characterized the alkaline lignin and suggested their possible use for phenol-formaldehyde resin due to the availability of huge OH groups. Peng et al. [182] reported the fabrication of phenol-formaldehyde thermoset resin with lignin fillers and investigated their chemo-rheological properties. They found that the curing rate of the resin system decreased with increasing lignin content. Guigo et al. [183] fabricated the poly(furfuryl alcohol)/lignin composite resins and reported their monophase behavior. This indicates the reactive monomer behavior of lignin in this thermoset system. Thielemans et al. [184] investigated the effect of kraft lignin on unsaturated thermosetting resin, which was a mixture of epoxidized soybean oil and styrene, for the fabrication of natural fibre reinforced thermoset composites. They found the complete solubility of lignin into the resin system and their result on natural fibre composites indicateed the compatibilizing effect of lignin. Nonaka et al. [185] reported the fabrication of a new resin system by aqueous mixing of alkaline kraft lignin with polyethylene glycol diglycidyl ether (PEGDGE), and a curing reagent. They identified the complete compatibility between lignin and PEGDGE though the studies on dynamic mechanical analysis.

3.3.1.3. Ligninin rubber blends

Although the history of lignin/rubber blend started in 1949, a very little work has been performed till date [186]. The role of lignin in rubber is identified as reinforcing filler and stabilizer or antioxidant. Kumaran et al. [186] performed an extensive research on the utilization of lignins in rubber compounding and identified the improvement of many properties. They reported that the addition of lignin into rubber improved their tear, abrasion and flexural crack resistances. Košíková et al. [167] investigated the reinforcement effect of sulfur-free lignin with styrene butadiene rubber (SBR). They identified that the lignin blending with SBR influenced their vulcanizing behavior and enhanced the various physicomechanical properties significantly. In addition to that, Wang et al. [187] investigated the fabrication of latex/modified lignin blend and identified their effective water barrier properties. Processing condition of the lignin/rubber is critical for the achievement of better properties. Tibenham et al. [188] reported the hot-milling of lignin/rubber precursors with hexamethylenetetramine, which yields a vulcanizate. They also found that the modulus, tensile strength, and hardness properties were in the same order as the rubber reinforced with carbon blacks.

3.3.1.4. Lignin in polyurethane

Polyurethanes are made of diisocyanate and polyol precursors, which have been used for the highly diversified applications. Traditionally, they were made from petroleum-based synthetic polyols and nowadays soy-based polyols are also widely used as the renewable feedstock. Nakamura and his co-researchers [189] investigated the lignin-based polyurethane (PU) films using polyethylene glycol (PEG) and diphenylmethane diisocyanate (MDI). They reported the thermal behavior of new polyurethane system, which indicates that the addition of lignin to PEG enhances their T_g proportionally. The combination of lignin and PEG for the formation of polyurethane resulted in various types of microstructure such as soft and flexible and hard. Their mechanical properties were highly dependent on their distribution as well as cross-linking ability between lignin-PEG-MDI segments [190]. Sarkar et al. [191] reported the synthesis of lignin–hydroxyl terminated polybutadiene (HTPB) co-polyurethanes using toluene diisocyanate as initiator. Their characterizations showed the better properties up to 3% lignin incorporation and further increment of lignin caused the reduction in their properties [191]. Saraf et al. [192-194] made an extensive research on various aspects of lignin-based polyurethane and suggested their suitable formulations for the enhanced performance. In addition to that, various types of lignin also investigated for the fabrication of polyurethane systems [195-196]. Thring et al. [195] reported the fabrication of polyurethanes from Alcell®. They found that the increasing lignin content decreases the degree of swelling and cross-linking and causes the formation of brittle and hard structures. Yoshida et al. [196] utilized the kraft lignin for the fabrication of polyurethanes. They reported that the increasing lignin content increases the cross-link density and generally causes a hard and brittle nature. They also fabricated the polyurethane from various kraft lignins with different molecular weight and found that the cross-link density has increased with increasing molecular weight [197]. These studies conclude that the higher loadings of lignin in polyurethane caused the formation of rigid structure due to higher cross-link density and resulted in poor mechanical properties. This can be overcome by employing suitable chemistry in controlling the order of cross-linking.

3.3.2. Lignin in adhesives

Phenolic structure of lignin offers possible substitution with phenol-formaldehyde (PF) resin, which exhibits a wide range of applications as adhesives. Lignin substitute in phenol-formaldehyde (PF) formulation can vary from 30 to 50%, which exhibits similar or better performance compared to virgin PF resin. Haars et al. [198] reported the fabrication of room-temperature curing adhesives using lignin and phenoloxidases as precursor chemicals. They reported the possible use of this new bioadhesive as thermosetting glue. They also indentified the increment of water resistance during the usage in particleboard production. Mansouri et al. [199] demonstrated the fabrication of lignin adhesives without formaldehyde for wood panel. Their synthesized lignin adhesives showed better internal bond strength, which also passed required international standard specifications. They found that the newly bioadhesive from lignin exhibits many properties comparable to formaldehyde-based commercial adhesives. Schneider et al. [200] patented the new technology for the fabrication of new kind of adhesives using furfuryl alcohol and lignin employing zinc chloride-based catalyst. Lignin isolated from bagasse was also experimented for the fabrication of biobased cost effective adhesives [201-202]. The obtained adhesives were used for the purpose of particleboard and wood adhesives.

3.3.3. Lignin based carbon nanostructures

Recently, carbonaceous nanomaterials that include carbon nanotubes, carbon nanofibres, graphene/graphite nanosheets and also particulate carbon nanostructures have received an extensive importance due to their possible commercial values in diversified areas like polymeric composites, sensors, energy storage/ conversion, catalysis, filters and biology [203-208]. Traditionally, carbonaceous materials were prepared from petroleum-based precursors (liquid/gaseous hydrocarbons and carbon rich polymers such as polyacrylonitrile-PAN). As the global demand for carbon materials (nano/micro) grows continually and also the conventional sources are finite there is a need to investigate for the alternate carbon source. Thus, renewable resource-based biomaterials such as seed, oil, dried fibres as well as stem have been explored for the development of various carbon materials [209-212]. The challenge in using plant-based materials as carbon feedstock is to control the carbonizing process as well as the usage of suitable catalysis in order to achieve nanostructured materials. In addition to the larger availability of biobased feedstocks for the carbonaceous materials, it also provides eco-friendliness with the reduced carbon footprint. The biofeedstock exhibits a diversified morphology with the various combination of chemical structures, which can result in the formation of varieties of carbon nanostructures. Lignin has been widely used for the fabrication of activated carbon, however synthesizing carbon nanostructures such as particles/fibres are very new, thus next section summaries the effective uses of lignin as precursor for the fabrication of carbon nanostructures and their emerging applications [213-214].

3.3.3.1. Carbon nanoparticles

Lignin can be used as an efficient precursor in synthesizing carbonaceous nanomaterials with different morphology not only due to their carbon rich phenolic structure, but also for their capable chemical modification. Synthesizing the carbon nanoparticles with different morphology is possible by adopting various order of chemical modification as well as the processing conditions. The challenge is to inhibit the nucleation of carbon structures during the carbonization process to avoid larger particles, which normally occurs at elevated temperatures. Chemical modification can result in the formation of cross-linked structure, which normally alters the carbonization mechanism and can cause the formation of carbon nanostructures with different morphology. Babel et al. [215] reported the synthesis of KOH activated lignin-based carbon nanoparticles and their effective hydrogen storage capability. Recently, Gonugunta et al. reported the fabrication of carbon nanoparticels from lignin by adopting freeze drying process [216-217].

3.3.3.2. Lignin-based carbon nanofibres

One dimensional (1D) nanostructures such as fibrous materials receives recent attention due to their unique physicochemical properties. A wide range of fabrication techniques have been used for the fabrication of fibrous nanomaterials, among them electrospinning has been found to be an efficient technique for the fabrication of various types of fibre nanostructures using polymeric solutions as precursors [218]. The fabrication of carbon nanofibres from lignin through electrospinning has three steps and they are (i) electrospinning of lignin fibres, (ii) thermal stabilization of lignin fibres and (iii) carbonization of thermo-stabilized lignin fibres [219]. Normally, lignin exhibits poor viscoelastic properties, which creates a lot of challenges during the electrospinning process. This can be overcome by blending the lignin with other kind of synthetic polymers such poly(ethylene oxide) [220]. Figure 9 shows the electrospun lignin fibre as the precursor for the fabrication of carbon nanofibres. Dallmeyer et al. [221] investigated seven different technical lignins (isolated lignin) for the fabrication of fibrous network. None of the lignins were able to be spun into fibres without a binding polymer such as PEO. In addition to PEO, utilization of polyacrylonitrile (PAN) as binding polymer was reported by Seo et al. for the fabrication of lignin-based carbon fibres [222]. The physicochemical properties and the morphology of lignin-based carbon nanofibres can be varied by manipulating the experimental parameters. Lallave et al. [219] reported the fabrication of various types of (filled and hollow) carbon nanofibers from Alcell lignins by coaxial electrospinning. Uniqueness of their process is the successful electrospinning of lignin without binder

Figure 9. Electrospun lignin fibres (reprinted the figure with permission) [219].

polymer. Recently, Spender et al. [223] reported the rapid freezing process for the fabrication of lignin fibres with nano dimension.

3.4. Proteineous meals

As discussed in the previous sections, the major proteineous meal coproducts of biofuel industry are corn gluten meal from wet mill bioethanol and soybean meal from biodiesel industries. Different types of soy protein including soy flour (48% protein), soy protein concentrate (64% protein) and soy protein isolate (92% protein) can be extracted from soybean meal after oil extraction of soybean powder with hexane [224]. Similarly, corn gluten meal can be used to extract zein, the major protein in corn [225]. These proteins can be plasticized to produce films and formable thermoplastics. The biomaterial application of the these proteins has been investigated and reviewed extensively [225-226]. Recently, the biomaterial application of the meals themselves has attracted attentions and been studied in the form of plasticized meals as well as reinforcing fillers used in polymeric biocomposites.

3.4.1. Corn gluten meal

Corn gluten meal (CGM) is much cheaper than zein protein, thus creating more attraction compared to zein in producing thermoplastic materials. In this context, several plasticizers have been tried by many researchers for plasticization of CGM. Lawton and coworkers [227] studied the effect plasticizers such as glycerol, triethylene glycol (TEG), dibutyl tartrate, and octanoic acid on melt processing and tensile properties of CGM. In another work, di Gioia et al. [228] plasticized CGM with different plasticizers including water, glycerol, polyethylene glycols (PEG), glucose, urea, diethanolamine, and triethanolamine, at concentrations of 10–30% (dwb). They implemented dynamic mechanical thermal analysis (DMTA) to investigate the change in glass transition temperature and rheological moduli of CGM. Similarly, the effect of "polar" plasticizers (such as water, glycerol) or "amphiphilic" plasticizers (such as octanoic and palmitic acids, dibutyl tartrate and phthalate, and diacetyl tartaric acid ester of mono-diglycerides) on the glass transition temperature of the CGM/plasticizer blends have been reported [229].

Plasticized CGM has been blended with several polymers. Corradini et al. [230] blended CGM with different plastics such as starch, polyvinyl alcohol (PVA) and poly(hydroxybutyrate-co-hydroxyvalerate), PHBV, using glycerol as plasticizer. After studying the glass transition temperature of the blends, they found that these blends are immiscible in the studied compositional range. Also in terms of mechanical properties, PVA improved the flexibility while PHBV enhanced the rigidity and starch caused slight changes in mechanical properties. CGM has also been blended with poly(ε-caprolactone), PCL [231]. In this work, CGM was first plasticized using glycerol/ethanol mixture, denatured by the addition of guanidine hydrochloride (GHCl), and then blended with PCL. They used twin screw extruder and injection molding for the processing. Their results showed that chemical modification of plasticized CGM with GHCl resulted in a high percent elongation. In another work CGM was blended with poly(lactic acid), PLA, plasticized with glycerol, water and ethanol using a single screw extruder followed by compression molding [212]. Their results showed that PLA enhanced the rigidity and improved the water resistance. CGM-wood fiber biocomposites have been the point of interest in several publications. CGM in these works has been used in the form of plasticized meal. Wu et al. [232] produced pellets of CGM-wood fiber, plasticized by glycerol, water and ethanol, to manufacture injection-molded plant pots for developing low cost, biodegradable containers used in agriculture. In another study, CGM plasticized with propylene glycol was blended with a biopolymer, poly(butylene succinate) (PBS), and wood fiber to produce a biodegradable material for plastic packaging applications [233]. The CGM content varied between 10–80 wt% and it was found that the produced biomaterial exhibited relatively high tensile strength, elongation at break and water resistance as long as the CGM content was less than 30 wt%. Similarly, CGM has been plasticized with different plasticizers such as glycerol, octanoic acid, polyethylene glycol and water, and reinforced with wood fiber using a twin screw extruder [234]. The best mechanical performance was achieved when a combination of 10 wt% octanoic acid and 30 wt% water was used as plasticizer with 20 wt% wood fibre as reinforcement. The mechanical properties were improved more when the CGM matrix was blended with polypropylene, coupling agent (maleated polypropylene) and cross-linking agent (benzoyl peroxide) with 50 wt% wood fibre [234].

3.4.2. *Soybean meal*

Soybean meal (SM) is still finding its way within scientific researches towards biobased material (biomaterial) applications. The number of publications on this topic is limited compared to corn gluten meal. SM has been characterized for its chemical composition, moisture content, thermal behavior and infrared spectrum and its potential as a particulate filler in value-added biocomposites was evaluated by compounding with polycaprolactone (PCL) [235]. The composite of PCL/SM (70/30) was prepared by extrusion and injection molding and then tested for mechanical properties such as tensile, flexural and impact. It has been observed that PCL/SM composite exhibited higher tensile/flexural modulus, but lower strength, elongation and impact strength compared to PCL. At the same time, the resulted biocomposite had relatively less cost than PCL itself. Thus, addition of SM to PCL increases the rigidity, but the particle-matrix adhesion needs to be improved.

SM has been plasticized and blended with other polymers such as polycaprolactone (PCL), poly(butylenes succinate) (PBS), poly(butylene adipate terephthalate) (PBAT) [236-237], and natural rubber [238-239]. SM was plasticized and destructurized successfully using glycerol and urea in a twin screw extruder and then blended with biodegradable polyesters, PCL and PBS [236]. As a result of destructurization phenomenon, improvement occurred in mechanical properties of the protein-based blends. In another work, SM was plasticized using glycerol in presence of two different denaturants (destructurizers), and the resulted thermoplastic SM was blended with different polyesters, PBS, PCL and PBAT [237]. Taguchi experimental design was adopted to investigate the effect of each constituent on the tensile properties of the final blend. Wu et al. [238] produced vulcanized blend of natural rubber and 50 wt% SM. The rubber phase was embedded by the SM matrix suggesting the interaction between phases, also approved by the increase in the glass transition temperature of the rubber phase. The produced blend exhibited good elasticity and water resistance.

Another area of research conducted on SM is producing edible films from defatted SM for food packaging applications [240-241]. For this purpose, SM was fermented in a soybean meal solution (15 g/100 ml of water) by inoculation with Bacillus subtilis bacteria fermentated under optimum conditions of 33°C and pH 7.0-7.5 for 33 h. Then, the fermented soybean solution was heated for 20 min at 75°C with 2-3 ml glycerol added to the solution to overcome film brittleness. The filtered solution was finally casted in a petri dish to produce films. Increasing amount of plasticizer in the fermented film led to a decrease in tensile strength and an increase in % elongation of the film compared to the ordinary soybean film. Moreover, the SM-based film exhibited higher water vapor permeability. On the other hand, experiments showed that growth inhibition of the produced SM-based film in the agar media containing E. coli was much higher than the ordinary soy protein film. These results indicated that the fermented SM-based films can be used as a new packaging material to extend the shelf-life of foods; however mechanical and physical properties need to be improved for more industrial applications.

3.5. Crude glycerol

Value-added uses of crude glycerol from biodesiel industries are highly diversified and it can be classified in to following categories, (i) chemicals/monomers: extraction and chemical/ biological conversion approach for the synthesis of various precursors, (ii) plasticizer: use as low cost plasticizing agent in various biomaterials applications, (iii) hydrogen generation: as the source for the generation of green hydrogen for energy applications, (iv) carbon source: carbon sole source for bacterial growth, which effectively used for the generation of bioplastics and (v) polyesters: combine with suitable organic acids and forms polyesters for various materials applications. In materials point of view, crude glycerol's application in plasticising as well as polyesters formation receives an immense attention and is discussed in this section.

As the global demand increases for the development of biodegradable materials, starch-based plastics receives an immense attention. The challenge is their unusual inter and intra molecular hydrogen bonds, which reduces their plastic performance [242]. This can be improved by incorporating plasticizers and manipulating processing conditions. Various plasticizers have been used on thermoplastic starch such as glycerol [243], glycol [244], sorbitol [245] and sugar [246]. Among them glycerol has been extensively used for the plasticization of starch-based materials. In addition to that, glycerol has also been used as an effective plasticizer in various other natural materials such as cellulose [247], chitosan [248], gelatin [249], DDG [250] and protein [233]. When these materials are plasticised with glycerol, their elasticity and toughness have been increased significantly with decreased brittleness. Numerous literatures are available for the glycerol plasticized natural products and most of the researchers have practiced pure glycerol as the precursor [233, 247-251]. Increasing trend of crude glycerol as plasticizer can be found in the literature for the development of various types of biomaterials [252-259].

Glycerol can be used to create polymeric materials by exploring their polyfunctional reactivity and the obtained polymers can have various applications as polyols substitutes, lubricant, raw materials to produce resins, polyesters and polyurethanes [260]. Among them, polyesters and polyurethanes receive more importance for their large volume biomaterial applications. The glycerol can easily react with carboxylic organic acids and forms polyesters, which can be used for various materials applications [260-261]. Carnahan et al. [262] demonstrated the synthesis of glycerol-based polyesters using succinic and adipic acids. Likewise, Tang et al. [263] synthesized the aliphatic polyesters from glycerol using sebacic acid with elastic properties. In both the cases they were using pure glycerol as a precursor and more publications are also available for fabrication of various other types of polyesters [264-265]. However, very few research works have been published based on the crude glycerol as feedstock for polyester synthesis. Brioude et al. [261] reported the fabrication of new polyester from crude glycerol and adipic acid through the bulk polymerization. They found that the newly developed crude glycerol-based polyester was amorphous in nature and has good mechanical and thermal stability. Similarly de Moura et al. [260] reported the polymerization between the glycerol and mono/bi-functional organic groups for the fabrication of new class of polyesters. They were able to synthesise two different classes of polymers with variable thermal stability. Their ultimate aim was to utilize these newly developed polyesters as matrix for the fabrication of natural fibre-reinforced composite materials. The obtained features (physicochemical) of these newly synthesised polyesters can also be used as modifiers for various types of thermosets (epoxy resins) as well as polyurethanes. Urea-formaldehyde (UF) has been widely known as an excellent adhesive used in wood panel fabrication. The challenge is to overcome their limited moisture resistance. The polyols made form glycerol can be used to enhance their moisture resistance properties. Recently, fungi-based biological transformation of crude glycerol in to various value-added products receives significant attention. Fungus exhibits more tolerance against the various impurities that exist in crude glycerol and found to be a suitable candidate for the biotransformations. Nicol et al. [266] reviewed the various research work on effective bioconversion of crude glycerol into many value added chemicals/materials employing fungus. They conclude that these techniques need to be further investigated for the extended applications. It is visible that glycerol will become as a high potential feedstock for various chemicals and tends to replace various existing petroleum derived products [267].

4. Conclusions

Growing energy demand for transportation, necessity of reducing the dependency on fossil fuel and the emerging environmental concerns about greenhouse gas emission have enhanced the growth of biofuel industries enormously and created new policy mandates by many countries. This caused the generation of a huge amount of under-valued coproduces of different types including distillers' grains, bagasse, lignin, protein-rich meals and crude glycerol. In order to claim biofuel production as the sustainable technology of future, it is necessary to create value-addition to these coproducts. This also provides solution to the emerging issues on the environmental impact of the accumulation of these coproducts. Moreover, the technological development related to the value-added uses of biofuel coproducts in biomaterials applications, is expected to (i) create new ecofriendly products, (ii) strengthen the bioeconomy and (iii) generate more rural job opportunity.

Among these coproducts, dried distillers' grains with solubles (DDGS), from dry mill ethanol industry, and the protein-rich meals such as corn gluten meal (CGM) and soybean meal (SM), respectively from wet mill ethanol and biodiesel industries, are traditionally used for animal feed applications. However, recent researches show the huge potential of these coproducts for biomaterial applications. DDGS can be used as filler or even as reinforcing phase in polymeric biocomposites by integrating pre-treatment, compatibilization and proper processing techniques. Moreover, it can be converted into a thermoplastic by chemical modifications such as acetylation and cyanoethylation. Also, it can be used for producing bioadhesive or extraction of fiber, oil and protein to be used for biomaterial usages.

Due to their high content of protein, CGM and SM can be plasticized using several plasticizers. Such plasticized meals have been blended successfully with several thermoplastics or been reinforced with wood fiber to create biomaterials with balanced properties. Besides, fermentation of these proteineous coproducts can provide new opportunities for the development of biobased films for food packaging applications.

It has been a long time since bagasse, the sugarcane ethanol coproduct, has been used in thermoset-fiber composites for particleboard production. In addition to that, bagasse is recently considered as the reinforcing filler compounded with different thermoplastics. Also, it has been shown that bagasse can turn into a thermo-formable material through esterification.

http://dx.doi.org/10.5772/55382

Lignin, the major coproduct of lignocellulosic ethanol, is being upgraded from its traditional utilization, i.e. burning, into various value-added commercial products such as polymer blends/composites, adhesives, and carbon fibres.

The industrial applications of pure glycerol in different sections are well established. Consequently, crude glycerol coproduced in biodiesel industry has huge opportunities in biomaterial applications including chemicals/monomers, plasticizer, hydrogen generation, carbon source for bacterial growth and polyesters production.

Novel innovative research in utilizing biofuel coproducts for biomaterials applications (biopolymers and biocomposites) impends the value-added high-end uses in the automotive, packaging and other structural/durable sectors. This is expected to create a fundamental change in materials point of view by utilizing renewable resources as feedstocks. In addition to that, these biofuel coproducts also create opportunity for the fabrication of various nano-structured materials for the high value applications.

Acknowledgements

Authors are thankful to the, Ontario Research Fund, Research Excellence, Round-4 (ORF RE04) from the Ontario Ministry of Economic Development and Innovation (MEDI); Ontario Ministry of Agriculture, Food, and Rural Affairs' (OMAFRA) New Directions and Alternative Renewable Fuels "Plus" Research Programs; OMAFRA-University of Guelph Bioeconomy-Industrial Uses Program; the Natural Sciences and Engineering Research Council of Canada (NSERC) Discovery grant individuals to M. Misra and A. K. Mohanty; NSERC Biomaterials and Chemicals Strategic Research Network; AUTO21-NCE (Canada's automotive R&D program); the Hannam Soybean Utilization Fund (HSUF) and the Grain Farmers of Ontario (GFO) fund for supporting research on various aspects of biobased materials, biofuel co-products and bionanotechnology at the University of Guelph's Bioproducts Discovery and Development Centre. N. Zarrinbakhsh acknowledges highly qualified personnel (HQP) scholarship from OMAFRA 2009.

Author details

S. Vivekanandhan[1,2], N. Zarrinbakhsh[1,2], M. Misra[1,2] and A. K. Mohanty[1,2]

1 Bioproducts Discovery and Development Centre, Department of Plant Agriculture, Crop Science Building, University of Guelph, Guelph, ON, N1G 2W1, Canada

2 School of Engineering, Thornbrough Building, University of Guelph, Guelph, ON, N1G 2W1, Canada

References

[1] 2012 The Outlook for Energy: A View to 2040 – Highlights. http://www.exxonmobil.com /corporate/files/news_pub_eo2012_highlights.pdf (Accessed on 30th July 2012).

[2] Weischer L, Wood D, Ballesteros A, Fu-Bertaux X. Grounding Green Power-Bottom-up Perspectives on Smart Renewable Energy Policy in Developing Countries, Climate & Energy Paper Series, The German Marshall Fund of the United States, Washington, 2011.

[3] Pereira MG, Camacho CF, Freitas MAV, da Silva NF. The Renewable Energy Market in Brazil: Current Status and Potential. Renewable and Sustainable Energy Reviews 2012;16 3786-3802.

[4] Rosegrant MW, Zhu TJ, Msangi S, Sulser T. Global Scenarios for Biofuels: Impacts and Implications. Applied Economic Perspectives and Policy 2008;30 495-505.

[5] Biodiesel Benefits, http://www.Biodiesel.com/index.php/biodiesel/biodiesel_benefits_why_use_biodiesel/. (accessed on 30th July 2012).

[6] Rajagopal D, Sexton SE, Roland-Holst D, Zilberman D. Challenge of Biofuel: Filling the Tank without Emptying the Stomach? Environmental Research Letters 2007;2 044004 (doi:10.1088/1748-9326/2/4/044004).

[7] Ragauskas AJ, Williams CK, Davison BH, Britovsek G, Cairney J, Eckert CA, et al. The Path Forward for Biofuels and Biomaterials. Science 2006;311 484-489.

[8] Walker GM. Bioethanol: Science and Technology of Fuel Alcohol, Graeme M. Wakler & Ventus Publishing ApS, Avilable online at http://bookboon.com/, 2010.

[9] Bhullar AS, Deo B, Sachdeva J. Transport Biofuel Production, Trade-Offs, and Promotion Policies in Canada—a Review. Agricultural Economics Research Review 2012;25 137-150.

[10] Letete T, Blottnitz H. Biofuel Policy in South Africa: A Critical Analysis, In: Rainer J, Dominik R, Bioenergy for Sustainable Development in Africa, Springer, 2012 191-199.

[11] Khanna M, Önal H, Crago CL, Mino K. Can India Meet Biofuel Policy Targets? Implications for Food and Fuel Prices. American Journal of Agricultural Economics 2012 doi: 10.1093/ajae/aas040.

[12] Wang Q, Tian Z. Biofuels and the Policy Implications for China. Asian Pacific Economic Literature 2011;25 161-168.

[13] Abdullah AZ, Salamatinia B, Mootabadi H, Bhatia S. Current Status and Policies on Biodiesel Industry in Malaysia as the World's Leading Producer of Palm Oil. Energy Policy 2009;37 5440-5448.

[14] Sparks GD, Ortmann GF. Global Biofuel Policies: A Review. Agrekon 2011;50 59-82.

[15] Lynd LR, Laser MS, Bransby D, Dale BE, Davison B, Hamilton R, et al. How Biotech Can Transform Biofuels. Nature Biotechnology 2008;26 169-172.

[16] Paulson ND, Ginder RG. The Growth and Direction of the Biodiesel Industry in the United States: Center for Agricultural and Rural Development, Iowa State University; 2007.

[17] Fernando S, Adhikari S, Chandrapal C, Murali N. Biorefineries: Current Status, Challenges, and Future Direction. Energy & Fuels 2006;20 1727-1737.

[18] OECD-FAO Agricultural Outlook 2011, Organisation for Economic Co-operation and Development; 2011. (DOI: 10.1787/agr_outlook-2011-en).

[19] OECD-FAO Agricultural Outlook 2011, Development of the World Ethanol Market. (http://dx.doi.org/10.1787/888932426467). Organisation for Economic Co-operation and Development; 2011.

[20] OECD-FAO Agricultural Outlook 2011, Evolution of Global Ethanol Production by Feedstocks Used (http://dx.doi.org/10.1787/888932426562). Organisation for Economic Co-operation and Development; 2011.

[21] OECD-FAO Agricultural Outlook 2011, Development of the World Biodiesel Market (http://dx.doi.org/10.1787/888932426486), Organisation for Economic Co-operation and Development; 2011.

[22] OECD-FAO Agricultural Outlook 2011, Evolution of Global Biodiesel Production by Feedstocks Used (http://dx.doi.org/10.1787/888932426581), Organisation for Economic Co-operation and Development; 2011.

[23] Rosentrater KA. Ethanol processing co-products: economics, impacts, sustainability. In: Eaglesham A, Hardy RWF, editors. Agricultural biofuels: technology, sustainability and profitability. Proceedings of the National Agricultural Biotechnology Council's 19th Annual Conference; 2007 May 22–24; Brookings (SD), USA. Ithaca (NY): National Agricultural Biotechnology Council; 2007. p. 105-26.

[24] RFA Renewable Fuels Association. Industry resources: co-products; Available from: http://www.ethanolrfa.org/pages/industry-resources-coproducts (Accessed on 1st October 2012).

[25] Yu P, Nuez-Ortin WG. Relationship of Protein Molecular Structure to Metabolisable Proteins in Different Types of Dried Distillers Grains with Solubles: A Novel Approach. British Journal of Nutrition 2010;104 1429-37.

[26] Kim Y, Mosier NS, Hendrickson R, Ezeji T, Blaschek H, Dien B, et al. Composition of Corn Dry-Grind Ethanol by-Products: DDGS, Wet Cake, and Thin Stillage. Bioresource Technology 2008;99 5165-5176.

[27] Dong FM, Rasco BA. The Neutral Detergent Fiber, Acid Detergent Fiber, Crude Fiber, and Lignin Contents of Distillers' Dried Grains with Solubles. Journal of Food Science 1987;52 403.

[28] Corn Refiners A. Corn Wet Milled Feed Products: http://www.corn.org/wp-content/uploads/2009/12/Feed2006.pdf; 2006. (Accessed on 17th September 2012)

[29] Getachew G, Robinson PH, DePeters EJ, Taylor SJ. Relationships between Chemical Composition, Dry Matter Degradation and in Vitro Gas Production of Several Ruminant Feeds. Animal Feed Science and Technology 2004;111 57-71.

[30] Bell JM, Keith MO. A Survey of Variation in the Chemical Composition of Commercial Canola Meal Produced in Western Canadian Crushing Plants. Canadian Journal of Animal Science 1991;71 469-480.

[31] Makkar HPS, Aderibigbe AO, Becker K. Comparative Evaluation of Non-Toxic and Toxic Varieties of Jatropha Curcas for Chemical Composition, Digestibility, Protein Degradability and Toxic Factors. Food Chemistry 1998;62 207-215.

[32] Anderson JL, Schingoethe DJ, Kalscheur KF, Hippen AR. Evaluation of Dried and Wet Distillers Grains Included at Two Concentrations in the Diets of Lactating Dairy Cows. Journal of Dairy Science 2006;89 3133-3142.

[33] Gill RK, VanOverbeke DL, Depenbusch B, Drouillard JS, DiCostanzo A. Impact of Beef Cattle Diets Containing Corn or Sorghum Distillers Grains on Beef Color, Fatty Acid Profiles, and Sensory Attributes. Journal of Animal Science 2008;86 923-935.

[34] Wahlstrom RC, German CS, Libal GW. Corn Distillers Dried Grains with Solubles in Growing-Finishing Swine Rations. Journal of Animal Science 1970;30 532-535.

[35] Oryschak M, Korver D, Zuidhof M, Meng X, Beltranena E. Comparative Feeding Value of Extruded and Nonextruded Wheat and Corn Distillers Dried Grains with Solubles for Broilers. Poultry Science 2010;89 2183-2196.

[36] Loar II RE, Schilling MW, McDaniel CD, Coufal CD, Rogers SF, Karges K, et al. Effect of Dietary Inclusion Level of Distillers Dried Grains with Solubles on Layer Performance, Egg Characteristics, and Consumer Acceptability. The Journal of Applied Poultry Research 2010;19 30-37.

[37] Cozannet P, Lessire M, Gady C, Métayer JP, Primot Y, Skiba F, et al. Energy Value of Wheat Dried Distillers Grains with Solubles in Roosters, Broilers, Layers, and Turkeys. Poultry Science 2010;89 2230-2241.

[38] Whitney TR, Lupton CJ. Evaluating Percentage of Roughage in Lamb Finishing Diets Containing 40% Dried Distillers Grains: Growth, Serum Urea Nitrogen, Nonesterified Fatty Acids, and Insulin Growth Factor-1 Concentrations and Wool, Carcass, and Fatty Acid Characteristics. Journal of Animal Science 2010;88 3030-3040.

[39] Webster CD, Tidwell JH, Yancey DH. Evaluation of Distillers' Grains with Solubles as a Protein Source in Diets for Channel Catfish. Aquaculture 1991;96 179-190.

[40] Shelby RA, Lim C, Yildrim-Aksoy M, Klesius PH. Effect of Distillers Dried Grains with Solubles incorporated Diets on Growth, Immune Function and Disease Resistance in Nile Tilapia (Oreochromis Niloticus L.). Aquaculture Research 2008;39 1351-1353.

[41] Ayadi FY, Muthukumarappan K, Rosentrater KA, Brown ML. Twin-Screw Extrusion Processing of Distillers Dried Grains with Solubles (DDGS)-Based Rainbow Trout (Oncorhynchus Mykiss) Feeds, ASABE Meeting Presentation, Pennsylvania, Paper Number: 1008337 2010.

[42] Tidwell JH, Webster CD, Clark JA, D'Abramo LR. Evaluation of Distillers Dried Grains with Solubles as an Ingredient in Diets for Pond Culture of the Freshwater Prawn Macrobrachium Rosenbergii. Journal of the World Aquaculture Society 1993;24 66-70.

[43] Babcock BA, Hayes DJ, Lawrence JD. Using Distillers Grains in the U.S. And International Livestock and Poultry Industries. USA: Midwest Agribusiness Trade Research and Information Center; 2008.

[44] Hoffman LA, Baker A. Market issues and prospects for U.S. distillers' grains supply, use, and price relationships. Washington, DC: United States Department of Agriculture, Economic Research Service; 2010 Dec. Outlook No.: FDS-10k-01.

[45] A Guide to Distiller's Dried Grains with Solubles (DDGS), U.S. Grains Council, http://www.ethanolrfa.org/page/-/rfa-association-site/studies/2012_DDGS_Handbook.pdf?nocdn=1 (Accessed on 2nd December 2012).

[46] Tokgoz S, Elobeid A, Fabiosa J, Hayes DJ, Babcock BA, Yu T-H, et al. Emerging Biofuels: Outlook of Effects on U.S. Grain, Oilseed, and Livestock Markets. Iowa State University: Center for Agricultural and Rural Development; 2007.

[47] Canadian Food Inspection Agency. Rg-6 Regulatory Guidance: Ethanol Distillers' Grains for Livestock Feed. Available Online at: http://www.inspection.gc.ca/animals/feeds/regulatory-guidance/rg-6/eng/1329275341920/1329275491608. (Accessed on 30th July 2012).

[48] AllAboutFeed (2011) US DDGS exports reach record high in 2010. http://www.allaboutfeed.net/news/us-ddgs-exports-reach-record-high-in-2010-5303.html. (Accessed on February 2012).

[49] Chevanan N, Muthukumarappan K, Rosentrater KA. Extrusion Studies of Aquaculture Feed Using Distillers Dried Grains with Solubles and Whey. Food and Bioprocess Technology 2009;2 177-185.

[50] Rasco BA, Rubenthaler G, Borhan M, Dong FM. Baking Properties of Bread and Cookies Incorporating Distillers' or Brewer's Grain from Wheat or Barley. Journal of Food Science 1990;55 424-429.

[51] Morey RV, Hatfield DL, Sears R, Tiffany DG. Characterization of Feed Streams and Emissions from Biomass Gasification/Combustion at Fuel Ethanol Plants. ASABE Meeting Presentation, Oregon, Paper Number: 064180 2006.

[52] Morey RV, Tiffany DG, Hatfield DL. Biomass for Electricity and Process Heat at Ethanol Plants. Applied Engineering in Agriculture 2006;22 723.

[53] Xu W, Reddy N, Yang Y. An Acidic Method of Zein Extraction from DDGS. Journal of Agricultural and Food Chemistry 2007;55 6279-6284.

[54] Xu W, Reddy N, Yang Y. Extraction, Characterization and Potential Applications of Cellulose in Corn Kernels and Distillers' Dried Grains with Solubles (DDGS). Carbohydrate Polymers 2009;76 521-527.

[55] Singh N, Cheryan M. Extraction of Oil from Corn Distillers Dried Grains with Solubles. Transactions of the ASAE 1998;41 1775-1777.

[56] Bruinsma K, Endres DL. Solvent Extraction of Oil from Distillers Dried Grains and Methods of Using Extraction Products. US Patent 2012; US 8,227,015 B2.

[57] Weber AD. The Relative Value of Cottonseed Meal, Linseed Meal, and Corn Gluten Meal in Fattening Cattle Rations. Journal of Animal Science 1934;1934 70-72.

[58] Umunna NN, Klopfenstein TJ, Hasimoglu S, Woods WR. Evaluation of Corn Gluten Meal with Urea as a Source of Supplementary Nitrogen for Growing Calves and Lambs. Animal Feed Science and Technology 1982;7 375-385.

[59] Fellner V, Belyea RL. Maximizing Gluten Feed in Corn Silage Diets for Dairy Cows. Journal of Dairy Science 1991;74 996-1005.

[60] Babidis V, Florou-Paneri P, Kufidis D, Christaki E, Spais AB, Vassilopoulos V. The Use of Corn Gluten Meal Instead of Herring and Meat Meal in Broiler Diets and Its Effect on Performance, Carcass Fatty Acid Composition and Othercarcass Characteristics. Archiv für Geflügelkunde 2002;66 145-150.

[61] Shurson G, Spiehs M, Whitney M. The Use of Maize Distiller's Dried Grains with Solubles in Pig Diets. Pig News and Information 2004;25 75N-83N.

[62] Funaba M, Oka Y, Kobayashi S, Kaneko M, Yamamoto H, Namikawa K, et al. Evaluation of Meat Meal, Chicken Meal, and Corn Gluten Meal as Dietary Sources of Protein in Dry Cat Food. Canadian Journal of Veterinary Research 2005;69 299-304.

[63] Wu YV, Rosati RR, Sessa DJ, Brown PB. Evaluation of Corn Gluten Meal as a Protein Source in Tilapia Diets. Journal of Agricultural and Food Chemistry 1995;43 1585-1588.

[64] Christians NE. Preemergence Weed Control Using Corn Gluten Meal. US Patent 1994; RE34594.

[65] Biopesticides registration action document, Corn Gluten Meal (PC Code 100137), US Environmental Protection Agency, Office of Pesticide Programs, Available online at: http://www.epa.gov/pesticides/chem_search/reg_actions/registration/decision_PC-100137_4-Mar-03.pdf. (Accessed: December 2012).

[66] Anselmo Filho P, Badr O. Biomass Resources for Energy in North-Eastern Brazil. Applied Energy 2004;77 51-67.

[67] Pandey A, Soccol CR, Nigam P, Soccol VT. Biotechnological Potential of Agro-Industrial Residues. I: Sugarcane Bagasse. Bioresource Technology 2000;74 69-80.

[68] Srinivasan VR, Han YW. Utilization of Bagasse. In: Hajny GJ, Reese ET, editors. Cellulases and Their Applications: American Chemical Society; 1969, 447-460.

[69] Mbohwa C, Fukuda S. Electricity from Bagasse in Zimbabwe. Biomass and Bioenergy 2003;25 197-207.

[70] van den Broek R, van den Burg T, van Wijk A, Turkenburg W. Electricity Generation from Eucalyptus and Bagasse by Sugar Mills in Nicaragua: A Comparison with Fuel Oil Electricity Generation on the Basis of Costs, Macro-Economic Impacts and Environmental Emissions. Biomass and Bioenergy 2000;19 311-335.

[71] Scaramucci JA, Perin C, Pulino P, Bordoni OFJG, da Cunha MP, Cortez LAB. Energy from Sugarcane Bagasse under Electricity Rationing in Brazil: A Computable General Equilibrium Model. Energy Policy 2006;34 986-992.

[72] Krüger H, Berndt W, Schwartzkopff U, Reitter FJ, Höpner T, Mühlig H-J. Production of Paper Pulp from Sugar Mill Bagasse. US Patent 1981; US 4260452.

[73] Shaikh AJ. Blending of Cotton Stalk Pulp with Bagasse Pulp for Paper Making. Biological Wastes 1990;31 37-43.

[74] Widyorini R, Xu J, Umemura K, Kawai S. Manufacture and Properties of Binderless Particleboard from Bagasse I: Effects of Raw Material Type, Storage Methods, and Manufacturing Process. Journal of Wood Science 2005;51 648-654.

[75] Zayed G, Abdel-Motaal H. Bio-Production of Compost with Low Ph and High Soluble Phosphorus from Sugar Cane Bagasse Enriched with Rock Phosphate. World Journal of Microbiology and Biotechnology 2005;21 747-752.

[76] Nigam P. Investigation of Some Factors Important for Solid-State Fermentation of Sugar Cane Bagasse for Animal Feed Production. Enzyme and Microbial Technology 1990;12 808-811.

[77] De Filippis P, Borgianni C, Paolucci M, Pochetti F. Gasification Process of Cuban Bagasse in a Two-Stage Reactor. Biomass and Bioenergy 2004;27 247-252.

[78] Laser M, Schulman D, Allen SG, Lichwa J, Antal Jr MJ, Lynd LR. A Comparison of Liquid Hot Water and Steam Pretreatments of Sugar Cane Bagasse for Bioconversion to Ethanol. Bioresource Technology 2002;81 33-44.

[79] Cuzens JC, Miller JR. Acid Hydrolysis of Bagasse for Ethanol Production. World Renewable Energy Congress IV Renewable Energy, Energy Efficiency and the Environment 1997;10 285-290.

[80] Teixeira LC, Linden JC, Schroeder HA. Optimizing Peracetic Acid Pretreatment Conditions for Improved Simultaneous Saccharification and Co-Fermentation (SSCF) of Sugar Cane Bagasse to Ethanol Fuel. Renewable Energy Energy Efficiency, Policy and the Environment 1999;16 1070-1073.

[81] Kim S, Dale BE. Global Potential Bioethanol Production from Wasted Crops and Crop Residues. Biomass and Bioenergy 2004;26 361-375.

[82] SugarCane.org. Brazilian Sugarcane Harvest. http://sugarcane.org/media-center/sugarcane-statistics/brazilian-sugarcane-harvest. (Accessed on 2nd December 2012).

[83] Taherzadeh MJ, Karimi K. Acid-Based Hydrolysis Processes for Ethanol from Lignocellulosic Materials: A Review. BioResources 2007; 2 472-499.

[84] Taherzadeh MJ, Karimi K. Enzymatic-Based Hydrolysis Processes for Ethanol from Lignocellulosic Materials: A Review. BioResources 2007; 2 707-738.

[85] Kumar S, Mohanty AK, Erickson L, Misra M. Lignin and Its Applications with Polymers. Journal of Biobased Materials and Bioenergy 2009;3 1-24.

[86] Mohanty AK, Misra M, Drzal LT. Sustainable Bio-Composites from Renewable Resources: Opportunities and Challenges in the Green Materials World. Journal of Polymers and the Environment 2002;10 19-26.

[87] Calvo-Flores FG, Dobado JA. Lignin as Renewable Raw Material. ChemSusChem 2010;3 1227-1235.

[88] Vishtal AG, Kraslawski A. Challenges in Industrial Applications of Technical Lignins. BioResources 2011;6 3547-3568.

[89] Stewart D. Lignin as a Base Material for Materials Applications: Chemistry, Application and Economics. Industrial Crops and Products 2008;27 202-207.

[90] Naik SN, Goud VV, Rout PK, Dalai AK. Production of First and Second Generation Biofuels: A Comprehensive Review. Renewable and Sustainable Energy Reviews 2010;14 578-597.

[91] Adler E. Lignin Chemistry—Past, Present and Future. Wood Science and Technology 1977;11 169-218.

[92] Sahoo S, Misra M, Mohanty AK. Enhanced Properties of Lignin-Based Biodegradable Polymer Composites Using Injection Moulding Process. Composites Part A: Applied Science and Manufacturing 2011, 42 1710–1718.

[93] Sahoo S, Misra M, Mohanty AK. Effect of Compatibilizer and Fillers on the Properties of Injection Molded Lignin Based Hybrid Green Composites. Journal of Applied Polymer Science 2012. DOI: 10.1002/APP.37667

[94] Hu TQ. Chemical Modification, Properties, and Usage of Lignin: Springer; 2002.

[95] Suhas, Carrott PJM, Ribeiro Carrott MML. Lignin–from Natural Adsorbent to Activated Carbon: A Review. Bioresource Technology 2007;98 2301-2312.

[96] Reisch MS. Getting the Steel Out. Chemical & Engineering News, 2011;89 10-14.

[97] United States Department of Agriculture, Economic Research Service. World Protein Meal Production. 2012. http://www.ers.usda.gov/dataproducts/oil-crops-yearbook.aspx. (Accesed on 1st December 2012).

[98] Faostat. http://faostat.fao.Org. (Accesed on 1st December 2012).

[99] United States National Biodiesel Board. US Biodiesel Production, http://www.biodiesel.org/production/production-statistics. (Accesed on 1st December 2012).

[100] Lang X, Dalai AK, Bakhshi NN, Reaney MJ, Hertz PB. Preparation and Characterization of Bio-Diesels from Various Bio-Oils. Bioresource Technology 2001;80 53-62.

[101] Sarin R, Sharma M, Sinharay S, Malhotra RK. Jatropha–Palm Biodiesel Blends: An Optimum Mix for Asia. Fuel 2007;86 1365-1371.

[102] Meher LC, Kulkarni MG, Dalai AK, Naik SN. Transesterification of Karanja (Pongamia Pinnata) Oil by Solid Basic Catalysts. European Journal of Lipid Science and Technology 2006;108 389-397.

[103] Pramanik K. Properties and Use of Jatropha Curcas Oil and Diesel Fuel Blends in Compression Ignition Engine. Renewable Energy 2003;28 239-248.

[104] Gerpen JV. Biodiesel Processing and Production. Fuel Processing Technology 2005;86 1097-1107.

[105] Thompson JC, He BB. Characterization of Crude Glycerol from Biodiesel Production from Multiple Feedstocks. Applied Engineering in Agriculture 2006;22 261-265.

[106] Meher LC, Vidya Sagar D, Naik SN. Technical Aspects of Biodiesel Production by Transesterification—a Review. Renewable and Sustainable Energy Reviews 2006;10 248-268.

[107] Pachauri N, He B. Value-Added Utilization of Crude Glycerol from Biodiesel Production: A Survey of Current Research Activities. ASABE Meeting Presentation, Oregon, Paper Number: 066223 2006.

[108] Mu Y, Teng H, Zhang D-J, Wang W, Xiu Z-L. Microbial Production of 1, 3-Propanediol by Klebsiella Pneumoniae Using Crude Glycerol from Biodiesel Preparations. Biotechnology Letters 2006;28 1755-1759.

[109] Soares RR, Simonetti DA, Dumesic JA. Glycerol as a Source for Fuels and Chemicals by Low Temperature Catalytic Processing. Angewandte Chemie 2006;118 4086-4089.

[110] Mothes G, Schnorpfeil C, Ackermann J-U. Production of PHB from Crude Glycerol. Engineering in Life Sciences 2007;7 475-479.

[111] Zhou C-HC, Beltramini JN, Fan Y-X, Lu GQM. Chemoselective Catalytic Conversion of Glycerol as a Biorenewable Source to Valuable Commodity Chemicals. Chemical Society Reviews, 2008;37 527-549.

[112] Cockerill S, Martin C. Are Biofuels Sustainable? The Eu Perspective. Biotechnology for Biofuels 2008;1. doi:10.1186/1754-6834-1-9.

[113] Bergeron C, Carrier DJ, Ramaswamy S. Biorefinery Co-Products: Phytochemicals, Primary Metabolites and Value-Added Biomass Processing: Wiley; 2012.

[114] Solomon BD. Biofuels and Sustainability. Annals of the New York Academy of Sciences 2010;1185 119-134.

[115] Biofuels and Sustainability Issues, http://www.biofuelstp.eu/sustainability.html (Accessed on 26th July 2012).

[116] Cook JH, Beyea J, Keeler KH. Potential Impacts of Biomass Production in the United States on Biological Diversity. Annual Review of Energy and the Environment 1991;16 401-431.

[117] Giampietro M, Ulgiati S, Pimentel D. Feasibility of Large-Scale Biofuel Production. BioScience 1997;47 587-600.

[118] Julson JL, Subbarao G, Stokke DD, Gieselman HH, Muthukumarappan K. Mechanical Properties of Biorenewable Fiber/Plastic Composites. Journal of Applied Polymer Science 2004;93 2484-2493.

[119] Tisserat BH, Reifschneider L, O'Kuru RH, Finkenstadt VL. Mechanical and Thermal Properties of High Density Polyethylene–Dried Distillers Grains with Solubles Composites. BioResources 2012;8 59-75.

[120] Wu Q, Mohanty AK. Renewable Resource Based Biocomposites from Coproduct of Dry Milling Corn Ethanol Industry and Castor Oil Based Biopolyurethanes. Journal of Biobased Materials and Bioenergy 2007;1 257-265.

[121] Tatara RA, Suraparaju S, Rosentrater KA. Compression Molding of Phenolic Resin and Corn-Based DDGS Blends. Journal of Polymers and the Environment 2007;15 89-95.

[122] Tatara RA, Rosentrater KA, Suraparaju S. Design Properties for Molded, Corn-Based DDGS-Filled Phenolic Resin. Industrial Crops and Products 2009;29 9-15.

[123] Cheesbrough V, Rosentrater KA, Visser J. Properties of Distillers Grains Composites: A Preliminary Investigation. Journal of Polymers and the Environment 2008;16 40-50.

[124] Li Y, Susan Sun X. Mechanical and Thermal Properties of Biocomposites from Poly(Lactic Acid) and DDGS. Journal of Applied Polymer Science 2011;121 589-597.

[125] Zarrinbakhsh N, Mohanty AK, Misra M. Fundamental Studies on Water-Washing of the Corn Ethanol Coproduct (DDGS) and Its Characterization for Biocomposite Applications. Biomass and Bioenergy 2013; http://dx.doi.org/10.1016/j.biombioe.2013.02.016.

[126] Zarrinbakhsh N, Misra M, Mohanty AK. Biodegradable Green Composites from Distiller's Dried Grains with Solubles (DDGS) and Polyhydroxy(Butyrate-Co-Valerate) (PHBV)-Based Bioplastic. Macromolecular Materials and Engineering 2011;296 1035-1045.

[127] Muniyasamy S, Reddy MM, Misra M, Mohanty AK. Biodegradable Green Composites from Bioethanol Co-Product and Poly(Butylene Adipate-Co-Terephatalate). Industrial Crops and Products 2013;43 812– 819.

[128] Saunders JA, Rosentrater KA. Properties of Solvent Extracted Low-Oil Corn Distillers Dried Grains with Solubles. Biomass and Bioenergy 2009;33 1486-1490.

[129] Srinivasan R, Singh V, Belyea RL, Rausch KD, Moreau RA, Tumbleson ME. Economics of Fiber Separation from Distillers Dried Grains with Solubles (DDGS) Using Sieving and Elutriation. Cereal Chemistry 2006;83 324-330.

[130] Singh A. Bioadhesives from Distiller's Dried Grains with Solubles (DDGS) and Studies on Sustainability Issues of Corn Ethanol Industries. Master Thesis, Michigan State University; 2008.

[131] Mohanty AK, Wu Q, Singh A. Bioadhesive from Distillers' Dried Grains with Solubles (DDGS) and the Methods of Making Those. US Patent 2009; US 7618660 B2.

[132] Hu C, Reddy N, Luo Y, Yan K, Yang Y. Thermoplastics from Acetylated Zein-and-Oil-Free Corn Distillers Dried Grains with Solubles. Biomass and Bioenergy 2011;35 884-892.

[133] Hu C, Reddy N, Yan K, Yang Y. Synthesis and Characterization of Highly Flexible Thermoplastic Films from Cyanoethylated Corn Distillers Dried Grains with Solubles. Journal of Agricultural and Food Chemistry 2011;59 1723-1728.

[134] Yu F, Le Z, Chen P, Liu Y, Lin X, Ruan R. Atmospheric Pressure Liquefaction of Dried Distillers Grains (DDG) and Making Polyurethane Foams from Liquefied DDG. Applied Biochemistry and Biotechnology 2008;148 235-243.

[135] Youngquist JA, Krzysik AM, English BW, Spelter HN, Chow P. Proceedings of 1996 symposium on The Use of Recycled Wood and Paper in Building Applications, Madison, WI, Forest products society 1996; p 123–134.

[136] Jústiz-Smith NG, Junior Virgo G, Buchanan VE. Potential of Jamaican Banana, Coconut Coir and Bagasse Fibres as Composite Materials. Materials Characterization 2008;59 1273-1278.

[137] Zárate CN, Aranguren MI, Reboredo MM. Resol-Vegetable Fibers Composites. Journal of Applied Polymer Science 2000;77 1832-1840.

[138] de Sousa MV, Monteiro SN, d'Almeida JRM. Evaluation of Pre-Treatment, Size and Molding Pressure on Flexural Mechanical Behavior of Chopped Bagasse–Polyester Composites. Polymer Testing 2004;23 253-258.

[139] Satyanarayana KG, Guimarães JL, Wypych F. Studies on Lignocellulosic Fibers of Brazil. Part I: Source, Production, Morphology, Properties and Applications. Composites Part A: Applied Science and Manufacturing 2007;38 1694-1709.

[140] Shibata S, Cao Y, Fukumoto I. Press Forming of Short Natural Fiber-Reinforced Biodegradable Resin: Effects of Fiber Volume and Length on Flexural Properties. Polymer Testing 2005;24 1005-1011.

[141] Cao Y, Shibata S, Fukumoto I. Mechanical Properties of Biodegradable Composites Reinforced with Bagasse Fibre before and after Alkali Treatments. Composites Part A: Applied Science and Manufacturing 2006;37 423-429.

[142] Vilay V, Mariatti M, Mat Taib R, Todo M. Effect of Fiber Surface Treatment and Fiber Loading on the Properties of Bagasse Fiber–Reinforced Unsaturated Polyester Composites. Composites Science and Technology 2008;68 631-638.

[143] McLaughlin EC. The Strength of Bagasse Fibre-Reinforced Composites. Journal of Materials Science 1980;15 886-890.

[144] Carvajal O, Valdés JL, Puig J. Bagasse Particleboards for Building Purpose. European Journal of Wood and Wood Products 1996;54 61-63.

[145] Aggarwal LK. Bagasse-Reinforced Cement Composites. Cement and Concrete Composites 1995;17 107-112.

[146] Bilba K, Arsene M-A, Ouensanga A. Sugar Cane Bagasse Fibre Reinforced Cement Composites. Part I. Influence of the Botanical Components of Bagasse on the Setting of Bagasse/Cement Composite. Cement and Concrete Composites 2003;25 91-96.

[147] Shen KC. Process for Manufacturing Composite Products from Lignocellulosic Materials. US Patent 1986; US 4627951.

[148] Mobarak F, Fahmy Y, Augustin H. Binderless Lignocellulose Composite from Bagasse and Mechanism of Self-Bonding. Holzforschung 1982;36 131-135.

[149] Patil YP, Gajre B, Dusane D, Chavan S, Mishra S. Effect of Maleic Anhydride Treatment on Steam and Water Absorption of Wood Polymer Composites Prepared from Wheat Straw, Cane Bagasse, and Teak Wood Sawdust Using Novolac as Matrix. Journal of Applied Polymer Science 2000;77 2963-2967.

[150] Paiva JMF, Frollini E. Sugarcane Bagasse Reinforced Phenolic and Lignophenolic Composites. Journal of Applied Polymer Science 2002;83 880-888.

[151] Salyer IO, Usmani AM. Utilization of Bagasse in New Composite Building Materials. Industrial & Engineering Chemistry Product Research and Development 1982;21 17-23.

[152] Stael GC, Tavares MIB, d'Almeida JRM. Evaluation of Sugar Cane Bagasse Waste as Reinforcement in Eva Matrix Composite Materials. Polymer-Plastics Technology and Engineering 2001;40 217-223.

[153] Stael GC, Tavares MIB, d'Almeida JRM. Impact Behavior of Sugarcane Bagasse Waste–EVA Composites. Polymer Testing 2001;20 869-872.

[154] Luz SM, Gonçalves AR, Del'Arco Jr AP. Mechanical Behavior and Microstructural Analysis of Sugarcane Bagasse Fibers Reinforced Polypropylene Composites. Composites Part A: Applied Science and Manufacturing 2007;38 1455-1461.

[155] Mulinari DR, Voorwald HJC, Cioffi MOH, da Silva MLCP, Luz SM. Preparation and Properties of Hdpe/Sugarcane Bagasse Cellulose Composites Obtained for Thermokinetic Mixer. Carbohydrate Polymers 2009;75 317-321.

[156] Chen Y, Ishikawa Y, Maekawa T, Zhang Z. Preparation of Acetylated Starch/Bagasse Fiber Composites by Extrusion. Transactions of the ASAE 2006;49 85-90.

[157] Chen Y, Zhang Z, Ishikawa Y, Maekawa T. Production of Starch-Based Biodegradable Plastics Reinforced with Bagasse Fiber. Journal of the Society of Agricultural Structures 2002;32 177-184.

[158] Hassan ML, Rowell RM, Fadl NA, Yacoub SF, Christainsen AW. Thermoplasticization of Bagasse. I. Preparation and Characterization of Esterified Bagasse Fibers. Journal of Applied Polymer Science 2000;76 561-574.

[159] Hassan ML, Rowell RM, Fadl NA, Yacoub SF, Christainsen AW. Thermoplasticization of Bagasse. II. Dimensional Stability and Mechanical Properties of Esterified Bagasse Composite. Journal of Applied Polymer Science 2000;76 575-586.

[160] Wang J, Manley RSJ, Feldman D. Synthetic Polymer-Lignin Copolymers and Blends. Progress in Polymer Science (UK) 1992;17 611-646.

[161] Lora JH, Glasser WG. Recent Industrial Applications of Lignin: A Sustainable Alternative to Nonrenewable Materials. Journal of Polymers and the Environment 2002;10 39-48.

[162] Thielemans W, Can E, Morye SS, Wool RP. Novel Applications of Lignin in Composite Materials. Journal of Applied Polymer Science 2002;83 323-331.

[163] Rodríguez-Mirasol J, Cordero T, Rodríguez JJ. Preparation and Characterization of Activated Carbons from Eucalyptus Kraft Lignin. Carbon 1993;31 87-95.

[164] Nitz H, Semke H, Mülhaupt R. Influence of Lignin Type on the Mechanical Properties of Lignin Based Compounds. Macromolecular Materials and Engineering 2001;286 737-743.

[165] Kadla JF, Kubo S. Lignin-Based Polymer Blends: Analysis of Intermolecular Interactions in Lignin–Synthetic Polymer Blends. Composites Part A: Applied Science and Manufacturing 2004;35 395-400.

[166] Alexy P, Košíková B, Podstránska G. The Effect of Blending Lignin with Polyethylene and Polypropylene on Physical Properties. Polymer 2000;41 4901-4908.

[167] Košíková B, Gregorová A. Sulfur Free Lignin as Reinforcing Component of Styrene–Butadiene Rubber. Journal of Applied Polymer Science 2005;97 924-929.

[168] Toriz G, Denes F, Young RA. Lignin Polypropylene Composites. Part 1: Composites from Unmodified Lignin and Polypropylene. Polymer Composites 2002;23 806-813.

[169] Canetti M, Bertini F, De Chirico A, Audisio G. Thermal Degradation Behaviour of Isotactic Polypropylene Blended with Lignin. Polymer Degradation and Stability 2006;91 494-498.

[170] Mikulášová M, Košíková B. Biodegradability of Lignin—Polypropylene Composite Films. Folia Microbiologica 1999;44 669-672.

[171] El Raghi S, Zahran RR, Gebril BE. Effect of Weathering on Some Properties of Polyvinyl Chloride/Lignin Blends. Materials Letters 2000;46 332-342.

[172] Banu D, El-Aghoury A, Feldman D. Contributions to Characterization of Poly (Vinyl Chloride)–Lignin Blends. Journal of Applied Polymer Science 2006;101 2732-2748.

[173] Kubo S, Kadla JF. The Formation of Strong Intermolecular Interactions in Immiscible Blends of Poly (Vinyl Alcohol)(PVA) and Lignin. Biomacromolecules 2003;4 561-567.

[174] Fernandes DM, Winkler Hechenleitner AA, Job AE, Radovanocic E, Gómez Pineda EA. Thermal and Photochemical Stability of Poly (Vinyl Alcohol)/Modified Lignin Blends. Polymer Degradation and Stability 2006;91 1192-1201.

[175] Kubo S, Kadla JF. Poly (Ethylene Oxide)/Organosolv Lignin Blends: Relationship between Thermal Properties, Chemical Structure, and Blend Behavior. Macromolecules 2004;37 6904-6911.

[176] Kubo S, Kadla JF. Kraft Lignin/Poly (Ethylene Oxide) Blends: Effect of Lignin Structure on Miscibility and Hydrogen Bonding. Journal of Applied Polymer Science 2005;98 1437-1444.

[177] Camargo FA, Innocentini-Mei LH, Lemes AP, Moraes SG, Durán N. Processing and Characterization of Composites of Poly (3-Hydroxybutyrate-Co-Hydroxyvalerate) and Lignin from Sugar Cane Bagasse. Journal of Composite Materials 2012;46 417-425.

[178] Mousavioun P, Doherty WOS, George G. Thermal Stability and Miscibility of Poly (Hydroxybutyrate) and Soda Lignin Blends. Industrial Crops and Products 2010;32 656-661.

[179] Vengal JC, Srikumar M. Processing and Study of Novel Lignin-Starch and Lignin-Gelatin Biodegradable Polymeric Films. Trends in Biomaterials and Artificial Organs 2005;18 237-241.

[180] Réti C, Casetta M, Duquesne S, Bourbigot S, Delobel R. Flammability Properties of Intumescent PLA Including Starch and Lignin. Polymers for Advanced Technologies 2008;19 628-635.

[181] El Mansouri NE, Yuan Q, Huang F. Characterization of Alkaline Lignins for Use in Phenol-Formaldehyde and Epoxy Resins. BioResources 2011;6 2647-2662.

[182] Peng W, Riedl B. The Chemorheology of Phenol-Formaldehyde Thermoset Resin and Mixtures of the Resin with Lignin Fillers. Polymer 1994;35 1280-1286.

[183] Guigo N, Mija A, Vincent L, Sbirrazzuoli N. Eco-Friendly Composite Resins Based on Renewable Biomass Resources: Polyfurfuryl Alcohol/Lignin Thermosets. European Polymer Journal 2010;46 1016-1023.

[184] Thielemans W, Wool RP. Butyrated Kraft Lignin as Compatibilizing Agent for Natural Fiber Reinforced Thermoset Composites. Composites Part A: Applied Science and Manufacturing 2004;35 327-338.

[185] Nonaka Y, Tomita B, Hatano Y. Synthesis of Lignin/Epoxy Resins in Aqueous Systems and Their Properties. Holzforschung-International Journal of the Biology, Chemistry, Physics and Technology of Wood 2009;51 183-187.

[186] Kumaran MG, De SK. Utilization of Lignins in Rubber Compounding. Journal of Applied Polymer Science 1978;22 1885-1893.

[187] Wang H, Easteal AJ, Edmonds N. Prevulcanized Natural Rubber Latex/Modified Lignin Dispersion for Water Vapour Barrier Coatings on Paperboard Packaging. Advanced Materials Research 2008;47-50 93-96.

[188] Tibenham FJ, Grace NS. Compounding Natural Rubber with Lignin and Humic Acid. Industrial & Engineering Chemistry 1954;46 824-828.

[189] Nakamura K, Hatakeyama T, Hatakeyama H. Thermal Properties of Solvolysis Lignin Derived Polyurethanes. Polymers for Advanced Technologies 1992;3 151-155.

[190] Nakamura K, Mörck R, Reimann A, Kringstad KP, Hatakeyama H. Mechanical Properties of Solvolysis Lignin Derived Polyurethanes. Polymers for Advanced Technologies 1991;2 41-47.

[191] Sarkar S, Adhikari B. Synthesis and Characterization of Lignin–HTPB Copolyurethane. European Polymer Journal 2001;37 1391-1401.

[192] Saraf VP, Glasser WG. Engineering Plastics from Lignin. III. Structure Property Relationships in Solution Cast Polyurethane Films. Journal of Applied Polymer Science 1984;29 1831-1841.

[193] Saraf VP, Glasser WG, Wilkes GL, McGrath JE. Engineering Plastics from Lignin. VI. Structure–Property Relationships of PEG Containing Polyurethane Networks. Journal of Applied Polymer Science 1985;30 2207-2224.

[194] Saraf VP, Glasser WG, Wilkes GL. Engineering Plastics from Lignin. VII. Structure Property Relationships of Poly (Butadiene Glycol) Containing Polyurethane Networks. Journal of Applied Polymer Science 1985;30 3809-3823.

[195] Thring RW, Vanderlaan MN, Griffin SL. Polyurethanes from Alcell® Lignin. Biomass and Bioenergy 1997;13 125-132.

[196] Yoshida H, Mörck R, Kringstad KP, Hatakeyama H. Kraft Lignin in Polyurethanes I. Mechanical Properties of Polyurethanes from a Kraft Lignin–Polyether Triol–Polymeric MDI System. Journal of Applied Polymer Science 1987;34 1187-1198.

[197] Yoshida H, Mörck R, Kringstad KP, Hatakeyama H. Kraft Lignin in Polyurethanes. II. Effects of the Molecular Weight of Kraft Lignin on the Properties of Polyurethanes from a Kraft Lignin–Polyether Triol–Polymeric Mdi System. Journal of Applied Polymer Science 1990;40 1819-1832.

[198] Haars A, Kharazipour A, Zanker H, Hüttermann A. Room-Temperature Curing Adhesives Based on Lignin and Phenoloxidases. In: Hemingway RW, Conner AH, Branham SJ (eds), Adhesives from Renewable Resources, American Chemical Society, ACS Symposium Series 1989;385 126-134 (DOI: 10.1021/bk-1989-0385.ch010).

[199] El Mansouri N, Pizzi A, Salvadó J. Lignin-Based Wood Panel Adhesives without Formaldehyde. European Journal of Wood and Wood Products 2007;65 65-70.

[200] Schneider MH, Phillips JG. Furfuryl Alcohol and Lignin Adhesive Composition. US Patents 2004; US 6747076 B2.

[201] Pizzi A, Cameron F-A, van der Klashorst GH. Soda Bagasse Lignin Adhesive for Particleboard. In: Hemingway RW, Conner AH, Branham SJ (eds), Adhesives from Renewable Resources, American Chemical Society, ACS Symposium Series 1989;385 82-95 (DOI: 10.1021/bk-1989-0385).

[202] Khan MA, Ashraf SM, Malhotra VP. Development and Characterization of a Wood Adhesive Using Bagasse Lignin. International Journal of Adhesion and Adhesives 2004;24 485-493.

[203] Allen MJ, Tung VC, Kaner RB. Honeycomb Carbon: A Review of Graphene. Chemical Reviews 2010;110 132.

[204] Niyogi S, Hamon MA, Hu H, Zhao B, Bhowmik P, Sen R, et al. Chemistry of Single-Walled Carbon Nanotubes. Accounts of Chemical Research 2002;35 1105-1113.

[205] Kim C, Yang KS. Electrochemical Properties of Carbon Nanofiber Web as an Electrode for Supercapacitor Prepared by Electrospinning. Applied Physics Letters 2003;83 1216-1218.

[206] Kim H, Abdala AA, Macosko CW. Graphene/Polymer Nanocomposites. Macromolecules 2010; 43 6515–6530.

[207] White RJ, Tauer K, Antonietti M, Titirici M-M. Functional Hollow Carbon Nanospheres by Latex Templating. Journal of the American Chemical Society 2010, 132, 17360–17363.

[208] Kang X, Wang J, Wu H, Liu J, Aksay IA, Lin Y. A Graphene-Based Electrochemical Sensor for Sensitive Detection of Paracetamol. Talanta 2010;81 754-759.

[209] Suriani AB, Azira AA, Nik SF, Md Nor R, Rusop M. Synthesis of Vertically Aligned Carbon Nanotubes Using Natural Palm Oil as Carbon Precursor. Materials Letters 2009;63 2704-2706.

[210] Sharon M, Sharon M. Carbon Nanomaterials and Their Synthesis from Plant Derived Precursors. Synthesis and Reactivity in Inorganic, Metal-Organic and Nano-Metal Chemistry 2006;36 265-279.

[211] Sharon M, Soga T, Afre R, Sathiyamoorthy D, Dasgupta K, Bhardwaj S, et al. Hydrogen Storage by Carbon Materials Synthesized from Oil Seeds and Fibrous Plant Materials. International Journal of Hydrogen Energy 2007;32 4238-4249.

[212] Yu SH, Cui XJ, Li LL, Li K, Yu B, Antonietti M, et al. From Starch to Metal/Carbon Hybrid Nanostructures: Hydrothermal Metal Catalyzed Carbonization. Advanced Materials 2004;16 1636-1640.

[213] Hayashi J, Kazehaya A, Muroyama K, Watkinson AP. Preparation of Activated Carbon from Lignin by Chemical Activation. Carbon 2000;38 1873-1878.

[214] Kubo S, Kadla JF. Lignin-Based Carbon Fibers: Effect of Synthetic Polymer Blending on Fiber Properties. Journal of Polymers and the Environment 2005;13 97-105.

[215] Babel K, Jurewicz K. KOH Activated Lignin Based Nanostructured Carbon Exhibiting High Hydrogen Electrosorption. Carbon 2008;46 1948-1956.

[216] Gonugunta P. Synthesis and Characterization of Biobased Carbon Nanoparticles from Lignin. Thesis, Master of Applied Science, University of Guelph, 2012.

[217] Gonugunta P, Vivekanandhan S, Mohanty AK, Misra M. A Study on Synthesis and Characterization of Biobased Carbon Nanoparticles from Lignin. World Journal of Nano Science and Engineering 2012;2 148-153.

[218] Ramesh Kumar P, Khan N, Vivekanandhan S, Satyanarayana N, Mohanty AK, Misra M. Nanofibers: Effective Generation by Electrospinning and Their Applications. Journal of Nanoscience and Nanotechnology 2012;12 1-25.

[219] Lallave M, Bedia J, Ruiz-Rosas R, Rodríguez-Mirasol J, Cordero T, Otero JC, et al. Filled and Hollow Carbon Nanofibers by Coaxial Electrospinning of Alcell Lignin without Binder Polymers. Advanced Materials 2007;19 4292-4296.

[220] Schreiber M, Vivekanandhan S, Mohanty AK, Misra M. A Study on the Electrospinning Behaviour and Nanofibre Morphology of Anionically Charged Lignin. Advanced Materials Letters 2012; 3(6), 476-480.

[221] Dallmeyer I, Ko F, Kadla JF. Electrospinning of Technical Lignins for the Production of Fibrous Networks. Journal of Wood Chemistry and Technology 2010;30 315-329.

[222] Seo DK, Jeun JP, Kim HB, Kang PH. Preparation and Characterization of the Carbon Nanofiber Mat Produced from Electrospun Pan/Lignin Precursors by Electron Beam Irradiation. Reviews on Advanced Materials Science 2011;28 31-34.

[223] Spender J, Demers AL, Xie X, Cline AE, Earle MA, Ellis LD, Neivandt DJ, Method for Production of Polymer and Carbon Nanofibers from Water-Soluble Polymers, Nano Letters, 2012; 12 3857-3860.

[224] Kumar R, Choudhary V, Mishra S, Varma IK, Mattiason B. Adhesives and Plastics Based on Soy Protein Products. Industrial Crops and Products 2002;16 155-172.

[225] Shukla R, Cheryan M. Zein: The Industrial Protein from Corn. Industrial Crops and Products 2001;13 171-192.

[226] Reddy M, Mohanty AK, Misra M. Thermoplastics from Soy Protein: A Review on Processing, Blends and Composites. Journal of Biobased Materials and Bioenergy 2010;4 298-316.

[227] Lawton JW, Selling GW, Willett JL. Corn Gluten Meal as a Thermoplastic Resin: Effect of Plasticizers and Water Content. Cereal Chemistry Journal 2008;85 102-108.

[228] di Gioia L, Cuq B, Guilbert S. Effect of Hydrophilic Plasticizers on Thermomechanical Properties of Corn Gluten Meal. Cereal Chemistry Journal 1998;75 514-519.

[229] Di Gioia L, Guilbert S. Corn Protein-Based Thermoplastic Resins: Effect of Some Polar and Amphiphilic Plasticizers. Journal of Agricultural and Food Chemistry 1999;47 1254-1261.

[230] Corradini E, Marconcini JM, Agnelli JAM, Mattoso LHC. Thermoplastic Blends of Corn Gluten Meal/Starch (CGM/Starch) and Corn Gluten Meal/Polyvinyl Alcohol and Corn Gluten Meal/Poly (Hydroxybutyrate-Co-Hydroxyvalerate) (CGM/PHB-V). Carbohydrate Polymers 2011;83 959-965.

[231] Aithani D, Mohanty AK. Value-Added New Materials from Byproduct of Corn Based Ethanol Industries: Blends of Plasticized Corn Gluten Meal and Poly(ε-Caprolactone). Industrial & Engineering Chemistry Research 2006;45 6147-6152.

[232] Wu Q, Sakabe H, Isobe S. Processing and Properties of Low Cost Corn Gluten Meal/ Wood Fiber Composite. Industrial & Engineering Chemistry Research 2003;42 6765-6773.

[233] Samarasinghe S, Easteal AJ, Edmonds NR. Biodegradable Plastic Composites from Corn Gluten Meal. Polymer International 2008;57 359-364.

[234] Beg MDH, Pickering KL, Weal SJ. Corn Gluten Meal as a Biodegradable Matrix Material in Wood Fibre Reinforced Composites. Materials Science and Engineering: A 2005;412 7-11.

[235] Diebel W, Reddy MM, Misra M, Mohanty A. Material Property Characterization of Co-Products from Biofuel Industries: Potential Uses in Value-Added Biocomposites. Biomass and Bioenergy, 2012; 37 88–96.

[236] Reddy MM, Mohanty AK, Misra M. Biodegradable Blends from Plasticized Soy Meal, Polycaprolactone, and Poly (Butylene Succinate). Macromolecular Materials and Engineering 2012; 297 455–463.

[237] Reddy MM, Mohanty AK, Misra M. Optimization of Tensile Properties Thermoplastic Blends from Soy and Biodegradable Polyesters: Taguchi Design of Experiments Approach. Journal of Materials Science 2012; 37 2591-2599.

[238] Wu Q, Selke S, Mohanty AK. Processing and Properties of Biobased Blends from Soy Meal and Natural Rubber. Macromolecular Materials and Engineering 2007;292 1149-1157.

[239] Mohanty AK, Wu Q, Selke S. "Green" Materials from Soy Meal and Natural Rubber Blends. US Patents 2010; US 7649036 B2.

[240] Kim H-W, Kim K-M, Ko E-J, Lee S-K, Ha S-D, Song K-B, et al. Development of Antimicrobial Edible Film from Defatted Soybean Meal Fermented by Bacillus Subtilis. Journal of Microbiology and Biotechnology 2004;14 1303-1309.

[241] Kim H-W, Ko E-J, Ha S-D, Song K-B, Park S-K, Chung D-H, et al. Physical, Mechanical, and Antimicrobial Properties of Edible Film Produced from Defatted Soybean Meal Fermented by Bacillus Subtilis. Journal of Microbiology and Biotechnology 2005;15 815-822.

[242] Jiugao Y, Ning W, Xiaofei M. The Effects of Citric Acid on the Properties of Thermoplastic Starch Plasticized by Glycerol. Starch Stärke 2005;57 494-504.

[243] Fishman ML, Coffin DR, Konstance RP, Onwulata CI. Extrusion of Pectin/Starch Blends Plasticized with Glycerol. Carbohydrate Polymers 2000;41 317-325.

[244] Yu J, Gao J, Lin T. Biodegradable Thermoplastic Starch. Journal of Applied Polymer Science 1996;62 1491-1494.

[245] Wang L, Shogren RL, Carriere C. Preparation and Properties of Thermoplastic Starch Polyester Laminate Sheets by Coextrusion. Polymer Engineering & Science 2000;40 499-506.

[246] Barrett A, Kaletunç G, Rosenburg S, Breslauer K. Effect of Sucrose on the Structure, Mechanical Strength and Thermal Properties of Corn Extrudates. Carbohydrate Polymers 1995;26 261-269.

[247] Kim J, Wang N, Chen Y, Yun G-Y. An Electro-Active Paper Actuator Made with Lithium Chloride/Cellulose Films: Effects of Glycerol Content and Film Thickness. Smart Materials and Structures 2007;16 1564-1569.

[248] Cervera MF, Heinämäki J, Krogars K, Jörgensen AC, Karjalainen M, Colarte AI, et al. Solid-State and Mechanical Properties of Aqueous Chitosan-Amylose Starch Films Plasticized with Polyols. AAPS Pharmscitech 2004;5 109-114.

[249] Bergo P, Sobral PJA. Effects of Plasticizer on Physical Properties of Pigskin Gelatin Films. Food Hydrocolloids 2007;21 1285-1289.

[250] Clarizio SC, Tatara RA. Tensile Strength, Elongation, Hardness, and Tensile and Flexural Moduli of PLA Filled with Glycerol-Plasticized DDGS. Journal of Polymers and the Environment, 2012; 20 638-646.

[251] Galietta G, Di Gioia L, Guilbert S, Cuq B. Mechanical and Thermomechanical Properties of Films Based on Whey Proteins as Affected by Plasticizer and Crosslinking Agents. Journal of Dairy Science 1998;81 3123-3130.

[252] Rosa DS, Bardi MAG, Machado LDB, Dias DB, Andrade e Silva LG, Kodama Y. Influence of Thermoplastic Starch Plasticized with Biodiesel Glycerol on Thermal Properties of PP Blends. Journal of Thermal Analysis and Calorimetry 2009;97 565-570.

[253] Rosa DS, Bardi MAG, Machado LDB, Dias D, Silva LGA, Kodama Y. Starch Plasticized with Glycerol from Biodiesel and Polypropylene Blends. Journal of Thermal Analysis and Calorimetry 2010;102 181-186.

[254] Dakka JM, Mozeleski EJ, Baugh LS, Process for making triglyceride plasticizer from crude glycerol, US Patent 2012; US 8,299,281 B2.

[255] Nobrega MM, Olivato JB, Bilck AP, Grossmann MVE, Yamashita F. Glycerol with Different Purity Grades Derived from Biodiesel: Effect on the Mechanical and Viscoe-

lastic Properties of Biodegradable Strands and Films. Materials Science and Engineering: C 2012; 32 2220–2222.

[256] Grazhulevichene V, Augulis L, Gumbite M, Makarevichene V. Biodegradable Composites of Polyvinyl Alcohol and Coproducts of Diesel Biofuel Production. Russian Journal of Applied Chemistry 2011;84 719-723.

[257] Stevens ES, Ashby RD, Solaiman DKY. Gelatin Plasticized with a Biodiesel Coproduct Stream. Journal of Biobased Materials and Bioenergy 2009;3 57-61.

[258] Bardi MAG, Machado LDB, Guedes CGF, Rosa DS. Influence of Processing on Properties of Aditived Polypropylene, Starch Plasticized with Glycerol from Biodiesel and Their Blends. http://www.ipen.br/biblioteca/cd/cbpol/2009/PDF/1450.pdf. (Accessed on 3rd December 2012)

[259] Singh M, Milano J, Stevens ES, Ashby RD, Solaiman DKY. Gelatin Films Plasticized with a Simulated Biodiesel Coproduct Stream. eXPRESS Polymer Letters; 2009; 3 201–206.

[260] de Moura CVR, Nunes ASL, Neto JMM, Neres HLS, de Carvalho LMG, de Moura EM. Synthesis and Characterization of Polyesters from Glycerol by-Product of Biodiesel Production. Journal of the Brazilian Chemical Society 2012;23 1226-1231.

[261] Brioude MM, Guimarães DH, Fiúza RP, Prado LASA, Boaventura JS, José NM. Synthesis and Characterization of Aliphatic Polyesters from Glycerol, by-Product of Biodiesel Production, and Adipic Acid. Materials Research 2007;10 335-339.

[262] Carnahan MA, Grinstaff MW. Synthesis of Generational Polyester Dendrimers Derived from Glycerol and Succinic or Adipic Acid. Macromolecules 2006;39 609-616.

[263] Tang J, Zhang Z, Song Z, Chen L, Hou X, Yao K. Synthesis and Characterization of Elastic Aliphatic Polyesters from Sebacic Acid, Glycol and Glycerol. European Polymer Journal 2006;42 3360-3366.

[264] Holser RA, Willett JL, Vaughn SF. Thermal and Physical Characterization of Glycerol Polyesters. Journal of Biobased Materials and Bioenergy 2008;2 94-96.

[265] Stumbé J-F, Bruchmann B. Hyperbranched Polyesters Based on Adipic Acid and Glycerol. Macromolecular Rapid Communications 2004;25 921-924.

[266] Nicol RW, Marchand K, Lubitz WD. Bioconversion of Crude Glycerol by Fungi. Applied Microbiology and Biotechnology 2012; 93 1865-1875.

[267] Johnson DT, Taconi KA. The Glycerin Glut: Options for the Value Added Conversion of Crude Glycerol Resulting from Biodiesel Production. Environmental Progress 2007;26 338-348.